Control Systems Engineering

Control Systems Engineering

William J. Palm III
University of Rhode Island

John Wiley & Sons
New York Chichester Brisbane Toronto Singapore

Library of Congress Cataloging-in-Publication Data

Palm, William J. (William John), 1944–
Control systems engineering.

Bibliography: p.
Includes index.
1. Automatic control. 2. Control theory.
I. Title
TJ213.P225 1986 629.8 85-26590
ISBN 0-471-81086-X

Printed in the United States of America

10 9 8 7 6 5 4 3

To my parents,
Lillian Hartmann Palm
and
William J. Palm

Preface

This text is an introduction to control systems engineering. Such a course is typically taken by undergraduates in mechanical, electrical, chemical, and aerospace engineering. It is assumed that the student has a background in calculus and college physics (mechanics, thermodynamics, and electrical circuits). Any other required material in physics and mathematics (e.g., differential equations, transforms, and matrices) is developed in the text and its appendices.

Teaching, and therefore writing a text in the control systems discipline present a great challenge for several reasons. The control engineer can be called on to develop mathematical models for a variety of applications involving hydraulic, pneumatic, thermal, mechanical, and electrical systems and therefore must be familiar with the modeling techniques appropriate to each area. Here, the challenge is to develop the student's ability to create a model that is sufficiently detailed to capture the dominant dynamic characteristics of the system yet simple enough to be useful for design purposes. Technological developments in the control field have been rapid, as evidenced by the increasing use of digital computers as controllers. Treatment of digital control necessitates covering sampled-data systems but without neglecting the fundamentals of continuous-time system analysis. In addition to controller hardware developments, the field has seen substantial improvement in computational techniques. Formulations using the state space approach provide a good basis for developing general algorithms that are well suited for computer-aided design purposes. On the other hand, the methods of "classical" control theory are still widely used and their utility has been significantly enhanced by modern computational techniques involving computer simulation and computer graphics. Today's engineer must be familiar with the classical methods and the new computational improvements.

We have attempted to satisfy these needs by introducing the required concepts and methods in a balanced and gradual way. First, a unified energy-based approach to modeling is presented. Examples from different fields are given to show the student how to develop relatively low-order models that are adequate for many applications. Next, there is a presentation of analysis techniques for such models. Emphasis is given to physical interpretations of the response patterns exhibited by these models and to the relationship between the form and complexity of the model and the resulting analytical requirements. Digital-simulation techniques are introduced to allow the student to deal with nonlinearities and other complexities that occur in many applications.

The basic principles of feedback control systems are then introduced. Commonly used controller hardware is introduced, and the analysis techniques developed earlier are applied to the design of such systems. Presentation of feedback control theory as merely a mathematical exercise is avoided, and the physical interpretations and practical considerations of the designs are emphasized. An in-depth coverage of a variety of design methods is included. The text concludes with a detailed study of digital control, including analysis of sampled-data systems and discrete-time models and the implementation of digital controllers.

Readers familiar with *Modeling, Analysis, and Control of Dynamic Systems*† will notice some additions and deletions as well as a substantial rearrangement of the material. This was done to better accommodate a particular control systems course sequence that is used at many schools. Most of the differences in existing sequences at various schools can be categorized by the relative emphasis given the following topics:

1. Modeling
2. Prediction and analysis of system response
3. Computer simulation methods
4. "Classical" control system design, emphasizing transfer functions and graphical methods, such as the root locus and frequency response plots
5. Digital control
6. "Modern" analysis and control techniques using matrix methods.

At the University of Rhode Island, the first five topics are covered in a two-semester sequence. The last topic, modern control methods, is covered at the graduate level and therefore not included in this text. This is one of the major differences from the previous text. The other differences are the addition of a new, separate chapter on computer simulation and the consolidation of the material on modeling, analytical techniques, and digital control into distinct chapters. This arrangement better reflects the course sequence offered at many schools.

It is my view that an undergraduate control systems sequence should have the following objectives:

1. To introduce the terminology and the basic principles of modeling, analysis, and feedback control of dynamic systems. A general understanding of these topics is important even for graduates who will not be actively involved in designing control systems, because such systems are encountered in many applications and their characteristics should be understood. This introduction should also prepare students for continuing their education in the controls area, either with self-study by reading the literature or attending short courses or graduate school.
2. To develop in students an ability to design control systems for applications likely to be encountered by a control systems specialist. This objective is important for those with possible career objectives in the controls area.

There is much disagreement in the academic community over how the topic of modern control theory fits into the preceding objectives. Having no delusions about settling this argument, nevertheless, we summarize the views of the classical school of thought on this subject, because it is for this group that the present text has been developed. Modern control theory refers to a collection of techniques developed over

---

†W. J. Palm, III, *Modeling, Analysis, and Control of Dynamic Systems*, John Wiley, New York, 1983.

the last 30 years, which deals with matrix methods, eigen analysis, and applications of such optimization methods as the calculus of variations. In the years since those methods were developed, there have been few significant implementations, and most of them have been in a single applications area, the aerospace industry. In addition, these applications are very complex and therefore not easily covered at the undergraduate level. It is interesting to note that the availability of low-cost, powerful microprocessor controllers has not led to a substantial increase in the applications of modern control theory. Such methods are useful for theoretical developments in control research and such related areas as signal processing. For this reason, these methods should be treated at the graduate level.

Modern control theory was developed primarily to deal with system models of high order. But most control systems applications encountered by a B.S. engineering graduate require controlling a physical system that can be represented by a differential equation of first or second order. With integral control action, the order is increased by one, so that the resulting system model is usually of third order or less. Thus, most applications can be handled with relatively low-order models, and the methods of classical control theory have proved to be well suited for these applications. It is unfortunate that the term *classical* suggests out-of-date, because this is not the case. In fact, classical design methods have been greatly enhanced by the availability and low cost of digital computers for systems analysis and simulation. The graphical tools of classical design can now be more easily used with computer graphics and the effects of nonlinearities and model approximations evaluated by computer simulation. The use of low-cost computers as digital controllers means that some of the limitations imposed on classical designs are eliminated and more flexibility now exists in the choice of the control algorithm. This text has been written in light of these new developments.

This is not to say that the control systems community has learned very little in the last 30 years. The state space and matrix *notation* introduced by modern control theory, as opposed to its *design* methods, has become widespread in the literature because of its ability to represent algorithms concisely. For this reason, many computer-aided design packages use this notation, and therefore it is helpful for control engineers to be familiar with it. Chapter 5, which deals with computer methods, contains an introduction to this notation.

Chapter 1 establishes the viewpoint of dynamic systems analysis and control and its associated modeling requirements. Terminology and an overview of analytical techniques are presented. Zero-order system examples provide a simple introduction to system analysis, because they require only algebraic models. Linearization, system diagrams, and the properties of feedback are introduced in this way.

The general structure of dynamic modeling based on energy principles and integral causality is developed in Chapter 2. Basic principles for modeling simple mechanical, electrical, fluid, and thermal systems are treated. Examples are given that introduce commonly used hardware.

The analysis of system response begins in Chapter 3. First- and second-order models are used as a vehicle for introducing free- and forced-linear system response to the common inputs: the step, ramp, and sine functions. Simple substitution methods are used to obtain the response. This approach, although less "automatic", serves to strengthen students' understanding of the basic response patterns of linear systems. Stability, dominant-root approximations, performance specifications, and linearization of dynamic models are also introduced.

The Laplace transform method for predicting system response is introduced in Chapter 4. This also provides a basis for introducing transfer functions and block diagram algebra. Such related concepts as impedance, frequency response plots, and the initial and final value theorems are presented. Treatment of impulse response, the effects of numerator dynamics, and signal flow graphs conclude the chapter.

Chapter 5 discusses computer simulation methods. The Euler, predictor-corrector, and Runge–Kutta methods are first developed for first-order models and then extended to include higher order cases. The utility of the state variable form is demonstrated, and methods for obtaining this form from the transfer function are given. Practical considerations for simulation are treated next, with a discussion of how linear or linearized analysis can be used to guide the simulation, and the use of step inputs with numerator dynamics. Finally, an introduction to matrix methods and their use in computer-aided design is given.

Chapter 6 presents the basics of feedback control. The common two-position and proportional-integral-derivative types of control laws are analyzed for first- and second-order systems. Typical hardware is described, and the implementation of the control laws with electronic, pneumatic, and hydraulic elements is developed.

A determined effort has been made throughout the text to justify and illustrate the analytical methods in terms of their practical applications. Book length prevents treating design problems in the fine detail needed in practice, and some design considerations, such as economics and reliability, are difficult to cover within the scope of this study. Nevertheless, the discussion and examples should give students as much of a feel for the design process as is possible in an academic environment. To this end, a number of topics that have not been given sufficient emphasis in the past are consolidated into Chapter 7. Methods for selecting the controller gains, and for interpreting the effects of modeling approximations, are among the design issues discussed. All physical elements are power-limited, and design methods are given for taking the associated saturation nonlinearities into account. The classical design approaches of feed-forward, feedback, and cascade compensation are presented, followed by some newer methods, such as state variable feedback, decoupling control, and pseudo-derivative feedback. The latter topic shows that control systems design is an exciting field with much to be discovered.

Control systems design has always relied heavily on graphical methods. Chapter 8 develops the root locus method. Although computer programs are readily available for generating these plots, designers must be able to sketch them first, at least roughly, in order to use the programs efficiently and interpret the results. The material in Chapter 8 is organized in this spirit. The chapter concludes with some examples of how the root locus plot is used for control systems design.

Chapter 9 uses the root locus plot, and other graphical methods—the Bode plot and the Nyquist criterion—to obtain a deeper understanding of control and compensation methods. The important topics of lead and lag compensation are covered in detail.

Chapter 10 introduces digital systems, and develops the necessary background for the analysis of sampled-data systems and discrete-time models. Sampling, quantization, and coding effects in analog-to-digital conversion are discussed. The Laplace transform of a sampled-time function provides a natural way of developing the z transform. This transform is then used to obtain the forced response for common input functions. Sampled-data systems result when a digital-to-analog converter couples a digital device to a continuous-time system, such as in computer control of a mechanical load. The zero-order hold model of this coupling is used with the z transform to develop methods for analyzing this important class of systems. In Chapter 11 digital equivalents of the analog control laws developed in previous chapters are discussed, and some examples are given of control algorithms that take advantage of the unique features of digital control. The concept of digital filtering is introduced, and analyzed with time and frequency domain techniques. The chapter concludes with a detailed development of a computer control system for regulating the speed of a dc motor.

Seven appendices are included for reference information and review purposes to fill in gaps in the reader's mathematical preparation, and for further study of some specialized topics. Appendix A deals with useful analytical techniques, such as the Taylor and Fourier series, matrix analysis, and complex numbers. A self-contained development of the Laplace transform appears in Appendix B. Appendix C gives a useful FORTRAN program implementing the Runge–Kutta integration scheme. Reduction of complicated systems diagrams is facilitated with Mason's rule (Appendix D). Analog simulation is covered in Appendix E. Appendix F presents the Routh–Hurwitz and Jury stability criteria. Appendix G contains tables of physical data, units, and conversion factors.

This text contains ample material for a two-semester course in control systems. The first course can introduce modeling, analysis, and control principles, while the second course is more specialized, dealing with control systems design. The text has been designed to be flexible, and can accommodate differences in courses because of differences in the students' mathematical preparation, previous training in modeling and system dynamics, and availability of computing resources for course work. The following chart shows how the various chapters can be covered, depending on the preceding factors.

For those wishing only a brief introduction or review of physical models, Sections 2.1–2.5 provide the necessary material, including the development of models of important mechanical and electrical control systems hardware. Because fluid and thermal systems are more difficult to model, Sections 2.6–2.9, which cover these systems, can be omitted if a briefer coverage of modeling is required.

Chapter 3 treats solution methods for differential equations and can be covered quickly if the students have had a prior course in this subject. The chapter should not be skipped, however, because it develops terminology and the students' understanding of how these methods are applied in engineering situations.

Chapter 4 introduces Laplace transform methods, to which some students might have had prior exposure in a mathematics course. However, transfer functions and block diagrams are rarely covered in such courses. Since these topics are essential in the succeeding chapters, its coverage should not be omitted.

Chapter 5 covers computer simulation methods. If computer facilities, especially those with xy-plotting capability, are available to the students, this chapter provides a

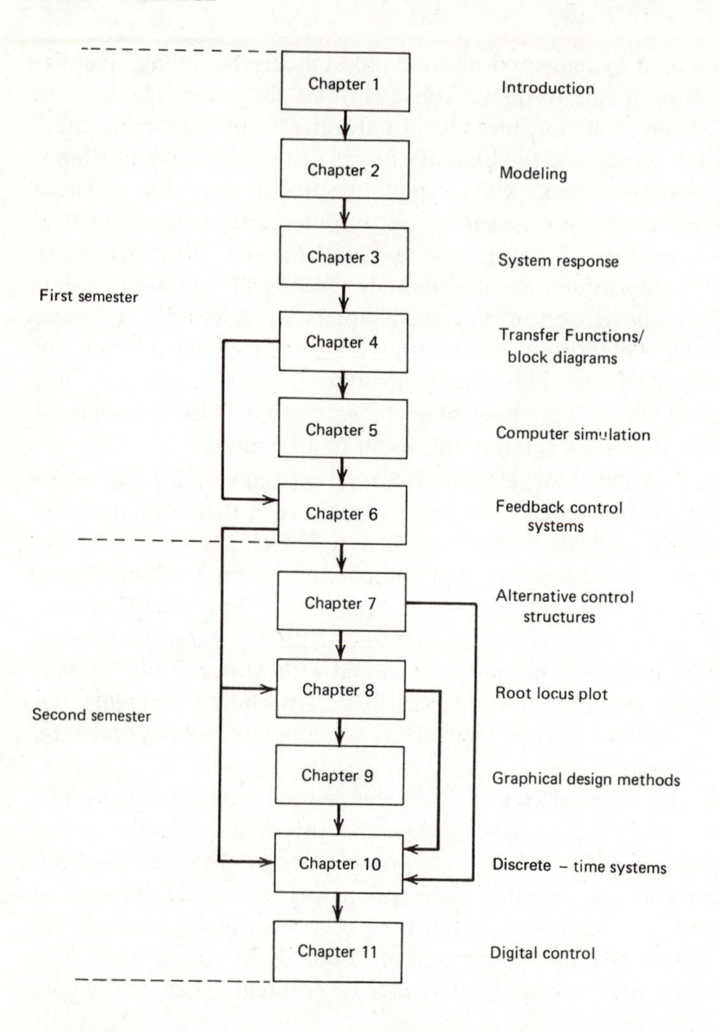

particularly instructive and motivating experience. However, if such facilities are not available, or if time is limited, this chapter can be omitted, because the following chapters do not depend on its results.

Chapter 6 is the basic chapter on feedback control systems and therefore is one of the core chapters that cannot be omitted. Beyond Chapter 6, there are several possible combinations if time is not available to cover all of the remaining chapters. These are shown in the chart shown earlier.

At the University of Rhode Island, second-semester junior students have already had a course in engineering analysis but no modeling course prior to beginning the control systems sequence. The analysis course covers ordinary differential equations, Laplace transforms, and matrices. Therefore, we can treat some of the material in Chapters 3 and 4 by way of a review. The required junior course thus covers the first six chapters; a senior professional elective covers Chapters 7–11. No formal laboratory accompanies the courses. If one is available, the text contains enough information on hardware to serve as a reference for the laboratory sessions.

With the exception of Appendix C, no specific computer or calculator language has been used to present the concepts, and no assumptions have been made concerning the availability of such equipment to the student. However, if it is available, there are many opportunities to use it throughout the text. Programmable calculators can be put to good use in response calculations for evaluating complicated functions, simulating nonlinear systems, and root finding (a good example is programming the solution of a cubic polynomial, since third-order systems are so common). Digital computers are also convenient for the applications mentioned earlier, especially with high-order systems. In addition, they can be used for matrix operations and generating frequency response and root locus plots. Sources for such programs are given at appropriate points in the text.

I am especially grateful to Bill Stenquist of John Wiley & Sons for suggesting this project and for his enthusiasm and support. Thanks are also due to Professors Thomas J. Kim, Frank M. White, and Charles D. Nash, Jr. of the University of Rhode Island, for useful discussions and encouragement. The suggestions of several anonymous reviewers were helpful. Marilyn Buckingham's editing of the manuscript was very thorough, and is much appreciated. My wife, Mary Louise, and our children. Aileen, Billy, and Andy, deserve my special gratitude for their interest and for providing an environment conducive to writing. Finally, I wish to dedicate this work to my parents, for their lifelong support.

Contents

CHAPTER 6
Feedback Control Systems **311**

CHAPTER 7
Control System Design: Modeling Considerations and Alternative Control Structures **389**

CHAPTER 8
The Root Locus Plot **441**

CHAPTER ONE
Introduction

Modeling, analysis, and control of dynamic systems have interested engineers for a long time. Within recent years, the subject has increased in importance for three reasons. Before the invention of the digital computer, calculations required for meaningful applications of the subject were often too time consuming and error prone to be seriously considered. Thus, gross simplifications were made, and only the simplest models of transient behavior were used, if at all. Now, of course, the widespread availability of computers, as well as pocket calculators, allows us to consider more detailed models and more complex algorithms for analysis and design.

Second, with this increased computational power, engineers have correspondingly increased the performance specifications required of their designs to make better use of limited materials and energy, for example, or to improve safety. This leads to the need for more detailed models, especially with regard to predicting transient behavior.

Finally, using computers as system elements for measurement and control now allows more complex algorithms to be employed for data analysis and decision making. For example, intelligent instruments with microprocessors can now calibrate themselves. This increased capability requires a better understanding of dynamic systems so that the full potential of these devices can be realized.

1.1 SYSTEMS

The term *system* has become widely used today, and as a result, its original meaning has been somewhat diluted. *A system is a combination of elements intended to act together to accomplish an objective.* For example, an electrical resistor is an element for impeding the flow of current, and it is usually not considered to be a system in the sense of our definition. However, when it is used in a network with other resistors, capacitors, inductors, etc., it becomes part of a system. Similarly, a car's engine is a system whose elements are the carburetor, the ignition, the crankshaft, etc. On a higher level, the car itself can be regarded as a system with the engine as an element. Since nothing in nature can be completely isolated from everything else, we see that our selection of the "boundaries" of the system depends on the purpose and the limitations of our study. This in part accounts for the widespread use of the term *system*, since almost everything can be considered a system at some level.

The Systems Approach

Automotive engineers interested in analyzing the car's overall performance would not have the need nor the time to study in detail the design of the gear train. They most likely would need to know only its gear ratio. Given this information, they would then consider the gear train as a "black box." This term is used to convey the fact that the details of the gear train are not important to the study (or at least constitute a luxury they cannot afford). They would be satisfied as long as they could compute the torque and speed at the axle, given the torque and speed at the drive shaft.

The black-box concept is essential to what has been called the "systems approach" to problem solving. With this approach, each element in the system is treated as a black box, and the analysis focuses on how connections between the elements influence the overall behavior of the system. Its viewpoint implies a willingness to accept a less detailed description of the operation of the individual elements in order to achieve this overall understanding. This viewpoint can be applied to the study of either artificial or natural systems. It reflects the belief that the behavior of complex systems is made up of basic behavior patterns that are contributed by each element and can be studied one at a time.

The behavior of a black-box element is specified by its *input–output relation.* An *input* is a *cause*; an *output* is an *effect* due to the input. Thus the input–output relation expresses the cause-and-effect behavior of the element. For example, a voltage v applied to a resistor R causes a current i to flow. The input–output or causal relation is $i = v/R$. Its input is v, its output is i, and its input–output relation is the preceding equation.

Block Diagrams

The black-box treatment of an element can be expressed graphically, as shown in Figure 1.1. The box represents the element, the arrow entering the box represents the input, and the one leaving the box stands for the output. Inside the box, we place the mathematical expression that relates the output to the input, if this expression is not too cumbersome and is known. This graphical representation is a *block diagram.*

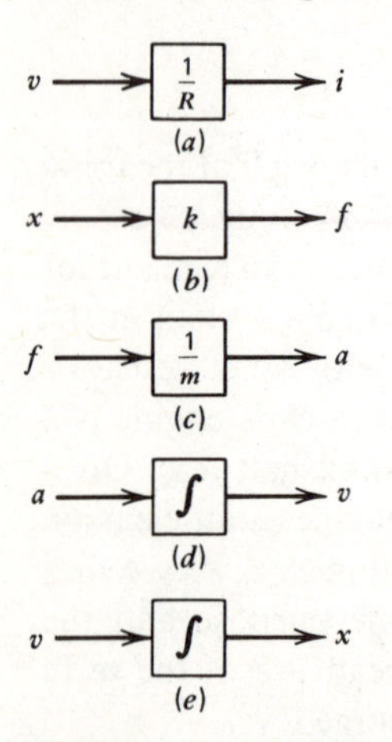

FIGURE 1.1 Block diagrams of input–output relations. (*a*) Voltage–current relation for a resistor. (*b*) Displacement–force relation for a spring. (*c*) Force–acceleration relation for a mass. (*d*) Velocity as the time integral of acceleration. (*e*) Displacement as the time integral of velocity.

The diagram in Figure 1.1*a* represents the resistor with input v and output i. They are related by the constant $1/R$. Figure 1.1*b* shows the representation of a spring whose resisting tensile force f is proportional to its extension x so that $f = kx$. Figure 1.1*c* shows how a force f applied to a mass m causes an acceleration a. The governing relation is Newton's law: $f = ma$. To obtain a from f, we must multiply the input f by the constant $1/m$. Thus, the symbol in the box represents the operation that must be performed on the input to obtain the output.

Not all black-box representations must refer to actual physical elements. Since they express cause-and-effect relations they can be used to display processes as well as components. Two examples of this are shown in figures 1.1*d* and 1.1*e*.

If we integrate the acceleration a over time, we obtain the velocity v, that is $v = \int a\,dt$. Thus, acceleration is the cause of velocity. Similarly, integrating velocity produces displacement x: $x = \int v\,dt$. The integration operator within each box in Figures 1.1*d* and 1.1*e* expresses these facts. Whenever an output is the time integral of the input, the element is said to exhibit *integral causality*. We will see that integral causality constitutes a basic form of causality for all physical systems.

The input–output relations for each element provide a means of specifying the connections between the elements. When connected together to form a system, the inputs to some elements will be the outputs from other elements. For example, the position of a speedometer needle is caused by the car's speed. Thus, for the speedometer element, the car's speed is an input. However, the speed is the result of action of the drive train element. The input–output relation can sometimes be reversed for an element, but not always. We can apply a current as input to a resistor and consider the voltage drop to be the output. On the other hand, the position of the speedometer needle can in no way physically influence the speed of the car.

The system itself can have inputs and outputs. These are determined by the selection of the system's boundary. Any causes acting on the system from the world external to this boundary are considered to be system inputs. Similarly, a system's outputs can be the outputs from any one or more of the elements, viewed in particular from outside the system's boundary. If we take the car engine to be the system, a system input would be the throttle position determined by the acceleration pedal, and a system output would be the torque delivered to the drive shaft. If the car is taken to be the system instead, the input would still be the pedal position, but the outputs might be taken to be the car's position, velocity, and acceleration. Usually, our choices for system outputs are a subset of the possible outputs, and are the variables in which we are interested. For example, a performance analysis of the car would normally focus on the acceleration or velocity, but not on the car's position.

A simple example of a system diagram is provided in Figure 1.2. Suppose a mass m is connected to one end of a spring. The other end of the spring is attached to a rigid support. In addition to the spring force f_s, another force f_o acts on the mass. This force is considered to be due to the external world and acts across the system "boundary"; that is, it is not generated by any action within the system itself. It might be due to gravity, for example.

The cause-and-effect relations can be summarized by the system diagram in Figure 1.2*b*. The net force on the mass in the direction of positive displacement x is

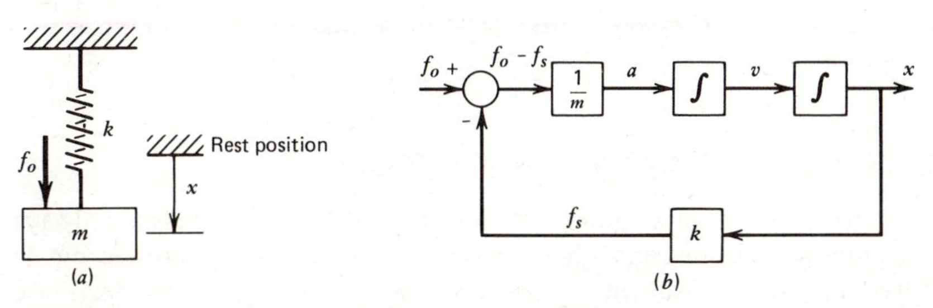

FIGURE 1.2 **(*a*) Mass-spring system with an external force f_o. (*b*) Block diagram of the causal relations.**

$f_o - f_s$, since the spring will pull up on the mass if the mass position is below the rest position ($x > 0$). The addition and subtraction of forces to produce the net force are represented by a new symbol, the *comparator*—a circle whose output is the signed sum of the inputs. The plus (+) sign indicates that f_o is to be added; the minus (−) sign indicates that f_s is to be subtracted to produce the net force.

The system input is f_o; its output could be any or all of the variables generated within the diagram. If we are interested in only the displacement x, then its arrow is shown leaving the system.

System diagrams such as this are a visually concise summary of the interplay between the causes and the effects. We will use them often.

Static and Dynamic Systems

In general, the present value of an element's output is the result of what has happened to the element in the past as well as what is currently affecting it. For example, the present position of a car depends on where it started and what its velocity has been from the start. We define a *dynamic* element to be one whose present output depends on past inputs. Conversely, a *static* element is one whose output at any given time depends on only the input at that time.

For the car considered as an element with an acceleration pedal position as input and car position as output, the preceding definition shows the element to be dynamic. On the other hand, we can consider a resistor to be a static element, because its present current depends on only the voltage applied at present, not on past voltages. This is an approximation, of course, because the resistor cannot respond instantaneously to voltage changes. This is true of all physical elements, and we therefore conclude that a static element is an approximation. Nevertheless, it is widely used, because it results in a simpler mathematical representation.

In popular usage, the terms *static* and *dynamic* distinguish situations in which no change occurs from those that are subject to changes over time. This usage conforms to the preceding definitions of these terms if the proper interpretation is made. A static element's output can change with time only if the input changes and will not change if the input is constant or absent. However, if the input is constant or removed from a dynamic element, its output can still change. For example, if the car's engine is turned off, the car's position will continue to change because of the car's velocity (because of past inputs). A similar statement cannot be made for the electrical resistor.

In the same way, we also speak of static and dynamic systems. A static system contains all static elements. Any system that contains at least one dynamic element is a dynamic system.

1.2 MODELING, ANALYSIS, AND CONTROL

We live in a universe that is undergoing continual change. This change is not always apparent if its time scale is long enough, such as with some geologic processes, but as engineers, we must often deal with situations where time-dependent effects are important. For example, designing high-speed production machinery for precise operation requires the vibrational motions due to high accelerations to be small in amplitude and die out quickly. Likewise, time-dependent behavior of fluid flow and heat transfer processes significantly affects the quality of the product of a chemical process. Even a relatively "stationary" object, such as a bridge, must be designed to accommodate the motions and forces produced by a heavy rolling load.

Modeling

In order to deal in a systematic and efficient way with problems involving time-dependent behavior, we must have a description of the objects or processes involved. We call such a description a *model.* A model for enhancing our understanding of the problem can take several forms. A physical model, like a scale model, helps us to visualize how the components of the design fit together and can provide insight not obtainable from a blueprint (which is another model form). Graphs or plots are still another type of model. They can often present time-dependent behavior in a concise way, and for that reason, we will rely on them heavily throughout this study. The model type we will use most frequently is the *mathematical model,* which is a description in terms of mathematical relations. These relations will consist of differential or difference equations if the model is to describe a dynamic system.

The concept of a mathematical model is undoubtedly familiar from elementary physics. Common examples include the voltage–current relation for a resistor $v = iR$, and the force–deflection relation for a spring $f = kx$. One of our aims here is to introduce a framework that allows the development of mathematical models for describing the time-dependent behavior of many types of phenomena: fluid flow, thermal processes, mechanical elements, and electrical systems, as well as some nonphysical applications. In this regard, it is important to remember that the precise nature of a mathematical model depends on its purpose. For example, an electrical resistor can be subjected to mechanical deformations if its mounting board is subjected to vibration. In this case, the force-deflection spring model could be used to describe the resistor's mechanical behavior.

Thus, we see that the nature of any object has many facets: thermal, mechanical, electrical, and so forth. No mathematical model can deal with all these facets. Even if it could, it would be useless, because its very complexity would render it cumbersome. We can draw an analogy from maps. A given region can be described by a road map, a terrain elevation map, a mineral resources map, a population density map, and so on. A single map containing all this information would be cluttered and therefore useless. Instead, we select the particular type of map required for the purpose at hand. In the same way, we select or construct a mathematical model to suit the requirements of the study.

The purpose of the model should guide the selection of the model's time scale, its length scale, and the particular facet of the object's nature to be described (thermal, mechanical, electrical, etc.). The time scale will in turn determine whether or not time-dependent effects should be included. (Tectonic plate motion constitutes time-dependent behavior on a geologic time scale but would not be considered by an engineer designing a bridge.) Similarly, the length scale partly dictates what details should or should not be included. The engineer analyzing the dynamics of high-speed machinery might treat a component as a point mass, whereas this approximation is useless to a metallurgist studying material properties at the molecular level.

Block diagrams are often used to display the mathematical model in a form that allows us to understand the interactions occurring between the system's elements. For example, the mathematical model of the mass–spring system shown in Figure 1.2*a* is

$$f_o - f_s = ma$$
$$f_s = kx$$
$$x = \int v\,dt$$
$$v = \int a\,dt$$

Each equation expresses a cause-and-effect relationship for one part of the system.

The diagram is a valuable aid, but if we wish to solve for the displacement $x(t)$, we need the model in equation form. If we differentiate the last two relations we obtain

$$\frac{dx}{dt} = v \qquad \frac{dv}{dt} = a$$

or

$$a = \frac{d^2x}{dt^2}$$

Substituting this and the second relation into the first gives

$$f_o - kx = m\frac{d^2x}{dt^2} \tag{1.2-1}$$

This differential equation is a quantitative description of the system. Given f_o, k, m, and the initial position $x(0)$ and velocity $v(0)$, we can use the methods of later chapters to solve the equation for $x(t)$.

Analysis

A mathematical model represents a concise statement of our hypotheses concerning the behavior of the system under study. We can deal with the verification of the model in two ways. Verification by experiment or testing is ultimately required of all serious design projects. This is not always done at the outset of a study, however, especially if we are dealing with component types whose behavior is known to be well described by a specific model on the basis of past experience. For example, we are on firm ground in using the resistor model $v = iR$ without verification as long as the operating conditions (voltage levels, temperatures, etc.) are not extreme. This is often the case for the types of problems we will be considering, since we will be concerned with the behavior of systems consisting of components whose individual behavior is often well understood.

Once we are satisfied with the validity of our chosen component models, they can be used to predict the performance of the system in question. Predicting the performance from a model is called *analysis*. For example, the current produced in a resistor by an applied voltage v can be predicted to be $i = v/R$ by solving the resistor model for the unknown variable i in terms of the given quantities v and R. Most of our mathematical models will describe dynamic behavior and will thus consist of differential or difference equations. They will not be so easy to solve as the algebraic model just seen. Nevertheless, the techniques for analyzing such models are straightforward. These are introduced in Chapter 3.

Just as the model's purpose partly determines its form, so also does the purpose influence the types of analytical techniques used to predict the system's behavior. It is not possible to discuss these concepts with the simple resistor model, because its solution is so simple. However, we will be developing many types of analytical techniques whose applicability depends on the purpose of the analysis. Not all of these techniques will be brought to bear on any one problem, but the engineer should be familiar with all of them. They are the tools of the trade—a means to an end. We will not study them, as a mathematician would, for their inherent interest. Instead, we will focus on how they can help us predict the performance of a proposed design before it is built. Thus, we can avoid a trial-and-error approach. This is especially important today, since most modern engineering endeavors are too complex and expensive to allow them to be built without a thorough analysis beforehand.

Control

The successful operation of a system under changing conditions often requires a control system. For example, a building's heating system requires a thermostat to turn the heating elements on or off as the room temperature rises and falls (Figure 1.3). Note that we have not shown the thermostat as one element, because it has two functions: (1) to measure the room temperature and compare it with the desired temperature and (2) to decide whether to turn the furnace on or off. The variation in the outdoor environment is the primary reason for the unpredictable change in the room temperature. If the outside conditions (temperature, wind, solar insolation, etc.) were predictable, we could design a heater that would operate continuously to supply heat at a predetermined rate just large enough to replace the heat lost to the outside environment. No controller would be necessary. Of course, the real world does not behave so nicely, so we must adjust the heat output rate of our system according to what the actual room temperature is.

We often wish to alter the desired operating conditions of a system. In our heating example, the thermostat allows the user to specify the desired room temperature, say, 68°F during the day and 60°F at night. When the thermostat setting is changed from 60°F to 68°F in the morning, the thermostat acts to bring the room

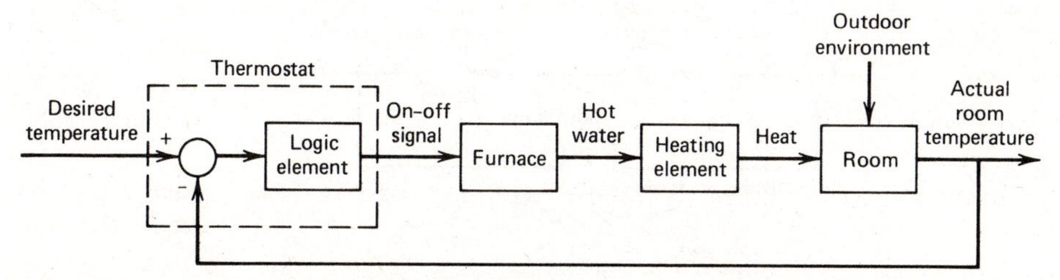

FIGURE 1.3 Block diagram of the thermostat system for temperature control.

temperature up to 68°F and to keep it near this value until the setting is changed again at night.

The term *control* refers to the process of deliberately influencing the behavior of an object in order to produce some desired result. The physical device inserted for this purpose is the *controller* or *control system.* Other common examples of controllers include:

1. An aircraft autopilot for maintaining desired altitude, orientation, and speed.
2. An automatic cruise control system for a car.
3. A pressure regulator for keeping constant pressure in a water supply system.

A cutaway view of a commonly used type of pressure regulator is shown in Figure 1.4 along with a block diagram of its operation. The desired pressure is set by turning a calibrated screw. This compresses the spring and sets up a force that opposes the upward motion of the diaphragm. The bottom side of the diaphragm is exposed to the water pressure that is to be controlled. Thus, the motion of the diaphragm is an indication of the pressure difference between the desired and the actual pressures. It acts like a comparator. The valve is connected to the diaphragm and moves according to the pressure difference until it reaches a position in which the difference is zero.

From the preceding examples, we see that the role of a controller is twofold.

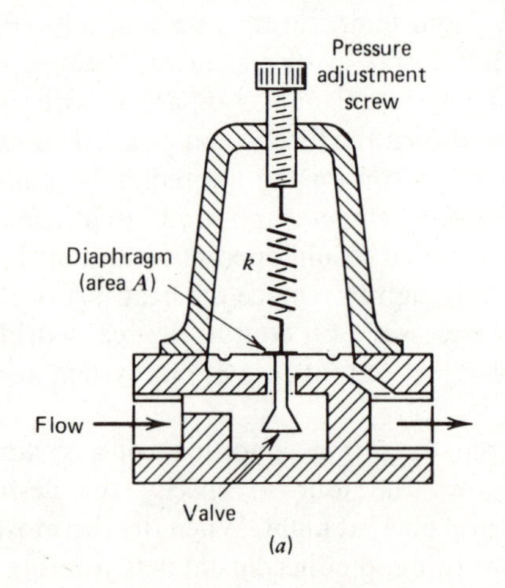

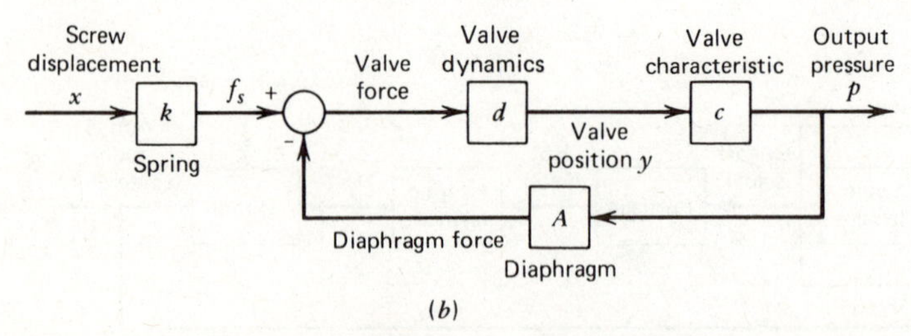

FIGURE 1.4 Pressure regulator. (*a*) Cutaway view. (*b*) Block diagram.

1. It must bring the system's operating condition to the desired value.
2. It must maintain the desired condition in the presence of variations caused by the external environment.

In the terminology of the control engineer, we say that the controller must respond satisfactorily to changes in *commands* and maintain system performance in the presence of *disturbances.*

One or more controllers are often required in complex dynamic systems in order to make the system elements act together to achieve the intended goal. So the design of dynamic systems quite naturally involves the study of control systems. On the other hand, the variations produced by command changes and disturbances tend to upset the system. Thus, control system design requires models that describe the dominant dynamic properties of the system to be controlled, and the analysis techniques must be capable of dealing with such a model. Modeling, analysis, and control of dynamic systems therefore constitute a unified area of study.

1.3 TYPES OF MODELS

As we have seen, a system model is a representation of the essential behavior of the system for the purposes at hand. In order to be useful, it must contain the minimum amount of information necessary to achieve its purpose and no more. This requirement is most immediately reflected in the choice of static- versus dynamic-element models. Those elements whose behavior is fast relative to other elements are often modeled as static elements in order to reduce the complexity of the model. For example, the switching time of a thermostat is fast compared to the time required for the room temperature to change appreciably. Thus, the room temperature in Figure 1.2 would most likely be modeled as a dynamic element and the thermostat as a static one.

Lumped- and Distributed-Parameter Models

We have implicitly assumed in the heating example that the temperature in the room can be described by a single number, a temperature that is average in some sense. In reality, the temperature varies according to location within the room, but if we did not choose to use a single representative temperature, the required room model would be much more complicated.

Many variables in nature are functions of location as well as time. The process of ignoring the spatial dependence by choosing a single representative value is called *lumping* (room air is considered to be one "lump" with a single temperature).

Lumping an element is a technique usually requiring experience. It reflects the judgment of the engineer about what is unimportant in terms of spatial variation. It can be described as the spatial equivalent of the process of dividing a system into static and dynamic elements. The model of a lumped element or system is called a *lumped-parameter model.* If it is dynamic, the only independent variable in the model will be time; that is, the model will be an ordinary differential equation, such as (1.2-1). Only time derivatives will appear, not spatial derivatives.

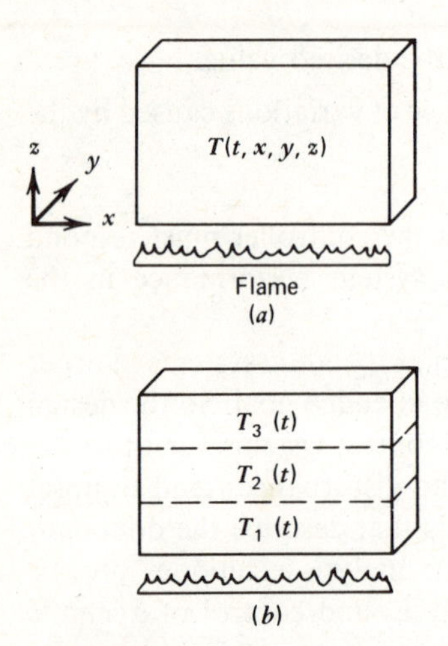

FIGURE 1.5 Temperature distribution in a plate. (*a*) Distributed-parameter representation. (*b*) Lumped-parameter representation using three elements.

When spatial dependence is included, the independent variables are the spatial coordinates as well as time. The resulting model is said to be a *distributed-parameter model.* It consists of one or more partial differential equations containing partial derivatives with respect to the independent variables. The difference is illustrated in Figure 1.5*a*, which shows the temperature T of a metal plate. If the plate is heated at one side, the temperature will be a function of location and time—$T = T(t, x, y, z)$—and the model will have the form

$$f\left(T, \frac{\partial T}{\partial t}, \frac{\partial^2 T}{\partial x^2}, \frac{\partial^2 T}{\partial y^2}, \frac{\partial^2 T}{\partial z^2}\right) = 0 \tag{1.3-1}$$

But if the plate temperature is lumped with a single value, the model will have the form

$$f\left(T, \frac{dT}{dt}\right) = 0 \tag{1.3-2}$$

which is mathematically easier to handle.

Lumping may be done at several levels. For example, we may take a single temperature to represent the entire house. In this case, a single differential equation would result. On the other hand, we may take a representative temperature for each room. In this case, the total model would consist of a differential equation for each room temperature.

Although choosing a temperature for each room leads to several equations, the model is usually more manageable than if the lumping were not performed.

There are applications in engineering where a detailed model like (1.3-1) is required, and we do not dismiss such models as useless. However, we will see that in analyzing systems with many elements, distributed-parameter models of elements are a luxury we usually cannot afford, because their complexity tends to prevent us from understanding the overall system behavior. We will therefore limit our treatment to lumped-parameter models. Note that if a more detailed model is needed, we can increase the number of lumped elements, as is shown in Figure 1.5*b*, where we have used three temperature lumps in an attempt to model the dependence of the plate temperature as a function of distance from the flame. The resulting model would have three coupled ordinary differential equations, one for each $T_i(t)$.

Linear and Nonlinear Models

We have seen that engineers should attempt to model elements as static rather than dynamic and as lumped rather than distributed. The reason is that engineers eventually have to analyze the resulting system model, and its complexity can easily get out of hand if there is too much detail in each element model. In a similar vein, we now discuss the distinction between model types based on *linearity*.

Let y be the output and x the input of an element that can be either static or dynamic. Its model is written as

$$y = f(x) \tag{1.3-3}$$

where the function $f(x)$ may include such operations as differentiation and integration. The model (or element) is said to be *linear* if, for an input $ax_1 + bx_2$, the output is

$$y = f(ax_1 + bx_2) = af(x_1) + bf(x_2) = ay_1 + by_2 \tag{1.3-4}$$

where a and b are arbitrary constants, x_1 and x_2 are arbitrary inputs, and

$$y_1 = f(x_1) \tag{1.3-5}$$

$$y_2 = f(x_2) \tag{1.3-6}$$

Thus, linearity implies that multiplicative constants and additive operations in the input can be factored out when considering the effects on the output. The linearity property (1.3-4) is sometimes called the *superposition principle*, because it states that a linear combination of inputs produces an output that is the superposition (linear combination) of the outputs that would be produced if each input term were applied separately. Any relation not satisfying (1.3-4) is nonlinear.

Let us consider some input–output relations to see if they are linear. The simple multiplicative relation $y = mx$ is linear, because

$$y = m(ax_1 + bx_2) = amx_1 + bmx_2 = ay_1 + by_2$$

where $y_1 = mx_1$ and $y_2 = mx_2$. The operation of differentiation, $y = dx/dt$, is linear, because

$$y = \frac{d}{dt}(ax_1 + bx_2) = a\frac{dx_1}{dt} + b\frac{dx_2}{dt} = ay_1 + by_2$$

Similarly, integration is a linear operation. If $y = \int x\, dt$, then

$$y = \int (ax_1 + bx_2)\, dt = a\int x_1\, dt + b\int x_2\, dt = ay_1 + by_2$$

Any relation involving a transcendental function or a power other than unity is nonlinear. For example, if $y = x^2$,

$$y = (ax_1 + bx_2)^2 = a^2x_1^2 + 2abx_1x_2 + b^2x_2^2 \neq ax_1^2 + bx_2^2$$

Similarly, if $y = \sin x$,

$$y = \sin(ax_1 + bx_2) \neq a \sin x_1 + b \sin x_2$$

The definition of linearity (1.3-4) can be extended to include functions of more than one variable, such as $f(x, z)$. This function is linear if and only if

$$f(ax_1 + bx_2, az_1 + bz_2) = af(x_1, z_1) + bf(x_2, z_2)$$

Differential equations represent input–output relations also and can be classified as linear or nonlinear. The outputs (solutions) of the model depend on the outputs' initial values and on the inputs. We will see in Chapter 3 that a superposition principle applies to linear differential equations. This is useful, because it allows us to separate the effects of more than one input and thus to consider each input one at a time. It also allows us to separate the effects of the initial values of the outputs from the effects of the inputs. For these reasons, we will always attempt to obtain a linear model for our systems provided that any approximations required to do so will not mask important features of the system's behavior.

A differential equation is easily recognized as nonlinear if it contains powers or transcendental functions of the dependent variable. For example, the following equation is nonlinear:

$$\frac{dy}{dt} = -\sqrt{y} + f$$

Time-Variant Models

The presence of a time-varying coefficient does not make a model nonlinear. For example, the model

$$\frac{dy}{dt} = c(t)y + f$$

is linear. Models with constant coefficients are called *time-invariant* or *stationary* models, while those with variable coefficients are *time variant* or *nonstationary*. As an example, suppose the mass m in Figure 1.1 represents a bucket of water with a leak. Its mass would then change with time, and $m = m(t)$ in (1.2-1).

Discrete- and Continuous-Time Models

Sometimes it is inconvenient to view the system's dynamics in terms of a continuous-time variable. In such cases, we use a discrete variable to measure time. Common examples of this practice include a person's age (we usually express it in integer years, with no fractions) and interest computations on savings accounts (compounded quarterly, annually, etc.). For engineers, the most important situation suggesting the use of discrete-time models occurs when a system contains a digital computer for measurement or control purposes. It is an inherently discrete-time device, because it is driven by an internal clock that allows activity to take place at only fixed intervals. Thus, a digital computer cannot take measurements continuously but must "sample" the measured variable at these instants.

If we choose to represent our system in terms of discrete time, the form of the model is a difference equation instead of a differential equation. For example, an amount of money x in a savings account drawing 5% interest compounded annually will grow according to the relation

$$x(k+1) = 1.05x(k) \tag{1.3-7}$$

The index k represents the number of years after the start of the investment. We will return to the analysis of discrete-time models in Chapter 10.

Model Order

Equation (1.2-1) is called a *second-order* differential equation, because its highest derivative is second order. It is equivalent to the relations

$$m\frac{dv}{dt} = f_o - kx \tag{1.3-8}$$

$$\frac{dx}{dt} = v \tag{1.3-9}$$

These two first-order equations are coupled to each other because of the x term in the first equation and the v term in the second. One cannot be solved without solving the other; they must be solved simultaneously. Taken together, they thus form a second-order model.

We will organize our study of dynamic systems partly according to the order of the model. In this way, we can begin simply and gradually progress to more difficult topics.

Stochastic Models

Sometimes, there is uncertainty in the values of the model's coefficients or inputs. If this uncertainty is great enough, it might justify using a *stochastic* model. In such a model, the coefficients and inputs are described in terms of probability distributions involving, for example, their means and variances. Such a model is useful for describing the effects of wind gusts on an aircraft autopilot. Although the wind is not random, presumably our knowledge of its behavior is poor enough to justify a probabilistic approach. However, the mathematics required to analyze such models is beyond the scope of this work, and we will not consider stochastic models further.

A Model Classification Tree

Figure 1.6 is a diagram of the relationship between the various model types. We have extended only the branches that lead to linear time-invariant models, since this is the type of most interest to us.

1.4 LINEARIZATION

Because of the usefulness of the superposition principle, we always attempt to obtain a linear model if possible. Sometimes, this can be done from the outset by neglecting effects that would lead to a nonlinear model. A common example of this is the small angle approximation. If we assume that the angle of rotation θ of the lever in Figure 1.7 is small, the rectilinear displacement of its ends is roughly proportional to θ such that $x = L\theta$. The same is not true for a large enough value of θ.

If such an approximation is not obvious, a systematic procedure based on the Taylor series expansion can be used (Appendix A). Let the input–output model for a static element be written as

$$w = f(y) \tag{1.4-1}$$

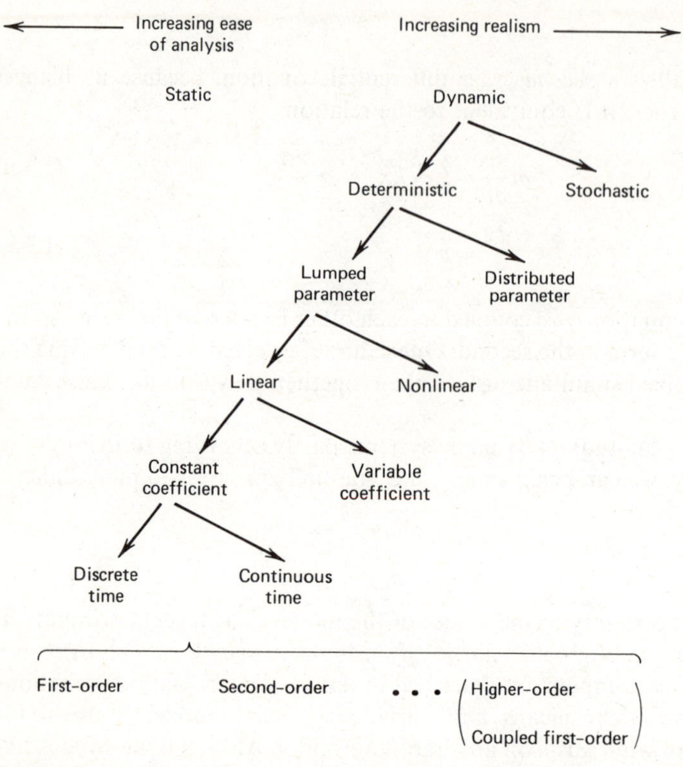

FIGURE 1.6 Classification of mathematical models. We have not continued the tree on every branch for simplicity. For example, nonlinear models can also be classified as discrete or continuous time.

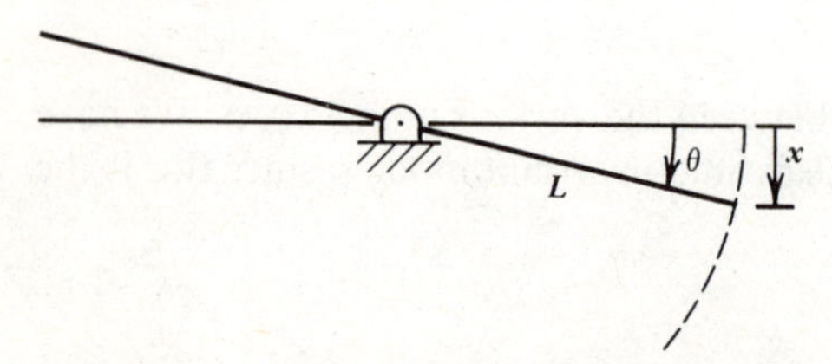

FIGURE 1.7 Small-angle approximation for the displacement of a lever endpoint. For θ small, $x \cong L\theta$.

Its form is sketched in a general way in Figure 1.8. A model that is approximately linear near the reference point (w_o, y_o) can be obtained by expanding $f(y)$ in a Taylor series near this point and truncating the series beyond the first-order term. The series is

$$w = f(y) = f(y_o) + \left(\frac{df}{dy}\right)_o (y - y_o) + \frac{1}{2!}\left(\frac{d^2f}{dy^2}\right)_o (y - y_o)^2 + \cdots \qquad \textbf{(1.4-2)}$$

where the subscript o on the derivatives means that they are evaluated at the reference

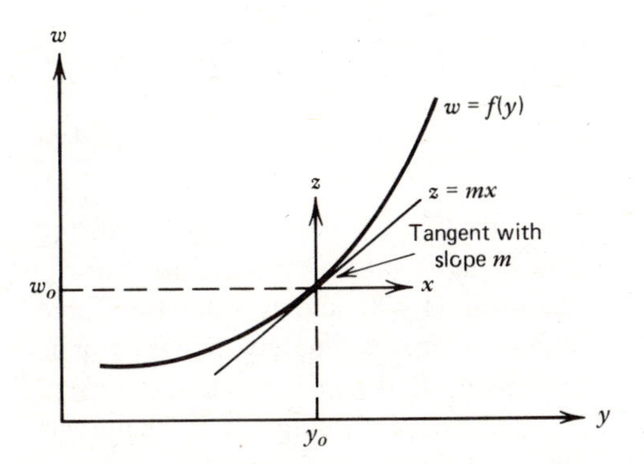

FIGURE 1.8 Linearization of the function $w = f(y)$ about the point (w_o, y_o).

point (w_o, y_o). If y is "close enough" to y_o, the terms involving $(y - y_o)^i$ for $i \geqslant 2$ are small compared to the first two terms. Ignoring these higher order terms gives

$$w = f(y) \cong f(y_o) + \left(\frac{df}{dy}\right)_o (y - y_o) \tag{1.4-3}$$

This is a linear relation. To put it into a simpler form, let

$$m = \left(\frac{df}{dy}\right)_o \tag{1.4-4}$$

$$z = w - w_o = w - f(y_o) \tag{1.4-5}$$

$$x = y - y_o \tag{1.4-6}$$

Then (1.4-3) becomes

$$z \cong mx \tag{1.4-7}$$

The geometric interpretation of this result is shown in Figure 1.8. We have replaced the original function with a straight line passing through the point (w_o, y_o) and having a slope equal to the slope of $f(y)$ at the reference point. With the (z, x) coordinates, a zero intercept occurs, and the relation is simplified.

A Nonlinear Spring Example

No spring is linear over an arbitrary range of extensions. Instead, the force will increase nonlinearly with extension beyond some point, and the linear model used to obtain (1.2-1) will no longer be valid. Suppose the correct relation for a particular spring is $f = y^2$, where y is the extension of the spring from its free length (Figure 1.9). Let it be attached to the mass as shown in Figure 1.2, and allow the mass to settle to its rest position y_o. At this position, the weight of the mass will equal the spring force, so that $mg = y_o^2$, or $y_o = \sqrt{mg}$. The Taylor series applied to the spring relation $f = y^2$ gives

$$\begin{aligned} f \cong y^2 &= y_o^2 + \left(\frac{dy^2}{dy}\right)_o (y - y_o) \\ &= y_o^2 + 2y_o(y - y_o) \end{aligned}$$

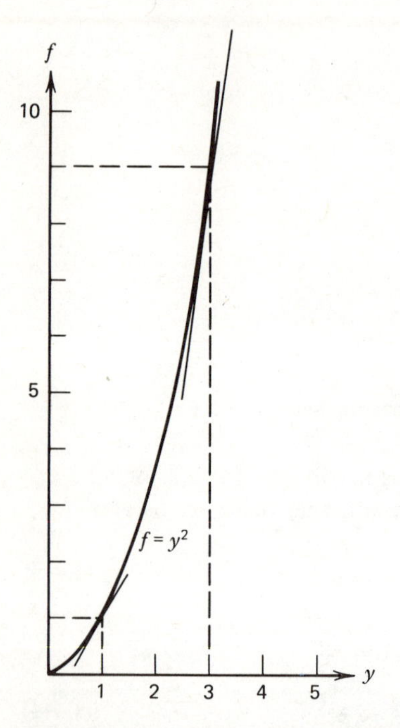

FIGURE 1.9 Linearization of the function $f=y^2$ about the points (1, 1) and (9, 3). Note the different slopes for each linearization.

Let $z=f-f_o=f-y_o^2$ and $x=y-y_o$. Thus,

$$z \cong 2y_o x \tag{1.4-8}$$

The variable z is the change in spring force from its value at the rest position y_o. Equation (1.4-8) shows that the force change is approximately linear if y is close to y_o; that is, if x is small.

The free-body diagram of the mass is shown in Figure 1.10, with the gravity force and spring force the only forces applied to the mass. From Newton's law,

$$m\frac{d^2x}{dt^2} = mg - (f_o + z)$$

Using (1.4-8) and the fact that $mg = y_o^2 = f_o$, we obtain

$$m\frac{d^2x}{dt^2} = -2y_o x \tag{1.4-9}$$

which is linear since y_o is constant If we had not approximated the spring force as a linear function, the system's differential equation would have been

$$m\frac{d^2y}{dt^2} = mg - y^2 \tag{1.4-10}$$

which is nonlinear because of the y^2 term.

From (1.4-8), we see that the "spring constant" k is $2y_o$. It depends on not only the spring's physical properties, which yield the constant 2, but also on the reference position y_o. The constant $2y_o$ is the slope of the force-extension curve of the spring at the position y_o. This position is determined by the spring's characteristics and the weight of the mass. If $mg=1$, then $y_o=1$ and $k=2$. For a larger weight, say $mg=9$, $y_o=3$ and $k=6$.

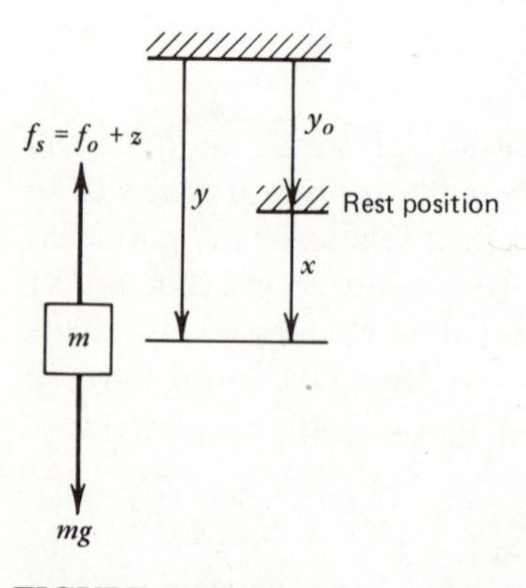

FIGURE 1.10 Free-body diagram of the mass-spring system.

The Multivariable Case

The Taylor series linearization technique can be extended to any number of variables. For two variables, the function is

$$w = f(y_1, y_2) \tag{1.4-11}$$

and the truncated series is

$$w \cong f(y_{1o}, y_{2o}) + \left(\frac{\partial f}{\partial y_1}\right)_o (y_1 - y_{1o}) + \left(\frac{\partial f}{\partial y_2}\right)_o (y_2 - y_{2o}) \tag{1.4-12}$$

Define

$$z = w - w_o = w - f(y_{1o}, y_{2o}) \tag{1.4-13}$$

$$x_1 = y_1 - y_{1o} \tag{1.4-14}$$

$$x_2 = y_2 - y_{2o} \tag{1.4-15}$$

The linearized approximation is

$$z \cong \left(\frac{\partial f}{\partial y_1}\right)_o x_1 + \left(\frac{\partial f}{\partial y_2}\right)_o x_2 \tag{1.4-16}$$

The partial derivatives are the slopes of the function f in the y_1 and y_2 directions at the reference point.

As an example, consider the perfect gas law

$$p = \frac{mRT}{V} \tag{1.4-17}$$

where p, V, T, and m are the gas pressure, volume, temperature, and mass, respectively. The universal gas constant is R. If the gas is isolated in a flexible chamber, its mass is constant, but its volume, temperature, and pressure can change. For given reference values T_o and V_o, a linearized expression for the pressure is

$$p \cong p_o + \left(\frac{\partial p}{\partial T}\right)_o (T - T_o) + \left(\frac{\partial p}{\partial V}\right)_o (V - V_o) \tag{1.4-18}$$

where $p_o = mRT_o/V_o$ and

$$\left(\frac{\partial p}{\partial T}\right)_o = \left(\frac{mR}{V}\right)_o = \frac{mR}{V_o} = a \tag{1.4-19}$$

$$\left(\frac{\partial p}{\partial V}\right)_o = \left(-\frac{mRT}{V^2}\right)_o = -\frac{mRT_o}{V_o^2} = -b \tag{1.4-20}$$

Sometimes, the following notation is used to represent variations from the reference values:

$$\delta p = p - p_o$$

$$\delta T = T - T_o$$

$$\delta V = V - V_o$$

In this case, (1.4-18)–(1.4-20) give

$$\delta p \cong a\,\delta T - b\,\delta V \tag{1.4-21}$$

Since a and b are positive, (1.4-21) shows that the pressure increases if T increases or V decreases.

Of course, we do not need the linearized form to calculate p given V, T, m, and R. However, if p were to appear in a differential equation with T or V as inputs or dependent variables, then the linearized form would be needed to obtain a linear differential equation.

Linearization of Operating Curves

An element's input–output relation is not always given in analytical form but might be available as an experimentally determined plot. For example, the operating curves of an electric motor might look something like those in Figure 1.11. For a fixed motor voltage v_2 and a fixed load torque T_2, the motor will eventually reach a fixed speed ω_2 some time after the motor is started. This steady-state speed can be found from the curve marked v_2. For another load torque, say, T_a, the resulting speed ω_a can be found from the same v_2 curve. However, if we fix the load torque but change the voltage, the resulting speed must be found by interpolating between the appropriate constant-voltage curves.

Suppose that the load torque T and the voltage v are inputs for a dynamic model that will have the speed ω as the output. Then we cannot use the curves as they are but must convert them to an analytical expression. This expression must be linear if the dynamic model is to be linear.

A linearized expression can be obtained from the operating curves by using (1.4-16) and calculating the required slopes numerically from the plot. The partial derivative $(\partial f/\partial y_1)_o$ is computed with y_2 held constant at the value y_{2_o}. It can be found

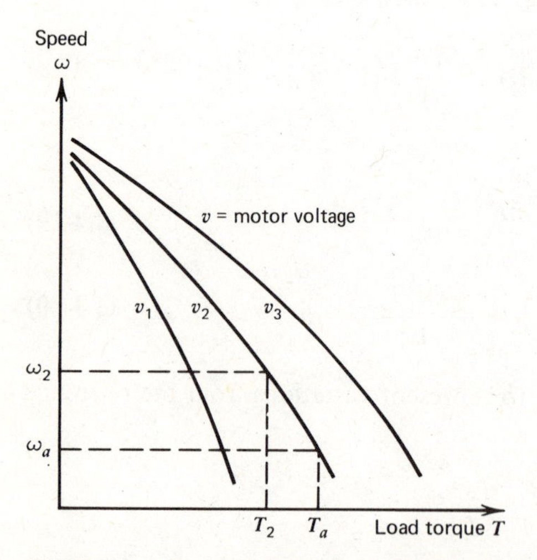

FIGURE 1.11 Steady-state operating curves for an electric motor.

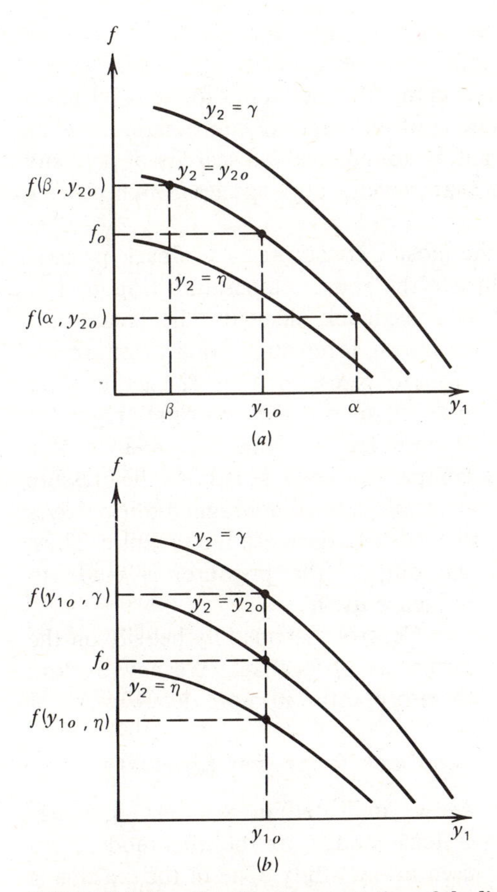

FIGURE 1.12 Graphical computation of the linearization derivatives for the function $f(y_1, y_2)$ near the point $f_o = f(y_{1o}, y_{2o})$.

approximately as follows (see Figure 1.12):

$$\left(\frac{\partial f}{\partial y_1}\right)_o \cong \frac{f(\alpha, y_{2o}) - f(\beta, y_{2o})}{\alpha - \beta} \tag{1.4-22}$$

where the y_1 range (β, α) straddles y_{1o}. Similarly,

$$\left(\frac{\partial f}{\partial y_2}\right)_o \cong \frac{f(y_{1o}, \gamma) - f(y_{1o}, \eta)}{\gamma - \eta} \tag{1.4-23}$$

The smaller $(\alpha - \beta)$ and $(\gamma - \eta)$ are made, the better the approximation, but this is limited by the ability to read the plot accurately for smaller increments.

1.5 FEEDBACK

A feature found in many static systems and in almost every dynamic system is one or more *feedback loops*, such as shown in Figures 1.2–1.4. *Feedback* is the process by which an element's input is altered by its output. It occurs frequently in physiological and ecological systems and is deliberately employed in engineering systems for several reasons, as we will see. The human body uses many feedback loops for such purposes as body temperature control, blood pressure control, hand-eye coordination, and so forth. In nature, we can describe the alternating abundance and scarcity of prey and predators as due to a feedback mechanism that prevents both species from becoming extinct.

Measuring room temperature by a thermostat constitutes a feedback process, because the measurement is used to influence the room temperature (Figure 1.3). Similarly, the spring in Figure 1.2 acts as a feedback element. The greater the mass displacement, the greater the spring's restoring force attempting to return the mass to its rest position. The action of the diaphragm in the pressure regulator (Figure 1.4) combines the actions of a comparator and a sensor. As the pressure p increases, the diaphragm motion moves the valve to decrease the pressure. Thus, the output (the pressure) is made to influence itself.

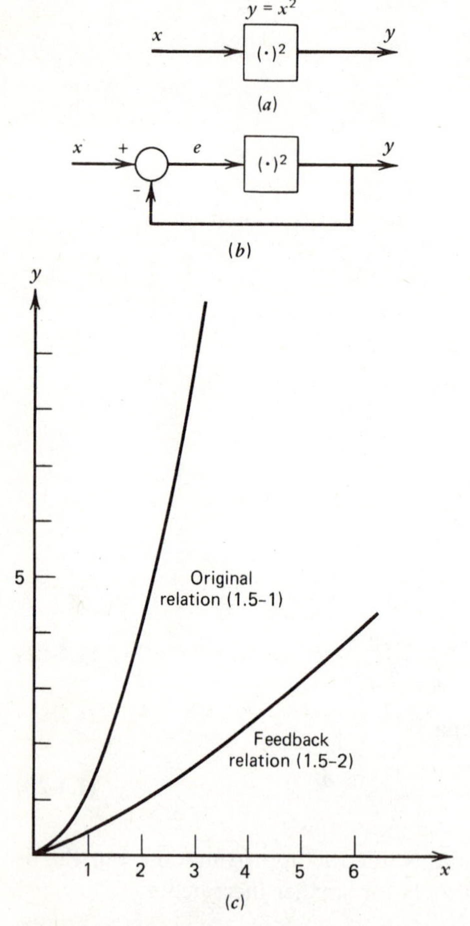

FIGURE 1.13 Improvement of an element's linearity with feedback. (*a*) Original nonlinear element. (*b*) Original element with a feedback loop added. (*c*) Plots of the two input–output relations.

Control systems rely heavily on the properties of feedback. We now explore these properties in more detail.

Feedback Improves Linearity

As we indicated in Section 1.3, linear systems models are the chief model form used in our study. One of the reasons is that using feedback often improves the linearity of the system. We can construct a system of elements whose individual behavior is nonlinear, but with proper use of feedback, the resulting system's behavior will be approximately linear.

To illustrate this effect, consider the nonlinear element shown in Figure 1.13*a*. Its input–output relation is

$$y = x^2 \tag{1.5-1}$$

If we introduce a feedback loop, as in Figure 1.13*b*, we can write the following relations.

$$y = e^2$$

$$e = x - y$$

Thus,

$$y = (x - y)^2 \tag{1.5-2}$$

The plot of y versus x for the original element and the feedback system is shown in Figure 1.13*c*. The feedback system's input–output relation is closer to being a straight line and therefore is approximately linear over a wider range of x than for the original nonlinear element. This wider range is what we mean by "improved linearity."

Feedback improves linearity in dynamic as well as static systems. Chapter 6 provides several examples of this effect.

Feedback Improves Robustness

The coefficient values and the form of a model are always approximations to reality and thus have some uncertainty associated with them. In addition, such factors as wear, heat, and pressure can cause the performance of a system to change with time. Thus, values of the design parameters that were optimal when the system was built might no longer give the desired performance. A vacuum tube amplifier is an example of this effect. The heat generated causes the amplification factor to change in time. In light of this, we should always investigate a prospective design to assess the sensitivity of its performance to uncertainties or variations in the system's parameters. Feedback can be used to improve the system's behavior in this respect.

We have seen that one purpose of the thermostat is to compensate for changes produced by variations in the outdoor environment—the system's "disturbances." This is another use for feedback, and systems that can maintain the output near its desired value in the presence of disturbances are said to have good *disturbance rejection*.

A system that has both good disturbance rejection and low sensitivity to parameter variations is said to be *robust*. We now illustrate how feedback can create or improve robustness.

Parameter Sensitivity

First, consider the reduction of parameter sensitivity. The element shown in Figure 1.14*a* has a proportional constant G (called the *gain*). We wish the gain value to be $G = 10$ and thus select an element that has this nominal value. However, suppose that because of heat, wear, or poor construction, the actual value of G can vary by $\pm 10\%$. In this case, the input–output relation will be somewhere between $y = 9x$ and $y = 11x$, and this is considered unacceptable.

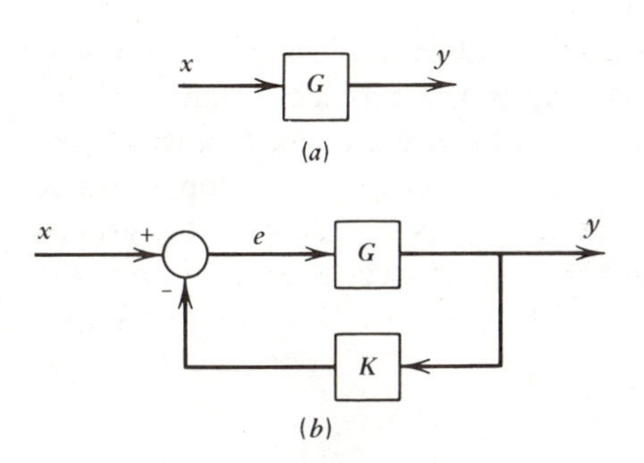

FIGURE 1.14 Reduction of parameter sensitivity with feedback. (*a*) Original element with an uncertain gain *G*. (*b*) Insertion of a feedback element with a reliable gain *K*. If $KG \gg 1$, $y \cong (1/K)x$.

To improve the situation, we place a feedback loop with its own gain K around the basic element G. Presumably the physical element we select to produce the gain K will be relatively insensitive, so that the value of K will be constant and predictable. This arrangement is shown in Figure 1.14*b*.

The governing relations are

$$y = Ge, \qquad e = x - Ky$$

or

$$y = \frac{G}{1 + GK} x \tag{1.5-3}$$

Note that if we pick G large enough so that $GK \gg 1$, then (1.5-3) becomes approximately

$$y \cong \frac{G}{GK} x = \frac{1}{K} x \tag{1.5-4}$$

The system's input–output relation becomes independent of G when GK is large! We now pick K to obtain our desired input–output relation, here, $y = 10x$. Thus, K must be $K = 0.1$, and we pick G such that $0.1G \gg 1$ or $G \gg 10$.

Let us use $G = 1000$ and see what happens. From (1.5-3),

$$y = \frac{1000}{1 + 100} x = 9.901x$$

which is very close to the desired relation. Now, if G varies by $\pm 10\%$ so that $G = 900$ and 1100, (1.5-3) gives

$$y = \frac{900}{1 + 90} x = 9.8901x$$

and

$$y = \frac{1100}{1 + 110} x = 9.91x$$

The sensitivity of the feedback system is much lower. This method of reducing the sensitivity is called *feedback compensation.*

A common engineering application of this approach is stabilizing the gain of an electronic amplifier (Figure 1.15). Engineers working on this problem in the early part of the twentieth century discovered the principle of feedback compensation. The amplifier gain A is large but subject to some uncertainty. The feedback loop is created by a resistor. Part of the voltage drop across the resistor is used to raise the ground level at the amplifier input. From the resistor's voltage–current relation we obtain

$$e_o = A\left(e_i - e_o \frac{R_2}{R_1}\right) \tag{1.5-5}$$

or

$$e_o = \frac{A}{1 + A\dfrac{R_2}{R_1}} e_i$$

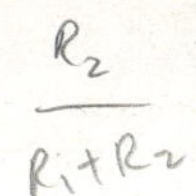

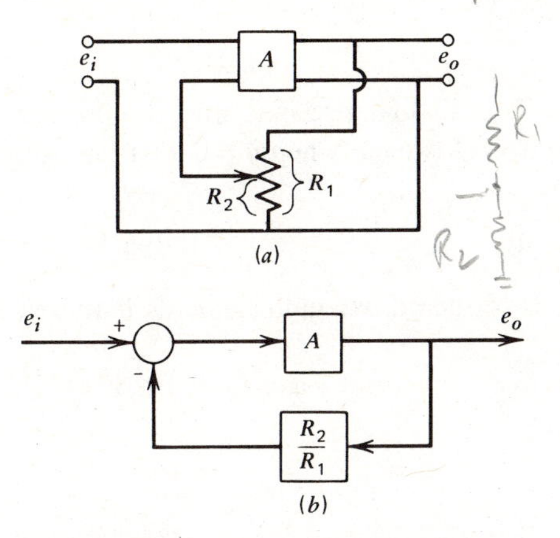

FIGURE 1.15 Feedback compensation of an amplifier. (a) Circuit diagram. (b) Block diagram.

If $AR_2/R_1 \gg 1$, then

$$e_o \cong \frac{R_1}{R_2} e_i \qquad \textbf{(1.5-6)}$$

Presumably, the resistor values are sufficiently accurate and constant enough to allow the system gain R_1/R_2 to be reliable.

Another version of this application uses the *operational amplifier* (op amp). This is a voltage amplifier with a very large gain ($A = 10^5$ to 10^9) that draws a negligible current. If resistors are placed in series and parallel around the op amp, as shown in Figure 1.16, the system's input–output relation will be

$$e_o \cong -\frac{R_2}{R_1} e_i \qquad \textbf{(1.5-7)}$$

This can be shown by writing the appropriate circuit equations. Since the current i_3 is negligible, the voltage e_1 is nearly zero, and $i_1 \cong i_2$. But

$$i_1 = \frac{e_i - e_1}{R_1}$$

$$i_2 = \frac{e_1 - e_o}{R_2}$$

Therefore,

$$\frac{e_i - e_1}{R_1} = \frac{e_1 - e_o}{R_2}$$

Since $e_1 \cong 0$, it follows that

$$\frac{e_i}{R_1} \cong -\frac{e_o}{R_2}$$

which is equivalent to (1.5-7).

Op amps appear in many designs, and we will see more of them.

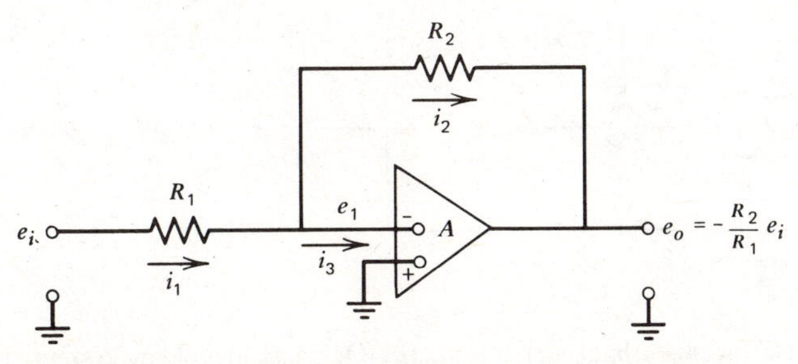

FIGURE 1.16 Op amp multiplier. Note the sign inversion.

Disturbance Rejection

Consider the system shown in Figure 1.17*a*. The disturbance is u, and the desired input–output relation between y and x is $y = 10x$. However, when $u \neq 0$, this relation is not obtained, because

$$y = 5(2x - u) = 10x - 5u$$

This can be remedied by introducing a feedback loop and two gain elements B and K, as shown in Figure 1.17*b*.

We can use the superposition principle to find the output y as a function of x and u. First, set $u = 0$ and solve for y as a function of x.

$$y\bigg|_{u=0} = \frac{10B}{1 + 10BK}x$$

Now replace u and set $x = 0$. Solve for y in terms of u to obtain

$$y\bigg|_{x=0} = \frac{-5}{1 + 10BK}u$$

Invoking superposition, we obtain the general expression for y by adding the preceding results.

$$y = \frac{10B}{1 + 10BK}x - \frac{5}{1 + 10BK}u \tag{1.5-8}$$

We desire u to have no effect on y, but this cannot be accomplished with finite values of B and K. Therefore, we relent somewhat and require instead that no more than 10% of the value of u appear in the output y; that is, we require that

$$\left|\frac{5}{1 + 10BK}u\right| \leqslant |0.1u|$$

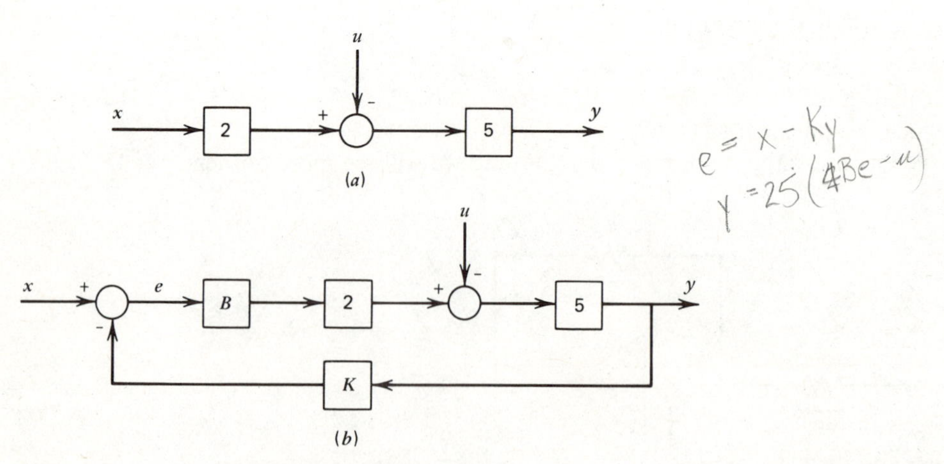

FIGURE 1.17 Use of feedback for disturbance rejection. (*a*) Original open-loop system. (*b*) Feedback system.

This is satisfied if $5/(1 + 10BK) = 0.1$. From the desired relation $y = 10x$, we also have

$$\frac{10B}{1 + 10BK} = 10$$

The last two conditions give $B = 50$ and $K = 49/500$. This design meets the specifications and can be implemented if the values of B and K can be obtained physically and are reliable. The price paid for this improvement is that we have a more expensive system (due to the added elements B and K) and possibly a less reliable system (since there are more elements to fail).

The Importance of Dynamic Models for Feedback Systems

We have used static systems to illustrate the properties of feedback, but its most important applications occur in dynamic systems. Consider the motor-operating curves shown in Figure 1.11. Suppose the motor speed, voltage, and load torque are ω_2, v_2, and T_2 initially and that we wish to maintain the speed at ω_2. If the load torque increases, the speed will decrease. However, the operating curves do not tell us *how long* it will take for the new speed to be established. The curves represent only the steady-state behavior of the system as a static element. The inertia of the motor obviously prevents the speed from changing instantaneously from the value ω_2 to its new steady-state value.

In order to keep the speed near its desired value, we would use a controller. A block diagram of the general situation is shown in Figure 1.18*a*. The controller would sense that the speed had decreased and would increase the motor voltage. Again, the curves give no information about the time it would take for the speed to return to its desired value ω_2.

At this point, it is not clear how we should design the controller to act. For example, should it change the voltage in proportion to the difference between the desired and actual speeds? Or should the voltage change be proportional to the *rate* of change of the speed difference? The fact that the motor and its load have inertia suggests that the time behavior of the speed depends to a great extent on the characteristics of the controller. For example, suppose we make the voltage change proportional to the *integral* of the speed difference such that

$$v = v_2 + K \int_0^t e \, dt \tag{1.5-9}$$

where e is the speed error $\omega_2 - \omega$. We will see in Chapter 6 that no steady-state error due to a constant load torque disturbance will exist if this control scheme is used. This is because the voltage will not stop increasing until the speed difference e becomes zero. However, if the proportionality constant K is made too large, the controller can "overcompensate." The result is an oscillation in speed about the desired value. This effect is shown in Figure 1.18*b*. The initial time in that plot is the time at which the load torque changes. On the other hand, if K is made smaller, the oscillation does not occur, but the time to return ω to ω_2 might be very long.

Obviously, we will need a dynamic model that includes the effect of the inertia in order to design the controller properly. Such a model can be obtained by applying

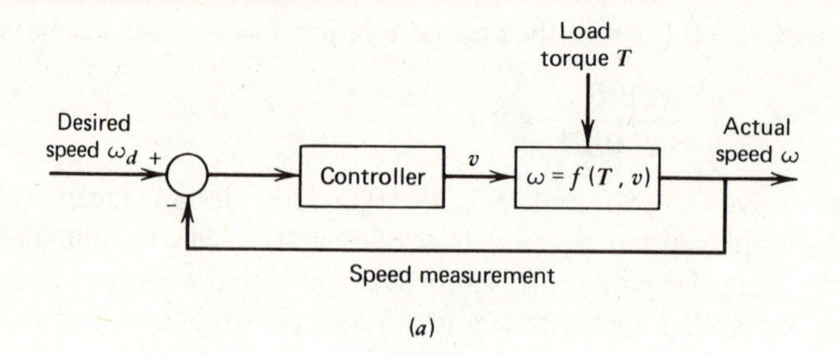

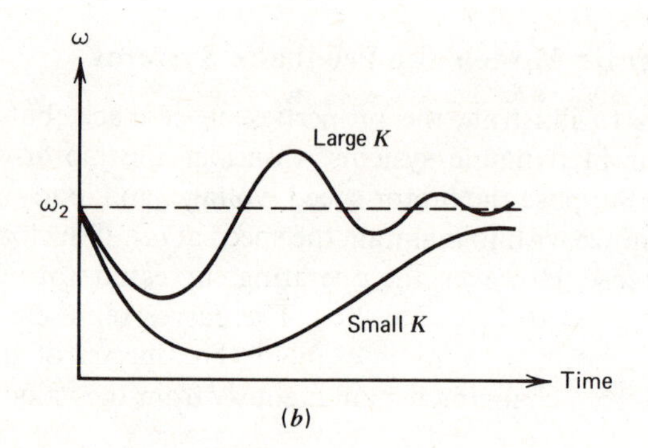

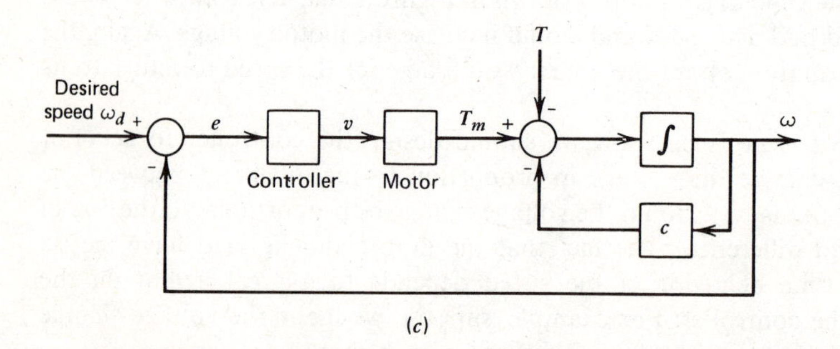

FIGURE 1.18 Motor speed control. (*a*) General block diagram. (*b*) Transient behavior of speed for different values of the controller gain *K*. (*c*) Block diagram using a dynamic model of the load.

Newton's law. We assume that the motor and load inertias can be lumped into one inertia I. The motor torque is T_m, and it is resisted by a friction torque T_f, which can often be modeled as proportional to the speed; that is, $T_f = c\omega$. By summing the torques on the inertia, we obtain the dynamic model for the load speed.

$$I\frac{d\omega}{dt} = T_m - T - c\omega \tag{1.5-10}$$

The block diagram of the system with the controller and motor is shown in Figure 1.18*c*. When used with a motor model that relates motor torque to motor voltage, the preceding equation is often sufficient for designing the controller—for example, for picking the proper value of K in (1.5-9).

The linearizing property of the feedback loop in the controller often allows us to model the system as a linear one. In addition, reducing the system's parameter sensitivity means that a lumped-parameter, low-order dynamic model is often satisfactory for the purpose of designing feedback systems. The model usually cannot be static but must describe at least the dominant dynamic behavior of the system. We take up procedures for developing such models in the next chapter.

PROBLEMS

1.1 What is the causal relation for the following elements with the given inputs and outputs?

(a) A capacitor (charge as input; voltage as output).
(b) An inertia (torque as input; angular acceleration as output).
(c) Angular acceleration as input; angular velocity as output.
(d) A water tank with vertical sides (water volume as input; water height as output).
(e) The heat energy stored in a body as input; the body temperature as output.

1.2 Draw a block diagram for the following models. The inputs are u and v; the output is y. The variables x and e are internal variables. Show them on the diagram.

(a) $y = 5x$
$x = v + 3e - 4y$
$e = u - 2y$

(b) $\dot{y} = 6x$
$\dot{e} = u - 2e$
$x = 5v + 3y + e$

1.3 Obtain the input–output relations for each of the diagrams shown in Figure P1.3. The inputs are u and v; the output is y.

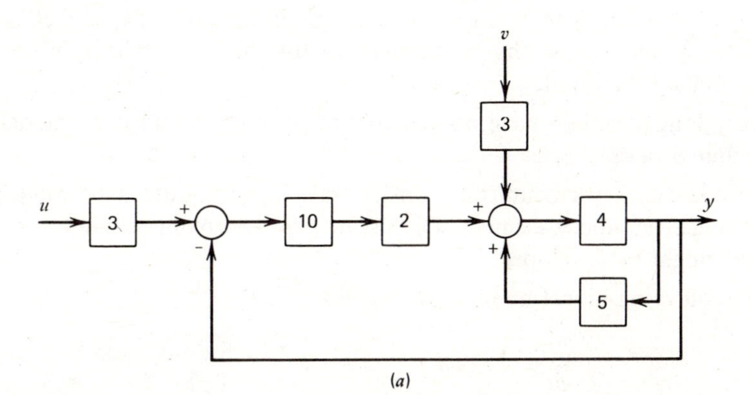

(a)

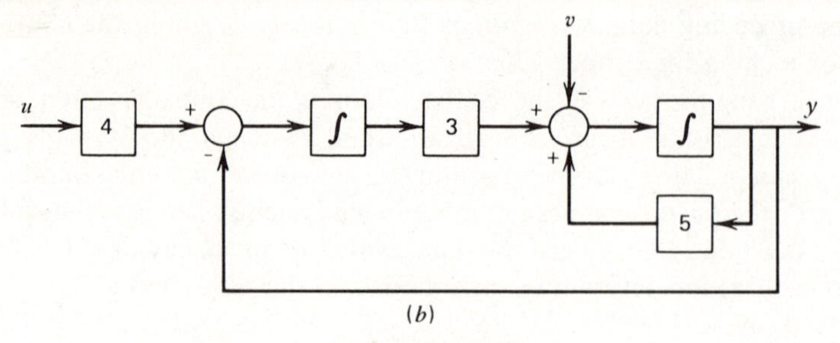

FIGURE P1.3

1.4 Water level controllers represent the earliest examples of control systems. A simple version using a float and lever is shown in Figure P1.4.

(a) Discuss the system's operation. How can we adjust the water level that the system will maintain?

(b) Draw the block diagram with the desired level as the command input, the actual level as the output, and the change in water supply pressure as the disturbance.

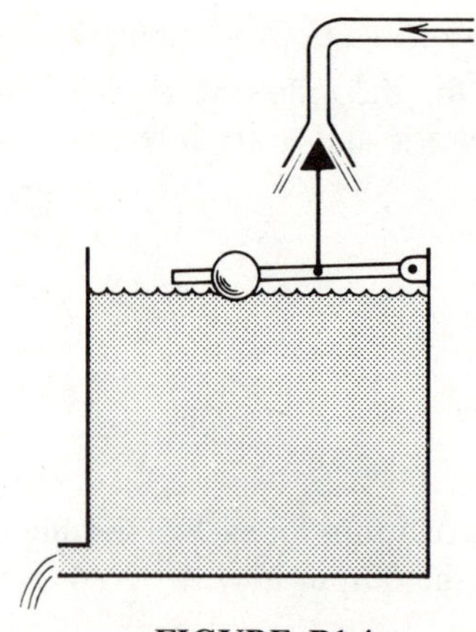

FIGURE P1.4

1.5 A person attempting to balance a stick on end in the palm of a hand constitutes a control system. Draw the block diagram of the system. Is more than one measurement involved?

1.6 A very long wire can have appreciable resistance. Should it be modeled as a distributed-parameter element?

1.7 A certain cantilever beam has considerable mass. Should it be modeled as a distributed-parameter element? Discuss how an approximate lumped-parameter model might be developed.

1.8 Is the following model nonlinear? Explain.

$$\frac{dx}{dt} = -3x + f, \qquad f = \begin{cases} +1 & \text{if} \quad x \geqslant 0 \\ -1 & \text{if} \quad x < 0 \end{cases}$$

1.9 Consider the savings growth model (1.3-7).

(a) Suppose that $x(0) = \$1.00$. Find $x(5)$, the amount of money at the end of five years.

(b) Generalize the results in (a) to find an expression for $x(k)$ in terms of $x(0)$ for any integer k.

1.10 Obtain a linearized expression for the following functions, valid near the given reference values:

(a) $w = \cos y, \quad y_o = 0$

(b) $w = \cos y, \quad y_o = \pi/4$

(c) $w = e^{3y}, \quad y_o = 1$

(d) $w = y_1^2 \sin y_2, \quad y_{1o} = 1, \quad y_{2o} = \pi/4$

(e) $w = y_1/y_2, \quad y_{1o} = 1, \quad y_{2o} = 3$

1.11 The area A of a rectangle is $A = y_1 y_2$, where y_1 and y_2 are the lengths of the sides.

(a) Obtain a linearized expression for A if $y_{1o} = 2$, $y_{2o} = 5$.

(b) Give a geometric interpretation of the error in the linearization.

1.12 Use the curves shown in Figure P1.12 to obtain a linearized expression for $w = f(y_1, y_2)$, valid near the point $y_{1o} = 1$, $y_{2o} = 10$.

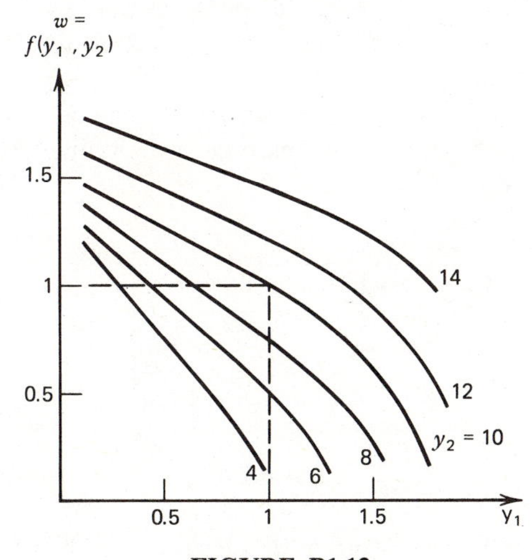

FIGURE P1.12

1.13 Shown in the figure is a "walking beam" linkage. The points whose displacements are labeled x, y, and z are to be attached to other parts of the mechanism.

(a) Derive a linear expression for z as a function of x and y. Hint: Consider two cases: (1) $x = 0$, and (2) $y = 0$, and find the expression for z in each case. Then use superposition to obtain the general solution for $x, y \neq 0$.

(b) Draw the block diagram with x and y as inputs, and z as output.

(c) In the derivation in part (a), how is the small angle approximation used?

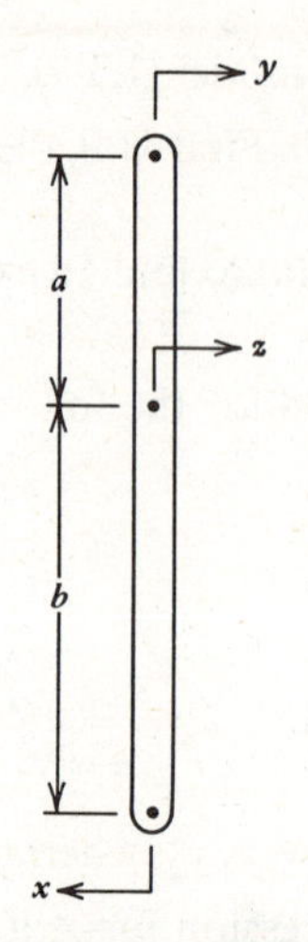

FIGURE P1.13

1.14 A certain component has the input–output relation $y = \sqrt{x}$. In an attempt to obtain somewhat more linear behavior over the range $0 \leqslant x \leqslant 4$, a feedback loop with a gain of K is placed around this component (see Figure P1.14).

(a) Suppose we use $K = 10$. Plot the input–output relation for the original and modified system over the range $0 \leqslant x \leqslant 4$. Which one more closely resembles a straight line?

(b) How should K be chosen to increase the range of positive x over which the input–output relation approximates a straight line? How is the overall gain (the slope) affected?

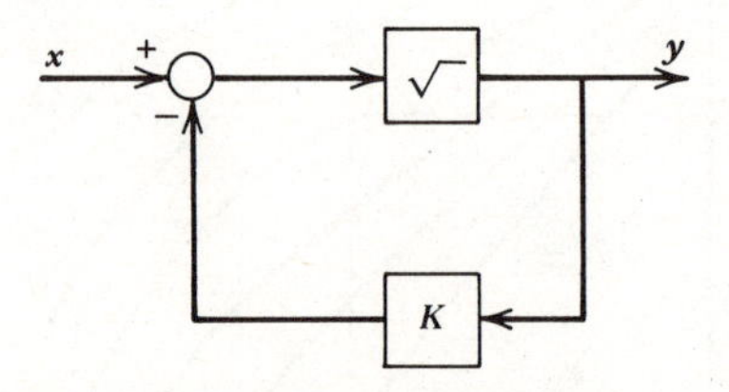

FIGURE P1.14

1.15 A certain component has a gain G that varies by $\pm 20\%$ but can be made very large. Select the feedback gain K to use with G so that the resulting system will have an overall gain of $100 \pm 10\%$ (see Figure 1.14).

1.16 For the system shown in Figure P1.16, the input u represents an unwanted effect. We desire that no more than 5% of the value of u show up in the output y. Also, if

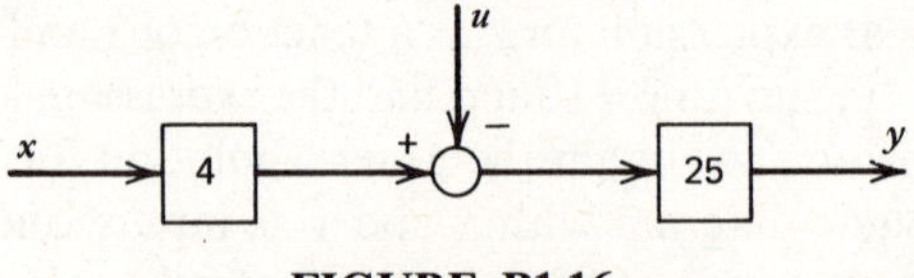

FIGURE P1.16

$u = 0$, the desired input–output relation is $y = 100x$. Design a feedback system and select its gain values to meet these objectives.

1.17 (a) Show that the two systems in Figure P1.17 have the same input–output relation when the gain K is 100.

(b) If the gain K is subject to a $\pm 10\%$ uncertainty, which system is the *least* sensitive?

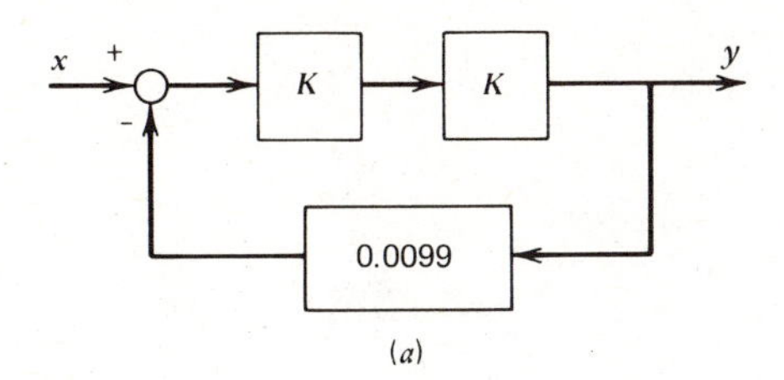

(*a*)

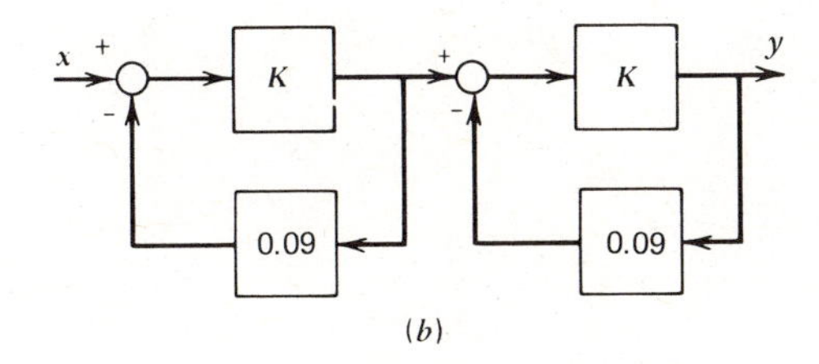

(*b*)

FIGURE P1.17

1.18 The following are common examples of control systems. Discuss their operation. Do they employ feedback?

(a) Toaster.
(b) Washing machine.
(c) Engine camshaft.
(d) Engine cooling system.
(e) Carburetor.
(f) Traffic light.

1.19 Explain the feedback action in the following systems:

(a) The law of supply and demand in economic systems.
(b) Temperature control in the human body.
(c) Predator–prey interactions.

1.20 This story circulates in several forms. A worker at a factory was in charge of activating the noon whistle at the plant. Being very conscientious and trusting, every day he phoned the research lab at the nearby university to set his clock according to their time. Eventually, he became curious and asked the university researcher how he managed to keep his clock accurate. "Why, we set our clock by the noon whistle at the local factory" came the reply.

Is this a feedback process? If the university clock gains 2 min per day, what is the error in the timing of the "noon" whistle at the end of 5 days?

CHAPTER TWO
Modeling Dynamic Systems

In Section 1.2, we saw that a mathematical model is like a map; it is useful only if it concisely describes that system aspect of interest to us. The system aspects of interest to control system engineers can be classified as electrical, mechanical, hydraulic, pneumatic, and thermal. Chapter 2 develops the ability to model these aspects for the purpose of control system design. We adopt an approach that emphasizes the similarity of modeling principles and techniques for various system types. We will see, for example, that the steps required to derive a thermal system model are very similar to those required for an electrical system model. Familiarity with modeling different system types will improve the engineer's ability to handle a variety of applications.

The control system engineer must employ a model that captures the essential dynamic behavior of the system, no more and no less. If the model is too simplistic, the control system will not function as desired. If the model is too detailed, its complexity will hinder the application of design methods and prevent the designer from understanding the essential behavior of the system. Developing such an appropriate model is made easier because the feedback loop of the control system improves the system's linearity, parameter sensitivity, and disturbance rejection, as discussed in Section 1.5. This means that a relatively simple model that describes the dominant time-dependent behavior is usually sufficient for control system design.

Time-dependent behavior involves both the rate of response and the pattern of the response. For example, a car's cruise control system should quickly return the car's speed to its desired value when the car begins to ascend a hill. But it should not overrespond and cause the speed to exhibit an oscillatory pattern about the desired value. The models and modeling techniques developed in Chapter 2 are usually sufficient to describe the dominant rate of response and response patterns of typical control system applications. In Chapter 3, we will develop methods for using these models to compute the dominant response rate and patterns.

Because we will be using models of realistic systems with typical parameter values as examples, we begin with a review of typical physical units. Then electrical and mechanical system models are introduced first, because their principles are perhaps the most familiar from statics, dynamics, and basic physics courses. These are followed by treatments of two types of fluid systems, hydraulic and pneumatic, and finally an introduction to thermal systems. Coverage of electrical and mechanical systems in Sections 2.2–2.5 is essential for understanding later chapters. The treatment of hydraulic, pneumatic, and thermal systems in Sections 2.6–2.9 has been greatly simplified through the generous use of empirical formulas and the easily understood principles of conservation of mass and energy. Although the material in these sections is not so vital as the material on electrical and mechanical systems, its coverage is nevertheless strongly recommended.

2.1 SYSTEMS OF UNITS

None of the modeling techniques we will develop depend on a specific system of units. In order to make quantitative statements based on the resulting models, a set of units must be employed. Most engineering work in the United States has been based on the British Engineering system. However, the metric Système Internationale (the SI system) is rapidly becoming the worldwide standard. Until the changeover is complete, engineers in the United States will have to be familiar with both systems. In our examples, we will use SI and British Engineering units in the hope that the student will become comfortable with both. Other systems are in use, such as the meter-kilogram-second (mks) and centimeter-gram-second (cgs) metric systems and the British system, in which the mass unit is a pound. We will not use these, because the British Engineering and SI systems are the most common in engineering applications. We now briefly summarize these two systems.

The British Engineering system is an example of a "gravitational" system. The primary variable is force, and the unit of mass is derived from Newton's second law. The pound is selected as the unit of force and the foot and second as units of length and time, respectively. From Newton's law, force equals mass × acceleration, so the unit of mass must be

$$\text{mass} = \frac{\text{force}}{\text{acceleration}} = \frac{\text{pound}}{\text{foot}/(\text{second})^2}$$

This mass unit is named the *slug*. Energy has the dimensions of mechanical work; namely, force × displacement. Therefore, the unit of energy in this system is the foot-pound (ft-lb). Another energy unit in common use for historical reasons is the British thermal unit (Btu). The relationship between the two is given in Table 2.1. Power is the rate of change of energy with time, and a common unit is *horsepower*. Finally, temperature in this system can be expressed in degrees Fahrenheit or in absolute units, degrees Rankine.

The SI metric system is an "absolute" system. This means that the mass is chosen as the primary variable, and the force unit is derived from Newton's law. Selecting the meter and the second as the length and time units and the kilogram as the mass unit, the derived force unit is the *newton*. The common energy unit is the newton-meter, also called the *joule*, while the power unit is the joule/second, or *watt*. The difference between the boiling and freezing temperatures of water is 100°C, with 0°C being the freezing point. The absolute temperature units are degrees Kelvin.

Table G.1 lists the most commonly needed factors for converting between the British Engineering and the SI systems. For example, if the elastic constant of a certain spring is given as 50 lb/ft, its SI equivalent is

$$50\frac{\text{lb}}{\text{ft}} = 50\frac{4.4482\ \text{N}}{0.3048\ \text{m}} = 729.7\frac{\text{N}}{\text{m}}$$

2.2 THE STRUCTURE OF DYNAMIC MODELS

In Chapter 1, we formulated several system models with the help of block diagrams and the principle of integral causality. We will now see how the modeling procedure can be generalized to treat a variety of system types.

TABLE 2.1 Systems of Units

Quantity	Usual Symbol	British Engineering	SI Metric
Time	t	second (sec)	second (sec or s)
Length	x	foot (ft)	meter (m)
Force	f	pound (lb)	newton (N) (1 N = 1 kg-m/sec^2)
Mass	m	slug (1 slug = 1 lb-sec^2/ft)	kilogram (kg)
Energy[a]	W or Q_h	foot-pound (ft-lb) or British thermal unit (Btu) (1 Btu = 778 ft-lb)	newton-meter (N-m) or joule (J) (1 J = 1 N-m)
Power	P	ft-lb/sec or horsepower (hp) (1 hp = 550 ft-lb/sec)	watt (W) (1 W = 1 N-m/sec)
Temperature	T	degrees Fahrenheit °F or degrees Rankine °R T°F = (T + 460)°R	degrees Celsius °C degrees Kelvin K T°C = (T + 273) K

[a] Energy in the form of mechanical work is usually denoted by W. Heat energy is typically represented by Q_h.

The dynamics of many physical systems results from the transfer, loss, and storage of mass or energy. Thus, one way of developing a model of such a system is to identify the flow paths and storage compartments of mass or energy and to describe quantitatively how these paths and compartments are connected. On the other hand, it is sometimes more appropriate to use another physical law, such as Newton's laws of motion. A basic law that is used to model electrical systems is conservation of charge. For a given system, we must decide what physical laws are appropriate to use and then develop the model using those laws. There are many other physical laws, but the systems we will treat in this book can be described by the following laws:

1. Conservation of mass
2. Conservation of energy
3. Conservation of charge
4. Newton's laws of motion

The physical laws alone do not provide enough information to write the equations that describe the system. Three more types of information must be provided; the four requirements are:

1. Appropriate basic physical laws.
2. Specific arrangement or way the system elements are interconnected.
3. Empirically based descriptions for some or all of the system elements.
4. Any relationships due to integral causality.

The need for the second type of information is easily understood, since it is possible to connect two elements in more than one way. For example, two resistors can be connected in either series or parallel to form two different systems (Figure 2.1); of course, the models are different for each system. We can use the voltage–current relation for a resistor along with conservation of charge and conservation of energy to obtain the models. For Figure 2.1*a*,

$$v = v_1 + v_2 = iR_1 + iR_2$$

$$v = (R_1 + R_2)i \tag{2.2-1}$$

Conservation of charge tells us that the current is the same through each resistor. In electrical systems, conservation of energy is commonly known as Kirchhoff's voltage law, which states that the algebraic sum of the voltages around a loop must be zero.

For Figure 2.1*b*, application of the same laws yields a different model.

$$i = i_1 + i_2 = \frac{v}{R_1} + \frac{v}{R_2}$$

$$i = \left(\frac{1}{R_1} + \frac{1}{R_2}\right)v \tag{2.2-2}$$

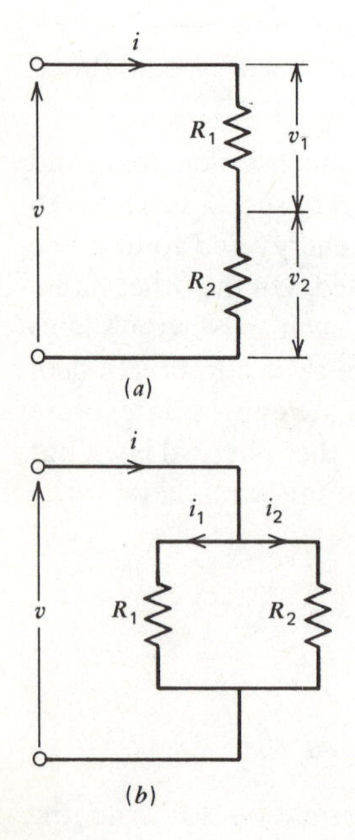

FIGURE 2.1 Series and parallel resistor arrangements.

The voltage–current relation for a resistor, $v = iR$, is an example of an empirically based description of a system element, the third type of required information. This type of description is called a *constitutive relation*, which is an algebraic (static) relation that is not derivable from a basic physical law. Rather, the relation is obtained from a series of measurements. For example, if we apply a range of currents to a resistor, then measure the resulting voltage drop for each current, we would find that the voltage is directly proportional to the applied current. The constant of proportionality is the resistance, and we can determine its value from the test results.

For the fourth type of required information, we must describe all the ways in which integral causality operates in the system being modeled. This description is needed to relate one type of variable, such as acceleration, to another type, such as velocity. If displacement must also appear in the model, it should be related to the velocity through another integral causality expression.

Model Structure and the Block Diagram

The system model is determined by the basic physical laws and the specific way the system elements are arranged. This arrangement also specifies how the integral causality relations and the constitutive relations for the system elements are used to express the appropriate physical laws. The block diagram displays all of these relationships.

Two electrical examples will illustrate these concepts. Figures 2.2*a* and 2.3*a* show a resistor and capacitor connected in two different ways. For the system in Figure 2.2, suppose we are given the value of the applied voltage v_1 and we want to find the voltage v_3. To do this, we first write all of the appropriate physical laws, constitutive

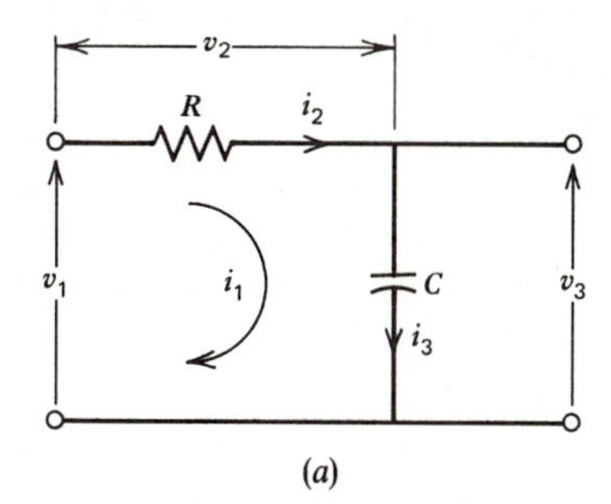

(*a*)

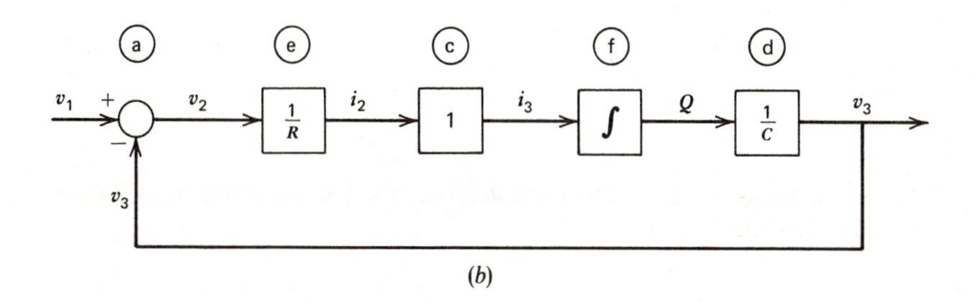

(*b*)

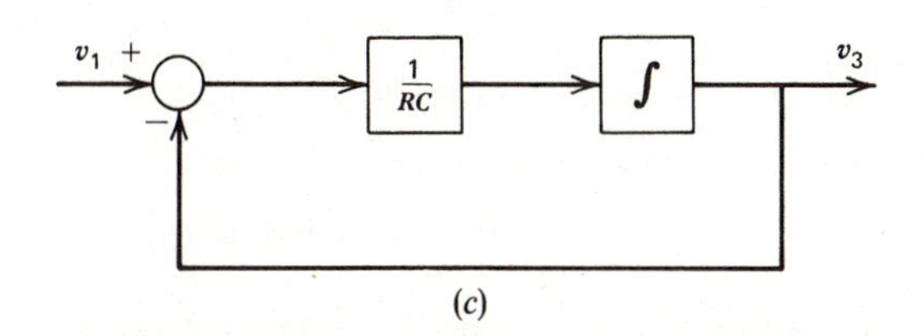

(*c*)

FIGURE 2.2 (*a*) A series *RC* circuit. (*b*) Its block diagram. Circled letters refer to equations in the text. (*c*) The simplified diagram.

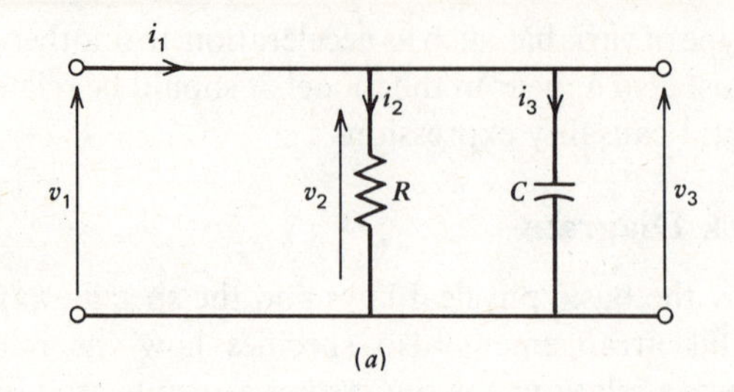

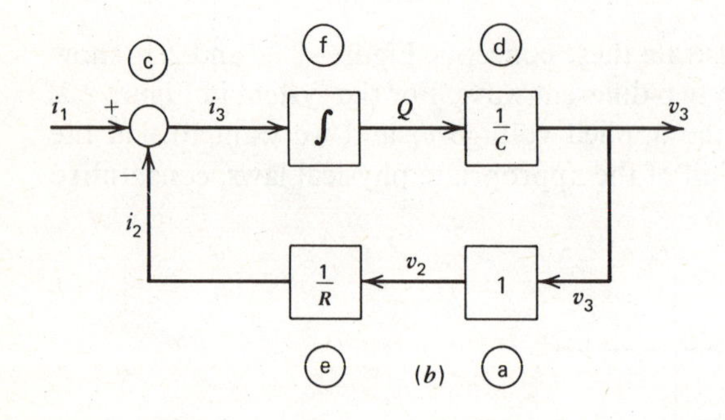

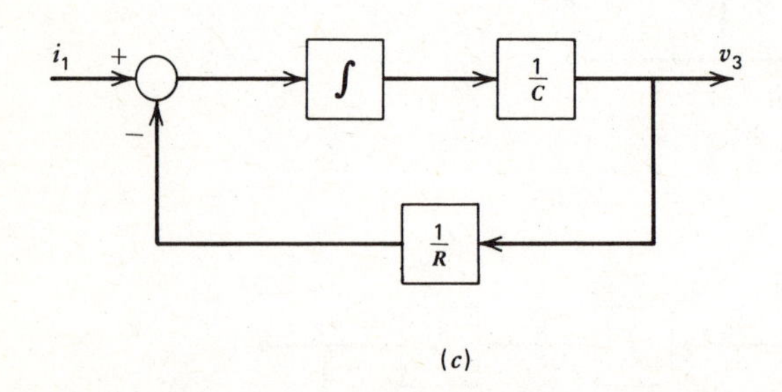

FIGURE 2.3 **(*a*) A parallel *RC* circuit. (*b*) Its block diagram. Circled letters refer to equations in the text. (*c*) The simplified diagram.**

relations, and integral causality relations in terms of the voltages, charges, and currents defined in the figure; these are:

Conservation of energy: $v_1 = v_2 + v_3$	(a)	Capacitor: $v_3 = Q/C$	(d)
Conservation of charge: $i_1 = i_2$	(b)	Resistor: $i_2 = v_2/R$	(e)
$i_2 = i_3$	(c)	Integral causality: $Q = \int i_3 \, dt$	(f)

Figure 2.2*b* shows the corresponding block diagram. Each block represents one of the preceding relations, and the comparator represents the conservation of energy statement. Note that relation (b) does not appear on the diagram, because we were not

given the value of i_1 and have no need to find it or relate it to i_2. Relation (b) is valid, but we do not need it to find v_3 in terms of v_1.

If we are interested in only v_1 and v_3, we can combine the multiplier blocks and obtain the simplified diagram shown in Figure 2.2*c*. The desired system model can be obtained from this diagram as follows:

$$v_3 = \frac{1}{RC}\int (v_1 - v_3)\, dt$$

or

$$RC\frac{dv_3}{dt} + v_3 = v_1 \qquad \textbf{(2.2-3)}$$

The block diagram is often useful for deriving the system model, but it is not necessary. Instead, equation (2.2-3) could have been obtained by using algebraic manipulation to eliminate all the variables except v_1 and v_3 from equations (a)–(f). It is a matter of personal preference.

For the system in Figure 2.3*a*, suppose we are given the value of the current i_1 and must find v_3. The descriptive relations are:

Conservation of energy: $v_1 = v_2$	(a)	Capacitor: $v_3 = Q/C$	(d)
$v_2 = v_3$	(b)	Resistor: $i_2 = v_2/R$	(e)
Conservation of charge: $i_1 = i_2 + i_3$	(c)	Integral causality: $Q = \int i_3\, dt$	(f)

The diagram of these relations is shown in Figure 2.3*b*. Note that relation (a) has not been used, because we have no need here to relate v_2 to v_1. The simplified diagram is shown in (c), and the system model can be derived from the diagram.

$$v_3 = \frac{1}{C}\int \left(i_1 - \frac{1}{R}v_3\right) dt$$

or

$$RC\frac{dv_3}{dt} + v_3 = Ri_1 \qquad \textbf{(2.2-4)}$$

The previous two examples show how the same components will result in different system models if they are connected in different arrangements or if different output or input variables are specified. Here, the outputs were the same (v_3), but the inputs were different (v_1 versus i_1). Different input–output specifications also mean that some of the basic relations (a)–(f) might not be needed to obtain the system description in terms of the desired variables.

The Three Constitutive Relations

There were two constitutive relations in the preceding examples—resistance and capacitance; inductance is the third type. A system with all three types is the series *RLC* circuit shown in Figure 2.4. Suppose we are given the value of the applied

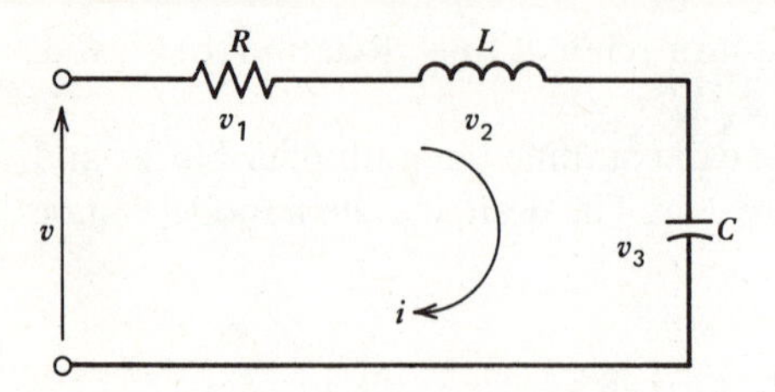

FIGURE 2.4 A series *RLC* circuit.

voltage v and we want to find the resulting current i. Conservation of charge implies that the current through all three elements is the same; namely, i. From Kirchhoff's voltage law (conservation of energy),

$$v = v_1 + v_2 + v_3$$

The constitutive relation for an inductance is

$$\phi = Li$$

where L is the inductance value and ϕ is the flux across the inductance. The integral causality relation between flux and voltage is

$$\phi = \int v\,dt$$

Combining the two preceding expressions gives the voltage–current relation for an inductance

$$v_2 = L\frac{di}{dt}$$

Similarly, we can combine the constitutive relation and the appropriate integral causality relation for the capacitor to obtain its voltage–current relation:

$$v_3 = \frac{1}{C}\int i\,dt$$

The resistor relation is $v_1 = iR$. Substituting these expressions into the voltage law gives the model

$$v = iR + L\frac{di}{dt} + \frac{1}{C}\int i\,dt$$

Since we do not know i as a function of time, we cannot evaluate the integral. Therefore, we differentiate both sides of the preceding equation to obtain a differential equation for which standard solution techniques are available.

$$L\frac{d^2i}{dt^2} + R\frac{di}{dt} + \frac{1}{C}i = \frac{dv}{dt} \tag{2.2-5}$$

A General View of Model Structure

The structure of electrical system models can be represented as shown in Figure 2.5. The electrical variables are charge, voltage, flux, and current; they are represented by the four oval compartments. The two unidirectional paths with the integral signs represent the two integral causality relations; one between flux and voltage, the other between charge and current. The remaining three paths represent the role of the three constitutive relations. The resistance path is the relationship between current and

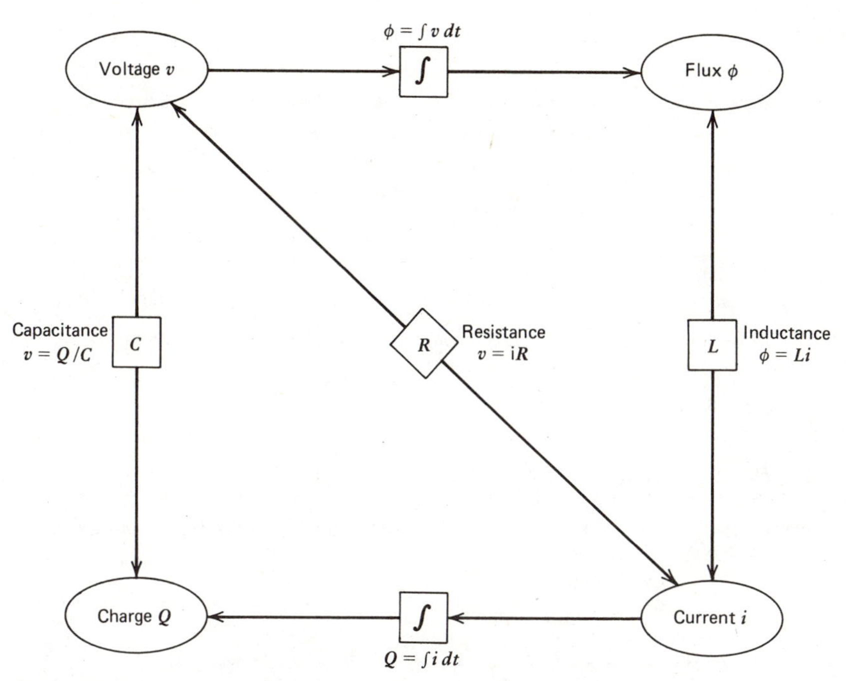

FIGURE 2.5 The general model structure for electrical systems.

voltage, and it is bidirectional. Voltage can produce current through a resistor, and current flow results in a voltage drop. Similar statements apply to the capacitance and inductance paths. The basic physical laws of energy and charge conservation are not shown on this diagram, because the specific arrangement of a given system determines how these laws can be used to relate the electrical variables to one another.

We will soon see that other system types, such as mechanical, fluid, and thermal systems, can be represented by similar diagrams. The four electrical variables—charge, voltage, flux, and current—are specific examples of the more general system variables called *quantity*, *effort*, *flux*, and *rate*. In mechanical systems, these variables are called *displacement*, *force*, *impulse*, and *velocity*. Table 2.2 lists the principal variables for various kinds of systems.

Figure 2.6 is a generalization of the electrical diagram given in Figure 2.5. *Quantity* is the term used to describe the accumulation of rate over time. Rate and quantity are related as cause and effect through integral causality. For example, displacement is the time integral of velocity. *Effort* is the variable produced when potential energy is stored. When charge accumulates in a capacitor, the voltage across the capacitor increases. When a spring is compressed, a force is built up. Voltage and force are effort variables.

Flux is the accumulation of effort over time through integral causality. The time integral of force is *impulse*, which is the flux variable in mechanical systems. *Rate* is the variable that is produced when kinetic energy is stored. When an impulse is applied to a mass, its kinetic energy and, therefore, its speed is increased.

The concepts of electrical resistance, capacitance, and inductance can be

TABLE 2.2 Primary System Variables and Their Usual Symbols

Type	Rate r	Quantity $Q = \int r\,dt$	Effort e	Flux $\phi = \int e\,dt$
Electrical	Current i	Charge Q	Voltage v	Flux ϕ
Mechanical (translation)	Velocity v	Displacement x	Force f	Impulse M_x
Mechanical (rotation)	Angular velocity ω	Angular displacement θ	Torque T	Angular impulse M_θ
Fluid (incompressible)[a]	Mass flow rate q_m or volume flow rate q	Mass Q (or m) or volume V	Pressure p	None
Fluid (compressible)	Mass flow rate q_m	Mass Q (or m)	Pressure p	None
Thermal	Heat flow rate q_h	Heat energy Q_h	Temperature T	None

[a]Mass and volume are readily interchangeable for incompressible fluids; frequently, the symbol Q denotes volume and q mass flow rate.

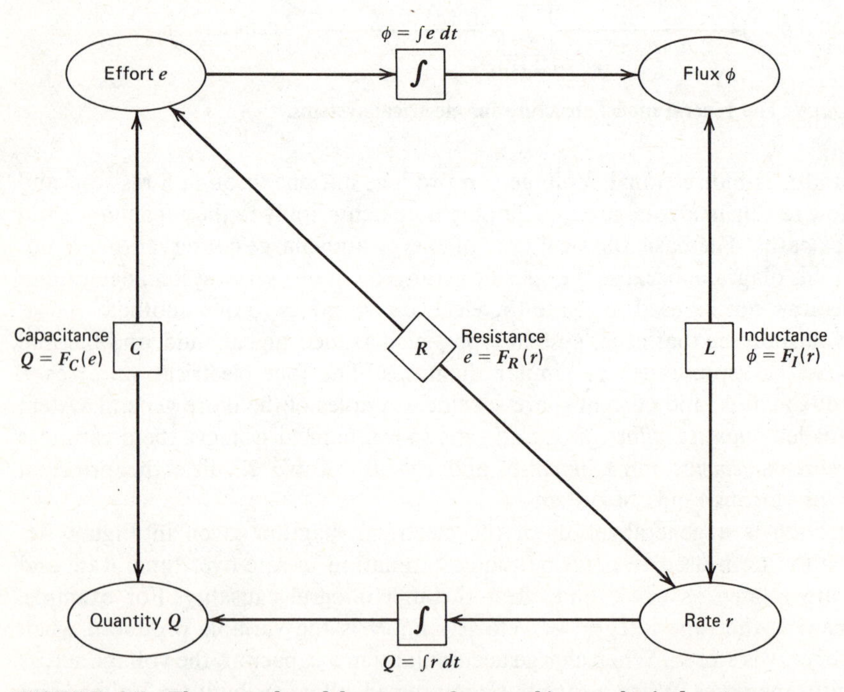

FIGURE 2.6 The general model structure for an arbitrary physical system.

extended to develop general constitutive relations for other types of systems. In a mechanical system, the capacitance relation is the relation between force and displacement, such as in a spring element, where $f = kx$. Inductance is the relation between velocity and impulse. For a mass m with zero initial velocity, this relation is

$$mv = \phi = \int f \, dt$$

which is the impulse–momentum form of Newton's laws. Finally, resistance is the relation between velocity and force, such as found in a shock absorber.

We can think of generalized resistance, capacitance, and inductance in terms of energy. Effort in physical systems is never completely free to act; it is always opposed in some manner. This occurs with resistance elements, and they dissipate energy. Capacitance represents the storage of potential energy, such as in a compressed spring. Inductance represents kinetic energy storage, as shown by the kinetic energy expression for a mass m moving at a speed v: K.E. $= mv^2/2$.

The resistance, capacitance, and inductance paths in Figure 2.6 are represented by the nonlinear functions $F_R(r)$, $F_C(e)$, and $F_I(r)$ for generality. If the functions are linear for specific components, then the resistance, capacitance, and inductance parameters are related to the constants of proportionality. For example, in the electrical resistance relation $v = iR$, the resistance function is $v = F_R(i) = Ri$.

The purpose of the general structure shown in Figure 2.6 is to help us to understand how the behavior of apparently different system types actually arises from very similar principles. This analogy can help the reader develop system models, just as the block diagram can help express the system equations. But it is not necessary to use either the analogy or the block diagram. The reader should use whatever is most convenient.

2.3 MODELING MECHANICAL ELEMENTS

In the terminology of systems engineering, *mechanical* refers to the analysis of forces and motions in solid bodies. In many situations, the internal forces in the body resulting from elastic deformation are small compared to the externally applied forces. In this case, the mass or moment of inertia can be considered to be lumped at a point, and its motion is analyzed by "rigid-body dynamics."

The units and common symbols are given in Table G.2. We assume the reader is familiar with the basic principles of rigid body dynamics found in introductory physics or engineering mechanics. Our purpose here is to relate these concepts to the modeling framework provided by the energy and causality concepts in Section 2.2 and to show how mechanical components are modeled as resistance, capacitance, or inductance elements.

Spring Elements

If in a mechanical system displacement acts against an elastic element, such as a spring, potential energy is stored, and a force is set up in the spring. The usual constitutive relation for an elastic element that undergoes small deflections is given by *Hooke's law*

$$f = kx \tag{2.3-1}$$

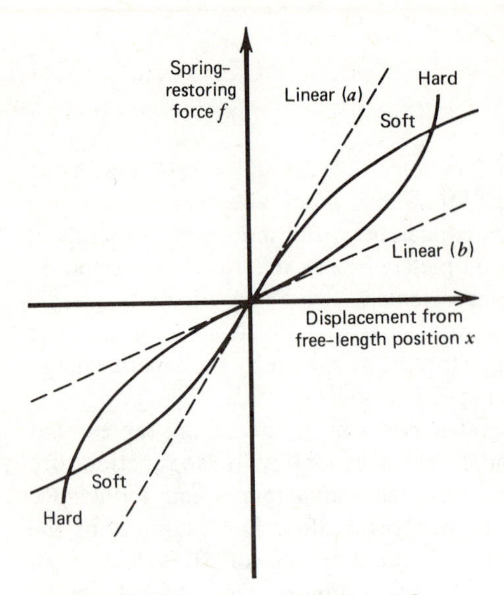

FIGURE 2.7 Some constitutive relations for spring elements. Linear model (*a*) is the small displacement model for the soft spring. Linear model (*b*) approximates the relation for the hard spring.

where f is the elastic force, x is the deflection, and k is the elastic constant (the spring constant).

In mechanical systems, elastic elements, such as beams, rods, and rubber mounts, are usually represented pictorially as "springs." The linear relation between force and displacement given by (2.3-1) for such elements becomes less accurate as the deflection becomes larger. In such cases, it is often replaced by

$$f = k_1 x + k_2 x^3 \tag{2.3-2}$$

where $k_1 > 0$. Again, x is the deflection of the spring from its free (unstressed) length and f is the restoring force set up in the spring by x. The spring is compressed if $x < 0$ and extended if $x > 0$. It is said to be *hard* if $k_2 > 0$ and *soft* if $k_2 < 0$. This relation is compared to the linear relation in Figure 2.7. Note that both the hard and soft springs have their own linear approximation tangent at the origin.

A similar process occurs when a rod is twisted. The torque so induced is proportional to the net angular deflection for small deflections. This type of element is called a "torsional spring."

Table 2.3 lists the expressions for the spring constant for several common elements. These formulas are derived from the force-displacement relations given in elementary texts on mechanics of materials. The moduli of elasticity E and G are given for steel and aluminum in Table G.3.

Damping Elements

Resistance in mechanical and fluid systems results from friction. Viscous friction in a damper or dashpot results from the fluid's viscosity (Figure 2.8). An example of such an element is the shock absorber on a car. The dashpot exerts a force f proportional to the velocity difference v across the element's endpoints for relatively small velocities. Thus,

$$f = cv \tag{2.3-3}$$

where c is a constant, the *damping coefficient.*

The damper force varies as the square of the velocity difference for larger velocities. This phenomenon results from the difference between laminar (smooth) and turbulent (rough) flow through the orifice as the piston moves. The square of the velocity does not change sign when the velocity difference is negative, but the resisting force does. Therefore the square law relation for the turbulent case must be written as

$$f = cv|v| \tag{2.3-4}$$

TABLE 2.3 Spring Constants for Common Elements

Rod in compression/tension

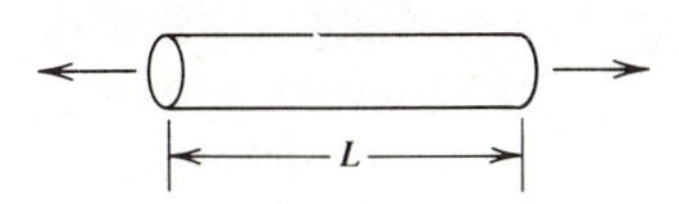

$$k = \frac{EA}{L}$$

A = cross-sectional area

Rod in torsion

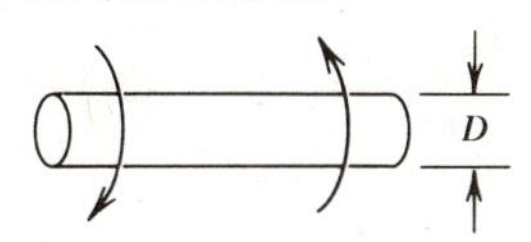

$$k = \frac{G\pi D^4}{32L}$$

L = length

Helical wire coil

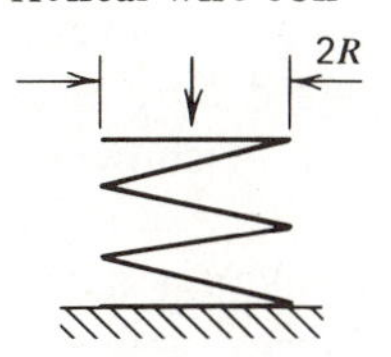

$$k = \frac{Gd^4}{64nR^3}$$

d = wire diameter

n = number of coils

Cantilever beam

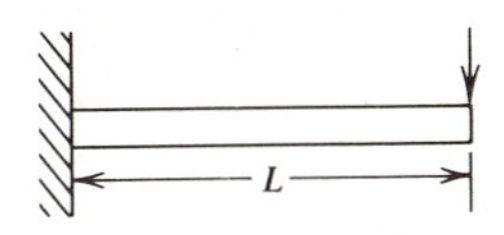

$$k = \frac{Ewh^3}{4L^3}$$

w = beam width

h = beam thickness

Doubly clamped beam

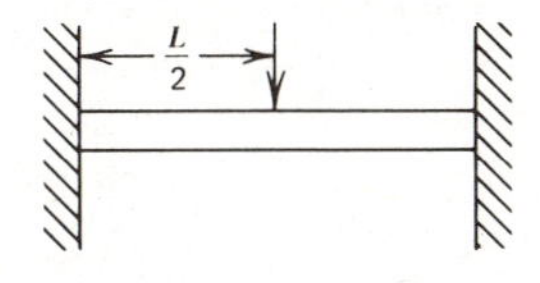

$$k = \frac{16Ewh^3}{L^3}$$

Air spring

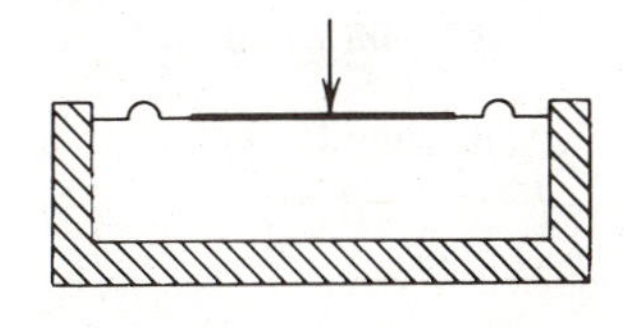

$$k = \frac{\gamma PA^2}{V}$$

A = diaphragm area

P, V = nominal pressure and volume

γ = ratio of specific heats

Damping constants for three common elements are given in Table 2.4. The first will be derived in Section 2.7. The dashpot relation for sliding viscous friction is a consequence of *Newton's law of viscosity*, which states that

$$\tau = \mu \frac{du}{dy} \tag{2.3-5}$$

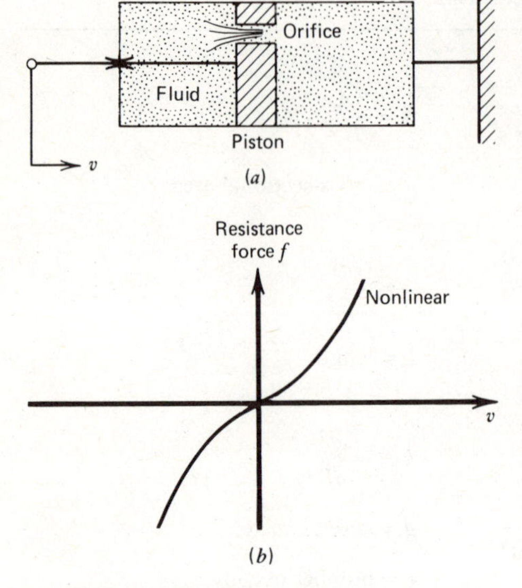

FIGURE 2.8 Viscous damper. (*a*) Construction. (*b*) Constitutive relationship.

where τ is the shear stress in the lubricating film, μ is the lubricant's viscosity coefficient, and du/dy is the velocity gradient in the film normal to the surface. If we assume that the gradient is constant and the fluid velocity near the surface is zero relative to the surface, it can be shown that the friction force is

$$f = \frac{A\mu}{d} v = cv \tag{2.3-6}$$

where A is the lubricated area under the mass and d is the film thickness. For high velocities, these assumptions are not valid, and (2.3-6) must be replaced by a nonlinear relation similar to that shown in Figure 2.8*b*.

The journal-bearing relation shown in Table 2.4 is known as Petrov's law and, it is a consequence of Newton's viscosity law applied to a rotating shaft.

TABLE 2.4 Damping Constants of Common Elements

Element	Damping constant
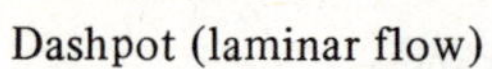 Dashpot (laminar flow) 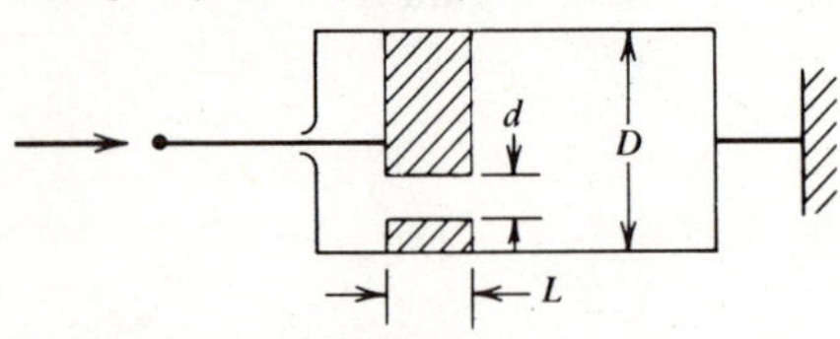	$c = 8\pi\mu L\left[\left(\frac{D}{d}\right)^2 - 1\right]^2$
Sliding viscous friction 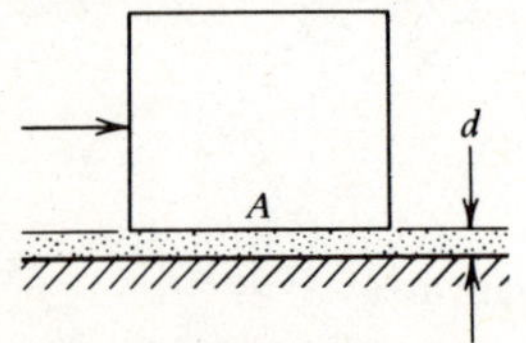	$c = \frac{A\mu}{d}$ A = lubricated area d = film thickness
Journal bearing 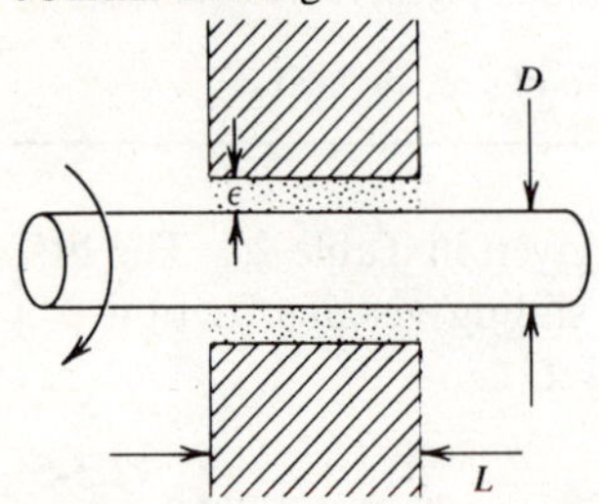	$c = \frac{\pi D^3 L\mu}{4\varepsilon}$ ε = radial clearance

The Laws of Motion

The Newtonian laws of motion for rigid bodies can be summarized as follows. The vector sum of external forces acting on the body must equal the time rate of change of momentum. Thus,

$$\frac{d\mathbf{P}}{dt} = \mathbf{f} \tag{2.3-7}$$

where $\mathbf{P}$ and $\mathbf{f}$ are the three-dimensional momentum and force vectors, respectively. If $\mathbf{v}$ is the vector velocity of the body's center of mass, and m is the mass, the momentum is $\mathbf{P} = m\mathbf{v}$, and (2.3-7) becomes

$$m\frac{d\mathbf{v}}{dt} = \mathbf{f} \tag{2.3-8}$$

In addition, the vector sum of the external torques acting on the body must equal the time rate of change of the angular momentum. Thus,

$$\frac{d\mathbf{\Omega}}{dt} = \mathbf{T} \tag{2.3-9}$$

where $\mathbf{\Omega}$ and $\mathbf{T}$ are the three-dimensional angular momentum and torque vectors.

Chasle's theorem states that the general motion of a rigid body can be considered as a translation plus a rotation about a suitable point. The center of mass is often taken to be this point. If the body's rotation and translation are not simply related to a convenient coordinate system, then (2.3-7) and (2.3-9) can result in complicated expressions. However, most of the systems of interest here involve rotation about a fixed axis. In this case, the angular momentum can be expressed as

$$\mathbf{\Omega} = I\boldsymbol{\omega} \tag{2.3-10}$$

where I is the body's mass moment of inertia about the axis, and $\boldsymbol{\omega}$ is the vector angular velocity. Thus, (2.3-9) becomes

$$I\frac{d\boldsymbol{\omega}}{dt} = \mathbf{T} \tag{2.3-11}$$

If the coordinate system used to express $\boldsymbol{\omega}$ has one direction aligned with the axis of rotation, the vector relation (2.3-11) becomes a scalar one.

$$I\frac{d\omega}{dt} = T \tag{2.3-12}$$

The mass moment of inertia I is given in Table 2.5 for several common objects.

For most of our problems involving rotation, (2.3-12) will suffice. In addition, the translational motion equation (2.3-8) can be written in scalar form as

$$m\frac{dv_i}{dt} = f_i \qquad i = 1, 2, 3 \tag{2.2-13}$$

where v_i and f_i are the velocity and force components along the rectilinear coordinates x_1, x_2, x_3. Further simplification results if the forces f_i are not functions of the

TABLE 2.5 Mass Moment of Inertia

Definition		
$I = \int r^2\, dm$		r = distance from reference axis to mass element dm
Units for I	*British*	*SI*
	slug-ft^2 =	1.356 kg-m^2
Values for common elements	(m = element mass)	

Cylinder

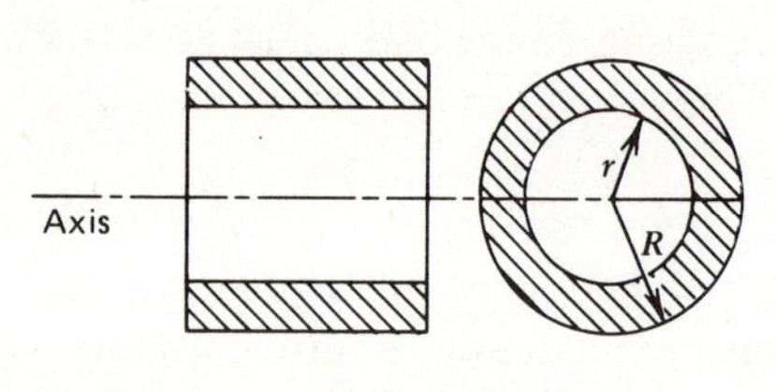

$$I = m\frac{R^2 + r^2}{2}$$

Point mass

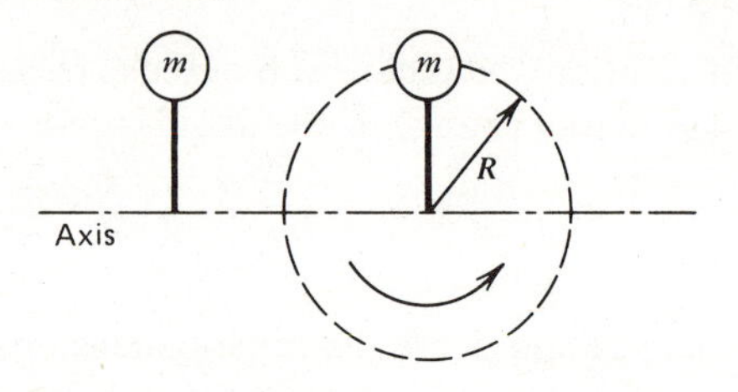

$$I = mR^2$$

Sphere

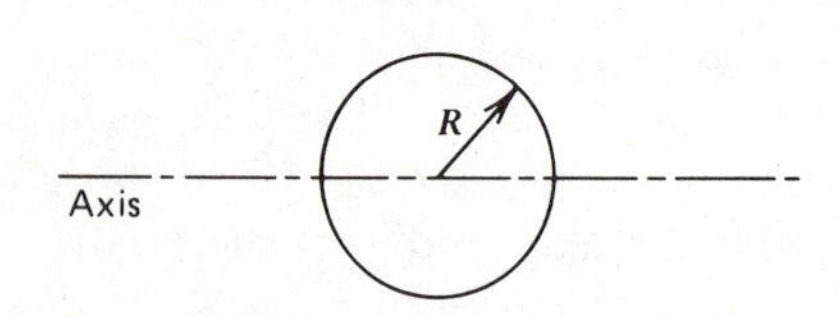

$$I = \frac{2}{5}mR^2$$

Lead screw

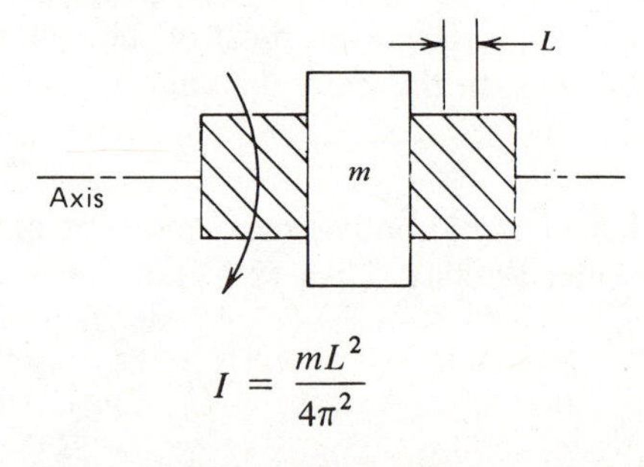

$$I = \frac{mL^2}{4\pi^2}$$

L = screw lead

displacements or velocities in the other coordinate directions. In this case, the set (2.3-13) becomes uncoupled. Each equation will be of the form

$$m\frac{dv_i}{dt} = f_i(v_i, x_i, t) \tag{2.3-14}$$

and can be solved independently of the other two.

Many problems involving both rotation and translation can be simplified by breaking them down into problems involving pure translation and pure rotation and by using (2.3-12) and (2.3-13).

Energy Storage and Dissipation

The product of the effort and rate variables e and r is power P. For mechanical systems, the expression is $P = fv$. This relation can be used to compute the power dissipated in a damping element. Since $f = cv$ in such an element, we have

$$P = cv^2 \tag{2.3-15}$$

For a rotational damper moving at an angular velocity ω, $P = c\omega^2$.

Because energy E is the time integral of power, we have

$$E = \int P\,dt = \int fv\,dt \tag{2.3-16}$$

We can use this expression to compute the energy that is stored in a spring or a moving mass. For a spring, $f = kx$, and the potential energy stored in the spring is

$$PE = \int kx\frac{dx}{dt}\,dt = k\int x\,dx = \frac{1}{2}kx^2 \tag{2.3-17}$$

For a torsional spring, $PE = k\theta^2/2$ where θ is the twist angle.

Because $f = m(dv/dt)$ from Newton's laws, the kinetic energy stored in a mass m moving with a velocity v is

$$KE = \int mv\frac{dv}{dt}\,dt = m\int v\,dv = \frac{1}{2}mv^2 \tag{2.3-18}$$

For an inertia I rotating with angular velocity ω, $KE = I\omega^2/2$.

Lumped-Parameter Equivalent Systems

In any real system, the mass will be distributed throughout and not concentrated at a point with infinitesimal volume. If the forces of interest are external to the system, the principles of rigid-body dynamics can be applied by considering the system's mass to be concentrated or "lumped" at the center of mass. Thus, the original distributed-mass system is modeled as an equivalent lumped-mass system.

If the forces of importance are internal to the system, such as bending stresses in a beam, the situation is more complicated. The application of Newton's second law requires that a displacement variable be assigned to each mass, so that a distributed-mass model requires an infinite number of displacement variables, one for each infinitesimal mass element. This leads to a description in terms of partial differential equations. On the other hand, a lumped-mass model results in a set of ordinary differential equations, one second-order equation for each dimension in which motion occurs (rotational and translational). Because ordinary differential equations are easier to solve, we usually prefer a lumped model.

Mass is not the only property we seek to represent in a lumped fashion. For example, the stress resulting from bending in a cantilever beam is distributed throughout the beam length. The beam's equivalent spring constant k is that value required of an ideal spring if it is to represent the force–deflection characteristics of the endpoint of the beam. Similarly, viscous friction forces are distributed throughout a

TABLE 2.6 Lumped-Mass Equivalent Systems

m_c = concentrated mass
m_d = distributed mass
m_e = equivalent lumped mass

	Actual System	Equivalent System
Helical spring or rod		$m_e = m_c + \frac{1}{3} m_d$
Cantilever beam		$m_e = m_c + 0.23 m_d$
Fixed-end beam		$m_e = m_c + 0.375 m_d$
Simply supported beam		$m_e = m_c + 0.50 m_d$

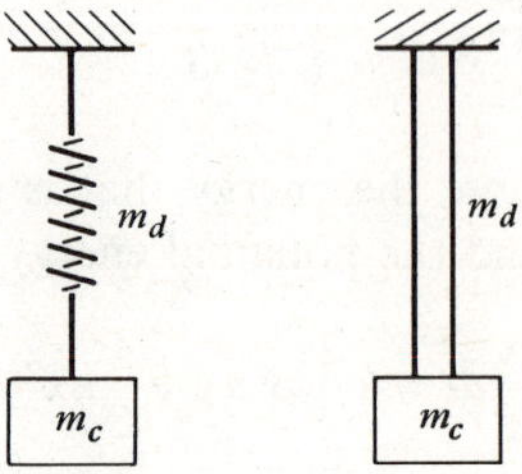

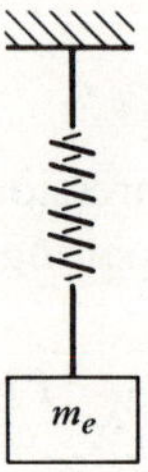

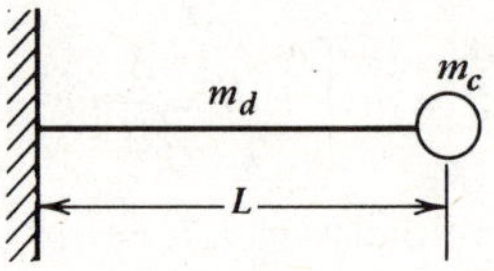

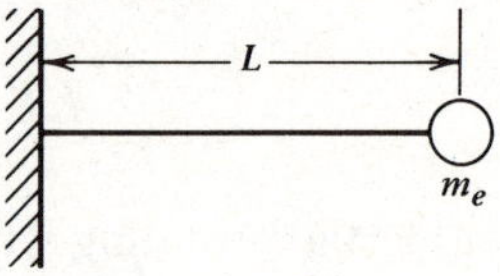

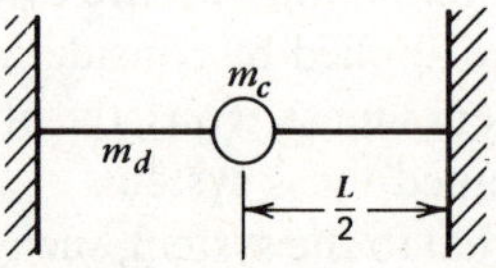

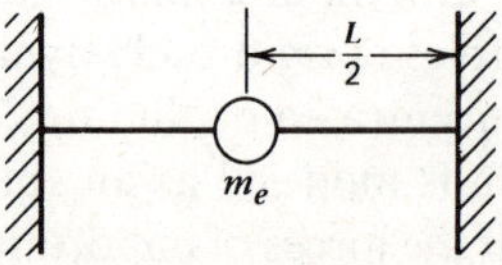

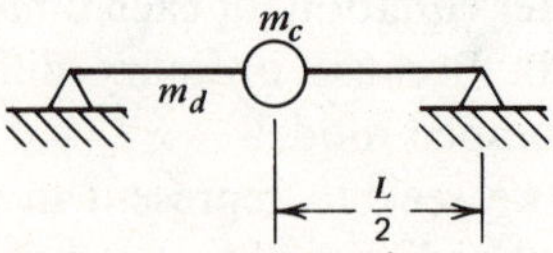

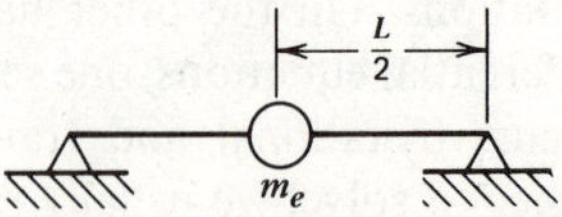

system. The equivalent damping constant c is that value required for an ideal damper to represent the force–velocity characteristics at a chosen point in the system.

There are several methods for modeling equivalent systems. None are exact, because no two systems can be identical in every respect. The equivalent masses in Table 2.6 were found by requiring the kinetic energy of the equivalent system to equal that of the original. The equivalent spring constants in Table 2.3 were obtained from the force–deflection characteristics of the original system at the point where the force was applied. An alternative means of computing the spring constant is to require equivalence of potential energy in the original and equivalent systems. Note that a spring constant found in this manner might not be the same as that found from the force–deflection characteristics. We will not go into the details of these methods, but will use the results for common elements as listed in the accompanying tables. Detailed discussion is given in Reference 3.

In the examples in Table 2.6, the origin of the coordinate system for the lumped model is taken to be at the location of the concentrated mass, and the expressions for the equivalent mass are valid for only that coordinate choice. If, for example, we wish to use a coordinate system located at the midpoint of the cantilever beam, we will need another expression for the equivalent mass m_e. Thus, in general, the values of the parameters in a lumped model depend on the location of the model's coordinate system relative to the original distributed-parameter system.

2.4 EXAMPLES OF MECHANICAL SYSTEMS MODELS

The modeling principles in the previous section can be applied to develop a system model by converting the system elasticities into equivalent springs, the system friction effects into equivalent dampers, and the system mass distribution into equivalent lumped masses. The simplest system models result from a single mass that can only translate in one direction or rotate about one axis. The required model becomes more complex when the system is capable of multiple motions or has multiple masses. The following examples illustrate the required modeling procedures for some typical applications.

A System in Pure Translation

A motor can be mounted on a cantilever beam as a support, as shown in Figure 2.9. The spring constant can be obtained from Table 2.3, entry 4, and the equivalent lumped mass from Table 2.6, entry 2. The effects of internal friction in the beam material while bending can be modeled as a damper, but the damping constant must be obtained experimentally. It is impossible to balance a motor perfectly, and the imbalance will result in a vertical force f that oscillates at the same frequency as the motor's rotational speed. The resulting equivalent system is shown in Figure 2.9*b*.

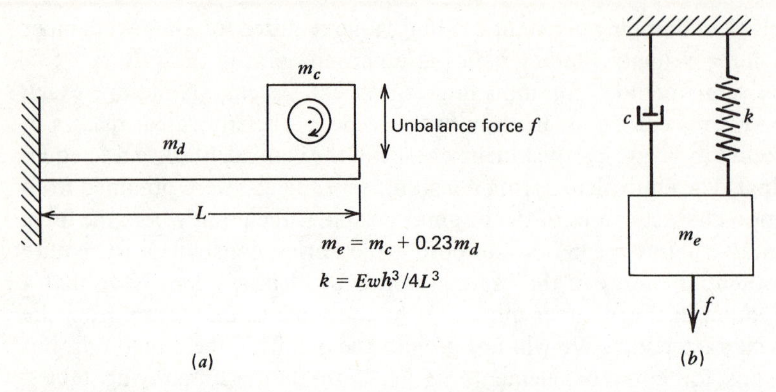

FIGURE 2.9 (*a*) A motor mounted on a cantilever beam. (*b*) Its equivalent representation.

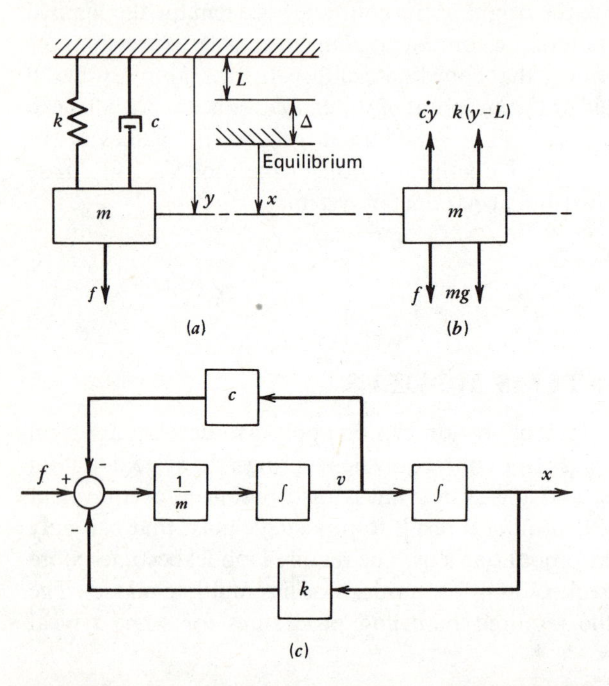

FIGURE 2.10 Second-order model for mechanical vibration. (*a*) Coordinates for mass, spring, and damper elements. (*b*) Free-body diagram. (*c*) Block diagram.

Example 2.1

Many mechanical systems can be represented as an equivalent mass–spring–damper system, such as in Figure 2.10*a*. Derive the equation of motion for the displacement, with the applied force f as the input.

Recall that the constitutive relation for a linear spring states that the spring's restoring force is directly proportional to the displacement from its free length L. If the mass m is attached to the spring and allowed to settle to its equilibrium position, the spring will be stretched by an amount Δ from its free length. The force balance at this equilibrium is (if $f=0$)

$$mg = k\Delta \tag{2.4-1}$$

where mg is the gravity force and $k\Delta$ is the spring force. This relation determines Δ.

Now consider the dynamic situation with $f \neq 0$. The damper force is $c\dot{y}$, and the spring force is $k(y-L)$. Newton's second law states that

$$m\ddot{y} = f - c\dot{y} - k(y-L) + mg \tag{2.4-2}$$

The model can be simplified by using the coordinate x, as defined in Figure 2.10*a*. The required relations are

$$y - L = x + \Delta$$
$$\dot{y} = \dot{x}$$
$$\ddot{y} = \ddot{x}$$

In terms of x, we have

$$m\ddot{x} = f - c\dot{x} - k(x + \Delta) + mg$$

In light of (2.4-1), the model becomes

$$m\ddot{x} = f - c\dot{x} - kx \tag{2.4-3}$$

In this form, the gravity term cancels the spring-force term $k\Delta$. This simplification is commonly used in vibration models; when the coordinate origin is located at the equilibrium position of the mass, the static-force terms do not appear in the resulting dynamic equation. This coordinate choice is often used but is sometimes not clearly stated. It is important to remember that the model form (2.4-3) applies only when x is measured from the equilibrium resulting when $f=0$.

Systems in Pure Rotation

When we study motor systems later on, we will see that motor shafts can behave like torsional springs when a torque is applied. Also, many automotive suspension designs depend on the restoring force generated by a rod in torsion for their performance. These are called torsion bar suspensions. The torsional spring model of a shaft is given in Table 2.3, entry 2. When the motor shaft is supported by bearings, damping often occurs due to the bearing friction. The damper model of a journal bearing is given in Table 2.4, entry 3.

Figure 2.11*a* represents a system with a shaft rigidly supported at one end. Usually, this end will be attached to another inertia. Including another inertia would produce a more complex model, so we will treat that case later. A torque T is applied to the cylinder with inertia I. This will cause the shaft to twist, and the twisting motion will be resisted by the bearing friction and the wall attachment. The inertias of the cylinders and the shaft can be found from Table 2.5, entry 1. Let us assume that the shaft's inertia is negligible, so that the total system inertia is I. Figure 2.11*b* shows the equivalent model, and Figure 2.11*c* shows the free-body diagram. From Newton's law

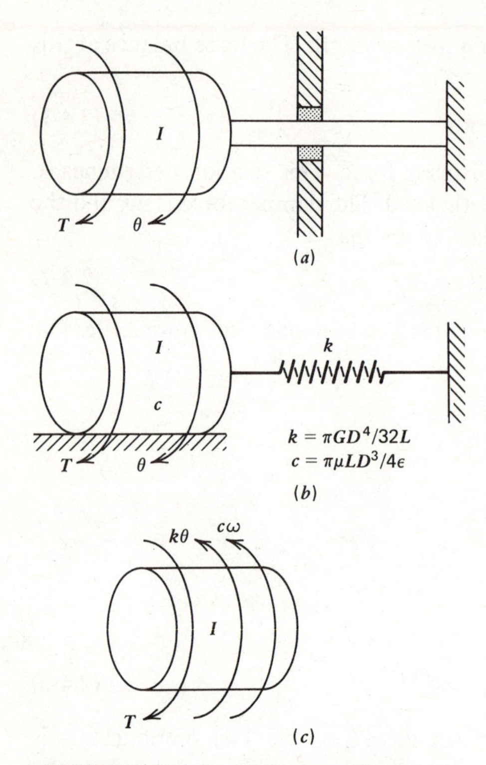

FIGURE 2.11 A torsional vibration example. (*a*) The shaft twists an amount θ. The applied torque is *T*. The shaft is also affected by viscous friction in the bearing. (*b*) The equivalent representation. (*c*) The free-body diagram.

for pure rotation, (2.3-12), with $\omega = \dot{\theta}$, we obtain

$$I\dot{\omega} = T - c\omega - k\theta$$

$$\dot{\omega} = \ddot{\theta}$$

So,

$$I\ddot{\theta} + c\dot{\theta} + k\theta = T \qquad \textbf{(2.4-4)}$$

Systems in Rotation Plus Translation

Systems whose motion consists of rotation and translation can often be modeled by the separate application of the appropriate form of Newton's law, (2.3-12) or (2.3-14). The equations will usually be coupled together by common variables. Algebraic manipulation can then be used to eliminate unwanted variables.

Example 2.2

A homogeneous cylinder of radius R and mass m is free to rotate about an axis that is connected to a support by a spring and damper (Figure 2.12*a*). Assume that the cylinder rolls without slipping on the surface of inclination α. Find the equation of motion.

Since the problem statement does not specify the coordinates to be used, we are free to make a convenient choice. Choose x_1 to represent the tangential displacement of the cylinder's center of mass from its equilibrium position. In this position, the

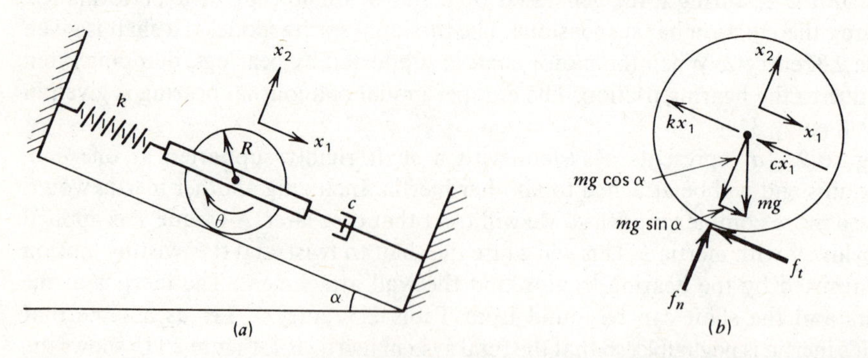

FIGURE 2.12 (*a*) Cylinder–spring–damper system. (*b*) Free-body diagram of the cylinder.

spring force is balanced by the tangential component of the cylinder's weight. Let x_2 be the direction perpendicular to x_1, normal to the inclined surface. Choose θ to be the positive angle of rotation when the cylinder moves in the positive x_1 direction.

The free-body diagram is given in Figure 2.12*b*. The spring force is kx_1, the damper force is cv_1, the normal component of the reaction force due to gravity is f_n, and the tangential reaction force is f_t. Since $f_n = mg \cos \alpha$, no motion occurs in the x_2 direction. For the x_1 direction, (2.3-14) gives

$$m\frac{dv_1}{dt} = -cv_1 - kx_1 + mg \sin \alpha - f_t - mg \sin \alpha \tag{2.4-5}$$

where $v_1 = \dot{x}_1$. The rotational motion is given by (2.3-12) as

$$I\frac{d\omega}{dt} = Rf_t \tag{2.4-6}$$

where $\omega = \dot{\theta}$.

The condition of no slipping is $R\theta = x_1$. Thus, $\dot{\omega} = \ddot{\theta} = \ddot{x}_1/R$, and (2.4-6) gives

$$f_t = \frac{I}{R^2}\ddot{x}_1$$

Substituting f_t into (2.4-5) results in

$$\left(m + \frac{I}{R^2}\right)\ddot{x}_1 + c\dot{x}_1 + kx_1 = 0$$

From Table 2.5, with $r = 0$, we have $I = \frac{1}{2}mR^2$, and the answer

$$1.5m\ddot{x}_1 + c\dot{x}_1 + kx_1 = 0 \tag{2.4-7}$$

Note that the effective mass for translation (1.5*m*) is greater than for pure translation because of the additional effort required to rotate the cylinder.

Multimass Systems

In situations where two or more masses are coupled together, make an assumption concerning the displacements and velocity of each mass relative to the other masses. This fixes the directions of the forces on the free-body diagram. Then apply Newton's law to each mass. The resulting equations will be independent of the relative motion assumptions, because the mathematics will take care of the sign reversals automatically.

Suppose that the cantilever-supported motor in Figure 2.9 is found to vibrate excessively after it is built. Rather than change the support, a *vibration absorber* can be added. This consists of a mass attached to the structure by means of an elastic element like a spring. This is shown in Figure 2.13*a*, where the absorber mass is m_2 and the absorber spring constant is k_2. The model for the resulting system is shown in Figure 2.13*b*, where m, c, k, and f are as defined for Figure 2.10, and $m = m_e$. The displacements x and x_2 are measured from the equilibrium positions of the masses.

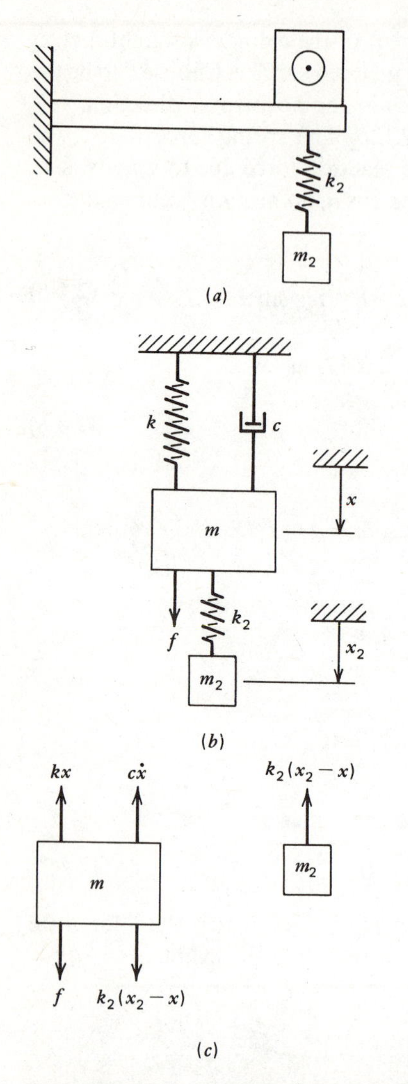

FIGURE 2.13 **(*a*) A vibration absorber of mass m_2 and elasticity k_2 attached to a motor. (*b*) The equivalent representation. (*c*) The free body diagrams for the assumption that $x > 0$, $\dot{x} > 0$, and $x_2 > x$.**

Assuming that $x > 0$, $\dot{x} > 0$, and $x_2 > x$, the free-body diagrams are as shown in Figure 2.13*c*. Newton's law applied to each mass gives

$$m\ddot{x} = f + k_2(x_2 - x) - c\dot{x} - kx$$
$$m_2\ddot{x}_2 = -k_2(x_2 - x)$$

Or

$$m\ddot{x} + c\dot{x} + (k + k_2)x = f + k_2x_2 \qquad \textbf{(2.4-8)}$$
$$m_2\ddot{x}_2 + k_2x_2 = k_2x \qquad \textbf{(2.4-9)}$$

In a later chapter, we will develop methods that can be used to select the values of m_2 and k_2 required to reduce the vibration of the motor.

Prescribed Motions as Inputs

The system in Figure 2.13 has a force as the input. In many applications, the input force is not known, but the device's motion is a specified function of time. In such cases, we may take the given motion to be the input if proper care is used in deriving the model. In Newtonian mechanics, force acting on a mass is the cause, and motion is its effect. The specified motion is caused by some force that might be unknown and might in fact be of little interest to us. In other words, when a motion input is specified, we are saying that the driven system has no choice but to move as specified by the input. When specifying a displacement or velocity input to a mass, we implicitly assume the existence of a force capable of producing the specified motion—the case for the system shown in Figure 2.14*a*. A rotating cam drives the mass, and the displacement $x(t)$ is determined by the cam profile and its rotational speed. No differential equation model is needed here. The designer might simply be interested in computing the force exerted on the support as a result of the motion. This force is $kx(t)$.

Motions also appear as inputs to spring and damper elements. Figures 2.14*b* and 2.14*c* show some examples. The vibration isolator (Figure 2.14*b*) is used to prevent the effects of the input displacement x from being transmitted to point 1. The arrangement

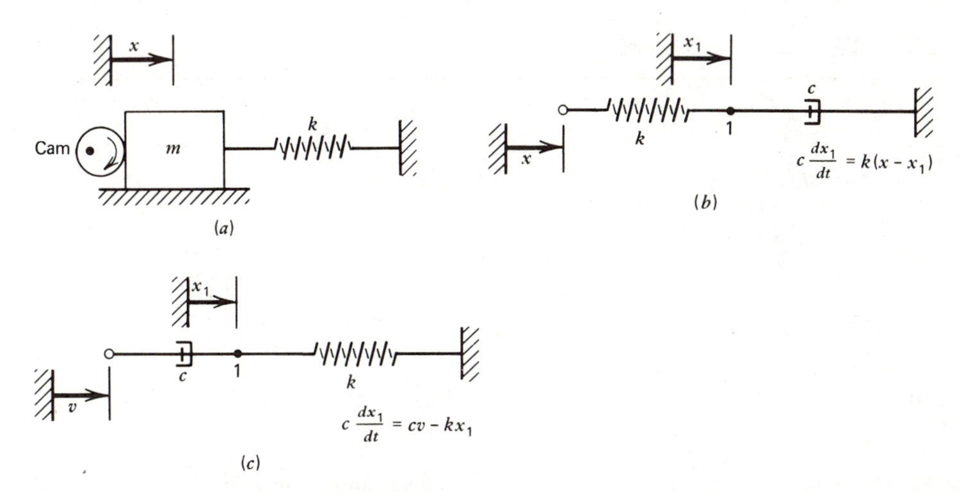

FIGURE 2.14 Some mechanical systems with motion inputs. (*a*) Cam-driven system. (*b*) Vibration isolator. (*c*) Velocity indicator.

in Figure 2.14*c* can be used in an instrument to provide a displacement at point 1 that is indicative of the input velocity magnitude.

Figures 2.14*b* and 2.14*c* imply that the mass of the system is considered to be negligible. Equations for such systems are easily derived by assuming the existence of a mass m at the point of interest, applying Newton's law, and then setting m to zero. For Figure 2.14*c*, we desire a model of the displacement x_1. For a mass m at point 1, its displacement is x_1, and Newton's law gives

$$m\ddot{x}_1 = c(v - \dot{x}_1) - kx_1$$

Set $m = 0$ to obtain

$$c\dot{x}_1 = cv - kx_1 \tag{2.4-10}$$

Geared Systems

Important examples of equivalent systems are given by devices called *mechanical transformers*. Such devices as levers, gears, cams, and chains transform the motion at the input into a related motion at the output. Their analysis is simplified by describing the effects of their mass, elasticity, and damping relative to a single location, such as the input location. The result is a lumped-parameter model whose coordinate system is centered at that location.

We can characterize a transformer by a single parameter b such that in terms of the general displacement variable Q,

$$Q_1 = bQ_2 \tag{2.4-11}$$

where the subscripts 1 and 2 refer to the input and output states, respectively. Since the rate variable r is related to Q by $\dot{Q} = r$, we also have

$$r_1 = br_2 \tag{2.4-12}$$

TABLE 2.7 Gear Pair Relations

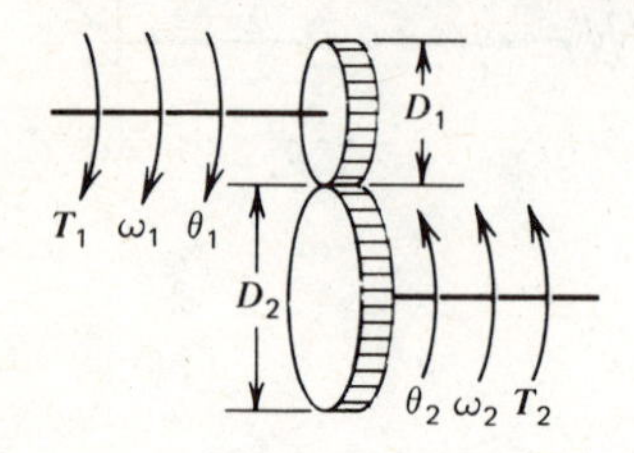

n = number of gear teeth

Gear ratio $\dfrac{D_2}{D_1} = \dfrac{n_2}{n_1} = N$

Velocity ratio $\dfrac{\omega_2}{\omega_1} = \dfrac{n_1}{n_2} = \dfrac{1}{N}$

Displacement ratio $\dfrac{\theta_2}{\theta_1} = \dfrac{n_1}{n_2}$

$= 1/N$

Torque ratio $\dfrac{T_2}{T_1} = \dfrac{n_2}{n_1} = N$

(neglects friction loss)

Original Geared System	**Equivalent Gearless System**

Inertia

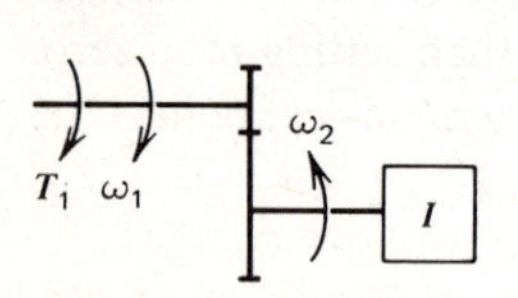

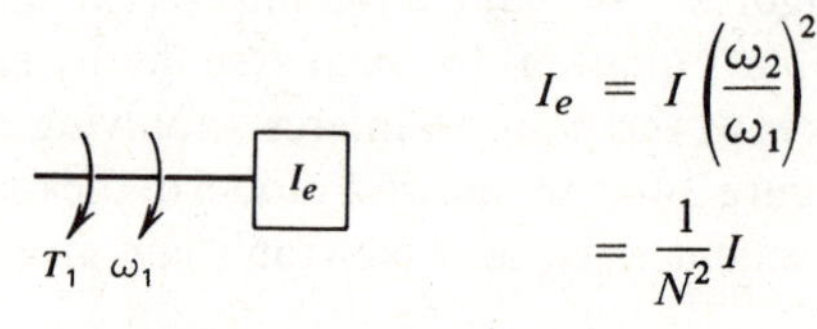

$$I_e = I\left(\frac{\omega_2}{\omega_1}\right)^2 = \frac{1}{N^2}I$$

(neglects gear inertias)

Spring constant

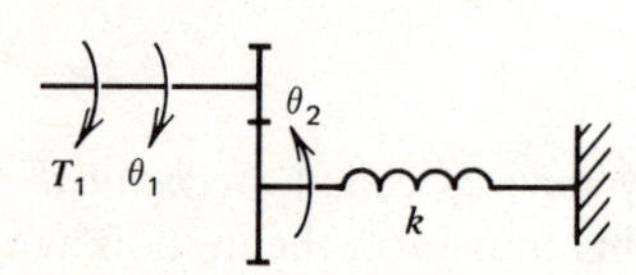

$$k_e = k\left(\frac{\theta_2}{\theta_1}\right)^2 = \frac{1}{N^2}k$$

(neglects gear-tooth elasticity)

Damping constant

$$c_e = c\left(\frac{\omega_2}{\omega_1}\right)^2 = \frac{1}{N^2}c$$

(neglects gear friction)

if b is constant. An ideal transformer neither stores nor dissipates energy. Therefore, the output power equals the input power, or $e_1 r_1 = e_2 r_2$. This shows the effort variables to be related as

$$e_1 = \frac{1}{b} e_2 \tag{2.4-13}$$

A specific example is the gear pair shown in Table 2.7, with $Q = \theta$, $e = T$, and $r = \omega$. The parameter b is the gear ratio N. If $N > 1$, the pair is a speed reducer whose output torque T_2 is greater than the input torque T_1.

Suppose an element possessing inertia, elasticity, or damping properties is connected to the output shaft of a gear pair, as shown in Table 2.7. How can we represent the geared system by an equivalent gearless system? If we wish to reference all motion to the input shaft, we can proceed as follows. We assume that the inertia, twisting, and damping of the input shaft are negligible. If not, they can be included in a manner similar to the following. We derive an equivalent system by requiring its kinetic energy, potential energy, and power dissipation terms to be identical to those of the original system. The expressions for these quantities in the equivalent system are

kinetic energy $= \frac{1}{2} I_e \omega_1^2$
potential energy $= \frac{1}{2} k_e \theta_1^2$
power dissipation $= T_1 \omega_1 = (c_e \omega_1)\omega_1 = c_e \omega_1^2$

If we neglect the gear inertias, gear tooth elasticity, and gear friction, the corresponding expressions for the original system are

kinetic energy $= \frac{1}{2} I \omega_2^2$
potential energy $= \frac{1}{2} k \theta_2^2$
power dissipation $= T_2 \omega_2 = c\omega_2^2$

Equating these expressions and using the gear ratio formulas, we obtain the parameters for the equivalent system

$$I_e = I\left(\frac{\omega_2}{\omega_1}\right)^2 = \frac{1}{N^2} I \tag{2.4-14}$$

$$k_e = k\left(\frac{\theta_2}{\theta_1}\right)^2 = \frac{1}{N^2} k \tag{2.4-15}$$

$$c_e = c\left(\frac{\omega_2}{\omega_1}\right)^2 = \frac{1}{N^2} c \tag{2.4-16}$$

The next two examples illustrate the lumping process.

Example 2.3

For systems with more than one inertia on a single shaft, it is frequently possible to add the inertias to produce an equivalent system. A pump is represented in Figure 2.15. The pump impeller I_2 is driven by the shaft I_3 connected to an electric

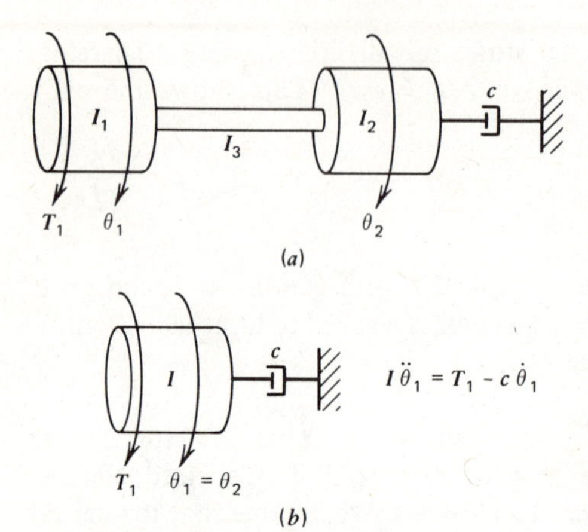

FIGURE 2.15 System with three inertias. (*a*) Original system. (*b*) Lumped-parameter equivalent system.

motor producing a torque T_1 on its rotor I_1. The viscous friction of the fluid acting on the impeller is represented by the damper. A model is desired for the behavior of the angular displacement θ_2 due to the applied torque T_1. Discuss how a lumped-parameter model can be developed (Figure 2.15*b*).

A good initial assumption might be to neglect the twist $\theta_1 - \theta_2$ between the inertias I_1 and I_2. This corresponds to neglecting the potential energy in the shaft and implies that $\theta_1 = \theta_2$ and $\omega_1 = \omega_2$. Equating the kinetic energy of the original and equivalent systems gives

$$\tfrac{1}{2}I\omega^2 = \tfrac{1}{2}I_1\omega_1^2 + \tfrac{1}{2}I_2\omega_1^2 + \tfrac{1}{2}I_3\omega_1^2$$

With a coordinate reference on the motor shaft, $\omega = \omega_1$, and thus,

$$I = I_1 + I_2 + I_3 \tag{2.4-17}$$

The equivalent system model is given by

$$I\ddot{\theta} = T_1 - c\dot{\theta}_1$$

or, since $\omega_1 = \dot{\theta}_1$,

$$I\dot{\omega}_1 = T_1 - c\omega_1 \tag{2.4-18}$$

If the damping force is large or if the torque T_1 is applied suddenly, the shaft twist will become important, and a model incorporating shaft elasticity will be needed.

Example 2.4

Suppose that the pump motor drives the impeller through a gear pair (Figure 2.16). Assume also that the gear and shaft inertias are significant. Neglect the elasticity of the gear teeth and shafts, and assume the friction and backlash in the gears are small. Develop a lumped-inertia model of the impeller velocity as a function of the motor torque.

The angular displacements of the gears are related through the gear ratio N such that

$$\theta_B = N\theta_C \tag{2.4-19}$$

$$N = \frac{D_C}{D_B}$$

Note that a positive displacement θ_B produces a positive displacement θ_C. Thus a negative sign is not needed in (2.4-19) even though the directions of rotation are opposite. From this, we also see that $\omega_B = N\omega_C$.

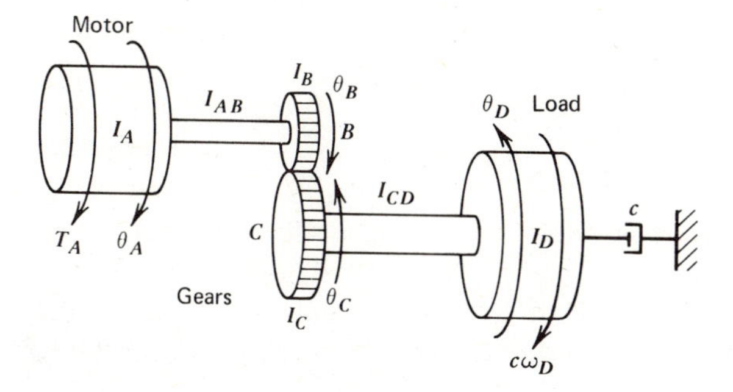

FIGURE 2.16 Geared system.

With negligible shaft elasticity, the twist is zero, and thus,

$$\theta_A = \theta_B$$
$$\omega_A = \omega_B$$
$$\theta_C = \theta_D$$
$$\omega_C = \omega_D$$

and

$$\omega_A = N\omega_D$$

Also, we can use the results from (2.4-17) to obtain I_1, the lumped inertia on shaft AB, as follows:

$$I_1 = I_A + I_{AB} + I_B$$

Similarly, the lumped inertia on shaft CD is

$$I_2 = I_C + I_{CD} + I_D$$

For a single-inertia model, we must reference all quantities to one shaft. Let us choose the motor shaft AB as the reference. Then, from Table 2.7, the inertia equivalent to I_2 referenced to shaft AB is

$$I_{e2} = \frac{1}{N^2} I_2$$

The total equivalent inertia I is the sum of all the inertias referenced to shaft AB, or

$$I = I_1 + I_{e2} \tag{2.4-20}$$

From Table 2.7, the viscous damping acting on shaft CD has an equivalent value on shaft AB of

$$c_e = \frac{1}{N^2} c \tag{2.4-21}$$

Therefore, the desired model is

$$I \frac{d\omega_A}{dt} = T_A - c_e \omega_A \tag{2.4-22}$$

with I and c_e given earlier.

Suppose that we had chosen shaft CD as the reference shaft. A good exercise is to use energy equivalence to derive the following model.

$$(I_1 N^2 + I_2)\frac{d\omega_D}{dt} = NT_A - c\omega_D \tag{2.4-23}$$

Of course, (2.4-22) and (2.4-23) are equivalent, as can be verified by substituting $\omega_A = N\omega_D$ into (2.4-22).

This model can be used to compute the motor torque required to maintain the load at some desired constant speed. For constant speed, (2.4-22) shows the required torque to be

$$T_A = c_e\omega_A = c_e N\omega_D \tag{2.4-24}$$

The time to achieve this speed from rest can also be estimated with this model using the techniques of Chapter 3. Lumping the two gears into one inertia prevents the model from giving information on the gear tooth forces, the effects of shaft elasticity, gear-bearing friction, and backlash. Such information must be obtained from multimass, higher order models. However, in many applications, the simple model (2.4-22) is adequate.

Levered Systems

Another example of a mechanical transformer is the lever shown in Table 2.8. The transformer parameter is the lever ratio a/b. For small angular displacement θ, geometry shows that $x = a\theta$ and $y = b\theta$. The lever is an ideal transformer (no power loss) if its mass, elasticity, and pivot friction are negligible. In this case, the input work $f_1 x$ equals the output work $f_2 y$. Thus, $f_2 = (a/b)f_1$.

We can choose to lump a levered system in terms of any one of the coordinates x, y, or θ. If the system's mass, elasticity, damping, and input force are concentrated at the left-hand end, the lever has no effect if we choose the coordinate x (Table 2.8). An equivalent model in terms of either y or θ can be found as for the gear pair by equating corresponding energy terms. The results are given in Table 2.8. They can also be used for systems whose mechanical properties are located at more than one point, as in the following example.

Example 2.5

Derive an equivalent model in terms of the coordinate x, and another in terms of θ, for the system shown in Figure 2.17.

Relative to the left-hand end, the equivalent values of k_2, m_b, and f_2 are

$$k_{2e} = k_2(b^2/a^2)$$
$$m_{be} = m_b(b^2/a^2)$$
$$f_{2e} = f_2(b/a)$$

Since c and m_a are located at the point of reference, they are unchanged. The parameters of the complete equivalent system are found by adding the values

TABLE 2.8 Lever Relations

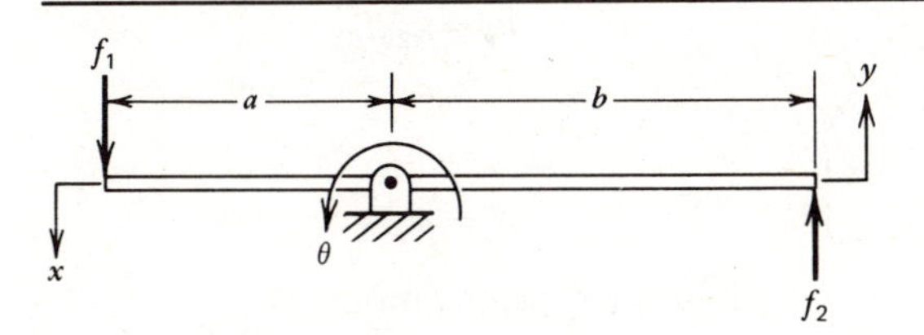

Lever ratio $\frac{x}{y} = \frac{a}{b}$ (for small displacements)

Force advantage $f_2 = \frac{a}{b} f_1$

Original levered system

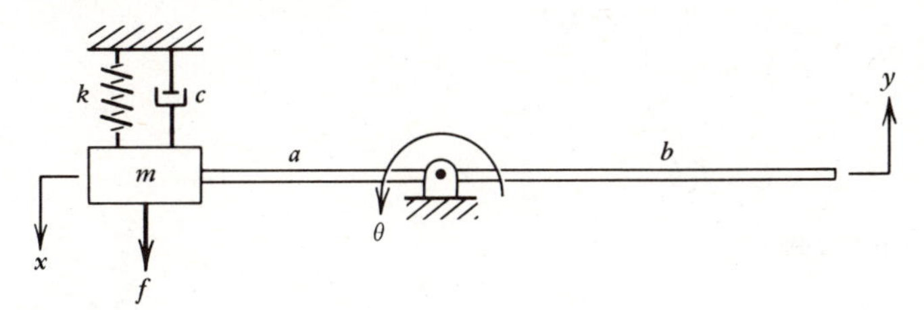

Equivalent Systems
(for small displacements and negligible lever mass, lever elasticity, and pivot friction)

Lumped for coordinate x

$$m_e = m$$
$$k_e = k$$
$$c_e = c$$
$$f_e = f$$

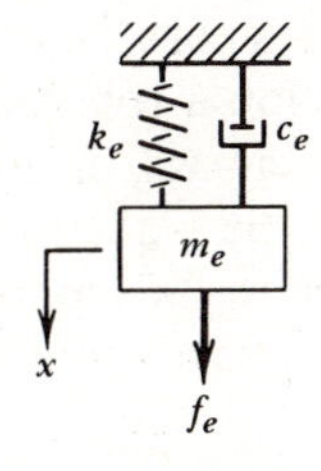

Lumped for coordinate y

$$m_e = m(a^2/b^2)$$
$$k_e = k(a^2/b^2)$$
$$c_e = c(a^2/b^2)$$
$$f_e = f(a/b)$$

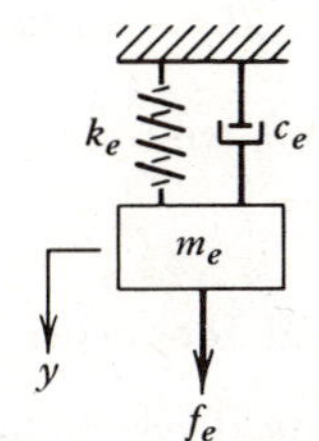

Lumped for coordinate θ

$I_e = ma^2 \qquad k_e = ka^2$

$c_e = ca^2 \qquad T_e = fa$

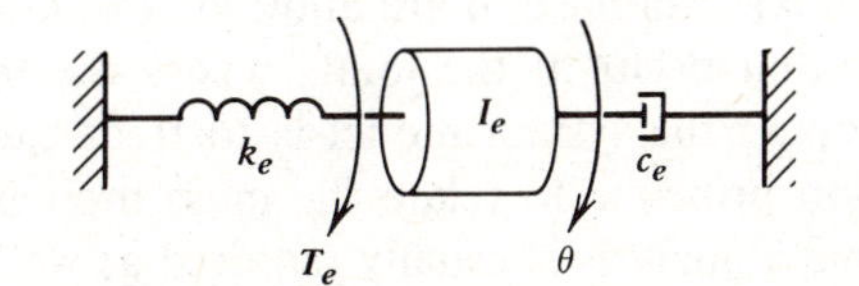

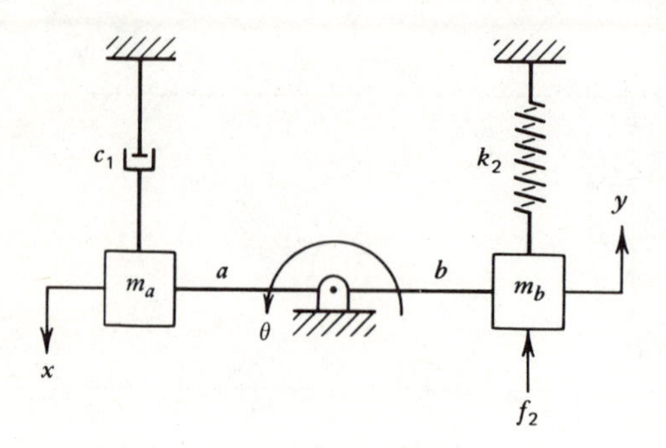

FIGURE 2.17 Levered system.

referenced to the left-hand end. They are

$$k_e = k_{2e}$$
$$c_e = c$$
$$f_e = f_{2e}$$
$$m_e = m_a + m_{be}$$

Thus, the dynamic model is

$$m_e \ddot{x} = -c_e \dot{x} - k_e x + f_e \quad \textbf{(2.4-25)}$$

For a rotational model, all the parameters must be changed, because none of them are located at the pivot. The equivalent values at the pivot are

$$I_e = m_a a^2 + m_b b^2$$
$$k_e = k_2 b^2$$
$$c_e = c_1 a^2$$
$$T_e = f_2 b$$

The equivalent dynamic model is

$$I_e \ddot{\theta} = -c_e \dot{\theta} - k_e \theta + T_e \quad \textbf{(2.4-26)}$$

Instead of energy equivalence, direct application of Newton's laws can be used to derive the models of levered systems. The choice depends on the specific problem and personal preference. To develop a rotational model for Figure 2.17 using Newton's law, note that from Table 2.5, entry 2, the inertias of m_a and m_b relative to the pivot are $m_a a^2$ and $m_b b^2$. The total inertia is the sum and given by I_e, shown previously. The torque about the pivot is due to the damper force $c_1 \dot{x}$, the spring force $k_2 y$, and the applied force f_2. These forces have moment arms of a, b, and b, respectively. Thus, Newton's law gives

$$I_e \ddot{\theta} = -c_1 a \dot{x} - k_2 b y + f_2 b$$

Substituting $\dot{x} = a\dot{\theta}$ and $y = b\theta$, we obtain (2.4-26).

Kinematic Constraints and Reaction Forces

Often the masses in a mechanical system are linked together in such a way that only certain types of motion are allowed. The kinematics are constrained by the reaction forces that occur at the joints where the masses are connected. One approach to developing the system model is to treat each mass separately, and use the action-reaction principle to relate the mass models to one another mathematically. Some kinematic analysis is usually required as well, in order to obtain the system model in terms of the desired coordinates. The following example illustrates the procedure.

Example 2.6

A simplified representation of a gantry robot is shown in Figure 2.18(*a*). The main body of mass M is propelled along a horizontal track by a traction force f. The main body contains an actuator for rotating the arm, which will have a grasping device like a hand at its end for picking up objects. The actuator applies a torque T to the arm. The arm, hand, and grasped object have a total mass m and moment of inertia I relative to its mass center at point C. Find the equations of motion for x_1 and θ in terms of the given quantities f and T.

The free body diagrams for the masses M and m are shown in Figure 2.18 (*b*) and (*c*). F_x and F_y are the reaction forces in the x and y directions due to the physical connection between the two masses. Note that they are applied in opposite directions on each mass. A similar statement applies to the torque T. R_1 and R_2 are the vertical reaction forces on the wheels. Newton's law applied to mass M in the x and y directions gives

$$M\ddot{x}_1 = F_x + f \tag{2.4-27}$$

$$M\ddot{y}_1 = F_y + R_1 + R_2 - Mg \tag{2.4-28}$$

Equation (2.4-28) is of no interest unless we need to find R_1 and R_2. For mass m,

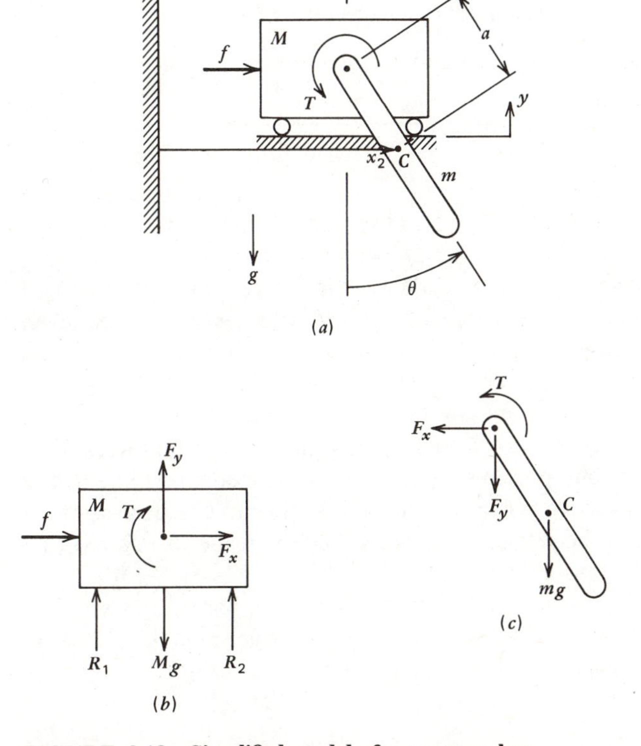

FIGURE 2.18 Simplified model of a gantry robot.

Newton's law applied to the center of mass gives

$$m\ddot{x}_2 = -F_x \tag{2.4-29}$$

$$m\ddot{y}_2 = -F_y - mg \tag{2.4-30}$$

$$I\ddot{\theta} = aF_x \cos\theta + aF_y \sin\theta + T \tag{2.4-31}$$

To eliminate F_x use (2.4-27) and (2.4-29) to obtain

$$M\ddot{x}_1 + m\ddot{x}_2 = f \tag{2.4-32}$$

Use (2.4-27) and (2.4-30) to eliminate F_x and F_y in (2.4-31). This gives

$$I\ddot{\theta} + m\ddot{x}_2\, a\cos\theta + ma(\ddot{y}_2 + g)\sin\theta = T \tag{2.4-33}$$

The displacements x_2 and y_2 are related to θ as follows.

$$x_2 = x_1 + a\sin\theta$$

$$y_2 = -a\cos\theta$$

Thus

$$\ddot{x}_2 = \ddot{x}_1 - a\sin\theta\dot{\theta}^2 + a\cos\theta\ddot{\theta} \tag{2.4-34}$$

$$\ddot{y}_2 = a\cos\theta\dot{\theta}^2 + a\sin\theta\ddot{\theta} \tag{2.4-35}$$

Substituting $\ddot{x}_2$, $\ddot{y}_2$ into (2.4-32) and (2.4-33) gives the desired result:

$$(I + ma^2)\ddot{\theta} + ma\cos\theta\ddot{x}_1 + mga\sin\theta = T \tag{2.4-36}$$

$$(M + m)\ddot{x}_1 - ma\sin\theta\dot{\theta}^2 + ma\cos\theta\ddot{\theta} = f \tag{2.4-37}$$

These equations can be used to find the $T(t)$ and $f(t)$ required to produce a specified path $x_1(t)$, $\theta(t)$.

Lagrange's Equations

As seen in Example 2.6, direct application of Newton's law to each mass in a multi-mass system requires that the reaction forces be explicitly accounted for. Then, once all the governing equations have been obtained, the reaction forces must be eliminated algebraically in order to obtain the equations of motion in terms of only the given input and output variables. This was not too difficult to accomplish in the previous example, but as the number of masses increases or the kinematic constraints become more complex, it becomes tedious to eliminate the reaction forces. There is however, another method available that avoids the need to deal with the reaction forces. This method, due to Lagrange, is briefly presented here so that the student will be aware that there is more than one way to solve a dynamics problem. The derivation of Lagrange's equations is beyond the scope of this book, and the reader is referred to a text on advanced dynamics for more discussion.

The Lagrangian method is energy based rather than force based, and requires that the user derive expressions for the system's potential and kinetic energies. Denote these energies by P and K respectively. Then Lagrange's equations are

$$\frac{d}{dt}\left(\frac{\partial K}{\partial \dot{q}_j}\right) - \frac{\partial K}{\partial q_j} + \frac{\partial P}{\partial q_j} = Q_j \qquad j = 1, 2, \ldots, n \tag{2.4-38}$$

where the variables q_j are a set of coordinates that completely describe the system's motion. The corresponding velocities are $\dot{q}_j$, and Q_j represents the forces or moments at the coordinate q_j. Q_j can represent either an externally applied force or moment, or one due to a dissipative effect such as a damping or friction force. The identification of q_j and Q_j might seem somewhat vague at this point, but is usually easily done in practice. The following example illustrates this.

Example 2.7

Consider the system of Example 2.6 (Figure 2.18). Derive the equations of motion using Lagrange's equations.

It is easily seen that x_1 and θ are convenient choices for q_1 and q_2 because it is then possible to choose $Q_1 = f$ and $Q_2 = T$. The next step is to develop the expressions for the kinetic and potential energies. The kinetic energy of M is due solely to horizontal translation, while that for m is due to translation of the mass center C in the horizontal and vertical directions, plus the rotation about the mass center. Thus the system's kinetic energy is

$$K = \tfrac{1}{2}M\dot{x}_1^2 + \tfrac{1}{2}mv_2^2 + \tfrac{1}{2}I\dot{\theta}^2 \tag{2.4-39}$$

where from the Pythagorean theorem,

$$v_2^2 = \dot{x}_2^2 + \dot{y}_2^2 = (\dot{x}_1 + a\cos\theta\dot{\theta})^2 + (a\sin\theta\dot{\theta})^2 \tag{2.4-40}$$

Thus, because $\sin^2\theta + \cos^2\theta = 1$,

$$K = \tfrac{1}{2}(M+m)\dot{x}_1^2 + ma\dot{x}_1\cos\theta\dot{\theta} + \tfrac{1}{2}(ma^2 + I)\dot{\theta}^2 \tag{2.4-41}$$

The system's potential energy is due entirely to the vertical displacement of the center of mass of I. Taking the energy to be zero when $y_2 = -a$, we obtain

$$P = mga(1 - \cos\theta) \tag{2.4-42}$$

Lagrange's equations (2.4-38) require the following derivatives:

$$\frac{\partial K}{\partial \dot{x}_1} = (M+m)\dot{x}_1 + ma\cos\theta\dot{\theta}$$

$$\frac{\partial K}{\partial x_1} = \frac{\partial P}{\partial x_1} = 0$$

$$\frac{\partial K}{\partial \dot{\theta}} = ma\dot{x}_1\cos\theta + (ma^2 + I)\dot{\theta}$$

$$\frac{\partial K}{\partial \theta} = -max_1\sin\theta\dot{\theta} \qquad \frac{\partial P}{\partial \theta} = mga\sin\theta$$

Thus for $j = 1$, (2.4-38) gives

$$\frac{d}{dt}[(M+m)\dot{x}_1 + ma\cos\theta\dot{\theta}] = f$$

or

$$(M+m)\ddot{x}_1 + ma\cos\theta\ddot{\theta} - ma\sin\theta\dot{\theta}^2 = f$$

which is identical to (2.4-37). For $j = 2$, we have

$$\frac{d}{dt}[ma\dot{x}_1 \cos\theta + (ma^2 + I)\dot{\theta}] + ma\dot{x}_1 \sin\theta\dot{\theta} + mga\sin\theta = T$$

or

$$ma\cos\theta\ddot{x}_1 - ma\dot{x}_1\sin\theta\dot{\theta} + (ma^2 + I)\ddot{\theta} + ma\dot{x}_1\sin\theta\dot{\theta} + mga\sin\theta = T$$

which gives (2.4-36) after the terms are combined.

In some applications the reaction forces must be found for mechanical design purposes. In such cases the Lagrangian approach does not directly give all the necessary information. In addition, although not the case in the previous example, Lagrange's method sometimes produces a set of equations that is not as computationally efficient as those derived with the Newton method. If the equations must be solved in real-time for control purposes, this can be a significant reason for choosing the Newton method.

2.5 ELECTROMECHANICAL SYSTEMS

Our principal interest in electrical systems here is concerned with those devices used for measurement and control instead of applications in power transmission or communications, for example. The theory of electrical devices can be divided into direct-current (dc) or alternating-current (ac) theory. We emphasize dc theory and applications here primarily, because advances in permanent magnet technology have made it possible to use dc motors in many control system applications that formerly required ac motors. Thus, the associated circuitry is now more often direct-current. Since a dc motor is an electromechanical system, the appropriate models are combinations of the rotational models in Section 2.4 and the electrical models in Section 2.2.

Circuit elements may also be classified as active or passive. Such passive elements as resistors, capacitors, or inductors are not sources of energy, although the last two can store it temporarily. The active elements are energy sources that drive the system. Their physical type can be chemical (batteries), mechanical (generators), thermal (thermocouples), and optical (solar cells).

Active elements are modeled as either ideal voltage sources or ideal current sources. Their circuit symbols are given in Table G.3. An ideal voltage source supplies the desired voltage no matter how much current is drawn by the loading circuit. The opposite is true for a current source. Obviously, no real source behaves exactly this way. For example, battery voltage drops because of heat generation as more current is drawn. If the current is small or constant in a given application, the battery can be treated as an ideal source. If not, the battery is frequently modeled as an ideal voltage source plus an internal resistance whose value can be obtained from the voltage–current curve for the battery.

A Direct-Current Motor System

Figure 2.19 shows an electromechanical system consisting of an armature-controlled dc motor driving a load inertia. The armature is the rotating part of the motor. It usually consists of a wire conductor wrapped around an iron core. This winding has an inductance L, shown in the figure. The resistance R in the figure represents the lumped value of the armature resistance and any external resistance deliberately introduced to change the motor's behavior.

The armature is surrounded by a magnetic field. The reaction of this field with the armature current produces a torque that causes the armature to rotate. If the armature voltage v is used to control the motor, the motor is said to be *armature-controlled.* In this case, the field is produced by an electromagnet supplied with a constant voltage or by a permanent magnet. It is now possible to manufacture permanent magnets of high field intensity and armatures of low inertia so that motors with a high torque-to-inertia ratio are available. This has opened up new areas of application for dc motors in control systems. This is fortunate, because the speed of a dc motor is easier to control than that of an ac motor. Also, the analysis and design of the associated dc circuitry is simpler than for ac systems.

Example 2.8

Develop a model for the armature-controlled system shown in Figure 2.19, with the armature voltage v as input and the load speed ω as output.

This motor type produces a torque T that is proportional to the armature

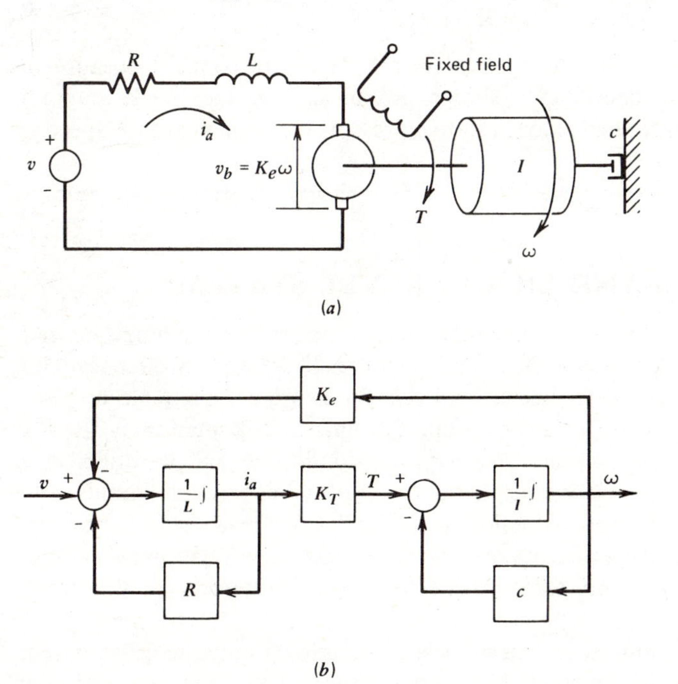

FIGURE 2.19 Armature-controlled dc motor with a load and its block diagram.

current i_a.

$$T = K_T i_a \tag{2.5-1}$$

The torque constant K_T depends on the strength of the field and other details of the motor's construction. The *user* of such motors (as opposed to the motor's *designer*) can obtain values of K_T for a specific motor from the manufacturer's literature.

The motion of a current-carrying conductor in a field produces a voltage in the conductor that opposes the current. This voltage is called the *back emf* (electromotive force). Its magnitude is proportional to the speed and is given by

$$e_b = K_e \omega \tag{2.5-2}$$

The value of the constant K_e is found with other product data (it is sometimes called the *voltage constant*). Equations (2.5-1) and (2.5-2) are constitutive relations for the motor.

The back emf is a voltage drop in the armature circuit. Thus, Kirchhoff's voltage law gives

$$v = i_a R + L\frac{di_a}{dt} + K_e \omega \tag{2.5-3}$$

From Newton's law applied to the inertia I,

$$\begin{aligned} I\frac{d\omega}{dt} &= T - c\omega \\ &= K_T i_a - c\omega \end{aligned} \tag{2.5-4}$$

These two equations constitute the system model. They are coupled because of the back emf term. This coupling is shown graphically by the block diagram (Figure 2.19*b*), in which the outer feedback loop represents the influence of the speed ω on the current i_a.

2.6 HYDRAULICS AND LIQUID LEVEL SYSTEMS

We can model fluid behavior as *incompressible*, which means that the fluid's density remains constant despite changes in the fluid pressure. If the density changes with pressure, the fluid is *compressible*. Because all real fluids are compressible to some extent, incompressibility is an approximation. But this approximation is usually sufficiently accurate for most liquids under typical conditions, and it results in a simpler model of the system. *Hydraulics* is the study of incompressible liquids, and hydraulic devices use an incompressible liquid, such as oil, for their working medium. Liquid level systems consisting of storage tanks and connecting pipes are a class of hydraulic systems whose driving force is due to relative differences in the liquid heights in the tanks.

Here, we will avoid complex system models by describing only the gross system behavior instead of the details of the fluid motion patterns. To do this, we need only one basic physical law: conservation of mass. For incompressible fluids, conservation of mass is equivalent to conservation of volume, since the fluid density is constant. If we know the density and the volume flow rate, we can compute the mass flow rate. That is, $q_m = \rho q$, where q_m and q are the mass and volume flow rates, and ρ is the fluid density. Typical units for these and other fluid quantities are given in Tables G.4 and G.5.

Basic Principles

The primary variables for hydraulic systems are *pressure*, *mass*, and *mass flow rate*. There is no variable equivalent to the general flux variable unless we choose to model the fluid forces in detail. The system must then be described as a mechanical system with distributed mass, and Newton's laws apply. We choose not to do that here, because such detailed models are usually not required for control system design.

Fluid resistance is the constitutive relation between pressure and mass flow rate. *Fluid capacitance* is the constitutive relation between pressure and mass. Since we choose not to use a flux variable, there will be no inductance relation. Integral causality relates mass m to mass flow rate, because

$$m = \int q_m \, dt \tag{2.6-1}$$

Conservation of mass can be stated as follows. For a container holding a mass of fluid m, the time rate of change $\dot{m}$ of mass in the container must equal the total mass inflow rate minus the total mass outflow rate. That is,

$$\dot{m} = q_i - q_o \tag{2.6-2}$$

where q_i is the inflow rate and q_o is the outflow rate. The fluid mass m is related to the container volume V by

$$m = \rho V \tag{2.6-3}$$

For an incompressible fluid, ρ is constant, and $\dot{m} = \rho \dot{V}$. Let q_1 and q_2 be the total *volume* inflow and outflow rates. Thus, $q_i = \rho q_1$ and $q_o = \rho q_2$. Substituting these relationships into (2.6-2) gives

$$\rho \dot{V} = \rho q_1 - \rho q_2$$

or

$$\dot{V} = q_1 - q_2 \tag{2.6-4}$$

This is a statement of conservation of *volume* for the fluid, and it is equivalent to conservation of mass, equation (2.6-2).

Fluid Capacitance and Resistance

The tank shown in Figure 2.20 illustrates these concepts. Let A be the surface area of the tank's bottom. If the tank's sides are vertical, the liquid height h is related to m by

$$m = \rho A h \tag{2.6-5}$$

In analogy with the electrical capacitance relation, $v = Q/C$, pressure plays the role of

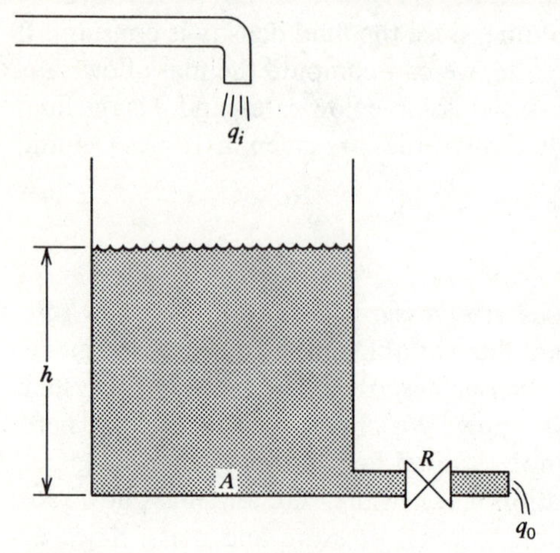

FIGURE 2.20 A liquid-level system.

voltage, and mass m plays the role of charge. Thus, the fluid capacitance relation is

$$p = \frac{m}{C} \tag{2.6-6}$$

where C is the fluid capacitance of the system. The hydrostatic pressure due to the height h is

$$p = \rho g h \tag{2.6-7}$$

where g is the acceleration due to gravity. Comparing (2.6-5)–(2.6-7), we see that

$$p = \frac{mg}{A} \tag{2.6-8}$$

Thus, the fluid capacitance of the tank in Figure 2.20 is $C = A/g$.

When the container does not have vertical sides, the relations between m and h, and between p and m are nonlinear. In this case, we define the capacitance to be the slope of the m versus p curve, and thus there is no single value for the container's capacitance. An example of this is given in the chapter problems.

With reference to Figure 2.20, if atmospheric pressure p_a exists at both the liquid's surface and the pipe outlet, the pressure difference across the pipe is $(p + p_a) - p_a = p$. The outflow rate q_o should obviously depend on p somehow. The greater the pressure p, the greater the outflow rate. For now, let us assume that the relation between q_o and p is linear. In analogy with the electrical resistance relation, $i = v/R$, the linear fluid resistance relation is written as

$$q_o = \frac{p}{R} \tag{2.6-9}$$

where R is the fluid resistance.

Liquid Height Model

We are now prepared to develop a complete model for Figure 2.20. Assume that we are given q_i as a function of time and that we want a model for the behavior of the height h. Using (2.6-2), (2.6-5), (2.6-7), and (2.6-9), we obtain

$$\rho A \dot{h} = q_i - \frac{\rho g}{R} h \tag{2.6-10}$$

If we write this equation in terms of the volume inflow rate q_1 instead of q_i, the density ρ can be divided out of the equation to obtain

$$A \dot{h} = q_1 - \frac{g}{R} h \tag{2.6-11}$$

General Resistance Relations

Fluid trying to flow out of a container, as in Figure 2.20, can meet with resistance in several ways. If the outlet is simply a hole, the fluid meets resistance merely because it cannot squeeze through the hole as fast as it would like. This is known as *orifice* flow. If the outlet is a pipe, the friction between the fluid and the pipe walls produces resistance to flow. Finally, the presence of a component, such as a valve in the pipe, increases the resistance. We now consider appropriate models for each type of resistance. For the derivations, more details on the resistance formulas, and some typical values, consult Reference 1 or a fluid mechanics text, such as Reference 2.

Pipe Flow:

Under certain conditions, the flow in the pipe will be smooth in the sense that the average fluid particle velocity equals the actual particle velocity. This condition is called *laminar* flow. Under other conditions, the flow will be rough, and the average particle velocity will be less than the actual particle velocity, because the fluid particles "meander" while moving downstream. This is *turbulent* flow. Determining the conditions under which laminar or turbulent flows exist is a more advanced problem in fluid mechanics and need not concern us here.

If the pipe flow is laminar, the linear relation (2.6-9) applies, and the *laminar resistance* R_L for a level pipe of diameter D and length L is given by the Hagen–Poiseuille formula

$$R_L = \frac{128\mu L}{\pi \rho D^4} \tag{2.6-12}$$

where μ is the fluid viscosity. Typical values for μ are given in Table G.6. The resistance relation for the turbulent pipe flow is nonlinear. In this case, (2.6-9) is replaced by

$$R_T q_o^2 = p \tag{2.6-13}$$

where R_T is the *turbulent resistance* whose value must be partly determined experimentally.

Component Resistance:

Components, such as valves, elbow bends, and couplings, resist flow and can also be described by (2.6-9) for small pressure drops and by (2.6-13) for higher pressure drops. In most cases, the component induces turbulent flow at typical pressures, and (2.6-13) is normally used. Experimentally determined values of R_T are available for common types of components (References 1, 2, and 4).

Orifice Flow:

If the size of the orifice is small enough so that the pressure variation over the orifice area is negligible compared to the average pressure at the orifice, it can be shown from conservation of energy that the mass flow rate through the orifice is

$$q_o = C_d A_o \sqrt{2p/\rho} \tag{2.6-14}$$

where A_o is the orifice area and C_d is an experimentally determined number called the *discharge coefficient*. For a liquid level system with a circular orifice, the preceding assumption implies that the liquid height above the orifice must be large compared to the orifice diameter.

The orifice relation (2.6-14) has the same form as the turbulent relation (2.6-13). To see this, square both sides of (2.6-14) and rearrange to obtain

$$R_o q_o^2 = p \tag{2.6-15}$$

where the *orifice resistance* is defined as

$$R_o = \frac{1}{2\rho C_d^2 A_o^2} \tag{2.6-16}$$

The restriction symbol in the outlet pipe in Figure 2.20 represents fluid resistance due to any of the preceding causes. If the resistance is due to laminar pipe flow, the system model is (2.6-11). If the pipe flow is turbulent, we use (2.6-13) to solve for q_o

$$q_o = \sqrt{\frac{p}{R_T}} \tag{2.6-17}$$

The positive square root is taken, because a positive value for p corresponds to an outflow ($q_o > 0$). From (2.6-2), (2.6-5), (2.6-7), and (2.6-17), we obtain the nonlinear model

$$\rho A\dot{h} = q_i - \sqrt{\frac{\rho g h}{R_T}} \tag{2.6-18}$$

These relations are summarized in Table 2.9.

TABLE 2.9 Hydraulic Resistance Relations

$q \rightarrow$

p_1 —⋈— p_2

$q =$ mass flow rate

$p =$ pressure drop $= p_1 - p_2$

Case	Relation
1. Laminar pipe flow.	$q = \frac{1}{R} p$
	$R = \frac{128\mu L}{\pi \rho D^4}$ [see (2.6-12)]
2. Turbulent pipe flow and flow through components.	$q = \sqrt{p/R}$ R found from experiments.
3. Orifice flow.	$q = \sqrt{p/R}$
	$R = \frac{1}{2\rho C_d^2 A_o^2}$ [see (2.6-16)]

Interconnected Storage Elements

When a hydraulic system contains more than one storage element, apply the conservation of mass equation (2.6-2) to each element. Then use the appropriate resistance relations to couple the resulting equations. It is necessary to assume that some pressures are greater than others and assign the positive-flow directions accordingly. If you are consistent, the mathematics will handle the reversals of flow direction automatically.

Example 2.9

Develop a model for the heights h_1 and h_2 in the liquid level system shown in Figure 2.21*a*. The input *volume* flow rate q is given. Assume that laminar flow exists in the pipes. The laminar resistances are R_1 and R_2, and the bottom areas of the tanks are A_1 and A_2. Also, draw the system block diagram.

Assume that $h_1 > h_2$ so that the mass flow rate q_1 is positive if going from tank 1 to tank 2. Conservation of mass applied to each tank gives

$$\rho A_1 \dot{h}_1 = \rho q - q_1$$

$$q_1 = \frac{\rho g}{R_1}(h_1 - h_2)$$

For tank 2,

$$\rho A_2 \dot{h}_2 = q_1 - q_o$$

$$q_o = \frac{\rho g}{R_2} h_2$$

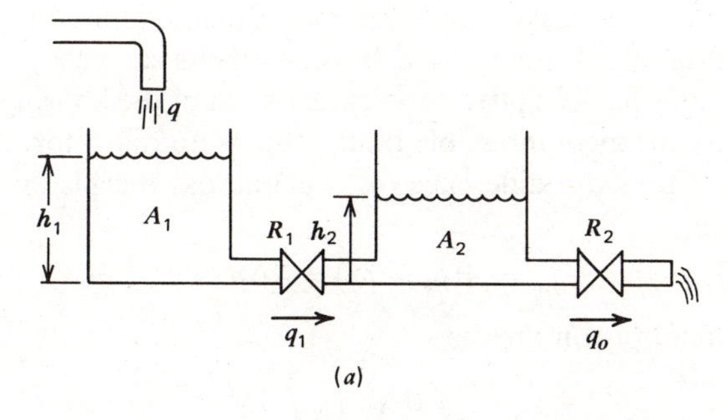

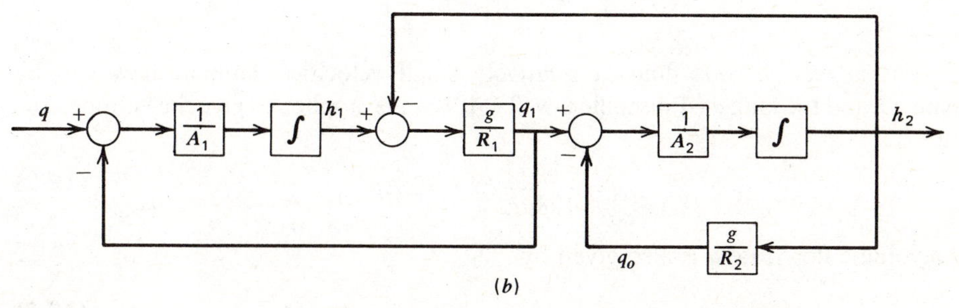

FIGURE 2.21 A system of two coupled tank elements and its block diagram.

Substituting for q_1 and q_o and dividing by ρ gives the desired model.

$$A_1 \dot{h}_1 = q - \frac{g}{R_1}(h_1 - h_2) \tag{2.6-19}$$

$$A_2 \dot{h}_2 = \frac{g}{R_1}(h_1 - h_2) - \frac{g}{R_2} h_2 \tag{2.6-20}$$

The density ρ does not appear in the final model because of the incompressibility assumption. The block diagram shown in Figure 2.21*b* is easily constructed from the preceding equations.

2.7 HYDRAULIC DEVICES

Hydraulic devices use an incompressible liquid, such as oil, for the working medium. They are widely used with high pressures to obtain large forces in control systems, allowing large loads to be moved or high accelerations to be obtained. Here, we analyze three devices commonly found in hydraulic systems: (1) dampers limit the velocity of mechanical elements; (2) hydraulic servomotors produce motion from a pressure source; (3) accumulators reduce fluctuations in pressure or flow rates.

A Fluid Damper

A *damper* or *dashpot* is a mechanical device that exerts a force as a result of a velocity difference across it. One common design relying on fluid friction is shown in Figure 2.22, and it is similar to that used in automotive shock absorbers. A piston of diameter D and thickness L has a cylindrical hole of diameter d drilled through it. The housing is assumed to be stationary here, but the results can easily be generalized to include a movable housing. The piston rod extends out of the housing, which is sealed and filled with a viscous incompressible fluid, such as an oil. A force f acting on the piston rod creates a pressure difference $(p_1 - p_2)$ across the piston such that if the acceleration is small,

$$f = A(p_1 - p_2) = A\Delta p \tag{2.7-1}$$

The net cross-sectional piston area is

$$A = \pi\left(\frac{D}{2}\right)^2 - \pi\left(\frac{d}{2}\right)^2$$

For a very viscous fluid at relatively small velocities, laminar flow can be assumed, and the Hagen–Poiseuille law (2.6-12) can be applied to give the volume flow rate

$$q = \frac{\pi d^4}{128\mu L}\Delta p = C_1 \Delta p \tag{2.7-2}$$

The volume flow rate q is also given by

$$q = A\frac{dx}{dt} \tag{2.7-3}$$

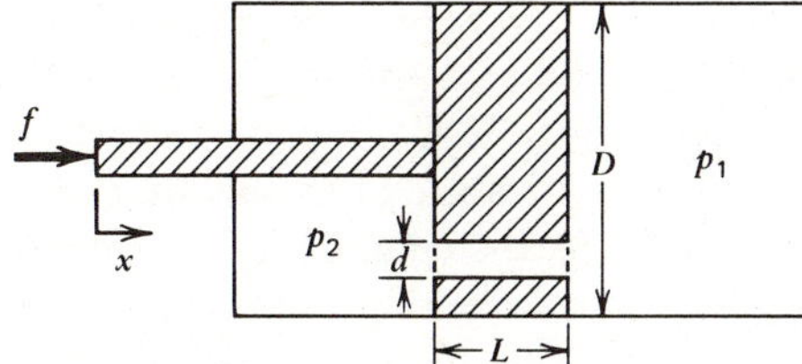

FIGURE 2.22 Hydraulic damper.

because the fluid is incompressible. Substituting (2.7-1) and (2.7-2) into (2.7-3) shows that

$$f = c\frac{dx}{dt} \tag{2.7-4}$$

where the *damping coefficient* (see Table 2.4) is

$$c = \frac{A^2}{C_1} = 8\pi\mu L\left[\left(\frac{D}{d}\right)^2 - 1\right]^2 \tag{2.7-5}$$

From the principle of action and reaction, f can also be considered to be the force exerted by the damper when moving at a velocity dx/dt. If the housing is also in motion, dx/dt is taken to be the relative velocity between housing and piston.

If the relative velocity is large enough to produce turbulent flow in the passage, the nonlinear resistance formulas should be used. In this case, the damper relation is the nonlinear expression given by (2.3-4) where

$$c = A^3 R_T \tag{2.7-6}$$

with R_T the turbulent resistance and $v = dx/dt$.

Note that the fluid inertia, or kinetic-energy storage compartment, has been assumed to be negligible and that potential energy resulting from pressure difference acts against the fluid resistance to create the system's dynamics. A similar analysis can be performed for *air dampers* in which the working fluid is air. The common storm door damper is such a device.

Accumulators

The compliance of fluid elements can have detrimental or beneficial effects. In hydraulic control systems operating with high force levels or at high speeds, the accuracy of the controller can be affected by the tendency of the nominally incompressible fluid medium to compress slightly. This compliance effect is greatly increased by the presence of bubbles in the hydraulic fluid, which makes the control action less positive and causes oscillations.

On the other hand, compliance can be useful in damping fluctuations or surges in pressure or flow rate, such as those resulting from reciprocating pumps and the addition or removal of other loads on a common pressure source. In such cases, a compliance element called an accumulator is added to the system. This general term includes such different elements as a surge tank for liquid level systems, such as tank 2 in Figure 2.21, a fluid column acting against an elastic element for hydraulic systems, and a bellows for gas systems. They all have the property of compliance, which allows them to store fluid during pressure peaks and release fluid during low pressure. The result is more efficient and safe operation by damping out pressure pulses the way a flywheel smooths the acceleration in a mechanical drive or an electrical capacitor smooths or "filters" voltage ripples.

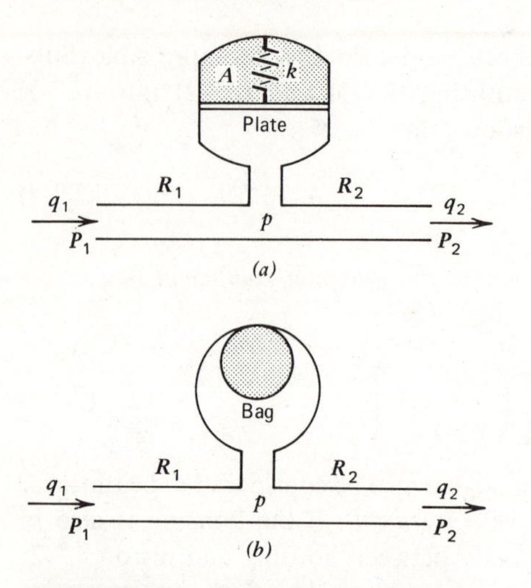

FIGURE 2.23 Hydraulic accumulators. (*a*) Mechanical spring principle. (*b*) Compressible gas principle.

Two typical arrangements for hydraulic systems are shown in Figure 2.23. In Figure 2.23*a*, a plate and spring oppose the pressure of the liquid, and in Figure 2.23*b*, the compliance is provided by a gas-filled bag that compresses as the liquid pressure increases. The simplest arrangement is a closed tube with air at the top. The air's compressibility acts against the liquid. This device is the common cure for the water-hammer effect in household plumbing.

All of the preceding devices use an equivalent elasticity to oppose the pressure, and thus they can be modeled similarly. If we take the displacement x of the interface to be proportional to the force f due to the liquid pressure, then from (2.3-1)

$$f = kx \tag{2.3-1}$$

where k is the elastic constant of the device. Let A be the area of the interface, p the liquid pressure, and q_1, q_2 the volume flow rates into and out of the section. Neglect the pressure drop between the flow junction and the plate. From mass (volume) conservation,

$$\frac{d}{dt}(Ax) = A\frac{dx}{dt} = q_1 - q_2$$

Neglecting the inertia of the interface gives the force balance

$$pA = kx$$

The model for the dynamic behavior of the pressure as a function of the flow rates is obtained by combining the last two expressions.

$$\frac{A^2}{k}\frac{dp}{dt} = q_1 - q_2 \tag{2.7-7}$$

When the resistances upstream and downstream are defined, the flow rates q_1 and q_2 can be expressed as functions of the upstream and downstream pressures.

Assume that the laminar resistances in the left- and right-hand pipe sections are R_1 and R_2. Then,

$$\rho q_1 = \frac{1}{R_1}(p_1 - p) \tag{2.7-8}$$

$$\rho q_2 = \frac{1}{R_2}(p - p_2) \tag{2.7-9}$$

and (2.7-7) becomes

$$\rho\frac{A^2}{k}\frac{dp}{dt}=\frac{1}{R_1}(p_1-p)-\frac{1}{R_2}(p-p_2) \tag{2.7-10}$$

This equation can be solved for p as a function of time if we are given p_1 and p_2 as time functions.

The Hydraulic Servomotor

A hydraulic servomotor is shown in Figure 2.24. The *pilot valve* controls the flow rate of the working medium to the receiving unit (a piston). The pilot valye shown is known as a spool-type valve because of its shape. Fluid under pressure is available at the supply port. For the pilot valve position shown (the "line-on-line" position), both cylinder ports are blocked and no motion occurs. When the pilot valve is moved to the right, the fluid enters the right-hand piston chamber and pushes the piston to the left. The fluid displaced by this motion exits through the left-hand drain port. The action is reversed for a valve displacement to the left. Both drain ports are connected to a tank (a sump) from which the pump draws fluid to deliver to the supply port. The filters and accumulators necessary to clean the recirculated fluid and dampen pressure fluctuations are not shown.

Let z denote the displacement of the pilot valve from its line-on-line position and x the displacement of the load and piston from their last position before the start of the motion. Note that a positive value of x (to the left) results from a positive value of z (to the right). The flow through the cylinder port uncovered by the pilot valve can be treated as flow through an orifice, and the orifice flow relation (2.6-14) can be applied. Let p be the pressure drop across the orifice. Thus,

$$q=C_dA_o\sqrt{2p/\rho} \tag{2.7-11}$$

where q is the *volume* flow rate through the cylinder port, A_o is the uncovered area of the port, C_d is the discharge coefficient (which usually lies between 0.6 and 0.8 for this application), and ρ is the mass density of the fluid. The area A_o is equal to wz, where w is the port width (into the page). If C_d, ρ, p, and w are taken to be constant, (2.7-11) can be written as

$$q=Cz \tag{2.7-12}$$

Conservation of mass requires the flow rate into the cylinder to equal the flow rate out. Therefore,

$$\rho q=\rho A\frac{dx}{dt}$$

Combining the last two equations gives

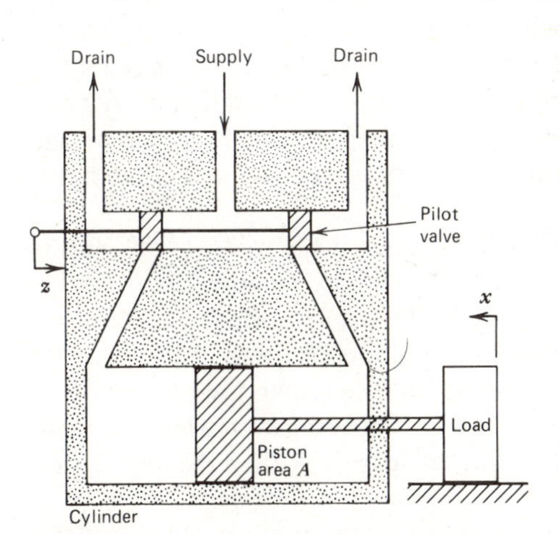

FIGURE 2.24 Hydraulic servomotor.

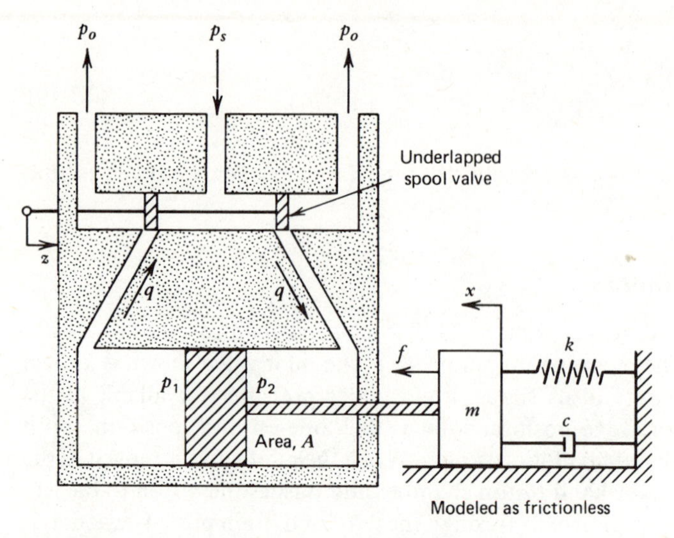

FIGURE 2.25 Hydraulic servomotor with a load.

the model for the servomotor.

$$\frac{dx}{dt} = \frac{C}{A} z \tag{2.7-13}$$

For high accelerations, this model must be modified, because it neglects the inertia of the load and piston. The piston force was assumed sufficient to move the load and is generated by the pressure difference across the piston.

Although (2.7-13) is accurate enough for many applications, we will now develop a more detailed model to account for the effects of the load. Figure 2.25 gives the necessary details. Machine tools for cutting metal are one application of such a system. The applied force f can be supplied by a hydraulic servomotor. The m represents the mass of a cutting tool and the power piston, while k represents the combined effects of the elasticity naturally present in the structure and that introduced by the designer to achieve proper performance. A similar statement applies to the damping c. The valve displacement z is generated by a control system in order to move the tool through its prescribed motion.

The spool valve shown in Figure 2.25 has two lands. If the width of the land is greater than the port width, the valve is said to be overlapped (Figure 2.26*a*). In this case, a dead zone exists in which a slight change in the displacement z produces no power piston motion. Such dead zones create control difficulties and are avoided by designing the valve to be underlapped (the land width is less than the port width—Figure 2.26*b*). For such valves there will be a small flow opening even when the valve is in the neutral position ($z = 0$). This gives it a higher sensitivity than the overlapped valve.

Let p_s and p_o denote the supply and outlet pressures, respectively. Then the pressure drop from inlet to outlet is $p_s - p_o$, and this must equal the sum of the drops

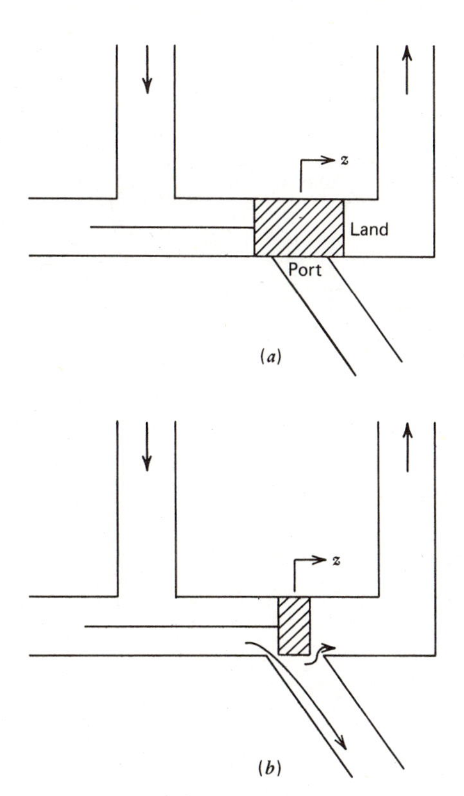

FIGURE 2.26 **Spool valve construction. (*a*) Overlapped. (*b*) Underlapped.**

across both valve openings and the power piston. That is,

$$p_s - p_o = 2\Delta p_v + \Delta p$$

where the drop across the valve openings is

$$\Delta p_v = p_s - p_2 = p_1 - p_o$$

and the drop across the piston is

$$\Delta p = p_2 - p_1$$

Therefore,

$$\Delta p_v = \tfrac{1}{2}(p_s - p_o - \Delta p)$$

If we take p_s and p_o to be constant, we see that Δp_v varies only if Δp changes. The derivation of (2.7-13) treated Δp_v as a constant.

Thus the variables z and Δp determine the volume flow rate, as

$$q = f(z, \Delta p)$$

This relation is nonlinear. For the reference equilibrium condition ($z = 0$, $\Delta p = 0$, $q = 0$), a linearization gives

$$q = C_1 z - C_2 \Delta p \qquad \textbf{(2.7-14)}$$

where the variations from equilibrium are simply z, Δp, and q. The linearization constants are available from theoretical and experimental results (Reference 6). The constant C_1 is identical to C in (2.7-13), and both C_1 and C_2 are positive. The effect of the underlapping shows up in the nonzero flow predicted from (2.7-14) when $z = 0$. The pressure drop Δp is caused by the reaction forces of the load mass, elasticity, and damping and by the supply pressure difference $(p_s - p_o)$.

Example 2.10

(a) Derive a model of the system shown in Figure 2.25. Take into account the inertia effects of the load and assume the valve is underlapped.

(b) Show that the model reduces to (2.7-13) when $c = k = 0$ and either of the following approximations is valid:

1. $m/A \to 0$ (large piston force compared to the load inertia)
2. $C_2 \to 0$ (q independent of Δp)

(a) Assume an incompressible fluid. Then conservation of mass and (2.7-14) give

$$A\dot{x} = q = C_1 z - C_2 \Delta p$$

The force generated by the piston is $A\Delta p$, and from Newton's law,

$$m\ddot{x} = -c\dot{x} - kx + A\Delta p$$

where x is measured from the equilibrium point of the mass m. Substitute Δp from the second equation to obtain

$$A\dot{x} = C_1 z - C_2\left(\frac{m}{A}\ddot{x} + \frac{c}{A}\dot{x} + \frac{k}{A}x\right)$$

or

$$\frac{C_2 m}{A}\ddot{x} + \left(\frac{cC_2}{A} + A\right)\dot{x} + \frac{C_2 k}{A}x = C_1 z \tag{2.7-15}$$

This is the desired model with z as the input and x as the output.

(b) For $c = k = 0$, (2.7-15) becomes

$$\frac{C_2 m}{A}\ddot{x} + A\dot{x} = C_1 z \tag{2.7-16}$$

When either $m/A \rightarrow 0$ or $C_2 \rightarrow 0$, we obtain

$$A\dot{x} = C_1 z \tag{2.7-17}$$

This is equivalent to (2.7-13), because $C_1 = C$.

2.8 PNEUMATIC ELEMENTS

While hydraulic devices use an incompressible liquid, the working medium in a *pneumatic* device is a compressible fluid, such as air. Industrial control systems frequently have pneumatic components to provide forces greater than those available from electrical devices. Also, they afford more safety from fire hazards. The availability of air is an advantage for these devices, because it may be exhausted to the atmosphere at the end of the device's work cycle, thus eliminating the need for return lines. On the other hand, the response of pneumatic systems is slower than that of hydraulic systems because of the compressibility of the working fluid.

The analysis of gas flow is more complicated than for liquid flow. For compressible fluids, the mass and volume flow rates are not readily interchangeable. Since mass is conserved, pneumatic systems analysis uses mass flow rate, here denoted q_m to distinguish from volume flow rate q. Because of the relatively low viscosity and density of gases, their flow is more likely to be turbulent. Finally, the possibility of supersonic (greater than the speed of sound) flow is more likely in gas systems.

Pressure is normally the only effort variable used for such systems. The kinetic energy of a gas is usually negligible, so the inductance relation is not employed. Typical units given in Table G.5 also apply here.

Thermodynamic Properties of Gases

Temperature, pressure, volume, and mass are functionally related for a gas. The model most often used to describe this relationship is the *perfect gas law* (Reference 5), which describes all gases if the pressure is low enough and the temperature high enough. The law states that

$$pV = mR_g T \tag{2.8-1}$$

where p is the absolute pressure of the gas with volume V, m is the mass, T its absolute temperature, and R_g the *gas constant* that depends on the particular type of gas. The values of this constant and properties of air at standard temperature and pressure are given in Table G.7.

The perfect gas law allows us to solve for one of the variables p, V, m, or T if the other three are given. We frequently do not know the values of three of the variables and need additional information. For a perfect gas, this information is usually available in the form of a pressure–volume or "process" relation. The following process models are commonly used: constant–pressure, constant–volume, constant–temperature (*isothermal*), and *reversible adiabatic* (*isentropic*). In the latter process, no heat is transferred between the gas and its surroundings. These four processes only approximate reality, but they allow some modeling simplifications to be made. A real process can be more accurately described by deriving or measuring an appropriate value for the exponent n in the *polytropic* process equation

$$p\left(\frac{V}{m}\right)^n = \text{constant} \tag{2.8-2}$$

If the mass is constant, the polytropic process describes the four previous processes if n is chosen as 0, ∞, 1, and γ, respectively, where γ is the ratio c_p/c_v of the specific heats of the gas (Table G.7).

Pneumatic Resistance

Here we develop methods for determining pneumatic resistances when the flow is compressible but subsonic. The subject of compressible fluid flow is complex, and here we present only enough theory to permit the development of models sufficiently accurate for control system design. By making generous use of such empirical constants as the orifice discharge coefficient, we can keep the necessary theory to a minimum. (More detailed coverage of compressible flow can be found in chapters 9 and 10 of Reference 2. More information on pneumatic as well as hydraulic systems is given in Reference 6.)

The theory of gas flow through an orifice provides the basis for modeling other pneumatic components. We assume that the perfect gas law applies and the effects of viscosity (friction) and heat transfer are negligible; hence, an isentropic process is assumed. Consider an orifice with a cross-sectional area A. The upstream absolute pressure is p_1, and the absolute back pressure p_b is that pressure eventually achieved downstream from the orifice. As p_b is lowered below p_1, the flow rate through the orifice increases. If the difference between p_1 and p_b becomes large enough, the gas flows through the orifice at the speed of sound. The value of p_b at which this occurs is the *critical pressure* p_c. For air, $\gamma = 1.4$ and the critical pressure is (Reference 2, chapter 10)

$$p_c = 0.528 p_1 \tag{2.8-3}$$

The results of chief interest to us are the expressions for the mass flow rate as functions of the pressures p_1 and p_b. Newton's laws, conservation of mass, and the isentropic process formula can be combined to show that the mass flow rate q_m in the subsonic case ($p_b > p_c$) is

$$q_m = C_d A \sqrt{\frac{2}{R_g T_1} p_b (p_1 - p_b)} \tag{2.8-4}$$

where C_d is an experimentally determined discharge coefficient that accounts for viscosity effects and T_1 is the absolute temperature upstream from the orifice. For the sonic case, q_m is independent of p_b as long as $p_b < p_c$. This case is treated in Reference 1, section 2.7.

A typical condition in pneumatic systems is such that a small pressure drop exists across the component. Hence, the flow is frequently subsonic. Also, the pressure changes often consist of small fluctuations about an average or steady-state constant pressure value. Under these circumstances, the compressible flow resistance of pneumatic components can be modeled in the form of the turbulent resistance relation (2.6-13), written here as

$$R_p q_m^2 = \Delta p \tag{2.8-5}$$

where Δp is the pressure drop across the component and R_p is the *pneumatic resistance.*

To show that (2.8-4) reduces to (2.8-5), we assume that the input pressure p_1 fluctuates about a constant pressure p_s by a small amount p_i. Thus, $p_1 = p_s + p_i$. Similarly, the back pressure can be written as $p_b = p_s + p_o$, and we assume that p_o is also small relative to p_s. Note that $p_1 - p_b = p_i - p_o$ and consider the following term from (2.8-4).

$$\begin{aligned}
\sqrt{p_b(p_1 - p_b)} &= \sqrt{(p_s + p_o)(p_i - p_o)} \\
&= \sqrt{p_s + p_o}\sqrt{p_i - p_o} \\
&= \sqrt{p_s\left(1 + \frac{p_o}{p_s}\right)}\sqrt{p_i - p_o} \\
&\approx \sqrt{p_s}\sqrt{p_i - p_o}
\end{aligned}$$

because $1 + p_o/p_s \approx 1$ from our assumptions. Thus, (2.8-4) becomes

$$q_m = C_d A \sqrt{\frac{2p_s}{R_g T_1}} \sqrt{p_i - p_o} \tag{2.8-6}$$

This has the same form as (2.8-5), where $\Delta p = p_i - p_o = p_1 - p_b$, and the pneumatic resistance is

$$R_p = \frac{R_g T_1}{2 p_s C_d^2 A^2} \tag{2.8-7}$$

Solve (2.8-5) for q_m assuming that $\Delta p > 0$.

$$q_m = \sqrt{\frac{\Delta p}{R_P}} \tag{2.8-8}$$

where we take the positive square root, because $q_m > 0$ if $\Delta p > 0$. For (2.8-8), the slope of q_m is infinite at $\Delta p = 0$. This physically unrealistic result is due to certain assumptions made to derive (2.8-4) that are not true as $p_b \rightarrow p_1$. For p_b very close to p_1, the orifice flow relation resembles that of laminar flow, and we will model it as

$$q_m = \frac{1}{R_L} \Delta p \tag{2.8-9}$$

where R_L is the equivalent resistance. If we match the solutions of (2.8-8) and (2.8-9) at some intermediate value of Δp, say, $\Delta p = B$, then R_L is related to R_P by

$$R_L = \sqrt{BR_P} \tag{2.8-10}$$

where the value of B must be determined by experiment. For this value of R_L, the slopes of (2.8-8) and (2.8-9) also match at $\Delta p = B$.

Thus, we have three models for subsonic orifice flow: (2.8-4) for Δp large; (2.8-8) for Δp small but greater than B; and (2.8-9) for Δp very small. These relations are summarized in Table 2.10.

TABLE 2.10 Pneumatic Resistance Relations

$q \rightarrow$

p_1 —⋈— p_b

q = mass flow rate

p = pressure drop

$= p_1 - p_b$

Air flow is subsonic if $p_b > 0.528 p_1$

Case

1. Subsonic with p large.

$$q = C_d A \sqrt{\frac{2}{R_g T_1} p_b (p_1 - p_b)} \qquad \text{[see (2.8-4)]}$$

2. Subsonic with p moderate.

$$q = \sqrt{\frac{p}{R}} \qquad R = \frac{R_g T_1}{2 p_s C_d^2 A^2} \qquad \text{[see (2.8-7)]}$$

3. Subsonic with p very small.

$$q = \frac{p}{R} \qquad R \text{ found from experiments. [see (2.8-10)]}$$

Pneumatic Capacitance

In pneumatic systems, mass is the quantity variable and pressure the effort variable. Thus, pneumatic capacitance is the relation between stored mass and pressure. Specifically, we take pneumatic capacitance to be the system's capacitance C (or compliance), defined as the ratio of the change in stored mass to the change in pressure, or

$$C = \frac{dm}{dp} \tag{2.8-11}$$

For a container of constant volume V with a gas density ρ, this expression may be written as

$$C = \frac{d(\rho V)}{dp} = V\frac{d\rho}{dp} \tag{2.8-12}$$

If the gas undergoes a polytropic process,

$$p\left(\frac{V}{m}\right)^n = \frac{p}{\rho^n} = \text{constant} \tag{2.8-13}$$

and

$$\frac{d\rho}{dp} = \frac{\rho}{np} = \frac{m}{npV}$$

For a perfect gas, this shows the capacitance of the container to be

$$C = \frac{mV}{npV} = \frac{V}{nR_g T} \tag{2.8-14}$$

Note that the same container can have a different capacitance for different expansion processes, temperatures, and gases, since C depends on n, T, and R_g.

Modeling Pneumatic Systems

Because fluid inertia is usually neglected in pneumatic analysis, the simplest model of such systems is a resistance–capacitance model for each mass storage element.

Example 2.11

Air passes through a valve (modeled as an orifice) into a rigid container, as shown in Figure 2.27. Develop a dynamic model of the pressure change p in the container as a function of the inlet pressure change p_i. Assume an isothermal process and that p and p_i are small variations from the steady-state pressure p_s.

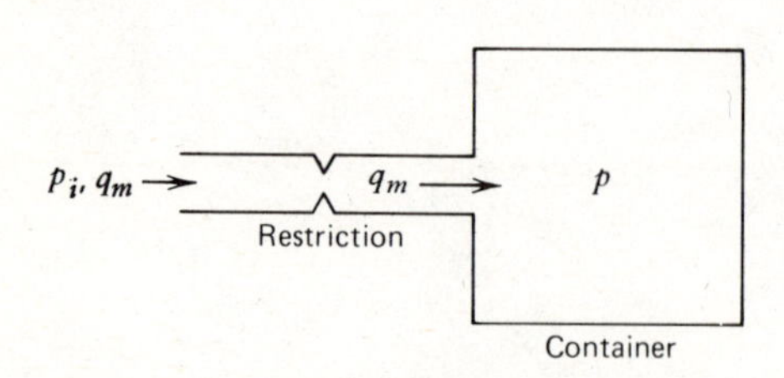

FIGURE 2.27 Gas flow into a rigid container.

From conservation of mass, the rate of mass increase in the container equals the mass flow rate through the valve. Thus, if $p_i - p > 0$,

$$\frac{dm}{dt} = q_m$$

$$\frac{dm}{dt} = \frac{dm}{dp}\frac{dp}{dt} = C\frac{dp}{dt}$$

$$q_m = \sqrt{\frac{p_i - p}{R_P}} \tag{2.8-15}$$

where C and R_P are given by (2.8-14) and (2.8-7), with $T = T_1$ and $n = 1$. The desired model is

$$C\frac{dp}{dt} = \sqrt{\frac{p_i - p}{R_P}} \tag{2.8-16}$$

If $p_i - p < 0$, the flow q_m is reversed, and the model is

$$C\frac{dp}{dt} = -\sqrt{\frac{p - p_i}{R_P}} \tag{2.8-17}$$

If $p_i - p$ is very close to zero, the linear resistance relation (2.8-9) can be used, and (2.8-15) is replaced by

$$q_m = \frac{1}{R_L}(p_i - p) \tag{2.8-18}$$

In this case, the system model is

$$C\frac{dp}{dt} = \frac{1}{R_L}(p_i - p) \tag{2.8-19}$$

and the flow reversal is accounted for automatically if $p_i - p$ changes sign.

Example 2.12

A pneumatic bellows, shown in Figure 2.28, is an expandable chamber usually made from corrugated copper because of this metal's good heat conduction and elastic properties. We model the elasticity of the bellows as a spring. The spring constant k for the bellows can be determined from vendor's data or standard formulas. This device, when employed with a variable resistance R_P is useful in pneumatic controllers as a feedback element to sense and control the pressure p_i. The displacement x is transmitted by a linkage or beam balance to a pneumatic valve regulating the air supply. Develop a model for the dynamic behavior of x with p_i a given input.

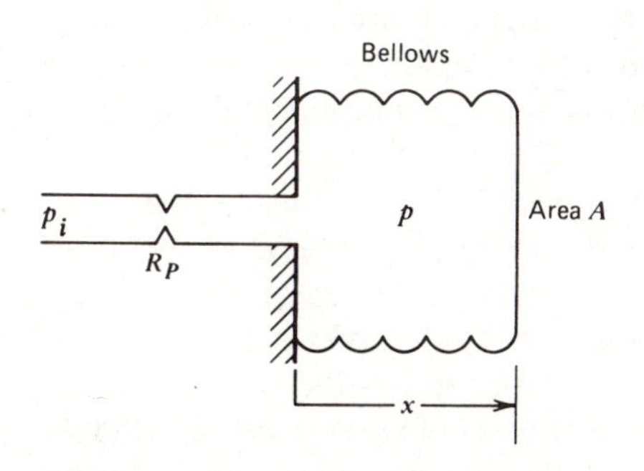

FIGURE 2.28 Pneumatic bellows.

We assume that the bellows expands

or contracts slowly. Thus, the product of its mass and acceleration is negligible, and a force balance gives

$$pA = kx \tag{2.8-20}$$

where the force exerted by the internal pressure is pA, A is the area of the bellows, and x is the displacement of the right-hand side (the left-hand side is fixed). The volume is $V = Ax$, and thus

$$pV = \frac{kx}{A} Ax = kx^2$$

The assumption of a slow process suggests an isothermal process. Thus, the time derivative of the perfect gas law gives

$$\frac{d}{dt}(pV) = R_g T \frac{dm}{dt}$$

From the law for resistance (2.8-9), assuming that $p_i - p$ is very small,

$$\frac{dm}{dt} = q_m = \frac{1}{R_L}(p_i - p)$$

Since

$$\frac{d(kx^2)}{dt} = 2kx\frac{dx}{dt}$$

these expressions give the following model:

$$2kx\frac{dx}{dt} = \frac{R_g T}{R_L}\left(p_i - \frac{kx}{A}\right) \tag{2.8-21}$$

This is nonlinear because of the cross-product between x and its derivative.

2.9 THERMAL SYSTEMS

In thermal systems, energy is stored and transferred as heat. For this reason, many energy conversion devices are examples of thermal systems, as are chemical processes where heat must be added or removed in order to maintain an optimal reaction temperature. Other examples occur in food processing and environmental control in buildings.

The primary system variables are given in Table 2.2. Typical units are given in Table G.8. The effort variable is temperature T, or more precisely, temperature difference ΔT. This can cause a heat transfer rate q_h by one or more of the following resistance modes: conduction, convection, or radiation. The transfer of heat causes a change in the system's temperature. Thermal capacitance relates the system temperature to the amount of heat energy stored. It is the product of the system mass and its specific heat. No physical elements are known to follow a thermal inductance causal path.

In order to have a single lumped-parameter model, we must be able to assign a single temperature that is representative of the system. Thermal systems analysis is sometimes complicated by the fact that it is difficult to assign such a representative temperature to a body or fluid because of its complex shape or motion and resulting complex distribution of temperature throughout the system. Here, we develop some guidelines for when such an assignment can be made. When it cannot, several coupled lumped-parameter models or even a distributed-parameter model will be required.

Unlike current in an electrical circuit, the flow of heat often does not occur along a single, simple path, but might involve more than one mode. If the convective mode is present, the analysis must also consider a fluid system with its attendant difficulties. While the analysis of such complex phenomena is beyond the scope of this work, fortunately many analytical and empirical results are available for common situations, and we can use them to obtain the necessary coefficients for our thermal resistances.

Heat Transfer

Heat can be transferred in three ways: by *conduction* (diffusion through a substance), *convection* (fluid transport), and *radiation* (mostly infrared waves). The effort variable causing a heat flow is a temperature difference. The constitutive relation takes a different form for each of the three heat transfer modes. The linear model for heat flow rate as a function of temperature difference is given by *Newton's law of cooling*, which is valid for both convection and conduction.

$$Rq_h = \Delta T \tag{2.9-1}$$

The thermal resistance for convection is

$$R = \frac{1}{hA} \tag{2.9-2}$$

where h is the *film coefficient* of the fluid–solid interface. For conduction, we will see that

$$R = \frac{d}{kA} \tag{2.9-3}$$

where k is the *thermal* conductivity of the material, A is the surface area, and d the material thickness. Table G.8 gives typical units for h and k. Table G.9 gives values of k for some common materials.

Convective heat transfer is divided into two categories: forced convection—due, for example, to fluid pumped past a surface—and free or natural convection, due to motion produced by density differences in the fluid. The heat transfer coefficient for convection is a complicated function of the fluid flow characteristics especially. For most cases of practical importance, the coefficient has been determined to acceptable accuracy, but a presentation of the results is lengthy and beyond the scope of this work. Standard references on heat transfer, such as Reference 7, contain this information for most cases.

Significant heat transfer can occur by radiation; the most notable example is solar energy. Thermal radiation produces heat when it strikes a surface capable of absorbing it. It can also be reflected or refracted, and all three mechanisms can occur at a single surface. When two bodies are in visual contact, a mutual exchange of energy occurs by emission and absorption. The *net* transfer of heat occurs from the warmer to the colder body. This rate depends on material properties, geometric factors affecting the portion of radiation emitted by one body and striking the other, and the amount of surface area involved. The net heat transfer rate can be shown to depend on the body temperatures raised to the fourth power (a consequence of the *Stefan–Boltzmann* law).

$$q_h = \beta(T_1^4 - T_2^4) \tag{2.9-4}$$

The absolute body temperatures are T_1 and T_2, and β is a factor incorporating the other effects. This factor is usually very small. Thus, the effect of radiation heat transfer is usually negligible compared to conduction and convection effects, unless the temperature of one body is much greater than that of the other. Determining β, like the convection coefficient, is too involved to consider here, but many results are available (see Reference 7).

Some Simplifying Approximations

The thermal resistance for conduction through a plane wall given by (2.9-3) is an approximation that is valid only under certain conditions. Let us take a closer look at the problem. If we consider the wall to extend to infinity in both directions, then the heat flow is one dimensional. If the wall material is homogeneous, the temperature gradient through the wall is constant under steady-state conditions (Figure 2.29*a*). *Fourier's law* of heat conduction states that the heat transfer rate per unit area within a homogeneous substance is directly proportional to the negative temperature gradient. The proportionality constant is the thermal conductivity k. For the case shown in Figure 2.29*a*, the gradient is $(T_2 - T_1)/d$, and the heat transfer rate is thus

$$q_h = -\frac{kA(T_2 - T_1)}{d} \tag{2.9-5}$$

where A is the area in question. Comparing this with (2.9-1) shows that the thermal resistance is given by (2.9-3), with $\Delta T = T_1 - T_2$.

In transient conditions where temperatures change with time, the temperature distribution is no longer a straight line. If the internal temperature gradients in the body are small, as would be the case for a large value of k, it is possible to treat the body as being at one uniform, average temperature and thus obtain a lumped- instead of distributed-parameter model. For solid bodies immersed in a fluid, a useful criterion for determining the validity of the uniform-temperature assumption is based on the *Biot* number, defined as

$$N_B = \frac{hL}{k} \tag{2.9-6}$$

where L is the ratio of the volume to surface area of the body and h is the film coefficient. If the shape of the body resembles a plate, cylinder, or sphere, it may be

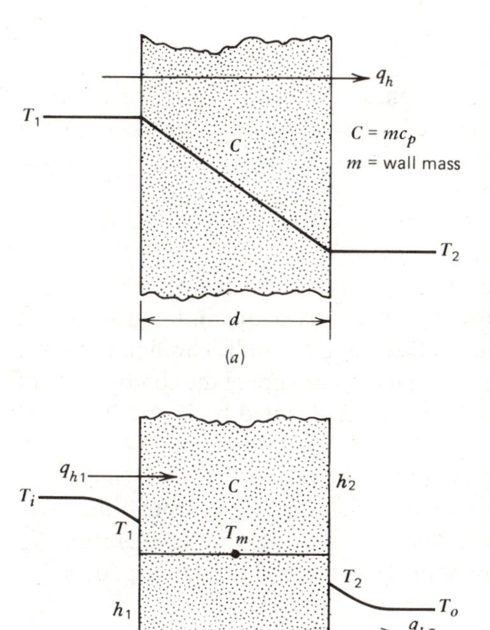

FIGURE 2.29 One-dimensional heat transfer through a wall. (*a*) Steady-state temperature gradient. (*b*) Lumped approximation.

considered to have a uniform temperature if N_B is small. If $N_B < 0.1$, the temperature is usually taken to be uniform, and the accuracy of this approximation improves if the inputs vary slowly.

Example 2.13

Consider a copper sphere 1 in. in diameter with $k = 212$ Btu/hr-ft-°F at 570°F and immersed in a fluid such that $h = 5$ Btu/hr-ft^2-°F. Show that its temperature can be considered uniform, and develop a model of the sphere's temperature as a function of the temperature T_o of the surrounding fluid.

The surface area and volume of the sphere are

$$A = 4\pi\left(\frac{1}{24}\right)^2 = 2.18 \times 10^{-2}$$

$$V = \left(\frac{4}{3}\right)\pi\left(\frac{1}{24}\right)^3 = 3.03 \times 10^{-4}$$

Thus, $L = V/A = 0.0139$, and $N_B = 3.28 \times 10^{-4}$, which is much less than 0.1. According to the Biot criterion, we may treat the sphere as a lumped-parameter system with a single uniform temperature T.

The amount of heat energy in the sphere is CT, where C is its thermal capacitance. If the outside temperature T_o is greater than T, heat is transferred into the sphere at the rate q_h given by (2.9-1), where $\Delta T = T_o - T$. Thus, from conservation of energy,

$$\frac{d}{dt}(CT) = q_h$$

Using (2.9-1) and noting that C is constant, we obtain

$$C\frac{dT}{dt} = \frac{1}{R}(T_o - T) \tag{2.9-7}$$

For the copper sphere, Table G.9 and (2.9-2) give

$$C = mc_p = \rho V c_p = 17.3(3.03 \times 10^{-4})(2.93) = 0.0154$$

$$R = \frac{1}{5(2.18 \times 10^{-2})} = 9.17$$

A Wall Model

Conductive and convective resistances to heat transfer frequently occur when one or more layers of solid materials are surrounded on both sides by fluid, such as shown in Figure 2.29*b*. We will use the figure to show how thermal resistance is computed when both conduction and convection are present.

Assume that the wall's material is homogeneous, and that it is to be treated with a lumped approximation. The wall temperature is considered to be constant throughout and is denoted by T_m. The temperatures T_1 and T_2 are the temperatures in the surrounding fluid just outside the wall surface. There is a temperature gradient on each side of the wall across a thin film of fluid adhering to the wall. This film is known as the *thermal boundary layer*. The film coefficient h is a measure of the conductivity of this layer. The temperatures outside of these layers are denoted by T_i and T_o.

The thermal capacitance C is the product of the wall mass times its specific heat. The capacitance of the boundary layer is generally negligible because of its small fluid mass. Thus, since no heat is stored in the layer, the heat flow rate through the layer must equal the heat flow rate from the surface to the mass at temperature T_m considered to be located at the center of the wall. The length of this latter path is $d/2$. For the left-hand side, this gives

$$q_{h1} = h_1 A(T_i - T_1) = \frac{kA}{d/2}(T_1 - T_m) \tag{2.9-8}$$

This allows us to solve for T_1 as a function of T_i and T_m.

$$T_1 = \frac{2kT_m + dh_1 T_i}{2k + dh_1} \tag{2.9-9}$$

Similar equations can be developed for the right-hand flow q_{h2} and temperature T_2.

An energy balance for the wall mass gives

$$C\frac{dT_m}{dt} = q_{h1} - q_{h2} \tag{2.9-10}$$

After T_1 and T_2 have been eliminated, this gives an equation for T_m with T_i and T_o as inputs. Define

$$a_j = \frac{2kh_j A}{2k + dh_j}, \qquad j = 1, 2 \tag{2.9-11}$$

and (2.9-10) becomes

$$C\frac{dT_m}{dt} = a_1(T_i - T_m) - a_2(T_m - T_o) \tag{2.9-12}$$

In (2.9-12), a_1 is the reciprocal of the resistance for the path from T_i to T_m. We can write this resistance as

$$R_1 = \frac{1}{a_1} = \frac{2k + dh_1}{2kh_1 A} = \frac{1}{h_1 A} + \frac{d}{2kA} \tag{2.9-13}$$

Comparing this with (2.9-2) and (2.9-3) shows that R_1 is the sum of the thermal boundary layer resistance and the wall's conductive resistance along the path of length $d/2$.

In many applications, the wall's thermal capacitance is small compared to the capacitance of the mass of fluid on either side of the wall. In this case, we model the wall as a pure resistance, and the total wall resistance is the sum of the path resistances. Equation (2.9-12) can be used to show this. If either the rate of change of T_m or the wall capacitance C is very small, the right-hand side of (2.9-12) can be taken to be zero and thus q_{h1} equals q_{h2}. This relation can be manipulated to show that the wall temperature is

$$T_m = \frac{a_1 T_i + a_2 T_o}{a_1 + a_2} \tag{2.9-14}$$

and that the heat flow rate is

$$q_h = \frac{a_1 a_2}{a_1 + a_2}(T_i - T_o) \tag{2.9-15}$$

Thus, the total resistance between T_i and T_o is $(a_1 + a_2)/a_1 a_2$. With (2.9-11) and some algebra, the total resistance can be shown to be

$$R = \left(\frac{1}{h_1} + \frac{d}{2k} + \frac{d}{2k} + \frac{1}{h_2}\right)\frac{1}{A} \tag{2.9-16}$$

When it is not possible to identify one representative temperature for a system, several temperatures can be chosen, one for each distinct mass. The resistance paths between the masses are then identified, and conservation of heat energy is applied to each mass. This results in a model whose order equals the number of representative temperatures.

Example 2.14

A simplified representation of the temperature dynamics of a room is shown in Figure 2.30. The room is perfectly insulated on all sides except one, which has a wall with a thermal capacitance C_2. The inner wall resistance is R_1, as given by (2.9-13). The outer resistance is R_2 and is found by a similar expression. The representative temperature of the room's air is designated T_i, that of the wall is T_m, and the outside air temperature is T_0. The room air capacitance is C_1. Develop a model of the behavior of T_i.

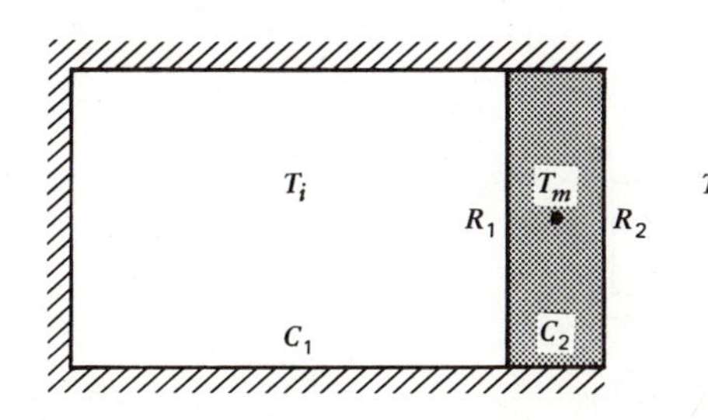

FIGURE 2.30 A model of room temperature.

Assume that $T_o > T_m > T_i$. Then the heat flow is from T_0 to T_m to T_i. Conservation of energy applied to the capacitances C_1 and C_2 gives

$$C_1 \dot{T}_i = \frac{1}{R_1}(T_m - T_i) \tag{2.9-17}$$

$$C_2 \dot{T}_m = \frac{1}{R_2}(T_o - T_m) - \frac{1}{R_1}(T_m - T_i) \tag{2.9-18}$$

This is the desired model. If the heat flow is not in the assumed direction, the mathematics takes care of the reversal automatically.

If the wall capacitance is negligible, then $C_2 = 0$, and we can substitute T_m from (2.9-18) into (2.9-17). Or, equivalently, we can use the principle of (2.9-16) to find the total wall resistance R. This is

$$R = R_1 + R_2 \tag{2.9-19}$$

The heat flow rate through the wall is $(T_o - T_i)/R$, and the model becomes

$$C_1 \dot{T}_i = \frac{1}{R}(T_o - T_i) \tag{2.9-20}$$

2.10 SUMMARY

For convenience, we have segregated our coverage of physical systems into fluid (compressible and incompressible), thermal, mechanical, and electrical systems. Many designs contain a mixture of system types, and these subsystems can interact, often in subtle ways.

For example, Figure 2.31 represents a common configuration for a chemical processing system. A dry chemical is fed into the tank of liquid where it is dissolved and reacts with the liquid. The proper ratio of the two chemicals must be maintained.

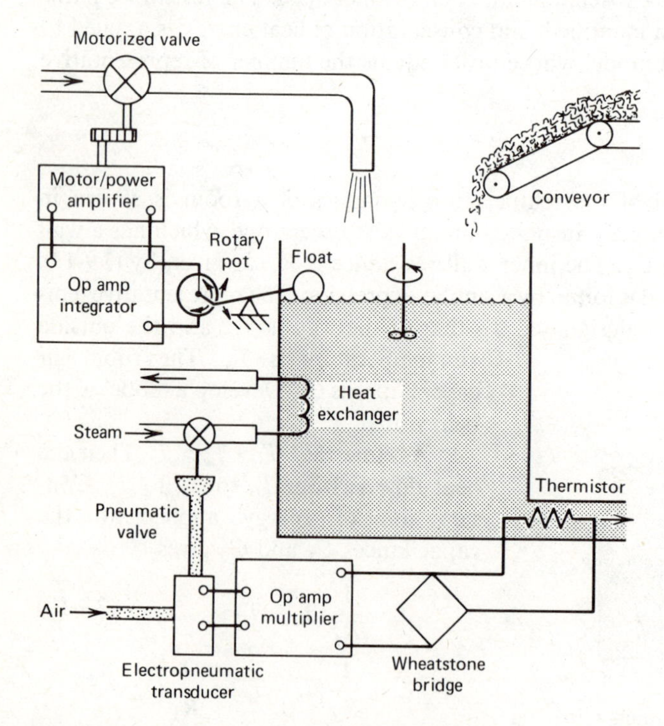

FIGURE 2.31 Chemical processing system.

Thus, the inflow rate of each is placed under feedback control. The volume in the tank is sensed by a float that moves a linkage proportionally to the tank height. The linkage in turn rotates a potentiometer (a variable resistance). This produces a voltage, which is compared to a reference voltage (not shown) indicative of the desired height. The operational amplifier (op amp) integrates this error signal voltage and opens or closes the motorized valve accordingly. The dynamics of the motor, gears, and load (valve) can be developed with the techniques in Sections 2.4 and 2.5. The device for controlling the feed rate of the dry chemical is not shown but would be a motor/amplifier connected to the conveyor drive rollers. The reasons for choosing an *integrating* op amp will be explored in Chapter 6.

The model for the tank height is linear if we assume that laminar flow exists and the dry chemical is completely dissolved. The mixture is kept uniform by stirring, and its reaction rate is a function of the chemical concentration and the temperature. This is controlled by allowing steam to flow through a heat exchanger coil. The model for the temperature of the tank mixture can be developed with an energy balance in the tank and the constitutive equation for the heat exchanger. The steam flow is controlled by a pneumatic valve whose input air pressure is produced by a transducer that makes the pressure proportional to the error signal voltage generated by a Wheatstone Bridge, which is a device for computing differences in voltages. Here, it is used to compare the voltage from a temperature-sensitive resistor (a thermistor) to a reference voltage indicative of the desired tank temperature.

The models of all these subsystems can be developed with the techniques of this chapter, and each would be a low-order model, probably a linear one. However, the total system model is of a higher order and is nonlinear. Let y be the concentration of the reactant in the tank, h the height, C the tank surface area, and T the reactant temperature. Then the total amount of reactant in the tank is Chy, and a material balance equation can be written for it. This equation is nonlinear because of the product of two of the state variables, h and y. Also, *Arrhenius's law* states that the rate of the reaction is proportional to Chy with the proportionality factor being a function of temperature. In this way, the heat-balance equation is coupled to the material-balance equation. The rate of heat generated by the reaction is proportional to the reaction rate and thus provides an additional coupling of the equations. If the thermistor or the pneumatic valve acts too slowly or too quickly, the action will result in additional coupling. The same can be said of the float, integrator, and motorized-valve dynamics. These interactions frequently require an overall supervisory control system to coordinate the individual controllers.

Even though coupling exists between the subsystems and the overall system model is of an order higher than one, it would be impossible to develop such a model without understanding how to describe each subsystem. Also, an analysis of the overall model is impossible without an ability to analyze a low-order model. We must start simply and work up the ladder of complexity. We will see that an analysis of each subsystem will enable us to decide whether or not its dynamics are such that they will cause significant interactions with the other subsystems. For example, if we can estimate the speed of response of the thermistor voltage to temperature changes, we can decide whether or not we must couple its behavior into the overall system model. Such analysis techniques are the subject of the next chapter.

REFERENCES

1. W. J. Palm III, *Modeling, Analysis, and Control of Dynamic Systems*, John Wiley & Sons, New York, 1983.
2. R. W. Fox and A. T. McDonald, *Introduction to Fluid Mechanics*, John Wiley & Sons, New York, 1978.
3. J. E. Shigley, *Stimulation of Mechanical Systems*, McGraw-Hill, New York, 1967.
4. *Chemical Engineers' Handbook*, 4th ed., McGraw-Hill, New York, 1963.
5. G. J. Van Wylen and R. E. Sonntag, *Fundamentals of Classical Thermodynamics*, John Wiley & Sons, New York, 1977.
6. D. McCloy and H. R. Martin, *The Control of Fluid Power*, Longman, London, 1973.
7. American Society of Heating, Refrigeration, and Air Conditioning Engineers, *ASHRAE Handbook of Fundamentals*, New York, 1972.

PROBLEMS

2.1 Derive models for the circuits shown in Figure P2.1. The input is the voltage e_1, and the output is the voltage e_2.

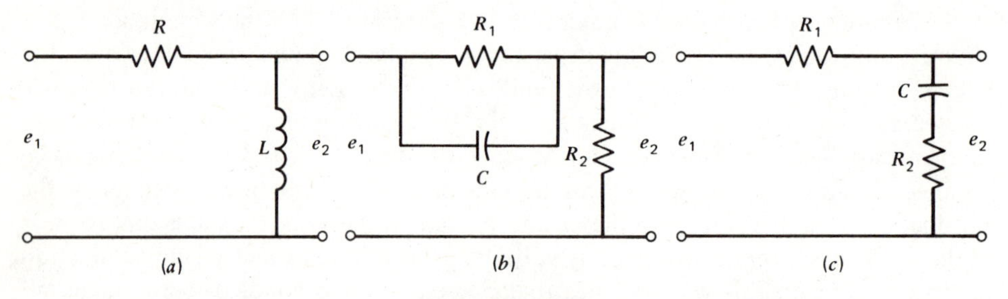

FIGURE P2.1

2.2 A circuit with significant applications is the op amp circuit shown in Figure P2.2. Find the output voltage v_o as a function of the input voltage v_i.

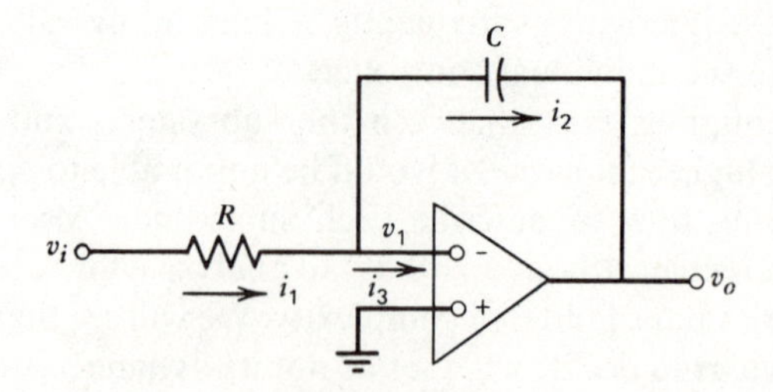

FIGURE P2.2

2.3 Derive the constitutive relations between the restoring force and displacement for the following systems (Figure P2.3). Do not assume small displacements.

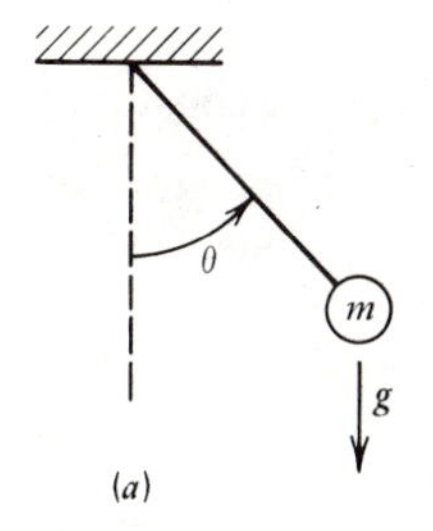

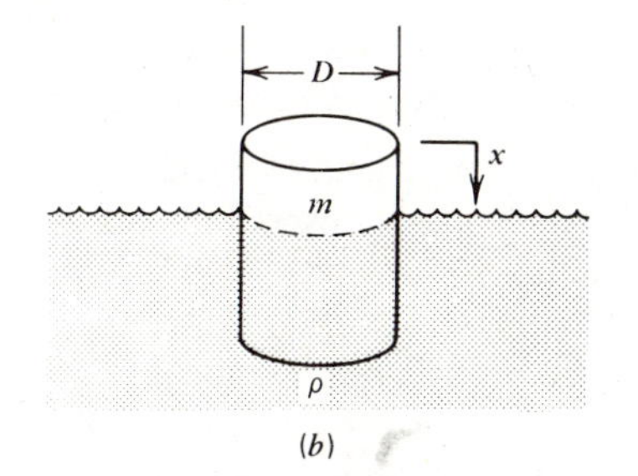

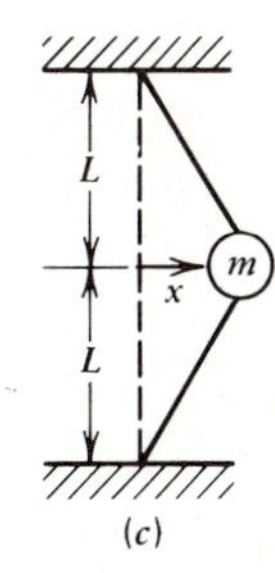

FIGURE P2.3

(a) A pendulum with a "gravity spring."
(b) A cylinder with a "buoyancy spring." (*Hint*: Use Archimede's principle.)
(c) A wire of length $2L$ with an initial tension T.

2.4 Plot the force exerted on the mass by the springs as a function of the displacement x (Figure P2.4). Neglect friction.

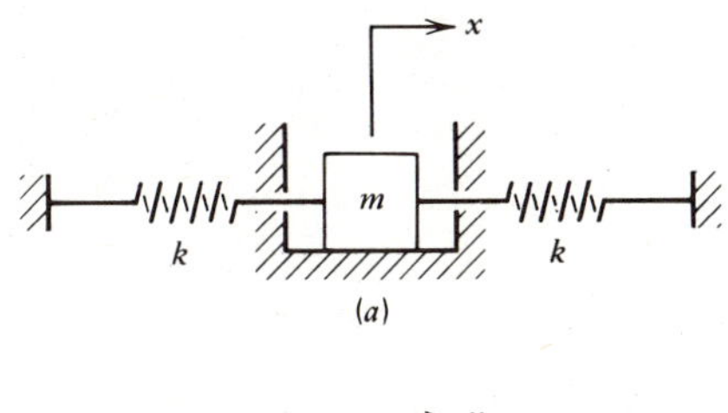

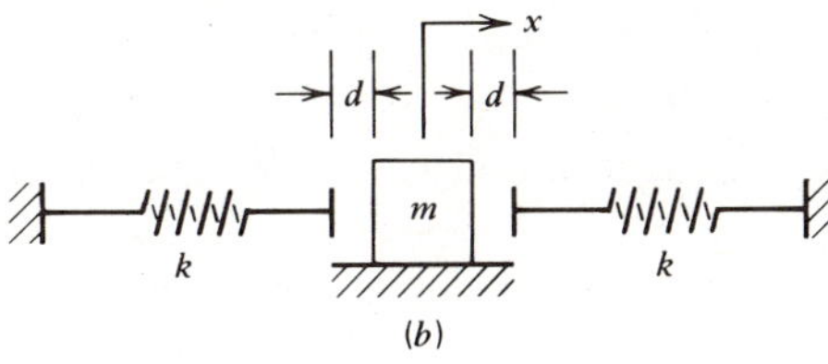

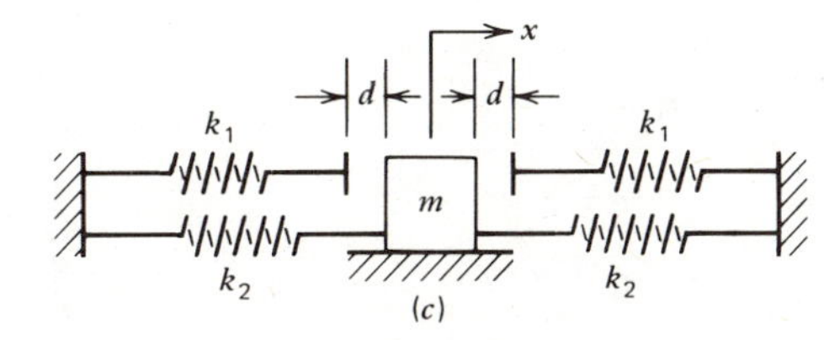

FIGURE P2.4

2.5 Gear teeth have an elasticity that sometimes must be accounted for in a dynamic model. They also exhibit the phenomenon of *backlash* if the gears do not mesh tightly. Sketch the equivalent spring force of a gear tooth under these conditions, as a function of angular displacement.

2.6 The vibration of a motor mounted on the end of a cantilever beam can be modeled as a mass–spring system. The motor mass is 2 slugs, and the beam mass is 1 slug. When the motor is placed on the beam, it causes an additional static deflection of 0.01 ft. Find the equivalent mass m and equivalent spring constant k.

2.7 A motor connected to a pinion gear of radius R drives a load of mass m on the rack (Figure P2.7). Neglect all masses except m. What is the energy-equivalent inertia due to m, as felt by the motor?

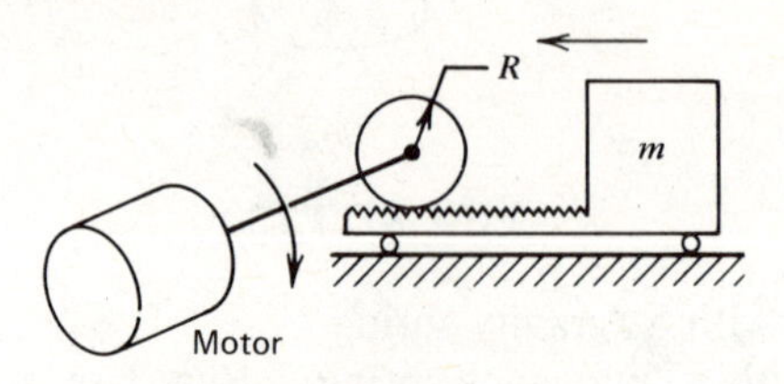

FIGURE P2.7

2.8 Find an expression for the torsional spring constant for the stepped shaft shown in Figure P2.8.

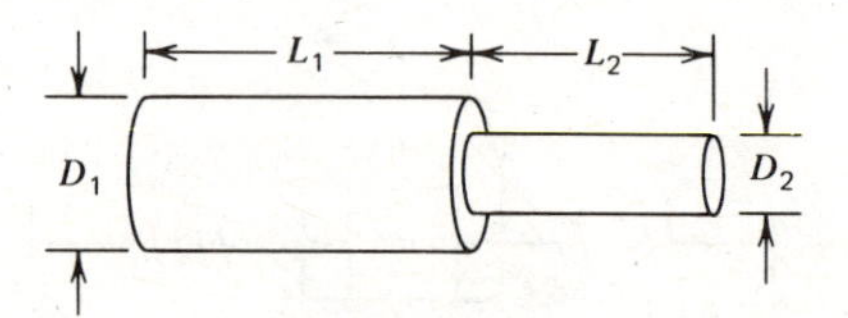

FIGURE P2.8

2.9 Derive the equivalent inertia formula for the lead screw given in Table 2.5. Use kinetic-energy equivalence to do this.

2.10 Find the differential equation models for the systems shown in Figure P2.10.

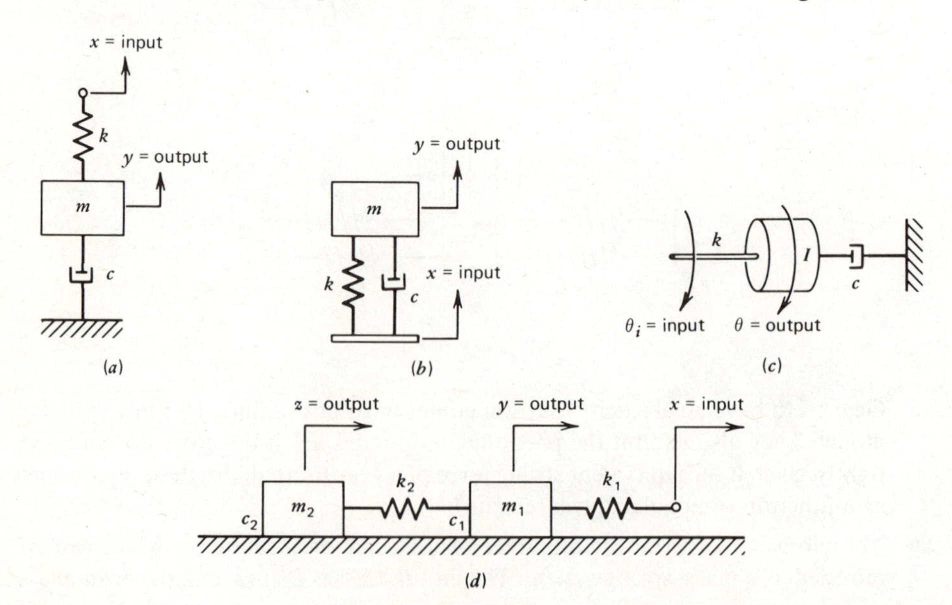

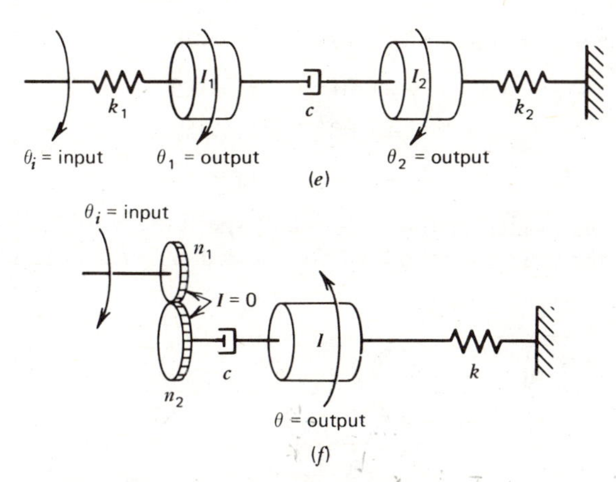

FIGURE P2.10

2.11 Derive the model for the mechanical system shown in Figure 2.14*b*.

2.12 The missile shown in Figure P2.12 is steered in the x-y plane by varying the thrust forces f_1 and f_2 from each engine. Treat the missile as a thin rod whose mass moment of inertia about the center of mass is $I = mL^2/12$, where $m = 7000$ slugs and $L = 20$ ft. Let x and y be the coordinates of the center of mass and θ the angular displacement of the missile from vertical. Derive the differential equation model of the missile for motion in the x-y plane. The inputs are f_1 and f_2. The outputs are x, y, and θ.

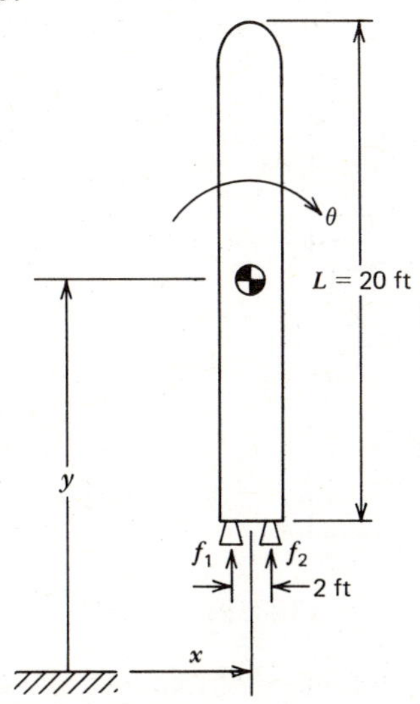

FIGURE P2.12

2.13 A load inertia I_2 is driven through gears by a motor with inertia I_1. The shaft inertias are I_3 and I_4; the gear inertias are I_5 and I_6. The gear ratio is 5:1 (the motor shaft has the greater speed). The motor torque is T_1, and the viscous damping coefficient is $c = 1.2$ lb-ft-sec/rad. Neglect elasticity in the system, and use the following inertia values (in sec²-ft-lb/rad):

$$I_1 = 0.01 \qquad I_2 = 0.5$$
$$I_3 = 0.001 \qquad I_4 = 0.005$$
$$I_5 = 0.02 \qquad I_6 = 0.03$$

(a) Derive the model for the motor shaft speed ω_1 with T_1 as input.
(b) Derive the model for the load shaft speed ω_2 with T_1 as input.

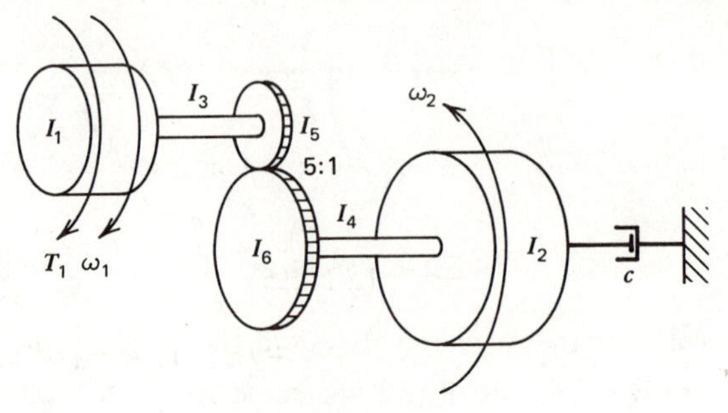

FIGURE P2.13

2.14 Determine a model for the levered systems shown in Figure P2.14, with the force f as input. Assume small displacements.

(a) The output is θ, and the lever is massless.
(b) The output is $\omega = \dot{\theta}$, and the lever has an inertia I relative to the pivot.

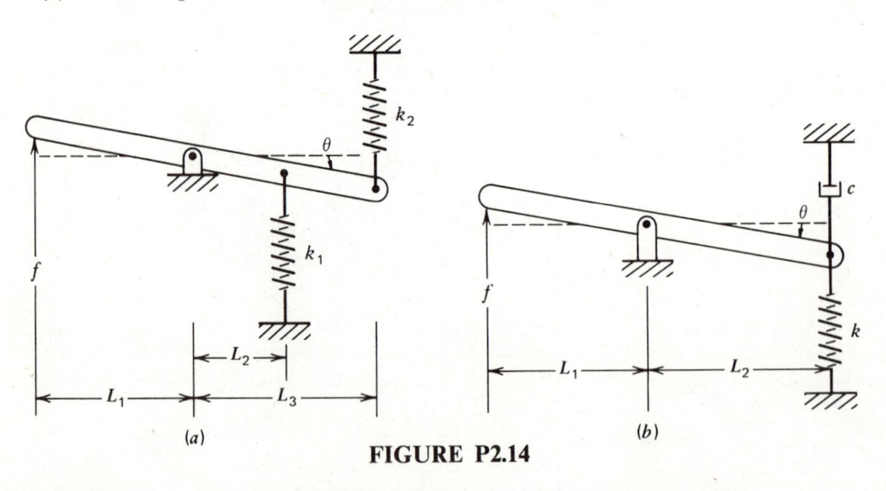

FIGURE P2.14

2.15 A simple robot is shown in Figure P2.15(a). An actuator at the base provides the torque T to rotate the robot through the angle θ. Another actuator in the arm provides the force f to extend the boom. The mass of the rotating base is m_1; its center of mass is at r_1, which is constant. The mass of the extendable boom is m_2, and its mass center is at r_2, which is variable. The lumped parameter representation is shown in Figure P2.15(b). Use Lagrange's equations to obtain the equations of motion for θ and r_2 as outputs, with T and f as given input functions of time.

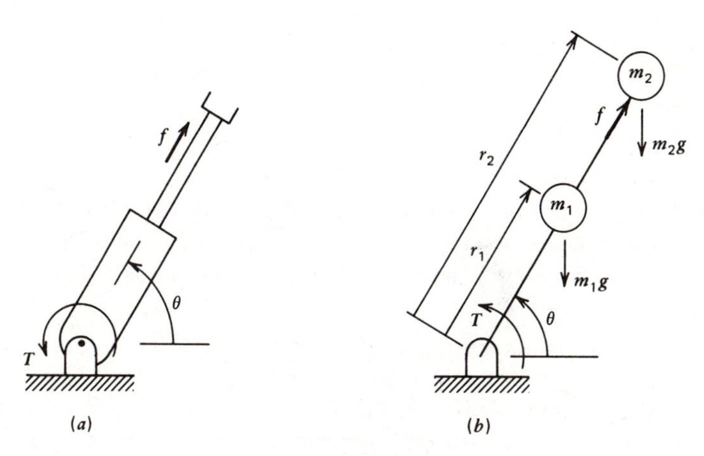

FIGURE P2.15

2.16 A double pendulum is shown in Figure P2.16. Use Lagrange's equations to obtain the equations of motion in terms of the variables θ_1 and θ_2.

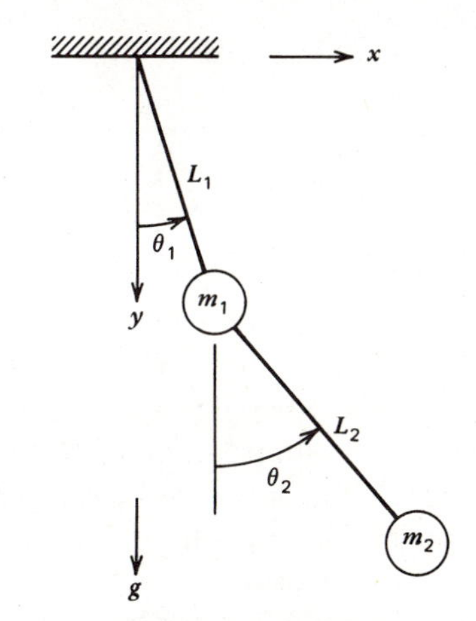

FIGURE P2.16

2.17 The derivation of the dc motor model in Example 2.8 neglected the elasticity of the motor–load shaft. Figure P2.17 shows a model that includes this elasticity,

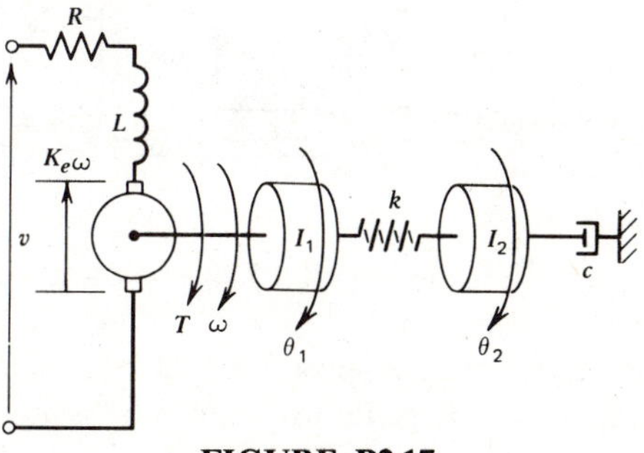

FIGURE P2.17

denoted by its equivalent spring constant k. The motor inertia is I_1, and the load inertia is I_2. Find the differential equation model with θ_2 as output and v as input.

2.18 The torque constant for an armature-controlled dc motor relates the armature current i to the motor torque T as $T = K_t i$. The constant K_e relates the back emf e_b to the motor speed ω as $e_b = K_e \omega$. Establish the relationship between K_t and K_e by equating the two expressions for the power developed by the motor. These expressions are $P = e_b i_a$ (W) and $P = \frac{1}{550} T\omega$ (hp). The units of K_t are lb-ft/A and those of K_e are V/rad/sec.

2.19 The dc motor treated in Example 2.8 is armature controlled. Motor control can also be obtained by varying the voltage applied to the field windings and keeping the armature current constant (see Figure P2.19). In this case, the motor torque is proportional to the field current, so that $T = K_m i_f$, where K_m is the motor constant.

Develop the equations for the field-controlled dc motor with e_f as input and ω as output. Explain why no back emf occurs.

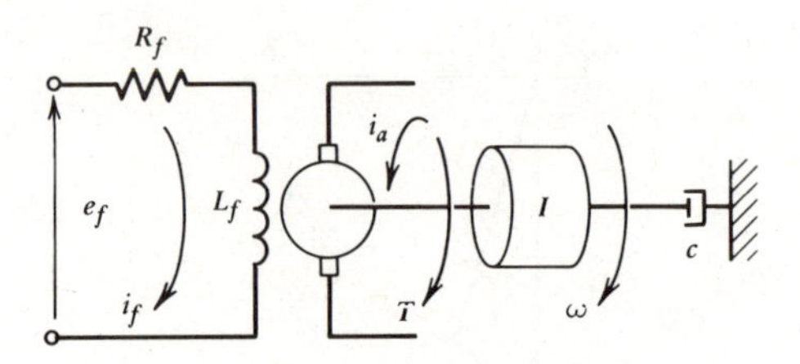

FIGURE P2.19

2.20 The system shown in Figure P2.20 might represent a machine tool slide driven through a rack and pinion by a motor, or an electrohydraulic actuator in which the mass m_1 represents the spool valve (see Problem 2.28).

The gear shaft is supported by frictionless bearings and thus can only rotate. Assume that the only significant masses in the system are the motor with inertia I and the slide with mass m_1. Obtain the system's differential equations with the motor torque T as the input.

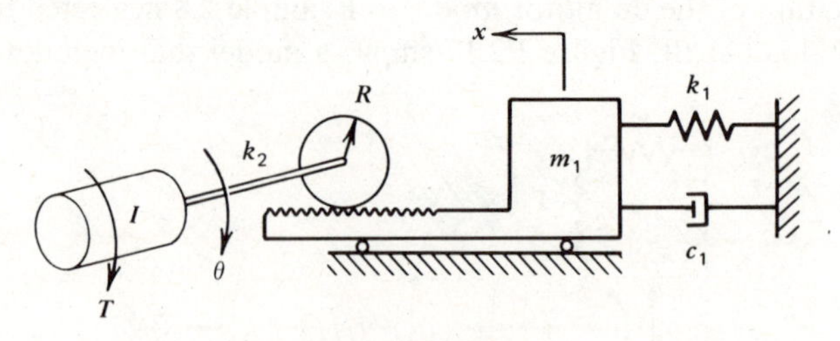

FIGURE P2.20

2.21 Figure P2.21 is the diagram of a speed-control system in which the dc motor voltage e_1 is generated by a dc generator driven by a prime mover. This system has been used on locomotives where the prime mover is a diesel engine that operates most efficiently at one speed. The efficiency of the electric motor is not so sensitive to speed and thus can be used to drive the locomotive. The motor voltage e_1 is varied by changing the generator input voltage e. The voltage e_1 is related to the generator field current i_f by $e_1 = K_1 i_f$.

Obtain the system's equations relating the speed ω to the voltage e.

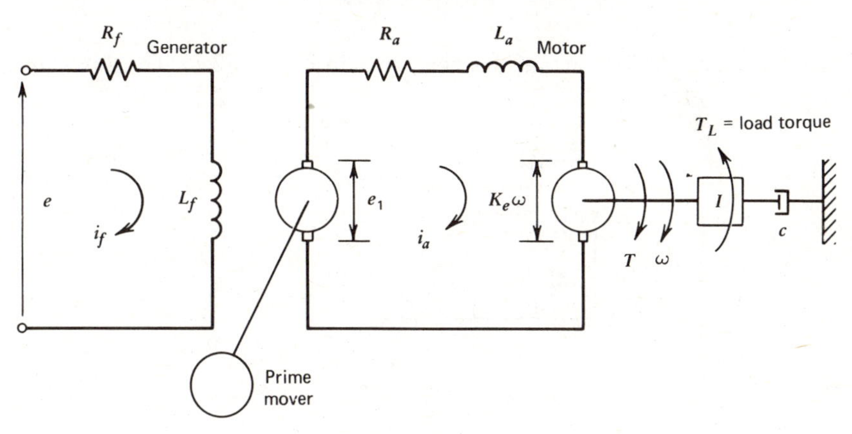

FIGURE P2.21

2.22 A tank initially holds v_o gallons of brine that contains w pounds of salt. The input flow rate to the tank is q_i gallons per minute of brine with a salt density of ρ pounds of salt per gallon. The solution in the tank is well stirred and is pumped out at the *constant* rate of q_o gal./min. Let $y(t)$ be the number of pounds of salt in the tank at time t. Develop the dynamic model for y with q_i and q_o as the input variables.

2.23 A spherical water tank has a radius r, an air vent on top, and an orifice at the bottom (Figure P2.23).

(a) Find the constitutive relation between the liquid height h and the liquid volume V.

(b) Develop a dynamic model for the height h.

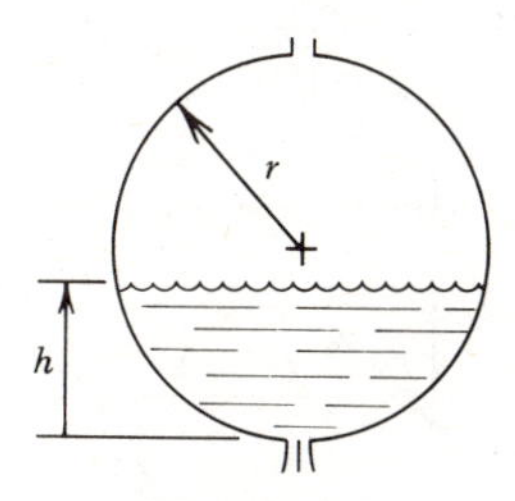

FIGURE P2.23

2.24 Derive a model for the liquid height h in a tank system whose input is the pressure p_i (Figure P2.24).

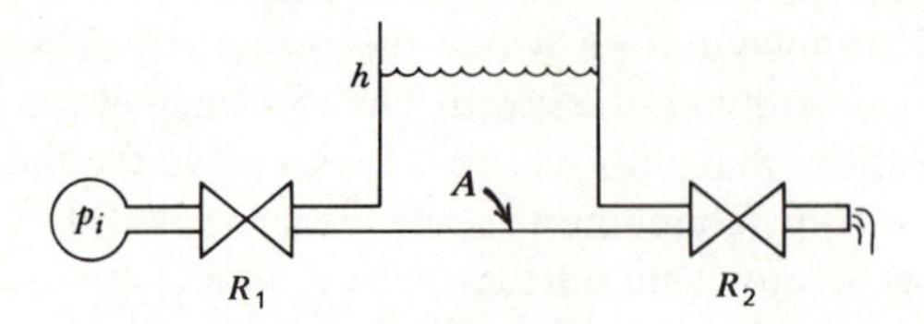

FIGURE P2.24

2.25 Derive the model for the tank system shown in Figure P2.25. The flow rate q_i is a mass flow rate.

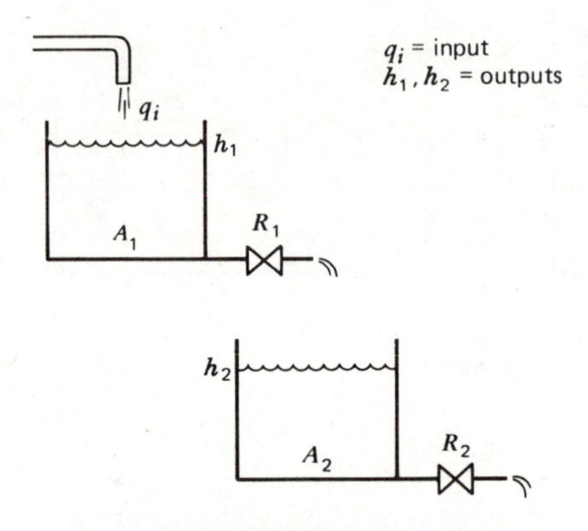

FIGURE P2.25

2.26 For the tank shown in Figure P2.26, brine containing w lb of salt/ft^3 flows in at a rate of p ft^3/min. The resulting mixture flows out at a rate of q ft^3/min. Assume that the tank mixture is kept uniform by stirring. Thus, the salt density in the outflow is the same as that in the tank; namely, v lb/ft^3. Let y be the total amount of salt in the tank in pounds. Neglect the volume of the outflow pipe. The input variables w, p and the parameters R, C are given. Find differential equations for y and h in terms of the given quantities.

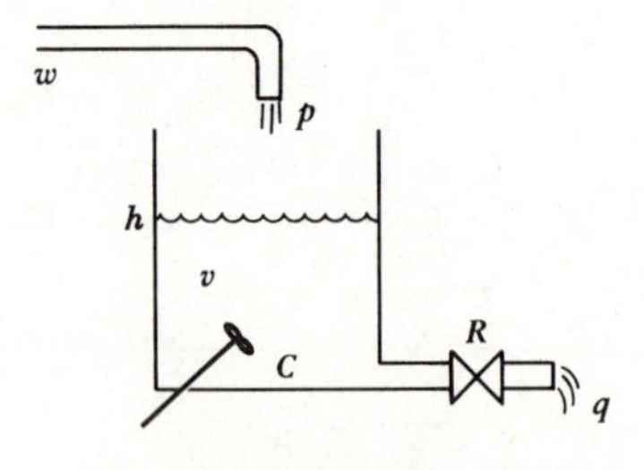

FIGURE P2.26

2.27 Derive the nonlinear damping relations (2.3-4) and (2.7-6) when turbulent flow exists in the flow passage in Figure 2.22.

2.28 The system in Problem 2.20 is now used to move the spool valve of a hydraulic servomotor, as shown in Figure P2.28. Develop a model of the system with the load displacement y as the output and the electric motor torque T as input. Use the underlapped valve model developed in Section 2.7 and the model developed in Problem 2.20 for the following special cases: (a) $I = 0$, and (b) $I = m_1 = 0$.

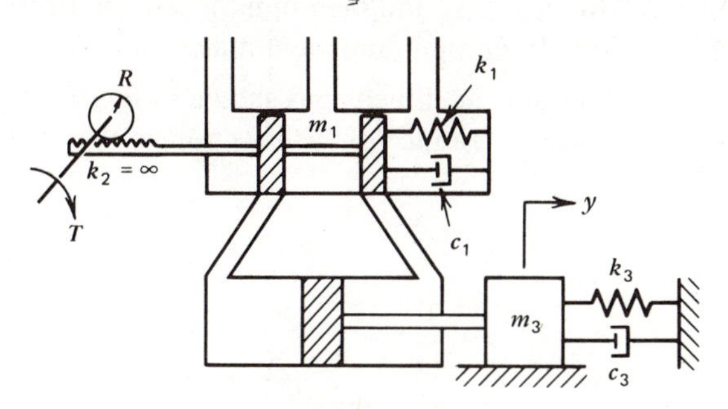

FIGURE P2.28

2.29 A pneumatic restriction was tested under the following conditions. The inlet and outlet pressures (absolute) were $p_1 = 20$ psia and $p_b = 14.7$ psia. The inlet air temperature was $T_1 = 68°F$, and the measured mass flow rate was $q_m = 0.01$ slug/sec.

(a) Estimate the effective cross-sectional area $C_d A$.

(b) Estimate the turbulent resistance R_p of the restriction at the reference pressure $p_s = 14.7$.

2.30 A rigid container has a volume of 10 ft^3. The air inside is initially at 68°F.

(a) Find the pneumatic capacitance of the container for an isothermal process.

(b) Develop a model for a system consisting of the container and the pneumatic inlet restriction analyzed in Problem 2.29.

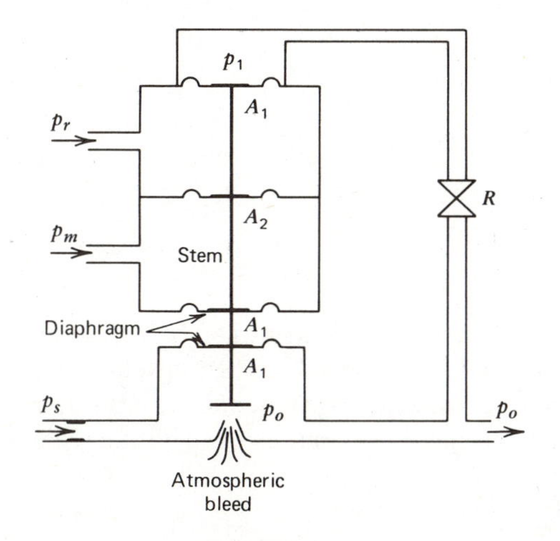

FIGURE P2.31

2.31 The pneumatic system shown in Figure P2.31 is a stack-type device used to control the output pressure p_o. The input pressures p_r and p_m represent reference and measurement pressures, respectively. The supply pressure is p_s.

Assume that the pressures are deviations from nominal values and develop an approximate model for the output pressure p_o. To do this, first develop a dynamic model relating p_1 to p_o, and then sum the pneumatic forces acting on the stem. Assume that the diaphragm and stem masses are negligible.

2.32 A stepped copper shaft has the dimensions shown in Figure P2.32. Compute the thermal resistance and capacitance of each section of the shaft and of the total shaft.

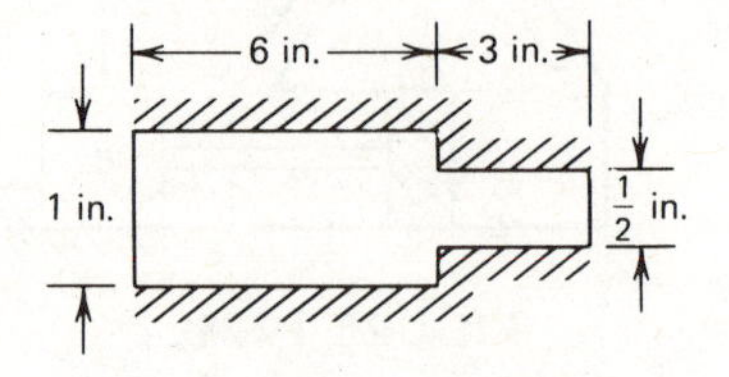

FIGURE P2.32

2.33 Liquid at a temperature T_i is pumped into a mixing tank at a volume flow rate q (Figure P2.33). The container walls are perfectly insulated. The container volume is V, and the liquid within is assumed to be mixed and at a uniform temperature T. The liquid's specific heat and mass density are c and ρ. Before the start of the process, the inlet and outlet temperatures were $T_i = T = T_R$. Then T_i is changed to $T_R + \Delta T_i$. Develop a model for $\Delta T = T - T_R$ with ΔT_i as the input.

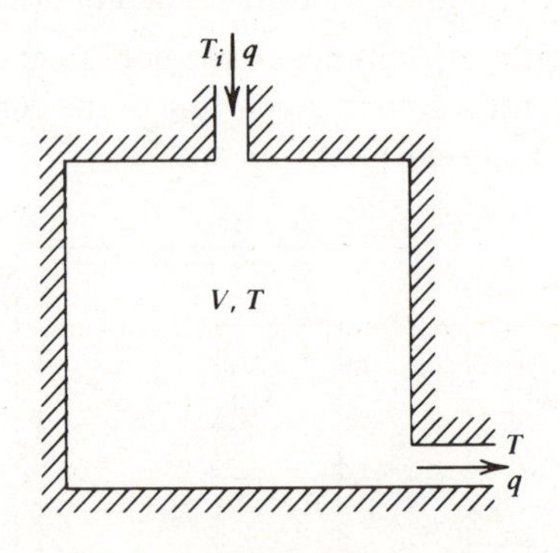

FIGURE P2.33

2.34 A hot metal part is to be quenched in a water bath. Assume that the Biot number is small enough to allow a lumped-temperature model of the part. Denote its temperature by T, its capacitance by C, and its surface resistance by R. Assume that the bath is large enough so that its temperature remains constant. Derive a model for T.

2.35 A thermometer can be represented as two thermal masses with thermal capacitances C_1 and C_2 (Figure P2.35). The total thermal resistances of the inner and outer glass surfaces are R_1 and R_2, respectively. Develop a model for the temperatures T_1 and T_2 with the external temperature T_o as input.

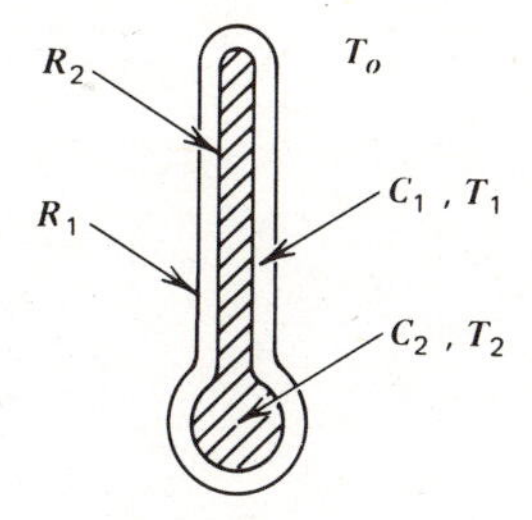

FIGURE P2.35

2.36 For the tank shown in Figure P2.36, liquid flows in at a temperature T_i and the volume flow rate q_i. Assume the outlet flow is laminar. Let T be the temperature of the liquid in the tank and outflow. Assume that the only way heat is lost from the tank liquid is by the outflow. The liquid's mass density is ρ, and its specific heat is c.

Derive the differential equations for the height h and the temperature T, with q_i and T_i as the input variables.

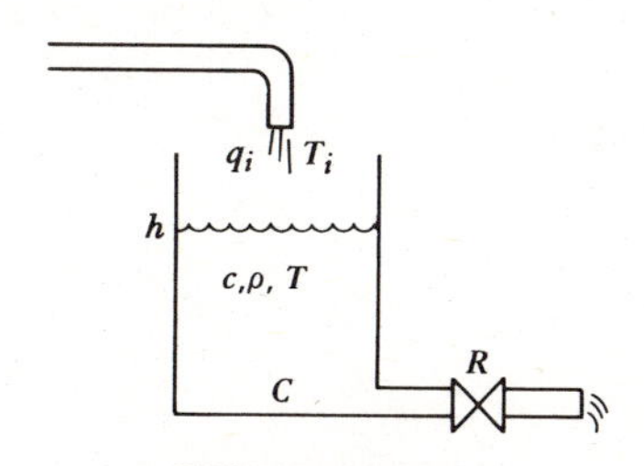

FIGURE P2.36

2.37 The oven shown in Figure P2.37 has a heater with appreciable capacitance C_1. The other capacitance is that of the oven air C_2. The corresponding temperatures are T_1 and T_2, and the outside temperature is T_o. The thermal resistance of the heater-air interface is R_1; that of the oven wall is R_2. Develop a model for T_1 and T_2, with input q, the heat flow rate delivered to the heater mass.

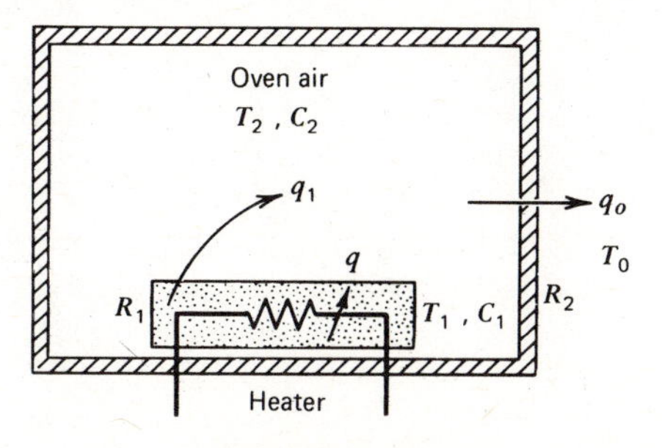

FIGURE P2.37

CHAPTER THREE

System Response

Once the system model has been developed, we can see what predictions it makes of the system's behavior under whatever conditions are of interest. In order to do this for a continuous-time, dynamic model, we must solve the model's differential equations. This chapter introduces the necessary solution methods for relatively low-order models with uncomplicated input functions. Techniques needed for more complex problems are treated in Chapters 4 and 5.

Most of the methods needed to analyze dynamic models can be developed in terms of first- and second-order models, the simplest cases. Most of the differences for higher order models lie in the increased algebraic complexity needed to analyze such models, and not in the basic solution concepts. By concentrating on first- and second-order models here, we do not obscure these fundamental concepts with algebraic manipulations.

Analytical methods for linear models are treated first, starting with simple input functions. As the complexity of the input and the resulting response are increased, we introduce appropriate methods to deal with the situation. We will see that linear models allow general solutions to be obtained for several types of input functions, without requiring numerical values for either the coefficients or the initial values. However, for nonlinear models, few closed-form solutions are available, and we must rely on either a linearized approximation or a numerical solution. The linearization approach is treated in this chapter, while numerical methods are treated in Chapter 5.

3.1 FREE RESPONSE OF A FIRST-ORDER MODEL

The simplest dynamic model to analyze is a linear, first-order, constant-coefficient, ordinary differential equation without inputs. This lengthy string of modifiers refers to the following simple equation:

$$\frac{dy}{dt} = ry \tag{3.1-1}$$

where r is a constant. There are several ways of solving this equation, but we wish to emphasize from the start the following useful fact.

> Any constant-coefficient, linear, ordinary differential equation or coupled set of such equations of any order, without inputs, can always be solved by assuming an exponential form for the solution.

With this in mind, the solution can easily be obtained. The general exponential form to be assumed for each unknown variable is

$$y(t) = Ae^{st} \tag{3.1-2}$$

where A and s are unknown constants. The time derivative of y in this form is

$$\frac{dy}{dt} = sAe^{st}$$

Substituting the last two relations into (3.1-1) gives

$$sAe^{st} = rAe^{st}$$

For the solution to be nontrivial (nonzero for arbitrary values of t), the constant A must be nonzero, and s must be finite. Thus, Ae^{st} does not equal zero, and we obtain

$$s = r \tag{3.1-3}$$

Equation (3.1-3) is known as the *characteristic* equation of the model and is a first-order polynomial equation in s. It has one solution, the *characteristic root*. Thus, the solution of (3.1-1) is

$$y(t) = Ae^{rt} \tag{3.1-4}$$

for any nonzero A. In order to determine a value for A, an additional condition must be stipulated. This condition usually specifies the value of y at time t_0, the start of the process and is the *initial condition*. Evaluating the preceding equation at $t = t_0$ gives

$$A = y(t_0)e^{-rt_0}$$

and accordingly the solution is

$$y(t) = y(t_0)e^{-rt_0}e^{rt}$$

or

$$y(t) = y(t_0)e^{r(t-t_0)} \tag{3.1-5}$$

For autonomous models,† the origin of the time axis may be shifted so that t_0 is taken to be zero without loss of generality. The solution is sketched in Figure 3.1 for positive, zero, and negative values of r. Because this solution describes the system's behavior when it is free of input's influence, this solution is called the *free response*.

The Time Constant

If r is negative, a new constant τ is usually introduced by the definition

$$\tau = -\frac{1}{r} \tag{3.1-6}$$

and the solution can be written as

$$y(t) = y(0)e^{-t/\tau} \tag{3.1-7}$$

The new parameter τ is the model's *time constant*, and it gives a convenient measure of the exponential decay curve. To see this, let $t = \tau$ in (3.1-7) to obtain

$$y(\tau) = y(0)e^{-1} \cong 0.37y(0) \tag{3.1-8}$$

---

†The model $dy/dt = f(y, t)$ is *autonomous* when $f(y, t)$ is independent of t.

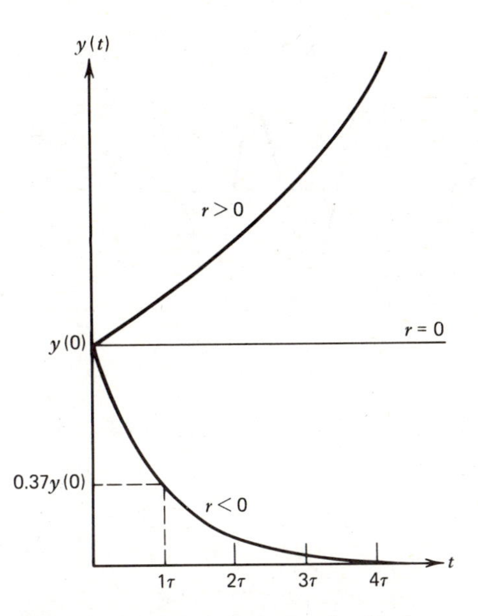

FIGURE 3.1 Free response of a linear first-order model.

After a time equal to one time constant has elapsed, y has decayed to 37% of its initial value. Alternatively, we can say that y has decayed by 63%. If $t = 4\tau$,

$$y(4\tau) = y(0)e^{-4} \cong 0.02y(0) \quad \textbf{(3.1-9)}$$

After a time equal to four time constants, y has decayed to 2% of its initial value (or by 98%). Other times, such as $t = 2\tau$, 3τ, etc., could be selected, but τ and 4τ are the common choices, because these times yield two points that, along with the initial value, provide sufficient spacing for sketching the curve. The results summarized by (3.1-8) and (3.1-9) are used frequently, and it is desirable to commit them to memory.

Point Equilibrium and Stability

Often the information desired from a dynamic model simply concerns the existence of a *point equilibrium* or *equilibrium state* (or simply *equilibrium* in common usage) and its *stability*. A point equilibrium is a condition of no change in the model's variables. Mathematically, it is determined by solving for the values of the variables that make *all* of the time derivatives identically zero. For (3.1-1), this implies that

$$0 = ry$$

If r is zero, an infinite number of equilibrium values for y exists. If r is nonzero, the only point equilibrium is $y = 0$. Thus, if y is initially zero, it will remain zero unless some cause (other than those already described by the model) displaces y from its equilibrium value.

If y is disturbed from equilibrium, a natural question to ask is, Does y return to its equilibrium? The answer is determined by the stability characteristics of the equilibrium. A *stable* equilibrium is one to which the model's variables return *and* remain if slightly disturbed. The significance of the last phrase will become apparent later. Conversely, an *unstable* equilibrium is one from which the model's variables continue to recede if slightly disturbed. There is a borderline situation, a *neutrally stable* equilibrium, defined to be one to which the model's variables do not return and remain but from which they do *not* continue to recede.

These three cases can be illustrated by a ball rolling on a surface (Figure 3.2). In Figure 3.2*a*, if the ball is displaced but still within the valley ("slightly displaced"), it will roll back and forth around the bottom, transferring energy between kinetic and potential. If there is friction, the ball finally comes to rest after the friction has dissipated all the energy. Thus, the equilibrium state is the state in which the ball rests at the bottom, and it is stable if friction is present. A frictionless surface will allow the

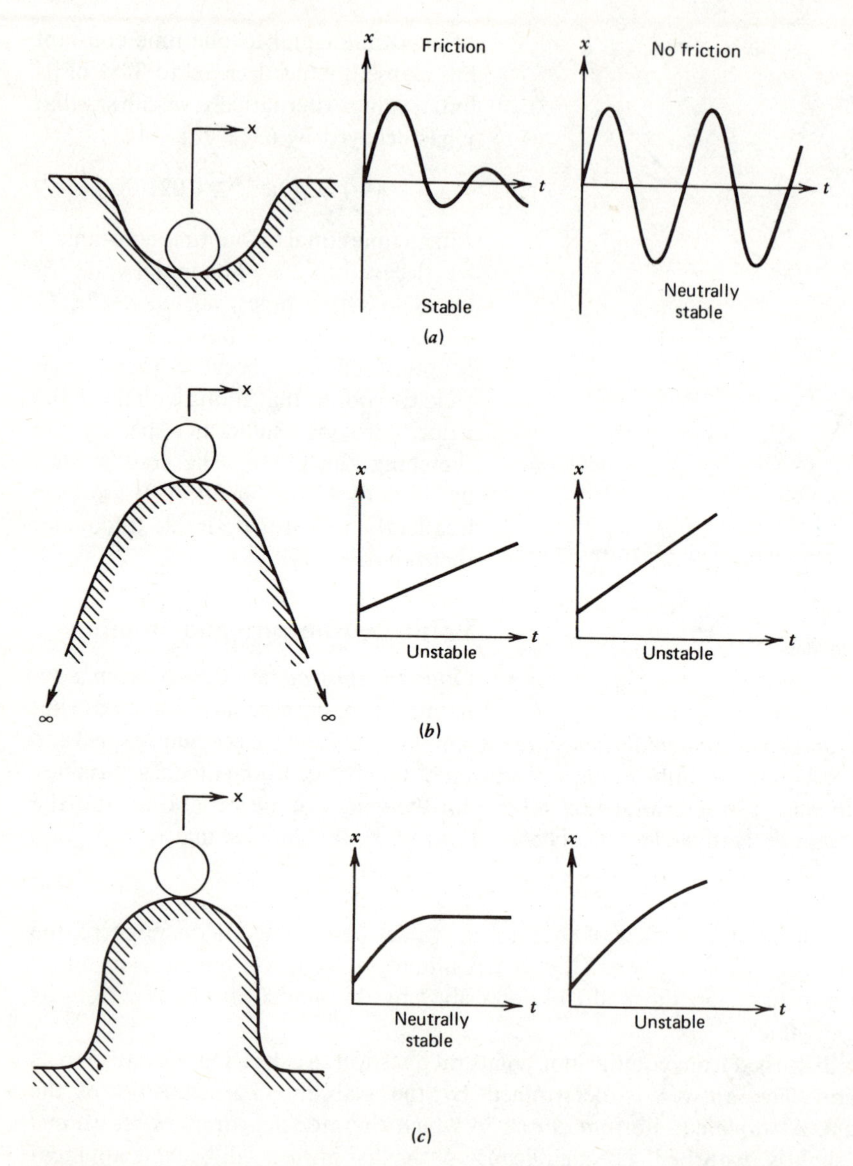

FIGURE 3.2 Stability examples. (*a*) Ball in a hole. (*b*) Ball on an infinite hill. (*c*) Ball on a finite hill.

ball to oscillate forever about the bottom, and in this case, the equilibrium is neutrally stable. If the ball is displaced so much that it now lies on the level surface, the equilibrium can no longer be considered stable even with friction. The term *local stability* is sometimes used to emphasize the restriction to small displacements, while *global stability* implies stability with respect to all displacements.

For the ball precariously balanced on top of the hill (Figure 3.2*b*), this equilibrium is globally unstable (and thus locally unstable as well) for a hypothetical hill without a bottom if the friction is not great enough to stop the ball. The absence of such a hill in reality points out that in interpreting instability, the assumptions of the model structure must be kept in mind. Thus, in a model of the form of (3.1-1), the equilibrium at $y = 0$ is globally unstable mathematically if $r > 0$, but this does not mean that the actual variable y will eventually become infinite, because the assumptions of the model will eventually be violated as y increases. For physical systems, a locally unstable equilibrium usually means either that the equipment will fail as it recedes from equilibrium or that the model's assumptions are no longer valid as the system enters a new mode of behavior requiring a new model.

If the hill has a bottom consisting of a level surface (Figure 3.2*c*), the equilibrium is still locally unstable but now becomes neutrally stable in a global sense, since the ball eventually comes to rest if there is friction. With no friction, it remains globally unstable. Comparing Figure 3.2*a* without friction and 3.2*c* with friction reveals that two types of neutral stability can occur. In the first type, the ball's position returns to equilibrium but does not remain. In the second type, the ball never returns but does not recede forever. Neutral stability of the first type (oscillatory behavior) is an abstraction that does not exist in physical systems unless a power source is present to sustain the oscillations. Otherwise, dissipative forces, such as friction, act to create a stable situation.

The first-order model has one characteristic root, and, as we will see, a second-order model has two roots, etc. The local stability properties of an equilibrium for a linear model are determined from the characteristic roots and by an equivalent but approximate process called *linearization* for a nonlinear model. A *linear* model is globally stable if it is locally stable, but determining global stability is frequently difficult for nonlinear models. This topic will be explored later in some detail, since the question of the stability of a control system is very important, because the introduction of feedback can cause instability if the designer is not careful.

The characteristic root of (3.1-1) is given by (3.1-3), and from Figure 3.1, we can extract the following general statement:

> An equilibrium of the first-order linear model is globally stable if and only if its characteristic root s is negative and is neutrally stable if and only if s is zero. Otherwise, the equilibrium is unstable.

This statement will be later extended, with some modifications, to higher order models. Thus, the characteristic roots play a key role in determining local stability.

Parameter Estimation

In many applications, a system can be described by (3.1-1), but the parameter r cannot be computed from basic principles. If past measurements of y at various times are available, a value for r can be estimated by the following technique. If the first data point is taken to be at $t_0 = 0$, the natural logarithm of both sides of (3.1-5) gives

$$\ln y(t) = \ln y(0) + rt \qquad \textbf{(3.1-10)}$$

If the logarithms of the measurements of y are plotted versus t, the transformed data should cluster about a straight line if (3.1-1) is an accurate model of the process (and if

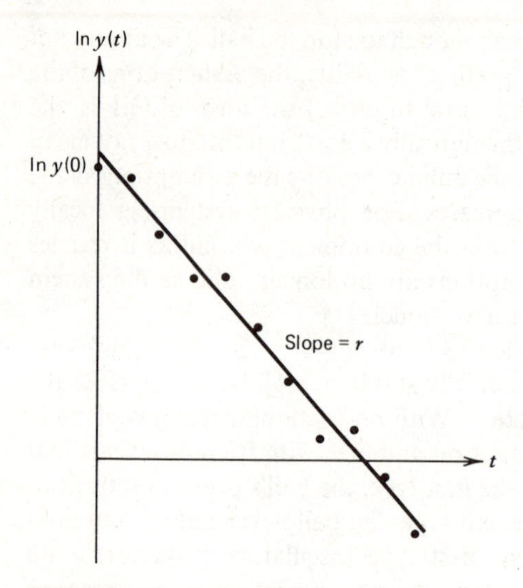

FIGURE 3.3 Logarithmic transformation of data.

the measurement error is small!). This situation is shown in Figure 3.3 for a negative value of r. A straight line can be fitted by eye if great accuracy is not required, and the value of r is estimated from the slope of this line. If the points have some scatter, the method of *least squares* can be used to fit a line to the data. This method is developed in the problems at the end of the chapter. In either case, the line giving the best fit does not necessarily pass through the initial data point and should not be constrained to do so unless some compelling reason exists. With the estimated value of r, predictions can then be made about future system behavior.

Example 3.1

For the series RC circuit model (2.2-3), let $y = v_3$ and $v_1 = 0$. Then,

$$RC\frac{dy}{dt} + y = 0$$

the characteristic root is $s = -1/RC$, and thus the time constant is $\tau = RC$. Typical values of R and C are 10,000 Ω and 0.1 μf, respectively. These give a time constant of 0.001 sec. If the input voltage becomes zero but the circuit remains closed, the capacitor voltage decays by 98% in 0.004 sec.

Example 3.2

Compute the time constant for a tank–pipe system as shown in Figure 2.20, assuming that the flow is laminar. The tank contains fuel oil at 70°F with a mass density ρ of 1.82 slugs/ft^3 and a viscosity μ of 0.02 lb-sec/ft^2. The outlet pipe diameter D is 1 in., and its length L is 2 ft. The tank is 2 ft in diameter.

From (2.6-11)

$$A\frac{dh}{dt} = q_1 - \frac{g}{R}h \tag{3.1-11}$$

where q_1 is a volume flow rate. Here, the tank area is $A = \pi$ ft^2. The laminar pipe resistance R can be obtained from (2.6-12). The resistance is

$$R = \frac{128\mu L}{\rho \pi D^4} = \frac{128(0.02)(2)}{(1.82)\pi(1/12)^4} = 18568.4\frac{\text{lb-sec}}{\text{ft}^5}$$

and

$$\frac{g}{R} = \frac{32.2}{18568.4} = 0.001734 \text{ ft}^2/\text{sec}$$

Thus,

$$\pi \frac{dh}{dt} = -0.001734h + q_1$$

and the time constant is $\tau = \pi/0.001734 = 1812$ sec $= 30.2$ min.

3.2 STEP RESPONSE OF THE FIRST-ORDER MODEL

Many of the models developed in Chapter 2 can be put into the following general form:

$$\dot{y} = ry + bv \tag{3.2-1}$$

where v is the input. The solution obtained in Section 3.1 applies when no input is present—that is, when $v = 0$—and this solution represents the intrinsic behavior of the system. Here, we begin the analysis of (3.2-1) for the commonly encountered input functions of time.

The simplest of these perhaps is the *step function* depicted in Figure 3.4. As its name implies, it has the appearance of a single stair step. Before the initial time $t = 0$, the step function has a constant value, usually zero. At $t = 0$, v jumps instantaneously to a new constant value, M, the magnitude, which it maintains thereafter. The step function is an approximate description of an input that can be switched on in a time interval that is very short compared to the time constant of the system. In the tank system in Figure 2.20, we may model the input flow rate q_i as a step if the inlet valve can be opened much faster than the time it takes for significant changes in liquid height to occur. In Example 3.2, since $\tau = 30.2$ min, a valve-opening time of less than, say, 1 min would seem to allow a step input approximation to be used with confidence. The magnitude M would be the value of the flow rate if it is held constant.

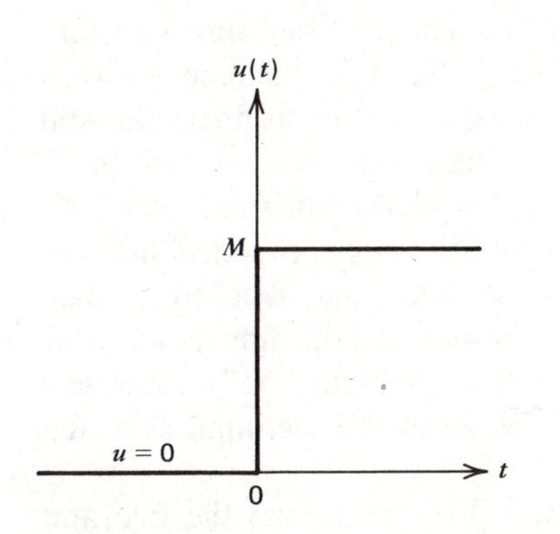

FIGURE 3.4 Step input of magnitude *M*.

A common notation for the *unit* step function ($M = 1$) is $u_s(t)$. Accordingly, the step with magnitude M is denoted by $Mu_s(t)$.

Solution for the Stable Case

There are several ways of obtaining the closed-form solution of (3.2-1) with a step input, if r and b are constants. The case where r and b are not constants is treated in Chapter 5. For now, we will restrict ourselves to the stable case ($r < 0$), in which case (3.2-1) can be written as

$$\dot{y} = -\frac{1}{\tau}y + bv \tag{3.2-2}$$

The simplest way of proceeding at this point is by employing some physical insight to obtain the form of the solution. If the input flow rate q_i in Figure 2.20 is a step input, the tank height and thus the outflow rate will increase until the outflow rate equals the input rate and the tank comes to an equilibrium state. Since this system is stable and its intrinsic behavior is given by (3.1-5), we assume that its response to a step input can be written in the form

$$y(t) = C_1 e^{-t/\tau} + C_2 \tag{3.2-3}$$

where C_1 and C_2 are undetermined constants. Note that as $t \to \infty$, $y \to C_2$ as we would expect. Substituting into (3.2-2) for $t > 0$ gives

$$-\frac{1}{\tau}C_1 e^{-t/\tau} = -\frac{1}{\tau}(C_1 e^{-t/\tau} + C_2) + bM$$

This yields

$$C_2 = bM\tau$$

The last condition to be satisfied is the initial condition

$$y(0) = C_1 e^0 + bM\tau$$

or

$$C_1 = y(0) - bM\tau$$

This gives the step response

$$y(t) = [y(0) - bM\tau]e^{-t/\tau} + bM\tau \tag{3.2-4}$$

which is plotted in Figure 3.5 for positive values of b, M, and $y(0)$. Note once again the usefulness of the time constant in supplying convenient points for plotting. After a time equal to one time constant has elapsed, 37% of the difference between the initial value of y and its final value still remains. After a four time constant interval, only 2% of this difference remains. Thus, for most cases of practical interest, we may say that y has reached its final value at $t = 4\tau$, and if the time required for the input to be "switched on" is much less than τ, we can model the input as a step function.

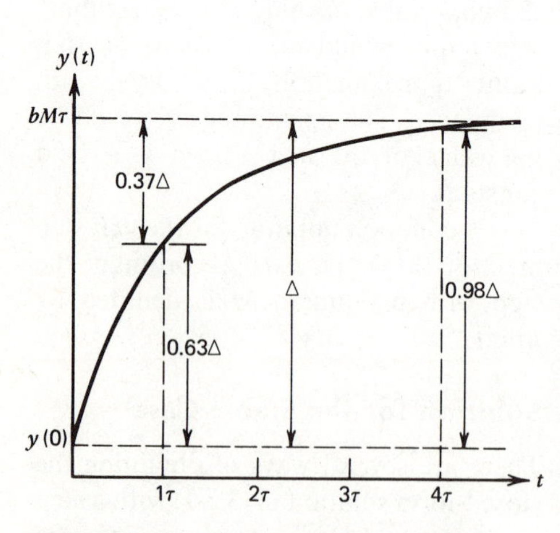

FIGURE 3.5 Step response of a first-order system.

Table 3.1 summarizes the free and step responses of the stable first-order model with $b = 1/\tau$.

TABLE 3.1 Free and Step Responses of the Stable First-order Model: $\tau\dot{y}+y=v(t)$

Free response $[v(t)=0]$:

$$y(t)=y(0)e^{-t/\tau} \qquad \tau = \text{time constant}$$

$$y(\tau)=0.37y(0)$$

$$y(4\tau)=0.02y(0)$$

Step response $[y(0)=0]$:

$$v(t)=\begin{cases} 0 & t\leqslant 0 \\ M & t>0 \end{cases}$$

$$y(t)=M[1-e^{-t/\tau}]$$

$$y_{ss}=M \text{ (steady-state response)}$$

$$y(\tau)=0.63y_{ss}$$

$$y(4\tau)=0.98y_{ss}$$

Complete step response:

$$y(t)=y(0)e^{-t/\tau}+M[1-e^{-t/\tau}]$$

Free and Forced Response

The solution given by (3.2-4) can be decomposed in several ways. We may think of the solution as being composed of two parts: one resulting from the "intrinsic" behavior of the system (the free response) and the other from the input (the forced response). Written this way, the solution is

$$y(t)=\underbrace{y(0)e^{-t/\tau}}_{\text{free}}+\underbrace{bM\tau(1-e^{-t/\tau})}_{\text{forced}}$$

The *principle of superposition* for a linear model implies that the *total* or *complete* response results from the sum of the free and the forced responses. This is borne out by the preceding equation. With this in mind, we could have found the complete solution by adding the free response from (3.1-5) to the forced response given by (3.2-4) with $y(0)=0$. This approach is sometimes useful for complicated (higher order) models. In summary, the free response represents that part of the system behavior stimulated by the initial value of y, and the forced response represents the effect of the input.

Transient and Steady-State Responses

We can also think of the response as being comprised of a term that eventually disappears (the *transient* response) and a term that remains (the *steady-state* response). For (3.2-4), the transient response is

$$[y(0)-bM\tau]e^{-t/\tau}$$

and the steady-state response is $bM\tau$. For a stable system, the free response is always part of the transient response, and some of the forced response might appear in the transient response.

Neutrally Stable and Unstable Cases

If r is zero in (3.2-1), the solution given by (3.2-4) does not apply, because τ is positive by definition. In this neutrally stable case, $\dot{y}$ is a constant for $t > 0$ and thus

$$y(t) = \int_0^t bM\, dt + y(0)$$

or

$$y(t) = bMt + y(0) \tag{3.2-5}$$

and y continues to increase indefinitely in magnitude. Contrast this behavior with the step response for the stable case, where y approaches a constant. This difference results from the self-regulating (feedback) nature of the term ry on the right-hand side of (3.2-1).

If r is positive (unstable case), an approach similar to that used for the stable case produces the following solution:

$$y(t) = \left[y(0) + \frac{bM}{r}\right]e^{rt} - \frac{bM}{r} \tag{3.2-6}$$

The response increases exponentially.

Example 3.3

(a) We want to find the motor torque necessary to drive a load inertia of 1.4 oz-in.-sec^2 at a constant speed of 150 rad/sec. The load's resisting torque is due entirely to dry friction and is measured to be 256 oz-in. The tentative motor choice has an armature inertia of 0.15 oz-in.-sec^2, a rotational loss coefficient of 7 oz-in./Krpm,† and a maximum internal dry friction torque of 15 oz-in. These data are available from manufacturer's catalogs (for example, see Reference 1).

(b) Once the motor torque is determined, find how long it will take to reach the desired speed starting from rest.

(a) The inertias of the connecting shaft and gears are assumed to have been lumped with the load inertia, using the methods in Section 2.4. If we neglect the twisting of the shaft, we can also lump the armature and load inertias to get a model similar in form to that shown in Figure 2.15*b*, where the applied torque T_1 is the motor torque T_m minus the total dry friction torque T_F. The damping coefficient c is given by the rotational-loss coefficient. Thus, the needed values are

$$I = 1.4 + 0.15 = 1.55 \text{ oz-in.-sec}^2$$

$$T_1 = T_m - 256 - 15 = T_m - 271 \text{ oz-in.}$$

$$c = 7 \text{ oz-in./Krpm} = 0.07 \text{ oz-in./rad/sec}$$

---

†Krpm = 1000 rpm (revolutions per minute).

At steady state, the torques must balance, so that

$$T_m - 271 = 0.07(150)$$

where the torque on the right is the damping torque experienced at the design speed of 150 rad/sec. The required motor torque therefore is $T_m = 281.5$ oz-in.

(b) If this torque is suddenly applied and maintained, the step response may be used. The system time constant is [see (2.4-18)]

$$\tau = \frac{I}{c} = \frac{1.55}{0.07} = 22 \text{ sec}$$

and it will take approximately 88 sec to reach the desired speed if the motor torque is held constant. Later, we will see that the torque of a dc motor does not remain constant when accelerating a load even if a constant motor voltage is applied. This is due to the coupling between the dynamics of the mechanical subsystem and the electrical subsystem. However, these results are accurate enough to be useful for finding the approximate motor size required if the response of the electrical subsystem is much faster than 22 sec.

Example 3.4

Consider Example 3.2 with inflow to a 2-ft diameter tank. If the input flow rate of fuel oil q_1 is 20 gal/min, how long will it take to raise the oil level from 1 to 10 ft if the 1-in. outlet line remains open?

Since 7.48 gal = 1 ft^3, the input flow rate is 0.0443 ft^3/sec, and this is the magnitude of the step. The tank model (3.1-11) when put into the standard form of (3.2-2) gives

$$y = h \qquad v = q_1$$

$$b = \frac{1}{A} = \frac{1}{\pi} \qquad \tau = \frac{RA}{g} = 1812$$

Equation (3.2-4) can be transformed logarithmically to find the time required to reach 10 ft if $h(0) = 1$ ft.

$$bM\tau = \left(\frac{1}{\pi}\right)(0.0443)(1812) = 25.55 \text{ ft}$$

$$\ln\left(\frac{10 - 25.55}{1 - 25.55}\right) = -\frac{t}{1812}$$

The time required is

$$t = 827 \text{ sec} = 13.8 \text{ min}$$

A by-product of this analysis is the final height, 25.55 ft, that would be attained if the input flow rate were maintained constant at 20 gal/min.

Example 3.5

For the previous example, how long would it take for the height to reach 10 ft if the outlet pipe were closed?

In this case, the outflow term gh/R in (3.1-11) is zero, and the neutrally stable case applies. From (3.2-5), the step response is

$$h(t) = \frac{1}{3.14}(0.0443)t + h(0)$$

With $h(t) = 10$ and $h(0) = 1$, the time required is

$$t = \frac{(10-1)(3.14)}{0.0443} = 638 \text{ sec} = 10.6 \text{ min}$$

3.3 FREE RESPONSE OF A SECOND-ORDER MODEL

The free response of second-order linear continuous-time models is more varied than that of first-order models, because oscillatory as well as exponential functions can occur.

Reduced and State Variable Forms

We saw in Chapter 2 that the application of physical laws can produce a second-order model in the form of either a single second-order equation or two coupled first-order equations. The mass–spring–damper model (2.4-3) $m\ddot{x} + c\dot{x} + kx = f$ is an example of the first form, which is called the *reduced form.* The dc motor model (2.5-3) and (2.5-4) is an example of the second form, which is the *state variable* form. The two forms are equivalent, but each has its own advantages for analyzing system behavior. The state variable form is more convenient for use with numerical-simulation techniques to be treated in Chapter 5. The reduced form is more convenient for finding the response analytically, for low-order models with relatively simple inputs.

It is easy to convert from one form to another. The model (2.4-3) can be converted to state variable form by defining as a new variable the time derivative of the variable x. Let the new variable be denoted v. Since $v = \dot{x}$, v is obviously the speed of the mass m. The two variables whose derivatives appear in the state variable form are called the *state variables* of the system. Generally, there is no unique choice for the state variables. Our current choices are x and v, but we could also have written the model in state variable form by choosing $x + v$ and $x - v$ as the state variables. But there is little incentive to do this, since the choice makes little sense physically.

Because $v = \dot{x}$, (2.4-3) can be written in state variable form as

$$\dot{x} = v \qquad \textbf{(3.3-1)}$$

$$m\dot{v} = f - cv - kx \qquad \textbf{(3.3-2)}$$

In order to solve these equations, it should be obvious that we need to know the starting value of the two state variables; that is, $x(0)$ and $v(0) = \dot{x}(0)$. In order to solve the reduced form (2.3-3), we need the same information.

Conversion from state variable form to reduced form requires one of the equations to be differentiated and substituted into the other to eliminate the unwanted variable. For the dc motor model, suppose we desire a reduced model in terms of the speed ω. To obtain this, differentiate (2.5-4) to yield

$$I\frac{d^2\omega}{dt^2} = K_T\frac{di_a}{dt} - c\frac{d\omega}{dt}$$

Solve (2.5-3) for di_a/dt and substitute into the preceding equation to obtain

$$I\frac{d^2\omega}{dt^2} = \frac{K_T}{L}(v - i_a R - K_e\omega) - c\frac{d\omega}{dt}$$

Finally, solve (2.5-4) for i_a, and substitute into the preceding equation to eliminate i_a. This gives

$$I\frac{d^2\omega}{dt^2} = \frac{K_T}{L}\left[v - \frac{R}{K_T}\left(I\frac{d\omega}{dt} + c\omega\right) - K_e\omega\right] - c\frac{d\omega}{dt}$$

or

$$LI\frac{d^2\omega}{dt^2} + (RI + cL)\frac{d\omega}{dt} + (cR + K_e K_T)\omega = K_T v \tag{3.3-3}$$

Table 3.2 gives the general relationship between the two forms for the second-order model.

Solution from the Reduced Form

The mass–spring–damper system shown in Figure 2.10 is often used to illustrate the free response of a second-order system. Its model (2.4-3) can be written as

$$m\ddot{x} + c\dot{x} + kx = f \tag{3.3-4}$$

TABLE 3.2 Conversion from State Variable Form to Reduced Form: a Second-order Case

State variable form:

$$\dot{x}_1 = a_{11}x_1 + a_{12}x_2 + b_1 f$$
$$\dot{x}_2 = a_{21}x_1 + a_{22}x_2 + b_2 f$$

Equivalent reduced form:

1. In terms of the variable x_1:

$$\ddot{x}_1 - (a_{11} + a_{22})\dot{x}_1 + (a_{11}a_{22} - a_{12}a_{21})x_1 = b_1\dot{f} + (a_{12}b_2 - a_{22}b_1)f$$

2. In terms of the variable x_2:

$$\ddot{x}_2 - (a_{11} + a_{22})\dot{x}_2 + (a_{11}a_{22} - a_{12}a_{21})x_2 = b_2\dot{f} + (a_{21}b_1 - a_{11}b_2)f$$

The free response can be obtained by using the same trial solution as in Section 3.1, namely,

$$x(t) = Ae^{st} \tag{3.3-5}$$

where A and s are constants to be determined. This approach is used to emphasize that the free response of any linear constant-coefficient equation can always be obtained by exponential substitution.

Differentiating (3.3-5) twice gives

$$\dot{x} = sAe^{st}$$
$$\ddot{x} = s^2 Ae^{st}$$

Substitute these into (3.3-4) with $f = 0$, and collect terms to get

$$(ms^2 + cs + k)Ae^{st} = 0$$

As with the first-order model in Section 3.1, a general solution is possible only if $Ae^{st} \neq 0$. Thus,

$$ms^2 + cs + k = 0 \tag{3.3-6}$$

This is the model's *characteristic equation.* It gives the following solution for the unknown constant s:

$$s = \frac{-c \pm \sqrt{c^2 - 4mk}}{2m} \tag{3.3-7}$$

We see immediately that two characteristic roots occur. Each root generates a solution of the form of (3.3-5). If the roots are distinct, the free response is a linear combination of these two forms. Let s_1 and s_2 denote the two roots. The free response is

$$x(t) = A_1 e^{s_1 t} + A_2 e^{s_2 t} \tag{3.3-8}$$

It is easy to show that this solves (3.3-4) with $f = 0$, by computing $\dot{x}$ and $\ddot{x}$, substituting these expressions into (3.3-4), and noting that both s_1 and s_2 satisfy (3.3-6).

Two state variables are required to describe this system's dynamics. Therefore, the initial values of these variables must be specified in order for the solution to be completely determined. This means that the two constants A_1 and A_2 are determined by the two initial conditions. If the values of $x(t)$ and $\dot{x}(t)$ are given at $t = 0$, then from (3.3-8),

$$x(0) = A_1 e^0 + A_2 e^0 = A_1 + A_2 \tag{3.3-9}$$

Differentiating (3.3-8) and evaluating at $t = 0$, we obtain

$$\dot{x}(0) = s_1 A_1 + s_2 A_2 \tag{3.3-10}$$

The solution for A_1 and A_2 in terms of $x(0)$ and $\dot{x}(0)$ is

$$A_1 = \frac{\dot{x}(0) - s_2 x(0)}{s_1 - s_2} \tag{3.3-11}$$

$$A_2 = \frac{s_1 x(0) - \dot{x}(0)}{s_1 - s_2} = x(0) - A_1 \tag{3.3-12}$$

Example 3.6

Find the free response of (3.3-4) for $m = 1$, $c = 5$, and $k = 4$. The initial conditions are $x(0) = 1$ and $\dot{x}(0) = 5$.

The characteristic roots from (3.3-7) are $s_1 = -1$ and $s_2 = -4$. From (3.3-11) and (3.3-12), $A_1 = 3$, $A_2 = -2$, and (3.3-8) gives

$$x(t) = 3e^{-t} - 2e^{-4t}$$

Oscillatory Solutions

The solution given by (3.3-8) is convenient to use only when the characteristic roots are real and distinct. If the roots are complex, the behavior of $x(t)$ is difficult to visualize from (3.3-8). We now rearrange the solution into a more useful form for the case of complex roots.

Complex roots of (3.3-6) occur if and only if $c^2 - 4mk < 0$. If so, the roots are complex conjugates. For now, assume that the real part of the roots is negative and write the roots as

$$s_1 = -a + ib \tag{3.3-13}$$

$$s_2 = -a - ib \tag{3.3-14}$$

where

$$a = \frac{c}{2m} \tag{3.3-15}$$

$$b = \frac{\sqrt{4mk - c^2}}{2m} \tag{3.3-16}$$

For this case, $b > 0$ if $m > 0$. The solution (3.3-8) can be written as

$$\begin{aligned} x(t) &= A_1 e^{(-a+ib)t} + A_2 e^{(-a-ib)t} \\ &= e^{-at}(A_1 e^{ibt} + A_2 e^{-ibt}) \end{aligned}$$

Euler's identity for the complex exponential shows that

$$e^{ibt} = \cos bt + i \sin bt \tag{3.3-17}$$

$$e^{-ibt} = \cos bt - i \sin bt \tag{3.3-18}$$

Thus,

$$x(t) = e^{-at}[(A_1 + A_2)\cos bt + i(A_1 - A_2)\sin bt] \tag{3.3-19}$$

Substituting s_1 and s_2 into the expressions for A_1 and A_2 shows that

$$A_1 = \frac{x(0)}{2} - \frac{\dot{x}(0) + ax(0)}{2b} i \tag{3.3-20}$$

and that A_2 is the complex conjugate of A_1. Write A_1 as $A_R + A_I i$. Then

$$A_1 + A_2 = 2A_R$$

$$i(A_1 - A_2) = -2A_I$$

and

$$x(t) = e^{-at}(2A_R \cos bt - 2A_I \sin bt) \tag{3.3-21}$$

where

$$A_R = \frac{x(0)}{2} \tag{3.3-22}$$

$$A_I = -\frac{\dot{x}(0) + ax(0)}{2b} \tag{3.3-23}$$

It is now apparent from (3.3-21) that the free response consists of exponentially decaying oscillations with a frequency of b radians per unit time. The precise nature of the oscillation can be displayed by writing (3.3-21) as a sine wave with a phase shift. Use the identity (see Section A.6)

$$B \sin (bt + \phi) = B \cos \phi \sin bt + B \sin \phi \cos bt \tag{3.3-24}$$

and compare with the term in parentheses in (3.3-21). This shows that

$$B \sin \phi = 2A_R \tag{3.3-25}$$

$$B \cos \phi = -2A_I \tag{3.3-26}$$

We can define B to be positive and absorb any negative signs with the phase angle ϕ. Thus,

$$B^2 \sin^2\phi + B^2 \cos^2\phi = 4(A_R^2 + A_I^2)$$

or

$$B = 2\sqrt{A_R^2 + A_I^2} \tag{3.3-27}$$

since $\sin^2\phi + \cos^2\phi = 1$. With B found from this expression, ϕ is computed from (3.3-25) with (3.3-26) used to determine the quadrant of ϕ.

The free response for complex roots is thus given by

$$x(t) = Be^{-at} \sin (bt + \phi) \tag{3.3-28}$$

This is illustrated in Figure 3.6. The sinusoidal oscillation has a frequency b (radians/unit time) and therefore a period of $2\pi/b$. The amplitude of oscillation decays exponentially; that is, the oscillation is bracketed on top and bottom by envelopes that are proportional to e^{-at}. These envelopes have a time constant of $\tau = 1/a$. Thus, the amplitude of the next oscillation occurring after $t = 1/a$ will be less than 37% of the peak amplitude. For $t > 4/a$, the amplitudes are less than 2% of the peak.

Example 3.7

Solve (3.3-4) for $m = 1$, $c = 3$, $k = 8.5$, and $f = 0$. The initial conditions are $x(0) = 1$ and $\dot{x}(0) = 5$.

The characteristic roots are

$$s = \frac{-3 \pm \sqrt{9 - 34}}{2} = -1.5 \pm 2.5i$$

and therefore $a = 1.5$ and $b = 2.5$. The constants A_R, A_I, B, and ϕ can be found from (3.3-22), (3.3-23), and (3.3-25) through (3.3-27). However, we now show that it is

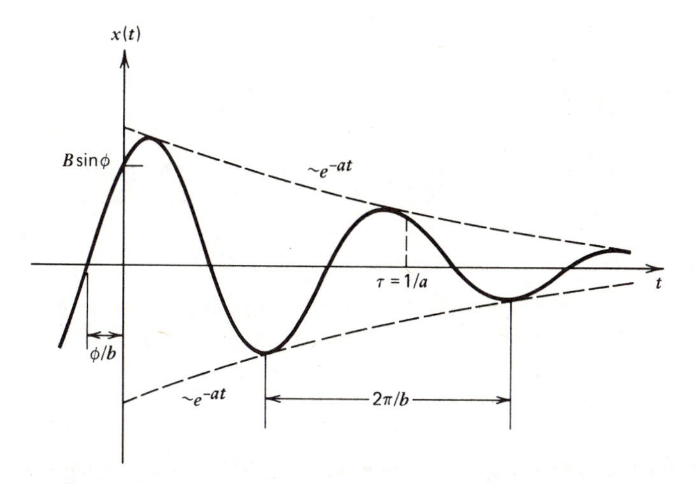

FIGURE 3.6 Free response of the second-order model with complex roots $x(t) = Be^{-at}\sin(bt + \phi)$.

sufficient simply to remember the solution form (3.3-28). From this, we see that

$$\dot{x}(t) = -aBe^{-at}\sin(bt + \phi) + bBe^{-at}\cos(bt + \phi) \tag{3.3-29}$$

and

$$x(0) = B\sin\phi \tag{3.3-30}$$

$$\dot{x}(0) = -aB\sin\phi + bB\cos\phi \tag{3.3-31}$$

Substitute $B\sin\phi$ into the last equation to obtain

$$B\cos\phi = \frac{\dot{x}(0) + ax(0)}{b} \tag{3.3-32}$$

The present numerical values give

$$B\sin\phi = 1$$

$$B\cos\phi = \frac{5 + 1.5}{2.5} = 2.6$$

Since B is defined to be positive, these relations imply that $\sin\phi > 0$ and $\cos\phi > 0$. Thus, ϕ is in the first quadrant and

$$\tan\phi = \frac{\sin\phi}{\cos\phi} = \frac{1}{2.6} = 0.3846$$

$$\phi = 21.04^\circ = 0.367 \text{ rad}$$

Finally,

$$B = \frac{1}{\sin\phi} = 2.786$$

and the solution is

$$x(t) = 2.786e^{-1.5t}\sin(2.5t + 0.367) \tag{3.3-33}$$

The time constant is

$$\tau = \frac{1}{1.5} = 0.667 \text{ time units}$$

and the oscillation will have essentially disappeared after $4(0.667) = 2.67$ time units.

Behavior with Repeated Roots

When the two roots of the characteristic equation are repeated (equal), the combination given by (3.3-8) does not consist of two linearly independent functions. In this case, it can be shown that the proper combination is

$$x(t) = A_1 e^{s_1 t} + t A_2 e^{s_2 t} \qquad \textbf{(3.3-34)}$$

where $s_1 = s_2$. This situation occurs for the vibration model (3.3-4) when $c^2 - 4mk = 0$. In this case, $s_1 = -c/2m$.

If the repeated roots are negative, the free response (3.3-34) decays with time despite the presence of t as a multiplier, because $e^{s_2 t}$ approaches zero faster than t becomes infinite. To see this, replace the exponential by its series representation. This series contains powers of t greater than one.

A peak can occur before the solution begins to decay, depending on the relative values of A_1 and A_2, which are found from the initial conditions as before.

$$A_1 = x(0) \qquad \textbf{(3.3-35)}$$

$$A_2 = \dot{x}(0) - s_1 x(0) \qquad \textbf{(3.3-36)}$$

Table 3.3 summarizes the free response for the three cases.

3.4 THE CHARACTERISTIC EQUATION

We will now present some convenient criteria to use in determining the system's stability, its time constant, oscillatory behavior, and frequency of oscillation, if any. All of these properties are determined by only the characteristic equation, and not by the initial conditions. The latter determine amplitudes and phase shifts. Because we have used the vibration model (3.3-4) as our example, we will use its form of the characteristic equation as the reference.

$$ms^2 + cs + k = 0 \qquad \textbf{(3.4-1)}$$

Stability

We have now treated all three situations that can arise for the free response of the linear second-order model with constant coefficients. With $f = 0$ in the vibration model (3.3-4), the only possible equilibrium is $x = 0$ if $k \neq 0$.† From the definition of

---

†If $k = 0$, there exists an infinite number of equilibrium positions, but in this case, the model (3.3-4) can be reduced to a first-order equation in terms of the speed $v = \dot{x}$.

TABLE 3.3 Free Response of the Second-order Model: $m\ddot{x} + c\dot{x} + kx = f$

Characteristic roots: $s = \dfrac{-c \pm \sqrt{c^2 - 4mk}}{2m} = s_1, s_2$

1. Real, distinct roots:

$$x(t) = A_1 e^{s_1 t} + A_2 e^{s_2 t}$$

$$A_1 = \frac{\dot{x}(0) - s_2 x(0)}{s_1 - s_2}$$

$$A_2 = \frac{s_1 x(0) - \dot{x}(0)}{s_1 - s_2} = x(0) - A_1$$

2. Real repeated roots: $s = s_1, s_1$

$$x(t) = A_1 e^{s_1 t} + t A_2 e^{s_1 t}$$

$$A_1 = x(0)$$

$$A_2 = \dot{x}(0) - s_1 x(0)$$

3. Complex conjugate roots: $s = -a \pm ib$

$$x(t) = Be^{-at} \sin(bt + \phi)$$

$$B = +2\sqrt{A_R^2 + A_I^2}$$

$$A_R = \frac{x(0)}{2} \qquad B \sin \phi = 2A_R$$

$$A_I = -\frac{\dot{x}(0) + ax(0)}{2b} \qquad B \cos \phi = -2A_I$$

Quadrant of ϕ		$\cos \phi$ $-$	$\cos \phi$ $+$
$\sin \phi$	$+$	2nd	1st
$\sin \phi$	$-$	3rd	4th

stability given in Section 3.1, it can be seen that the solution $x(t)$ must approach zero as t becomes infinite, in order for the equilibrium to be stable. This occurs in the real, distinct-roots and the repeated-roots cases only if both roots are negative. In the case of complex roots, (3.3-28) shows that the real part of the roots must be negative. We can summarize all three cases by stating that the system is stable if and only if both roots have negative real parts. (For the real-root cases, the imaginary parts are considered to be zero.)

Neutral, or limited, stability occurs for this second-order system in the complex-roots case if the real part is zero. The solution is then a constant-amplitude sinusoidal oscillation about the equilibrium. Neutral stability also occurs if one root is negative and the other zero. Since two repeated zero roots are impossible here, we can generalize this result to state that the equilibrium is neutrally stable when at least one root has a zero real part and the other root does not have a positive real part.

Conversely, it can be seen from (3.3-8) that only one root need be positive for $x(t)$ to become infinite. For the complex-roots case, this happens if the real part is positive. Thus, for all root cases, if at least one root has a positive real part, the equilibrium is unstable. These observations hold true for systems of higher order.

Discussion of a second-order system automatically implies that the coefficient m in (3.3-4) is nonzero. With this restriction, an analysis of the root solution (3.3-7) will show that both roots have negative real parts if and only if m, c, and k are positive. If either c or k is zero, the equilibrium is neutrally stable.

The Damping Ratio

The behavior of the preceding free response for the stable case can be conveniently characterized by the *damping ratio* ζ (sometimes called the *damping factor*). For the characteristic equation (3.4-1), this is defined as

$$\zeta = \frac{c}{2\sqrt{mk}} \tag{3.4-2}$$

Repeated roots occur if $c^2 - 4mk = 0$; that is, if $c = 2\sqrt{mk}$. This value of the damping constant is the *critical damping constant*, and when c has this value, the system is said to be critically damped. If the actual damping constant is greater than $2\sqrt{mk}$, two real distinct roots exist, and the system is *overdamped*. If $c < 2\sqrt{mk}$, complex roots occur, and the system is *underdamped*. The damping ratio is thus seen to be the ratio of the actual damping constant c to the critical value. For a critically damped system, $\zeta = 1$. Exponential behavior occurs if $\zeta > 1$, and oscillations exist for $\zeta < 1$.

The damping ratio is used with the reduced model form as a quick check for oscillatory behavior. In Example 3.6, $m = 1$, $c = 5$, and $k = 4$. Thus, $\zeta = 5/4 > 1$, and therefore the roots are real and distinct. There will be no oscillatory free response. In Example 3.7, $m = 1$, $c = 3$, and $k = 8.5$. Since the damping ratio is $\zeta = 0.51 < 1$, the roots are complex, and the free response will oscillate.

Natural and Damped Frequencies of Oscillation

When there is no damping, the characteristic roots are purely imaginary. The imaginary part, and therefore the frequency of oscillation, for this case is $b = \sqrt{k/m}$. This frequency is termed the *undamped natural frequency*, or simply the *natural frequency*, and is used as a reference value denoted by ω_n. Thus,

$$\omega_n = \sqrt{\frac{k}{m}} \tag{3.4-3}$$

We can write the characteristic equation in terms of the parameters ζ and ω_n. First, divide (3.4-1) by m and use the fact that $2\zeta\omega_n = c/m$. The equation becomes

$$s^2 + 2\zeta\omega_n s + \omega_n^2 = 0 \quad \textbf{(3.4-4)}$$

and the roots are

$$s = -\zeta\omega_n \pm i\omega_n\sqrt{1-\zeta^2} \quad \textbf{(3.4-5)}$$

Comparison with (3.3-13) shows that $a = \zeta\omega_n$ and $b = \omega_n\sqrt{1-\zeta^2}$. Since the time constant τ is $1/a$,

$$\tau = \frac{1}{\zeta\omega_n} \quad \textbf{(3.4-6)}$$

The frequency of oscillation is sometimes called the *damped natural frequency*, or simply *damped frequency* ω_d, to distinguish it from ω_n.

$$\omega_d = \omega_n\sqrt{1-\zeta^2} \quad \textbf{(3.4-7)}$$

For the underdamped case ($\zeta < 1$), we see that $\omega_d < \omega_n$.

Graphical Interpretation

The preceding relationships can be represented graphically by plotting the location of the roots (3.4-5) in the complex plane (Figure 3.7), in which the imaginary part is plotted versus the real part. This plot is said to be in the s plane. Since the roots are conjugate, we consider only the upper root.

From the stability criterion, we see that a root lying in the right-half plane results in an unstable system. The parameters ζ, ω_n, ω_d, and τ are normally used to describe stable systems only, and so we will assume for now that all roots lie in the left-hand plane.

The lengths of two sides of the right triangle shown in Figure 3.7 are $\zeta\omega_n$ and $\omega_n\sqrt{1-\zeta^2}$. Thus, the hypoteneuse has length ω_n. It makes an angle β with the negative real axis, and

$$\cos\beta = \zeta \quad \textbf{(3.4-8)}$$

Therefore, all roots lying on the circumference of a given circle centered on the origin are associated with the same undamped natural frequency ω_n. Figure 3.8*a* illustrates this for two different frequencies, ω_{n1} and ω_{n2}. From (3.4-8), we see that all roots lying on the same line passing through the origin are associated with the same damping ratio (Figure 3.8*b*). The limiting values of ζ correspond to the imaginary axis ($\zeta = 0$) and the negative real axis ($\zeta = 1$). Roots lying on a given line parallel to the real axis all give the same damped natural frequency (Figure 3.8*c*).

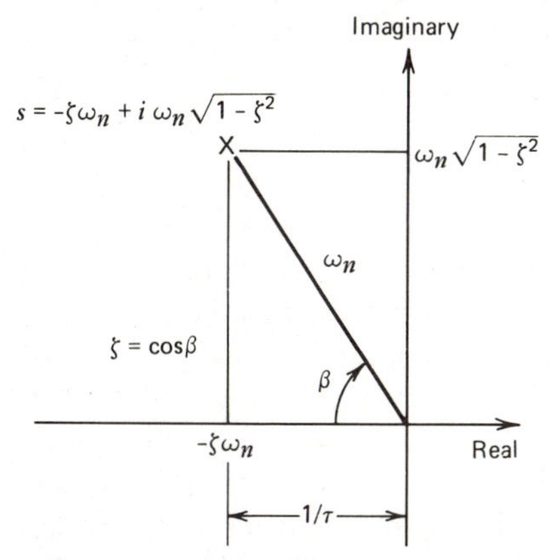

FIGURE 3.7 Location of the upper complex root in terms of the parameters ζ, τ, ω_n, ω_d.

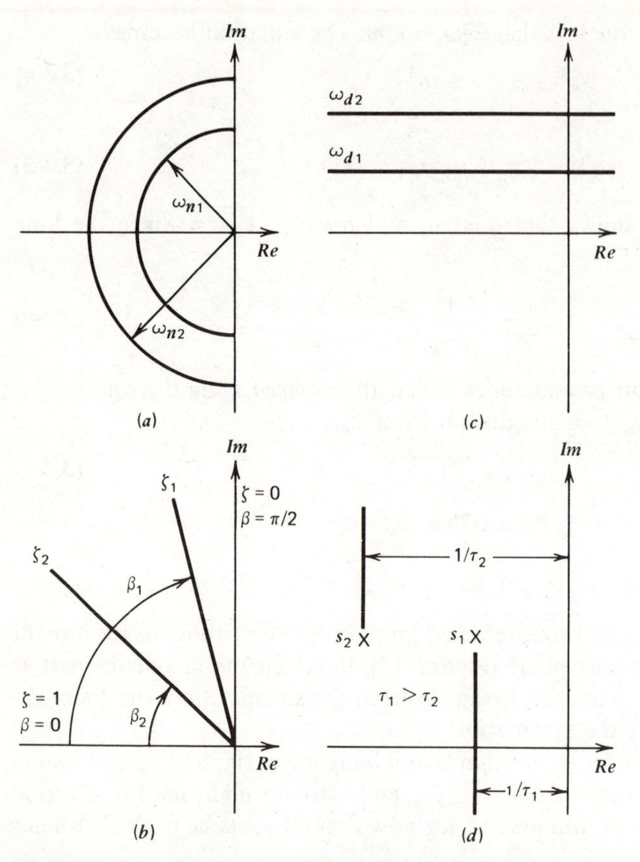

FIGURE 3.8 Graphical representation of the parameters ζ, τ, ω_n, and ω_d in the complex plane. (*a*) Roots with the same natural frequency ω_n lie on the same circle. (*b*) Roots with the same damping ratio ζ lie on the same line through the origin. (*c*) All roots on a given line parallel to the real axis have the same damped natural frequency ω_d. (*d*) All roots on a given line parallel to the imaginary axis have the same time constant τ. The root lying the farthest to the right is the dominant root.

The Dominant-Root Concept

Since $\tau = 1/\zeta\omega_n$, the distance from the root to the imaginary axis equals the reciprocal of the time constant for that root. All roots lying on a given vertical line have the same time constant, and the greater the distance of the line from the imaginary axis, the smaller the time constant (Figure 3.8*d*); this leads to the concept of the *dominant root.* For a given characteristic equation, this is the root that lies the farthest to the right in the s plane. Therefore, if the system is stable, the dominant root is the root with the largest time constant.

The dominant-root concept allows us to simplify the analysis of a given system by focusing on the root that plays the most important role in the system's dynamics. For example, if the two roots are $s_1 = -20$ and $s_2 = -2$, the free response is of the form

$$x(t) = A_1 e^{-20t} + A_2 e^{-2t}$$

For time measured in seconds, the first exponential has a time constant $\tau_1 = 1/20$ sec, while the second exponential's time constant is $\tau_2 = 1/2$ sec. The first exponential will have essentially disappeared after 4/20 sec, while the second exponential takes about $4/2 = 2$ sec to disappear. Unless the constant A_1 is very much larger than A_2, after about 0.2 sec, the solution is given approximately by

$$x(t) \simeq A_2 e^{-2t}$$

The root $s_2 = -2$ is the dominant root. The term in the free response corresponding to the dominant root remains nonzero longer than the other terms.

The usefulness of the dominant root in the preceding example is that it allows us to estimate the response time for the system. This is the time constant of the dominant root. Obviously, the greater the separation between the dominant root and the other roots, the better the dominant-root approximation.

For a second-order system with two real roots, the damping ratio ζ is a useful indicator of how much separation exists between the roots. Therefore, ζ is an indicator of the accuracy of the dominant-root approximation for such systems. Suppose we wish the dominant root to be $s = -b$, where b is a specified number. Write the secondary root as $s = -nb$, where n is the root separation factor. To see how n depends on ζ, write the polynomial corresponding to these two roots as

$$(s + b)(s + nb) = s^2 + (n + 1)bs + nb^2 = 0 \tag{3.4-9}$$

TABLE 3.4 Root Separation Factor as a Function of the Damping Ratio ζ

Dominant root: $s = -b$
Secondary root: $s = -nb$

$$\zeta = \frac{n + 1}{2\sqrt{n}}$$

n	ζ
1	1
2	1.06
5	1.34
10	1.74
13.93	2
20	2.35
33.97	3
50	3.61
61.98	4
97.99	5
100	5.05

The damping ratio is

$$\zeta = \frac{b(n+1)}{2\sqrt{nb^2}} = \frac{n+1}{2\sqrt{n}} \tag{3.4-10}$$

Table 3.4 shows the relation between ζ and n for some representative values.

3.5 STEP RESPONSE OF A SECOND-ORDER MODEL

The step function models any rapid change in the input from one constant level to another. In this section, we treat the step response of second-order systems.

Trial Solution Method

Consider the following model:

$$m\ddot{x} + c\dot{x} + kx = u(t) \tag{3.5-1}$$

where $u(t)$ is the input and m, c, k, are constants. If this model is stable, any constant input u will produce a steady-state response such that $\ddot{x} = \dot{x} = 0$ and

$$kx = u \tag{3.5-2}$$

Let $u(t)$ be a unit-step input, so that $u(t) = 0$, $t < 0$, and $u(t) = 1$, $t \geqslant 0$. Then from (3.5-2), the response at steady state is $x = 1/k$.

Before steady state is reached, a transient term exists that is of the same form as the free response. If the model is underdamped, this term is of the form $Be^{-at} \sin(bt + \phi)$, where the characteristic roots have been written as $s = -a \pm ib$. We therefore try a solution in the form of the sum of the transient and steady-state forms; that is,

$$x(t) = Be^{-at} \sin(bt + \phi) + \frac{1}{k} \tag{3.5-3}$$

Substituting this trial form into (3.5-1) with $u(t) = 1$ will verify its correctness. The constants B and ϕ are determined from the initial conditions *after* the steady-state term $1/k$ has been included. Because

$$\dot{x}(t) = -aBe^{-at} \sin(bt + \phi) + bBe^{-at} \cos(bt + \phi)$$

we obtain

$$x(0) = B \sin \phi + \frac{1}{k}$$

$$\dot{x}(0) = -Ba \sin \phi + bB \cos \phi$$

or

$$B \sin \phi = x(0) - \frac{1}{k} \tag{3.5-4}$$

$$B \cos \phi = \frac{a}{bk}[kx(0) - 1] + \frac{1}{b}\dot{x}(0) \tag{3.5-5}$$

These equations can be solved for B and ϕ in a manner similar to that used to find the free response. For the case of zero initial conditions, we obtain

$$x(t) = \frac{1}{k}\left[\frac{\sqrt{a^2+b^2}}{b} e^{-at} \sin(bt+\phi) + 1\right] \tag{3.5-6}$$

$$\tan\phi = \frac{b}{a} \tag{3.5-7}$$

where ϕ is in the third quadrant (remember, we are assuming $a > 0, b > 0$). In terms of the parameters ζ, ω_n, this solution becomes

$$x(t) = \frac{1}{k}\left[\frac{1}{\sqrt{1-\zeta^2}} e^{-\zeta\omega_n t} \sin(\omega_n\sqrt{1-\zeta^2}\,t+\phi) + 1\right] \tag{3.5-8}$$

$$\tan\phi = \frac{\sqrt{1-\zeta^2}}{\zeta} \tag{3.5-9}$$

where ϕ is in the third quadrant. If the initial conditions are not zero, we can simply add the free response to the result.

For an overdamped or critically damped system, the method is essentially the same. A function of the form of the free response is added to the steady-state term. For an overdamped case, this gives

$$x(t) = A_1 e^{s_1 t} + A_2 e^{s_2 t} + \frac{1}{k}$$

where s_1 and s_2 are the distinct negative roots. For zero initial conditions, this gives

$$x(t) = \frac{1}{k}\left(\frac{s_2}{s_1 - s_2} e^{s_1 t} - \frac{s_1}{s_1 - s_2} e^{s_2 t} + 1\right) \tag{3.5-10}$$

The result for the critically damped case is found in a similar way. For zero initial conditions, the unit-step response is

$$x(t) = \frac{1}{k}[(s_1 t - 1)e^{s_1 t}] + 1 \tag{3.5-11}$$

The preceding step responses are summarized in Table 3.5.

The preceding solutions are for a *unit*-step input. If the magnitude of the step were M instead, it is easy to see that the resulting response can be obtained by multiplying by M. This result is due to the linearity properties of the differential equation (3.5-1).

The step response is illustrated in Figure 3.9, with a plot of the normalized response variable kx as a function of the normalized time variable $\omega_n t$. The plot gives the response for several values of the damping ratio, with ω_n held constant. When $\zeta > 1$, the response is sluggish and does not overshoot the steady-state value. As ζ is decreased, the speed of response increases. The critically damped case $\zeta = 1$ is the case in which the steady-state value is reached most quickly but without oscillation. Instrument mechanisms are frequently designed to be critically damped for this reason. When a voltage to be measured is applied to a voltmeter, it acts like a step

TABLE 3.5 Unit Step Response of the Second-order Model: $m\ddot{x} + c\dot{x} + kx = du(t)$

$$u(t) = 1, \qquad t > 0$$
$$x(0) = \dot{x}(0) = 0$$

1. Real, distinct roots: $s = s_1, s_2$

$$x(t) = \frac{d}{k}\left(\frac{s_2}{s_1 - s_2} e^{s_1 t} - \frac{s_1}{s_1 - s_2} e^{s_2 t} + 1\right)$$

2. Real, repeated roots: $s = s_1, s_1$

$$x(t) = \frac{d}{k}\,[s_1 t e^{s_1 t} - e^{s_1 t} + 1]$$

3. Complex conjugate roots: $s = -a \pm ib,\ b > 0$

$$x(t) = \frac{d}{k}\left[\frac{\sqrt{a^2 + b^2}}{b} e^{-at} \sin(bt + \phi) + 1\right]$$

$$\tan\phi = \frac{b}{a} \qquad (\phi \text{ in third quadrant if } a > 0)$$

Alternate form for: $s = -\zeta\omega_n \pm i\omega_n\sqrt{1 - \zeta^2}$

$$x(t) = \frac{d}{k}\left[\frac{1}{\sqrt{1 - \zeta^2}} e^{-\zeta\omega_n t} \sin(\omega_n\sqrt{1 - \zeta^2}\,t + \phi) + 1\right]$$

$$\tan\phi = \frac{\sqrt{1 - \zeta^2}}{\zeta} \qquad (\phi \text{ in third quadrant})$$

input to the mechanism. We would like the needle of the meter to approach the indicated position as quickly as possible but without oscillating around it. This is the critically damped case.

As ζ is decreased below 1, the response overshoots and oscillates about the final value. The smaller ζ is, the larger the overshoot, and the longer it takes for the oscillations to die out. There are design applications in which we wish the response to be near its final value as quickly as possible, with some oscillation tolerated. (For example, a radar antenna might need to be pointed quickly in the general direction of a target.) As ζ is decreased to zero (no damping), the oscillations never die out.

Because the axes of Figure 3.9 have been normalized by k and ω_n, respectively, the plot shows only the variation in the response as ζ is varied, with k and ω_n held constant. Let us see how each of the parameters affects the step response. Figure 3.10 shows what happens when ζ and k are held constant while ω_n is changed. The effect of increasing ω_n is to speed up the response and make the overshoot occur earlier. Figure 3.11 shows two cases in which ζ and ω_n are held constant while k is changed. The effect of increasing k is to decrease the magnitude of the response.

Finally, Figure 3.12 shows three cases for which c and k are fixed while ζ is varied by changing m. The plot shows that as ζ is decreased, the response becomes more oscillatory, the overshoot becomes larger and occurs later. This occurs even though

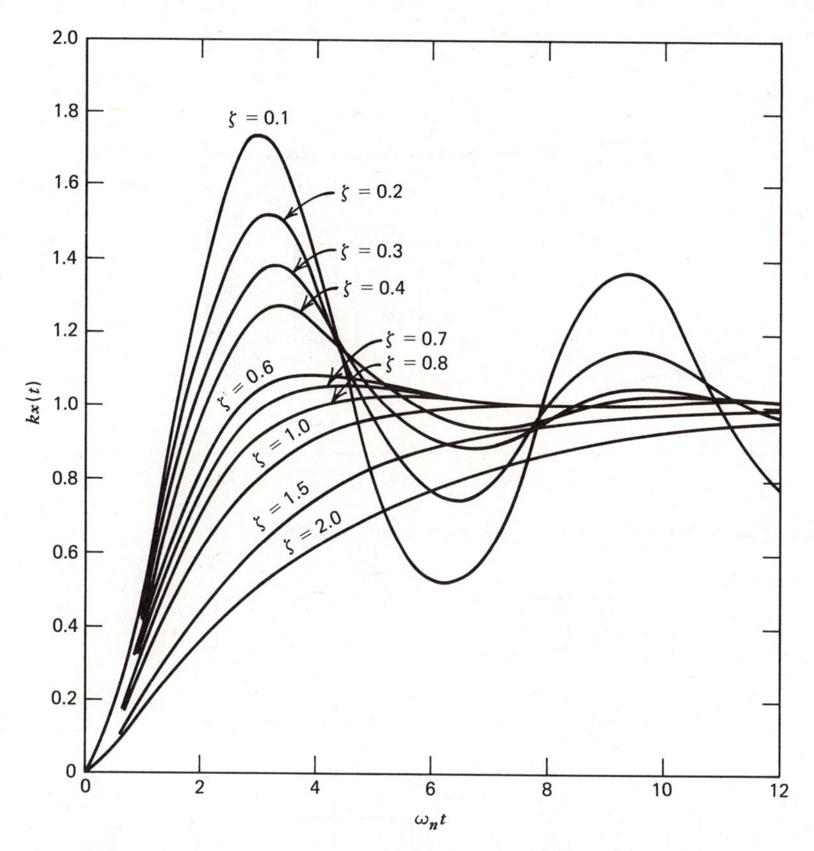

FIGURE 3.9 Normalized unit step response of the model $m\ddot{x} + c\dot{x} + kx = u(t)$.

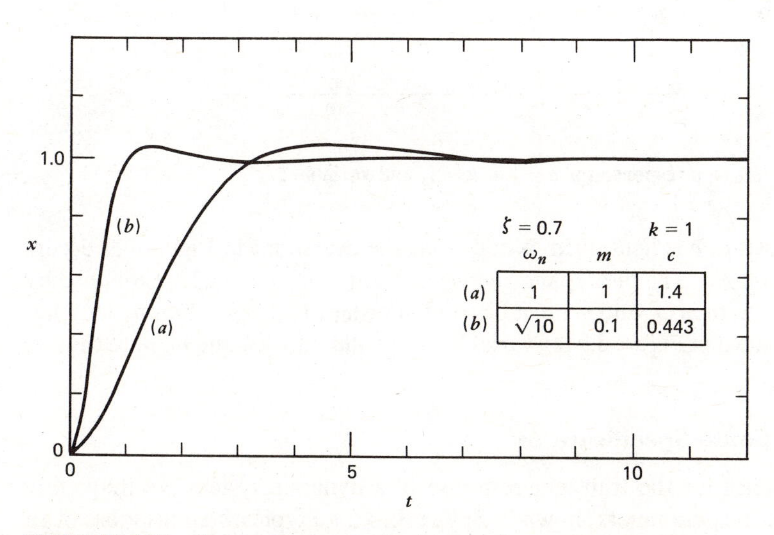

FIGURE 3.10 Unit-step response for $\zeta = 0.7$, $k = 1$, and variable ω_n.

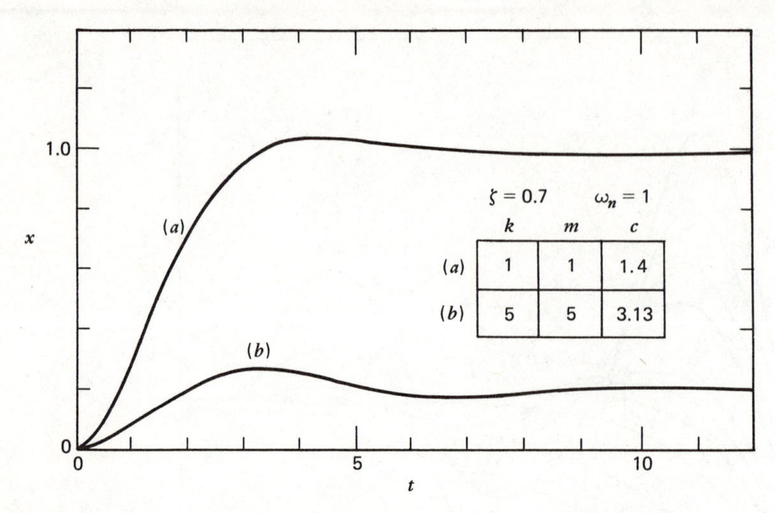

FIGURE 3.11 Unit-step response for $\zeta = 0.7$, $\omega_n = 1$, and variable k.

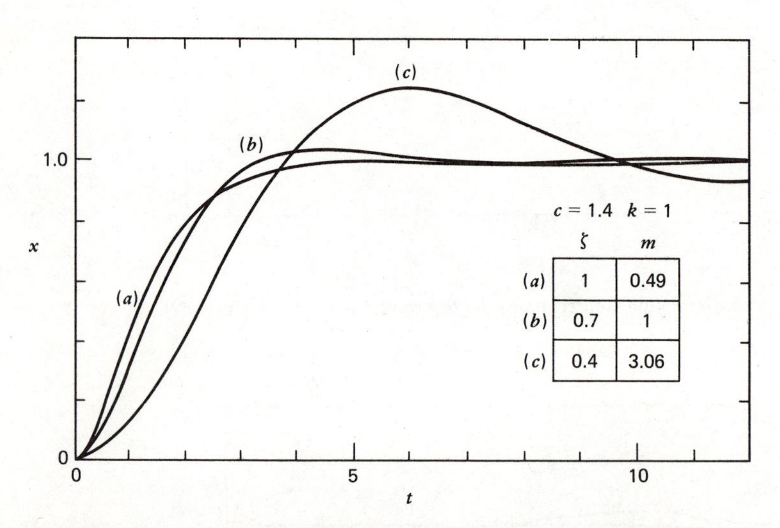

FIGURE 3.12 Unit step response for $c = 1.4$, $k = 1$, and variable ζ.

the damping constant c is held fixed. Notice that the overshoot in Figure 3.9 occurs *earlier* as ζ is decreased. The discrepancy between Figures 3.9 and 3.12 is explained by the fact that if $k = 1$, then ω_n and m must be fixed in order to interpret Figure 3.9. That is, the patterns shown in Figure 3.9 for variable ζ are valid only if k and ω_n (and thus m) are fixed values.

Transient-Response Specifications

Performance criteria for the transient response of a dynamic system are frequently stated in terms of the parameters shown in Figure 3.13 for a typical step response of an underdamped system (not necessarily a second-order system). The *maximum overshoot*

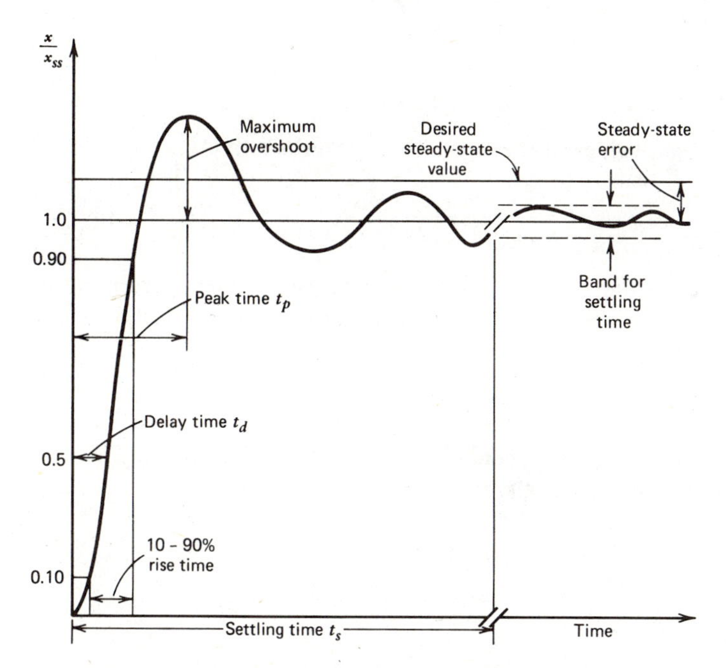

FIGURE 3.13 Transient performance specifications based on step response.

is the maximum deviation of the output x above its steady-state value x_{ss}. It is sometimes expressed as a percentage of the final value. Since the maximum overshoot increases with decreasing ζ, it is sometimes used as an indicator of the relative stability of the system. The *peak time* t_p is the time at which the maximum overshoot occurs. The *settling time* t_s is the time required for the oscillations to stay within some specified small percentage of the final value. The most common values used are 2% and 5%; the latter choice is shown in the figure. If the final value of the response differs from some desired value, a steady-state error exists.

The *rise time* t_r can be defined as the time required for the output to rise from 10% to 90% of its final value. However, no agreement exists on this definition. Sometimes, the rise time is taken to be the time required for the response to reach the final value for the first time. Other definitions are also in use. Finally, the *delay time* t_d is the time required for the response to reach 50% of its final value.

These parameters are relatively easy to obtain from an experimentally determined step-response plot. However, if they are to be determined in analytical form from a differential equation model, the task is difficult for models of order greater than two. Here, we obtain expressions for these quantities from the second-order step response given by (3.5-8).

Setting the derivative of (3.5-8) equal to zero gives expressions for both the maximum overshoot and the peak time t_p. After some trigonometric manipulation, the result is

$$\frac{dx}{dt} = \frac{1}{k}\left(\frac{\omega_n}{\sqrt{1-\zeta^2}} e^{-\zeta\omega_n t} \sin \omega_n \sqrt{1-\zeta^2}\, t\right) = 0$$

For $t < \infty$, this gives

$$\omega_n\sqrt{1-\zeta^2}\,t = n\pi, \qquad n = 0, 1, 2, \ldots$$

The times at which extreme values of the oscillations occur are thus

$$t = \frac{n\pi}{\omega_n\sqrt{1-\zeta^2}} \tag{3.5-12}$$

The odd values of n give the times of overshoots, and the even values correspond to the times of undershoots. The maximum overshoot occurs when $n = 1$. Thus,

$$t_p = \frac{\pi}{\omega_n\sqrt{1-\zeta^2}} \tag{3.5-13}$$

The magnitudes of the overshoots and undershoots are found by substituting (3.5-12) into (3.5-8). After some manipulation, the result is

$$x(t)|_{\text{extremum}} = \frac{1}{k}[1 + (-1)^{n-1}e^{-n\pi\zeta/\sqrt{1-\zeta^2}}] \tag{3.5-14}$$

The maximum overshoot is found when $n = 1$.

$$\text{maximum overshoot} = x_{\max} - x_{ss} = \frac{1}{k}e^{-\pi\zeta/\sqrt{1-\zeta^2}} \tag{3.5-15}$$

The preceding expressions show that the maximum overshoot and the peak time are functions of only the damping ratio ζ for a second-order system. The percent overshoot is

$$\begin{aligned}\text{percent maximum overshoot} &= \frac{x_{\max} - x_{ss}}{x_{ss}}100 \\ &= 100e^{-\pi\zeta/\sqrt{1-\zeta^2}}\end{aligned} \tag{3.5-16}$$

This is shown graphically in Figure 3.14*a*. The normalized peak time $\omega_n t_p$ is plotted versus ζ in Figure 3.14*b*.

Analytical expressions for the delay time, the rise time, and the settling time are difficult to obtain. For the delay time, set $x = 0.5x_{ss} = 0.5/k$ in (3.5-8) to obtain

$$e^{-\zeta\omega_n t_d}\sin(\omega_n\sqrt{1-\zeta^2}\,t_d + \phi) = -0.5\sqrt{1-\zeta^2} \tag{3.5-17}$$

where ϕ is given by (3.5-9). For a given ζ and ω_n, t_d can be obtained by a numerical procedure, such as Newton's method (Appendix A). This is easily done on a calculator especially if the following straight-line approximation is used as a starting guess:

$$t_d \simeq \frac{1 + 0.7\zeta}{\omega_n}, \qquad 0 \leqslant \zeta \leqslant 1 \tag{3.5-18}$$

A similar procedure can be applied to find the rise time t_r. In this case, two equations must be solved, one for the 10% time and one for the 90% time. The difference between these times is the rise time. These calculations are made easier by using the

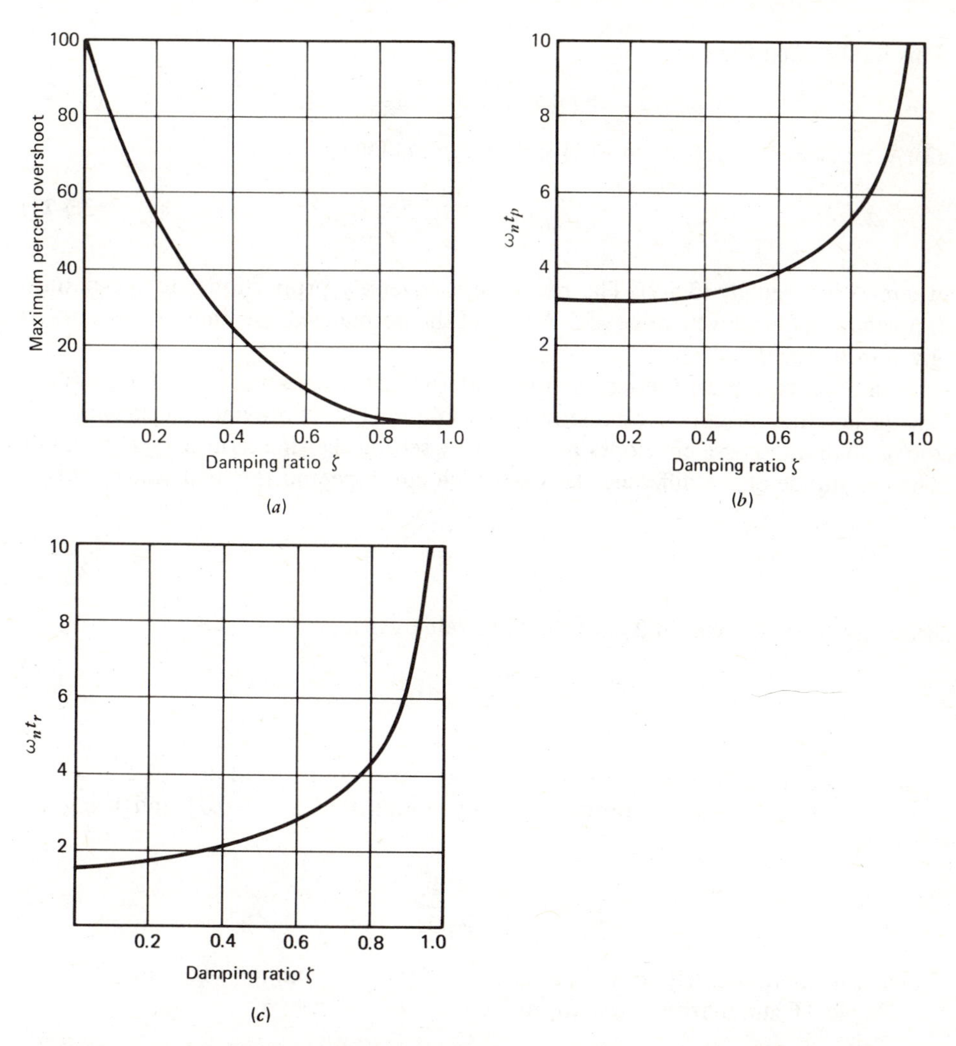

FIGURE 3.14 Transient response specifications as functions of the damping ratio ζ. (*a*) Maximum percent overshoot. (*b*) Normalized peak time $\omega_n t_p$. (*c*) Normalized 100% rise time $\omega_n t_r$.

following straight-line approximation:

$$t_r|_{10,90\%} \simeq \frac{0.8 + 2.5\zeta}{\omega_n}, \qquad 0 \leqslant \zeta \leqslant 1 \tag{3.5-19}$$

This approximation was obtained by plotting the results of many such computer solutions.

If we choose the alternative definition of rise time—that is, the first time at which the final value is crossed—then the solution for t_r is easier to obtain. Set $x = x_{ss} = 1/k$ in (3.5-8) to obtain

$$e^{-\zeta\omega_n t} \sin\left(\omega_n\sqrt{1-\zeta^2}\,t + \phi\right) = 0$$

This implies that for $t < \infty$,

$$\omega_n\sqrt{1-\zeta^2}\,t + \phi = n\pi, \qquad n = 0, 1, 2, \ldots \tag{3.5-20}$$

For $t_r > 0$, $n = 2$, since ϕ is in the third quadrant. Thus,

$$t_r|_{100\%} = \frac{2\pi - \phi}{\omega_n\sqrt{1-\zeta^2}} \tag{3.5-21}$$

where ϕ is given by (3.5-9). The rise time is inversely proportional to the natural frequency ω_n for a given value of ζ. A plot of the normalized rise time $\omega_n t_r$ versus ζ is given in Figure 3.14c.

In order to express the settling time in terms of the parameters ζ and ω_n, we can use the fact that the exponential term in the solution (3.5-8) provides the envelopes of the oscillations. These envelopes are found by setting the sine term to ± 1 in (3.5-8). The magnitude of the difference between each envelope and the final value $1/k$ is

$$\frac{1}{k}\frac{e^{-\zeta\omega_n t}}{\sqrt{1-\zeta^2}}$$

Both envelopes are within 2% of the final value when

$$\frac{e^{-\zeta\omega_n t}}{\sqrt{1-\zeta^2}} \leqslant 0.02$$

The 2% settling time can be found from the preceding expression. For ζ less than about 0.7, t_s can also be approximated by noting that $e^{-4} \cong 0.02$ and using the formula

$$t_s|_{2\%} \cong \frac{4}{\zeta\omega_n} \tag{3.5-22}$$

Thus, t_s is approximately four time constants for ζ less than approximately 0.7.

Table 3.6 summarizes these formulas.

Example 3.8

Typical parameter values for an armature-controlled dc motor of the type shown in Figure 2.19 are (Reference 1):

$K_e = 0.1$ V/rad/sec
$K_T = 26.9$ oz-in./A
$L = 0.0021$ H

$R = 0.43\ \Omega$
$c = 0.07$ oz-in./rad/sec
$I = 0.1$ oz-in.-sec^2

(a) Find the differential equation relating the load speed ω to the applied voltage v.

(b) Find ζ, τ, ω_n, and ω_d for this system.

(c) A step input voltage of $v = 10$ V is applied when the system is initially at rest. Evaluate the step response.

TABLE 3.6 Useful Formulas for Second-order Model: $m\ddot{x} + c\dot{x} + kx = f$

1. Characteristic roots: $s = \dfrac{-c \pm \sqrt{c^2 - 4mk}}{2m} = s_1, s_2$

2. Stability: Stable if m, c, and k have the same sign.

3. Damping ratio or damping factor: $\zeta = \dfrac{c}{2\sqrt{mk}}$

4. Undamped natural frequency: $\omega_n = \sqrt{\dfrac{k}{m}}$

5. Damped natural frequency: $\omega_d = \omega_n \sqrt{1 - \zeta^2}$

6. Time constant: $\tau = 2m/c = 1/\zeta\omega_n$ if $\zeta \leqslant 1$.
   $\tau = -1/s_1$ if $\zeta > 1$ (s_1 = dominant root)

Step Response Characteristics

7. Maximum percent overshoot: $100e^{-\pi\zeta/\sqrt{1-\zeta^2}}$

8. Peak time: $t_p = \dfrac{\pi}{\omega_n \sqrt{1 - \zeta^2}}$

9. Delay time: $t_d \simeq \dfrac{1 + 0.7\zeta}{\omega_n}$, $0 \leqslant \zeta \leqslant 1$

10. 100% rise time: $t_r|_{100\%} = \dfrac{2\pi - \phi}{\omega_n \sqrt{1 - \zeta^2}}$ $\tan \phi = \dfrac{\sqrt{1 - \zeta^2}}{\zeta}$ (third quadrant)

11. 2% settling time: $t_s = 4\tau$ for $\zeta < 0.7$.

(a) The governing state equations for the state variables ω and i_a (the armature current) were obtained in Section 2.5 and converted to reduced form in Section 3.3. Differential equation (3.3-3) is repeated here.

$$LI\ddot{\omega} + (RI + cL)\dot{\omega} + (cR + K_e K_T)\omega = K_T v(t) \tag{3.5-23}$$

The electrical units (Ω, V, H, A, sec) are independent of the mechanical units (oz, in., rad, sec) except for the time unit. Thus, because seconds are used in both systems, we need not convert any of the units; they are all compatible. From the given data, we obtain

$$0.00021\ddot{\omega} + 0.04315\dot{\omega} + 2.7201\omega = 26.9v(t) \tag{3.5-24}$$

where ω is in rad/sec and $v(t)$ is in V.

(b) Comparing the preceding with the left-hand side of the general form (3.5-1) and using the formulas in Table 3.6, we obtain

$$\zeta = 0.902 \qquad \omega_n = 113.8 \text{ rad/sec}$$
$$\omega_d = 49.2 \text{ rad/sec} \qquad \tau = 0.0097 \text{ sec}$$

(c) Since $\zeta < 1$, the system is underdamped, and oscillations will occur when a step voltage is applied. The solutions (3.5-8) and (3.5-9) must be multiplied by $26.9(10) = 269$. (Recall that (3.5-8) is for a unit-step input). This gives the following response:

$$\omega(t) = 98.893[2.313e^{-102.7t} \sin(49.2t + 3.589) + 1] \tag{3.5-25}$$

where the phase angle is in radians. The steady-state speed of the load is 98.893 rad/sec. From (3.5-13), the time of the maximum overshoot is $t_p = 0.0639$ sec. The maximum speed is 99.03 rad/sec. Thus, the percent overshoot is approximately 0.14%.

The delay time can be found approximately from (3.5-18). This gives

$$t_d = \frac{1 + 0.7(0.902)}{113.8} = 0.014 \text{ sec}$$

The delay time here is approximately 50% greater than the time constant. The 100% rise time is found from (3.5-21) to be

$$t_r|_{100\%} = \frac{2\pi - 3.589}{49.13} = 0.054 \text{ sec}$$

The 2% settling time is

$$t_s|_{2\%} = 0.046 \text{ sec} = 4.77\tau$$

The solution for the armature current i_a can be found by using the solution for $\omega(t)$ in (2.5-4), rewritten here as

$$\begin{aligned} i_a &= \frac{1}{K_T}(I\dot{\omega} + c\omega) \\ &= 0.003717\dot{\omega} + 0.0026\omega \end{aligned}$$

Differentiate (3.5-25) to obtain $\dot{\omega}$ and substitute in the preceding equation to obtain

$$\begin{aligned} i_a(t) = 41.8237e^{-102.7t} &\cos(49.2t + 3.589) \\ -86.62e^{-102.7t} &\sin(49.2t + 3.589) + 0.257 \end{aligned} \tag{3.5-26}$$

The solutions are plotted in Figure 3.15. The peak current that must be supplied by the power source can be seen to be 16.49 A.

This example shows the importance of plotting the solution as an aid to interpretation. What is striking about the plots in Figure 3.15 is that they are simple to interpret compared to the rather complex expressions given in (3.5-25) and (3.5-26). For example, it is much easier to compute the peak value from the plot. Also, no oscillations are apparent in the plots, although the solutions for ω and i_a contain sines and cosines. This discrepancy is explained by the fact that the time constant τ is much smaller than the period of the oscillation, which is $T = 2\pi/\omega_d = 0.128$ sec. Thus, the

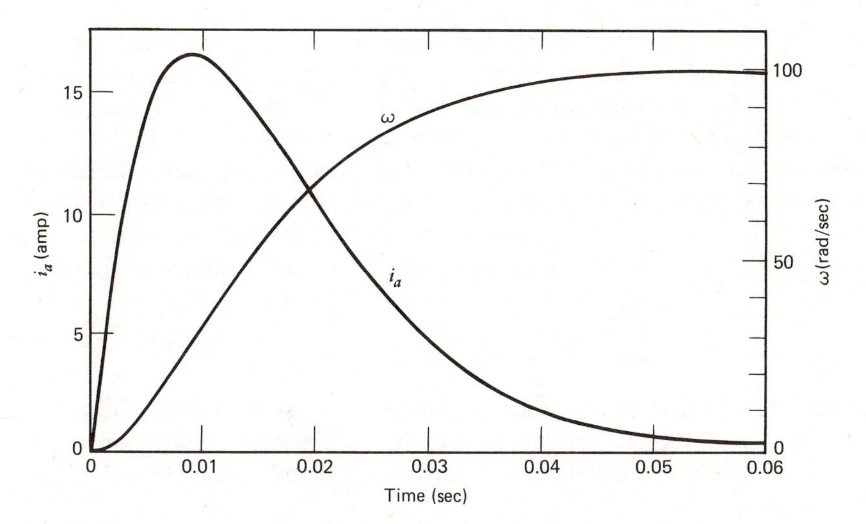

FIGURE 3.15 Step response of the dc motor system of Example 3.8.

envelopes of the sines and cosines decay to negligible values before the oscillations can become apparent.

3.6 RAMP AND SINUSOIDAL RESPONSE PATTERNS

The step input is perhaps the most commonly used input, and we have therefore devoted much attention to the step response. Two other input types commonly encountered are the ramp and sinusoidal inputs, which we now treat. Our purpose in this section is twofold: first, to show how an understanding of the free response can be used along with the trial solution method to obtain a quick estimate of the general behavior pattern of the forced response; and second, to develop an understanding of the ramp and sinusoidal responses for first- and second-order systems.

Trial Solution Method

The trial solution method used in Sections 3.2 and 3.5 for the step response can be generalized as follows. If $x(t)$ is the forced response to be found, try a solution of the form

$$x(t) = (\text{free response form}) + (\text{particular solution form}) \tag{3.6-1}$$

where the free-response form has the same functional form as the free response but with arbitrary constants replacing the initial conditions. For example, the free-response form corresponding to differential equation (3.2-1) is like (3.1-4); that is, Ae^{rt}. For differential equation (3.3-4), the free-response form is $Be^{-at}\sin(bt + \phi)$ for $\zeta < 1$. Equations (3.3-8) and (3.3-34) give the forms for $\zeta > 1$ and $\zeta = 1$.

The particular solution form is determined as follows. If the order of the differential equation is n, compute all the time derivatives of the given input function up to and including the nth-order derivative or until the derivatives become zero or begin to repeat their functional form. The particular solution form is the linear combination of the input function and these derivatives. The total trial solution $x(t)$ given by (3.6-1) is then substituted into the differential equation, and the coefficients of identical functional forms are compared. Form (3.6-1) is also used with the initial conditions to find n of the constants. If only the forced response is desired, set the initial conditions to zero. The preceding steps will generate as many linear algebraic equations to solve as there are arbitrary constants.

Ramp Response of a First-Order System

The ramp input is used to describe a start-up process for a system initially at rest. The ramp response can be used to determine how well the system's output follows the start-up command. Consider the first-order model (3.2-1) and assume that $b = 1$ for simplicity. Thus,

$$\dot{y} = ry + v \tag{3.6-2}$$

If the input $v(t)$ is a ramp function with a slope m, then

$$v = mt \tag{3.6-3}$$

The free response form is Ae^{rt}. Because (3.6-2) is first order, we need to compute only $\dot{v}$, which is a constant, $\dot{v} = m$. Thus, the particular solution form is $B + Ct$. The total trial solution is

$$y(t) = Ae^{rt} + B + Ct \tag{3.6-4}$$

Substituting this and (3.6-3) into (3.6-2) gives

$$rAe^{rt} + C = rAe^{rt} + rB + rCt + mt$$

Collect terms to obtain

$$(rA - rA)e^{rt} + (rC + m)t + rB - C = 0$$

For this equation to be true, the coefficient of each functional form must be zero. Thus,

$$rC + m = 0$$
$$rB - C = 0$$

Therefore, $C = -m/r$ and $B = C/r = -m/r^2$. The constant A is found by setting $t = 0$ in (3.6-4).

$$y(0) = A + B$$

Thus, $A = y(0) - B = y(0) + m/r^2$. To obtain the forced response, set $y(0) = 0$. The solution is

$$y(t) = \frac{m}{r^2}(e^{rt} - 1 - rt) \tag{3.6-5}$$

For the stable case, $r = -1/\tau$ and

$$y(t) = m\tau(t - \tau + \tau e^{-t/\tau}) \tag{3.6-6}$$

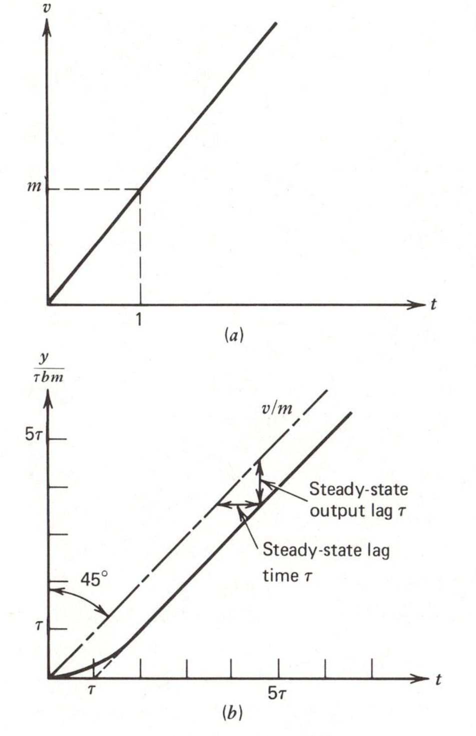

FIGURE 3.16 Ramp response of a first-order linear system. (*a*) Ramp input. (*b*) Normalized ramp input and ramp response.

Note that we could have used a ramp input with a unit slope and multiplied the resulting response by the slope m to obtain the general solution (3.6-6). This is shown in normalized form in Figure 3.16 along with the ramp input. The response approaches a straight line, which passes through the point $(0, \tau)$ and is parallel to the normalized ramp input v/m, a unit ramp. In the steady-state condition, the normalized output $y/\tau bm$, at any given time, is less than the normalized input by an amount τ. Also, in this condition, the normalized output at time t equals what the normalized input was at time $(t-\tau)$. This phenomenon is the steady-state *lag time* and another interpretation of the time constant.

The time constant also indicates how long it takes for the steady-state lag to become established. The difference between the normalized input and output is

$$\frac{v(t)}{m} - \frac{y(t)}{\tau m} = \tau(1 - e^{-t/\tau})$$

From this, it can be seen that the steady-state difference is τ (as $t \to \infty$). At $t = \tau$, the difference is 0.63τ; at $t = 4\tau$, the difference is 0.98τ.

Ramp Response of a Second-Order System

The steady-state part of the solution (3.6-6) could have been obtained without specifying the initial conditions. Let us use this observation to develop a quick way of determining the steady-state ramp response of the second-order model (3.3-4) with a ramp function; that is, let

$$f(t) = pt \tag{3.6-7}$$

The second-order and higher derivatives of this function are zero, so the particular solution form is $B_1 + B_2 t$. Assume the system is overdamped, with characteristic roots s_1 and s_2. Then, the total trial solution is

$$x(t) = A_1 e^{s_1 t} + A_2 e^{s_2 t} + B_1 + B_2 t \tag{3.6-8}$$

Substituting this into (3.3-4) gives

$$\sum_{i=1}^{2} (ms_i^2 + cs_i + k)A_i e^{s_i t} + cB_2 + kB_1 + kB_2 t = pt$$

Because the roots s_1 and s_2 satisfy the characteristic equation (3.3-6), the preceding equation reduces to

$$cB_2 + kB_1 + (kB_2 - p)t = 0 \tag{3.6-9}$$

Thus,

$$kB_2 - p = 0$$
$$cB_2 + kB_1 = 0$$

Or

$$B_2 = \frac{p}{k}$$

$$B_1 = -\frac{cp}{k^2}$$

The free response form is a transient, so the steady-state solution is

$$x(t) = \frac{p}{k}\left(t - \frac{c}{k}\right) \tag{3.6-10}$$

which passes through the point $x = 0$ at $t = c/k$. From this, we see that if k does not equal 1, the solution diverges from the ramp input as t increases.

If the system is underdamped, the trial solution (3.6-8) will be replaced with

$$x(t) = Be^{-at}\sin(bt + \phi) + B_1 + B_2 t \tag{3.6-11}$$

and (3.6-9) would still result. Thus, (3.6-10) is the steady-state ramp response for all values of ζ as long as the system is stable ($\zeta > 0$). One advantage of the trial solution method is that it can provide this sort of insight.

The unit-ramp response is illustrated in Figure 3.17. The steady state is reached

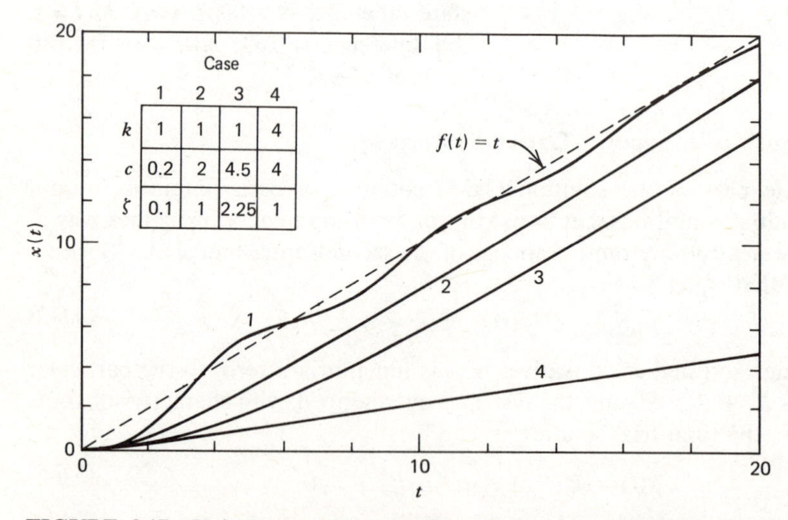

FIGURE 3.17 Unit ramp response of the second-order system: $m\ddot{x} + c\dot{x} + kx = f(t)$. For all cases, $m = 1$. The dominant time constants are $\tau = 10, 1, 2.13$, and 0.5 respectively, so all cases have reached steady-state except case 1. The steady-state response is parallel to the input ramp only when $k = 1$.

TABLE 3.7 Unit Ramp Response of the Second-order Model: $m\ddot{x} + c\dot{x} + kx = du(t)$

$$u(t) = t, \qquad t > 0$$
$$x(0) = \dot{x}(0) = 0$$

1. Real, distinct roots: $s = s_1, s_2$

$$x(t) = C_1 e^{s_1 t} + C_2 e^{s_2 t} + C_3 t + C_4$$

$$C_1 = \frac{d}{m s_1^2 (s_1 - s_2)} \qquad C_3 = \frac{d}{m s_1 s_2}$$

$$C_2 = \frac{d}{m s_2^2 (s_2 - s_1)} \qquad C_4 = \frac{d(s_1 + s_2)}{m s_1 s_2}$$

2. Real, repeated roots: $s = s_1, s_1$

$$x(t) = \frac{d}{m s_1^2}\left[e^{s_1 t}\left(t - \frac{2}{s_1}\right) + t + \frac{2}{s_1}\right]$$

3. Complex conjugate roots: $s = \dfrac{-c \pm \sqrt{c^2 - 4mk}}{2m}$

$$x(t) = \frac{d}{k}\left[t - \frac{2\zeta}{\omega_n} + \frac{e^{-\zeta\omega_n t}}{\omega_n\sqrt{1-\zeta^2}} \sin\left(\omega_n\sqrt{1-\zeta^2}\, t + \phi\right)\right]$$

$$\phi = \begin{cases} \tan^{-1}\dfrac{2\zeta\sqrt{1-\zeta^2}}{2\zeta^2 - 1}, & \text{if } 2\zeta^2 - 1 \geq 0 \\ \tan^{-1}\dfrac{2\zeta\sqrt{1-\zeta^2}}{2\zeta^2 - 1} + \pi, & \text{if } 2\zeta^2 - 1 < 0 \end{cases}$$

when the free-response form has disappeared. This occurs after approximately four time constants, unless the system has two real roots spaced closely together. No general statement can be made in that case. Note that when $k = 1$, the term c/k $(= c)$ plays the same role here as τ played in Figure 3.16; that is, c is the steady-state output lag as well as the lag time. Table 3.7 summarizes the ramp responses for the second-order model's three root types.

Sinusoidal Response of a First-Order System

Periodic inputs occur extensively in both natural and artificial systems. An input $v(t)$ is periodic with a period p if $v(t + p) = v(t)$ for all values of time t, where p is a constant called the *period*. The earth's rotation and revolution about the sun produce periodic forcing functions in nature. Rotating unbalanced machinery produces such a force on the supporting structure. Internal combustion engines produce a periodic torque resulting from the engine's cycle, and reciprocating fluid pumps produce hydraulic and pneumatic pressures with one or more periodic components. Perhaps the most familiar example is the ac voltage. Many of these periodicities are sinusoidal, or nearly so, and we will begin the analysis with this type. In Chapter 4, it will be shown how results for a sinusoidal input can be used to obtain the system response to a general periodic input by means of the *Fourier series* representation (Appendix A).

Consider again (3.2-1) where now the input is

$$v(t) = A \sin \omega t \tag{3.6-12}$$

where $A > 0$ and $\dot{v}(t) = \omega A \cos \omega t$. The derivatives of $v(t)$ repeat after the first one, so we use the following trial solution form:

$$y(t) = C_1 e^{rt} + C_2 \sin \omega t + C_3 \cos \omega t \tag{3.6-13}$$

Substituting this into (3.2-1), we obtain

$$(rC_1 - rC_1)e^{rt} - (rC_2 + \omega C_3 + bA) \sin \omega t + (\omega C_2 - rC_3) \cos \omega t = 0$$

Setting the coefficients of $\sin \omega t$ and $\cos \omega t$ to zero gives

$$\omega C_2 - rC_3 = 0$$
$$rC_2 + \omega C_3 = -Ab$$

or

$$C_2 = \frac{-rAb}{\omega^2 + r^2}$$
$$C_3 = \frac{-\omega Ab}{\omega^2 + r^2}$$

For $y(0) = 0$, $C_1 + C_3 = 0$ and

$$C_1 = \frac{\omega Ab}{\omega^2 + r^2}$$

For the stable case, $r = -1/\tau$, and the response is

$$y(t) = \frac{bA\tau}{\tau^2\omega^2 + 1}(\sin \omega t - \omega\tau \cos \omega t + \omega\tau e^{-t/\tau}) \tag{3.6-14}$$

At $t = 4\tau$, the transient term has essentially disappeared, and the steady-state response is

$$\begin{aligned} y(t) &= C(\sin \omega t - \omega\tau \cos \omega t) \\ &= B \sin (\omega t + \phi) \end{aligned} \tag{3.6-15}$$

where

$$C = \frac{bA\tau}{\tau^2\omega^2 + 1} \tag{3.6-16}$$

$$B = \frac{|b|A\tau}{\sqrt{\omega^2\tau^2 + 1}} > 0 \tag{3.6-17}$$

$$\tan \phi = -\omega\tau \qquad \begin{cases} -\pi/2 \leqslant \phi \leqslant 0, & b > 0 \\ \pi/2 \leqslant \phi \leqslant \pi, & b < 0 \end{cases} \tag{3.6-18}$$

We define B to be positive for convenience, and absorb the required sign into ϕ (see Section A.6). Thus, the steady-state response is sinusoidal with the same frequency as the input, and its amplitude decreases as ω increases. It is retarded in phase relative to the input, and this retardation increases with the frequency. An example of this effect is shown in Figure 3.18, where $\tau = 2$, $b = 1$, and $A = 1$. Note that the transient response also has a greater magnitude at the lower frequency.

These results can be used to find the response for related input types. For example, if $v = A \sin(\omega t + \psi)$, then the phase angle ϕ is simply increased by ψ. An important case in this category is the cosine input, because $\cos \omega t = \sin(\omega t + \pi/2)$.

The ramp and sine responses of the first-order model are summarized in Table 3.8.

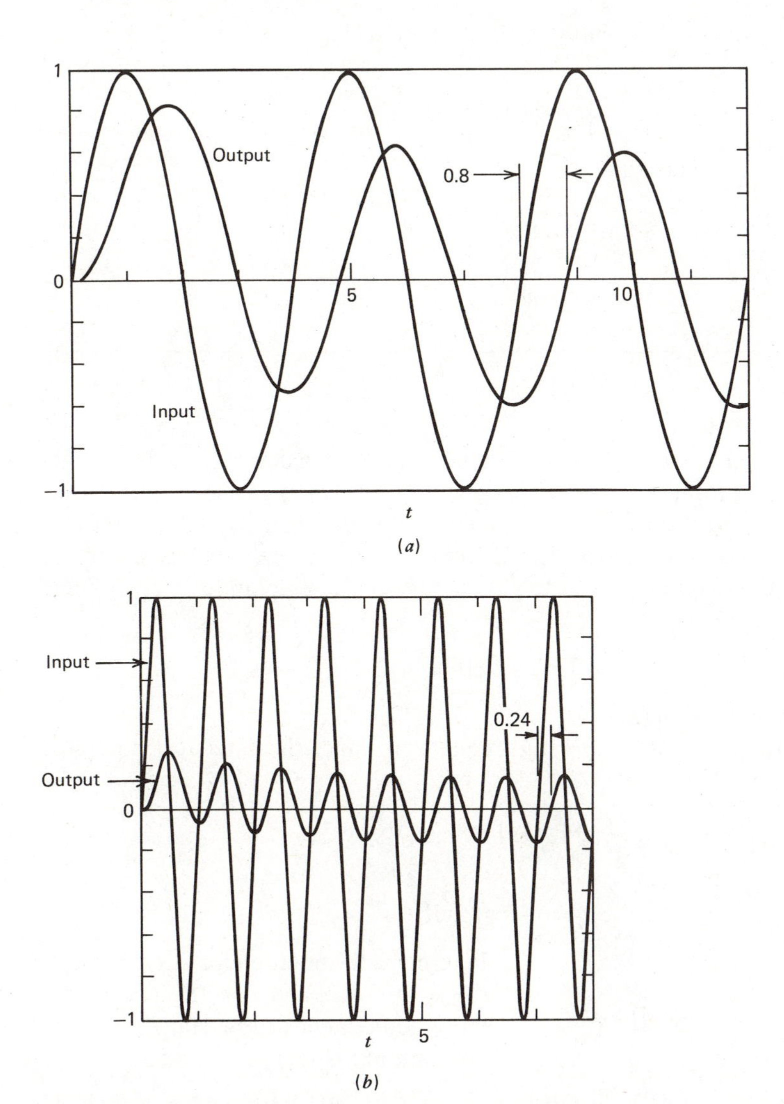

FIGURE 3.18 Sinusoidal response of the first-order system: $\dot{y} = -0.5y + v(t)$ for $v(t) = \sin \omega t$. The steady-state is reached at approximately $t = 8$. (*a*) $\omega = \pi/2$. The phase shift is $\phi = -1.26$ and the time shift is $t = \phi/\omega = -0.8$. (*b*) $\omega = 2\pi$. Here $\phi = -1.49$ rad and the time shift is -0.24.

TABLE 3.8 Ramp and Sine Responses of the First-order Model: $\tau\dot{y}+y=av(t)$

Ramp response $[y(0)=0]$:

$$v(t)=mt$$

$$y(t)=am(t-\tau+\tau e^{-t/\tau})$$

Sine response $[y(0)=0]$:

$$v(t)=A\sin\omega t$$

$$y(t)=\frac{aA}{\tau^2\omega^2+1}(\sin\omega t-\omega\tau\cos\omega t+\omega\tau e^{-t/\tau})$$

$$=\frac{A}{\sqrt{\tau^2\omega^2+1}}\left[|a|\sin(\omega t+\phi)+\frac{a\omega\tau}{\sqrt{\tau^2\omega^2+1}}e^{-t/\tau}\right]$$

$$\phi=-\tan^{-1}(\omega\tau)$$

$$-\frac{\pi}{2}\leqslant\phi\leqslant 0 \qquad \text{if } a>0$$

$$\frac{\pi}{2}\leqslant\phi\leqslant\pi \qquad \text{if } a<0$$

Example 3.9

A model of a *vibration isolator* is shown in Figure 3.19. Such a device is used to isolate the motion of point 1 from the input displacement y_2, which might be produced by a mechanism driving a lightweight tool at point 2. Point 1 represents a lightweight tool support that cannot be subjected to large deflections. Determine the values of c and K required to keep the amplitude of y_1 less than 10% of the amplitude of y_2. Assume that y_2 is

$$y_2(t)=2\sin(100t)\text{ in.}$$

with time measured in seconds.

The sum of the forces at point 1 must equal zero, since the mass at this point is assumed to be zero. Thus,

$$K(y_2-y_1)-c\dot{y}_1=0$$

or

$$c\dot{y}_1=-Ky_1+Ky_2$$

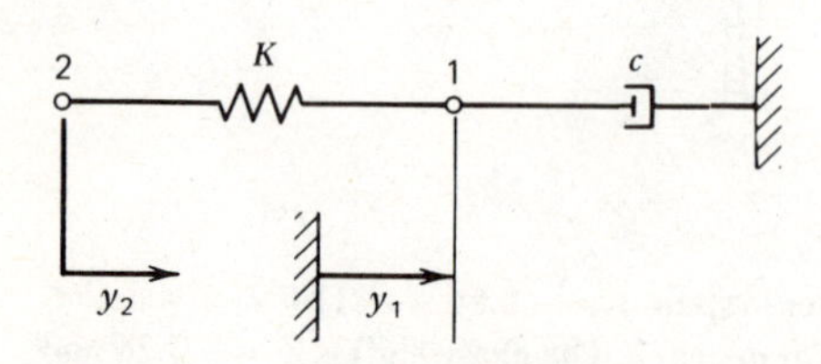

FIGURE 3.19 Vibration isolator.

In terms of the notation for (3.2-2), $y=y_1$, $v=y_2$, $b=K/c$, and $\tau=c/K$. The amplitude requirement can be stated with the notation in (3.6-15) as

$$B\leqslant 0.1A=0.2$$

The requirement is met exactly if $B=0.2$ at $\omega=100$ rad/sec. From (3.6-17), since here $b\tau = 1$ and $A = 2$,

$$B = \frac{2}{\sqrt{100^2\tau^2 + 1}} = 0.2$$

The result is $\tau = 0.1$ sec. The isolation requirement is met for any K and c such that $c/K = 0.1$ sec.

Other constraints must be used to specify the values of both K and c. One such constraint might be the maximum force allowed to be transmitted to the mechanism causing the displacement y_2. Since we have designed the system so that point 1 is relatively motionless, the deflection of the spring K is approximately equal to y_2, and the spring force is the force transmitted to the mechanism. This is $Ky_2 = K(2 \sin 100t)$. If this force must be less than, say, 50 lb, then K must be less than 25 lb/in. This means that c must be less than $K\tau = 25(0.1) = 2.5$ lb/in./sec. With this information, formulas such as the damper equation (Table 2.4) can be used to design or select the proper elements.

3.7 RESPONSE OF HIGHER ORDER SYSTEMS

The primary difference between first-order and second-order linear models is that the characteristic roots of second-order models can be complex numbers. This leads to types of behavior not seen in first-order models and sometimes requires modified algorithms in order to handle quantities with real and imaginary parts. The second-order case contains all of the possible types of behavior that can occur in linear systems. This is because a characteristic equation with real coefficients must have roots that fall into one of the three categories.

1. Real and distinct.
2. Distinct but complex conjugate pairs.
3. Repeated.

Once we understand the behavior generated by each of the three root cases, the superposition principle will allow us to apply the results to a linear system of any order. For example, the response of a sixth-order system with a complex conjugate pair of roots, two real repeated roots, and two real distinct roots can be analyzed by determining that part of the response resulting from each root type. The linear combination of each response pattern will give the total system response.

For example, consider the third-order model

$$b_3\dddot{y} + b_2\ddot{y} + b_2\dot{y} + b_0 y = f(t) \quad \textbf{(3.7-1)}$$

Its characteristic equation is

$$b_3 s^3 + b_2 s^2 + b_1 s + b_0 = 0 \quad \textbf{(3.7-2)}$$

If the three roots s_1, s_2, and s_3 are real and distinct, the free response is an extension of the form given by (3.3-8),

$$y(t) = A_1e^{s_1t} + A_2e^{s_2t} + A_3e^{s_3t} \tag{3.7-3}$$

If s_1 and s_2 are complex conjugates but s_3 is real, the free response is an extension of (3.3-28); namely,

$$y(t) = Be^{-at}\sin(bt + \phi) + A_3e^{s_3t} \tag{3.7-4}$$

where $s_1 = -a + ib$, and $s_2 = -a - ib$. Finally, if all three roots are repeated, then the extension of (3.3-34) gives

$$y(t) = A_1e^{s_1t} + tA_2e^{s_2t} + t^2A_3e^{s_3t} \tag{3.7-5}$$

where $s_1 = s_2 = s_3$. These cover all the cases for the third-order model. The forced response can be obtained using the trial solution method using equation (3.6-1), where the free-response form is given by one of the preceding solutions, and the particular solution form is obtained as outlined in Section 3.6.

The only situation requiring further algebraic manipulation is the case where any of the repeated roots are complex. This can occur in only fourth-order models or higher. The free response is an extension of the form given by (3.3-34), where there are now two pairs of repeated roots, $s_1 = s_2 = -a + ib$ and $s_3 = s_4 = -a - ib$. The free response is

$$y(t) = A_1e^{s_1t} + tA_2e^{s_1t} + A_3e^{s_3t} + tA_4e^{s_3t} \tag{3.7-6}$$

The algebra required to convert $y(t)$ into a useful form containing sine functions is similar to the procedures used following (3.3-16).

The trial solution method can be used to solve a linear differential equation of any order. However, the algebra required to solve for the solution's constants becomes tedious as the system's order increases. For example, the step response of a third-order model requires four constants to be found, one from the differential equation and three from the initial conditions $y(0)$, $\dot{y}(0)$, and $\ddot{y}(0)$. Two alternatives to the trial solution method are: (1) the *Laplace transform* method, presented in Chapter 4, and (2) *digital simulation*, covered in Chapter 5. The Laplace transform method gives the solution in closed form but also becomes tedious as the model's order increases. It can be applied only to linear models, but it has some advantages over the trial solution method in certain cases. These are discussed in Chapter 4. On the other hand, digital simulation does not give the solution in closed form, but rather as a table or graph of the response versus time for specific coefficient values and initial conditions. This at first seems like a disadvantage, but if the solution must eventually be plotted for ease of interpretation, as in Example 3.8, then it might be easier to use digital simulation. Also, digital simulation can be used to solve nonlinear models. The choice of solution method depends on the ultimate use of the solution and on the willingness of the analyst to perform algebraic manipulations. These choices will be discussed in more detail throughout Chapters 4 and 5.

Dominant-Root Approximation

Often, the model is needed only to obtain a general idea of the system's response pattern. In this case, we often do not need to solve for the coefficients in the solution, and the concept of the dominant root is useful. As an example, consider the fourth-order system

$$\frac{d^4y}{dt^4} + 24\frac{d^3y}{dt^3} + 225\frac{d^2y}{dt^2} + 900\frac{d^2y}{dt^2} + 2500y = f \tag{3.7-7}$$

where the roots are $s = -10 \pm i5$ and $-2 \pm i4$. The step response is of the form

$$y(t) = B_1e^{-10t}\sin(5t + \phi_1) + B_2e^{-2t}\sin(4t + \phi_2) + B_3 \tag{3.7-8}$$

where $B_3 = f/2500$. From the plot of the roots in Figure 3.20, we see that the dominant-root pair is $-2 + i4$, and after about 4/10 of a time unit, $y(t)$ is given approximately by

$$y(t) = B_2e^{-2t}\sin(4t + \phi_2) + B_3 \tag{3.7-9}$$

The period of the sine wave is 1.57, or about three time constants. Thus, the oscillation is not very apparent in the response, because it dies out before it can repeat itself. From the formulas in Section 3.5, we can see that a second-order system with the roots $s = -2 \pm i4$ has $\zeta = 0.447$ and $\omega_d = 4.47$. Its step response will have a maximum overshoot of 20.5% at $t = 0.79$ and will have nearly reached its steady-state value at $t = 2.0$. The step response of the fourth-order system (3.7-7) for $f = 2500$ is shown in Figure 3.21. The maximum overshoot is 18.5% at $t = 0.95$, and the steady-state value is nearly reached at $t = 2.0$. Thus, the qualitative features of the response as predicted

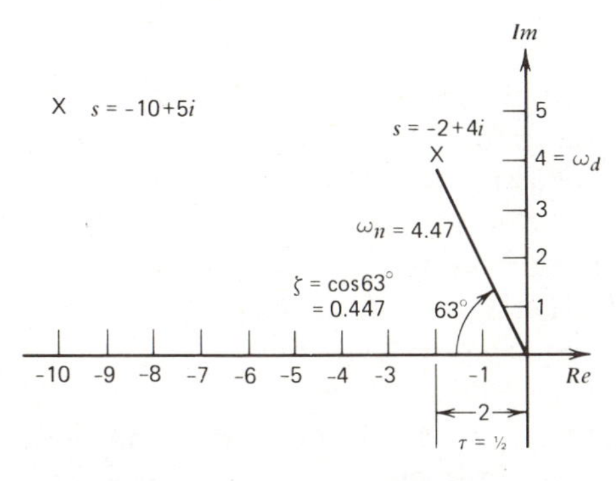

FIGURE 3.20 Dominant-root concept applied to a fourth-order system. An estimate of the system's behavior is given by the parameters ζ, τ, ω_n, and ω_d of the dominant root $s = -2 + 4i$. (The two conjugate roots lying in the lower half plane are not shown.)

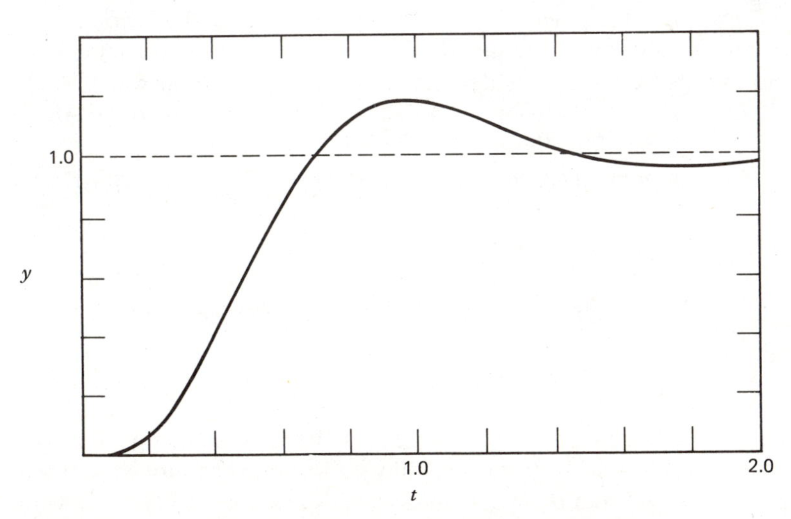

FIGURE 3.21 Unit-step response of the fourth-order model (3.7-7).

by the second-order dominant-root approximation are close to those of the fourth-order step response in this example.

The agreement between the actual response and that predicted by the dominant-root approximation becomes better as the separation between the dominant and minor roots increases. The greatest difference will occur early in the response when the terms due to the minor roots have not yet died out.

3.8 THE ROUTH–HURWITZ STABILITY CRITERION

Perhaps the most important question to be answered by a dynamic model concerns the stability of the system under study. If the proposed system is predicted to be unstable, the designer immediately seeks to modify the system to make it stable. This will be especially true in the design of feedback control systems.

Thus, it is convenient to have available a criterion that gives a yes-or-no answer to the stability question. Slightly different criteria were developed separately by Hurwitz in 1895 and by Routh in 1905 for the purpose of predicting the stability of rotation of a body about a given axis. Their criteria furnish the same information and are summarized here for the second- and third-order cases. In Appendix F, we state the results for the general case without proof (which is lengthy).

Recall that the stability properties of a linear system are determined by the roots of its characteristic equation. If any root lies in the right half of the complex plane, the system is unstable. The Routh–Hurwitz criterion tells us how many roots lie in the right-half plane. By shifting the origin of the complex plane to the left, we will see that the criterion can also be used to determine the number of roots that lie to the right of any given vertical line. This enables us to estimate the location and type of the dominant root or root pair and thus to obtain the approximate time constant, oscillation frequency, and damping ratio for the system without the necessity of solving for the characteristic roots. Finally, the Routh–Hurwitz criterion sometimes gives enough information to enable us to determine the location of all the roots. While closed-form solutions exist for the roots of polynomials up to fourth order, these formulas are cumbersome for third- and fourth-order equations, and the Routh–Hurwitz criterion often gives enough information with much less effort.

The Second-Order Case

Since the quadratic formula is relatively simple, it is not difficult to use it to show that the characteristic equation

$$b_2 s^2 + b_1 s + b_0 = 0 \tag{3.8-1}$$

represents a stable system if and only if b_2, b_1, and b_0 all have the same sign. If $b_0 = 0$ and b_2 has the same sign as b_1, the system is neutrally stable, since one root is $s = 0$ and the other is negative. If $b_1 = 0$ and $b_0/b_2 > 0$, the system is again neutrally stable with two purely imaginary roots. (Of course, if $b_2 = 0$, the system is no longer second order.)

The Third-Order Case

Third-order models occur quite frequently, and the Routh–Hurwitz results are simply stated. Consider the equation

$$b_3 s^3 + b_2 s^2 + b_1 s + b_0 = 0 \tag{3.8-2}$$

For simplicity, assume that we have normalized the equation so that $b_3 > 0$. The system is stable if and only if all of the following conditions are satisfied:

$$\left.\begin{aligned} b_2 > 0, \quad b_1 > 0, \quad b_0 > 0 \\ b_1 b_2 > b_0 b_3 \end{aligned}\right\} \tag{3.8-3}$$

The neutrally stable cases occur if (3.8-3) is satisfied except for the following:

1. $b_0 = 0$ (a root exists at $s = 0$).
2. $b_2 = b_0 = 0$ (roots are $s = 0, \pm i\sqrt{b_1}$).
3. $b_1 b_2 - b_0 b_3 = 0$ (roots are $s = \pm\sqrt{b_1}, -b_2/b_3$).

Example 3.10

A certain system has the characteristic equation

$$s^3 + 9s^2 + 26s + K = 0 \tag{3.8-4}$$

Find the range of K values for which the system will be stable.

From (3.8-3) it is necessary and sufficient that

$$K > 0$$

$$9(26) - K > 0$$

Thus, the system is stable if and only if $0 < K < 234$.

Example 3.11

For the characteristic equation (3.8-4), find the value of K required so that the dominant time constant is no larger than 1/2.

The time constant requirement means that no root can lie to the right of $s = -2$. Translate the origin of the s plane to $s = -2$ by substituting $s = p - 2$ into (3.8-4). This gives

$$p_3 + 3p^2 + 2p + K - 24 = 0 \tag{3.8-5}$$

We can apply the Routh–Hurwitz criterion to this equation to determine when all roots p have negative real parts (and thus when all s roots lie to the left of $s = -2$). From (3.8-3), this occurs when

$$K - 24 > 0$$

$$3(2) - (K - 24) > 0$$

Or $24 < K < 30$.

TABLE 3.9 Routh–Hurwitz Stability Criteria

The highest-numbered coefficients have been normalized to be positive.

1. First-order model: $b_1 \dot{y} + b_0 y = 0$, $b_1 > 0$
   Stable only if $b_0 > 0$.
   Neutrally stable if $b_0 = 0$.

2. Second-order model: $b_2 \ddot{y} + b_1 \dot{y} + b_0 y = 0$, $b_2 > 0$
   Stable only if $b_1 > 0$ and $b_0 > 0$.
   Neutrally stable if either
   (a) $b_0 = 0$ and $b_1 > 0$ (roots are $s = 0, -b_1/b_2$)
   or (b) $b_1 = 0$ and $b_0 > 0$ (roots are $s = \pm i\sqrt{b_0/b_2}$).

3. Third-order model: $b_3 \dddot{y} + b_2 \ddot{y} + b_1 \dot{y} + b_0 y = 0$, $b_3 > 0$
   Stable only if $b_2 > 0$, $b_1 > 0$, $b_0 > 0$ and $b_1 b_2 > b_0 b_3$.
   Neutrally stable if the stability conditions are satisfied except for
   (a) $b_0 = 0$ (one root is $s = 0$)
   (b) $b_2 = b_0 = 0$ (roots are $s = 0, \pm i\sqrt{b_1/b_3}$)
   (c) $b_1 b_2 - b_0 b_3 = 0$ (roots are $s = \pm i\sqrt{b_1/b_3}, -b_2/b_3$).

Even more information about the system's behavior (damping ratio, etc.) can be extracted from the Routh–Hurwitz criterion (this is explored in more detail in Appendix F, Example F.3). The stability criteria for first, second, and third-order systems are summarized in Table 3.9.

3.9 NONLINEAR MODELS AND LINEARIZATION

Most of the linear models encountered thus far have resulted from some simplifying assumption. For example, a closer examination of the viscous damper model in Section 2.7 reveals that the underlying physical process is inherently nonlinear and can be approximated by a linear relation only when the appropriate velocity is small. It is solely for reasons of mathematical simplicity that such approximations are made; as we will see, no general solution technique is available for nonlinear models.

The general form of a first-order *linear* ordinary differential equation is

$$a(t)\dot{y} + b(t)y + c(t) = 0 \tag{3.9-1}$$

Any first-order model not fitting this form is *nonlinear*. Thus, the class of nonlinear models can be seen to be much larger than the linear class, and this indicates the difficulty in obtaining general nonlinear techniques. Unhappily for the analyst, predicting the response of a nonlinear system is neither easy nor generally understood. Happily for the modeler, nonlinear models are not so restricted in their behavioral possibilities.

An important characteristic of nonlinear-model response is that its form depends on the magnitude of the input and the initial conditions. For example, step inputs of different magnitudes might produce completely different forms of response for the same model. Also, the response might be oscillatory for one initial condition, whereas another might produce exponential behavior. This cannot occur in a linear model, whose response form (sinusoidal, exponential, etc.) is completely determined by the input form and the system's characteristic roots.

Some techniques are available for obtaining the closed-form solution of relatively simple nonlinear equations (see Section 3.11 of Reference 2 for examples). In Chapter 5, we develop digital-simulation techniques for solving nonlinear equations, but few control system design techniques are available for nonlinear models. In this section, we are primarily concerned with how to approximate a nonlinear differential equation with a linear one so that we can use the numerous design techniques available for linear models. We can then take advantage of the linear model's generality, in the sense that its response form is independent of the specific values of the initial conditions. This feature makes linear models extremely useful in system design, because with them, the designer can focus on the structure of the system and need not worry about the effects of different initial conditions that might be encountered when the system is in operation. This is especially true in the design of control systems, which must produce stable system behavior for a variety of initial conditions. The sacrifice incurred by a linear approximation is that its validity is limited to a certain range of operating conditions. However, we will see that if a feedback control system is operating properly, it preserves the validity of the assumption on which the linearization is based. We will attempt to shed some light on these concepts as well as the procedures for developing linear approximations.

Nonlinear Response

An illustration of how nonlinear stability and response characteristics can depend on the initial conditions is given by the following equation, which is a model for some types of growth processes (Reference 2, Section 2.9).

$$\dot{y} = ry\left(1 - \frac{y}{K}\right) \tag{3.9-2}$$

The transformation given by $y = z^n$ gives

$$\dot{z} = \frac{r}{n}z - \frac{r}{nK}z^{n+1}$$

This is linear if $n = -1$, and the solution is given by the step response of the linear model (3.2-4). In the present notation, this is

$$z(t) = \left[z(0) - \frac{1}{K}\right]e^{-rt} + \frac{1}{K} \tag{3.9-3}$$

Inverting the transformation produces the solution in terms of the original variable.

$$y(t) = \frac{Ky(0)}{y(0) + [K - y(0)]e^{-rt}} \tag{3.9-4}$$

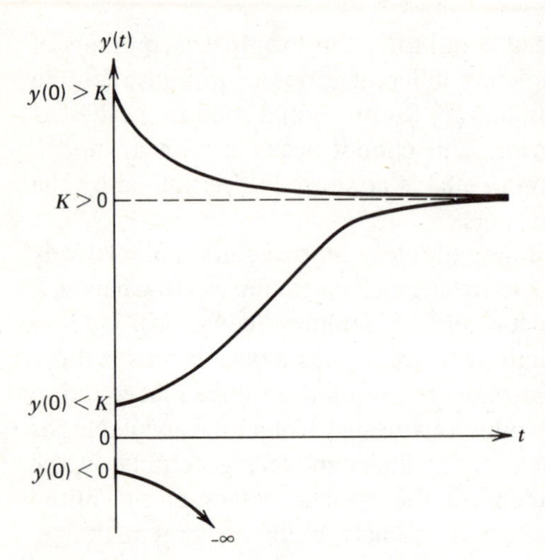

FIGURE 3.22 Solutions of the logistic growth model.

The solution's behavior for various initial conditions is illustrated in Figure 3.22 for r and K positive. The value $y = K$ represents the stable equilibrium to which all solutions are attracted provided that $y(0) > 0$.

If $y(0) < 0$, the solution approaches $-\infty$ and the system is unstable. The solution for $y(0) < 0$ cannot be obtained from (3.9-4), because the transformation from z to y becomes undefined when the denominator in this equation becomes zero. An examination of the original model (3.9-2) shows that y approaches $-\infty$ for negative initial conditions. Thus, the system is *locally* stable for $y(0) > 0$ and *globally* unstable for $y(0) < 0$.

If $y(0) > 0$, the system is stable, but the response characteristics still depend on the specific values of $y(0)$. To see this, let $r = K = 1$, and consider two initial-condition values equally spaced about the equilibrium at $y = 1$. Let the values be $y(0) = 1.9$ and $y(0) = 0.1$, and find the time required in each case to traverse half the distance to the equilibrium. For $y(0) = 1.9$, this time is $t = 0.423$ [$y(0.423) = 1.45$]. For $y(0) = 0.1$, the time is $t = 2.4$ [$y(2.4) = 0.55$]. This shows that the approach to equilibrium is faster for the larger initial condition.

Coordinate Translations in Dynamic Models

At this point, it is appropriate to consider the effect of a coordinate system translation in a linear dynamic system. Suppose that the model

$$\dot{y} = ay + bv \tag{3.9-5}$$

is subjected to a constant input v_e. Then

$$\dot{y} = ay + bv_e$$

At steady state, this will produce an output, denoted by y_e, such that $\dot{y}_e = 0$ and thus

$$y_e = -\frac{bv_e}{a} \tag{3.9-6}$$

The equilibrium state of the model is y_e when the input is the constant v_e.

In many situations, the input may change slightly from the reference or nominal value v_e, or the system may be subjected to a disturbance that suddenly changes y from y_e to a new value. In order to investigate the behavior of the system under these conditions, it is convenient to define a new set of variables x and u such that $x = y - y_e$ and $u = v - v_e$. To obtain the differential equation describing x in terms of u, note that

$\dot{x} = \dot{y} - \dot{y}_e = \dot{y}$, since y_e is a constant. Substituting for y and v in (3.9-5) results in

$$\begin{aligned} \dot{y} = \dot{x} &= a(x + y_e) + b(u + v_e) \\ &= ax + bu + ay_e + bv_e \end{aligned}$$

Using (3.9-6) gives

$$\dot{x} = ax + bu \tag{3.9-7}$$

and we see that the form of the equation for x and u is the same as that for y and v, because the equation is linear. Equation (3.9-7) is more convenient for analyzing deviations from the reference condition, since the values of y_e and v_e do not appear explicitly. Note that the system's equilibrium state is now described by $x = 0$ when $u = 0$.

Linearization by Taylor Series

The linearization method presented here is based on the Taylor series expansion of a function of one or more variables (see Appendix A). As demonstrated in Chapter 1, the series can be used to approximate a general (differentiable) nonlinear function in the vicinity of a reference point of interest. By truncating the series after the first-order derivative terms, a straight line or plane is obtained that passes through the reference point. To be more specific, consider a static function of two variables $f(y, v)$. Denote the reference point by (y_e, v_e). Then the Taylor series expansion of the function around the reference point is

$$\begin{aligned} f(y, v) = f(y_e, v_e) &+ \left(\frac{\partial f}{\partial y}\right)_e (y - y_e) + \left(\frac{\partial f}{\partial v}\right)_e (v - v_e) \\ &+ \frac{1}{2!}\left(\frac{\partial^2 f}{\partial y^2}\right)_e (y - y_e)^2 + \left(\frac{\partial^2 f}{\partial y \partial v}\right)_e (y - y_e)(v - v_e) \\ &+ \frac{1}{2!}\left(\frac{\partial^2 f}{\partial v^2}\right)_e (v - v_e)^2 + \cdots \end{aligned} \tag{3.9-8}$$

where $(\ )_e$ indicates that the term within the parentheses is evaluated at the point (y_e, v_e). If y and v are sufficiently close to y_e and v_e, then the series converges, and the second- and higher order terms are smaller than the first-order terms. Thus, the approximation is

$$\Delta f = ax + bu \tag{3.9-9}$$

where we have defined the following.

$$\Delta f = f(y, v) - f(y_e, v_e) \tag{3.9-10}$$

$$x = y - y_e \tag{3.9-11}$$

$$u = v - v_e \tag{3.9-12}$$

$$a = \left(\frac{\partial f}{\partial y}\right)_e \tag{3.9-13}$$

$$b = \left(\frac{\partial f}{\partial v}\right)_e \tag{3.9-14}$$

The quantities Δf, x, and u represent the deviations or perturbations of f, y, and v from their reference values. In terms of these deviations, the original nonlinear function $f(y, v)$ has been approximated by a plane in the (x, u) coordinate system. Note that the slopes of this plane in the x and u directions equal the slopes of $f(y, v)$ in the y and v directions at the reference point and that the values of these slopes depend on the specific reference point chosen. Different reference points will produce different planes, in general.

Now consider the nonlinear dynamic model

$$\dot{y} = f(y, v) \tag{3.9-15}$$

with the equilibrium values y_e, v_e. From the definitions given by (3.9-10)–(3.9-14), we can write the left- and right-hand sides of (3.9-15) as follows:

$$\dot{y} = \dot{x} + \dot{y}_e = \dot{x}$$
$$f(y, v) = f(y_e, v_e) + ax + bu$$

The latter equation is an approximation, of course. Since $\dot{y}_e = f(y_e, v_e) = 0$, we obtain from (3.9-15)

$$\dot{x} = ax + bu \tag{3.9-16}$$

This is the linearized approximation to the original equation (3.9-15). This procedure is summarized in Table 3.10.

The translation of the origin of a linear model to the equilibrium point results in the form (3.9-7), which is the same form as (3.9-16). However, in the first case, the resulting model is exact, and in the second case, it is only approximately correct for small values of x and u.

TABLE 3.10 Taylor Series Linearization of a Dynamic Model

Nonlinear model	$\dot{y} = f(y, v), \quad v = v(t)$
Reference solution	$\dot{y}_e = f(y_e, v_e)$
	$\dot{y}_e = 0$ for a point equilibrium
Linearized model	$\dot{x} = ax + bu$
	$x = y - y_e$
	$u = v - v_e$
	$a = \left(\dfrac{\partial f}{\partial y}\right)_e$
	$b = \left(\dfrac{\partial f}{\partial v}\right)_e$

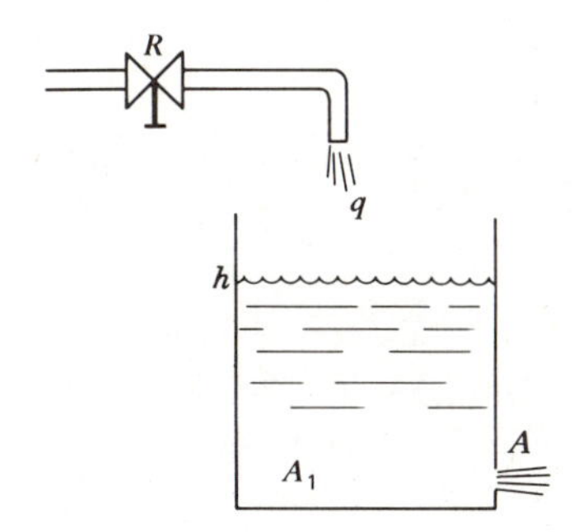

FIGURE 3.23 Tank with orifice.

Example 3.12

In Section 2.6, we saw that the height h of liquid in a tank with an orifice as in Figure 3.23 can be described by a model of the form

$$\dot{h} = \frac{1}{A_1} q - \frac{\alpha}{A_1}\sqrt{h} \tag{3.9-17}$$

$$\alpha = C_d A\sqrt{2g} \tag{3.9-18}$$

where q is a volume rate. Suppose the given parameter values are $A_1 = 1 = \alpha$. Obtain a linearized model in terms of the given equilibria at (a) $q_e = h_e = 1$, and (b) $q_e = 3$, $h_e = 9$.

For the given values, the model is

$$\dot{h} = q - \sqrt{h} = f(h, q) \tag{3.9-19}$$

If the inflow rate is held constant at q_e, the height will eventually come to equilibrium at the value h_e where

$$h_e = \left(\frac{q_e}{\alpha}\right)^2 \tag{3.9-20}$$

From Table 3.10, the linearization coefficients are

$$a = \left(\frac{\partial f}{\partial h}\right)_e = -\tfrac{1}{2}h_e^{-1/2}$$

$$b = \left(\frac{\partial f}{\partial q}\right)_e = 1$$

and the variations are $x = h - h_e$ and $u = q - q_e$.

(a) For $h_e = q_e = 1$, $a = -1/2$, $b = 1$, and the linearized model is

$$\dot{x} = -\tfrac{1}{2}x + u$$

The time constant of the linearized model is 2.

(b) For $h_e = 9$, $q_e = 3$, we have $a = -1/6$, $b = 1$, and

$$\dot{x} = -\tfrac{1}{6}x + u$$

The time constant is 6. Note that the linearized model's time constant depends on the particular equilibrium chosen for the linearization.

An equation can be linearized around one of its solutions other than a point equilibrium. The subscript e usually denotes an equilibrium value, but to avoid additional notation, let $y_e(t)$ and $v_e(t)$ denote the functions about which the linearization is to be performed. The function $y_e(t)$ is the solution of $\dot{y} = f(y, v)$ when $v = v_e(t)$. Then $f(y_e, v_e, t)$ is no longer zero but a known function of time and equals $\dot{y}_e$. The same procedure as before results in (3.9-16) where the coefficients a and b are now known functions of time.

Linearization Accuracy

The question that inevitably arises in a discussion of linearization is how small must x and u be for the truncated Taylor series to be acceptable? There are two ways of answering this question.

Often the linearized model is not obtained for solution purposes, but to indicate the local stability of an equilibrium. If the linearized equation is stable, then any slight perturbation will die out and the equilibrium is locally stable. On the other hand, linearization cannot be used to investigate the global stability of the equilibrium, because the series truncation is not valid for large perturbations. If the linearized equation is unstable, the equilibrium is locally unstable at least. However, if the linearized equation is neutrally stable, no information is obtained about local stability, because the second-order terms in the expansion are now critical. To see this, note that the coefficient a in (3.9-16) is zero for neutral stability, and thus the first-order derivative term in the expansion is also zero. The remaining terms are thus significant and cannot be truncated. This invalidates the linearization, and a nonlinear analysis is required to resolve the issue.

Linearized models are widely used in designing feedback control systems and are successful for the following reason. If the controller is designed to keep the system output near the desired equilibrium value and if the design is based on a linearization around this value, then the controller will tend to keep the linearization accurate. This interesting property partly explains why models for controller design need not be so detailed as those required for other purposes.

The Phase Plane

Phase-plane analysis was developed as a convenient graphical way of presenting the dynamics of a mechanical system. The *phase variables* used in this plot were originally taken to be the displacement and velocity of the mass in question. However, any convenient choice of a pair of state variables can be employed. The *phase plot* is a plot of one state variable versus the other, with time as a parameter on the resulting curve. For given initial values of the state variables, one curve is generated. This is the *trajectory* of the system's behavior. When the trajectories are sketched for many different sets of initial conditions, the resulting plot provides a compact and easily interpreted summary of the system's response.

The phase-plane concept can be applied to both linear and nonlinear models and first- and second-order models. Consider the nonlinear logistic growth model (3.9-2), repeated here.

$$\dot{y} = ry\left(1 - \frac{y}{K}\right) \tag{3.9-2}$$

A plot of $\dot{y}$ versus y is shown in Figure 3.24a for $y \geq 0$. For first-order models with

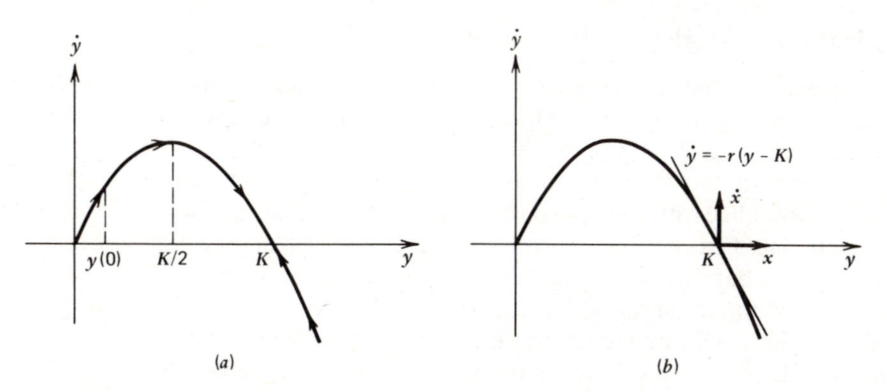

FIGURE 3.24 **(*a*) Phase plane plot of the logistic equation. (*b*) Graphical representation of linearization. The straight line is tangent to the phase plot at $y = K$.**

unique solutions, only one trajectory is possible. The maximum value of $\dot{y}$ is seen to occur at $y = K/2$. The plot is interpreted as follows. For a given initial value $y(0)$, the arrows on the plot indicate the behavior of the system for increasing time. The direction of the arrows can be found from the sign of $\dot{y}$ in each quadrant. From (3.9-2) we see that $\dot{y} > 0$ if $0 < y < K$. Thus, if $y(0) < K$, $y(t)$ increases until it reaches the equilibrium at $y = K$. If $y(0) > K$, $y(t)$ decreases until the same equilibrium is reached. This result is clearly shown by the plot.

The linearized analysis can be demonstrated on this plot. We approximate the plot near the equilibrium $y = K$ by a straight line passing through the point $y = K$, $\dot{y} = 0$, with a slope equal to the slope of the phase plot at that point (Figure 3.24*b*). This slope is

$$\left.\frac{\partial \dot{y}}{\partial y}\right|_{y=K} = \left.\left(r - 2\frac{r}{K}y\right)\right|_{y=K} = -r$$

The straight line is given by

$$\dot{y} = -r(y - K)$$

If the coordinate system is shifted to the point $\dot{y} = 0$, $y = K$, and the new coordinates denoted $(\dot{x}, x)$, this line can be expressed as

$$\dot{x} = -rx \tag{3.9-21}$$

since $x = y - K$ and $\dot{x} = \dot{y}$. Equation (3.9-21) is the linearized approximation to (3.9-2). It tells us that the equilibrium is stable if $r > 0$. This result is also plain from the phase plot. Equation (3.9-21) also gives us an estimate of the time constant near equilibrium.

Phase-plane plots for second-order systems can be developed in a similar manner by plotting one state variable x_1 versus the other, x_2. Because two initial conditions are required to generate a solution, the plot consists of a family of curves, one for each initial-condition pair. Techniques for obtaining the plots, along with examples, can be found in References 2 and 3.

Linearization of Higher Order Models

The linearization technique presented earlier in this section for first-order models is based on a truncated Taylor series representation of the model's nonlinearity and is equivalent to the straight-line approximation used in Figure 3.24*b*. The method can be easily extended to higher order models, by either first linearizing the function producing the nonlinearity and then developing the differential equations or developing the equations first and then linearizing them. The first approach is sometimes more convenient, but we must be careful that the linearization is performed for values of the state and input variables that correspond to an equilibrium solution of the system's equations. The second approach automatically accounts for this.

We consider now a second-order example. The pendulum shown in Figure 3.25*a* can be considered as a representation of an unbalanced rotating load shown in Figure 3.25*b*. The mass m represents the unbalanced mass lumped at its center of mass. We assume that a torque T is applied to the system about the fixed axis of rotation. This torque would be produced by a control motor trying to position the load at some desired angle.

The mass moment of inertia of the pendulum as given in Table 2.5 is $I = mL^2$ for a massless rod. The gravity moment about the axis is $mgL \sin \theta$. This is the nonlinear term. If we are interested in motion about $\theta = 0$, the Taylor series expansion of $\sin \theta$ near $\theta = 0$ gives $\sin \theta \cong \theta$ for θ in radians. Thus, the gravity moment is approximately $mgL\theta$, and Newton's law gives the model

$$mL^2\ddot{\theta} = T - mgL\theta \tag{3.9-22}$$

The reference point $\theta = 0$ is also an equilibrium solution of the model (3.9-22) if $T = 0$. The free response is like that of the mass–spring system, with a natural frequency of $\omega_n = \sqrt{g/L}$ for small amplitudes.

If we are interested in keeping the mass near the vertical position, we use the expansion of $\sin \theta$ near $\theta = \pi$; that is,

$$\sin \theta \cong \sin \pi + \left.\frac{\partial \sin \theta}{\partial \theta}\right|_{\theta=\pi} (\theta - \pi) = \pi - \theta$$

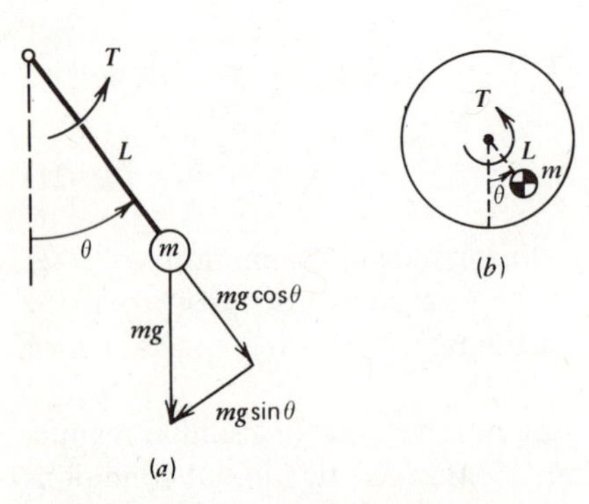

FIGURE 3.25 **(*a*) Pendulum with an applied torque *T*. (*b*) Equivalent model of an unbalanced load driven by a torque *T*.**

The gravity moment is approximately $mgL(\pi - \theta)$, and

$$mL^2\ddot{\theta} = T - mgL(\pi - \theta) \tag{3.9-23}$$

Again, the linearization reference point $\theta = \pi$ is also an equilibrium solution of the model (3.9-23) if $T = 0$. Here, it is convenient to introduce the deviation variable $x = \theta - \pi$ to obtain

$$mL^2\ddot{x} = T + mgLx \tag{3.9-24}$$

The equilibrium at $x = 0$ is seen to be unstable.

The second approach to linearization

requires that the describing differential equation be written first, then the equilibrium solutions obtained, and finally the linearization performed. This process is more easily stated in terms of the state-variable form of the model, and is easily demonstrated with an example.

Example 3.13

Obtain a linearized model for the system shown in Figure 3.25 for the equilibrium $\theta_e = 3\pi/4$, $T_e = 0.707mgL$. Evaluate the stability.

Newton's law gives the reduced form

$$mL^2\ddot{\theta} = T - mgL \sin \theta \tag{3.9-25}$$

To put this into state-variable form, let $z_1 = \theta$, $z_2 = \dot{\theta}$, and $v = T$. Then

$$\dot{z}_1 = z_2 \tag{3.9-26}$$

$$\dot{z}_2 = \frac{1}{mL^2} v - \frac{g}{L} \sin z_1 \tag{3.9-27}$$

The first equation is linear; the second is nonlinear. The equilibrium values of the new variables are: $z_{1e} = 3\pi/4$, $z_{2e} = 0$, and $v_e = 0.707mgL$, so we define the deviation variables as $x_1 = z_1 - 3\pi/4$, $x_2 = z_2$, and $u = v - 0.707mgL$. The state equations in terms of the deviation variables are

$$\dot{x}_1 = x_2 \tag{3.9-28}$$

$$\dot{x}_2 = \frac{u + 0.707mgL}{mL^2} - \frac{g}{L} \sin\left(x_1 + \frac{3\pi}{4}\right) \tag{3.9-29}$$

For x_1 near 0,

$$\sin\left(x_1 + \frac{3\pi}{4}\right) \approx \sin\frac{3\pi}{4} + \left(\cos\frac{3\pi}{4}\right)x_1$$
$$= 0.707 - 0.707x_1$$

and (3.9-29) becomes

$$\dot{x}_2 = \frac{1}{mL^2}u + 0.707\frac{g}{L} - 0.707\frac{g}{L} + 0.707\frac{g}{L}x_1$$
$$= \frac{1}{mL^2}u + 0.707\frac{g}{L}x_1 \tag{3.9-30}$$

Equations (3.9-28) and (3.9-30) constitute the linearized model in state-variable form.

We can convert the linearized model to reduced form by eliminating x_2.

$$\ddot{x}_1 = \dot{x}_2 = \frac{1}{mL^2}u + 0.707\frac{g}{L}x_1$$

or

$$mL^2\ddot{x}_1 - 0.707mgLx_1 = u \tag{3.9-31}$$

Its characteristic equation is $mL^2s^2 - 0.707mgL = 0$ and the equilibrium is locally unstable because of the positive root $s = \sqrt{0.707g/L}$.

The linearization procedure in Example 3.13 can be generalized for the following model with state variables z_1 and z_2, and input v.

$$\dot{z}_1 = f_1(z_1, z_2, v) \tag{3.9-32}$$

$$\dot{z}_2 = f_2(z_1, z_2, v) \tag{3.9-33}$$

The linearized model in terms of the deviation variables x_1, x_2, and u is

$$\dot{x}_1 = a_{11}x_1 + a_{12}x_2 + b_1 u \tag{3.9-34}$$

$$\dot{x}_2 = a_{21}x_1 + a_{22}x_2 + b_2 u \tag{3.9-35}$$

where the following derivatives are evaluated at the equilibrium or other reference solution of the original nonlinear equation.

$$a_{ij} = \frac{\partial f_i}{\partial z_j} \tag{3.9-36}$$

$$b_i = \frac{\partial f_i}{\partial v} \tag{3.9-37}$$

3.10 SUMMARY

Chapter 3 deals with the free- and forced-response patterns of continuous-time models. The trial solution method was used to obtain the responses, because this technique is probably the easiest to use for first and second-order models with step, ramp, and sine inputs. The following is a summary of the chapter's concepts by section.

(3.1) The free response of the linear model $\dot{y} = ry + bv$ is the solution for a zero input and can be obtained by substituting an assumed exponential time function into the equation. The two unknown constants are found from the resulting characteristic (algebraic) equation and the initial condition. The characteristic root $s = r$ determines the stability of the solution: $r > 0$ (unstable), $r = 0$ (neutrally stable), and $r < 0$ (stable). For the stable case, the time constant is $\tau = -1/r$ and a measure of the rate of exponential decay of the solution. At $t = t_0 + \tau$, the solution is 37% of the initial value; at $t = t_0 + 4\tau$, it is 2% of the initial value.

(3.2) The response of $\dot{y} = ry + bv$ to a step input can be obtained by an assumed exponential-plus-constant solution form. The time constant is a measure of the time required to reach the final output value (63% after one time constant; 98% after four time constants). The superposition principle says that the total solution is the sum of the free response (dependent on only the initial condition) and the forced response (dependent on only the input). The solution can also be decomposed into a transient part (which disappears with time) and a steady-state part (which remains).

(3.3) The primary difference between first- and second-order linear models is that the characteristic roots of second-order models can be complex numbers. This leads to types of behavior not seen in first-order models and sometimes requires modified algorithms in order to handle quantities with real and imaginary parts. The second-order case contains all of the possible types of behavior that can occur in linear systems. This is because a characteristic equation with real coefficients must have roots that fall into one of the three categories:

1. real and distinct,
2. repeated, and
3. distinct but complex conjugate pairs.

The free response for the first case is the sum of two exponentials, and the behavior is governed by the term possessing the dominant root if the roots are far apart. The dominant time constant is the negative reciprocal of the dominant root. Similar behavior occurs in the case of repeated real roots. For the third case, where the roots are complex, the free response is oscillatory, with a decaying amplitude if the model is stable. The rate of decay is given by the time constant, which is the negative reciprocal of the real part of the root. The oscillation frequency is given by the imaginary part. When the model's order is greater than three, it is possible to have repeated complex pairs, and the response generated by these roots is similar to that just described.

(3.4) The characteristic roots provide useful indicators of the response pattern; namely, stability, time constant, damping ratio, and oscillation frequency. A linear model is stable if and only if all of its roots lie in the left-half plane; that is, if they are negative or have negative real parts. If one root or complex root pair is far to the right of the others, it can be used to find the dominant response pattern, which will be that of a first- or second-order system.

(3.5) The step response of the stable second-order model (3.5-1) can be categorized according to the three root cases. For two distinct roots ($\zeta > 1$), the response does not overshoot its final value. For two real repeated roots ($\zeta = 1$), the response is similar to that for $\zeta > 1$. For complex roots ($\zeta < 1$), the response overshoots and oscillates about the final value. Several performance specifications have been defined with reference to the second-order step response, but they also apply to the step response of any system (Figure 3.13). These include the maximum percent overshoot, peak time, rise time, delay time, steady-state error, and settling time.

(3.6) The ramp response is a good description of a system's start-up behavior. The time constant is a measure of the difference between the input and output, as well as how long it takes for the steady-state lag to become established. The frequency response of a system describes its steady-state behavior resulting from periodic inputs. Later, we will see that it is also useful in predicting transient behavior and stability properties. A sinusoidal input applied to a linear system produces a steady-state sinusoidal output of the same frequency but with a different amplitude and a phase shift, which are functions of the input frequency.

(3.7) Once we understand the behavior generated by each of the three root cases, the superposition principle will allow us to apply the results to a linear system of any order. For example, the response of a sixth-order system with a complex conjugate pair of roots, two real repeated roots, and two real distinct roots can be analyzed by determining that part of the response resulting from each root type. The linear combination of each response pattern will give the total system response. For an approximate analysis, the dominant-root concept can be used to reduce the system to an equivalent first- or second-order system.

(3.8) Most control systems can be designed using a model of third order or less. For these models, the Routh–Hurwitz criterion provides a quick test of the system's stability in terms of the coefficients of the characteristic polynomial. Similar tests exist for higher order models, and are given in Appendix F.

(3.9) Most of the preceding does not apply to nonlinear models. The solution form of nonlinear models can change drastically with the initial conditions or the magnitude of the input. If a nonlinear equation cannot be solved in closed form, numerical methods or reduction to an approximate linear form must be used. Linearization by Taylor series produces an approximate model in terms of the deviations of the state variables and inputs from their reference values. It is useful for stability analysis, when an approximate closed-form solution is needed, and for design of feedback control systems (whose performance reinforces the linearization's accuracy).

In this chapter we have found the responses of first and second-order models for a variety of conditions, and these are summarized in the chapter's tables. The following chart is a guide to these tables.

	First-Order	**Second-Order**
Free response	Table 3.1	Table 3.3
Stability	Table 3.9	Table 3.9
State variable form	—	Table 3.2
Step response	Table 3.1	Table 3.5
Useful parameters	—	Table 3.6
Ramp response	Table 3.8	Table 3.7
Sine response	Table 3.8	See Section 4.5

REFERENCES

1. Electrocraft Corporation, *DC Motors, Speed Controls, Servo Systems*, 5th ed., Hopkins, Minn., 1980.
2. W. J. Palm III, *Modeling, Analysis, and Control of Dynamic Systems*, John Wiley & Sons, New York, 1983.

3. Y. Takahashi, M. Rabins, and D. Auslander, *Control*, Addison-Wesley, Reading, Mass., 1970.

PROBLEMS

3.1 Given a set of measurements $x(t_i)$, t_i, we can select the two parameters a and b to obtain a straight line $x(t) = at + b$ that best fits the data according to the following least-squares criterion. We choose a and b to minimize J, the sum of the squares of the differences between the predicted and the measured values; that is,

$$J = \sum_{i=1}^{n} [at_i + b - x(t_i)]^2$$

where n is the number of measurements. The criterion J obviously can never be negative, and its minimum occurs when $\partial J/\partial a = \partial J/\partial b = 0$.

(a) Obtain the expressions for the values of a and b that minimize J.

(b) Find the straight line $x(t) = at + b$ that best fits the following data in the least-squares sense.

t_i	$x(t_i)$
0	0.1
1	1.8
2	3.9
3	6.1

3.2 **(a)** Apply the results from Problem 3.1(a) to find the parameters r and $y(0)$ that give the best fit for the free response $y(t) = y(0)e^{rt}$ of a first-order linear model.

(b) The liquid height $y(t)$ in a certain tank was measured at several times with no inflow applied. The data are given in the following table. For a linear outlet resistance, the model is $\dot{y} = -gy/RA$. Assume the measurements have some error, and estimate the value of the resistance R with a least-squares approach. The tank area A is 3 ft^2.

(c) Now suppose that $y(0)$ is known to be exactly 10.1 ft. How does this change the results in (b)?

t_i (sec)	$y(t_i)$ (ft)
0	10.1
500	7.1
1000	5.0
1500	3.8
2000	2.6

3.3 A tank has the model $RA\dot{h} = -h + Rq_1$, with $A = 3\ \text{ft}^2$ and $R = 700\ \text{sec/ft}^2$. Suppose that the height h is initially 5 ft when the inflow rate q_1 is suddenly turned on at a constant value of 0.05 ft^3/sec. Find $h(t)$.

3.4 **(a)** A motor torque increases with time until it reaches a constant value after 0.1 sec. When this torque is applied to a load with inertia $I = 0.007$ lb-ft-sec^2 and damping $c = 0.001$ lb-ft-sec/rad, is a step function an adequate representation of the torque input? Explain.

(b) If at $t = 0.2$, the torque is suddenly removed, can it be modeled as an impulse? Explain.

3.5 Consider the model

$$2\dot{y} + y = 10v(t)$$

(a) If $y(0) = 0$ and $v(t) = 1$ (a unit step), what is the steady-state value of the output y_{ss}? How long does it take before 98% of the difference between $y(0)$ and y_{ss} is eliminated?

(b) Repeat (a) with $y(0) = 5$ and $v(t) = 1$.

(c) Repeat (a) with $y(0) = 0$ and $v(t) = 20$.

3.6 Find the free response and the step response of the following models. If the solution is oscillatory, it should be expressed in the form of a sine function with a phase shift.

(a) $\ddot{x} + 4\dot{x} + 8x = 2,\ x(0) = 0,\ \dot{x}(0) = 1$

(b) $\ddot{x} + 8\dot{x} + 12x = 2,\ x(0) = 0,\ \dot{x}(0) = 1$

(c) $\ddot{x} + 4\dot{x} + 4x = 2,\ x(0) = 0,\ \dot{x}(0) = 1$

3.7 Put the models given in Problem 3.6 into state-variable form.

3.8 Convert the following models into reduced form in terms of the variable x_1. The input is f.

(a) $\dot{x}_1 = x_2$

$\dot{x}_2 = -3x_1 - 5x_2 + f$

(b) $\dot{x}_1 = -x_1 + x_2$

$\dot{x}_2 = -3x_1 - 5x_2 + f$

3.9 Determine ζ, ω_n, τ, and ω_d for the dominant root in each of the following sets of characteristic roots.

(a) $s = -2,\ -3 \pm i$

(b) $s = -3,\ -2 \pm 2i$

3.10 Find ζ, ω_n, ω_d, and τ for the models given in Problem 3.6.

3.11 Given the model

$$\ddot{x} - (r + 2)\dot{x} + 2r + 5 = 0$$

(a) Find the values of the parameter r for which the system is

(1) Stable.

(2) Neutrally stable.

(3) Unstable.

(b) For the stable case, for what values of r is the system
  (1) Underdamped?
  (2) Overdamped?

3.12 Write a computer or calculator program to obtain the exact roots for the following:

(a) Quadratic polynomial.
(b) Cubic polynomial (see a mathematical handbook for the solution formulas).

3.13 For the model given in Problem 3.6(a), compute the maximum percent overshoot, the actual maximum overshoot, the peak time, the 100% rise time, the delay time, and the 2% settling time.

3.14 A certain system is described by the model

$$\ddot{x} + c\dot{x} + 4x = f$$

Set the value of the damping constant c so that both of the following specifications are satisfied. Give priority to the overshoot specification. If both cannot be satisfied, state the reason.

1. Maximum percent overshoot as small as possible and no greater than 20%.
2. 100% rise time as small as possible and no greater than 3.

3.15 Repeat Problem 3.14 for the model

$$9\ddot{x} + c\dot{x} + 4x = 0$$

3.16 Find the forced response of the following models:

(a) $\dot{y} + 2y = e^{-t}$
(b) $\dot{y} + 2y = t^2$

3.17 Find the steady-state part of the ramp response of the following model where $f(t) = 6t$. At steady-state, what is the difference between $f(t)$ and $x(t)$? Approximately how long will it take to reach steady-state?

$$\ddot{x} + 8\dot{x} + x = f$$

3.18 A thermometer is immersed in a liquid of temperature $v(t)$ °F. The temperature of the thermometer fluid is $y(t)$ °F. Time t is in seconds. The thermal time constant of the thermometer wall is 100 sec, so that

$$100\dot{y} + y = v(t)$$

Suppose that $v(t)$ varies sinusoidally with an amplitude of 10°F and a frequency of 0.5 cycles per min.

(a) Find the steady-state amplitude of y.
(b) Find the time lag in seconds between the occurrence of a peak in v and a peak in y, at steady state.

3.19 For the following model:

$$\dddot{x} + 22\ddot{x} + 131\dot{x} + 110x = f(t)$$

use the fact that

$$s^3 + 22s^2 + 131s + 110 = (s+1)(s+10)(s+11)$$

to

(a) Find the unit step response.
(b) Compare the answer in (a) with the results predicted by the dominant-root approximation.

3.20 Use the Routh–Hurwitz criterion to determine the number of roots that have positive real parts.

(a) $s^3 + 2s^2 + 5s + 24 = 0$
(b) $s^3 + 2s^2 + 4s + 8 = 0$

3.21 Use the Routh–Hurwitz criterion to determine the allowable range of K so that the systems with the following characteristic equations will be stable.

(a) $s^3 + as^2 + Ks + b = 0$ (a, b known constants)
(b) $s^3 + as^2 + bs + K = 0$ (a, b known constants)
(c) $s^3 + 3s^2 + 3s + 1 + K = 0$

3.22 A certain system has the characteristic equation

$$s(Ts + 1) + K = 0$$

We desire that all the roots lie to the left of the line $s = -b$ to guarantee a time constant no larger than $\tau = 1/b$. Use the Routh–Hurwitz criterion to determine the value of K and T so that both roots meet this requirement.

3.23 Determine the range of K values that will give a time constant of less than 1/2 for the following characteristic equation.

$$s^3 + 10s^2 + 31s + K = 0$$

3.24 Sketch the phase-plane plot for the following model of a liquid level system with an orifice and no inflow:

$$\dot{h} = -\sqrt{h}$$

3.25 Find the equilibria of the following model and obtain a linearized model for each of the first two nonnegative equilibria. Determine their stability properties.

$$\dot{y} = \sin y$$

3.26 Obtain a linearized set of equations for the model developed in Problem 2.26. Assume the nominal values of y, h, w, and p are constants.

3.27 Obtain a linearized approximation for the bellows model (2.8-21).

3.28 In (a) and (b), develop a linearized approximation of the pendulum model (3.9-25) for the equilibrium values given.

(a) $\theta_e = \pi/3$
(b) $\theta_e = \pi/4$
(c) Compare the values of τ, ζ, and ω_d for the two linearized models found in (a) and (b).

3.29 Suppose that the two-tank system shown in Figure 2.21 has nonlinear resistances. Then the model will take the following form for particular coefficient values:

$$\dot{h}_1 = q - \sqrt{h_1 - h_2}$$
$$\dot{h}_2 = \sqrt{h_1 - h_2} - 0.5\sqrt{h_2}$$

Obtain a linearized model valid near the reference equilibrium that results when $q_e = 1$.

CHAPTER FOUR

Transfer Functions and System Diagrams

It is helpful at this point to review where we stand in our study of control systems. Chapter 1 provided an overview of the major concepts, while Chapter 2 covered the principles for developing differential equation models of common systems. Chapter 3 treated methods for solving the equations to predict the system's response. These methods are helpful for understanding the dominant response patterns for relatively low-order systems with simple input functions. But there are other methods that can improve our understanding of how to design control systems that consist of several interconnected elements. These methods are based on the transfer function concept that treats each element in terms of its input–output properties. The Laplace transform allows us to extend the transfer function concept to dynamic elements and enables us to use an algebraic description of the system. This, in turn, lets us portray the system structure graphically by means of simplified block diagrams. It has been said that graphics is the primary means of communication in the engineering profession, and we will see that block diagrams play a prominent role in control systems design. These concepts are introduced in Sections 4.1–4.4.

The transfer function is also helpful in producing a concise graphical description of a system's frequency response characteristics. This consists of plots of the phase shift and the logarithm of the magnitude ratio versus the logarithm of the input frequency. Construction of these plots from the transfer function and their application are covered in Sections 4.5–4.7.

Sections 4.8 and 4.9 present some useful methods for evaluating system response with the transfer function and Laplace transform. The *final value theorem* in Section 4.8 gives a quick method for predicting the steady-state response. Section 4.9 introduces some new types of input functions, along with methods for dealing with them. It also presents an approach for analyzing models somewhat more complicated than those we have seen so far.

Section 4.10 concludes the chapter with the introduction of an alternative form of system diagrams, the signal flow graph.

4.1 THE LAPLACE TRANSFORM AND SYSTEM RESPONSE

We will make extensive use of the Laplace transform in this and succeeding chapters. The advantage of the transform is that it converts linear differential equations into algebraic relations. With proper algebraic manipulation of the resulting quantities, the solution of the differential equation can be recovered in an orderly fashion by inverting the transformation process to obtain a function of time. Here, we introduce this technique as applied to some simple cases. Later in the chapter, we will extend the method to other cases.

In addition to providing a systematic solution procedure for differential equations, the Laplace transform allows us to develop a graphical representation of the system's dynamics in terms of block diagrams. Manipulation of such diagrams in Chapter 1 was limited to static elements. With the Laplace transform, the diagram concept can be extended to include dynamic elements with the use of the *transfer function* concept. These topics are introduced in Sections 4.2 and 4.3.

The Laplace Transform

Appendix B is a self-contained presentation of the basic theory of the Laplace transform. A more detailed treatment is given by Kreysig (Reference 1). Here, we review only those properties of the transform that are needed to deal with some simple applications.

The Laplace transform $\mathscr{L}[y(t)]$ of a function $y(t)$ is defined to be

$$\mathscr{L}[y(t)] = \int_0^\infty y(t)e^{-st}\,dt \tag{4.1-1}$$

The integration removes t as a variable, and the transform is thus a function of only the Laplace variable s, which may be a complex number. This integral with an infinite limit exists for most of the commonly encountered functions if suitable restrictions are placed on s. An alternative notation is the use of the uppercase symbol to represent the transform of the corresponding lowercase symbol; that is,

$$Y(s) = \mathscr{L}[y(t)] \tag{4.1-2}$$

The variable $y(t)$ is assumed to be zero for negative time $t < 0$. For example, the unit-step function $u_s(t)$ is such a function. If $y(t) = Mu_s(t)$, then its transform is

$$\mathscr{L}[y(t)] = \int_0^\infty Mu_s(t)e^{-st}\,dt = M\int_0^\infty e^{-st}\,dt = M\frac{e^{-st}}{s}\bigg|_0^\infty = \frac{M}{s}$$

where we have assumed that the real part of s is greater than zero, so that the limit of e^{-st} exists as $t \to \infty$. Similar considerations of the region of convergence of the integral apply for other functions of time. However, we need not concern ourselves with this here, since the transforms of all the common functions have been calculated and tabulated. Some other examples of the transform calculation are given in Appendix B. Table 4.1 is a short table of transforms; a more extensive table is given in Appendix B.

The transform of a derivative is of use. Applying integration by parts to the definition of the transform, we obtain

$$\begin{aligned}\mathscr{L}\left(\frac{dy}{dt}\right) &= \int_0^\infty \frac{dy}{dt}e^{-st}\,dt = y(t)e^{-st}\bigg|_0^\infty + s\int_0^\infty y(t)e^{-st}\,dt \\ &= s\mathscr{L}[y(t)] - y(0)\end{aligned} \tag{4.1-3}$$

We have just seen how a multiplicative constant can be factored out of the integral. Also, the integral of a sum equals the sum of the integrals. These facts point out the *linearity property* of the transform; namely, that

$$\mathscr{L}[af_1(t) + bf_2(t)] = a\mathscr{L}[f_1(t)] + b\mathscr{L}[f_2(t)] \tag{4.1-4}$$

TABLE 4.1 Laplace Transform Pairs

$f(t), t \geqslant 0$	$\mathscr{L}[f(t)] = F(s)$
1. $u_s(t)$ (unit step)	$\dfrac{1}{s}$
2. t	$\dfrac{1}{s^2}$
3. t^n $(n > -1)$	$\dfrac{n!}{s^{n+1}}$
4. e^{-at}	$\dfrac{1}{s+a}$
5. $\sin bt$	$\dfrac{b}{s^2+b^2}$
6. $\cos bt$	$\dfrac{s}{s^2+b^2}$
7. $\dfrac{df}{dt}$	$sF(s) - f(0)$
8. $\dfrac{d^2f}{dt^2}$	$s^2F(s) - sf(0) - \left.\dfrac{df}{dt}\right\|_{t=0}$
9. $\int_0^t f(t)\,dt$	$\dfrac{1}{s}F(s)$
10. $f(t) = y(t-D)$	$F(s) = e^{-sD}Y(s)$

Application to Differential Equations

The derivative and linearity properties can be used to solve the differential equation

$$\dot{y} = ry + bv \tag{4.1-5}$$

If we multiply both sides of (4.1-5) by exp $(-st)$ and then integrate over time from $t = 0$ to $t = \infty$, we obtain

$$\begin{aligned}\mathscr{L}(\dot{y}) &= \mathscr{L}(ry + bv)\\ &= r\mathscr{L}(y) + b\mathscr{L}(v)\end{aligned}$$

where the second equality results from the linearity property of the transform. Using (4.1-3) and the alternative transform notation, we obtain

$$sY(s) - y(0) = rY(s) + bV(s) \tag{4.1-6}$$

where $V(s)$ is the transform of v. This equation is an algebraic equation for $Y(s)$ in terms of $V(s)$ and $y(0)$. The solution is

$$Y(s) = \frac{y(0)}{s-r} + \frac{b}{s-r} V(s) \tag{4.1-7}$$

The *inverse Laplace transform* $\mathscr{L}^{-1}[Y(s)]$ is that time function $y(t)$ whose transform is $Y(s)$. The inverse operation is also linear, and when applied to (4.1-7), it gives

$$y(t) = \mathscr{L}^{-1}\left[\frac{y(0)}{s-r}\right] + \mathscr{L}^{-1}\left[\frac{b}{s-r} V(s)\right] \tag{4.1-8}$$

From Table 4.1, it is seen that

$$\mathscr{L}^{-1}\left[\frac{y(0)}{s-r}\right] = y(0)e^{rt} \tag{4.1-9}$$

which is the free response. The forced response must therefore be given by

$$\mathscr{L}^{-1}\left[\frac{b}{s-r} V(s)\right] \tag{4.1-10}$$

This cannot be evaluated until $V(s)$ is specified.

Example 4.1

Compute the step response of (4.1-5).

If v is a step of magnitude M, $V(s) = M/s$ from Table 4.1. For this case, the forced response is

$$\mathscr{L}^{-1}\left(\frac{b}{s-r}\frac{M}{s}\right)$$

This transform can be converted into a sum of simple transforms by a partial fraction expansion.

$$\frac{bM}{(s-r)s} = \frac{C_1}{s-r} + \frac{C_2}{s}$$

This is true only if

$$C_1 s + C_2(s-r) = bM$$

for arbitrary values of s. This implies that

$$C_1 + C_2 = 0$$
$$-rC_2 = bM$$

or

$$C_1 = -C_2 = \frac{bM}{r}$$

Thus, the forced response is

$$\frac{bM}{r}e^{rt} - \frac{bM}{r}$$

The addition of the forced and free responses gives the previous results (3.2-6) or (3.2-4) for $r = -1/\tau$.

The form of a partial fraction expansion of a function depends on the roots of the function's denominator. When there are only a few roots, cross multiplication by the lowest common denominator quickly produces a solution for the expansion's coefficients $C_1, C_2, \ldots$. For more complicated functions, the coefficients can be determined by the general-purpose formulas given in Appendix B.

The Second-Order Case

It is useful to see how the preceding concepts extend to the second-order model

$$m\ddot{x} + c\dot{x} + kx = f(t) \tag{4.1-11}$$

To transform this equation, we need a new property (8) from Table 4.1. In terms of x, this gives

$$\mathscr{L}[\ddot{x}] = s^2 X(s) - sx(0) - \dot{x}(0) \tag{4.1-12}$$

Transforming (4.1-11) gives

$$m[s^2 X(s) - sx(0) - \dot{x}(0)] + c[sX(s) - x(0)] + kX(s) = F(s)$$

or

$$(ms^2 + cs + k)X(s) = mx(0)s + m\dot{x}(0) + cx(0) + F(s) \tag{4.1-13}$$

The free response is

$$x(t) = \mathscr{L}^{-1}\left[\frac{mx(0)s + m\dot{x}(0) + cx(0)}{ms^2 + cs + k}\right] \tag{4.1-14}$$

The forced response is

$$x(t) = \mathscr{L}^{-1}\left[\frac{F(s)}{ms^2 + cs + k}\right] \tag{4.1-15}$$

In order to evaluate either response, we must expand the transforms within the square brackets into a series of transforms that appear in our table. To perform this expansion, we need to know the values of m, c, and k to determine whether the roots of $ms^2 + cs + k$ are (1) real and distinct, (2) real and repeated, or (3) complex, because the form of the expansion is different for each case. In addition, we also need to know $f(t)$, so that we can find $F(s)$ for (4.1-15). The free, step, and ramp responses were found in Chapter 3 with the trial solution method, so we will not dwell on these here. In Sections 4.9 and 4.10, we develop techniques for using (4.1-15) to deal with other types of input functions. For now, two examples will suffice to illustrate the approach.

Example 4.2

Find the free response of (4.1-11) for $m = 1$, $c = 5$, and $k = 4$. The initial conditions are $x(0) = 1$ and $\dot{x}(0) = 5$.

The characteristic roots are $s_1 = -1$ and $s_2 = -4$. From (4.1-14),

$$x(t) = \mathscr{L}^{-1}\left[\frac{s+10}{(s+1)(s+4)}\right]$$
$$= \mathscr{L}^{-1}\left(\frac{3}{s+1}\right) + \mathscr{L}^{-1}\left(\frac{-2}{s+4}\right)$$
$$= 3e^{-t} - 2e^{-4t}$$

Example 4.3

Find the unit-step response of (4.1-11) for $m = 2$, $c = 8$, and $k = 8$.

The roots are repeated: $s = -2, -2$. From (4.1-15), with $F(s) = 1/s$, we have

$$x(t) = \mathscr{L}^{-1}\left[\frac{1}{s(2s^2+8s+8)}\right]$$
$$= \mathscr{L}^{-1}\left[\frac{0.5}{s(s+2)^2}\right]$$
$$= \mathscr{L}^{-1}\left[\frac{-0.25}{(s+2)^2}\right] + \mathscr{L}^{-1}\left(\frac{-0.125}{s+2}\right) + \mathscr{L}^{-1}\left(\frac{0.125}{s}\right)$$
$$= -0.25te^{-2t} - 0.125e^{-2t} + 0.125$$

The General Case

If the coefficients of the following equation are constants, we can transform it.

$$a_n\frac{d^n x}{dt^n} + a_{n-1}\frac{d^{n-1}x}{dt^{n-1}} + \cdots + a_1\frac{dx}{dt} + a_0 x = f(t) \quad \textbf{(4.1-16)}$$

The result is

$$(a_n s^n + a_{n-1}s^{n-1} + \cdots + a_1 s + a_0)X(s) = I(s) + F(s) \quad \textbf{(4.1-17)}$$

where $I(s)$ is a polynomial in s whose coefficients depend on the a_i and the initial conditions $x(0)$, $\dot{x}(0)$, $\ddot{x}(0)$, ...

The free response is

$$x(t) = \mathscr{L}^{-1}\left[\frac{I(s)}{P(s)}\right] \quad \textbf{(4.1-18)}$$

and the forced response is

$$x(t) = \mathscr{L}^{-1}\left[\frac{F(s)}{P(s)}\right] \quad \textbf{(4.1-19)}$$

where the characteristic polynomial is

$$P(s) = a_n s^n + a_{n-1}s^{n-1} + \cdots + a_1 s + a_0 \quad \textbf{(4.1-20)}$$

Thus, the Laplace transform provides a systematic way of finding the solution for any constant-coefficient linear differential equation. But for high-order equations, the algebra required to compute the inverse transforms in (4.1-18) and (4.1-19) is prohibitive, and in practice, other methods are used for those cases. We will return to this subject in the next chapter.

4.2 TRANSFER FUNCTIONS

Equation (4.1-7) contains much information about the system's behavior, as we have seen. Because the free and forced responses add to produce the total response, it is frequently convenient to assume that the initial condition $y(0)$ is zero. This allows us to focus our attention on the effects of the inputs. When the input analysis is completed, any free response due to a nonzero initial condition can be added to the result. This is a direct consequence of the superposition property of linear models.

We will be deeply involved in the analysis of systems made up of several components. Each component can influence the others by generating inputs to them. For example, the input flow rate to a tank might be the outlet flow rate from another tank. We therefore seek to establish an "input–output" format for analyzing the effect of each component on the overall behavior of the system. If this format can be developed in terms of algebraic relations, it will be to our advantage, because the analysis task will be simplified. This algebraic framework is provided by (4.1-7).

Assume that $y(0) = 0$. Then (4.1-7) gives

$$\frac{Y(s)}{V(s)} = \frac{b}{s - r} = T(s) \tag{4.2-1}$$

Equation (4.2-1) describes the *transfer function* $T(s)$ of the system (4.1-5). The transfer function of a linear system is the ratio of the transform of the output to the transform of the input, with the initial conditions assumed to be zero. In other words, the transfer function is the ratio of the transforms of the forced response and the input.

Equation (4.1-13) provides similar information about the response of the second-order model (4.1-11). Its transfer function is seen to be

$$T(s) = \frac{X(s)}{F(s)} = \frac{1}{ms^2 + cs + k} \tag{4.2-2}$$

The transfer function for the nth-order model (4.1-16) is

$$T(s) = \frac{X(s)}{F(s)} = \frac{1}{P(s)} \tag{4.2-3}$$

or

$$T(s) = \frac{1}{a_n s^n + a_{n-1} s^{n-1} + \cdots + a_1 s + a_0} \tag{4.2-4}$$

since $\mathscr{L}(d^n x/dt^n) = s^n X(s)$ for zero initial conditions (see Table B.2, entry 4).

From this, we see that the denominator of the transfer function is the system's characteristic polynomial. Thus, if we are given only the transfer function, we could still make some assessment of the system's behavior, because the roots of the characteristic polynomial give us stability information, time constants, damping ratios, and oscillation frequencies (see Section 3.4). In the preceding examples, the numerator of the transfer function does not give any particularly useful information, but we will soon see examples where the numerator also contains information about the system's response.

Transfer Functions from the State-Variable Form

The differential equations in the preceding examples were all in reduced form (see Section 3.3). If the system model is in state-variable form, some algebra is required to obtain the transfer function. The general procedure is to transform the state equations using zero initial conditions (this gives the forced response, which is required for the transfer function). Then solve for the desired output variable in terms of the input. To do this, all the other variables must be eliminated algebraically. The following simple example illustrates this procedure.

Example 4.4

The state-variable form of the model for the mass–spring–damper system is given by (3.3-1) and (3.3-2), repeated here.

$$\dot{x} = v \tag{4.2-5}$$

$$m\dot{v} = f - cv - kx \tag{4.2-6}$$

Find the transfer function between the input f and the output x.

Transform the state equations with zero initial conditions to obtain

$$sX(s) = V(s) \tag{4.2-7}$$

$$msV(s) = F(s) - cV(s) - kX(s) \tag{4.2-8}$$

Eliminate $V(s)$ algebraically by substituting $V(s) = sX(s)$ into (4.2-8).

$$ms^2X(s) = F(s) - csX(s) - kX(s)$$

or

$$(ms^2 + cs + k)X(s) = F(s)$$

Solve for $X(s)/F(s)$ to obtain the transfer function given by (4.2-2).

Multiple Inputs

Because a transfer function defines a relationship between an input and an output, there can be more than one transfer function for a system if it has more than one input or more than one output. In fact, there is a transfer function for each input–output pair. To find a particular transfer function, set all inputs to zero except the one of interest, and solve the transformed equations for the desired output as before.

Example 4.5

The accumulator model (2.7-10) is of the form

$$a_3\dot{p} = a_1(p_1 - p) - a_2(p - p_2) \tag{4.2-9}$$

where the output is the accumulator pressure p and the inputs are the upstream and downstream pressures p_1 and p_2. Find the system's transfer function.

Transform (4.2-9) and collect terms to obtain

$$(a_3 s + a_1 + a_2)P(s) = a_1 P_1(s) + a_2 P_2(s) \tag{4.2-10}$$

To find the transfer function $P(s)/P_1(s)$, set $P_2(s) = 0$ and solve (4.2-10) to obtain

$$\frac{P(s)}{P_1(s)} = \frac{a_1}{a_3 s + a_1 + a_2} \tag{4.2-11}$$

Similarly, setting $P_1(s) = 0$, we obtain the second transfer function.

$$\frac{P(s)}{P_2(s)} = \frac{a_2}{a_3 s + a_1 + a_2} \tag{4.2-12}$$

Note that the transfer functions in Example 4.5 have the same denominator. All transfer functions of any given system will always have the same denominator, because it is the system's characteristic polynomial.

Multiple Outputs

A single input can generate more than one transfer function if we are interested in more than one variable as output. Each transfer function is found by algebraically eliminating the other variables, as shown in Example 4.6.

Example 4.6

Find the transfer functions for the speed ω and the current i_a of the dc motor system in Figure 2.19. The input is the voltage v.

The models are (2.5-3) and (2.5-4), repeated here.

$$v = i_a R + L\frac{di_a}{dt} + K_e \omega \tag{4.2-13}$$

$$I\frac{d\omega}{dt} = K_T i_a - c\omega \tag{4.2-14}$$

Transforming and collecting terms, we obtain

$$(Ls + R)I_a(s) + K_e \Omega(s) = V(s) \tag{4.2-15}$$

$$-K_T I_a(s) + (Is + c)\Omega(s) = 0 \tag{4.2-16}$$

These equations can be solved for $I_a(s)$ and $\Omega(s)$ by standard algebraic means. Cramer's determinant for this system is

$$P(s) = \begin{vmatrix} (Ls + R) & K_e \\ -K_T & (Is + c) \end{vmatrix}$$

$$= (Ls + R)(Is + c) + K_e K_T$$

$$= LIs^2 + (RI + cL)s + Rc + K_e K_T \tag{4.2-17}$$

The solutions for $I_a(s)$ and $\Omega(s)$ are

$$I_a(s) = \frac{\begin{vmatrix} V(s) & K_e \\ 0 & (Is+c) \end{vmatrix}}{P(s)}$$

$$= \frac{Is+c}{P(s)} V(s) \tag{4.2-18}$$

$$\Omega(s) = \frac{\begin{vmatrix} (Ls+R) & V(s) \\ -K_T & 0 \end{vmatrix}}{P(s)}$$

$$= \frac{K_T}{P(s)} V(s) \tag{4.2-19}$$

The two transfer functions are the coefficients of $V(s)$ in (4.2-18) and (4.2-19). Note that Cramer's determinant is the characteristic polynomial of the system.

Recovering the Differential Equation Model

There will be applications where we are given a component's transfer function, but the differential equation model is required for analysis. The equation is easily recovered from the transfer function by cross multiplying by the denominators, using the following identity, which is valid for zero initial conditions.

$$\mathscr{L}\left(\frac{d^n x}{dt^n}\right) = s^n X(s) \tag{4.2-20}$$

The general form for a transfer function $T(s)$ between an input $F(s)$ and an output $Y(s)$ is

$$\frac{Y(s)}{F(s)} = T(s) = \frac{N(s)}{D(s)} \tag{4.2-21}$$

where $N(s)$ and $D(s)$ are the numerator and denominator polynomials of $T(s)$. This gives

$$D(s)Y(s) = N(s)F(s) \tag{4.2-22}$$

Equation (4.2-1) in this form is

$$(s-r)Y(s) = bV(s)$$

which implies

$$\dot{y} - ry = bv$$

To recover the differential equation for the current i_a in Example 4.6, use (4.2-17) and (4.2-18) to obtain

$$\frac{I_a(s)}{V(s)} = \frac{Is+c}{LIs^2 + (RI+cL)s + Rc + K_e K_T} \tag{4.2-23}$$

or

$$[LIs^2 + (RI + cL)s + Rc + K_e K_T] I_a(s) = (Is + c)V(s)$$

This gives

$$LI\frac{d^2 i_a}{dt^2} + (RI + cL)\frac{di_a}{dt} + (Rc + K_e K_T)i_a = I\frac{dv}{dt} + cv \tag{4.2-24}$$

Numerator Dynamics

The transfer function given by (4.2-23) is the first example thus far of a transfer function whose numerator is a polynomial in s. All the numerators previously seen were constants. The numerator's s term is seen from (4.2-24) to have the effect of computing the time derivative of the input v. Because differentiation is a dynamic operation, a transfer function having s terms in the numerator is said to have *numerator dynamics.*

Numerator dynamics can drastically alter the response of a system. When numerator dynamics are present, we cannot rely on only the characteristic polynomial (the denominator) to predict system response. This is illustrated by the response of the current i_a shown in Figure 3.15. With the parameter values given in Example 3.8, the maximum overshoot predicted for (4.2-24), neglecting the effect of numerator dynamics, is 0.14% using (3.5-16). The speed ω displays such a response [note that the transfer function for ω in (4.2-19) has no numerator dynamics]. But the current i_a from (3.5-26) and Figure 3.15 has a maximum overshoot of $(16.49 - 0.257)/0.257 = 6316\%$! The numerator dynamics can obviously have a great influence on the response. We will return to this topic in Section 4.9 to develop methods for predicting the effects of the numerator dynamics.

Table 4.2 summarizes the transfer functions of the differential equation models seen thus far.

4.3 BLOCK DIAGRAMS

It is apparent that the transfer function describes that part of the forced response that results from the characteristics of the system itself. For example, the denominator of $T(s)$ in (3.4-1) is the characteristic polynomial of the system. This fact is of great significance in analyzing and designing linear systems.

The remainder of the forced response results from characteristics of the input. The transform of the forced response is the product of the input transform and the transfer function; that is,

$$Y(s) = T(s)V(s) \tag{4.3-1}$$

This allows an algebraic and graphical representation of the cause-and-effect relationships in a given system. The algebraic representation of the system's equations permits easier manipulation for analysis and design purposes, and the graphical representation allows the analyst to see the interaction between the system's components. The principal graphical representation to be employed here is the block diagram, and it is a visual display of the algebraic relations.

In Chapter 1, we saw a simple block diagram consisting of a multiplier block,

TABLE 4.2 Transfer Functions for Common Models

1. Definition of the transfer function $T(s)$:

$y(t) = \text{output}$ $\qquad v(t) = \text{input}$

$$T(s) = \frac{\text{transform of the forced response}}{\text{transform of the input}} = \frac{Y(s)}{V(s)}$$

Time Domain Model	Transfer Function
2. Multiplication by a constant:	
$y(t) = Kv(t)$	$T(s) = K$
3. Integrator:	
$\dot{y} = v$	$T(s) = \dfrac{1}{s}$
4. Double integrator:	
$\ddot{y} = v$	$T(s) = \dfrac{1}{s^2}$
5. First-order model:	
$\tau\dot{y} + y = av$	$T(s) = \dfrac{a}{\tau s + 1}$
6. First-order model with numerator dynamics:	
$\tau\dot{y} + y = K(c\dot{v} + v)$	$T(s) = K\dfrac{cs + 1}{\tau s + 1}$
7. Second-order model, reduced form:	
$m\ddot{y} + c\dot{y} + ky = v$	$T(s) = \dfrac{1}{ms^2 + cs + k} = \dfrac{1}{m}\dfrac{1}{s^2 + 2\zeta\omega_n s + \omega_n^2}$
8. Second-order model, state variable form:	
$\dot{y}_1 = a_{11}y_1 + a_{12}y_2 + b_1 v,$	$T(s) = \dfrac{b_1 s + (a_{12}b_2 - a_{22}b_1)}{D(s)}$
$\dot{y}_2 = a_{21}y_1 + a_{22}y_2 + b_2 v,$	$T(s) = \dfrac{b_2 s + (a_{21}b_1 - a_{11}b_2)}{D(s)}$

$$D(s) = s^2 - (a_{11} + a_{22})s + (a_{11}a_{22} - a_{12}a_{21})$$

repeated in Figure 4.1*a*. The block represents the algebraic relation between the Laplace transforms of the input v and the output y. The arrows represent the transforms, and their direction indicates the cause-and-effect relationship between the input and output. The *summer* and *comparator* symbols are used to represent addition and subtraction. For these elements, as well as the multiplier, the diagramed variables need not be transforms, but may be functions of time, because the mathematical operations of addition, subtraction, and multiplication are the same in the time domain and the Laplace transform domain.

The concept of a transfer function is used to extend the notion of the multiplier block. If the input and output are related by a dynamic operation (integration or differentiation with respect to time), the time domain and transform domain relations are no longer identical. Only in the transform domain can such operations be represented by a multiplicative relation. The transfer function is the multiplication factor.

We must be careful to specify whether time domain or Laplace domain variables are used in the diagram. The reduction techniques to be developed for obtaining the overall system transfer function cannot be employed with time domain diagrams and can only be applied to block diagrams where all the variables are transformed variables.

The *takeoff point* in Figure 4.1*e* is used to connect a variable with another part of the diagram and does *not* imply subtraction. A variable is taken to have the same value along the entire length of a line until it is modified by one of the elements shown in (*a*)–(*d*) in Figure 4.1.

Type	Input-Output Relations: Time Domain	Input-Output Relations: Transform Domain	Symbol
(*a*) Multiplier	$y(t) = Kv(t)$	$Y(s) = KV(s)$	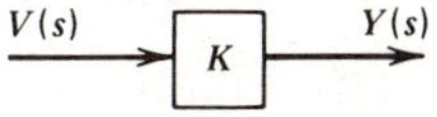 $V(s)$ → [K] → $Y(s)$
(*b*) General transfer function	$y(t) = \mathscr{L}^{-1}[T(s)V(s)]$	$Y(s) = T(s)V(s)$	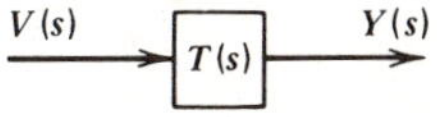 $V(s)$ → [$T(s)$] → $Y(s)$
(*c*) Summer	$y(t) = v_1(t) + v_2(t)$	$Y(s) = V_1(s) + V_2(s)$	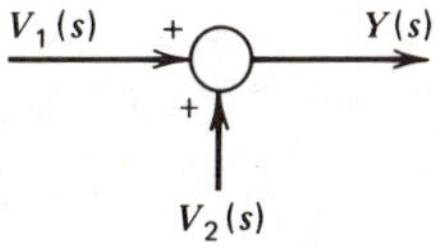 $V_1(s)$ (+), $V_2(s)$ (+) → ○ → $Y(s)$
(*d*) Comparator	$y(t) = v_1(t) - v_2(t)$	$Y(s) = V_1(s) - V_2(s)$	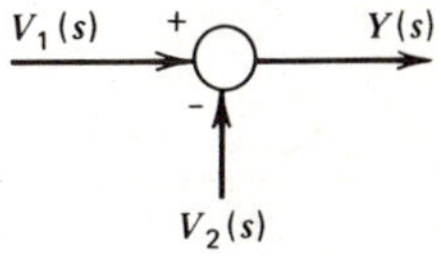 $V_1(s)$ (+), $V_2(s)$ (−) → ○ → $Y(s)$
(*e*) Takeoff point	$y(t) = v(t)$	$Y(s) = V(s)$	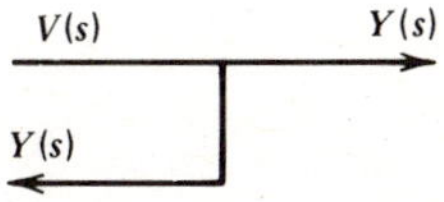 $V(s)$ → $Y(s)$; $Y(s)$

FIGURE 4.1 Basic block diagram elements.

Constructing a System Diagram

These concepts are more easily understood with the aid of an example. Equation (4.1-5) when transformed, gives (4.1-6), which for $y(0) = 0$ can be rearranged as

$$Y(s) = \frac{1}{s}[rY(s) + bV(s)]$$

Thus, $Y(s)$ can be treated as the output of a block with a transfer function $1/s$, with the input to the block being the terms within the brackets. Figure 4.2 shows the step-by-step decomposition to obtain the final diagram. We now have the results of step (*a*). At this point, it may seem strange to have $Y(s)$ in both the input and output. This is resolved by the use of a takeoff point, a multiplier, and a summer to create a *feedback* loop in step (*b*).

In step (*c*), the final form is obtained by using a multiplier. Sometimes it is preferable to add or subtract more than two variables at a summer or comparator. This is permissible if a convention is established for assigning the (+) and (−) signs to the proper variables. *The convention used here is that the sign should appear clockwise from the variable's arrowhead, if possible.* Because of this more general usage, the terms *summer* and *comparator* are sometimes used interchangeably.

Because block diagrams are intended primarily to display the system's structure, the common practice with these diagrams is to treat the initial conditions as zero and add in their effect (the free response) at the conclusion of the analysis. The emphasis of the diagram analysis is on the effect of the inputs on the system's behavior (the forced response).

Block Diagram Reduction

A system with two inputs is the accumulator model (4.2-9) in Example 4.5. A block diagram representation is shown in Figure 4.3*a*. However, if we had rearranged (4.2-9) as

$$a_3\dot{p} = a_1 p_1 + a_2 p_2 - (a_1 + a_2)p$$

we probably would have drawn the diagram shown in Figure 4.3*b*. Or, if (4.2-10) had

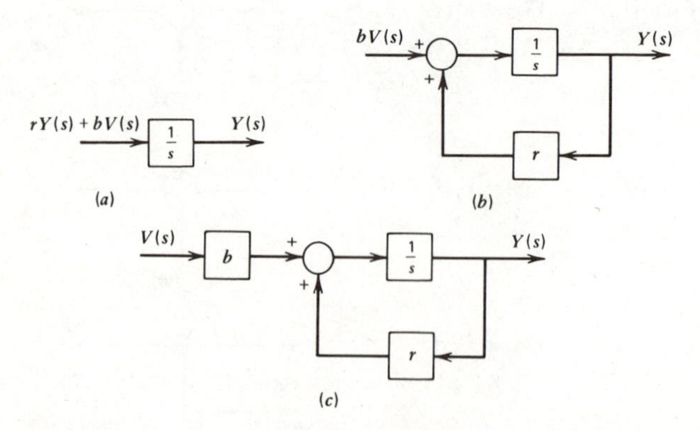

FIGURE 4.2 Construction of a block diagram. (*a*) Integrator representation. (*b*) Intermediate step. (*c*) Final result.

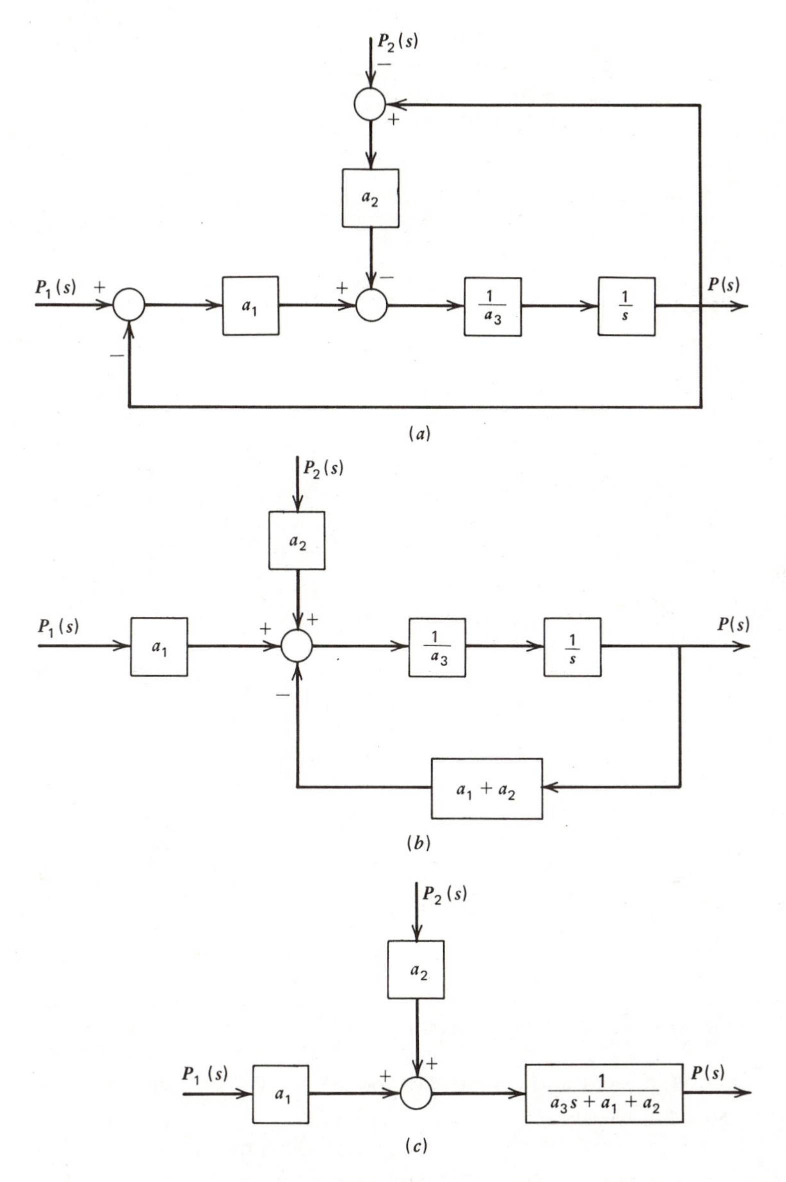

FIGURE 4.3 Several possible block diagrams for the accumulator model (4.2-9).

been used, the diagram shown in Figure 4.3*c* might have resulted. Since the equations are equivalent, the diagrams must be equivalent as well. We can reduce the diagrams in Figure 4.3*a* and 4.3*b* to that in Figure 4.3*c* by using the rules of block diagram algebra. The form of the diagram obtained depends on how the equation is arranged. The diagram manipulation rules simply require that the algebraic relations defined by the transformed equations be maintained. Any two diagram arrangements are equivalent if they correctly express the algebraic system

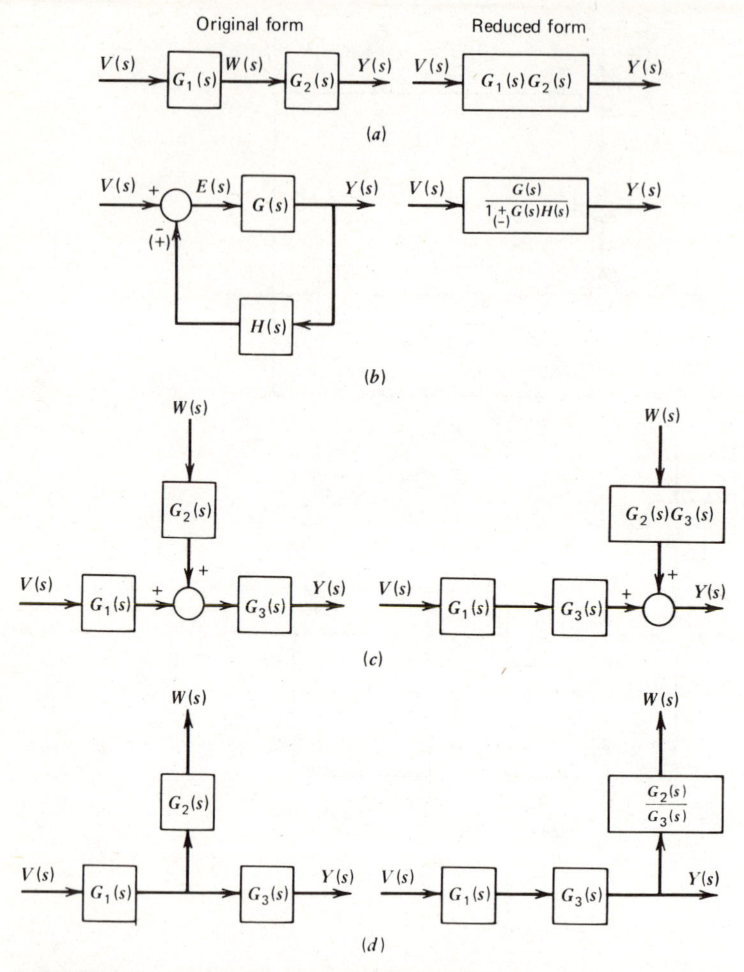

FIGURE 4.4 Some common reduction formulas. (*a*) Series or cascaded elements. (*b*) Feedback loop. (*c*) Relocated summer. (*d*) Relocated takeoff point.

equations and have the same input and output variables. However, one form may be more convenient for some purposes.

Figure 4.4 summarizes some of the common arrangements, along with an equivalent arrangement. The equivalence can easily be shown by working with the describing equations. In Figure 4.4*a* these are

$$W(s) = G_1(s)V(s)$$

$$Y(s) = G_2(s)W(s)$$

By eliminating the intermediate variable $W(s)$, we obtain

$$Y(s) = G_2(s)G_1(s)V(s)$$

The transfer function between v and y is $G_1(s)G_2(s)$. This reduction, called the *series* law, implies that we are not interested in analyzing the variable w per se.

It is important to note that the series (or *cascade*) representation of two elements implies that the elements are *nonloading*; that is, the output w of the first element is not changed by connecting it to the second element. In electrical terminology, this requires the input impedance of the second element to be infinite, which means that no power is being withdrawn from the first element. Of course, this situation is an idealization never achieved in practice. But the degree of loading should be determined before accepting the series representation. In electrical systems, nonloading is achieved by inserting an *isolation amplifier* with high impedance between the two circuits in question. In cases where significant loading exists, the overall transfer function for the combined elements must be obtained from their equations rather than from the series law.

A simple example is shown in Figure 4.5*a*. An amplifier with a gain K (the voltage amplification factor) is connected to a series RC circuit. We assume that the amplifier is designed so that it is not loaded by the circuit. This means that the amplifier is capable of supplying the current drawn by the circuit, and the amplifier is said to have a low-output impedance. The external power supply to the amplifier is not shown. The input voltage v_i is a signal voltage applied to the amplifier.

With the nonloading assumption, we can use the schematic diagram to draw the block diagram (Figure 4.5*b*). When combined, the series blocks give the transfer function between v_i and the output voltage y.

$$\frac{Y(s)}{V_i(s)} = \frac{K}{RCs+1}$$

Another important reduction formula is illustrated in Figure 4.4*b*. In order to obtain the overall transfer function of a system, the feedback loops must each be reduced to an equivalent single block. The formula can be derived by writing the equations for the original diagram.

$$E(s) = V(s) \underset{(+)}{-} H(s)Y(s)$$

$$Y(s) = G(s)E(s)$$

where the sign in parentheses refers to the situation shown by parentheses in the diagram. Eliminating $E(s)$ gives

$$Y(s) = G(s)[V(s) \underset{(+)}{-} H(s)Y(s)]$$

or

$$Y(s) = \frac{G(s)}{1 \underset{(-)}{+} G(s)H(s)} V(s) \qquad \textbf{(4.3-2)}$$

Note that if a comparator is used, $G(s)H(s)$ is added to 1 in the denominator. If a summer is used, $G(s)H(s)$ is subtracted from 1. Thus, the sign in the denominator is

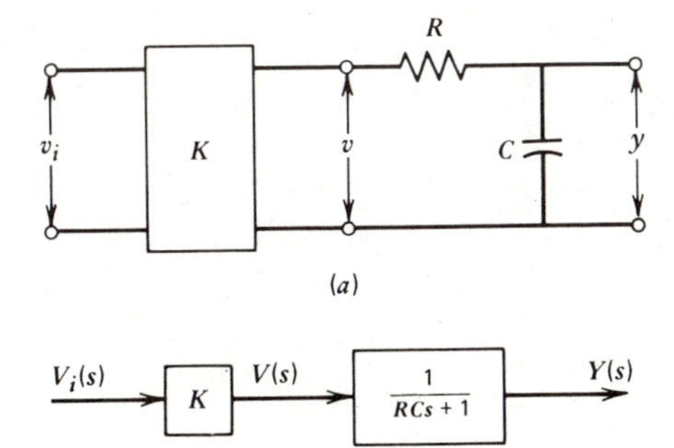

FIGURE 4.5 Series *RC* circuit with an amplifier. (*a*) Circuit diagram. (*b*) Block diagram.

always opposite to that in the feedback loop. The transfer function between y and v given in (4.3-2) is called the *closed-loop* transfer function, because it incorporates the loop effects.

Figure 4.4*c* provides some more insight into the process of modifying diagrams. Follow the "flow" of the variable $V(s)$ through the original diagram. By the time its effect is felt on the output $Y(s)$, $V(s)$ has been multiplied by $G_1(s)$ and $G_2(s)$, and its sign has been preserved in passing through the summer. In the equivalent diagram, exactly the same operations act on $V(s)$ before reaching the output, although in a different order. The same can be said of the flow of $W(s)$ through both diagrams. While this equivalence could have been shown from the system's equations, it seems easier to establish equivalence by tracing the variable's flow through the diagram. Figure 4.4*d* has two outputs, but the same technique can be applied in order to trace $V(s)$ through to each output.

Example 4.7

Consider the diagram shown in Figure 4.6*a*, and find the transfer function between y and v.

The first step in reducing the diagram to a single block is separating the feedback

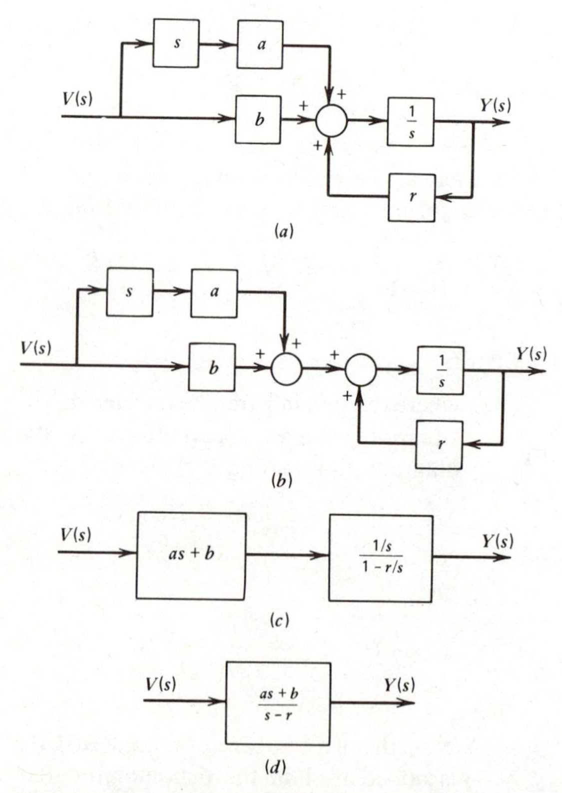

FIGURE 4.6 Diagram reduction example. (*a*) Original diagram. (*b*) Relocated summer. (*c*) Loop reduction. (*d*) Final result.

loop from the paths containing the multipliers with the factors a and b (Figure 4.6*b*). Next, reduce the loop to a single transfer function and do the same for the paths with the multipliers (Figure 4.6*c*). The last step is simply the combination of the two series elements (Figure 4.6*d*). It is standard practice to present transfer functions in the form of a numerator polynomial in s divided by a denominator polynomial. This is the normalized form given in Figure 4.6*d*.

Example 4.8

Apply the diagram reduction principles to reduce Figures 4.3*a* and 4.3*b* to 4.3*c*.

For Figure 4.3*b*, applying the loop reduction formula is all that is needed. For Figure 4.3*a*, one method involves moving a_1 and a_2 as shown in Figure 4.7*a*. Then combine the comparators as shown Figure 4.7*b*. Finally, move the comparator for $P_2(s)$ outside of the loop (Figure 4.7*c*). Note how the plus sign, denoted by ③ in Figure 4.7*c*, results from the two minus signs, denoted by ① and ② in Figure 4.7*a*. The diagram given in Figure 4.3*c* can be obtained from Figure 4.7*c* by applying the loop reduction formula to the upper and the lower loops.

System Structure and Block Diagrams

In general, diagrams in the time domain can be converted into s-domain diagrams by replacing the integrator blocks with blocks containing $1/s$. Of course, all the variables in the s-domain diagram must represent Laplace-transformed quantities. The advantage of s-domain diagrams is that they can be reduced to find the transfer functions. But the main purpose of block diagrams is to provide a visual representation of the system's structure, as it affects the response characteristics. The diagrams are especially useful for understanding the feedback processes that exist in the system.

4.4 IMPEDANCE AND TRANSFER FUNCTIONS

The transfer function is a useful vehicle for discussing the concept of impedance in more detail. In general, any transfer function between a rate variable as input and an effort variable as output is considered an impedance. In electrical systems, of course, this is the ratio of a voltage transform to a current transform and thus implies a current source. If the voltage output is measured at the terminals to which the driving current is applied, the impedance so obtained is the *driving-point* or *input impedance*. If the voltage is measured at another place in the circuit, the impedance obtained is a *transfer impedance* (because the effect of the input current has been transferred to another point). Impedance may be thought of as a generalized resistance, because it *impedes* current flow (ac or dc) just as a resistor *resists* dc current. Sometimes the term *admittance* is used. This is the reciprocal of impedance, and it is an indication of to what extent a circuit *admits* current flow.

The transfer functions (impedances) of individual elements can be combined with series and parallel laws to find the transfer function or impedance at any point in the

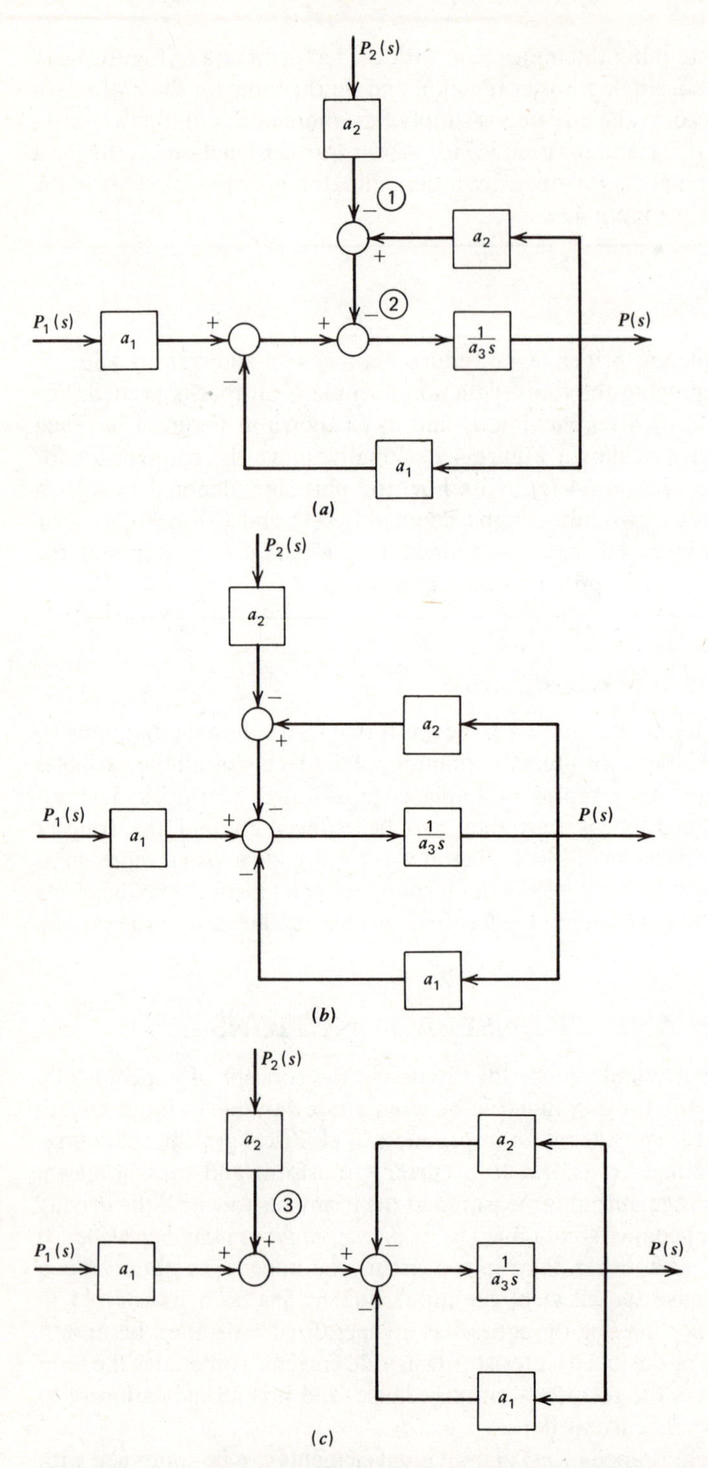

FIGURE 4.7 Reduction of the accumulator block diagram to a simpler form.

system. The impedances of three common electrical elements are found as follows. For the capacitor,

$$v(t) = \frac{1}{C}\int_0^t i\,dt$$

or

$$V(s) = \frac{1}{Cs}I(s) \tag{4.4-1}$$

and the impedance is

$$T(s) = \frac{1}{Cs}$$

For the inductor,

$$v(t) = L\frac{di}{dt}$$

or

$$V(s) = LsI(s) \tag{4.4-2}$$

and its impedance is $T(s) = Ls$. Of course, the impedance of a resistor is $T(s) = R$.

Example 4.9

The *lead compensator* network shown in Figure 4.8a is widely used to improve the performance of instruments and controllers. We wish to find the transfer function between the input voltage v_i and the output voltage v.

Two impedances are in series if they have the same rate variable. If so, the total impedance is the sum of the individual impedances. If the impedances have the same effort difference across them, they are in parallel, and their impedances combine by the reciprocal rule

$$\frac{1}{T} = \frac{1}{T_1} + \frac{1}{T_2}$$

where T is the total equivalent impedance. These two laws are extensions to the dynamic case of the laws governing series and parallel static resistance elements.

It can be seen that R_1 and C are in parallel. Thus, their equivalent impedance is found from

$$\frac{1}{Z(s)} = \frac{1}{1/Cs} + \frac{1}{R_1}$$

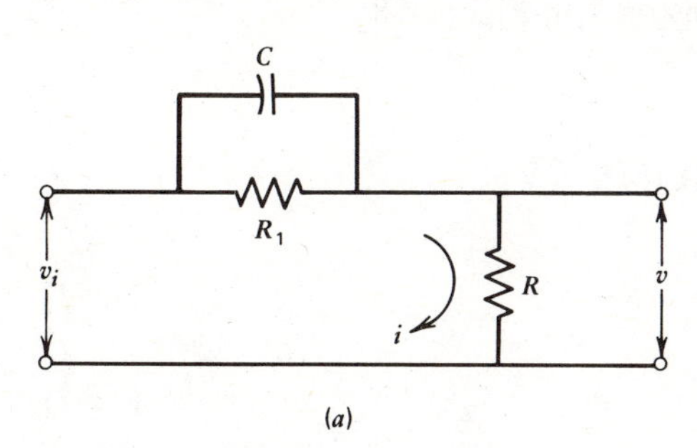

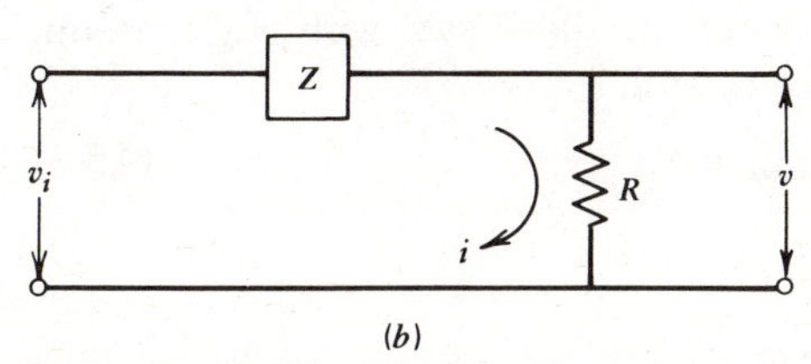

FIGURE 4.8 Lead compensator. (*a*) Circuit. (*b*) Equivalent impedance representation.

or

$$Z(s) = \frac{R_1}{R_1 C s + 1}$$

The equivalent circuit is shown in Figure 4.8*b*. From this, it can be seen that Z and R are in series. Thus,

$$\frac{V_i(s)}{I(s)} = Z(s) + R$$

and

$$V(s) = RI(s)$$

Eliminating $I(s)$ from the last two relations yields the desired transfer function.

$$V_i(s) = \frac{V(s)}{R}[Z(s) + R]$$

or

$$T(s) = \frac{V(s)}{V_i(s)} = \frac{1}{\frac{Z(s)}{R} + 1}$$

$$= \frac{RR_1 Cs + R}{RR_1 Cs + R_1 + R} \tag{4.4-3}$$

The network is seen to be a first-order system with numerator dynamics.

Impedances and transfer functions in general can be used to develop a systematic algebraic procedure for determining network transfer functions for most system types when the network is a ladderlike arrangement of series impedances alternating with impedances *shunted* to ground like the resistor R in Figure 4.8.

4.5 FREQUENCY RESPONSE AND TRANSFER FUNCTIONS

Consider the linear model

$$\dot{y} = ry + bv \tag{4.5-1}$$

where

$$v(t) = A \sin \omega t \tag{4.5-2}$$

with constant amplitude A and frequency ω. For the stable case with b a constant, the steady-state solution found in Section 3.6 is, with $r = -1/\tau$,

$$y(t) = B \sin(\omega t + \phi) \tag{4.5-3}$$

where

$$B = \frac{\tau|b|A}{\sqrt{1 + \omega^2\tau^2}} \tag{4.5-4}$$

$$\phi = \tan^{-1}(-\omega\tau) \tag{4.5-5}$$

with $-\pi/2 \leqslant \phi \leqslant 0$ for $b > 0$ and $\pi/2 \leqslant \phi \leqslant \pi$ for $b < 0$.

The Frequency Transfer Function

The solution given by (4.5-3)–(4.5-5) points out a useful application of the transfer function. Assume that the system is stable and write the transfer function $T(s) = b/(s - r)$ in terms of $\tau = -1/r$. This is

$$T(s) = \frac{\tau b}{\tau s + 1} \tag{4.5-6}$$

The transform of a sinusoidal steady-state output of frequency ω has the denominator $s^2 + \omega^2$ (see entry 5, Table 4.1). The roots are $s = \pm i\omega$; that is, the values of s corresponding to sinusoidal oscillation of frequency ω are $s = \pm i\omega$. With this as a guide, let us see what the transfer function reveals when s is replaced with $i\omega$.

$$T(i\omega) = \frac{\tau b}{\tau\omega i + 1} \tag{4.5-7}$$

This complex number is a function of only ω for fixed τ and b. It can be expressed in a more convenient form by multiplying through by the complex conjugate of the denominator.

$$T(i\omega) = \frac{\tau b}{\tau\omega i + 1}\,\frac{1 - \tau\omega i}{1 - \tau\omega i}$$

$$= \frac{\tau b(1 - \tau\omega i)}{1 + \tau^2\omega^2} \tag{4.5-8}$$

$T(i\omega)$ can be thought of as a *vector* in the complex plane where the vector's components are the real and imaginary parts of $T(i\omega)$. Figure 4.9 shows the case for $b > 0$. The vector rotates, and its length changes as ω varies from 0 to ∞. The locus of the vector's tip is shown, and it is a semicircle in the fourth quadrant for positive ω.

The transfer function with s replaced by $i\omega$ is called the *frequency transfer function*, and its plot in vector form is the *polar plot*. The vector's magnitude M and angle ϕ relative to the positive real axis are easily found from trigonometry to be

$$M(\omega) = \frac{\tau|b|}{\sqrt{1 + \tau^2\omega^2}} \tag{4.5-9}$$

$$\phi(\omega) = \tan^{-1}(-\omega\tau) \tag{4.5-10}$$

where M and ϕ are explicitly written as functions of ω. Comparing these ex-

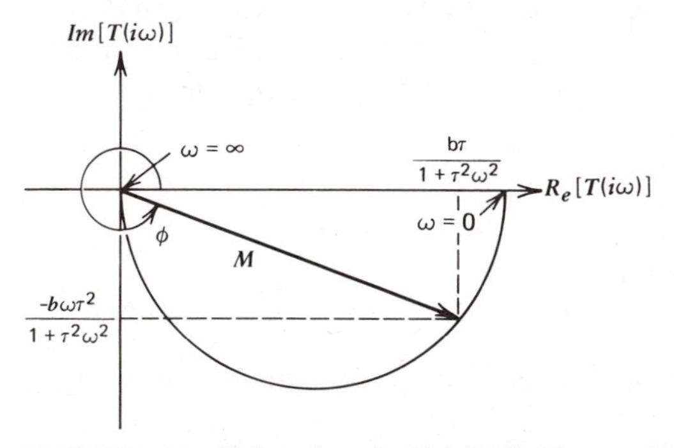

FIGURE 4.9 Polar plot of a first-order linear system.

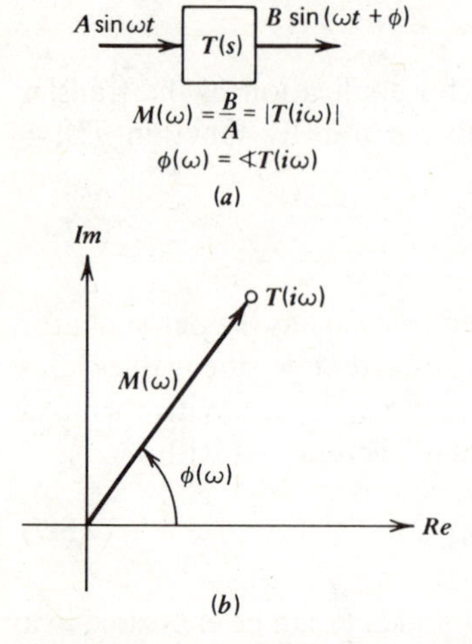

FIGURE 4.10 Steady-state sinusoidal response of a linear system. (*a*) Input–output relation. (*b*) Polar representation of the frequency transfer function *T*(*i*ω).

pressions for $M(\omega)$ and $\phi(\omega)$ with those of B and ϕ in (4.5-4) and (4.5-5) reveals the following useful fact (see Figure 4.10). The magnitude M of the frequency transfer function $T(i\omega)$ is the ratio of the sinusoidal steady-state output amplitude to the sinusoidal input amplitude. The phase shift of the output relative to the input is ϕ, the angle of $T(i\omega)$ [the *argument* of $T(i\omega)$]. Both M and ϕ are functions of the input frequency ω. In an alternative notation, this means that

$$M(\omega) = |T(i\omega)| \quad \textbf{(4.5-11)}$$

$$\phi(\omega) = \measuredangle\, T(i\omega) \quad \textbf{(4.5-12)}$$

$$B = AM(\omega) \quad \textbf{(4.5-13)}$$

where $|T(i\omega)|$ and $\measuredangle\, T(i\omega)$ denote the magnitude and angle of the complex number $T(i\omega)$. Because of (4.5-13), M is sometimes called the *amplitude ratio*.

The Logarithmic Plots

Inspecting (4.5-9 reveals that the steady-state amplitude of the output decreases as the frequency of the input increases for $\tau \neq 0$. The larger τ is, the faster the output amplitude decreases with frequency. At a high frequency, the system's "inertia" prevents it from closely following the input. The larger τ is, the more sluggish the system response is. This also produces an increasing phase lag as ω increases.

The curves of M and ϕ versus ω are difficult to sketch accurately. For this reason, logarithmic plots are usually employed. Here, the amplitude ratio M is specified in *decibel* units, denoted db. The relationship between a number M and its decibel equivalent m is

$$m = 20 \log M \quad \text{db} \quad \textbf{(4.5-14)}$$

where the logarithm is to the base 10. For example, the number 10 corresponds to 20 db; the number 1 corresponds to 0 db; numbers less than 1 have negative decibel values. It is common practice to plot $m(\omega)$ in decibels versus $\log \omega$. Thus, if M is first converted to decibel units, semilog graph paper may be used. For easy reference, $\phi(\omega)$ is also plotted versus $\log \omega$.

Referring to (4.5-9), we have

$$\begin{aligned} m(\omega) &= 20 \log \frac{\tau|b|}{\sqrt{1+\tau^2\omega^2}} \\ &= 20 \log (\tau|b|) - 10 \log (1 + \tau^2\omega^2) \end{aligned} \quad \textbf{(4.5-15)}$$

To sketch the curve, we approximate $m(\omega)$ in three frequency ranges. For $\omega\tau \ll 1$, $1 + \tau^2\omega^2 \cong 1$, and

$$\begin{aligned} m(\omega) &\cong 20 \log (\tau|b|) - 10 \log 1 \\ &= 20 \log (\tau|b|) \end{aligned} \quad \textbf{(4.5-16)}$$

Thus, for $\omega \ll 1/\tau$, $m(\omega)$ is approximately constant. For $\omega\tau \gg 1$, $1 + \tau^2\omega^2 \cong \tau^2\omega^2$, and

$$\begin{aligned} m(\omega) &\cong 20 \log (\tau|b|) - 10 \log \tau^2\omega^2 \\ &= 20 \log (\tau|b|) - 20 \log \tau - 20 \log \omega \end{aligned} \tag{4.5-17}$$

This gives a straight line versus log ω. Its slope is -20 db/decade, where a *decade* is any 10:1 frequency range. At $\omega = 1/\tau$, this line gives $m(\omega) = 20 \log (\tau|b|)$. This is useful for plotting purposes but does not represent the true value of m at that point. For $\omega = 1/\tau$, (4.5-15) gives

$$\begin{aligned} m(\omega) &= 20 \log (\tau|b|) - 10 \log 2 \\ &= 20 \log (\tau|b|) - 3.01 \end{aligned} \tag{4.5-18}$$

Thus, at $\omega = 1/\tau$, $m(\omega)$ is 3.01 db below the low-frequency asymptote given by (4.5-16). The low-frequency and high-frequency asymptotes meet at $\omega = 1/\tau$, which is the *breakpoint* frequency. It is also called the *corner* frequency.

The curve of ϕ versus ω is constructed as follows. For $\omega \ll 1/\tau$, (4.5-10) gives

$$\phi(\omega) \cong \tan^{-1}(0) = 0°$$

For $\omega = 1/\tau$,

$$\phi(\omega) = \tan^{-1}(-1) = -45°$$

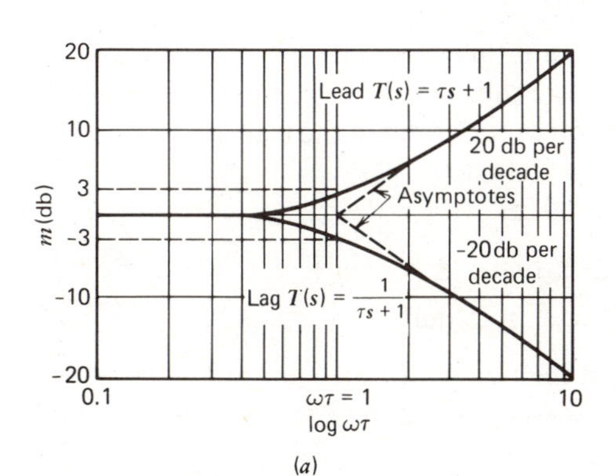

(a)

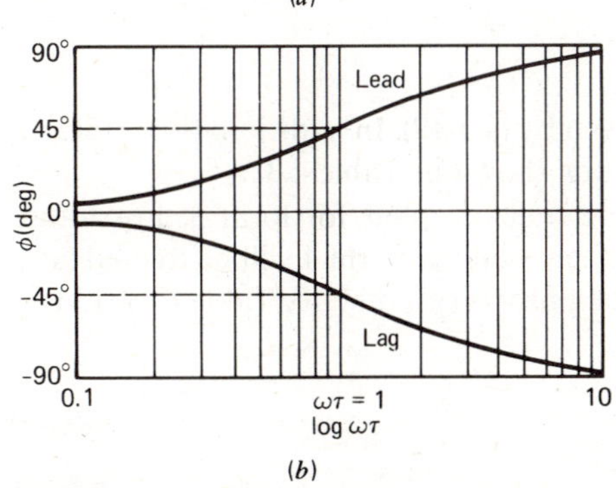

(b)

FIGURE 4.11 Logarithmic frequency response plots for two first-order transfer functions. (*a*) Magnitude ratio *m* in decibels. (*b*) Phase angle.

and for $\omega \gg 1/\tau$,

$$\phi(\omega) \cong \tan^{-1}(-\infty) = -90°$$

Using these facts, the curve is easily sketched.

The curves for $T(s) = 1/(\tau s + 1)$ are shown in Figure 4.11. This corresponds to $\tau b = 1$ in (4.5-6). If $\tau b \neq 1$, the scale for the m curve shown in the figure should be shifted by $20 \log (\tau|b|)$. The phase angle curve remains unchanged.

Also shown in the figure are the curves for $T(s) = \tau s + 1$. It is easy to show that these are the mirror images of the curves for $T(s) = 1/(\tau s + 1)$. We note that

$$\left|\frac{1}{1 + i\omega\tau}\right| = \frac{1}{|1 + i\omega\tau|}$$

Thus,

$$20 \log \left|\frac{1}{1 + i\omega\tau}\right| = -20 \log |1 + i\omega\tau|$$

Also,

$$\measuredangle(1 + i\omega\tau) = \tan^{-1} \omega\tau$$

This angle is the negative of that given in (4.5-10).

General Formulation

The previous results are for first-order models without numerator dynamics, but they can be easily extended to a stable linear system of any order. The general form of a transfer function is

$$T(s) = K\frac{N_1(s)N_2(s)\ldots}{D_1(s)D_2(s)\ldots} \tag{4.5-19}$$

where K is a constant real number. In general, if a complex number $T(i\omega)$ consists of products and ratios of complex factors, such that

$$T(i\omega) = K\frac{N_1(i\omega)N_2(i\omega)\ldots}{D_1(i\omega)D_2(i\omega)\ldots} \tag{4.5-20}$$

where K is a real constant, then from the properties of complex numbers

$$|T(i\omega)| = \frac{|K||N_1(i\omega)||N_2(i\omega)|\ldots}{|D_1(i\omega)||D_2(i\omega)|\ldots} \tag{4.5-21}$$

In decibel units, this implies that

$$\begin{aligned} m(\omega) = 20\log|T(i\omega)| &= 20\log|K| + 20\log|N_1(i\omega)| + 20\log|N_2(i\omega)| + \cdots \\ &\quad - 20\log|D_1(i\omega)| - 20\log|D_2(i\omega)| - \cdots \end{aligned} \tag{4.5-22}$$

That is, when expressed in logarithmic units, multiplicative factors in the numerator are summed, while those in the denominator are subtracted. We can use this principle graphically to add or subtract the contribution of each term in the transfer function to obtain the plot for the overall system transfer function.

Also, for the form (4.5-20), the phase angle is

$$\begin{aligned} \phi(\omega) = \measuredangle T(i\omega) &= \measuredangle K + \measuredangle N_1(i\omega) + \measuredangle N_2(i\omega) + \cdots \\ &\quad - \measuredangle D_1(i\omega) - \measuredangle D_2(i\omega) - \cdots \end{aligned} \tag{4.5-23}$$

The phase angles of multiplicative factors in the numerator are summed, while those in the denominator are subtracted. This allows us to build the composite phase angle plot from the plots for each factor.

Common Building Blocks

Most transfer functions occur in the form given by (4.5-19). In addition, the "building blocks," $N_j(s)$ and $D_j(s)$, usually take the forms shown in Table 4.3.

We have already obtained the frequency response plots for form 3. For future reference, we will now develop the plots for forms 1 and 2. Form 4 will be treated later in this section. Form 5 results from the shifting property (entry 10, Table 4.1), and its plots will be obtained in Chapter 9.

For

$$T(s) = K \tag{4.5-24}$$

we have

$$m = 20\log|K| \tag{4.5-25}$$

$$\phi = \measuredangle K = \begin{cases} 0^\circ & K > 0 \\ -180^\circ & K < 0 \end{cases} \tag{4.5-26}$$

TABLE 4.3 Factors Commonly Found in Transfer Functions of the Form:

$$T(s) = K\frac{N_1(s)N_2(s)\ldots}{D_1(s)D_2(s)\ldots}$$

Factor $N_j(s)$ or $D_j(s)$
1. Constant, K
2. s^n
3. $\tau s + 1$
4. $s^2 + 2\zeta\omega_n s + \omega_n^2 = \left[\left(\frac{s}{\omega_n}\right)^2 + \frac{2\zeta}{\omega_n}s + 1\right]\omega_n^2, \qquad \zeta < 1$
5. e^{-Ds}

The plots are shown in Figure 4.12.

For

$$T(s) = s^n \tag{4.5-27}$$

we have

$$m = 20 \log |i\omega|^n = 20n \log |i\omega| = 20n \log \omega \tag{4.5-28}$$

$$\phi = \measuredangle (i\omega)^n = n \measuredangle (i\omega) = n90° \tag{4.5-29}$$

The most common case is $n = 1$, and these plots are given in Figure 4.13.

Example 4.10

The presence of numerator dynamics in a first-order system can significantly alter the system's frequency response. Consider the series RC circuit, where the output voltage is now taken to be across the resistor (Figure 4.14). The two amplifiers serve to isolate the circuit from the loading effects of adjacent elements. The input impedance between

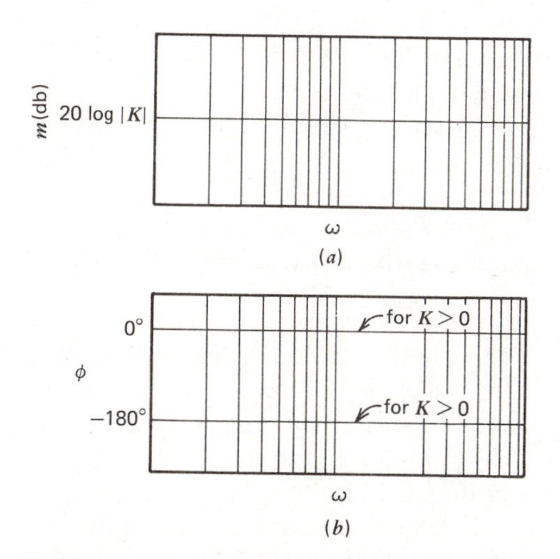

FIGURE 4.12 Frequency response plots for $T(s) = K$.

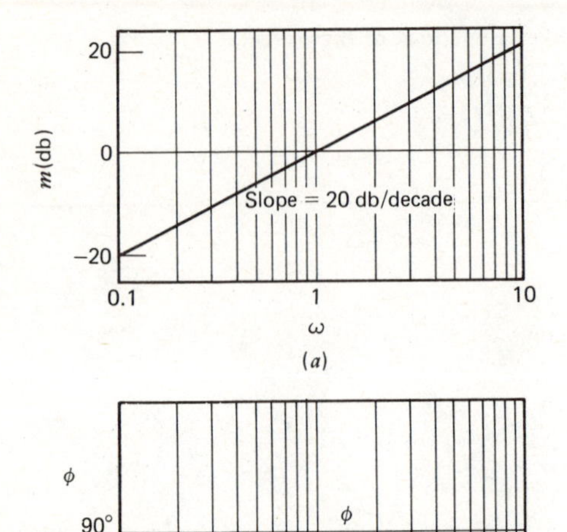

FIGURE 4.13 Frequency response plots for $T(s) = s$.

the voltage v and the current i is found from the series law.

$$\frac{V(s)}{I(s)} = R + \frac{1}{Cs}$$

Thus,

$$Y(s) = I(s)R = \frac{V(s)}{R + \dfrac{1}{Cs}} R$$

Determine the circuit's frequency response, and interpret its effect on the input.

The transfer function is

$$T(s) = \frac{Y(s)}{V(s)} = \frac{RCs}{RCs+1} \qquad \textbf{(4.5-30)}$$

Let $\tau = RC$ to obtain

$$T(s) = \tau \frac{s}{\tau s + 1}$$

This is of the form (4.5-19) with $K = \tau$, $N_1(s) = s$, and $D_1(s) = \tau s + 1$. Therefore, from (4.5-22),

$$m(\omega) = 20 \log |\tau| + 20 \log |i\omega| - 20 \log |\tau\omega i + 1| \qquad \textbf{(4.5-31)}$$

For the low-frequency asymptote, $\omega \ll 1/\tau$ and the term due to $\tau s + 1$ is negligible. Thus, the asymptote is described by

$$m(\omega) = 20 \log |\tau| + 20 \log |\omega|, \qquad \omega \ll \frac{1}{\tau} \qquad \textbf{(4.5-32)}$$

It can be sketched by noting that it has a slope of 20 db/decade and it passes through the point $m = 0$, $\omega = 1/\tau$. The asymptote is shown by the dotted line in Figure 4.15. At high frequencies ($\omega \gg 1/\tau$), the slope of -20 due to the term $\tau s + 1$ in the denominator cancels the slope of $+20$ due to the term s in the numerator. Therefore, at high frequencies, $m(\omega)$ has a slope of approximately zero and also $\sqrt{\tau^2\omega^2 + 1} \approx \tau\omega$. Thus,

$$\begin{aligned} m(\omega) &\approx 20 \log |\tau| + 20 \log |\omega| - 20 \log |\tau\omega| \\ &= 20 \log |\tau| + 20 \log |\omega| - 20 \log |\tau| - 20 \log |\omega| = 0 \end{aligned}$$

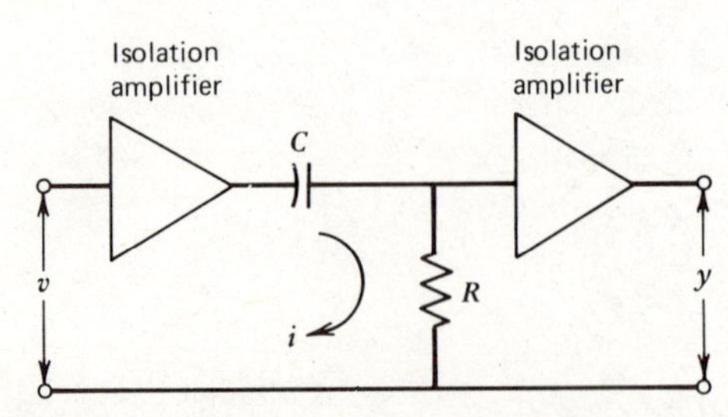

FIGURE 4.14 High-pass filter circuit.

for $\omega \gg 1/\tau$. At $\omega = 1/\tau$, the term $\tau s + 1$ contributes -3 db. The composite curve for $m(\omega)$ is obtained by "blending" the low-frequency and high-frequency asymptotes through this point, as shown in Figure 4.15*a*.

A similar technique can be used to sketch $\phi(\omega)$. From (4.5-23),

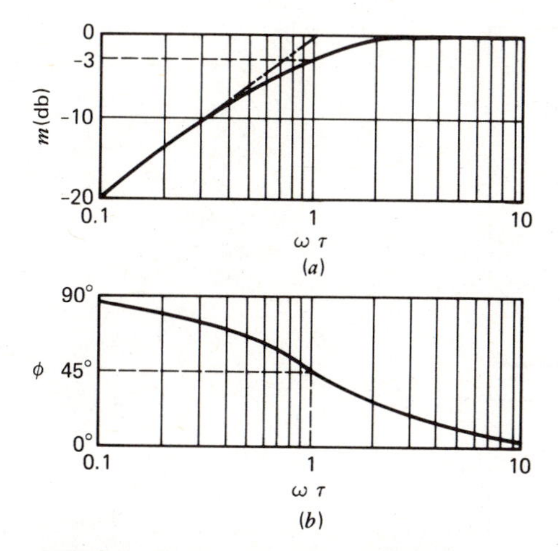

FIGURE 4.15 Frequency response of the high-pass filter. (*a*) Magnitude plot. (*b*) Phase plot.

$$\phi(\omega) = \measuredangle\, \tau + \measuredangle\, (i\omega) - \measuredangle\, (\tau\omega i + 1)$$
$$= 0° + 90° - \tan^{-1}(\omega\tau)$$

For $\omega \ll 1/\tau$, the low-frequency asymptote is

$$\phi(\omega) \approx 90° - \tan^{-1}(0) = 90°$$

For $\omega = 1/\tau$,

$$\phi(\omega) = 90° - \tan^{-1}(1) = 45°$$

For $\omega \gg 1/\tau$, the high-frequency asymptote is

$$\phi(\omega) \approx 90° - \tan^{-1}(\infty) = 90° - 90° = 0°$$

The result is sketched in Figure 4.15*b*.

The log magnitude plot shows that the circuit passes signals with frequencies above $\omega = 1/\tau$ with little attenuation, and thus it is called a *high-pass filter*. It is used to remove dc and low-frequency components from a signal. This is desirable when we wish to study high-frequency components whose small amplitudes would be undiscernible in the presence of large-amplitude, low-frequency components. Such a circuit is incorporated into oscilloscopes for this reason. The voltage scale can then be selected so that the signal components to be studied will fill the screen.

Another example of numerator dynamics is the transfer function

$$T(s) = K\frac{\tau_1 s + 1}{\tau_2 s + 1} \tag{4.5-33}$$

An example of this form is the transfer function (4.4-3) of the lead compensator circuit. Thus,

$$T(i\omega) = K\frac{\tau_1 \omega i + 1}{\tau_2 \omega i + 1} \tag{4.5-34}$$

From (4.5-22),

$$m(\omega) = 20 \log |K| + 20 \log |\tau_1 \omega i + 1| - 20 \log |\tau_2 \omega i + 1| \tag{4.5-35}$$

Thus, the plot of $m(\omega)$ can be obtained by subtracting the plot of $\tau_2 s + 1$ from that of $\tau_1 s + 1$. The scale is then adjusted by $20 \log |K|$. The term $\tau_1 s + 1$ causes the curve to break upward at $\omega = 1/\tau_1$. The term $\tau_2 s + 1$ causes the curve to break downward at $\omega = 1/\tau_2$. If $1/\tau_1 > 1/\tau_2$, the composite curve looks like Figure 4.16*a*, and the system is a *low-pass filter*. If $1/\tau_1 < 1/\tau_2$, it is a high-pass filter (Figure 4.16*b*). The plots of $m(\omega)$ in Figure 4.16 were obtained by using only the asymptotes of the terms $\tau_1 s + 1$ and $\tau_2 s + 1$, without using the 3-db corrections at the corner frequencies $1/\tau_1$ and $1/\tau_2$. This sketching technique allows the designer to understand the system's general behavior quickly.

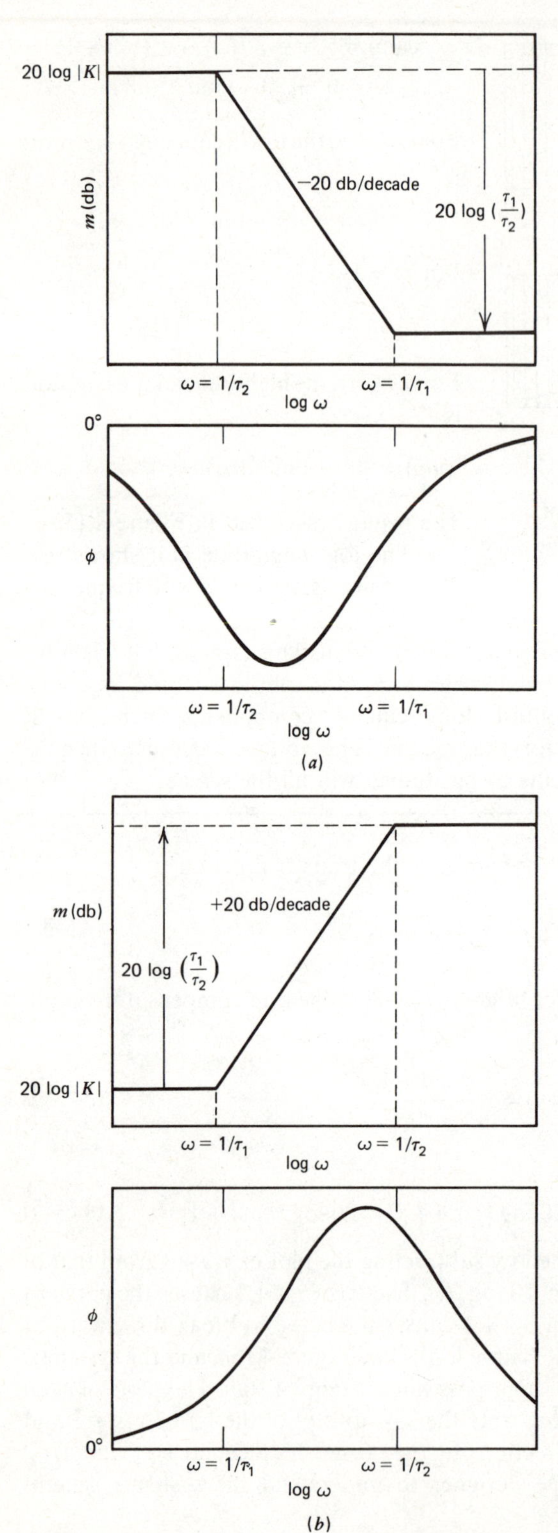

FIGURE 4.16 Approximate sketches of the frequency response plots for

$$T(s) = K\frac{\tau_1 s + 1}{\tau_2 s + 1}$$

For the low-pass filter in (*a*), $\tau_1 < \tau_2$. For the high-pass filter in (b), $\tau_1 > \tau_2$.

From (4.5-23), the phase angle is

$$\begin{aligned}\phi(\omega) &= \measuredangle K + \measuredangle(\tau_1\omega i + 1) - \measuredangle(\tau_2\omega i + 1)\\ &= 0^\circ + \tan^{-1}(\tau_1\omega) - \tan^{-1}(\tau_2\omega)\end{aligned} \tag{4.5-36}$$

Its plot can also be found by combining the plots of $\phi(\omega)$ for K, $\tau_1 s + 1$, and $\tau_2 s + 1$. The sketches shown in Figure 4.16 can be obtained by using the low-frequency, corner frequency, and high-frequency values of 0°, 45°, and 90° for the term $\tau s + 1$, supplemented by evaluations of (4.5-36) in the region between the corner frequencies. This sketching technique is more accurate when the corner frequencies are far apart.

More accurate plots can be obtained by using (4.5-35) and (4.5-36) with a calculator or computer. Some examples are shown in Figure 4.17 for various values of τ_1 and τ_2.

Second-Order Systems

The denominator of a second-order transfer function with $\zeta < 1$ can be expressed as form 4, Table 4.3. If $\zeta \geqslant 1$, the denominator can be written as the product of two first-order factors like form 3. A special case of a second-order system that does not fit into

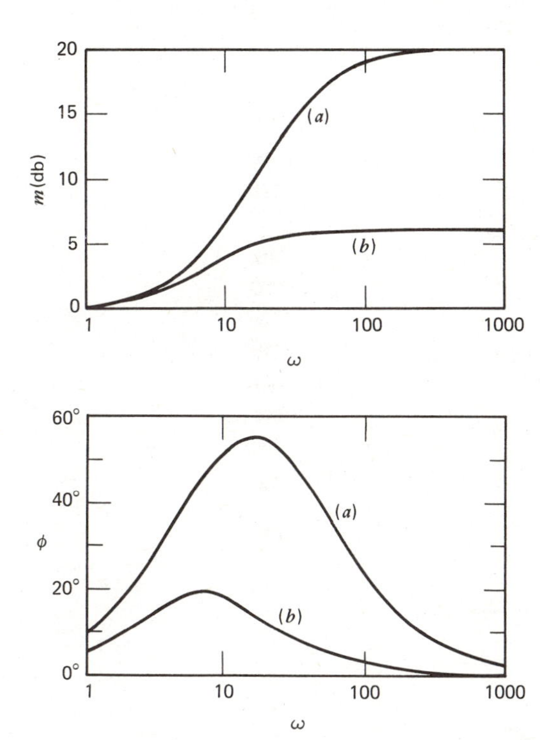

FIGURE 4.17 Representative plots for

$$T(s) = K\frac{\tau_1 s + 1}{\tau_2 s + 1}$$

with $K = 1$. For a, $\tau_1 = 0.2$, $\tau_2 = 0.02$. For b, $\tau_1 = 0.2$, $\tau_2 = 0.1$.

the previous cases occurs when $\zeta = \infty$. The form is

$$T(s) = \frac{K}{s(\tau s + 1)} \tag{4.5-37}$$

An example is a mass with a damper but no spring ($k = 0$). The three building blocks are K, s, and $\tau s + 1$. Because it is in the denominator, the s term shifts the composite m curve upward for $\omega < 1$ and shifts it down for $\omega > 1$. The composite m curve follows that of the s term until $\omega \approx 1/\tau$, when the $(\tau s + 1)$ term begins to have an effect. For $\omega \gg 1/\tau$, the composite slope is -40 db/decade. The s term contributes a constant $-90°$ to the ϕ curve. The result is to shift the first-order lag curve (Figure 4.11*b*) down by 90°. The results are shown in Figure 4.18.

An Overdamped System

Consider the second-order model

$$m\ddot{x} + c\dot{x} + kx = du(t)$$

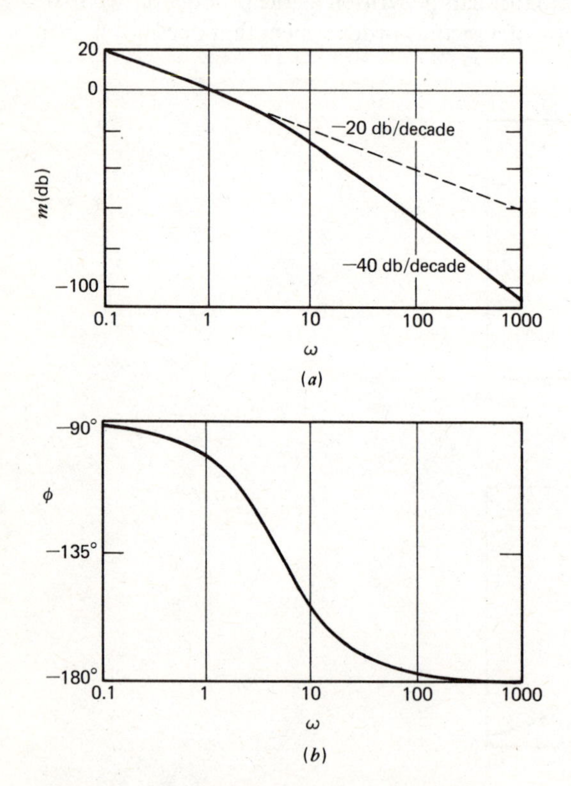

FIGURE 4.18 Frequency response plots for

$$T(s) = \frac{K}{s(\tau s + 1)}$$

with $K = 1$ and $\tau = 0.2$.

Its transfer function is

$$T(s) = \frac{d}{ms^2 + cs + k} \tag{4.5-38}$$

If the system is overdamped, both roots are real and distinct, and we can write $T(s)$ as

$$T(s) = \frac{d/k}{\frac{m}{k}s^2 + \frac{c}{k}s + 1} = \frac{d/k}{(\tau_1 s + 1)(\tau_2 s + 1)} \tag{4.5-39}$$

where τ_1 and τ_2 are the time constants of the roots.

Returning to (4.5-22) and (4.5-23), we see that

$$m(\omega) = 20 \log |T(i\omega)| = 20 \log \left|\frac{d}{k}\right| - 20 \log |\tau_1 \omega i + 1| - 20 \log |\tau_2 \omega i + 1| \tag{4.5-40}$$

$$\phi(\omega) = \measuredangle \frac{d}{k} - \measuredangle(\tau_1 \omega i + 1) - \measuredangle(\tau_2 \omega i + 1) \tag{4.5-41}$$

where $K = d/k$, $D_1(i\omega) = \tau_1 \omega i + 1$ and $D_2(i\omega) = \tau_2 \omega i + 1$. Thus, the magnitude ratio plot in db consists of a constant term, $20 \log |d/k|$, minus the sum of the plots for two first-order lead terms. Assume that $\tau_1 > \tau_2$. Then for $1/\tau_1 < \omega < 1/\tau_2$, the slope is approximately -20 db/decade. For $\omega > 1/\tau_2$, the contribution of the term $(\tau_2 \omega i + 1)$ is significant. This causes the slope to decrease by an additional 20 db/decade, to produce a net slope of -40 db/decade for $\omega > 1/\tau_2$. The rest of the plot can be sketched as before. The result is shown in Figure 4.19*a* for $d > k$. The phase angle plot shown in Figure 4.19*b* is produced in a similar manner by using (4.5-41). Note that if $d/k > 0$, $\measuredangle(d/k) = 0°$.

An Underdamped System

If the transfer function given by (4.5-38) has complex conjugate roots, it can be expressed as form 4 in Table 4.3.

$$T(s) = \frac{d/k}{\frac{m}{k}s^2 + \frac{c}{k}s + 1} = \frac{d/k}{\left(\frac{s}{\omega_n}\right)^2 + 2\zeta\frac{s}{\omega_n} + 1} \tag{4.5-42}$$

We have seen that the constant term d/k merely shifts the magnitude ratio plot up or down by a fixed amount and adds either 0° or $-180°$ to the phase angle plot. Therefore, for now, let us take $d/k = 1$ and consider the following quadratic factor, obtained from (4.5-42) by replacing s with $i\omega$.

$$T(i\omega) = \frac{1}{\left(\frac{i\omega}{\omega_n}\right)^2 + \frac{2\zeta}{\omega_n}\omega i + 1} = \frac{1}{1 - \left(\frac{\omega}{\omega_n}\right)^2 + \frac{2\zeta\omega}{\omega_n}i} \tag{4.5-43}$$

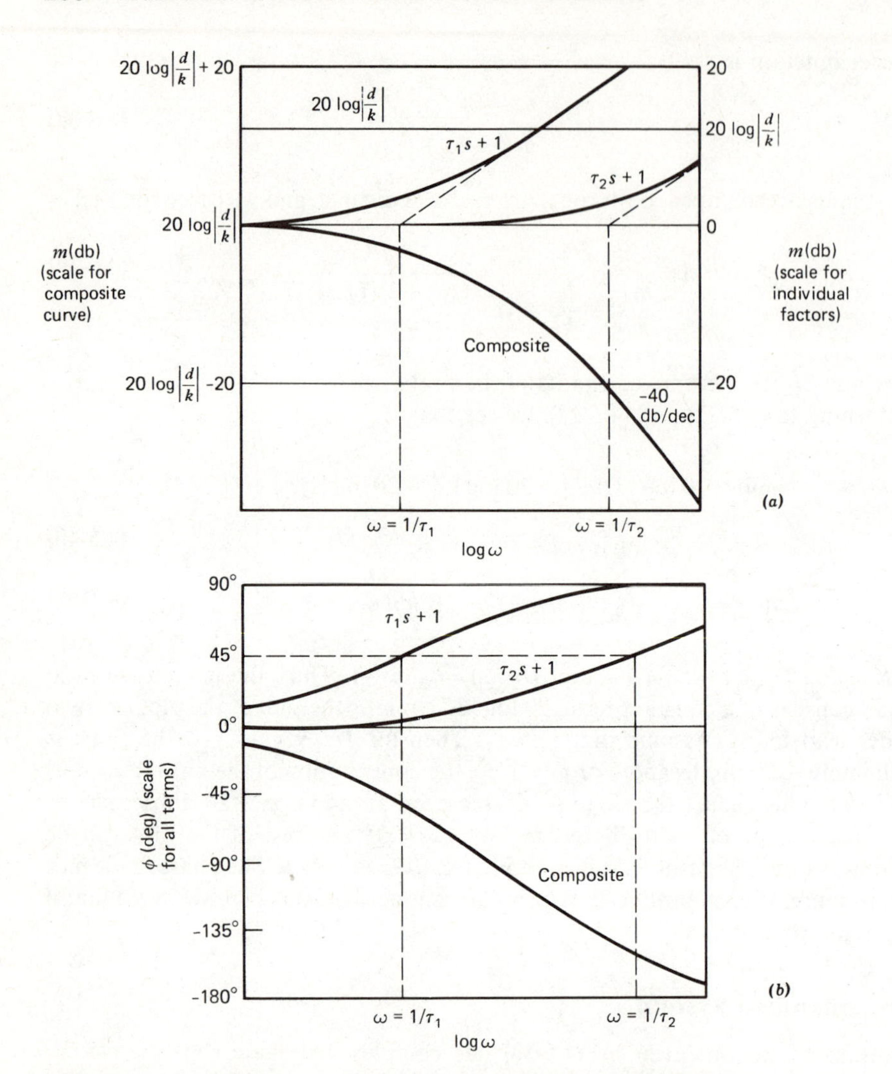

FIGURE 4.19 Frequency response plots for the overdamped system

$$T(s) = \frac{d/k}{(\tau_1 s + 1)(\tau_2 s + 1)}.$$

The magnitude ratio is

$$
\begin{aligned}
m(\omega) &= 20 \log \left| \frac{1}{1 - \left(\dfrac{\omega}{\omega_n}\right)^2 + \dfrac{2\zeta\omega}{\omega_n} i} \right| \\
&= -20 \log \sqrt{\left(1 - \frac{\omega^2}{\omega_n^2}\right)^2 + \left(\frac{2\zeta\omega}{\omega_n}\right)^2} \\
&= -10 \log \left[\left(1 - \frac{\omega^2}{\omega_n^2}\right)^2 + \left(\frac{2\zeta\omega}{\omega_n}\right)^2\right] \qquad \textbf{(4.5-44)}
\end{aligned}
$$

The asymptotic approximations are as follows. For $\omega \ll \omega_n$,

$$m(\omega) \cong -20 \log 1 = 0$$

For $\omega \gg \omega_n$,

$$m(\omega) \cong -20 \log \sqrt{\frac{\omega^4}{\omega_n^4} + 4\zeta^2 \frac{\omega^2}{\omega_n^2}}$$

$$\cong -20 \log \sqrt{\frac{\omega^4}{\omega_n^4}}$$

$$= -40 \log \frac{\omega}{\omega_n}$$

Thus, for low frequencies, the curve is horizontal at $m = 0$, while for high frequencies, it has a slope of -40 db/decade, just as in the overdamped case. The high-frequency and low-frequency asymptotes intersect at the corner frequency $\omega = \omega_n$.

The underdamped case differs from the overdamped case in the vicinity of the corner frequency. To see this, examine $M(\omega)$.

$$M(\omega) = \frac{1}{\sqrt{\left(1 - \dfrac{\omega^2}{\omega_n^2}\right)^2 + \left(\dfrac{2\zeta\omega}{\omega_n}\right)^2}} \qquad \textbf{(4.5-45)}$$

This has a maximum value when the denominator has a minimum. Setting the derivative of the denominator with respect to ω equal to zero shows that the maximum $M(\omega)$ occurs at $\omega = \omega_n\sqrt{1 - 2\zeta^2}$. This frequency is the *resonant frequency* ω_r. The peak of $M(\omega)$ exists only when the term under the radical is positive; that is, when $\zeta \leqslant 0.707$. Thus,

$$\omega_r = \omega_n\sqrt{1 - 2\zeta^2} \qquad 0 \leqslant \zeta \leqslant 0.707 \qquad \textbf{(4.5-46)}$$

The value of the peak M_p is found by substituting ω_r into $M(\omega)$. This gives

$$M_p = M(\omega_r) = \frac{1}{2\zeta\sqrt{1 - \zeta^2}} \qquad 0 \leqslant \zeta \leqslant 0.707 \qquad \textbf{(4.5-47)}$$

If $\zeta > 0.707$, no peak exists, and the maximum value of M occurs at $\omega = 0$ where $M = 1$. Note that as $\zeta \to 0$, $\omega_r \to \omega_n$, and $M_p \to \infty$. For an undamped system, the resonant frequency is the natural frequency ω_n.

A plot of $m(\omega)$ versus $\log \omega$ is shown in Figure 4.20*a* for several values of ζ. Note that the correction to the asymptotic approximations in the vicinity of the corner frequency depends on the value of ζ. The peak value in decibels is

$$m_p = m(\omega_r) = -20 \log (2\zeta\sqrt{1 - \zeta^2}) \qquad \textbf{(4.5-48)}$$

At $\omega = \omega_n$,

$$m(\omega_n) = -20 \log 2\zeta \qquad \textbf{(4.5-49)}$$

The curve can be sketched more accurately by repeated evaluation of (4.5-44) for values of ω near ω_n.

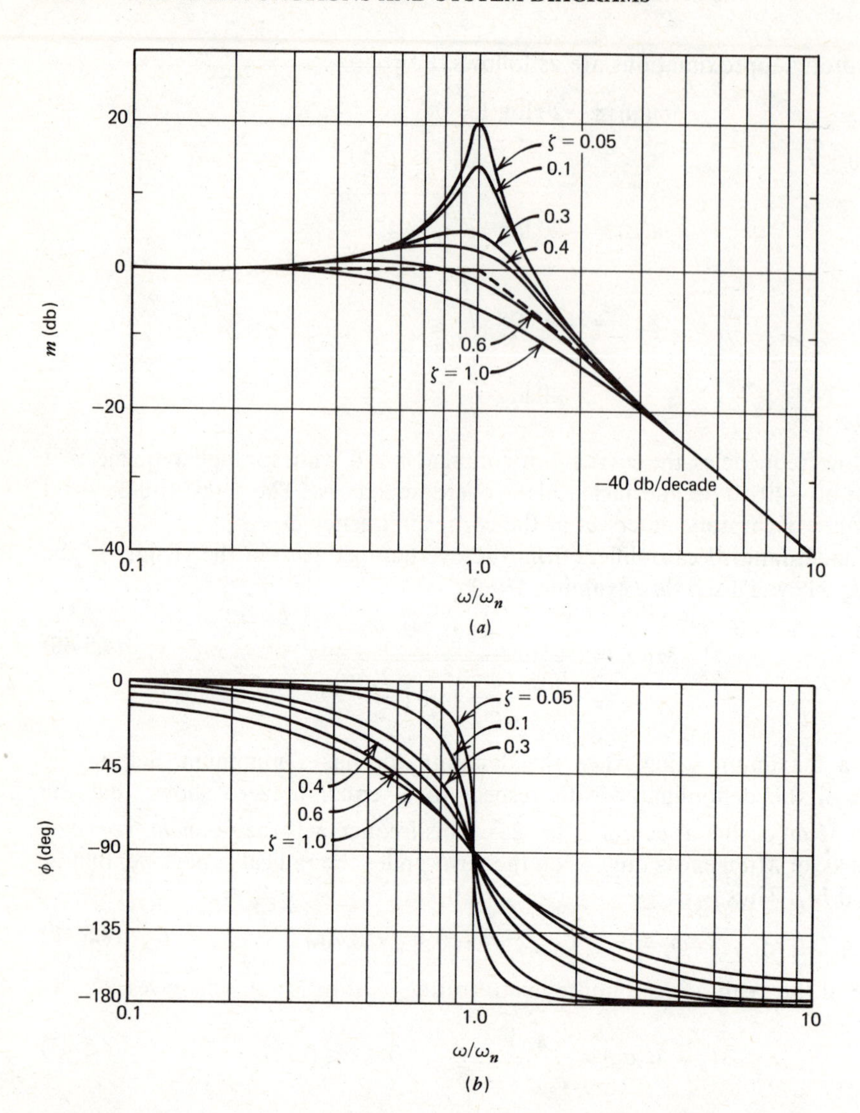

FIGURE 4.20 Frequency response plots for the underdamped system

$$T(s) = \frac{1}{\left(\dfrac{s}{\omega_n}\right)^2 + \dfrac{2\zeta s}{\omega_n} + 1}$$

The phase angle plot is obtained in a similar manner. From the additive property for angles (4.5-23), we see that for (4.5-43),

$$\phi(\omega) = -\measuredangle\left[1 - \left(\frac{\omega}{\omega_n}\right)^2 + \frac{2\zeta\omega}{\omega_n}i\right]$$

Thus

$$\tan \phi(\omega) = -\left[\frac{\dfrac{2\zeta\omega}{\omega_n}}{1-\left(\dfrac{\omega}{\omega_n}\right)^2}\right] \tag{4.5-50}$$

where $\phi(\omega)$ is in the 3rd or 4th quadrant. For $\omega \ll \omega_n$,

$$\phi(\omega) \cong -\tan^{-1} 0 = 0°$$

For $\omega \gg \omega_n$,

$$\phi(\omega) \cong -180°$$

At the corner frequency,

$$\phi(\omega_n) = -\tan^{-1} \infty = -90°$$

This result is independent of ζ. The curve is skew-symmetric about the inflection point at $\phi = -90°$ for all values of ζ. The rest of the plot can be sketched by evaluating (4.5-50) at various values of ω. The plot is shown for several values of ζ in Figure 4.20*b*. At the resonant frequency.

$$\phi(\omega_r) = -\tan^{-1} \frac{\sqrt{1-2\zeta^2}}{\zeta} \tag{4.5-51}$$

For our applications, the quadratic factor given by form 4 in Table 4.3 almost always occurs in the denominator; therefore, we have developed the results assuming this will be the case. If a quadratic factor is found in the numerator, its values of $m(\omega)$ and $\phi(\omega)$ are the negative of those given by (4.5-44) and (4.5-50).

Experimental Determination of a Transfer Function

The frequency response curves are an aid to developing a system model from test data. If a suitable apparatus can be devised to provide a sinusoidal input of adjustable frequency, then the system's output amplitude and phase shift relative to the input can be measured for various input frequencies. When these data are plotted on the logarithmic plot for a sufficient frequency range, it will be obvious whether or not the system can be modeled as a first-order linear system. If *both* the m and ϕ data plots are of the forms shown in Figure 4.11, the parameters τ and b can be estimated. This procedure is easiest for systems with electrical inputs and outputs, because for these systems, variable-frequency oscillators and frequency response analyzers are commonly available. Some of these can automatically sweep through a range of frequencies and plot the decibel and phase angle data. For nonelectrical outputs, such as a displacement, a suitable transducer can be used to produce an electrical measurement that can be analyzed. For nonelectrical inputs, the task is more difficult but is frequently used nonetheless.

An advantage of frequency response tests is that they can often be used on a device without interrupting its normal operations. The small-amplitude sinusoidal test signal is superimposed on the operating inputs, and the sinusoidal component of the output with the input frequency is subtracted from the output measurements. Computer algorithms are available to do this.

4.6 INTERPRETATION OF FREQUENCY RESPONSE†

Section 4.5 focused on techniques for obtaining the frequency response plots from the transfer function. Here, we provide three examples of the interpretation of these plots. Example 4.11 shows how the response is affected by the frequency and amplitude of the input. Example 4.12 shows how useful calculations can be made in some cases, even if the damping constant is not known accurately. The third example illustrates how the frequency response plots can be used to design instruments for measuring displacement, velocity, and acceleration. These are helpful for control systems applications, and we will see more uses for frequency response plots when we treat control systems design in Chapters 6, 7, and 9.

A Vehicle Suspension System

Example 4.11

A simplified representation of a vehicle's front-wheel system consisting of the tire, springs, and shock absorber is shown in Figure 4.21*a*. The spring k_1 represents the elasticity of the tire, while k represents that of the shock absorber. Determine the frequency response of the tire displacement x to a road surface displacement y. Assume that the tire does not leave the road and the inertia of the car body is so large that the body can be considered as a fixed support. The tire weighs 30 lb, and the other constants are $k = 12{,}000$ lb/ft, $k_1 = 2400$ lb/ft, and $c = 360$ lb-sec/ft.

Let the tire displacement x be measured from a suitable static equilibrium position. Then Newton's law gives the following model:

$$m\ddot{x} + c\dot{x} + kx = k_1(y - x) \tag{4.6-1}$$

For the given values, $m = 30/g = 0.933$ slugs, and

$$0.933\ddot{x} + 360\dot{x} + 14400x = 2400y$$

or

$$\ddot{x} + 386\dot{x} + 15434x = 2572y$$

The transfer function between y as input and x as output is

$$T(s) = \frac{2572}{s^2 + 386s + 15434}$$

---

†This section contains additional material to develop the reader's understanding of frequency response by using examples from mechanical systems and instrumentation. No new theory is introduced, and the section can be skipped if necessary.

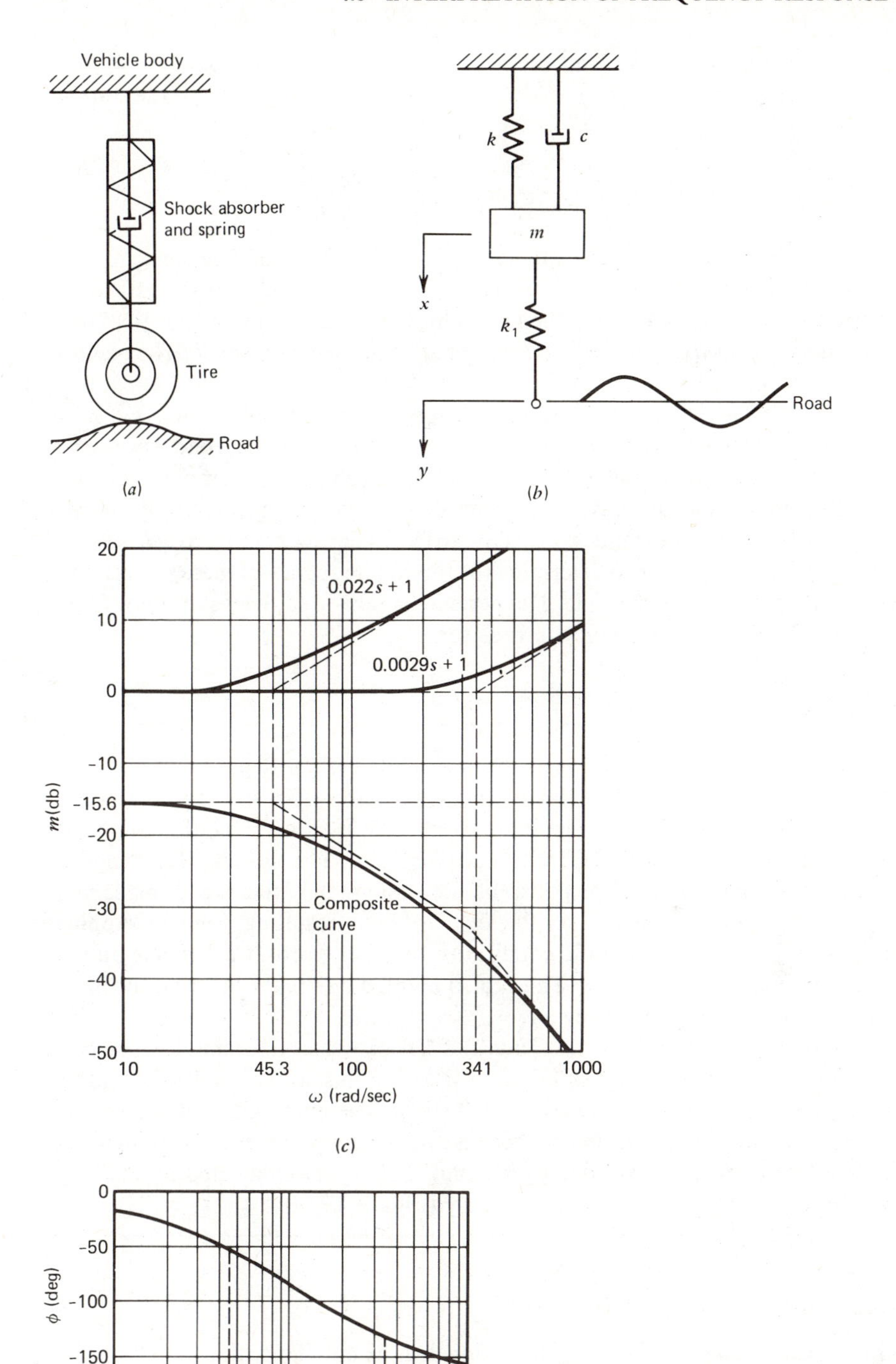

FIGURE 4.21 Model of a wheel suspension system. (*a*) System components. (*b*) Lumped-parameter representation. (*c*) Attenuation curve. (*d*) Phase angle curve.

The damping factor is $\zeta = 1.55$, and the roots are $s_1 = -45.3$ and $s_2 = -341$. The corresponding time constants are $\tau_1 = 0.022$ sec and $\tau_2 = 0.0029$ sec. Thus, the transfer function can be written as

$$T(s) = \frac{2572}{(s+45.3)(s+341)} = \frac{0.1665}{(0.022s+1)(0.0029s+1)}$$

The corner frequencies are $\omega_1 = 45.3$ rad/sec and $\omega_2 = 341$ rad/sec. The constant term 0.1665 shifts the $m(\omega)$ curve by 20 log 0.1665 = −15.6 db, and contributes 0° to the phase angle plot. It is best to sketch the $m(\omega)$ plot first without the shift of −15.6 db and then simply adjust the scale by this amount after the sketch is made. The result is shown in Figure 4.21*c*.

Measurements made on vehicles traveling on typical roads have shown that some construction techniques result in periodic wave patterns in the road surface. A representative wavelength is 50 ft, and a representative road surface amplitude is 3 in. Ripples with a much shorter wavelength can also appear but usually only on dirt roads, where they give the road a "washboard" appearance and have a smaller amplitude, say, 0.5 in. In order to determine the forcing frequency of these patterns, we must know the speed V of the vehicle. The period τ is related to the wavelength L and V by $\tau = L/V$. The forcing frequency in rads/sec is

$$\omega = 2\pi\frac{1}{\tau} = \frac{2\pi V}{L}$$

For a speed of 60 mph = 88 ft/sec, the wavelength of 50 ft produces a forcing frequency of $\omega = 11.06$ rad/sec. At this frequency, Figure 4.21*c* shows that $m(11) \cong -15.6$ db. Thus, $M(11) = 0.1665$. The input amplitude is 3 in. Thus, the amplitude of the tire displacement is $x = 0.1665(3) = 0.4995$ in. The forcing frequency is so low relative to the speed of response of the system that the tire displacement amplitude is essentially the same as the displacement resulting from an applied displacement of $y = 3$ in. under static conditions. In other words, the tire has time to follow the input motion very closely. This is also pointed out by the fact that the phase shift is almost zero at this frequency.

The washboard effect with a wavelength of 1 ft produces a forcing frequency of $\omega = 553$ rad/sec. At this frequency, $m(553) = -42.5$ db. Thus, $M(553) = 0.0075$, and the resulting tire amplitude is $x = 0.0075(0.5) = 0.0038$ in. The amplitude is much smaller than for the 50-ft wavelength, primarily because the forcing frequency is so high that the wheel system does not have time to respond. This is also indicated by the large negative phase shift of −143° at this frequency (Figure 4.21*d*).

Rotating Unbalance in Machinery

A common cause of sinusoidal forcing in machines is the unbalance that exists to some extent in every rotating machine. The unbalance is caused by the fact that the center of mass of the rotating part does not coincide with the center of rotation. Let M be the total mass of the machine and m the rotating mass causing the unbalance. Consider the entire unbalanced mass m to be lumped at its center of mass, a distance R from the

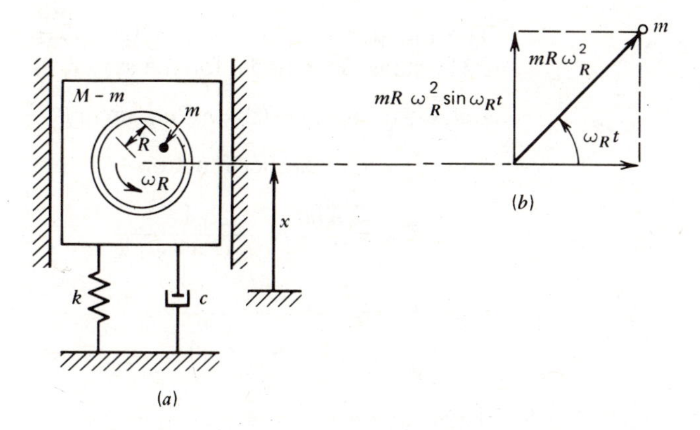

FIGURE 4.22 Machine with rotating unbalance. (*a*) System components. (*b*) Force diagram.

center of rotation. This distance is the *eccentricity*. Figure 4.22*a* shows this situation. The main mass is thus $(M-m)$ and assumed to be constrained to allow only vertical motion.

The motion of the unbalanced mass m will consist of the vector combination of its motion relative to the main mass $(M-m)$ and the motion of the main mass. For a constant speed of rotation ω_R, the rotation produces a radial acceleration of m equal to $R\omega_R^2$. This causes a force to be exerted on the bearings at the center of rotation. This force has a magnitude $mR\omega_R^2$ and is directed radially outward. The vertical component of this unbalance force is, from Figure 4.22*b*.

$$f = mR\omega_R^2 \sin \omega_R t \tag{4.6-2}$$

Example 4.12

Alternating-current motors are usually designed to run at a constant speed, typically either 1750 or 3500 rpm. One such motor for a power tool weighs 20 lb and is to be mounted on a steel cantilever beam as shown in Figure 4.23*a*. Static-force calculations and space considerations suggest that a beam 6 in. long, 4 in. wide, and 3/8 in. thick would be suitable. The rotating part of the motor weighs 10 lb and has an eccentricity of 0.001 ft. (This distance can be determined by standard balancing methods or applying the analysis in this example in reverse, using a vibration test.) The damping ratio for such beams is difficult to determine but is usually very small, say, $\zeta \leqslant 0.1$. Estimate the amplitude of vibration of the beam at steady state.

For the equivalent system shown in Figure 4.23*b*, we have from Tables 2.3, G.3, and 2.6, the following spring constant and equivalent mass (after converting inches to feet):

$$k = \frac{Ewh^3}{4L^3} = \frac{(4.32 \times 10^9)(0.333)(0.03125)^3}{4(0.5)^3}$$
$$= 8.78 \times 10^4 \text{ lb/ft}$$
$$m_e = \frac{20}{32.17} + 0.23(15.2)(0.333)(0.5) = 0.640 \text{ slugs}$$

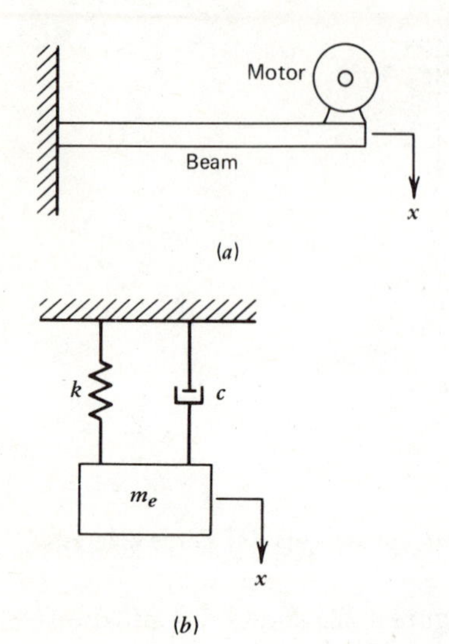

FIGURE 4.23 Motor vibration problem in Example 4.12. (*a*) Motor supported by a cantilever beam. (*b*) Lumped-parameter model.

The unbalanced mass is $m = 10/32.17 = 0.311$ slugs. The model for the system is

$$m_e \ddot{x} + c\dot{x} + kx = f(t) = mR\omega_R^2 \sin \omega_R t$$

This gives the transfer function

$$T(s) = \frac{X(s)}{F(s)} = \frac{1}{m_e s^2 + cs + k}$$

$$= \frac{1/k}{\dfrac{m_e}{k}s^2 + \dfrac{c}{k}s + 1}$$

Thus,

$$T(i\omega) = \frac{1/k}{1 - \left(\dfrac{\omega}{\omega_n}\right)^2 + \dfrac{2\zeta\omega}{\omega_n}i}$$

where $1/k = 1.139 \times 10^{-5}$ and $\omega_n = \sqrt{k/m_e} = 370$ rad/sec $= 3537$ rpm. Comparing this transfer function with (4.5-42), we see that $d/k \neq 1$, and we must include this factor in our calculations. This factor shifts the $m(\omega)$ curve by -98.9 db. Alternatively, we may use the response plot shown in Figure 4.20*a* and multiply the result by $1/k$.

The motor speed of 1750 rpm is 183 rad/sec. This gives $\omega_R/\omega_n = 0.485$. From the plot [or from (4.5-44)], we see that $m = 2.37$ db, which corresponds to a magnitude ratio of 1.314. Multiplying by the $1/k$ factor gives $(1.314)1.139 \times 10^{-5} = 1.496 \times 10^{-5}$. The amplitude of the forcing function is $mR\omega_R^2 = 0.311(0.001)(183)^2 = 10.4$ lb. Thus, the steady-state amplitude is $10.4(1.496 \times 10^{-5}) = 15.6 \times 10^{-5}$ ft.

On the downward oscillation, the total amplitude as measured from horizontal is the preceding value plus the static deflection, or $15.6 \times 10^{-5} + 20/87800 = 3.84 \times 10^{-4}$ ft. If this deflection would cause too much stress in the beam, one or more of the following changes could be made in the design:

1. Add a damper to the system.
2. Reduce the unbalance in the motor.
3. Increase the separation between the forcing frequency and the natural frequency either by selecting a beam with more stiffness (a larger k), by reducing the mass of the system, or by adding mass to shift ω_n far to the left of ω_R on the plot.

The preceding results are not very sensitive to the assumed value of $\zeta = 0.1$. If $\zeta = 0.05$ or 0.2, the calculated amplitude of vibration would be 15.7×10^{-5} ft or 15.2×10^{-5} ft, respectively. However, if we had used a motor with a speed of 3500 rpm $= 366$ rad/sec. this choice would put the forcing frequency very close to the natural frequency. In this region, the assumed value of ζ would be critical in the amplitude calculation. In practice, such a design would be avoided. That is, in vibration analysis, the most important quantity to know is the natural frequency ω_n. If the damping is slight, the resonant frequency is near ω_n. If ω_n is designed so that it is not close to the forcing frequency, it is not necessary to know the precise amount of damping.

Instrument Design

The second-order models we have seen so far have not had the s operator in the numerator of their transfer functions. Consequently, the magnitude ratio is small at high frequencies. However, introducing numerator dynamics can produce a large magnitude ratio at high frequencies. This effect can be used to advantage—in instrument design, for example. The instrument shown in Figure 4.24 illustrates this point. With proper selection of the natural frequency of the device, it can be used either as a *vibrometer* to measure the amplitude of a sinusoidal displacement y or an *accelerometer* to measure the amplitude of the acceleration $\ddot{y}$, which is also sinusoidal. When used to measure ground motion from an earthquake, for example, the instrument is commonly referred to as a seismograph.

The mass displacement x and the support displacement y are relative to an inertial reference, with $x = 0$ corresponding to the equilibrium position of m when $y = 0$. With the potentiometer arrangement shown, the voltage v is proportional to the relative displacement z between the support and the mass m, where $z = x - y$. Newton's law gives

$$m\ddot{x} = -c(\dot{x} - \dot{y}) - k(x - y) = 0$$

In terms of z, this becomes

$$m\ddot{z} + c\dot{z} + kz = -m\ddot{y} \tag{4.6-3}$$

The transfer function between the input y and the output z is

$$T(s) = \frac{Z(s)}{Y(s)} = \frac{-ms^2}{ms^2 + cs + k} = \frac{-s^2/\omega_n^2}{\dfrac{s^2}{\omega_n^2} + \dfrac{2\zeta s}{\omega_n} + 1} \tag{4.6-4}$$

Substituting $s = i\omega$ and referring to (4.5-22), we see that the numerator gives the following contribution to the log magnitude ratio:

$$20 \log |N_i(i\omega)| = 20 \log \left| \left(\frac{i\omega}{\omega_n} \right)^2 \right|$$

$$= 40 \log \frac{\omega}{\omega_n}$$

This term contributes 0 db to the net curve at the corner frequency $\omega = \omega_n$, and it increases the slope by 40 db/decade over all frequencies. Thus, at low frequencies,

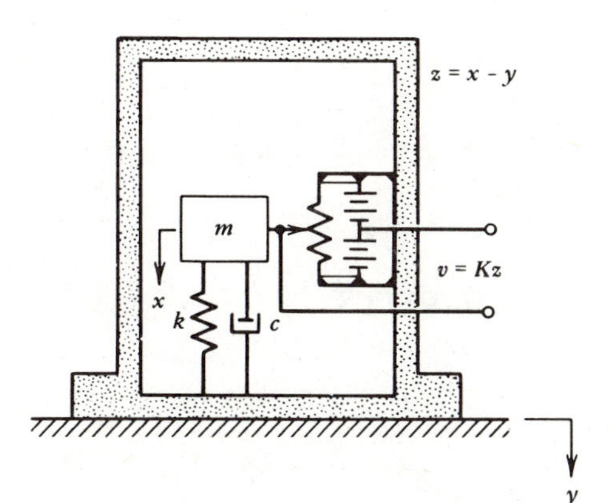

FIGURE 4.24 Vibration instrument.

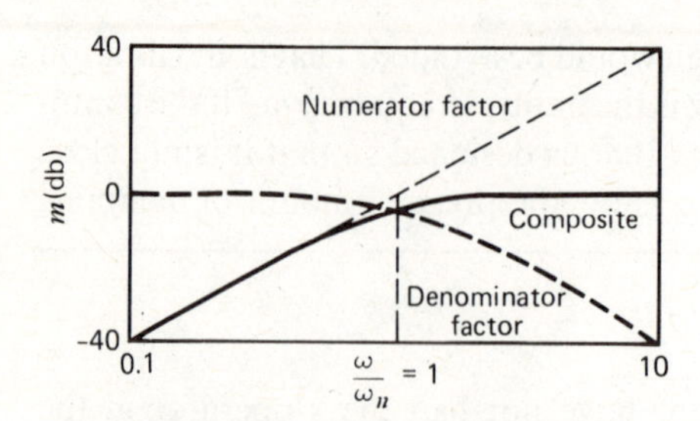

FIGURE 4.25 Attenuation curve for the vibration instrument shown in Figure 4.24 (assuming that $\zeta = 1$).

the slope of $m(\omega)$ is 40 db/decade, and at high frequencies, the slope is zero. The plot is sketched in Figure 4.25.

For a vibrometer, this plot shows that the device's natural frequency ω_n must be selected so that $\omega \gg \omega_n$, where ω is the oscillation frequency of the displacement to be measured $[y(t) = A \sin \omega t]$. For $\omega \gg \omega_n$,

$$|T(i\omega)| \cong 40 \log \frac{\omega}{\omega_n} - 40 \log \frac{\omega}{\omega_n} = 0 \text{ db}$$

and thus $|z| \cong |y| = A$, as desired. The voltage v is directly proportional to A in this case. The physical explanation for this result is the fact that the mass m cannot respond to high-frequency input displacements. Its displacement x therefore remains fixed, and the motion z directly indicates the motion y.

To design a specific vibrometer, we must know the lower bound of the input displacement frequency ω. The frequency $\omega_n = \sqrt{k/m}$ is then made much smaller than this bound by selecting a large mass and a "soft" spring (small k). However, these choices are governed by constraints on the allowable deflections. For example, a very soft spring will have a large distance between the free length and the equilibrium positions.

An accelerometer can be obtained by using the lower end of the frequency range; that is, selecting $\omega_n \gg \omega$, or equivalently, for s near zero, (4.6-4) gives

$$T(s) \cong -\frac{s^2}{\omega_n^2} = \frac{Z(s)}{Y(s)}$$

or

$$Z(s) \cong -\frac{1}{\omega_n^2} s^2 Y(s)$$

The term $s^2 Y(s)$ represents the transform of $\ddot{y}$, so the output of the accelerometer is

$$|z| \cong \frac{1}{\omega_n^2} |\ddot{y}| = \frac{\omega^2}{\omega_n^2} A$$

With ω_n chosen large (using a small mass and a "stiff" spring), the input acceleration amplitude $\omega^2 A$ can be determined from z (or v).

4.7 BANDWIDTH AND PERIODIC INPUTS

The instantaneous power in an electrical circuit is

$$p(t) = i(t)\, v(t) \tag{4.7-1}$$

If the input current is

$$i(t) = I \sin \omega t$$

we can write the steady-state power as

$$p(t) = I^2|T(i\omega)| \sin \omega t \sin(\omega t + \phi) \tag{4.7-2}$$

since

$$v(t) = |T(i\omega)|I \sin(\omega t + \phi) \tag{4.7-3}$$

$$\phi = \angle T(i\omega) \tag{4.7-4}$$

where $T(s)$ is the circuit's impedance $V(s)/I(s)$.

Bandwidth

From (4.7-2), we see that the magnitude of the instantaneous power is a function of the input frequency ω. This fact is used to establish another measure of system performance called *bandwidth*. Consider the series RL circuit in Figure 4.26. Its impedance is

$$T(s) = \frac{V(s)}{I(s)} = R + Ls \tag{4.7-5}$$

Thus,

$$I(i\omega) = \frac{V(i\omega)/R}{1 + i\omega L/R}$$

At zero input frequency $\omega = 0$, the current attains its maximum value of

$$I_o = \frac{V(0)}{R}$$

Form the ratio

$$\left|\frac{I(i\omega)}{I_o}\right| = \left|\frac{1}{1 + i\omega L/R}\right|$$

$$= \frac{1}{\sqrt{1 + (\omega L/R)^2}}$$

$$= \frac{|T(0)|}{|T(i\omega)|}$$

When $\omega = R/L = \omega_c$, the current is 0.707 of its maximum value I_o, and the power consumed by the circuit is one-half the maximum value at $\omega = 0$. At $\omega = \omega_c$, the preceding equation shows that

$$\frac{|T(0)|}{|T(i\omega_c)|} = 0.707$$

or

$$|T(i\omega_c)|_{\text{db}} = |T(0)|_{\text{db}} - 3.01 \text{ db} \tag{4.7-6}$$

That is, for input frequencies greater than ω_c (the *cutoff frequency*), the power de-

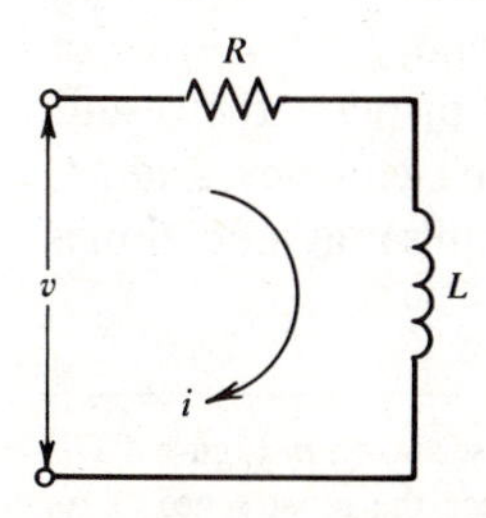

FIGURE 4.26 Series RL circuit.

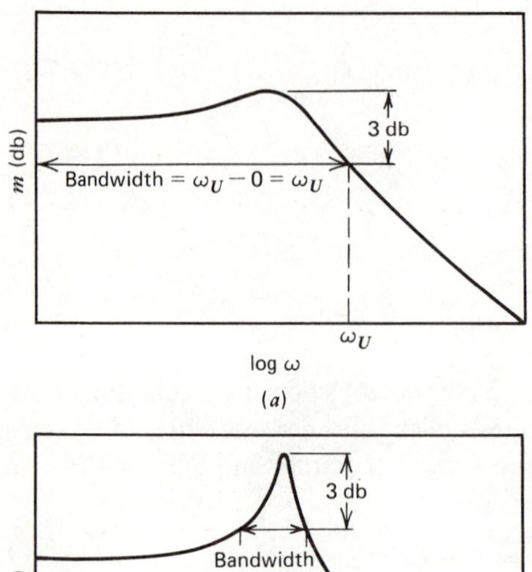

FIGURE 4.27 Bandwidth definition for (*a*) One cutoff frequency, and (*b*) Two cutoff frequencies.

livered to the circuit is less than one-half its maximum value. Thus, ω_c is termed the *half-power* frequency.

Equation (4.7-6) can be taken as the defining relation for the cutoff frequency ω_c, the frequency at which the log magnitude curve drops 3.01 db below its peak value. In this example, the peak occurred at $\omega = 0$, but this is not true for all systems (for example, see Figure 4.15). Also, there may be two cutoff frequencies; one to the left and one to the right of a peak in $T(i\omega)$.

The *bandwidth* is the frequency spread between the cutoff frequencies, or between $\omega = 0$ and the cutoff frequency, if the peak power occurs at $\omega = 0$. Figure 4.11 shows that ω_c is the corner frequency $1/\tau$ for $T(s) = 1/(\tau s + 1)$. Its bandwidth is $1/\tau$. If the peak in the response plot of the underdamped second-order system (4.5-42) is not high enough, the bandwidth begins at $\omega = 0$ (Figure 4.27*a*). For ζ small enough, two cutoff frequencies occur, and the bandwidth is as shown in Figure 4.27*b*.

Bandwidth is a measure of the range of frequencies for which a significant portion of the system's input is felt by the output. The system filters out to a greater extent those input components whose frequencies lie outside the bandwidth. The bandwidth definition is arbitrarily based on the half-power points. Other definitions can be used, but this is the most common.† Since the first-order system's bandwidth is the reciprocal of its time constant, the bandwidth is seen to be an indication of the speed of response. The larger τ is, more sluggish is the response. A small bandwidth indicates a sluggish system, which poses a problem for the designer. For example, if the time constant of a pneumatic device is picked to be large enough to filter out unwanted pressure fluctuations, its response might be too slow for other purposes.

Table 4.4 shows how the magnitude ratio rapidly decreases for frequencies outside the bandwidth of the first-order system. The cutoff frequency is $\omega_c = 1/\tau$. For $\omega > 10\omega_c$, the system's magnification is less than 10%. Table 4.5 displays the results for a second-order system, where ω_U is the upper bandwidth cutoff frequency, and M_p is the peak value of the magnitude ratio M. For this case, the system's magnification is

---

†Some authors define the bandwidth to be from $\omega = 0$ to $\omega = \omega_U$ even for the case shown in Figure 4.27*b*. The definition we have used seems to be the most common one. It also makes the most sense as an indication of the system's filtering characteristics.

TABLE 4.4 The Magnitude Ratio at Selected Frequencies for $T(s) = 1/(\tau s + 1)$

$\omega/\omega_c = \omega\tau$	M
0	1
1	0.707
2	0.447
3	0.316
4	0.243
5	0.196
10	0.100

TABLE 4.5 The Magnitude Ratio at Selected Frequencies for $T(s) = \omega_n^2/(s^2 + 2\zeta\omega_n s + \omega_n^2)$

ζ	**0.2**	**0.5**	**0.7**
ω_U/ω_n	**1.145**	**1.168**	**1.009**
M_p	**2.5576**	**1.1547**	**1.0002**
ω/ω_U		M/M_p	
1	0.707	0.707	0.707
2	0.090	0.172	0.240
3	0.036	0.073	0.109
4	0.020	0.041	0.061
5	0.012	0.026	0.039
10	0.003	0.006	0.010

less than 10% of its peak value when ω is two or three times greater than ω_U. This is due to the steeper high-frequency slope of the second-order system (−40 db/decade).

Response to General Periodic Inputs

The application of the sinusoidal input response is not limited to cases involving a single sinusoidal input. A basic theorem of analysis states that under some assumptions, which are generally satisfied in most practical applications, any periodic function can be expressed by a constant term plus an infinite series of sines and cosines with increasing frequencies. This theorem is the *Fourier theorem*, and its associated series is the *Fourier series*. It has the form

$$v(t) = a_0 + a_1 \cos\left(\frac{\pi t}{p}\right) + a_2 \cos\left(\frac{2\pi t}{p}\right) + \cdots + b_1 \sin\left(\frac{\pi t}{p}\right) + b_2 \sin\left(\frac{2\pi t}{p}\right) + \cdots \tag{4.7-7}$$

where $v(t)$ is the periodic function and p is the *half-period* of $v(t)$. The constants a_i and b_i are determined by integration formulas applied to $v(t)$. Appendix A contains these formulas and summarizes the procedure required to construct the series.

In previous sections, we have seen how to determine the steady-state response of a linear system when subjected to an input that is a constant (a step), a sine, or a cosine. When the input v is periodic and expressed in the form of (4.7-7), the superposition principle states that the complete steady-state response is the sum of the steady-state responses due to each term in (4.7-7). Although this is an infinite series, in practice we have to deal with only a few of its terms, because those terms whose frequencies lie outside the system's bandwidth can be neglected as a result of the filtering property of the system.

Example 4.13

For the vibration isolator shown in Figure 3.19, suppose that $c/K = \tau = 0.1$ sec and the displacement y_2 is no longer purely sinusoidal but consists of a sine for 0.5 sec followed by zero displacement for the second half of the period (see Figure 4.28). Such a motion might be produced by a rotating cam with a *dwell* (the zero displacement portion). Determine the steady-state response of y_1.

The Fourier series' representation of $y_2(t)$ is, from Appendix A,

$$y_2(t) = \frac{1}{\pi} + \frac{1}{2}\sin 2\pi t - \frac{2}{\pi}\left(\frac{\cos 4\pi t}{1(3)} + \frac{\cos 8\pi t}{3(5)} + \frac{\cos 12\pi t}{5(7)} + \cdots\right) \qquad \textbf{(4.7-8)}$$

For the system transfer function with $\tau = 0.1$ and $b = 1/\tau = 10$, the magnitude ratio M from (4.5-9) is

$$M(\omega) = \frac{1}{\sqrt{1 + 0.01\omega^2}}$$

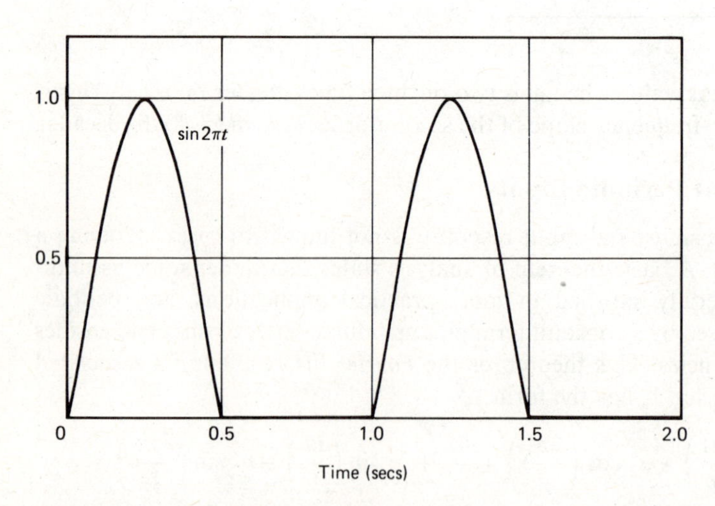

FIGURE 4.28 Half-sine function.

The decibel equivalent $m(\omega)$ may also be used, and it is convenient when working from a logarithmic plot. From (4.5-10), the phase shift is

$$\phi(\omega) = \tan^{-1}(-0.1\omega)$$

We desire to find the steady-state motion y of point 1 due to y_2. The first term in the series expansion for y_2 is a constant $1/\pi = 0.318$. This corresponds to an input with zero frequency; at steady state, it produces a response of $y_1 = 0.318$, since $M(0) = 1$. The second term in the series is a sine with an amplitude of 0.5 and a frequency of 2π rad/sec. Thus,

$$M(2\pi) = \frac{1}{\sqrt{1 + 0.01(2\pi)^2}} = 0.847$$

It produces a sinusoidal response of the same frequency with an amplitude of $0.847(0.5) = 0.424$, with a phase shift of $\phi(2\pi) = \tan^{-1}(-0.2\pi) = -0.56$ rad. These are the only terms within the bandwidth of the isolator (10 rad/sec). To demonstrate the filtering property and illustrate how cosine terms are handled, we will treat the third term in the series.

Since $\cos 4\pi t = \sin(4\pi t + \pi/2)$, this term can be handled as a sine wave. The phase shift of $\pi/2$ rad is then added to $\phi(4\pi)$. Here,

$$M(4\pi) = \frac{1}{\sqrt{1 + 0.01(4\pi)^2}} = 0.623$$

The amplitude of the response is $0.623(2/3\pi) = 0.132$, and the phase shift is $\phi(4\pi) + \pi/2 = -0.90 + \pi/2 = 0.67$ rad. From superposition, the total steady-state response resulting

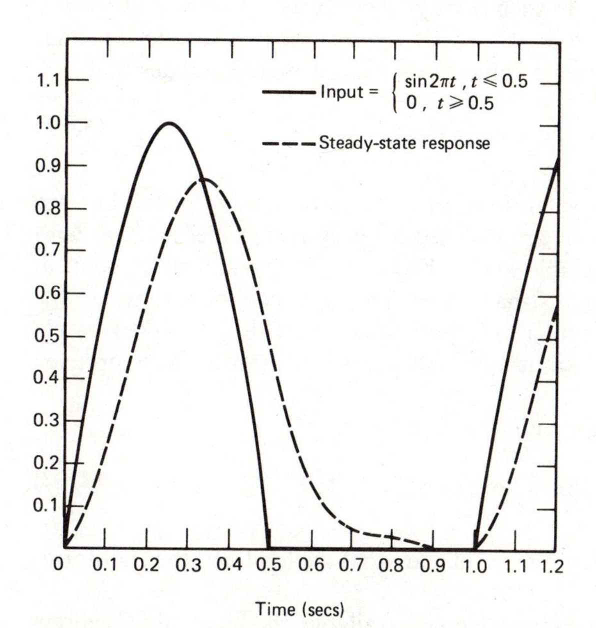

FIGURE 4.29 Steady-state response of the isolator to the half-sine input.

from the three series terms retained is

$$y_1(t) = 0.318 + 0.423 \sin(2\pi t - 0.56) - 0.132 \sin(4\pi t + 0.67)$$

The input and the response are plotted in Figure 4.29. The difference between the input and output wave shapes results from the resistive or lag effect of the system, not from the omission of the higher order terms in the series. To see this, we compute that part of the response amplitude resulting from the fourth term in the series. The amplitude contribution is $2M(8\pi)/15\pi = 0.016$, which is about 10% of the amplitude of the third term. The decreasing amplitude of the higher order terms in the series for $y_2(t)$, when combined with the filtering property of the system, allows us to truncate the series when the desired accuracy has been achieved.

Before leaving this example, we point out that the maximum displacement of point 1 is 85% of the maximum displacement of the input. Thus, the system's isolation characteristics for this input are quite different from those for the previous application (Example 3.9) in which the input was a pure sine with a frequency of 100 rad/sec. Whenever the input frequency or wave shape is altered very much from its design value, the isolator cannot be expected to perform as well.

4.8 THE INITIAL AND FINAL VALUE THEOREMS

In systems analysis, the investigator often needs to know only the steady-state value of the output for a preliminary study. In such cases, the *final value theorem* is frequently useful (Appendix B). If $y(t)$ and dy/dt possess Laplace transforms, if the following limit exists, and if $y(t)$ approaches a definite value as $t \to \infty$, this theorem states that

$$y(\infty) = \lim_{s \to 0} [sY(s)] \tag{4.8-1}$$

The derivation of this theorem is given in most texts on the theory of the Laplace transform. The theorem's conditions are satisfied if the system model is linear with constant coefficients [i.e., transformable so that $Y(s) = T(s)V(s)$] and if all the roots of the denominator of $sY(s)$ lie in the left-half plane. The most common failure of the theorem is the case where the input v is a pure sinusoid. This introduces purely imaginary roots and a steady-state sinusoidal response. Thus, $y(t)$ does not approach a definite value as $t \to \infty$.

The companion theorem is the *initial value theorem.*

$$y(0) = \lim_{s \to \infty} [sY(s)] \tag{4.8-2}$$

The conditions under which the theorem is valid are that the limit and the transforms of y and dy/dt exist.

If the ramp response of $\dot{y} = ry + bv$ were not available, the final value theorem could have been used to find the steady-state difference between the input and output quickly. This is because from (4.3-1) and (4.5-6) with $v(t) = mt$, we obtain

$$
\begin{aligned}
v(\infty) - y(\infty) &= \lim_{s\to 0} s[V(s) - Y(s)] \\
&= \lim_{s\to 0} s\left(\frac{m}{s^2} - \frac{\tau b}{\tau s + 1}\frac{m}{s^2}\right) \\
&= \lim_{s\to 0} \frac{m}{s}\left(\frac{\tau s + 1 - \tau b}{\tau s + 1}\right) \\
&= \begin{cases} \infty, & \tau b \neq 1 \\ m\tau & \tau b = 1 \end{cases}
\end{aligned}
\tag{4.8-3}
$$

Thus, the output does not follow the input unless $\tau b = 1$.

The initial and final value theorems are especially useful when the system is represented only in terms of its transfer function. The theorems can thus be directly applied without working through the solution of a differential equation.

Example 4.14

(a) Derive formulas for the steady-state difference between the input $V(s)$ and the output $Y(s)$ for a unit-step input and a unit-ramp input. The system diagram is given in Figure 4.30.

(b) Apply the results to the case where $G(s) = K/s$.

(a) The difference between $V(s)$ and $Y(s)$ is $E(s)$ in the figure, where

$$E(s) = V(s) - Y(s) = V(s) - G(s)E(s) \tag{4.8-4}$$

Solve for $E(s)$ in terms of $V(s)$.

$$E(s) = \frac{V(s)}{1 + G(s)}$$

The steady-state difference from the final value theorem is

$$e_{ss} = \lim_{s\to 0} sE(s) = \lim_{s\to 0} \frac{sV(s)}{1 + G(s)} \tag{4.8-5}$$

For a unit-step input, $V(s) = 1/s$, and (4.8-5) gives

$$e_{ss} = \lim_{s\to 0} \frac{1}{1 + G(s)} = \frac{1}{1 + G(0)} \tag{4.8-6}$$

where $G(0) = \lim_{s\to 0} G(s)$.

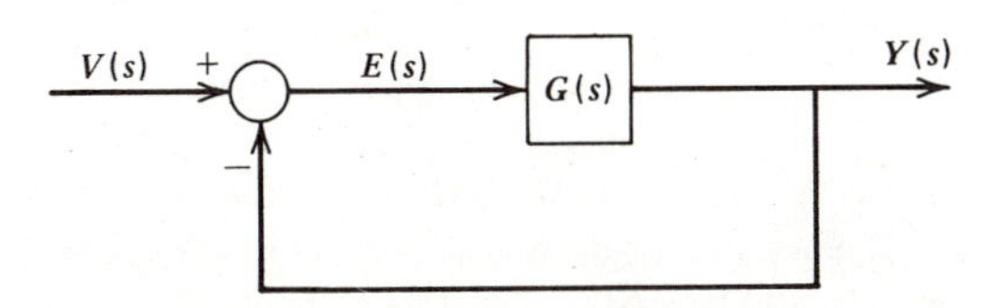

FIGURE 4.30 A unity feedback system. The error is $e(t) = v(t) - y(t)$.

For a unit-ramp input, $V(s) = 1/s^2$ and

$$e_{ss} = \lim_{s \to 0} \frac{1}{s + sG(s)} = \frac{1}{\lim_{s \to 0} sG(s)} \tag{4.8-7}$$

(b) For $G(s) = K/s$, $G(0) = \lim_{s \to 0} K/s \to \infty$, and $e_{ss} \to 0$. Thus, this particular system has a zero steady-state difference between the output and the step input.

For a unit-ramp input, $\lim_{s \to 0} sG(s) = K$, and $e_{ss} = 1/K$, which is a finite but nonzero difference.

4.9 RESPONSE CALCULATIONS WITH THE LAPLACE TRANSFORM†

The trial solution method presented in Chapter 3 allows the system response to be determined quickly in some cases. However, for higher order systems, more complicated input functions, or systems with numerator dynamics, this method relies too heavily on memorized solution forms to be generally useful. In these situations, the Laplace transform is often useful in computing the response. Here, we demonstrate its usefulness in dealing with complex roots, systems having numerator dynamics, and two new input types: the pulse and the impulse.

Response Calculations with Complex Roots

The Laplace transform can be used to obtain the response directly in the form of a sine function with a phase shift when a complex root pair occurs. Assume that the Laplace transformation of the system model results in the following expression for the transform of the response $X(s)$:

$$X(s) = \frac{P(s)}{Q(s)} \tag{4.9-1}$$

where $Q(s)$ is a quadratic polynomial of the form

$$Q(s) = (s + a)^2 + b^2$$

Thus, the roots are $s = -a \pm ib$. The term $P(s)$ represents the remaining terms in $X(s)$. A partial fraction expansion of $X(s)$ for the quadratic roots gives

$$X(s) = \frac{C_1}{s + a - ib} + \frac{C_2}{s + a + ib} \tag{4.9-2}$$

---

†This section deals with more general analytical methods for evaluating the system response. In some cases, the simulation methods to be presented in Chapter 5 can also be used to obtain the response. If time is limited for covering system response, this section can be omitted.

where

$$C_1 = \lim_{s \to -a+ib} \left[\frac{P(s)}{Q(s)}(s+a-ib) \right] \tag{4.9-3}$$

$$C_2 = \lim_{s \to -a-ib} \left[\frac{P(s)}{Q(s)}(s+a+ib) \right] \tag{4.9-4}$$

After the factor is canceled, C_1 has the form

$$C_1 = \lim_{s \to -a+ib} \left[\frac{P(s)}{s+a+ib} \right]$$

$$= \frac{1}{2ib} R(-a+ib) \tag{4.9-5}$$

where R is a function of the complex number $-a+ib$, and

$$R(-a+ib) = \lim_{s \to -a+ib} [P(s)] \tag{4.9-6}$$

Since $R(-a+ib)$ can be complex in general, it can be expressed as a magnitude and a phase angle, as shown in Figure 4.31.

$$R(-a+ib) = |R(-a+ib)|e^{i\phi}$$

Thus,

$$C_1 = \frac{1}{2ib} |R(-a+ib)|e^{i\phi}$$

and since C_2 is the conjugate of C_1,

$$C_2 = -\frac{1}{2ib} |R(-a+ib)|e^{-i\phi}$$

Reverting to the time domain from (4.9-2), we obtain

$$x(t) = C_1 e^{-at} e^{ibt} + C_2 e^{-at} e^{-ibt}$$

$$= \frac{1}{b} |R(-a+ib)| e^{-at} \frac{e^{ibt} e^{i\phi} - e^{-ibt} e^{-i\phi}}{2i}$$

$$= \frac{1}{b} |R(-a+ib)| e^{-at} \sin(bt+\phi) \tag{4.9-7}$$

where we have used the identity

$$\sin\theta = \frac{e^{i\theta} - e^{-i\theta}}{2i}$$

$$\theta = bt + \phi$$

The phase angle is given by

$$\phi = \measuredangle R(-a+ib) \tag{4.9-8}$$

The preceding expressions can be used to determine that part of the free or forced response resulting from a pair of complex roots. If the system is of order greater than

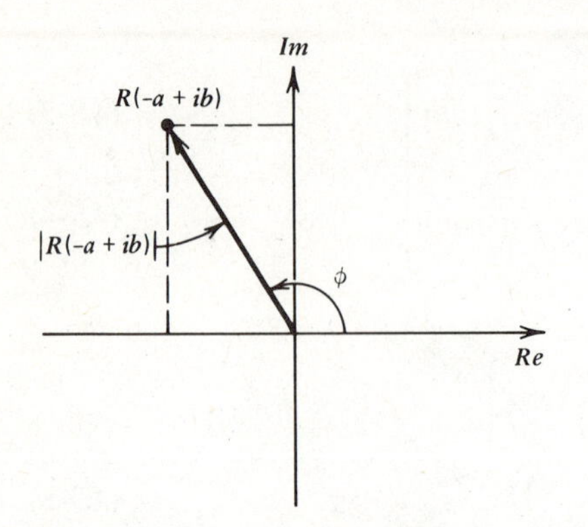

FIGURE 4.31 Polar representation of the complex factor $R(-a+ib)$.

two, or if the input function introduces more roots, then (4.9-7) gives only part of the response. The rest of the response is found by doing a complete partial-fraction expansion for all the roots and adding all of the resulting response terms. The method is summarized in Table 4.6, and illustrated by Example 4.15.

Example 4.15

Two similar mechanical systems are shown in Figure 4.32. Their transfer functions are

$$\frac{X(s)}{U(s)} = \frac{k+cs}{ms^2+cs+k} \tag{4.9-9}$$

and

$$\frac{X(s)}{U(s)} = \frac{k}{ms^2+cs+k} \tag{4.9-10}$$

The system in Figure 4.32*a* has numerator dynamics. Both systems can be represented by the general differential equation

$$m\ddot{x} + c\dot{x} + kx = du + g\dot{u} \tag{4.9-11}$$

where $u(t)$ is the input, and $g = 0$ for the system in Figure 4.32*b*.

(a) Obtain the unit-step response of (4.9-11). Assume that $g \neq 0$ and that $\zeta < 1$.

(b) Compare the response found in (a) with that for $g = 0$.

(a) With zero-initial conditions, the transformation gives

$$(ms^2 + cs + k)X(s) = (d + gs)U(s) = \frac{d}{s} + g \tag{4.9-12}$$

TABLE 4.6 Response Calculation for Complex Roots

1. Response transform $= X(s)$

$$X(s) = \frac{P(s)}{Q(s)} \qquad Q(s) = (s+a)^2 + b^2$$
$P(s)$ absorbs all other terms.

2. Solution form:

$$x(t) = \frac{1}{b}|R(-a+ib)|e^{-at}\sin(bt+\phi)$$
$$+ \text{ terms due to the roots of } P(s)$$
$$R(-a+ib) = \lim_{s\to -a+ib} [P(s)]$$
$$\phi = \measuredangle R(-a+ib) \qquad \text{(see Figure 4.31)}$$

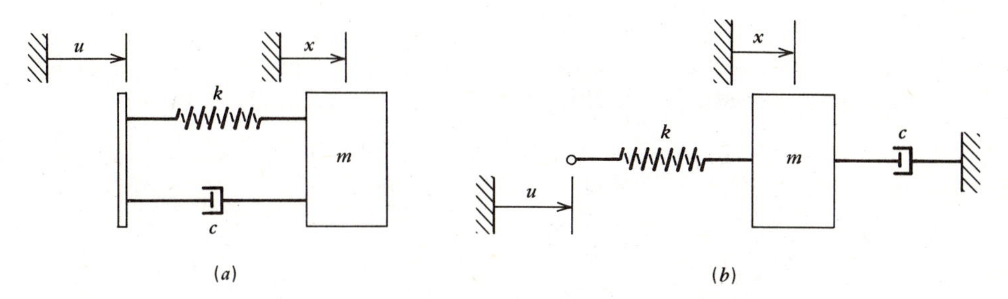

FIGURE 4.32 Two mechanical systems. The model for *a* has numerator dynamics.

Solving for $X(s)$ and using the parameters ζ and ω_n, we obtain

$$X(s) = \frac{d\omega_n^2}{zk} \frac{s+z}{(s^2 + 2\zeta\omega_n s + \omega_n^2)s}$$

where we have defined $z = d/g$. Since $P(s)$ is defined to contain everything in $X(s)$ except the quadratic factors,

$$P(s) = \frac{d\omega_n^2}{zk} \frac{s+z}{s}$$

Therefore, with $a = \zeta\omega_n$ and $b = \omega_n\sqrt{1-\zeta^2}$,

$$\begin{aligned} R(-a+ib) &= \lim_{s \to -a+ib} \left[\frac{d\omega_n^2}{zk} \frac{(s+z)}{s} \right] \\ &= \frac{d\omega_n^2}{zk} \left(\frac{z-a+ib}{-a+ib} \right) \left(\frac{-a-ib}{-a-ib} \right) \\ &= \frac{d\omega_n^2}{zk} \frac{a^2+b^2-az-ibz}{a^2+b^2} \\ &= \frac{d\omega_n}{zk} (\omega_n - \zeta z - iz\sqrt{1-\zeta^2}) \end{aligned}$$

The magnitude and phase angle are

$$\begin{aligned} |R(-a+ib)| &= \frac{d\omega_n}{zk} \sqrt{(\omega_n - \zeta z)^2 + z^2(1-\zeta^2)} \\ &= \frac{d\omega_n}{zk} \sqrt{\omega_n^2 - 2\zeta\omega_n z + z^2} \end{aligned}$$

$$\phi = \measuredangle R(-a+ib) = -\tan^{-1}\left(\frac{z\sqrt{1-\zeta^2}}{\omega_n - \zeta z} \right), \qquad \omega_n - \zeta z > 0 \qquad \textbf{(4.9-13a)}$$

or

$$\phi = \pi + \tan^{-1}\left(\frac{z\sqrt{1-\zeta^2}}{\zeta z - \omega_n} \right), \qquad \omega_n - \zeta z < 0 \qquad \textbf{(4.9-13b)}$$

or

$$\phi = \frac{\pi}{2}, \qquad \omega_n - \zeta z = 0 \qquad \textbf{(4.9-13c)}$$

That part of the response resulting from the quadratic roots is found from (4.9-7), using the preceding expressions for $R(-a+ib)$ and ϕ. The complete expansion of $X(s)$ here must also include the effect of the third root $s=0$. The expansion is

$$X(s)=\frac{C_1}{s+a-ib}+\frac{C_2}{s+a+ib}+\frac{C_3}{s}$$

where

$$C_3=\lim_{s\to 0}[X(s)s]=\frac{d}{k}$$

The root $s=0$ contributes the term d/k to the response. Adding the contributions for each root gives the total solution

$$x(t)=\frac{d}{k}\left[\frac{1}{z}\sqrt{\frac{\omega_n^2-2\zeta\omega_n z+z^2}{1-\zeta^2}}\,e^{-\zeta\omega_n t}\sin\left(\omega_n\sqrt{1-\zeta^2}\,t+\phi\right)+1\right] \quad \textbf{(4.9-14)}$$

where ϕ was given previously.

(b) The larger g is, the more significant is the effect of the derivative of the input. To see exactly what the effect is, compare the solution for (a) to (3.5-8), which is the solution for the case $g=0$. Both solutions have the same frequency and steady-state value; however, the amplitude of oscillation and the phase angle are both different. The ratio of the amplitude of (4.9-14) to that of (3.5-8) is

$$\frac{1}{z}\sqrt{\omega_n^2-2\omega_n z+z^2}=\sqrt{\left(\frac{\omega_n}{z}\right)^2-\frac{2\omega_n}{z}+1}$$

As g increases, $z=d/g\to 0$, and the amplitude for the case with numerator dynamics becomes very large compared to the case $g=0$. As $g\to 0$, $z\to\infty$, and the amplitudes become equal, as expected.

As the numerator zero approaches the imaginary axis (i.e., as $z\to 0$), the phase angle $\phi\to 0$. Physically this means that g is so large that the response follows the step input very quickly. For $g\neq 0$, a plot of the response versus $\omega_n t$ shows that it passes the final value earlier than when $g=0$ and the overshoot is greater. An example of the response is shown in Figure 4.33 for the values: $m=1$, $k=1$, $c=1.4$, $\zeta=0.7$. The case where $g=0$ corresponds to the absence of numerator dynamics.

Table 4.7 summarizes the response for the three root cases.

Initial Conditions and Numerator Dynamics

The Laplace transform can be used to provide some insight into how numerator dynamics affects the response. The transform of (4.9-12) with $g=0$, for a unit-step input and nonzero initial conditions gives

$$(ms^2+cs+k)X(s)=\frac{d}{s}+msx(0)+cx(0)+m\dot{x}(0) \quad \textbf{(4.9-15)}$$

Comparing the right-hand sides of (4.9-12) and (4.9-15) shows that the solution (4.9-14) can also be obtained from the unit-step response with $g=x(0)=0$, and

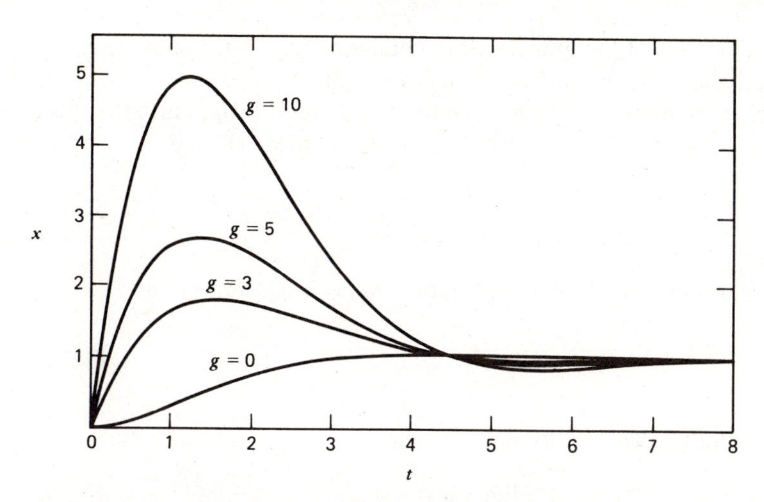

FIGURE 4.33 The effects of numerator dynamics on the unit step response. The model is: $\ddot{x} + 1.4\dot{x} + x = u + g\dot{u}$.

TABLE 4.7 Step Response for the Numerator-Dynamics Model: $m\ddot{x} + c\dot{x} + kx = du + g\dot{u}$

$$u(t) = 1, \qquad t > 0$$

$$x(0) = \dot{x}(0) = 0$$

1. Real, distinct roots: $s = s_1, s_2$

$$x(t) = C_1 e^{s_1 t} + C_2 e^{s_2 t} + C_3$$

$$C_1 = \frac{d + gs_1}{ms_1(s_1 - s_2)} \qquad C_2 = \frac{d + gs_2}{ms_2(s_2 - s_1)}$$

$$C_3 = \frac{d}{ms_1 s_2}$$

2. Real, repeated roots: $s = s_1, s_1$

$$x(t) = C_1 t e^{s_1 t} + C_2 e^{s_1 t} + C_3$$

$$C_1 = \frac{d + gs_1}{ms_1} \qquad C_2 = -\frac{d}{ms_1^2} \qquad C_3 = \frac{C_1}{s_1}$$

3. Complex conjugate roots:

$$x(t) = \frac{d}{k}\left[\frac{1}{z}\sqrt{\frac{\omega_n^2 - 2\zeta\omega_n z + z^2}{1 - \zeta^2}}\, e^{-\zeta\omega_n t} \sin\left(\omega_n\sqrt{1 - \zeta^2}\, t + \phi\right) + 1\right], \quad z = \frac{d}{g}$$

$$\phi = \measuredangle R(-a + ib) = -\tan^{-1}\left(\frac{z\sqrt{1 - \zeta^2}}{\omega_n - \zeta^2}\right), \qquad \omega_n - \zeta z > 0$$

$$\phi = \pi + \tan^{-1}\left(\frac{z\sqrt{1 - \zeta^2}}{\zeta z - \omega_n}\right), \qquad \omega_n - \zeta z < 0$$

$$\phi = \frac{\pi}{2}, \qquad \omega_n - \zeta z = 0$$

$\dot{x}(0) = g/m$. Thus, the effect of the numerator dynamics acting on the step input is to create an "artificial" nonzero initial velocity of value g/m.

The initial value theorem can also be used to provide some insight into the effects of numerator dynamics. Consider the transfer function of (4.9-11)

$$\frac{X(s)}{U(s)} = \frac{d + gs}{ms^2 + cs + k} \tag{4.9-16}$$

For zero initial conditions and a unit-step input $[U(s) = 1/s]$, we obtain

$$X(s) = \frac{d + gs}{s(ms^2 + cs + k)} \tag{4.9-17}$$

Applying the initial value theorem to (4.9-17) gives

$$x(0) = \lim_{s \to \infty} sX(s) = 0$$

for any value of g. However, suppose that we solve for the initial velocity using $V(s) = sX(s)$. Then we obtain

$$V(s) = \frac{d + gs}{ms^2 + cs + k} \tag{4.9-18}$$

The initial value theorem gives

$$v(0) = \lim_{s \to \infty} \frac{ds + gs^2}{ms^2 + cs + k} = \frac{g}{m} \tag{4.9-19}$$

Thus, $v(0) = 0$ for $g = 0$, but $v(0) \neq 0$ if $g \neq 0$. This confirms the result obtained from (4.9-15).

The discrepancy is explained as follows. We can think of the input as being applied just slightly before $t = 0$ (this time is denoted as $t = 0-$). The true initial conditions are given at this time. Thus, in this example, $v(0-) = 0$. Just slightly after applying the input, at $t = 0+$, the velocity has the value $v(0+)$. For $g \neq 0$, the effect of the numerator dynamics is to differentiate the input. Since the step input is changing very quickly (in fact, instantaneously) at $t = 0$, the numerator dynamics term gs provides an extra but short-lived "kick" to the system at $t = 0+$. This is reflected in an instantaneous change in the velocity from $v(0-) = 0$ to $v(0+) = g/m$. This shows up as a discontinuous slope on the plot of the response $x(t)$ (see Figure 4.34). In the figure, curve a corresponds to $g \neq 0$, and curve b corresponds to $g = 0$.

The possibility of a physical system behaving in this manner is another question, but since numerator dynamics often appear in approximate models of systems, we must be aware of their implications. A true step input is also physically impossible, because no physical variable can change value instantaneously. Thus, if we are interested in predicting the response of a model with numerator dynamics and a step input, we should examine more closely either the model's assumptions that led to the numerator dynamics or the assumptions that led us to describe the input as a step function. This is explored in more detail in later chapters.

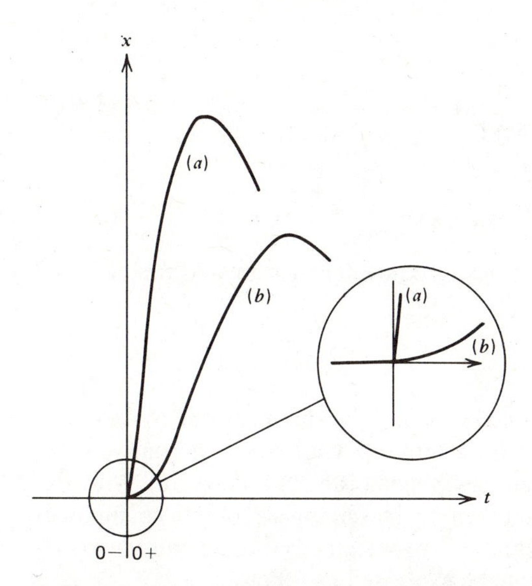

FIGURE 4.34 The effect of numerator dynamics on the response at $t = 0+$. The discontinuous slope in curve a is caused by the numerator dynamics.

Impulse Response

Besides the step function, the *pulse* function and its approximation, the *impulse*, appear quite often in the analysis and design of dynamic systems. In addition to being an analytically convenient approximation of an input applied for only a very short time, the impulse is also useful for estimating the system's parameters experimentally. The impulse is an abstraction that does not exist in the physical world but can be thought of as the limit of a rectangular pulse whose duration T approaches zero while maintaining its *strength* A. The strength of an impulse or pulse is the area under its time curve (Figure 4.35).

The impulse response of the first-order model $\dot{y} = ry + bv$ can be obtained by the Laplace transform method. From Table B.1, the transform of an impulse $v(t)$ of strength A is

$$V(s) = A$$

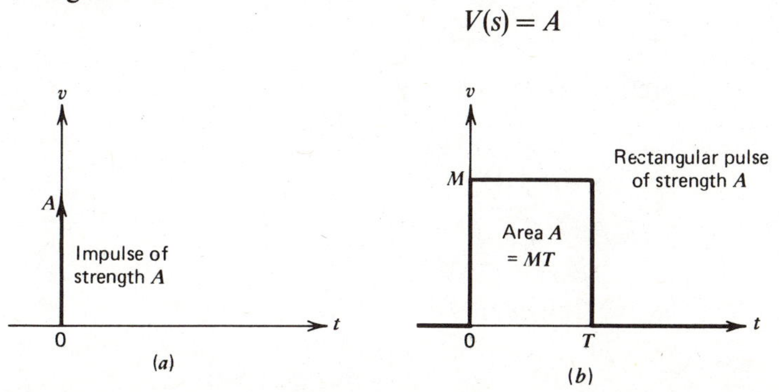

FIGURE 4.35 Impulse and rectangular pulse. (*a*) Impulse of strength A. (*b*) Rectangular pulse. The impulse is the limit of the pulse as $T \to 0$ with A held constant.

From (4.1-7),

$$Y(s) = \frac{y(0)}{s-r} + \frac{b}{s-r}A = \frac{y(0)+bA}{s-r} \tag{4.9-20}$$

In the time domain, this becomes

$$y(t) = [y(0) + bA]e^{rt} \tag{4.9-21}$$

Thus, the impulse can be thought of as being equivalent to an additional initial condition of magnitude bA.

Pulse Response

The response due to a pulse can be obtained by using the step response to find $y(T)$, which is then used as the initial condition for a zero input solution. Alternatively, the *shifting property* of Laplace transforms can be applied (Table 4.1, entry 10). With this viewpoint, the pulse in Figure 4.35*b* is taken to be composed of a step input of magnitude M starting at $t = 0$, followed at $t = T$ by a step input of magnitude $(-M)$ (see Figure 4.36). The pulse input can now be expressed as follows.

$$v(t) = Mu_s(t) - Mu_s(t-T) \tag{4.9-22}$$

Its transform is

$$\begin{aligned} V(s) &= M\mathscr{L}[u_s(t)] - M\mathscr{L}[u_s(t-T)] \\ &= M\frac{1}{s} - Me^{-sT}\frac{1}{s} \\ &= \frac{M}{s}(1 - e^{-sT}) \end{aligned} \tag{4.9-23}$$

Assume that the system is stable and the initial condition is zero. The pulse response is found from (4.1-7) with $r = -1/\tau$ and a partial fraction expansion.

$$\begin{aligned} Y(s) &= \frac{\tau bM}{\tau s + 1}\left(\frac{1 - e^{-sT}}{s}\right) \\ &= \frac{C_1}{\tau s + 1} + \frac{C_2}{s} - \frac{C_1}{\tau s + 1}e^{-sT} - \frac{C_2}{s}e^{-sT} \end{aligned}$$

The transform has been expressed as the sum of elementary transforms. Cross multiplication gives

$$C_1 s + C_2(\tau s + 1) = \tau bM$$

or

$$C_1 = -\tau^2 bM$$
$$C_2 = \tau bM$$

In the time domain, we obtain

$$y(t) = \frac{C_1}{\tau}e^{-t/\tau} + C_2 u_s(t) - \frac{C_1}{\tau}e^{-(t-T)/\tau}u_s(t-T) - C_2 u_s(t-T)$$

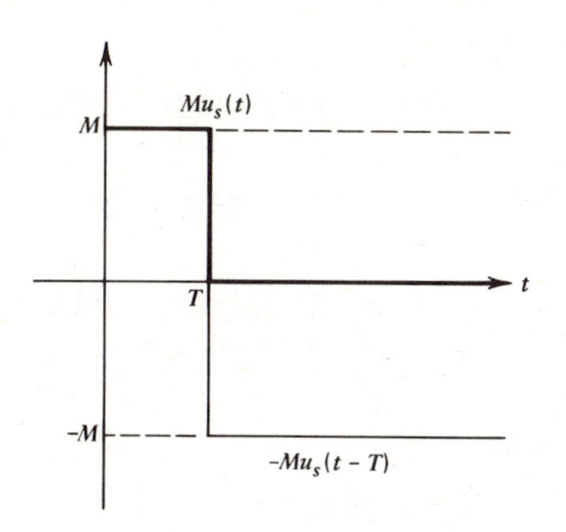

FIGURE 4.36 Rectangular pulse as the superposition of two step functions.

For $0 < t < T$,

$$y(t) = \frac{C_1}{\tau} e^{-t/\tau} + C_2$$
$$= \tau bM - \tau bMe^{-t/\tau} \quad \textbf{(4.9-24)}$$

For $t \geqslant T$,

$$y(t) = \frac{C_1}{\tau} e^{-t/\tau} - \frac{C_1}{\tau} e^{-(t-T)/\tau} + C_2 - C_2 .$$
$$= -\tau bM(1 - e^{T/\tau})e^{-t/\tau} \quad \textbf{(4.9-25)}$$

This response is shown in Figure 4.37. The previous equation, when written in terms of the pulse strength $A = MT$, is

$$y(t) = -\frac{\tau bA}{T}(1 - e^{T/\tau})e^{-t/\tau} \quad \textbf{(4.9-26)}$$

If the strength A is kept constant as T approaches zero, L'Hôpital's rule gives

$$\lim_{T \to 0} y(t) = bAe^{-t/\tau}$$

This is the same as the impulse response when $y(0)$ is zero.

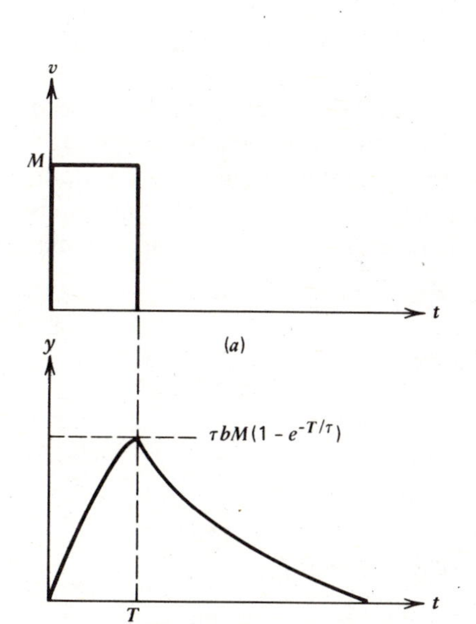

FIGURE 4.37 Pulse response of a linear first-order system. (*a*) Pulse input. (*b*) Pulse response.

Approximation of Pulse Response

A rule of thumb for determining when a pulse can be approximated by an impulse is obtained as follows. The Taylor series expansion for $\exp(T/\tau)$ is

$$e^{T/\tau} = 1 + \frac{T}{\tau} + \frac{1}{2}\left(\frac{T}{\tau}\right)^2 + \cdots$$
$$+ \frac{1}{n!}\left(\frac{T}{\tau}\right)^n + \cdots$$

If the first two terms in the series are retained and substituted into (4.9-26), the result is

$$y(t) = -\frac{\tau bA}{T}\left(1 - 1 - \frac{T}{\tau} - \cdots\right)e^{-t/T}$$
$$\cong bAe^{-t/\tau}$$

This is identical to the impulse response.

To see when it is justifiable to truncate the series in this way, consider the case where $T/\tau = 0.1$. Thus,

$$e^{T/\tau} = 1 + 0.1 + 0.005 + 0.00017 + \cdots$$

For accuracy to the second decimal place, only the first two terms need be kept. This leads to the following guide, which can also be shown true for an *arbitrarily shaped* pulse.

If the pulse duration is of the order of $\tau/10$ or less, the first-order system response is nearly the same as the impulse response.

If we are willing to accept the dominant-root approximation, then the preceding guide can be extended to higher order systems, where τ is the dominant-time constant.

The Transfer Function and Impulse Response

Equation (4.9-20) shows that if the initial condition $y(0) = 0$, the unit-impulse response is given by

$$Y(s) = \frac{b}{s - r}$$

or

$$y(t) = be^{rt}$$

But from (4.2-1),

$$T(s) = \frac{b}{s - r}$$

This shows that the Laplace transform of the unit-impulse response is identical to the system's transfer function. In general, since $Y(s) = T(s)V(s)$, if $V(s) = 1$ (a unit impulse), then $Y(s) = T(s)$, as stated.

This important property allows the transfer function of a system to be determined experimentally by placing a pulse input on the system and measuring the output as a function of time. If the pulse duration is small compared to the system's time constant, the measured response will approximate the impulse response. The transfer function is found by computing the Laplace transform of the measured function by approximating it with a convenient functional form.

Impulse Response of a Second-Order System

Consider the second-order model

$$m\ddot{x} + c\dot{x} + kx = du(t) \tag{4.9-27}$$

If the input $u(t)$ is zero, the transform gives

$$(ms^2 + cs + k)X(s) = (ms + c)x(0) + m\dot{x}(0)$$

However, if the initial conditions are zero and the input is a unit impulse $[U(s) = 1]$, then

$$(ms^2 + cs + k)X(s) = d$$

Comparing the last two expressions shows that the response to a unit impulse is equivalent to the free response for the special set of initial conditions such that

$$d = (ms + c)x(0) + m\dot{x}(0)$$

or

$$x(0) = 0$$
$$m\dot{x}(0) = d$$

For a mass–spring–damper system, this means that if the mass is started at the equilibrium $x(0) = 0$ with a velocity $\dot{x}(0) = d/m$, the resulting free response is equivalent to the unit-impulse response for zero initial conditions. This corresponds to the impulse momentum principle of Newtonian mechanics, which says that the change in momentum, here $m\dot{x}(0)$, must equal the impulse, which is the time integral of the applied force.

We have now seen two situations that act to create the effect of fictitious initial conditions at $t = 0+$; namely, an impulse input, and a step input with numerator dynamics. Neither situation changes $x(0+)$ for the second-order system. The unit impulse gives $\dot{x}(0+) = d/m$; the unit step with numerator dynamics gives $\dot{x}(0+) = g/m$. Therefore, numerator dynamics acting on a unit-step input is equivalent to applying an impulse input of magnitude g/d to a system without numerator dynamics.

4.10 SIGNAL FLOW GRAPHS†

An alternative graphical representation of the system's transformed equations is the *signal flow graph* in which variables are represented as *nodes* and the operations on these variables are represented by directed line segments between the nodes. This contrasts with the block diagram's representation of variables as lines and operations as blocks. Because of this difference, it is easier to find the transfer function of the system with a signal flow graph if the system has many loops. A general formula useful for reducing complicated graphs is *Mason's gain formula* (Appendix D). Here, we will confine ourselves to a brief introduction to signal flow graphs.

The principal advantage of a block diagram is that the system's physical components (the elements performing the operations) are themselves represented by blocks indicating their operations rather than by line segments. Thus, the block diagram more closely resembles the physical structure of the system. The choice between the two representations is made on the basis of convenience.

The value of a variable at a node is obtained by *summing* the incoming signals at that node. This sum is transmitted to all outgoing line segments from the node. The directed line segments show the relationship between the variables whose nodes are connected by the line, with the arrow indicating the direction of causality. Subtraction is indicated by a negative sign with the operation denoted by the proper line segment. An example of a signal flow graph is shown in Figure 4.38. The variables are $V(s)$, $E(s)$,

---

†This section presents an alternative to the block diagram for graphically representing a system's structure. It can be omitted without jeopardizing comprehension of material in succeeding chapters.

and $Y(s)$ and are shown as nodes (circles). The variable $E(s)$ is obtained by summing the incoming signals to obtain

$$\begin{aligned} E(s) &= 1V(s) + [-H(s)Y(s)] \\ &= V(s) - H(s)Y(s) \end{aligned}$$

The output $Y(s)$ is

$$Y(s) = G(s)E(s)$$

Combining the last two equations results in the same transfer function obtained from the block diagram. With the basic properties of nodes and line segments, we can develop rules similar to those given in Figure 4.4, with Mason's gain formula as one result.

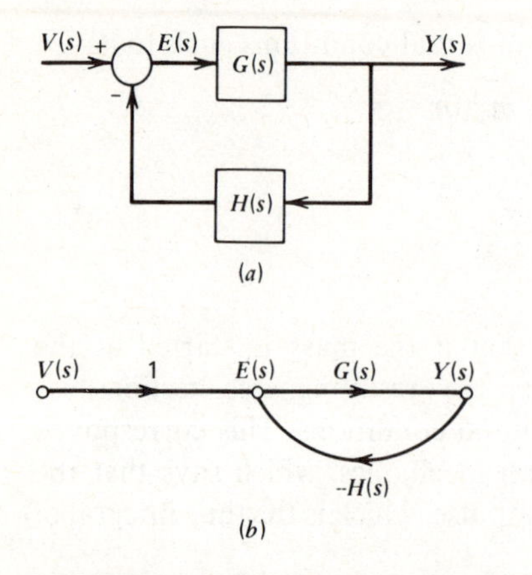

FIGURE 4.38 Comparison of (*a*) Block diagram and (*b*) Signal flow graph.

As an additional example, we display in Figure 4.39 the block diagram and signal flow graph for the mass–spring–damper system shown in Figure 2.10.

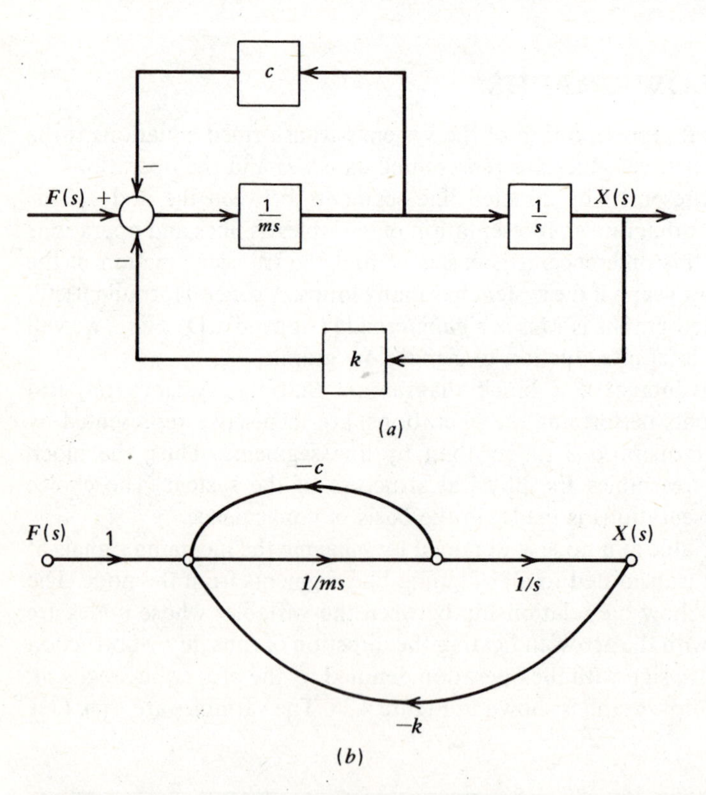

FIGURE 4.39 Comparison of the block diagram and signal flow graph for the mass–spring–damper system in Figure 2.10.

4.11 SUMMARY

Chapter 4 explores using the Laplace transform to provide algebraic and graphical representations of system structure and response. These methods will be especially useful in studying and designing control systems, starting in Chapter 6. The following is a summary of Chapter 4, arranged by section.

(4.1)–(4.4) Operator notation is introduced to provide algebraic representation of system models. The Laplace transform operator provides a general algebraic solution method for linear constant-coefficient models. A simple partial-fraction expansion technique enables a systematic table look-up procedure to be used to recover the time domain solution. The system's transfer function is the ratio of the output transform to the input transform for zero initial conditions. Block diagram representations of the system are possible using transfer function blocks for each subsystem. In addition to providing visualization of the system's dynamics, these diagrams can be reduced algebraically to find the overall system transfer function and differential equation. Thus, the effect of new system components on the overall behavior is easily determined. An alternative representation, the signal flow graph, is presented, and it has some advantages when dealing with systems composed of many elements. The transfer function between a rate variable as input and an effort variable as output is an impedance, which may be thought of as a generalized resistance. It is useful in determining the power-loading effects of connected elements and in combining dynamic elements in a model.

(4.5)–(4.7) The frequency response of a system describes its steady-state behavior resulting from periodic inputs. Later, we will see that it is also useful in predicting transient behavior and stability properties. A sinusoidal input applied to a linear system produces a steady-state sinusoidal output of the same frequency, but with a different amplitude and a phase shift. The frequency transfer function is the transfer function with the Laplace variable s replaced by $i\omega$, where ω is the input frequency. The magnitude M of the frequency transfer function is the amplitude ratio between the input and output, and its argument is the phase shift ϕ. Both are functions of ω and usually plotted against $\log \omega$, with M expressed in decibels. A polar plot is also useful sometimes. The logarithmic plots are easily sketched for the first-order system by using low- and high-frequency asymptotes, with the corner frequency given by $1/\tau$. The logarithmic plots for an underdamped second-order system can also be sketched with low- and high-frequency asymptotes, but near $\omega = \omega_n$, the plots depend heavily on the value of ζ. Because most transfer functions consist of terms like K, s, $\tau s + 1$, and $s^2 + 2\zeta\omega_n s + \omega_n^2$, sketching the plot is simplified by adding or subtracting the contribution of each term, depending on whether the term is in the numerator or denominator.

The bandwidth measures the filtering property of the system. The Fourier series representation of a periodic function can be used with the system's filtering property to compute the general periodic response in terms of the sinusoidal response.

(4.8) The ramp response provides a good example of the use of the initial and final value theorems. Given the transfer function and the input transform, these theorems provide a quick indication of the initial and steady-state values of the output under mild restrictions.

(4.9) The Laplace transform is more useful than the trial solution method for higher order systems, complicated input functions, and models with numerator dynamics. The Laplace transform can be used to obtain the response directly in the form of a sine function when complex roots occur. The transform's utility was demonstrated by obtaining the step response of a second-order system with numerator dynamics. The effects of numerator dynamics are to increase the overshoot and make it occur earlier. The initial value theorem shows how numerator dynamics act to create fictitious initial conditions at $t=0+$.

The impulse is an analytically convenient approximation of an input that is applied for only a very short time. If the input duration is less than one-tenth of the system's time constant, the impulse response may be used with considerable accuracy. The impulse response is equivalent to the free response with adjusted initial conditions. The transfer function is the Laplace transform of the unit-impulse response. For longer duration inputs, the pulse response can be obtained with the shifting property of the Laplace transform.

REFERENCES

1. E. Kreysig, *Advanced Engineering Mathematics*, 5th ed., John Wiley & Sons, New York, 1983.

2. W. J. Palm III, *Modeling, Analysis, and Control of Dynamic Systems*, John Wiley & Sons, New York, 1983.

PROBLEMS

4.1 Use the Laplace transform method to obtain the forced response of the system $\dot{y} = -2y + v(t)$ for $t \geqslant 0$, for the following cases:

(a) $v(t) = t$
(b) $v(t) = t^2$
(c) $v(t) = te^{-t}$

4.2 Use the Laplace transform to find the free response and the step response (for zero initial conditions) of the following models:

(a) $\ddot{x} + 8\dot{x} + 12x = 2,\ x(0) = 0,\ \dot{x}(0) = 1$
(b) $\ddot{x} + 6\dot{x} + 9x = 2,\ x(0) = 0,\ \dot{x}(0) = 1$

4.3 Find the transfer function, and draw the block diagram for the following model. The input is u; the output is x_1. Show the variables x_2 and x_3 on the diagram.

$$\dot{x}_1 = x_2$$
$$\dot{x}_2 = x_3$$
$$\dot{x}_3 = x_2 - 2x_3 + u$$

4.4 Find the transfer functions, and draw the block diagrams of the following models for the given inputs and outputs. Show both variables x_1 and x_2 on each diagram.

(a) $\dot{x}_1 = -5x_1 + 3x_2$
$\dot{x}_2 = x_1 - 4x_2 + 5u$
outputs $= x_1, x_2$ input $= u$

(b) $\dot{x}_1 = -5x_1 + 3x_2 + 4u_1$
$\dot{x}_2 = x_1 - 4x_2 + 5u_2$
output $= x_1$ inputs $= u_1, u_2$

4.5 Find the second-order differential equation for x_2 with u as input for the equations given in Problem 4.4*a*.

4.6 **(a)** Use the block diagram in Figure P4.6 to write the system's differential equation model directly in state variable form. The state variables are x_1 and x_2.

(b) Use block diagram reduction to derive the transfer function $X_1(s)/U(s)$.

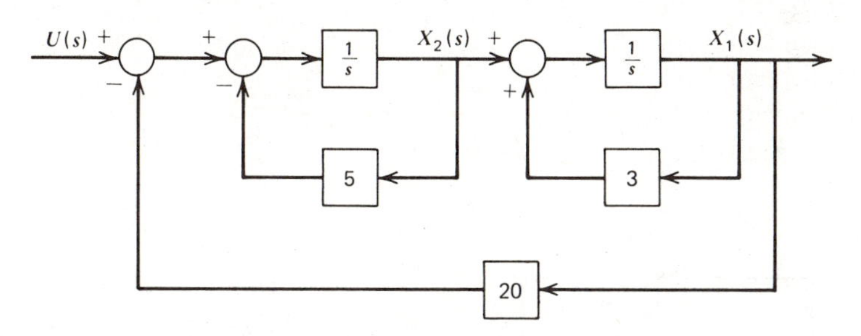

FIGURE P4.6

4.7 **(a)** The forced response of a linear system is the sum of the responses resulting from each input. Use this fact to reduce the diagram shown in Figure P4.7 in order to find the transfer functions relating y to v_1 and y to v_2.

(b) Find the differential equation model of the system.

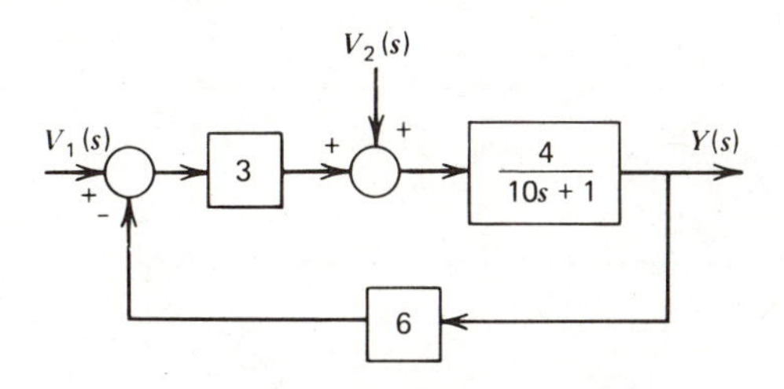

FIGURE P4.7

4.8 Find the transfer functions $Y(s)/U_1(s)$ and $Y(s)/U_2(s)$ for the block diagram shown. [*Hint:* Write the transformed equations using the intermediate variable $M(s)$. Then eliminate $M(s)$ algebraically.]

4.9 Differential equation models were derived for the following systems in Chapter 2's problems. Use these models to fill in the missing transfer functions in each

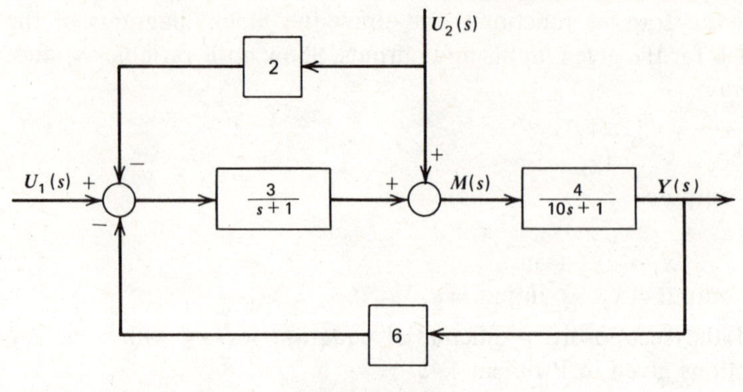

FIGURE P4.8

numbered block on the given diagram. Then find the overall transfer function for the system.

(a) Problem 2.10*b*
(b) Problem 2.10*d*
(c) Problem 2.10*f*
(d) Problem 2.21
(e) Problem 2.25

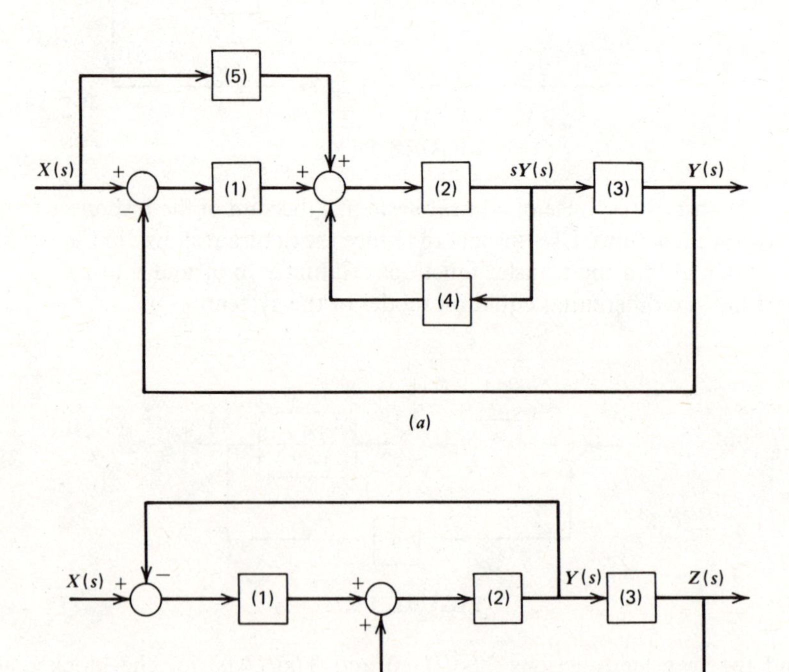

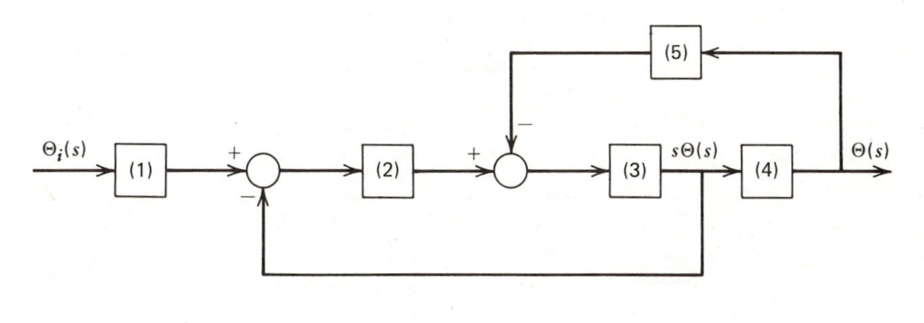

(c)

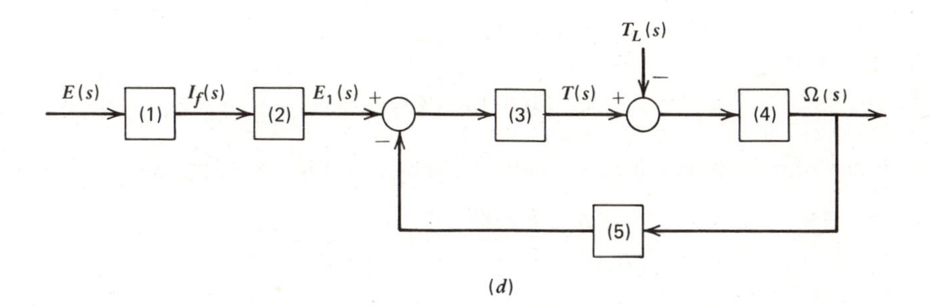

(d)

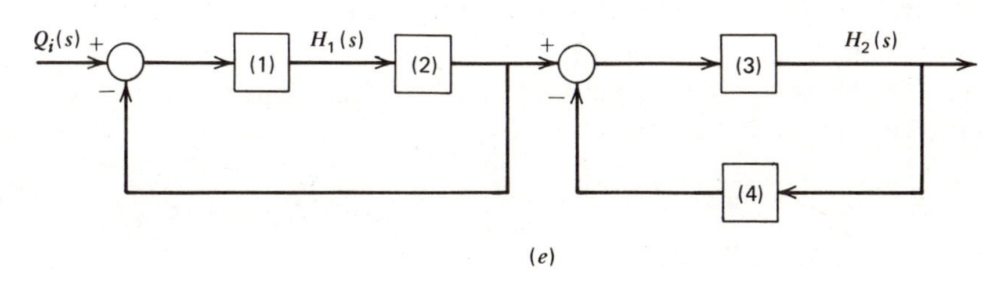

(e)

FIGURE P4.9

4.10 Differential equation models were derived for the following systems in Chapter 2's problems. Use these models to find the transfer function for the system. The input and output are specified in Chapter 2's problem statement.

(a) Problem 2.17
(b) Problem 2.19
(c) Problem 2.20
(d) Problem 2.28

4.11 Reduce the results of Problem 4.10*a* to the rigid-shaft case ($k \to \infty$, $I_1 + I_2 \to I$), and compare with the transfer function obtained in Example 2.8.

4.12 The transfer function of the series *RC* circuit shown in Figure P4.12*a* is

$$\frac{E_2(s)}{E_1(s)} = \frac{1}{RCs + 1}$$

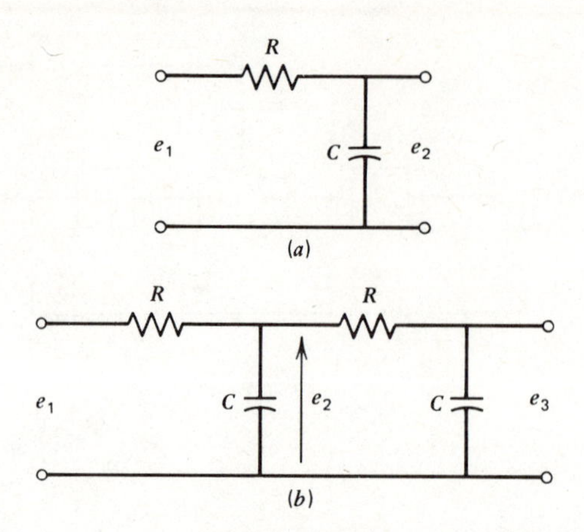

FIGURE P4.12

Is the transfer function of the network shown in Figure P4.12*b* equal to

$$\frac{E_3(s)}{E_1(s)} = \frac{E_3(s)E_2(s)}{E_2(s)E_1(s)} = \left(\frac{1}{RCs+1}\right)^2$$

Explain.

4.13 Use the impedance method to find the transfer functions for the circuits shown in Figure P4.13 for the indicated inputs and outputs.

(a) The input is e; the output is i.

(b) The input is e_i; the output is e_o. (This is a *lag* compensator.)

(c) The input is e_i; the output is e_o.

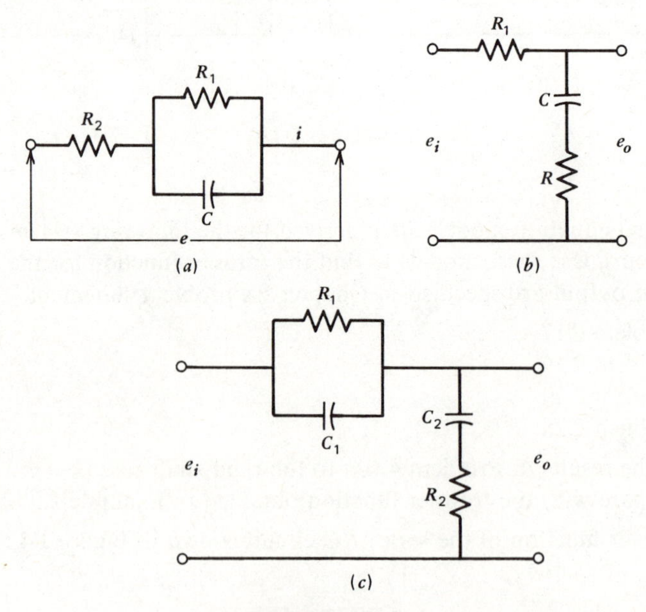

FIGURE P4.13

4.14 A torque T is applied to a load of inertia I. The damping is negligible so that $Is\Omega(s) = T(s)$, where ω is the speed of the load. For a sinusoidal torque $T(t) = A \sin \omega_f t$, plot the frequency response curves with $A/I = 2$ for a two-decade range centered at $\omega_f = 1$ rad/unit time.

4.15 Construct the frequency response plots for the following transfer functions:

(a) $T(s) = \dfrac{1}{3} \dfrac{1 + 0.02s}{1 + 0.00667s}$

(b) $T(s) = \dfrac{1 + 0.01s}{1 + 0.03s}$

(c) $T(s) = \dfrac{5}{(10s + 1)(4s + 1)}$

(d) $T(s) = \dfrac{4}{s^2 + 10s + 100}$

(e) $T(s) = \dfrac{4}{s(s^2 + 10s + 100)}$

(f) $T(s) = \dfrac{10}{s^2(s + 1)}$

(g) $T(s) = \dfrac{s}{(2s + 1)(5s + 1)}$

(h) $T(s) = \dfrac{s^2}{(2s + 1)(5s + 1)}$

4.16 Figure P4.16 shows an application of a *dynamic vibration absorber*. The absorber consists of a mass m_2 and elastic element k_2 that are connected to the main mass m_1. If the forcing function f is sinusoidal, proper selection of m_2 and k_2 can reduce the amplitude of motion of m_1 to zero. The technique is very effective and has been widely used.

(a) Derive the frequency transfer function for the displacements x_1 and x_2 with

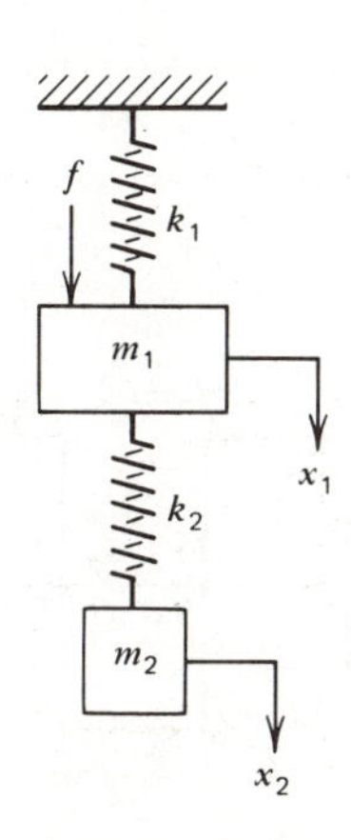

FIGURE P4.16

the force f as the input. The expressions are simplified if we let $r_1 = \omega/\omega_{n1}$, $r_2 = \omega/\omega_{n2}$, $b = \omega_{n2}/\omega_{n1}$, and $\mu = m_2/m_1$, where

$$\omega_{n1} = \sqrt{k_1/m_1}, \quad \omega_{n2} = \sqrt{k_2/m_2}$$

(b) What value of r_2 will make $|X_1(i\omega)| = 0$? This result gives the design condition for the absorber.

(c) For the value of r_2 found in (b), what is $|X_2(i\omega)|$? Assume that the forcing function is $f(t) = F_0 \sin \omega t$.

4.17 A certain machine with supports has an experimentally determined natural frequency of 3.43 Hz. It will be subjected to a rotating unbalanced force with an amplitude of 3 lb and a frequency of 3 Hz. Design a vibration absorber for this machine. The available clearance for the absorber's motion is 1 in.

4.18 For the plots found in Problem 4.15, find the resonance frequency and bandwidth if applicable.

4.19 Given the model

$$0.1\dot{y} + y = v(t)$$

with the following Fourier series representation of the input

$$v(t) = 0.5 \sin 4t + 2 \sin 8t + 0.02 \sin 12t$$
$$+ 0.03 \sin 16t + \cdots$$

(a) What is the bandwidth of the system?

(b) Find an approximate description of the output $y(t)$ at steady state, using only those input components that lie within the bandwidth.

4.20 A mass–spring–damper system is described by the model

$$m\ddot{x} + c\dot{x} + kx = f(t)$$

Where

$m = 0.05$ slugs

$c = 0.4$ lb/ft/sec

$k = 5$ lb/ft

$f(t) =$ externally applied force (lb), shown in Figure P4.20.

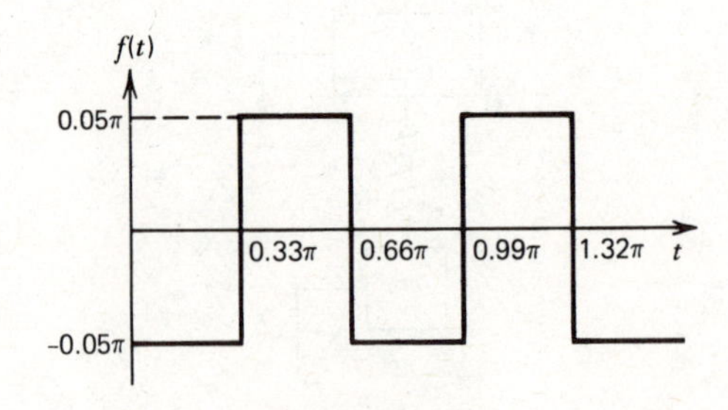

FIGURE P4.20

The forcing function can be expanded in a Fourier series as follows (see Appendix A).

$$f(t) = -0.2\left(\sin 3t + \tfrac{1}{3}\sin 9t + \tfrac{1}{5}\sin 15t + \tfrac{1}{7}\sin 21t + \cdots \right.$$
$$\left. + \frac{1}{n}\sin 3nt + \cdots\right) \qquad n \text{ odd}$$

(a) Find the bandwidth of this system.
(b) Find the steady-state response $x(t)$ by considering only those components of the $f(t)$ expansion that lie within the bandwidth of the system.

4.21 Use the final value theorem to compute the steady-state error e between the input and the output for the system shown in Figure P4.21, with the input functions as follows:

(a) $R(s) = 1/s$
(b) $R(s) = 1/s^2$

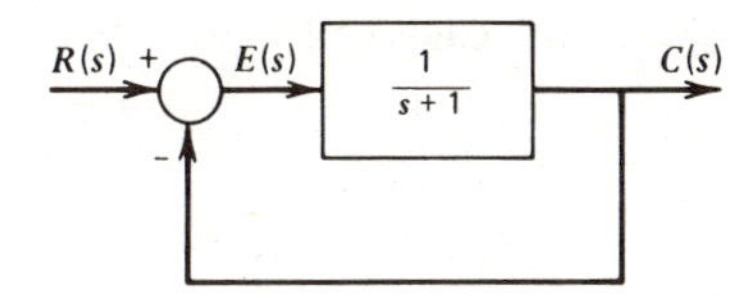

FIGURE P4.21

4.22 Compute the unit-step response for the models that follow. Assume they are stable ($d > 0$). Consider three cases: $c = 0$, $c > d$, and $c < d$. Compare the response with the step response of the first-order model without numerator dynamics.

(a) $T(s) = \dfrac{Ks}{s + d}$

(b) $T(s) = \dfrac{K(s + c)}{s + d}$

4.23 Apply the initial value theorem to the two models with numerator dynamics given in Problem 4.22, for a unit-step input and zero initial conditions. Explain the results.

4.24 Use the Laplace transform to compute the free response and the step response (for zero initial conditions) of the following model. Obtain the solution in the form of a sine function with a phase shift.

$$\ddot{x} + 4\dot{x} + 8x = 2, \quad x(0) = 0, \quad \dot{x}(0) = 1$$

4.25 Use the Laplace transform to find the ramp response for the following models. For (b) and (c) obtain the solution in the form of a sine function with a phase shift.

(a) $\ddot{x} + 8\dot{x} + 12x = 6t$, $x(0) = 1$, $\dot{x}(0) = 0$
(b) $\ddot{x} + 4\dot{x} + 8x = t$, $x(0) = \dot{x}(0) = 0$

(c) $m\ddot{x} + c\dot{x} + kx = dt$, $x(0) = \dot{x}(0) = 0$ (assume $\zeta < 1$)
(d) $m\ddot{x} + c\dot{x} + kx = dt$, $x(0) = \dot{x}(0) = 0$ (assume $\zeta = 1$)

4.26 Use the Laplace transform to find the response of the following models for a unit-step input (all initial conditions are zero).

(a) $\ddot{x} + 7\dot{x} + 12x = \dot{u} + 6u$
(b) $\ddot{x} + 6\dot{x} + 9x = \dot{u} + 9u$
(c) $\ddot{x} + 2\dot{x} + 5x = \dot{u} + u$
(d) $\dddot{x} + 6\ddot{x} + 11\dot{x} + 6x = u$ [*Hint:* $s^3 + 6s^2 + 11s + 6 = (s+1)(s+2)(s+3)$]
(e) $\ddot{x} + 3\dot{x} + 2x = \dot{u}$

4.27 Use the Laplace transform to obtain the forced response of $\dot{y} = -2y + v(t)$ for

$$v(t) = \begin{cases} 4t, & 0 \leqslant t \leqslant 2 \\ -4t + 16, & 2 \leqslant t \leqslant 4 \\ 0, & t \geqslant 4 \end{cases}$$

(*Hint:* Use the ramp response and the shifting theorem.)

4.28 Draw the signal flow graph for each of the following problems:

(a) Problem 4.3
(b) Problem 4.4
(c) Problem 4.9*b*

CHAPTER FIVE

Computer Simulation Methods†

Chapters 3 and 4 presented some mathematical methods for obtaining solutions of differential equations in what is known as *closed form.* In a closed-form solution, the response variable is given by a mathematical expression in terms of the independent variable (which is usually time t). These solution methods work well when the system model consists of linear, time-invariant differential equations of relatively low order, with relatively simple input functions. If the model does not have all of the preceding characteristics, a closed-form solution is usually difficult if not impossible to find. In such cases, it is necessary to resort to *numerical-solution* methods. With such a method, the differential equation model is converted into *difference equation* form, which can then be solved in an iterative way with a calculator or computer. The disadvantage of this type of solution is that generality is lost, because the numerical values of the initial conditions and coefficients of the model and its input function must be specified. But often, there is no alternative.‡

Using numerical methods to predict the response of a system is called *simulation.* In Sections 5.1–5.3, we present the simulation methods most commonly used in control system engineering. Section 5.1 introduces simple methods for first-order models, while Section 5.2 develops some more advanced methods. Section 5.3 extends them to deal with higher order models. The state variable form allows general algorithms to be written, and Section 5.4 discusses techniques for converting models into state variable form. These techniques are particularly useful for models with numerator dynamics. When the model to be simulated is nonlinear, or linear but of high order, some insight is needed to set up the simulation properly. This insight can be obtained by studying the characteristic roots of the linear or linearized model. Section 5.5 treats this topic. Converting some model forms into the discrete form needed for simulation can introduce errors due to the approximations used. These errors are often greatest when step inputs are used. Section 5.6 discusses how to deal with these difficulties and introduces an input function called the *practical step function* that alleviates this problem. Sections 5.7 and 5.8 introduce the vector matrix notation for representing the system model. This notation provides a compact way of

†The mathematical techniques in Chapters 3 and 4 are well suited to handle linear, time-invariant first-, second-, and third-order models with simple input functions. Because most of the models used in the following chapters are of this form, Chapter 5 can be omitted if time is limited. However, if computers are available to the student, Chapter 5 provides a motivating educational experience because the problems and examples in the following chapters can be worked using computer methods. This will give the student a better understanding of how control systems design is done in practice.

‡The *analog computer* method provides another way of solving differential equations. This method is rarely used now because of the advantages of digital computer methods. However, it still provides an excellent means of developing control system prototypes. See Appendix E for details.

describing both analytical and numerical techniques. An overview of these techniques is given to familiarize the reader with the notation and terminology, since vector matrix notation is often used to describe computer-aided design packages for control systems engineering. The application of this software is described in Section 5.9.

5.1 NUMERICAL METHODS: AN INTRODUCTION

Despite the relatively powerful methods presented in previous chapters, it is not always possible to obtain the closed-form solution of a dynamic model. In this section, we introduce numerical methods for solving differential equations. Our discussion will focus on first-order models, but the techniques are easily generalized to higher order models for which the analytical methods of the previous chapters become difficult to apply because of the cumbersome algebra involved.

The essence of a numerical method is to convert the differential equation into an equivalent model composed of difference equations. In this form, the model can be programmed on a calculator or digital computer. Numerical algorithms differ mostly as a result of the specific procedure used to obtain the difference equations. In general, as the accuracy of the approximation is increased, so is the complexity of the programming involved.

The Euler Method

Consider the model

$$\frac{dy}{dt} = ry, \; r = \text{constant} \tag{5.1-1}$$

From the definition of the derivative,

$$\frac{dy}{dt} = \lim_{\Delta t \to 0} \frac{y(t + \Delta t) - y(t)}{\Delta t}$$

If the time increment Δt is chosen small enough, the derivative can be replaced by the approximate expression

$$\frac{dy}{dt} \cong \frac{y(t + \Delta t) - y(t)}{\Delta t} \tag{5.1-2}$$

Assume that the right-hand side of (5.1-1) remains constant over the time interval $(t, t + \Delta t)$, and replace (5.1-1) by the following approximation:

$$\frac{y(t + \Delta t) - y(t)}{\Delta t} = ry(t)$$

or

$$y(t + \Delta t) = y(t) + ry(t)\Delta t \tag{5.1-3}$$

The smaller Δt is made, the more accurate are our two assumptions leading to (5.1-3). This technique for replacing a differential equation with a difference equation is the *Euler method*.

Equation (5.1-3) can be written in more convenient form as follows. At the initial time t_0, (5.1-3) gives

$$y(t_0 + \Delta t) = y(t_0) + ry(t_0)\Delta t$$

Let $t_1 = t_0 + \Delta t$. Then

$$y(t_1 + \Delta t) = y(t_1) + ry(t_1)\Delta t$$

That is, (5.1-3) is applied successively at the instants t_k, where $t_{k+1} = t_k + \Delta t$, and we can write in general

$$y(t_{k+1}) = y(t_k) + ry(t_k)\Delta t \tag{5.1-4}$$

It is convenient to introduce a time index k such that

$$y_k \triangleq y(t_k)$$

Thus, (5.1-4) becomes

$$y_{k+1} = y_k + ry_k\Delta t = (1 + r\Delta t)y_k, \quad k = 0, 1, 2, \ldots \tag{5.1-5}$$

In this form, it is easily seen that the continuous-time variable $y(t)$ has been represented by the discrete-time variable y_k. Equation (5.1-5) is a *recursion relation* or *difference equation.* It can be solved recursively (sequentially) at the instants $k = 0$, 1, 2, ..., starting with the initial value $y_0 = y(t_0)$.

In cases where the subscript notation is inconvenient, the variable y_k is written as $y(k)$. Care must be taken not to confuse $y(k)$ with $y(t_k)$. For $k = 0, 1, 2, 3, \ldots, y(k)$ represents $y(t)$ at the values of $t = t_0, t_1, t_2, \ldots$ In this notation, (5.1-5) becomes

$$y(k + 1) = (1 + r\Delta t)y(k), \quad k = 0, 1, 2, \ldots \tag{5.1-6}$$

With r a constant, we can easily solve the original differential equation (5.1-1), so we really have no need for (5.1-6). However, this is not the case for a general equation of the form

$$\dot{y} = f(y, v) \tag{5.1-7}$$

Assume the right-hand side to be constant over the interval (t_k, t_{k+1}) and equal to the value $f[y(t_k), v(t_k)]$. With the approximation (5.1-2), we have the Euler method for (5.1-7).

$$y(t_{k+1}) = y(t_k) + \Delta t\, f[y(t_k), v(t_k)] \tag{5.1-8}$$

This is easily programmed on a computer or calculator. The values of t_0, $y(t_0)$, Δt, and any parameters in the functions f and v must be read into the program along with a stopping criterion. To be specific, suppose the model to be solved is

$$\dot{y} = 3ty^2 + v(t)$$

where $v(t)$ is a ramp function with a slope of 5: $v(t) = 5t$. The structure of a FORTRAN program to implement the Euler method for this equation would look like the following:

```
read in: DT, Y0, T0, TMAX
T = T0
DEL = TMAX − T0
Y = Y0
IMAX = INT(DEL/DT) + 1
DO 1 I = 1, IMAX
    T = T + DT
    V = 5*T
    Y = Y + DT*(3*T*Y*Y + V)
    print: T, Y
1 CONTINUE
  STOP
```

The statements for reading and printing depend on the particular machine. The variables DT, Y0, T0, and TMAX represent Δt, $y(t_0)$, t_0, and t_{max}, respectively. The time t_{max} is the largest time at which the solution is desired.

The Euler method is the simplest algorithm for numerical solution of a differential equation, but it usually gives the least accurate results. We will therefore consider some more accurate algorithms later.

Example 5.1

Compare the exact solution and the Euler method using a step size $\Delta t = 0.1$ for

$$\dot{y} = ty, \quad y(0) = 1$$

where $0 \leqslant t \leqslant 1$.

The Euler algorithm (5.1-8) for this model is

$$y(t_{k+1}) = (1 + 0.1t_k)y(t_k) \tag{5.1-9}$$

For this case, the exact solution can be obtained by separating variables and direct integration. First, we write

$$\frac{dy}{y} = t\,dt$$

and integrate to obtain

$$\int_{y(0)}^{y(t)} \frac{dy}{y} = \int_0^t t\,dt$$

or

$$\ln y \Big|_{y(0)}^{y(t)} = \frac{t^2}{2}$$

This can be rearranged as

$$y(t) = y(0)\exp(t^2/2)$$

The following table compares the results to four significant figures.

	$y(t_k)$	
t_k	**Exact**	**Euler**
0.1	1.005	1.000
0.2	1.020	1.010
0.3	1.046	1.030
0.4	1.083	1.061
0.5	1.133	1.104
0.6	1.197	1.159
0.7	1.278	1.228
0.8	1.377	1.314
0.9	1.499	1.419
1.0	1.649	1.547

The error in the Euler method increases with time. It is in error by 0.5% in the first step; by the tenth step, the error is 6.19%. The accuracy can be improved by using a smaller value of Δt. The step size $\Delta t = 0.1$ was chosen arbitrarily here to illustrate the mechanics of the method.

As with this example, other examples will present the results of several iterations so the student can test the methods. A hand-held calculator is usually sufficient for this purpose.

Selecting a Step Size

Thus far, we have not considered the effect of the "run time" of a numerical method. This is the time required for the software to perform the calculations, and it is frequently an important factor in selecting an algorithm. It is difficult to make a prior determination of the run time for a given algorithm. The complexity and length of a program are indicators. Also, obviously the smaller Δt is, the more iterations will be required and the longer the total computer time required.

The process of selecting a proper size Δt begins by considering the dynamics of the model and how fast the input changes with time. A trial value of Δt is then selected that is small in relation to these considerations. The trial solution for the response is obtained with this value. The step size is then reduced significantly (say, by a factor of two), and the responses are compared. If they agree to the number of significant figures specified by the analyst, the correct answer is taken to be the last solution. If not, Δt is again reduced by a significant factor, and the process is repeated until the solution converges. Note that two values of Δt that are close to each other will give similar solutions. Thus, the method fails unless the successive Δt values differ significantly.

In strictly analytical terms, as Δt approaches zero, the solution of the difference equation approaches that of the corresponding differential equation. This is the reason for using a small Δt. The error produced when Δt is not small enough to represent the differential equation accurately is termed *truncation error*. However, numerical considerations as well as computer time require not making Δt too small. The reason is that the effect of *round-off error* is greater for smaller Δt due to the larger number of iterations required. The round-off error results from the inability of any digital device to represent a number by more than a finite number of binary bits or significant figures. For example, suppose we have a hypothetical device capable of representing numbers to only two significant figures. If we solve the equation

$$\frac{dy}{dt} = -y, \quad y(0) = 1$$

with the Euler method, the algorithm is

$$y(k+1) = (1 - \Delta t)y(k)$$

The time constant τ is unity, and so we might try $\Delta t = 0.1$. The results are shown in Table 5.1 along with the true solution for comparisons. Now, suppose we use $\Delta t = 0.01$ to improve the accuracy. The displayed results indicate that instead of decreasing, the error has increased drastically due to the large number of iterations required relative to the number of significant figures carried by the machine.

For large Δt, the truncation error dominates, while for small Δt, the round-off error is most important. Our suggested method for selecting Δt usually results in starting above the optimum Δt and moving down. On most large machines in use today, the number of available significant figures is great enough so that round-off error is usually not a problem unless we are solving high-order equations over very

TABLE 5.1 Illustration of Round-off Error Effects

Model $\frac{dy}{dt} = -y, \quad y(0) = 1$

Solution $y(t) = e^{-t}$

	Euler[a]		True Solution[a]
	$\Delta t = 0.1$	$\Delta t = 0.01$	
t	y	y	y
0	1	1	1
0.1	0.90	0.90	0.90
0.2	0.81	0.80	0.82
0.3	0.73	0.70	0.74
0.4	0.66	0.59	0.67
0.5	0.59	0.50	0.61

[a]Euler results were computed using only two significant figures. True solution has been rounded off to two figures.

long times. However, with many personal computers, the number of available significant figures is quite limited, and round-off error can be a problem. Sometimes, a double-precision representation is available, which can be used to reduce the round-off error.

5.2 ADVANCED NUMERICAL METHODS

In light of the large variety of equation types that are nonlinear, it is no wonder that many different numerical methods exist for solving them. Some methods work well for only special classes of problems. Here, we consider two methods that are generally useful. The first is a so-called predictor–corrector method based on the Euler method but with greater accuracy. It is easy to program and thus suitable for calculators. The second method is the Runge–Kutta family of algorithms. Of the more advanced techniques, these are perhaps the most widely used in engineering applications. Runge–Kutta algorithms are widely available at computer installations and in software packages for more advanced calculators. In this section, we apply these techniques to first-order models. We then extend them to higher order models in the next section.

Trapezoidal Integration

Some useful results can be obtained from the trapezoidal rule for integration. Consider the equation

$$\frac{dy}{dt} = v \tag{5.2-1}$$

The variable y is the integral of v, or

$$y(t_{k+1}) = y(t_k) + \int_{t_k}^{t_{k+1}} v(\lambda)d\lambda \tag{5.2-2}$$

The trapezoidal formula is

$$\int_{t_k}^{t_{k+1}} v(\lambda)d\lambda = \frac{\Delta t}{2}[v(t_{k+1}) + v(t_k)] \tag{5.2-3}$$

where $\Delta t = t_{k+1} - t_k$. Substituting this into (5.2-2) and omitting the subscript notation gives

$$y(k+1) = y(k) + \frac{\Delta t}{2}[v(k+1) + v(k)] \tag{5.2-4}$$

where $y(k)$ stands for $y(t_k)$, etc. Equation (5.2-4) is the recursive algorithm that implements the integration in (5.2-2) via the trapezoidal formula and therefore solves (5.2-1) numerically.

The difference between the Euler and trapezoidal integration methods is shown in Figure 5.1. The shaded area represents the approximation to the integral

$$\int_{t_k}^{t_{k+1}} v(\lambda)\, d\lambda$$

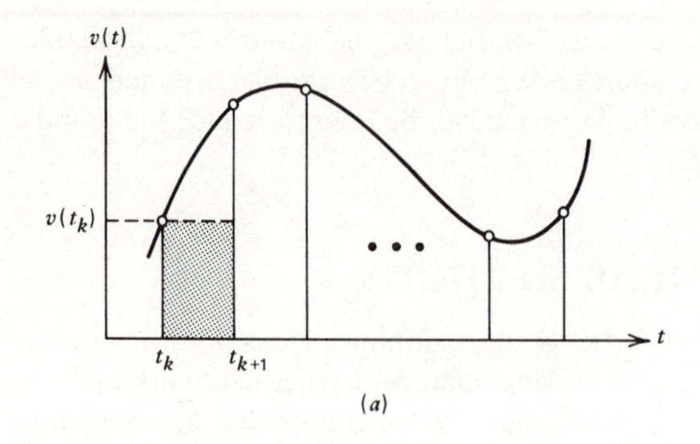

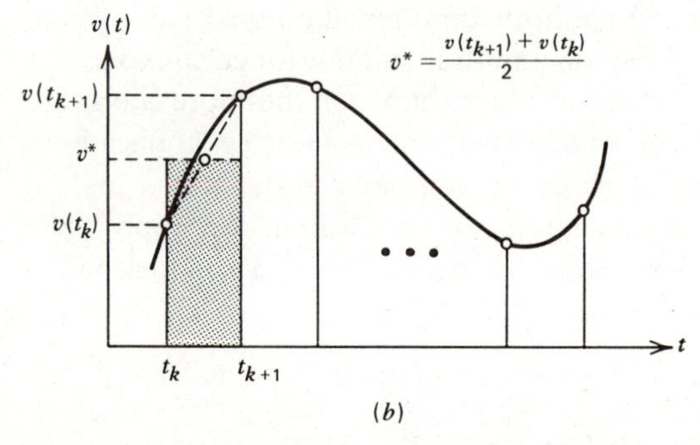

FIGURE 5.1 Comparison of the Euler and trapezoidal integration approximations. The shaded area represents the approximation to the integral. (*a*) Euler. (*b*) Trapezoidal.

Both methods use a rectangle of width Δt. The Euler method takes the height of the rectangle to be $v(t_k)$ and approximates the integral as

$$\int_{t_k}^{t_{k+1}} v(\lambda)\, d\lambda = \Delta t\, v(t_k)$$

On the other hand, the trapezoidal method takes the rectangle's height as the average value v^* of $v(t_k)$ and $v(t_{k+1})$ and approximates the integral with (5.2-3).

Predictor–Corrector Methods

As can be seen from Figure 5.1, the Euler method can have a serious deficiency in problems where the variables are rapidly changing, because the method assumes the variables are constant over the time interval Δt. As a result, its truncation error can be shown to be of the order of $(\Delta t)^2$ (see Reference 1).

One way of improving the method would be by using a better approximation to the right-hand side of the model

$$\frac{dy}{dt} = f(y, v, t) \tag{5.2-5}$$

where v is the input function. The Euler approximation is

$$y(t_{k+1}) = y(t_k) + \Delta t\, f[y(t_k), v(t_k), t_k] \tag{5.2-6}$$

Suppose instead we use the average of the right-hand side of (5.2-5) on the interval (t_k, t_{k+1}). This gives

$$y(t_{k+1}) = y(t_k) + \frac{\Delta t}{2}(f_k + f_{k+1}) \tag{5.2-7}$$

where

$$f_k = f[y(t_k), v(t_k), t_k] \tag{5.2-8}$$

with a similar definition for f_{k+1}. Equation (5.2-7) is equivalent to integrating (5.2-5) with the trapezoidal rule.

The difficulty with (5.2-7) is that f_{k+1} cannot be evaluated until $y(t_{k+1})$ is known, but this is precisely the quantity being sought. A way out of this difficulty is by using the Euler formula (5.2-6) to obtain a preliminary estimate of $y(t_{k+1})$. This estimate is then used to compute f_{k+1} for the use in (5.2-7) to obtain the required value of $y(t_{k+1})$.

The notation can be changed to clarify the method. Let x_{k+1} be the estimate of $y(t_{k+1})$ obtained from the Euler formula (5.2-6). Then, by omitting the t_k notation from the other equations, we obtain the following description of the *predictor–corrector* process.

Euler predictor: $$x_{k+1} = y_k + \Delta t\, f(y_k, v_k, t_k) \tag{5.2-9}$$

Trapezoidal corrector: $$y_{k+1} = y_k + \frac{\Delta t}{2}[f(y_k, v_k, t_k) + f(x_{k+1}, v_{k+1}, t_{k+1})] \tag{5.2-10}$$

This algorithm is sometimes called the *modified Euler method*. However, note that any algorithm can be tried as a predictor or a corrector. Thus many methods can be classified as predictor–corrector, but we will limit our treatment to the modified Euler method. The truncation error for this method is of the order of $(\Delta t)^3$, a significant improvement over the $(\Delta t)^2$ error for the basic Euler method.

For purposes of comparison with the Runge–Kutta methods to follow, we can express the modified Euler method as

$$g_1 = \Delta t\, f(y_k, v_k, t_k) \tag{5.2-11}$$

$$g_2 = \Delta t\, f(y_k + g_1, v_{k+1}, t_k + \Delta t) \tag{5.2-12}$$

$$y_{k+1} = y_k + \tfrac{1}{2}(g_1 + g_2) \tag{5.2-13}$$

Example 5.2

Use the modified Euler method with $\Delta t = 0.1$ to solve the equation

$$\frac{dy}{dt} = ty, \quad y(0) = 1$$

up to time $t = 1.0$.

Here, the equations for the algorithm become

$$x_{k+1} = y_k + \Delta t\, t_k\, y_k$$

$$y_{k+1} = y_k + \frac{\Delta t}{2}(t_k\, y_k + t_{k+1} x_{k+1})$$

On the first step, $t_0 = 0$, $k = 0$, $y_0 = 1$, and

$$x_1 = 1 + 0.1(0) = 1$$
$$y_1 = 1 + 0.05(0 + 0.1) = 1.005$$

On the second step, $t_1 = 0.1$, $k = 1$, and

$$x_2 = 1.005 + 0.1[0.1(1.005)] = 1.01505$$
$$y_2 = 1.005 + 0.05[0.1(1.005) + 0.2(1.01505)]$$
$$= 1.0201755$$

The results are shown in the following table rounded to six figures along with the analytical solution (see Example 5.1). It can be seen that the modified Euler method gives very accurate results for this problem. Its performance here is much superior to the basic Euler method.

y_k

t_k	Exact	Modified Euler
0	1	1
0.1	1.00501	1.00500
0.2	1.02020	1.02018
0.3	1.04603	1.04599
0.4	1.08329	1.08322
0.5	1.13315	1.13305
0.6	1.19722	1.19707
0.7	1.27762	1.27739
0.8	1.37713	1.37677
0.9	1.49930	1.49876
1.0	1.64872	1.64788

Corrector equation (5.2-10) can also be used more than once on each Δt interval in order to increase the accuracy. For example, the refined estimate y_{k+1} from (5.2-10) can be used to replace x_{k+1} on the right-hand side of that equation in order to get an improved estimate of y_{k+1}. This is repeated until the desired accuracy is achieved; then the next Δt interval is treated the same way. However, this modification obviously increases the complexity of the programming.

Runge–Kutta Methods†

The Taylor series representation forms the basis of several methods for solving differential equations, including the Runge–Kutta methods. The Taylor series may be

†If an abbreviated treatment of numerical methods is desired, the remainder of Section 5.2 can be omitted. The examples in the rest of the chapter can be solved, using the modified Euler method instead of the Runge–Kutta method.

used to represent the solution $y(t + \Delta t)$ in terms of $y(t)$ and its derivatives as follows.

$$y(t + \Delta t) = y(t) + \Delta t\, \dot{y}(t) + \tfrac{1}{2}(\Delta t)^2 \ddot{y}(t) + \tfrac{1}{6}(\Delta t)^3 \dddot{y}(t) + \cdots \tag{5.2-14}$$

The required derivatives are calculated from the differential equation. For an equation of the form

$$\frac{dy}{dt} = f(t, y) \tag{5.2-15}$$

these derivatives are

$$\begin{aligned} \dot{y}(t) &= f(t, y) \\ \ddot{y}(t) &= \frac{df}{dt} \\ \dddot{y}(t) &= \frac{d^2 f}{dt^2}, \text{ etc.} \end{aligned} \tag{5.2-16}$$

where (5.2-15) is to be used to express the derivatives in terms of only $y(t)$. If these derivatives can be found, (5.2-14) can be used to march forward in time. In practice, the high order derivatives can be difficult to calculate, and the series (5.2-14) is truncated at some term. The number of terms kept in the series thus determines its accuracy. If terms up to and including the nth derivative of y are retained, the truncation error at each step is of the order of the first term dropped – namely,

$$\frac{(\Delta t)^{n+1}}{(n+1)!} \frac{d^{n+1} y}{dt^{n+1}} \tag{5.2-17}$$

Sometimes the closed-form expression for the series can be recognized. For example, consider the equation

$$\dot{y} = ay^2, \; a < 0 \tag{5.2-18}$$

The derivatives required for the series are

$$\begin{aligned} \ddot{y} &= 2ay\dot{y} = 2a^2 y^3 \\ \dddot{y} &= 6a^2 y^2 \dot{y} = 6a^3 y^4 \\ &\vdots \\ \frac{d^n y}{dt^n} &= (n+1)a^n y^{n+1} \end{aligned}$$

Substituting these into (5.2-14) gives

$$\begin{aligned} y(t + \Delta t) = y(t) &+ a\, \Delta t\, y^2(t) + (a\, \Delta t)^2 y^3(t) \\ &+ (a\, \Delta t)^3 y^4(t) + \cdots + (a\, \Delta t)^n y^{n+1}(t) + \cdots \end{aligned}$$

The closed-form expression can be recognized by multiplying by $a\, \Delta t$ to obtain

$$\begin{aligned} a\, \Delta t\, y(t + \Delta t) &= \sum_{i=1}^{\infty} [a\, \Delta t\, y(t)]^i \\ &= \frac{a\, \Delta t\, y(t)}{1 - a\, \Delta t\, y(t)} \end{aligned}$$

Thus,

$$y(t+\Delta t) = \frac{y(t)}{1 - a\,\Delta t\, y(t)} \tag{5.2-19}$$

Since we have not truncated the series, no approximation errors have been introduced, and (5.2-19) is exact for any Δt.

While few problems can be solved exactly with the Taylor series, the preceding example serves to illustrate the power of the method. Its chief difficulty is the need to compute higher order derivatives. For this reason, the Runge–Kutta methods were developed. These methods use several evaluations of the function $f(t, y)$ in a way that approximates the Taylor series. The number of terms in the series that is duplicated determines the order of the Runge–Kutta method. Thus, a fourth-order Runge–Kutta algorithm duplicates the Taylor series through the term involving $(\Delta t)^4$.

Development of the Second-Order Algorithm

We now demonstrate the development of the Runge–Kutta algorithms for the second-order case. For notational compactness, we replace the step size Δt with h. Thus, $t_{k+1} = t_k + h$. Also, let the subscript k denote evaluation at time t_k. Truncating (5.2-14) beyond the second-order term and using (5.2-5) gives

$$\begin{aligned} y_{k+1} &= y_k + h\dot{y}_k + \tfrac{1}{2}h^2 \ddot{y}_k \\ &= y_k + hf_k + \tfrac{1}{2}h^2 \left(\frac{df}{dt}\right)_k \end{aligned}$$

But

$$\frac{df}{dt} = \frac{\partial f}{\partial t} + \frac{\partial f}{\partial y}\frac{dy}{dt} = \frac{\partial f}{\partial t} + \frac{\partial f}{\partial y} f$$

In simpler notation, f_y denotes $\partial f/\partial y$, and f_t denotes $\partial f/\partial t$. Thus,

$$\frac{df}{dt} = f_t + f_y f$$

and

$$y_{k+1} = y_k + hf_k + \tfrac{1}{2}h^2(f_t + f_y f)_k \tag{5.2-20}$$

The second-order Runge–Kutta methods express y_{k+1} as

$$y_{k+1} = y_k + w_1 g_1 + w_2 g_2 \tag{5.2-21}$$

where w_1 and w_2 are constant weighting factors, and

$$g_1 = hf(t_k, y_k) \tag{5.2-22}$$

$$g_2 = hf(t_k + \alpha h, y_k + \beta h f_k) \tag{5.2-23}$$

for constant α and β. We now compare (5.2-21) with (5.2-20). First, expand f in g_2 in a two-variable Taylor series.

$$g_2 = h[f_k + (f_t)_k \alpha h + (f_y)_k \beta h f_k + \cdots] \tag{5.2-24}$$

where the omitted terms are of the order of h^2. Substituting this and g_1 into (5.2-21) and collecting terms give

$$y_{k+1} = y_k + h[w_1 f + w_2 f]_k + h^2[w_1 \alpha f_t + w_2 \beta f f_y]_k$$

Comparison with (5.2-20) for like powers of h shows that

$$w_1 + w_2 = 1 \tag{5.2-25}$$

$$w_1 \alpha = \tfrac{1}{2} \tag{5.2-26}$$

$$w_2 \beta = \tfrac{1}{2} \tag{5.2-27}$$

Thus, the family of second-order Runge–Kutta algorithms is categorized by the parameters (α, β, w_1, w_2), one of which can be chosen independently. The choice $\alpha = 2/3$ minimizes the truncation error term (Reference 1). For $\alpha = 1$, the Runge–Kutta algorithm (5.2-21)–(5.2-23) corresponds to the trapezoidal integration rule if f is a function of only t, and is the same as the predictor–corrector algorithm (5.2-9)–(5.2-10) for a general $f(y, t)$.

Fourth-Order Algorithms

Second- and third-order Runge–Kutta algorithms have found extensive applications in the past. However, the fourth-order algorithm is now the most commonly used. Its derivation follows the same pattern as earlier, with all terms up to h^4 retained. The algebra required is extensive and given in Reference 2. Here, we present the results. The algorithm is

$$y_{k+1} = y_k + w_1 g_1 + w_2 g_2 + w_3 g_3 + w_4 g_4 \tag{5.2-28}$$

$$\begin{aligned} g_1 &= hf(t_k, y_k) \\ g_2 &= hf(t_k + \alpha_1 h, y_k + \alpha_1 g_1) \\ g_3 &= hf[t_k + \alpha_2 h, y_k + \beta_2 g_2 + (\alpha_2 - \beta_2) g_1] \\ g_4 &= hf[t_k + \alpha_3 h, y_k + \beta_3 g_2 + \gamma_3 g_3 + (\alpha_3 - \beta_3 - \gamma_3) g_1] \end{aligned} \tag{5.2-29}$$

Comparison with the Taylor series yields eight equations for the ten parameters. Thus, two parameters can be chosen in light of other considerations. Three common choices are as follows.

1. *Gill's Method.*† This choice minimizes the number of memory locations required to implement the algorithm and thus is well suited for programmable calculators.

$$\begin{aligned} w_1 &= w_4 = 1/6 \\ w_2 &= (1 - 1/\sqrt{2})/3 \\ w_3 &= (1 + 1/\sqrt{2})/3 \\ \alpha_1 &= \alpha_2 = 1/2, \quad \alpha_3 = 1 \\ \beta_2 &= 1 - 1/\sqrt{2}, \quad \beta_3 = -1/\sqrt{2} \\ \gamma_3 &= 1 + 1/\sqrt{2} \end{aligned} \tag{5.2-30}$$

†Appendix C contains a FORTRAN subroutine that implements Gill's version of the fourth-order Runge–Kutta algorithm.

2. *Ralston's Method.* This choice minimizes a bound on the truncation error.

$$w_1 = 0.17476028, \quad w_2 = -0.55148066$$
$$w_3 = 1.20553560, \quad w_4 = 0.17118478$$
$$\alpha_1 = 0.4, \quad \alpha_2 = 0.45573725$$
$$\alpha_3 = 1, \quad \gamma_3 = 3.83286476$$
$$\beta_2 = 0.15875964, \quad \beta_3 = -3.05096516 \qquad \textbf{(5.2-31)}$$

3. *Classical Method.* This method reduces to Simpson's rule for integration if f is a function of only t. Because the coefficients are computed from whole numbers, this method also requires less computer storage than the Ralston method. It is commercially available for programmable calculators.

$$w_1 = w_4 = 1/6$$
$$w_2 = w_3 = 1/3$$
$$\alpha_1 = \alpha_2 = 1/2$$
$$\beta_2 = 1/2, \quad \beta_3 = 0$$
$$\gamma_3 = \alpha_3 = 1 \qquad \textbf{(5.2-32)}$$

Example 5.3

Use the classical Runge–Kutta parameter values (5.2-32) with a step size $h = \Delta t = 0.1$ to solve

$$\dot{y} = ty, \quad y(0) = 1$$

up to $t = 1.0$.

The algorithm for the classical parameter values is

$$y_{k+1} = y_k + \tfrac{1}{6}g_1 + \tfrac{1}{3}g_2 + \tfrac{1}{3}g_3 + \tfrac{1}{6}g_4$$
$$g_1 = hf(t_k, y_k)$$
$$g_2 = hf(t_k + \tfrac{1}{2}h, y_k + \tfrac{1}{2}g_1)$$
$$g_3 = hf[t_k + \tfrac{1}{2}h, y_k + \tfrac{1}{2}g_2 + (\tfrac{1}{2} - \tfrac{1}{2})g_1]$$
$$g_4 = hf[t_k + h, y_k + 0g_2 + g_3 + (1 - 0 - 1)g_1]$$

For the given problem, $f = ty$. For the first iteration, $k = 0$, $t_0 = 0$, $y_0 = 1$, and

$$g_1 = 0.1(0) = 0$$
$$g_2 = 0.1(0 + 0.05)(1 + \tfrac{1}{2}0) = 0.005$$
$$g_3 = 0.1(0 + 0.05)(1 + \tfrac{1}{2}0.005) = 0.0050125$$
$$g_4 = 0.1(0 + 0.1)(1 + 0.0050125) = 0.010050125$$

Thus,

$$y(0.1) = y_1 = 1 + \tfrac{1}{6}0 + \tfrac{1}{3}(0.005) + \tfrac{1}{3}(0.0050125) + \tfrac{1}{6}(0.010050125)$$
$$= 1.005012521$$

This answer agrees with the exact solution $\exp(t^2/2)$ out to the number of significant figures available on most pocket calculators. The result demonstrates the power of the Runge–Kutta method. The rest of the solution is as follows.

t	$y(k)$
0.1	1.005012521
0.2	1.02020134
0.3	1.046027859
0.4	1.08387065
0.5	1.133148446
0.6	1.197217347
0.7	1.277621279
0.8	1.377127695
0.9	1.499302362
1.0	1.648721007

The exact solution at $t = 1.0$, to ten figures, is 1.648721271. The numerical solution is correct to seven figures.

Numerical methods must be used carefully when the true solution changes rapidly. This can occur when either the free response or the forcing function is oscillatory. This also arises when the time span for significant changes to occur in the free response is different from the time span for the forcing function. An example of this effect is given by the equation

$$\dot{y} = -y + 0.001e^{10t}, \quad y(0) = 10 \tag{5.2-33}$$

Since the solution can be found with analytical methods from Chapter 3, a numerical method is not needed. The problem is illustrative only. Its solution is

$$y(t) = 10e^{-t} + \frac{0.001}{11}(e^{10t} - e^{-t}) \tag{5.2-34}$$

The time constant is $\tau = 1$, and thus the free response decreases by 63% in one time unit. On the other hand, the forcing function $\exp(10t)$ increases by 22,026% in the same time. The existence of more than one time scale in the problem means that the step size Δt must be selected with care. Also, for $0 \leqslant t \leqslant 0.9$, the free response dominates, and y decreases. For $t > 0.9$, the forcing function dominates, and y increases. For $t < 0.9$, a step size of one-tenth of the time constant might be sufficient but would be too large to use for $t > 0.9$. One approach would be to decrease the step size for $t > 0.9$. However, the Runge–Kutta method can deal with this solution by using a constant step size if it is chosen small enough.

Example 5.4

Solve (5.2-33) using $\Delta t = 0.01$.

The classical Runge–Kutta algorithm gives the following solution. The exact solution is shown for comparison. The table presents the results of every twentieth iteration.

t	Runge–Kutta	Exact
0.2	8.187904833	8.187904833
0.4	6.708102991	6.708102991
0.6	5.524741816	5.524741818
0.8	4.764244988	4.764245001
1.0	5.681167053	5.68116715
1.2	17.80780563	17.80780635
1.4	111.793615	111.7936204
1.6	809.8472175	809.8472567
1.8	5970.741384	5970.741674

The characteristics of the Runge–Kutta algorithms can be summarized as follows.

1. The Runge–Kutta algorithms do not require calculating the higher derivatives of $y(t)$, as is required by the Taylor method. Calculating these derivatives can be quite tedious, especially for sets of coupled differential equations of high order. Instead, the Runge–Kutta algorithms use the computation of $f(t, y)$ at various points. These algorithms can be easily generalized to handle higher order equations and equation sets, as we will see in the next section. On the other hand, error propagation in the Runge–Kutta method is not so easy to keep track of as in the Taylor method. It is advisable to watch the g_i values. If they differ drastically from each other, reduce the step size.
2. The Runge–Kutta algorithms are self-starting, and the step size can be changed easily between iterations. While we have not seen any methods for which this is not true, there are more accurate and faster methods that suffer those disadvantages. For example, higher order predictor–corrector methods are not self-starting, and a Runge–Kutta method is frequently used to provide a starting solution for the predictor–corrector (Reference 2).
3. For some numerical methods, the difference equations used to approximate the original differential equation have more than one solution. One of these solutions is the true one, while the others result from round-off errors. If these extraneous solutions grow with time (thus swamping the true solution), the method is said to be unstable. The Runge–Kutta methods have good stability characteristics in this respect.
4. The fourth-order Runge–Kutta method is probably the most widely used method for engineering applications. Its software is available for advanced calculators and most computers. Faster, more accurate methods are available, but these are most difficult to program (Reference 2). If great efficiency is required, such as with production codes to be run many times, one might consider a more advanced method. However, the Runge–Kutta method is suitable for most problems that involve a small number of runs.

5.3 EXTENSION TO HIGHER ORDER MODELS

The numerical methods in the two previous sections were applied to only first-order equations. However, they can easily be generalized to handle higher order models if the model is first converted into state variable form. There are several ways of making this conversion, as we will see in the next section.

The Euler Method

To see how numerical methods are applied to higher order models, consider the following second-order model:

$$\dot{y}_1 = f_1(y_1, y_2, t) \tag{5.3-1}$$

$$\dot{y}_2 = f_2(y_1, y_2, t) \tag{5.3-2}$$

First, consider the Euler method. Apply the Euler approximation (5.1-8) to each equation to obtain

$$y_1(t_{k+1}) = y_1(t_k) + \Delta t\ f_1[y_1(t_k), y_2(t_k), t_k] \tag{5.3-3}$$

$$y_2(t_{k+1}) = y_2(t_k) + \Delta t\ f_2[y_1(t_k), y_2(t_k), t_k] \tag{5.3-4}$$

It is easy to see how this procedure can be extended to higher order models in a straightforward manner.

Consider the specific model

$$\dot{y}_1 = y_2 \tag{5.3-5}$$

$$\dot{y}_2 = y_1 y_2 + t \tag{5.3-6}$$

Then (5.3-3) and (5.3-4) give

$$y_1(t_{k+1}) = y_1(t_k) + \Delta t\ y_2(t_k) \tag{5.3-7}$$

$$y_2(t_{k+1}) = y_2(t_k) + \Delta t\ [y_1(t_k)y_2(t_k) + t_k] \tag{5.3-8}$$

A partial listing of the FORTRAN code to implement the algorithm follows. The input routine and initializations have been omitted (see the listing in Section 5.1).

```
DO 1 I = 1, IMAX
      T = T + DT   (t_k = t_k-1 + Δt)
   W1 = Y1 + DT*Y2
   Y2 = Y2 + DT*(Y1*Y2 + T)
   Y1 = W1
   (print T, Y1, Y2)
1 CONTINUE
```

Note the use of the intermediate variable $W1$. If the third line of the listing were instead

```
Y1 = Y1 + DT*Y2
```

then the statement in the fourth line would use the updated value of y_1 to compute y_2. That is, the value of $y_1(t_{k+1})$ would have been used instead of $y_1(t_k)$. This would be equivalent to the following algorithm:

$$y_2(t_{k+1}) = y_2(t_k) + \Delta t \, [y_1(t_{k+1}) y_2(t_k) + t_k] \tag{5.3-9}$$

which is not correct according to (5.3-8). Using $W1$ preserves the time indexing specified in (5.3-7) and (5.3-8). The line $Y1 = W1$ is then used to insert the value of $y_1(t_{k+1})$ into the variable $Y1$.

The Predictor–Corrector Method

The predictor–corrector method given by the modified Euler algorithm (5.2-9) and (5.2-10) can be applied in turn to each equation in the set (5.3-1) and (5.3-2) as follows.

Euler Predictor

$$x_1(t_{k+1}) = y_1(t_k) + \Delta t \, f_1[y_1(t_k), y_2(t_k), t_k] \tag{5.3-10}$$

$$x_2(t_{k+1}) = y_2(t_k) + \Delta t \, f_2[y_1(t_k), y_2(t_k), t_k] \tag{5.3-11}$$

Trapezoidal Corrector

$$y_1(t_{k+1}) = y_1(t_k) + 0.5 \, \Delta t \, \{f_1[y_1(t_k), y_2(t_k), t_k] + f_1[x_1(t_{k+1}), x_2(t_{k+1}), t_{k+1}]\} \tag{5.3-12}$$

$$y_2(t_{k+1}) = y_2(t_k) + 0.5 \, \Delta t \, \{f_2[y_1(t_k), y_2(t_k), t_k] + f_2[x_1(t_{k+1}), x_2(t_{k+1}), t_{k+1}]\} \tag{5.3-13}$$

The code for the example given by (5.3-5) and (5.3-6) is

```
DO 1 I = 1, IMAX
     T = T + DT        (t_k = t_{k-1} + Δt)
   X1 = Y1 + DT*Y2
   X2 = Y2 + DT*(Y1*Y2 + T)
   W1 = Y1 + 0.5*DT*(Y2 + X2)
   Y2 = Y2 + 0.5*DT*(Y1*Y2 + T + X1*X2 + T + DT)
   Y1 = W1
   (print T, Y1, Y2)
1 CONTINUE
```

Note that in the expressions for $X2$ and $Y2$, the variable T represents t_k; $T + DT$ represents t_{k+1}.

The pattern displayed by (5.3-10)–(5.3–13) can be readily extended to models of third order or higher.

The Runge–Kutta Method

From the preceding discussion, it should be clear that numerical methods developed for first-order models are easily applied to higher order models if they are in state variable form. This is because the state variable form consists of a series of coupled first-order equations. The Runge–Kutta method given by (5.2-28) and (5.2-29) can be extended in this manner. First, all the g_i terms for every state equation are evaluated with (5.2-29) using the same numerical values for h, α_i, β_i, and γ_i. After this is done, then all the updated values of the $y_i(t_{k+1})$ are calculated from (5.2-28). Care must be taken to observe the proper time indexing and update the state variables for time t_{k+1} only after *all* of the intermediate calculations have been made. We will not write out all the equations here, because they are lengthy, but the algorithm can be seen in the program given in Appendix C.

5.4 CONVERSION TO STATE VARIABLE FORM

There are several ways of converting a model to state variable form. If the variables representing the energy storage in the system are chosen as the state variables, the model must then be manipulated algebraically to produce the standard state variable form. The required algebra is not always obvious. However, we note here that the model

$$\frac{d^n y}{dt^n} = f\left[t, y, \frac{dy}{dt}, \ldots, \frac{d^{n-1}y}{dt^{n-1}}, v(t)\right] \tag{5.4-1}$$

with the input $v(t)$ can always be put into state variable form by the following choice of state variables:

$$\begin{aligned} x_1 &= y \\ x_2 &= \frac{dy}{dt} \\ &\vdots \\ x_n &= \frac{d^{n-1}y}{dt^{n-1}} \end{aligned} \tag{5.4-2}$$

The resulting state model is

$$\begin{aligned} \dot{x}_1 &= x_2 \\ \dot{x}_2 &= x_3 \\ &\vdots \\ \dot{x}_{n-1} &= x_n \\ \dot{x}_n &= f[t, x_1, x_2, \ldots, x_n, v(t)] \end{aligned} \tag{5.4-3}$$

The numbering scheme is not unique; in fact, we often find the following state variable choice:

$$\begin{aligned} y_1 &= \frac{d^{n-1}y}{dt^{n-1}} \\ y_2 &= \frac{d^{n-2}y}{dt^{n-2}} \\ &\vdots \\ y_{n-1} &= \frac{dy}{dt} \\ y_n &= y \end{aligned} \tag{5.4-4}$$

which gives the model

$$\begin{aligned}
\dot{y}_1 &= f[t, y_n, y_{n-1}, \ldots, y_1, v(t)] \\
\dot{y}_2 &= y_1 \\
\dot{y}_3 &= y_2 \\
&\vdots \\
\dot{y}_n &= y_{n-1}
\end{aligned} \tag{5.4-5}$$

Coupled Higher Order Models

Equation (5.4-1) is somewhat general in that it can be nonlinear, but it does not cover all cases. For example, consider the coupled higher order model

$$\ddot{y} = f_1(t, y, \dot{y}, z, \dot{z}) \tag{5.4-6}$$

$$\ddot{z} = f_2(t, y, \dot{y}, z, \dot{z}) \tag{5.4-7}$$

Choose the state variables as

$$\begin{aligned}
x_1 &= y \\
x_2 &= \dot{y} \\
x_3 &= z \\
x_4 &= \dot{z}
\end{aligned} \tag{5.4-8}$$

Then the state model is

$$\begin{aligned}
\dot{x}_1 &= x_2 \\
\dot{x}_2 &= f_1(t, x_1, x_2, x_3, x_4) \\
\dot{x}_3 &= x_4 \\
\dot{x}_4 &= f_2(t, x_1, x_2, x_3, x_4)
\end{aligned} \tag{5.4-9}$$

State Equations From Transfer Functions

We now illustrate a method for using the transfer function to rearrange the diagram so that state variables can be identified. The order of the system, and therefore the number of state variables required, can be found by examining the denominator of the transfer functions. If the denominator polynomial is of order n, then n state variables are required. Physical considerations (integral causality) can be used to show that the outputs of integration processes ($1/s$) can be chosen as state variables.

To illustrate how a state model can be derived from a transfer function, consider the following first-order model with numerator dynamics:

$$T(s) = \frac{Y(s)}{U(s)} = \frac{b_0 s + b_1}{s + a_1} \tag{5.4-10}$$

where the input is u and the output is y. Cross multiplying and reverting to the time domain, we obtain

$$\dot{y} + a_1 y = b_0 \dot{u} + b_1 u \tag{5.4-11}$$

An alternative way of obtaining the differential equation is by dividing the numerator and denominator of (5.4-10) by s.

$$T(s) = \frac{b_0 + b_1/s}{1 + a_1/s} = \frac{Y(s)}{U(s)} \tag{5.4-12}$$

The essence of the technique is to obtain a 1 in the denominator, which is then used to isolate $Y(s)$. Cross multiplication gives

$$\begin{aligned} Y(s) &= -\frac{a_1}{s}Y(s) + b_0U(s) + \frac{b_1}{s}U(s) \\ &= \frac{1}{s}[b_1U(s) - a_1Y(s)] + b_0U(s) \end{aligned}$$

The term multiplying $1/s$ is the input to an integrator; the integrator's output can be selected as a state variable x. Thus,

$$\begin{aligned} Y(s) &= X(s) + b_0U(s) \\ X(s) &= \frac{1}{s}[b_1U(s) - a_1Y(s)] \\ &= \frac{1}{s}[b_1U(s) - a_1X(s) - a_1b_0U(s)] \end{aligned}$$

or

$$\dot{x} = -a_1x + (b_1 - a_1b_0)u \tag{5.4-13}$$

$$y = x + b_0u \tag{5.4-14}$$

The block diagram is shown in Figure 5.2. Compare (5.4-11) with (5.4-13), and note that the derivative of the input does not appear in the latter form.

Now, consider the second-order model for the mechanical vibration problem. Its transfer function is

$$T(s) = \frac{X(s)}{F(s)} = \frac{1}{ms^2 + cs + k} \tag{5.4-15}$$

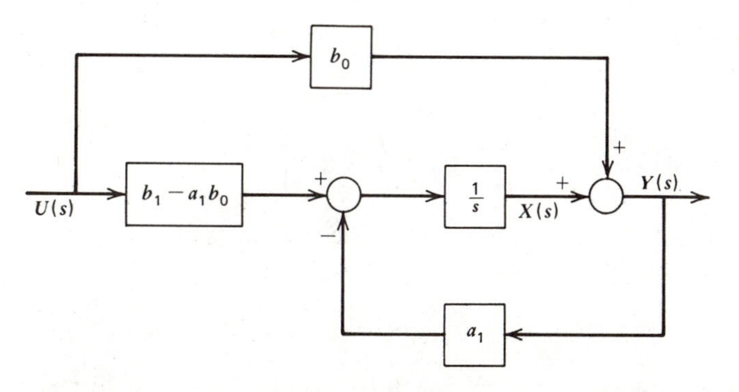

FIGURE 5.2 The block diagram for (5.4-13) and (5.4-14).

Divide by ms^2 to obtain a unity term.

$$T(s) = \frac{X(s)}{F(s)} = \frac{1/ms^2}{1 + c/ms + k/ms^2}$$

Cross multiply and rearrange to identify the integration operators.

$$X(s) = -\frac{c}{ms}X(s) - \frac{k}{ms^2}X(s) + \frac{1}{ms^2}F(s)$$

$$= \frac{1}{s}\left\{-\frac{c}{m}X(s) + \frac{1}{s}\left[\frac{1}{m}F(s) - \frac{k}{m}X(s)\right]\right\} \tag{5.4-16}$$

Defining the state variables as the outputs of the integrators, we see that

$$X_1(s) = X(s) = \frac{1}{s}\left[-\frac{c}{m}X(s) + X_2(s)\right]$$

$$X_2(s) = \frac{1}{s}\left[\frac{1}{m}F(s) - \frac{k}{m}X(s)\right]$$

or

$$X_1(s) = \frac{1}{s}\left[-\frac{c}{m}X_1(s) + X_2(s)\right]$$

$$X_2(s) = \frac{1}{s}\left[\frac{1}{m}F(s) - \frac{k}{m}X_1(s)\right]$$

Figure 5.3a is the block diagram derived from these relations. The corresponding state equations in standard form are

$$\dot{x}_1 = -\frac{c}{m}x_1 + x_2 \tag{5.4-17}$$

$$\dot{x}_2 = -\frac{k}{m}x_1 + \frac{1}{m}f \tag{5.4-18}$$

where $x_1 = x$.

Equation (5.4-16) can also be arranged as follows. Let

$$X_1(s) = mX(s)$$

$$X_2(s) = sX_1(s)$$

Then

$$X_2(s) = \frac{1}{s}\left[F(s) - \frac{c}{m}X_2(s) - \frac{k}{m}X_1(s)\right]$$

These relations give the diagram shown in Figure 5.3*b*. The resulting state model with $x_1 = mx$ is

$$\dot{x}_1 = x_2 \tag{5.4-19}$$

$$\dot{x}_2 = -\frac{k}{m}x_1 - \frac{c}{m}x_2 + f \tag{5.4-20}$$

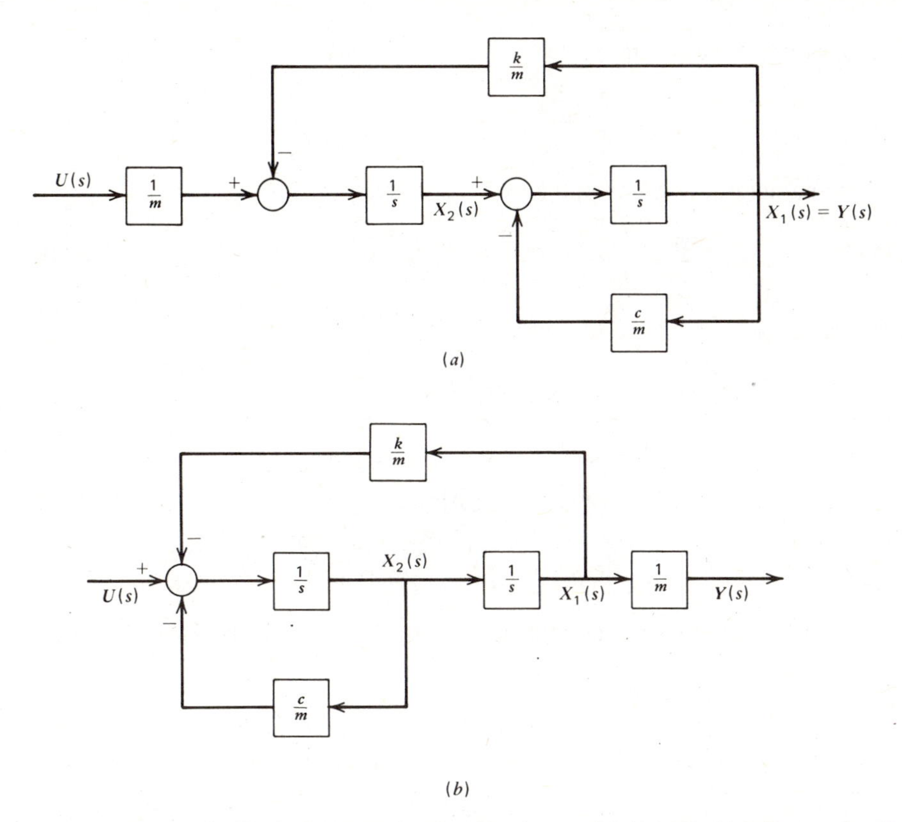

FIGURE 5.3 The block diagrams for the vibration model (5.4-15). (*a*) Diagram for the state model (5.4-17) and (5.4-18). (*b*) Diagram for the state model (5.4-19) and (5.4-20).

Note that the state variables obtained by this technique do not always have straightforward physical interpretations. With the original choice of state variables prescribed by (5.4-2) for (5.4-15), x_1 is the displacement x, and x_2 is the velocity $\dot{x}$. For (5.4-17) and (5.4-18), x_1 is the displacement x, but x_2 is the integral of the difference between the applied force f and the spring force kx, divided by the mass m (x_2 is the acceleration due to this force difference). In (5.4-19) and (5.4-20), x_2 is the momentum $m\dot{x}$, while x_1 is an artificial variable that is proportional to displacement ($x_1 = mx$). Sometimes convenient physical interpretations of the state variables are sacrificed to obtain special forms of the state equations that are useful for analytical purposes. However, this is not apparent in this example, because there is not much difference between the various state models.

State Variable Models and Numerator Dynamics

The case given by (5.4-1) does not include models containing time derivatives of the input v. Linear models with numerator dynamics are examples of this class of models (see Sections 4.2 and 4.9 for discussions of numerator dynamics). We now generalize the transfer function method applied to the first-order model (5.4-10) to show how we can obtain a state variable model that does not contain derivatives of the input.

The general time-invariant model with numerator dynamics can be written as

$$\frac{Y(s)}{V(s)} = \frac{b_n s^n + b_{n-1}s^{n-1} + \cdots + b_1 s + b_0}{a_n s^n + a_{n-1}s^{n-1} + \cdots + a_1 s + a_0} \tag{5.4-21}$$

Dividing the numerator and denominator by $a_n s^n$ gives

$$\frac{Y(s)}{V(s)} = \frac{\beta_n + \beta_{n-1}s^{-1} + \cdots + \beta_1 s^{1-n} + \beta_0 s^{-n}}{1 + \alpha_{n-1}s^{-1} + \cdots + \alpha_1 s^{1-n} + \alpha_0 s^{-n}} \tag{5.4-22}$$

where $\beta_i = b_i/a_n$ and $\alpha_i = a_i/a_n$. This gives

$$\begin{aligned}
Y(s) &= \beta_n V(s) + [\beta_{n-1}V(s) - \alpha_{n-1}Y(s)]s^{-1} \\
&\quad + [\beta_{n-2}V(s) - \alpha_{n-2}Y(s)]s^{-2} \\
&\quad + \cdots + [\beta_1 V(s) - \alpha_1 Y(s)]s^{1-n} \\
&\quad + [\beta_0 V(s) - \alpha_0 Y(s)]s^{-n} \\
&= \beta_n V(s) + s^{-1}\{\beta_{n-1}V(s) - \alpha_{n-1}Y(s) \\
&\quad + s^{-1}[\beta_{n-2}V(s) - \alpha_{n-2}Y(s)] + \cdots\}
\end{aligned}$$

We choose the output of each integrator $(1/s)$ to be a state variable. Thus,

$$\begin{aligned}
Y(s) &= \beta_n V(s) + X_1(s) \\
X_1(s) &= s^{-1}[\beta_{n-1}V(s) - \alpha_{n-1}Y(s) + X_2(s)] \\
X_2(s) &= s^{-1}[\beta_{n-2}V(s) - \alpha_{n-2}Y(s) + X_3(s)] \\
&\vdots \\
X_n(s) &= s^{-1}[\beta_0 V(s) - \alpha_0 Y(s)]
\end{aligned} \tag{5.4-23}$$

Use the first equation in (5.4-23) to eliminate $Y(s)$ in the remaining equations. Thus,

$$\begin{aligned}
sX_1(s) &= (\beta_{n-1} - \alpha_{n-1}\beta_n)V(s) - \alpha_{n-1}X_1(s) + X_2(s) \\
sX_2(s) &= (\beta_{n-2} - \alpha_{n-2}\beta_n)V(s) - \alpha_{n-2}X_1(s) + X_3(s) \\
&\vdots \\
sX_n(s) &= (\beta_0 - \alpha_0\beta_n)V(s) - \alpha_0 X_1(s)
\end{aligned}$$

This gives the differential equation set

$$\begin{aligned}
\dot{x}_1 &= \gamma_{n-1}v - \alpha_{n-1}x_1 + x_2 \\
\dot{x}_2 &= \gamma_{n-2}v - \alpha_{n-2}x_1 + x_3 \\
&\vdots \\
\dot{x}_n &= \gamma_0 v - \alpha_0 x_1
\end{aligned} \tag{5.4-24}$$

where

$$\gamma_i = \beta_i - \alpha_i\beta_n \tag{5.4-25}$$

$$y = \beta_n v + x_1 \tag{5.4-26}$$

Note that not only is the model (5.4-24) in state variable form, but it does not contain derivatives of the input function v. The original initial conditions of the problem are given in terms of y, as $y(0)$, $\dot{y}(0)$, $\ddot{y}(0)$, etc. Equations (5.4-24) and (5.4-26) can be used to find the relation between $x_i(0)$ and $v(t)$, $y(0)$, $\dot{y}(0)$, ... From (5.4-26), at $t = 0$:

$$x_1(0) = y(0) - \beta_n v(0) \tag{5.4-27}$$

Also, from the first equation in (5.4-24), we have

$$x_2 = \dot{x}_1 + \alpha_{n-1} x_1 - \gamma_{n-1} v \tag{5.4-28}$$

Differentiating (5.4-26) and substituting gives

$$\begin{aligned} x_2 &= \dot{y} - \beta_n \dot{v} + \alpha_{n-1} x_1 - \gamma_{n-1} v \\ &= \dot{y} - \beta_n \dot{v} + \alpha_{n-1} y - \alpha_{n-1} \beta_n v - \gamma_{n-1} v \\ &= \dot{y} + \alpha_{n-1} y - \beta_n(\dot{v} + \alpha_{n-1} v) - \gamma_{n-1} v \end{aligned}$$

Generalizing this procedure and denoting $d^i y/dt^i$ by $y^{(i)}$ gives the result:

$$\begin{aligned} x_i &= y^{(i-1)} + \alpha_{n-1} y^{(i-2)} + \cdots + \alpha_{n-i+1} y \\ &\quad - \beta_n[v^{(i-1)} + \alpha_{n-1} v^{(i-2)} + \cdots + \alpha_{n-i+1} v] - \gamma_{n-i+1} v \\ i &= 1, 2, \ldots, n \end{aligned} \tag{5.4-29}$$

Thus the initial values $x_i(0)$ can be found from (5.4-29) given the initial values of $y(t)$, $v(t)$, and their derivatives. Interpreting $t = 0$ as equivalent to $t = 0-$, we can take $v(0) = \dot{v}(0) = \cdots = 0$. In this case (5.4-29) gives

$$x_i(0) = y^{(i-1)}(0) + \alpha_{n-1} y^{(i-2)}(0) + \cdots + \alpha_{n-i+1} y(0) \tag{5.4-30}$$

The transformation to state variable form is summarized in Table 5.2.

TABLE 5.2 A State Variable Form for Numerator Dynamics

Transfer function model:

$$\frac{Y(s)}{V(s)} = \frac{\beta_n s^n + \beta_{n-1} s^{n-1} + \cdots + \beta_1 s + \beta_0}{s^n + \alpha_{n-1} s^{n-1} + \cdots + \alpha_1 s + \alpha_0}$$

State variable model:

$$\begin{aligned} \dot{x}_1 &= \gamma_{n-1} v - \alpha_{n-1} x_1 + x_2 \\ \dot{x}_2 &= \gamma_{n-2} v - \alpha_{n-2} x_1 + x_3 \\ &\vdots \\ \dot{x}_j &= \gamma_{n-j} v - \alpha_{n-j} x_1 + x_{j+1}, \quad j = 1, 2, \ldots, n-1 \\ &\vdots \\ \dot{x}_n &= \gamma_0 v - \alpha_0 x_1 \\ y &= \beta_n v + x_1 \end{aligned}$$

where

$$\gamma_i = \beta_i - \alpha_i \beta_n$$

If $v(0) = \dot{v}(0) = \cdots = 0$ (the usual case), then

$$\begin{aligned} x_i(0) &= y^{(i-1)}(0) + \alpha_{n-1} y^{(i-2)}(0) + \cdots + \alpha_{n-i+1} y(0) \\ i &= 1, 2, \ldots, n \end{aligned}$$

For example, consider

$$\frac{Y(s)}{V(s)} = \frac{s^2 + 3s + 2}{10s^2 + 5s + 6} \tag{5.4-31}$$

where $y(0) = 4$, $\dot{y}(0) = 7$, and $v(t) = 0$ for $t \leqslant 0$. The state model (5.4-24) and (5.4-26) is

$$\dot{x}_1 = 0.25v - 0.5x_1 + x_2 \tag{5.4-32}$$

$$\dot{x}_2 = 0.14v - 0.6x_1 \tag{5.4-33}$$

$$y = 0.1v + x_1 \tag{5.4-34}$$

The initial conditions are, from (5.4-30),

$$x_1(0) = y(0) = 4 \tag{5.4-35}$$

$$x_2(0) = \dot{y}(0) + 0.5y(0) = 7 + 2 = 9 \tag{5.4-36}$$

Equations (5.4-32)–(5.4-34) are interpreted for $t > 0$. Thus, from (5.4-34)–(5.4-35), we see that $y(0+) = 0.1v(0+) + 4$. If $v(t)$ is a unit step, then $v(0+) = 1$ and $y(0+) = 4.1$.

5.5 LINEAR ANALYSIS AS AN AID TO SIMULATION

Proper choice of the step size Δt is essential for successfully applying numerical simulation methods. When there is insufficient physical insight to choose an appropriate value for Δt, we can sometimes use a linear analysis of the model to estimate the required value of Δt. The analysis techniques for linear and linearized models given in the previous chapters can be useful when:

1. The model to be simulated is linear, but the order is high enough or the input functions are complicated enough to discourage a closed-form solution from being attempted.
2. The model to be simulated is nonlinear. In this case, the model can be linearized for nominal values of the system's variables that represent the simulation conditions.

In either case, a linear analysis can be used to estimate the smallest time constant and the smallest oscillation period of the model. The initial choice for the time step Δt is then taken to be a small fraction of the smallest of these. The situation is illustrated in Figure 5.4. In 5.4*a* the smallest oscillation period P is less than the smallest system time constant τ. In this case, a reasonable initial choice for $\Delta t = P/100$, assuming that 100 points are sufficient to identify the form of the oscillatory function for one period. In 5.4*b*, the opposite is true, and we might choose $\Delta t = \tau/10$, since a decaying curve with a time constant τ can be plotted well with 10 points. The reason for these choices is that the time step Δt must be small compared to the time span over which significant changes occur in the system variables.

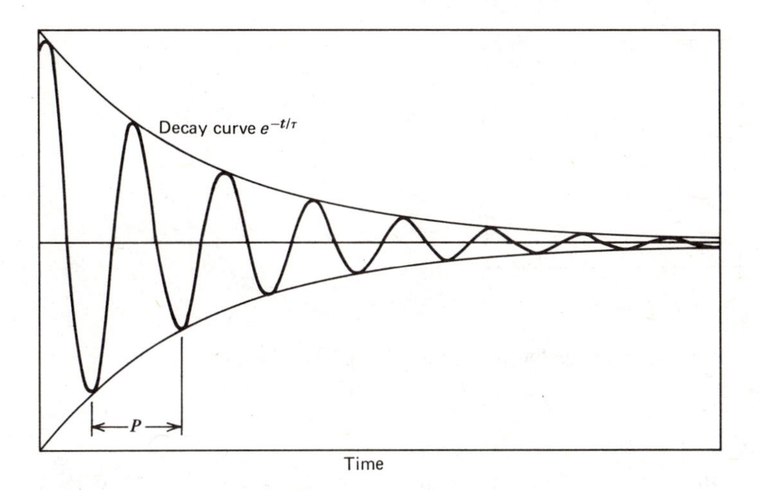

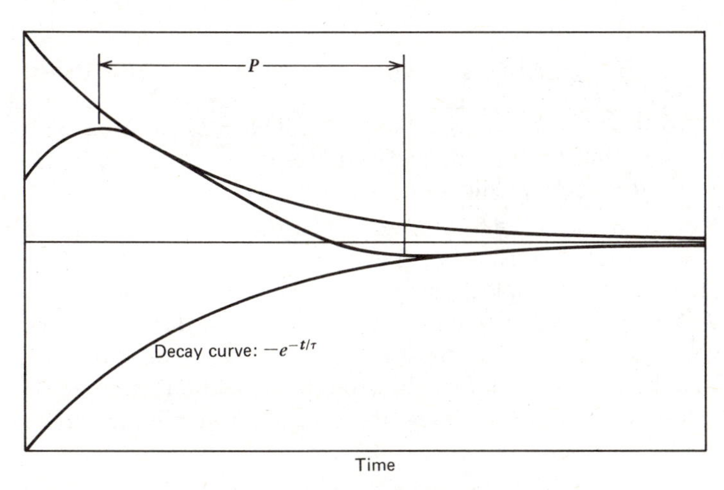

FIGURE 5.4 Using the linear response to choose the step size Δt. (*a*) The oscillation period *P* determines Δt. (*b*) The time constant τ determines Δt.

Example 5.5

Use simulation to compute the time it will take to fill a tank with liquid to a height of 9 ft if the input flow rate is $q = 3t$ ft^3/min. The initial height is $h(0) = 1$ ft. The tank's model, given in Example 3.12, is

$$\dot{h} = q - \sqrt{h} \tag{5.5-1}$$

From the linearized analysis of Example 3.12, the tank's time constant is $\tau = 2$ when $h = 1$ and $\tau = 6$ when $h = 9$. We can choose Δt to be 1/10 of the smallest τ; that is, $\Delta t = 0.1(2) = 0.2$. The results using the Runge–Kutta method are shown in Figure 5.5, and they were checked for accuracy by using one-half the original value of Δt. The results were essentially the same, so we can take $\Delta t = 0.2$ to be an appropriate value. The time to raise the height to 9 ft can be seen from Figure 5.5 to be approximately 2.9 min.

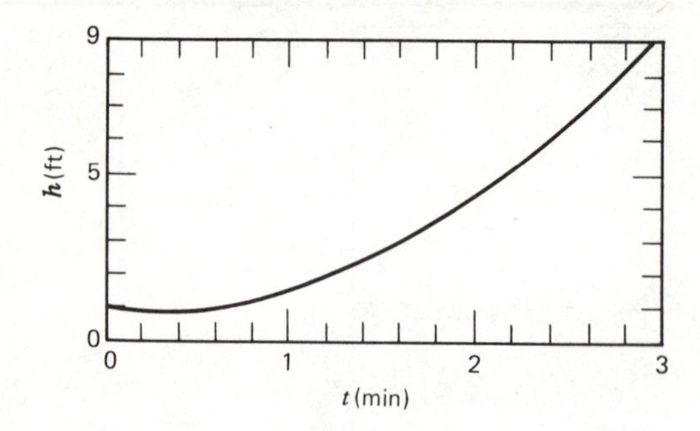

FIGURE 5.5 Ramp response of the nonlinear tank model (5.5-1).

Example 5.6

The linearized equation of motion for the pendulum shown in Figure 3.25, with a reference angle $\theta = 0$ is

$$mL^2\ddot{\theta} + mgL\theta = 0 \tag{5.5-2}$$

The linearized natural frequency is $\omega_n = \sqrt{g/L}$. Suppose that $L = 2$ ft, and $g = 32.2$. Use numerical simulation to find $\theta(t)$ for one half-period if $\theta(0) = \pi/2$ and $\dot{\theta}(0) = 0$. Compare with the linearized model's predictions.

For the given values, $\omega_n = \sqrt{16.1} = 4.012$ rad/sec. Thus, the predicted oscillation period is $P = 2\pi/4.012 = 1.57$ sec. Therefore, Δt might be initially selected as $\Delta t = 0.01(1.57) \approx 0.016$ sec. (Note that this model has no damping and therefore no time constant. The period P is the only meaningful time scale provided by the model.) The results with the Runge–Kutta method are shown in Figure 5.6. A check with $\Delta t = 0.008$ gave essentially the same results. Another check on the simulation's accuracy is provided by physical insight. With no damping, the oscillation amplitude should

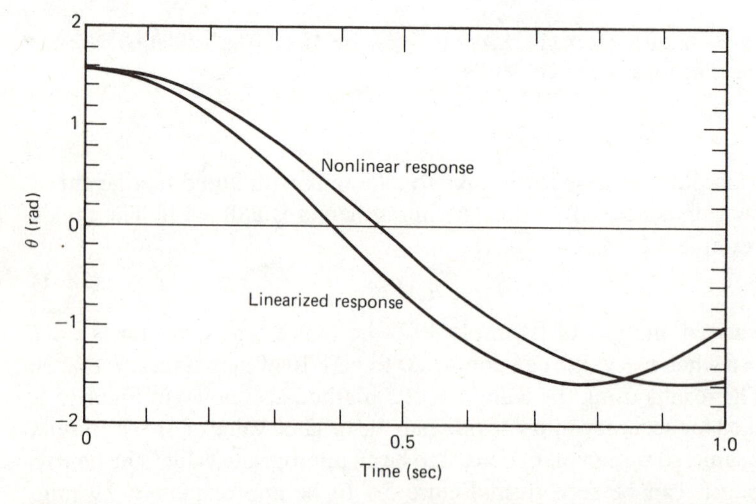

FIGURE 5.6 Free response of the nonlinear and linearized pendulum models about $\theta = 0$.

remain constant. Thus, θ should be $-\pi/2$ at the end of the first half-period. With the resolution of the plot shown in Figure 5.6, there is no discernible error for $\Delta t = 0.016$. However, if $\Delta t = 0.1P = 0.16$ were used, the amplitude error would be 4%.

A Design Example

Simulation methods must often be used in situations where we do not know the values of some of the model's coefficients, because they must be determined as part of the design process. In these cases, a linearized analysis can often be used to pick approximate values for the desired coefficients in order to satisfy such design criteria as stability, time constants, damping ratio, etc. The nonlinear model is then simulated using these coefficient values to verify the predictions of the linearized analysis.

As an illustration, consider the equations of motion for the rotation of a rigid body. These are the *Euler* equations

$$I_1\dot{\omega}_1 + (I_3 - I_2)\omega_2\omega_3 = T_1 \tag{5.5-3}$$

$$I_2\dot{\omega}_2 + (I_1 - I_3)\omega_3\omega_1 = T_2 \tag{5.5-4}$$

$$I_3\dot{\omega}_3 + (I_2 - I_1)\omega_1\omega_2 = T_3 \tag{5.5-5}$$

where I_i, ω_i, and T_i are the inertia, angular velocity, and applied torque about the ith principal axis (Reference 3). Suppose the rigid body is an artificial satellite that is cylindrical in shape with mass m, radius r, and length L (see Figure 5.7). Then from Reference 3,

$$I_1 = I_3 = \frac{m(3r^2 + L^2)}{12}$$

$$I_2 = \frac{mr^2}{2}$$

Suppose that $m = 100$ slugs, $r = 3$ ft, and $L = 20$ ft. Then the Euler equations become

$$\dot{\omega} = -0.8735\omega_2\omega_3 + 0.00028T_1 \tag{5.5-6}$$

$$\dot{\omega}_2 = 0.00222T_2 \tag{5.5-7}$$

$$\dot{\omega}_3 = 0.8735\omega_1\omega_2 + 0.00028T_3 \tag{5.5-8}$$

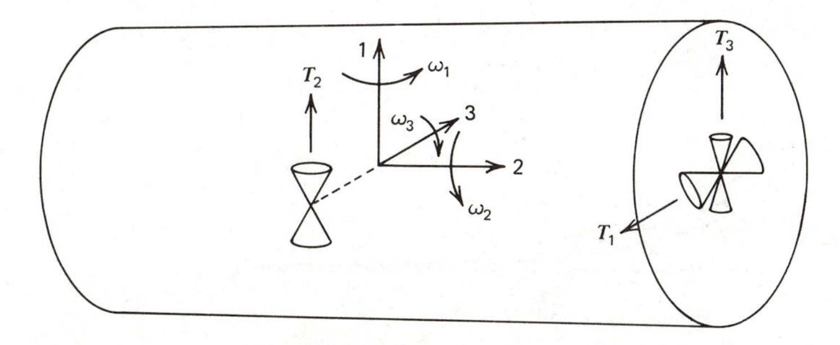

FIGURE 5.7 A cylindrical satellite and its principal axes.

Our design problem is to find a way of controlling the thruster torques T_1, T_2, T_3 in order to keep the satellite from spinning; that is, to keep ω_1, ω_2, and ω_3 near zero. A logical way would be to have the controller adjust the torques so that they oppose the rotation and their magnitudes are directly proportional to the angular velocity about their respective axes. That is, make

$$T_i = -K_i\omega_i \qquad i = 1, 2, 3$$

This makes sense, because the larger the angular velocity, the more torque applied by the thruster in the direction opposite to the rotation. Our analysis must answer the following questions:

1. Will this scheme stabilize the rotation? [Will $\omega_i(t) \to 0$ for $i = 1, 2, 3$?]
2. If the answer is yes, what values should be used for the K_i so that the angular velocities are made nearly zero in a specified manner within a specified time?

To address the first question, we linearize the model (5.5-3)–(5.5-5) about $\omega_i = 0$, using $T_i = K_i\omega_i$. This gives

$$\begin{aligned} \dot{\omega}_1 &= -0.00028K_1\omega_1 \\ \dot{\omega}_2 &= -0.00222K_2\omega_2 \\ \dot{\omega}_3 &= -0.00028K_3\omega_3 \end{aligned} \qquad \textbf{(5.5-9)}$$

From these, we see that $\omega_i(t) \to 0$ if the K_i are chosen to be positive.

To answer the second question, suppose we wish to have the angular velocity decay to zero with a time constant of 1 sec for ω_1 and ω_3 and 10 sec for ω_2. From

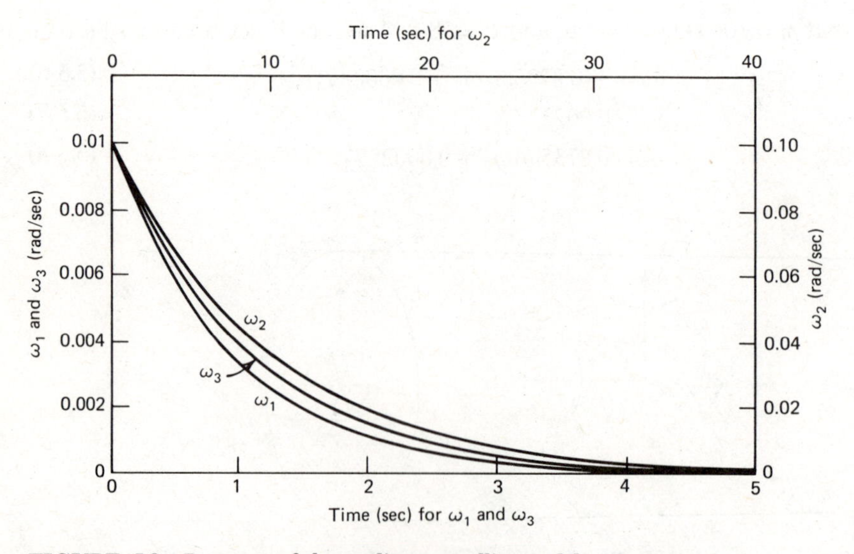

FIGURE 5.8 Response of the nonlinear satellite model.

(5.5-9), we find that the K_i must satisfy

$$0.00028K_1 = \frac{1}{\tau_1} = 1$$

$$0.00222K_2 = \frac{1}{\tau_2} = 0.1$$

$$0.00028K_3 = \frac{1}{\tau_3} = 1$$

or $K_1 = K_3 = 3571.4$ and $K_2 = 45.045$.

To see if our design performs as desired, we now simulate the original nonlinear model (5.5-6)–(5.5-8) using $T_1 = -3571.4\omega_1$, $T_2 = -45.045\omega_2$, and $T_3 = -3571.4\omega_3$. The smallest specified time constant is 1 sec, so $\Delta t = 0.1$ was chosen to use with the Runge–Kutta method. Then Δt was reduced by one-half to 0.05. This gave results that were close to those obtained with $\Delta t = 0.1$. These are shown in Figure 5.8 for initial velocities of $\omega_1(0) = \omega_3(0) = 0.01$, and $\omega_2(0) = 0.1$ rad/sec. The simulation shows that each angular velocity decays to 37% of its initial value in a time approximately equal to that predicted by the time constants of the linearized model. Thus, our scheme for controlling the thrusters works well.

5.6 SIMULATION AND STEP INPUTS

The step function models a sudden change in the system's input and as such represents perhaps the most severe test of system performance. It is commonly used for that reason, but because it signifies an instantaneous change in the input, we must be sure that our system model is valid for such an input. Models containing time derivatives of the input function are prime examples of this problem. We derived the step response of one such model in Section 4.9 in order to generate some insight about effects of numerator dynamics. Because the time derivative of a step function is infinite at $t = 0$, the input generates an impulse at $t = 0$, and thus a step input applied to a model with input derivatives is meaningless from a physical point of view. In spite of this, the step response of such models can be useful if viewed as the limiting case of a rapidly changing input function.

The difficulty occurs when we cannot either obtain an analytical solution for the step response of models containing input derivatives, or transform the model to eliminate input derivatives as done in Section 5.4. Such cases must then be solved numerically, and the question arises of how we can numerically represent the rapid change in the step input at $t = 0$. Of course, the effective impulse generated by the input derivatives acting on the step function cannot be represented numerically, so we will take the approach of modeling the step function with a similar function that makes more sense physically.

Here are four ways of finding the step response of a model with input derivatives. The first two will work only with linear models.

1. Find the response with the Laplace transform. This is the most direct approach, but care must be used to interpret the results realistically from a physical viewpoint.

2. Use artificial initial conditions to allow elimination of the input derivatives from the model (see Section 4.8). The resulting model can then be handled by any conventional solution method.
3. Redefine the system's state variables to convert the model into one without input derivatives. This was done in Section 5.4 with (5.4-10) and more generally with (5.4-21). Conventional solution methods can then be used. This approach is particularly useful for a linear model whose order is high enough to make the Laplace transform solution impractical. Instead, a numerical solution method can be applied to the model after its variables have been redefined to eliminate the input derivatives.
4. Model the step input with a more realistic function that does not possess discontinuities. This approach allows the use of numerical solution methods for cases that cannot be treated with the preceding methods. We will develop such a function in this section and refer to it as the "practical" step function to distinguish it from the "pure" step function we have used thus far.

The following examples illustrate the four approaches.

Laplace Transform Solution

Consider the following model:

$$\dot{y} + 10y = b\dot{v} + v \tag{5.6-1}$$

$$y(0) = 0 \tag{5.6-2}$$

If $v(t)$ is a unit-step input and $b = 1$, then

$$Y(s) = \frac{s+1}{s+10}\frac{1}{s} = \frac{0.1}{s} + \frac{0.9}{s+10} \tag{5.6-3}$$

and

$$y(t) = 0.1 + 0.9e^{-10t} \tag{5.6-4}$$

This is plotted in Figure 5.9. Notice that (5.6-4) implies that $y(0+) = 1$. This represents an instantaneous jump in the value of y at $t = 0$, which is impossible if y represents a physical variable. Since this behavior results from combining the $\dot{v}$ term in (5.6-1) with a step input, assumptions that lead to such a combined representation should be carefully examined from a physical point of view.

Using Artificial Initial Conditions

The Laplace transform of (5.6-1) with a step input of magnitude m and initial condition $y(0) = c$ gives

$$\begin{aligned}(s+10)Y(s) &= (bs+1)V(s) + c \\ &= (bs+1)\frac{m}{s} + c \\ &= bm + \frac{m}{s} + c\end{aligned} \tag{5.6-5}$$

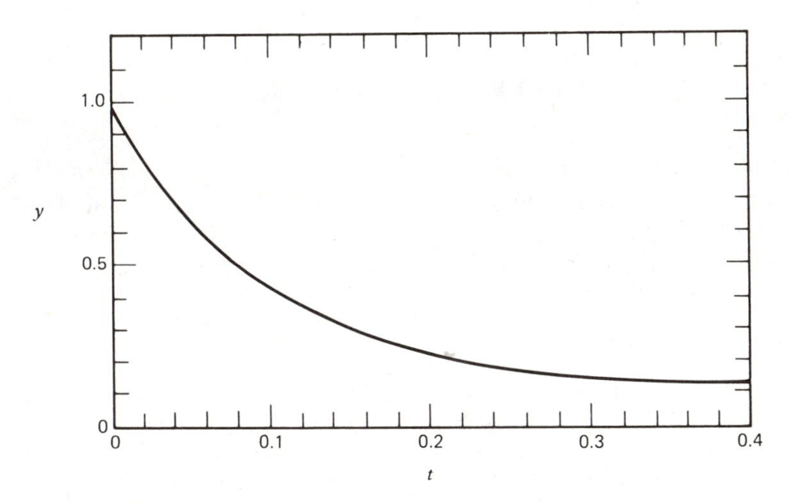

FIGURE 5.9 Unit-step response of equation (5.6-1).

From this, we see that the effect of the derivative term $b\dot{v}$ for a step input can be represented by a proper choice of initial value $y(0)$. That is, if we pick $y(0) = bm + c$, then the following model can be used to replace (5.6-1) as long as the input is a step function.

$$\dot{y} + 10y = v \tag{5.6-6}$$

$$y(0) = bm + c \tag{5.6-7}$$

Equation (5.6-6) can then be solved with any convenient method. There is no great advantage in using this approach for the simple model given here, but the example illustrates the general approach. It is useful for higher order models that can be solved numerically after the input derivatives have been eliminated.

Redefining the State Variables

With the approach taken to derive (5.4-10), the model (5.6-1) can be converted into the equivalent form

$$\dot{x} + 10x = (1 - 10b)v \tag{5.6-8}$$

$$y = x + bv \tag{5.6-9}$$

$$x(0) = y(0-) - bv(0-) \tag{5.6-10}$$

First, we solve (5.6-8) by any method to obtain the intermediate variable $x(t)$. Then we use (5.6-9) to obtain the variable of interest $y(t)$. For $b = 1$, $y(0-) = 0$ and a unit-step input, (5.6-10) gives $x(0) = 0$. Equation (5.6-8) gives

$$X(s) = \frac{-9}{s+10}\frac{1}{s} = \frac{-0.9}{s} + \frac{0.9}{s+10}$$

or

$$x(t) = -0.9 + 0.9e^{-10t}$$

From (5.6-9) for $t \geqslant 0+$,

$$y(t) = 0.1 + 0.9e^{-10t}$$

which agrees with (5.6-4).

Equations (5.4-24) give the general form for converting the linear model (5.4-21) into one without input derivatives. Numerical-solution methods can then be applied to (5.4-24) if necessary. No general procedure is available for redefining the state variables for a nonlinear model. Each case must be dealt with individually.

The Practical Step Function

The fourth approach to dealing with the step response is replacing the pure step function with a more realistic function, which we will call the practical step function. This replacement function should satisfy at least the following three conditions.

1. The function and its time derivatives should be continuous (so that we can use the function with models possessing input derivatives); and
2. The function should start at zero and approach a constant value in a time span that is very short compared to the response of the model. This is required in order for the function to represent a rapidly applied input; and
3. The function should possess a simple analytic form so that we may easily compute its derivatives, and use it to find the response analytically in some cases.

The following function fits the preceding requirements, and we will use it as our *practical step function.*

$$v(t) = m(1 - e^{-at}) \tag{5.6-11}$$

where m is the magnitude of the function and a is a constant that is selected to satisfy the second requirement. The first requirement is met because

$$\begin{aligned} \dot{v} &= mae^{-at} \\ \ddot{v} &= -ma^2e^{-at} \\ &\vdots \\ \frac{d^nv}{dt^n} &= (-1)^{n+1}ma^ne^{-at} \end{aligned} \tag{5.6-12}$$

Thus, $v(t)$ and all its derivatives are continuous in t. The third requirement is also satisfied, because $v(t)$ and its derivatives have the following simple Laplace transforms:

$$\mathscr{L}[v(t)] = \frac{ma}{s(s+a)} \tag{5.6-13}$$

$$\mathscr{L}\left[\frac{d^nv}{dt^n}\right] = \frac{(-1)^{n+1}ma^n}{s+a} \tag{5.6-14}$$

Our choice (5.6-11) also makes sense physically. Suppose that $v(t)$ is the voltage input to a motor. The amplifier producing $v(t)$ is itself a dynamic system, and the most reasonable dynamic model for the amplifier is a first-order model. When the amplifier is turned on, the first-order model predicts that the voltage $v(t)$ will increase according to (5.6-11). In general, it is not unreasonable to assume that the device producing the input $v(t)$ behaves like a first-order system and its output is described by (5.6-11).

The larger the constant a is in (5.6-11), the closer $v(t)$ resembles a pure step function. This is illustrated in Figure 5.10, where $m = 1$. If we wish to use (5.6-11) to approximate a pure step function, we should choose the constant a so that $1/a$ is small compared to the smallest characteristic response time of the model being analyzed. For example, if we know that the smallest time constant of the model is τ, then it is suggested that a be selected so that

$$a \geqslant 10/\tau \qquad \textbf{(5.6-15)}$$

The use of the practical step function is illustrated with the model (5.6-1), whose transfer function is

$$\frac{Y(s)}{V(s)} = \frac{bs+1}{s+10} \qquad \textbf{(5.6-16)}$$

From (5.6-13),

$$V(s) = \frac{ma}{s(s+a)} \qquad \textbf{(5.6-17)}$$

For $b = m = 1$ and $y(0) = 0$, we have

$$Y(s) = \frac{s+1}{s+10}\frac{a}{s(s+a)}$$
$$= \frac{C_1}{s} + \frac{C_2}{s+a} + \frac{C_3}{s+10}$$

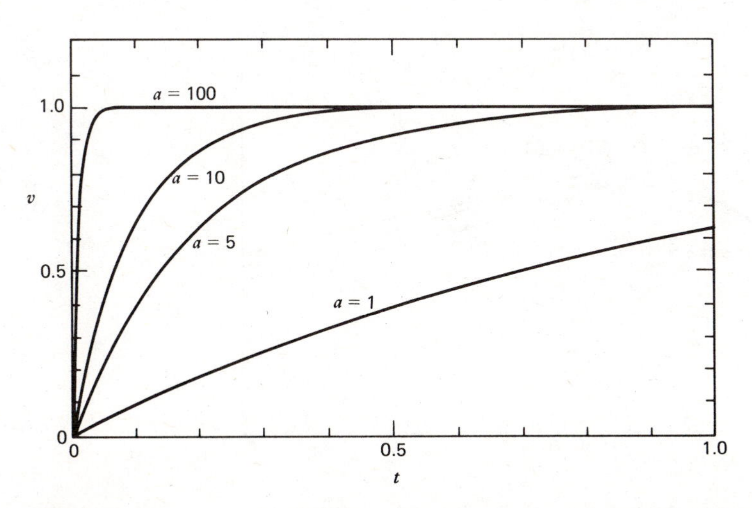

FIGURE 5.10 The practical step function (5.6-11) for $m = 1$ and various values of the parameter a.

and

$$y(t) = C_1 + C_2 e^{-at} + C_3 e^{-10t} \tag{5.6-18}$$

where $C_1 = 0.1$, $C_2 = (1 - a)/(a - 10)$, and $C_3 = 0.9a/(a - 10)$. The solution is shown in Figure 5.11 for $a = 5/\tau = 50$ and $a = 10/\tau = 100$ along with the pure step response (5.6-4). From the figure, we can see that the response to the practical step input follows (the pure step response rather closely except for the latter's unrealistic discontinuity at $t = 0$. In fact, for $t \geqslant 0.1$, the response curves are very close together (recall that the time constant of the model is $\tau = 0.1$).

The preceding example merely illustrates the response produced by the practical step function; the pure step response can easily be obtained analytically in this case. However, in the following example, the model is nonlinear, an analytical solution is not available, and artificial initial conditions cannot be used. Also, the input derivatives cannot be eliminated using the method in Section 5.4 because the model is nonlinear. Therefore, a numerical solution is called for. The example to be used is given in Figure 4.32*a*, where we now assume that the spring's force displacement relation is the nonlinear function

$$f = k_1(y + k_2 y^3) \tag{5.6-19}$$

where

$$y = u - x$$
$$k_1 = 1$$
$$k_2 = 5$$

For $m = 1$ and $c = 2$, the system model becomes

$$\ddot{x} = u - x + 5(u - x)^3 + 2(\dot{u} - \dot{x}) \tag{5.6-20}$$

For simulation purposes, we convert to state variable form by defining $x_1 = x$, $x_2 = \dot{x}$,

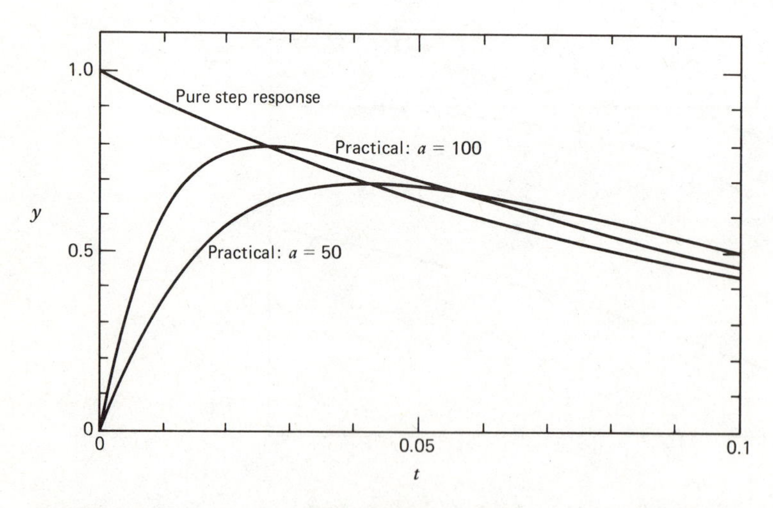

FIGURE 5.11 Response of the model (5.6-11) to a pure step input, and to the practical step function for two values of the parameter *a*.

to obtain

$$\dot{x}_1 = x_2 \tag{5.6-21}$$

$$\dot{x}_2 = u - x_1 + 5(u - x_1)^3 + 2(\dot{u} - x_2) \tag{5.6-22}$$

Suppose the applied displacement u changes rapidly from $u = 0$ to $u = 1$. The practical step function representation is

$$u(t) = 1 - e^{-at}$$

Equation (5.6-22) requires $\dot{u}$, which is easily found to be $\dot{u} = ae^{-at}$. We must now estimate an appropriate value to use for the constant a. One way to do this is to obtain an estimate of the model's time constants from the linearized model. Here, for $u \approx x \approx 0$, the linearized model is

$$\ddot{x} + 2\dot{x} + x = 2\dot{u} + u$$

The characteristic roots are $s = -1, -1$; so the model's smallest time constant is $\tau = 1$. Therefore, we choose $a \geqslant 10/\tau = 10$. To determine the simulation's sensitivity to the chosen value of a, it is helpful to use some larger values as well. Here, we used $a = 30$ and $a = 100$. The Runge–Kutta results are shown in Figure 5.12 for $x_1(0) = x_2(0) = 0$. For an increase in a from 30 to 100, not much change occurred in the response curve, so we conclude that $a = 100$ is large enough and accept that response curve as our final answer.

The size of the time step to be used in numerical simulations with the practical step function must be much smaller than $1/a$, because $1/a$ is the time constant of the practical step function. In general, a good rule of thumb is to choose $\Delta t \leqslant 0.2/a$. For the results shown in Figure 5.12, Δt was chosen to be 0.02, 0.007, and 0.002 for $a = 10$, 30, and 100. A large value for a requires using a small time step. Since this can result in a large number of iterations, care should be taken to select both a and Δt with this in mind.

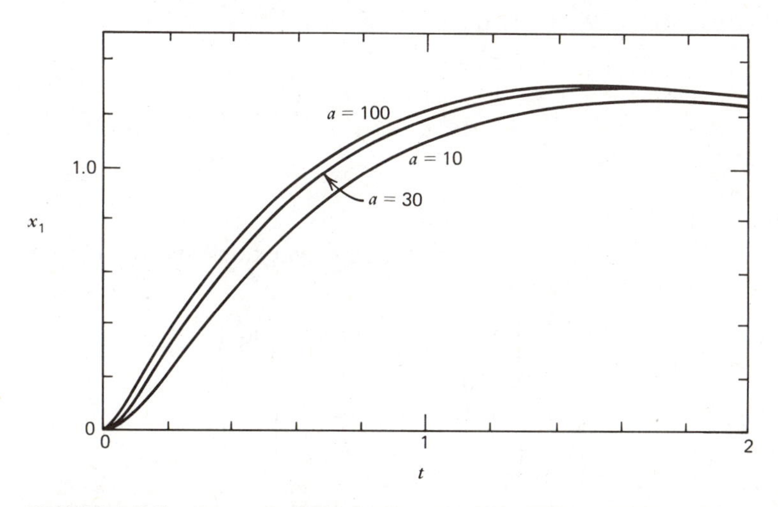

FIGURE 5.12 Response of the nonlinear model (5.6-21) and (5.6-22) to a practical step input for several values of the parameter *a*.

5.7 VECTOR–MATRIX METHODS†

The state variable form of a linear model can be expressed compactly using vector–matrix notation. This compactness can be exploited by simulation and analysis techniques. For example, a single computer program can be written for simulating the response of linear model of arbitrary order. Similarly, algorithms for determining stability and response characteristics can be developed independently of the system order. This section introduces vector–matrix notation and indicates how such notation can be used. A review of vector–matrix algebra is given in Appendix A, Section A.4. More detailed coverage of these topics can be found in Reference 4, chapters 5, 9, and 10.

Vector–Matrix Form of the State Equations

The general form of the second-order linear model in terms of the state variables x_1 and x_2 is

$$\dot{x}_1 = a_{11}x_1 + a_{12}x_2 + b_{11}u_1 + b_{12}u_2 \quad \textbf{(5.7-1)}$$

$$\dot{x}_2 = a_{21}x_1 + a_{22}x_2 + b_{21}u_1 + b_{22}u_2 \quad \textbf{(5.7-2)}$$

where we have assumed that two inputs (u_1, u_2) act on the system. This is the *state variable* form of the model. Assume also that there are two outputs (y_1, y_2), which are linear combinations of the state variables and the inputs. Then

$$y_1 = c_{11}x_1 + c_{12}x_2 + d_{11}u_1 + d_{12}u_2 \quad \textbf{(5.7-3)}$$

$$y_2 = c_{21}x_1 + c_{22}x_2 + d_{21}u_1 + d_{22}u_2 \quad \textbf{(5.7-4)}$$

We can form a two-dimensional column vector x with the two state variables x_1 and x_2; that is,

$$\mathbf{x} = \begin{bmatrix} x_1 \\ x_2 \end{bmatrix} \quad \textbf{(5.7-5)}$$

The vector $\mathbf{x}$ is the *state vector.* Similarly the *input vector* is

$$\mathbf{u} = \begin{bmatrix} u_1 \\ u_2 \end{bmatrix} \quad \textbf{(5.7-6)}$$

and the *output vector* is

$$\mathbf{y} = \begin{bmatrix} y_1 \\ y_2 \end{bmatrix} \quad \textbf{(5.7-7)}$$

From the rules of matrix–vector multiplication, the state equations and the output equations (5.7-1)–(5.7-4) can be written as

$$\begin{bmatrix} \dot{x}_1 \\ \dot{x}_2 \end{bmatrix} = \begin{bmatrix} a_{11} & a_{12} \\ a_{21} & a_{22} \end{bmatrix} \begin{bmatrix} x_1 \\ x_2 \end{bmatrix} + \begin{bmatrix} b_{11} & b_{12} \\ b_{21} & b_{22} \end{bmatrix} \begin{bmatrix} u_1 \\ u_2 \end{bmatrix}$$

$$\begin{bmatrix} y_1 \\ y_2 \end{bmatrix} = \begin{bmatrix} c_{11} & c_{12} \\ c_{21} & c_{22} \end{bmatrix} \begin{bmatrix} x_1 \\ x_2 \end{bmatrix} + \begin{bmatrix} d_{11} & d_{12} \\ d_{21} & d_{22} \end{bmatrix} \begin{bmatrix} u_1 \\ u_2 \end{bmatrix}$$

---

†This section introduces the student to state vector terminology, which appears more frequently in the technical literature. However, the following chapters do not depend on this material, and Sections 5.7 and 5.8 may be omitted if necessary.

or

$$\dot{\mathbf{x}} = \mathbf{Ax} + \mathbf{Bu} \tag{5.7-8}$$

$$\mathbf{y} = \mathbf{Cx} + \mathbf{Du} \tag{5.7-9}$$

where

$$\dot{\mathbf{x}} = \begin{bmatrix} \dot{\mathbf{x}}_1 \\ \dot{\mathbf{x}}_2 \end{bmatrix} \tag{5.7-10}$$

$$\mathbf{A} = [a_{ij}] \tag{5.7-11}$$

$$\mathbf{B} = [b_{ij}] \tag{5.7-12}$$

$$\mathbf{C} = [c_{ij}] \tag{5.7-13}$$

$$\mathbf{D} = [d_{ij}] \tag{5.7-14}$$

for $i = 1, 2$ and $j = 1, 2$. The matrix **A** is the *system matrix*; **B** is the *control* or *input matrix*; **C** and **D** are the *output matrices.*

If the coefficients are constant, these equations can be Laplace transformed. The system diagram can then be readily drawn using two integrator elements whose outputs are x_1 and x_2 and whose inputs are the right-hand sides of (5.7-1) and (5.7-2). The system outputs are then constructed from (5.7-3) and (5.7-4). This diagram is called the *state diagram,* because it is constructed from the state variable form of the model and thus graphically shows how the state variables are generated.

Example 5.7

Draw the state diagram for the following model:

$$\dot{x}_1 = -4x_1 + 2x_2 + 4u_1$$

$$\dot{x}_2 = -3x_2 + 5u_2$$

The output variables are

$$y_1 = x_1 + x_2 + 3u_1$$

$$y_2 = x_2$$

Assuming zero initial conditions, we transform the model and divide the first two equations by s to obtain

$$X_1(s) = \frac{1}{s}[-4X_1(s) + 2X_2(s) + 4U_1(s)]$$

$$X_2(s) = \frac{1}{s}[-3X_2(s) + 5U_2(s)]$$

$$Y_2(s) = X_2(s)$$

Using the guidelines for the inputs and outputs of the two integrators, we obtain the diagram in Figure 5.13.

The state equations and output equations need not be linear to be represented in vector form. The general vector form is

$$\dot{\mathbf{x}} = \mathbf{f}(\mathbf{x}, \mathbf{u}, t) \tag{5.7-15}$$

$$\mathbf{y} = \mathbf{g}(\mathbf{x}, \mathbf{u}) \tag{5.7-16}$$

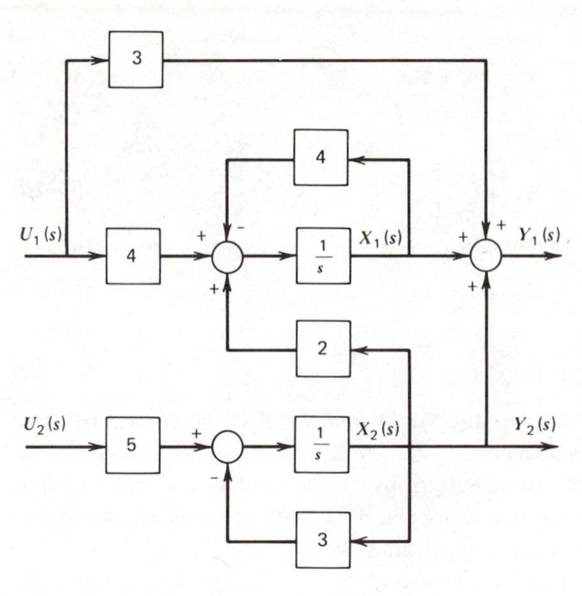

FIGURE 5.13 State diagram for Example 5.7.

Numerical Simulation Methods in Vector Form

The compactness of (5.7-15) and (5.7-16) makes it easy to describe a numerical simulation method in a completely general way. This allows a computer code to be written more easily. For example, the Euler method applied to (5.7-15) yields

$$\mathbf{x}(t_{k+1}) = \mathbf{x}(t_k) + \Delta t\, \mathbf{f}[\mathbf{x}(t_k), \mathbf{u}(t_k), t_k] \qquad \textbf{(5.7-17)}$$

Note the similarity to the scalar case given by (5.1-7) and (5.1-8), with $\mathbf{x}$ and $\mathbf{u}$ now playing the roles of y and v. Compare the conciseness of (5.7-17) with that of (5.3-3) and (5.3-4) to appreciate the advantages of vector notation.

The predictor–corrector method given by the modified Euler algorithm can be stated in general as:

Euler Predictor

$$\mathbf{z}_{k+1} = \mathbf{x}_k + \Delta t\, \mathbf{f}(\mathbf{x}_k, \mathbf{u}_k, t_k) \qquad \textbf{(5.7-18)}$$

Trapezoidal Corrector

$$\mathbf{x}_{k+1} = \mathbf{x}_k + 0.5\Delta t\, [\mathbf{f}(\mathbf{x}_k, \mathbf{u}_k, t_k) + \mathbf{f}(\mathbf{z}_{k+1}, \mathbf{u}_{k+1}, t_{k+1})] \qquad \textbf{(5.7-19)}$$

where we have suppressed the t_k notation by letting $\mathbf{x}_k = \mathbf{x}(t_k)$.

The Runge–Kutta algorithm can be expressed in vector form in like manner. The resulting vector equations would resemble those of the scalar form (5.2-28) and (5.2-29), with the scalars y and f replaced by $\mathbf{x}$ and $\mathbf{f}$. A computer program implementing this method is given in Appendix C. The reader should study this program to see how the algorithm is programmed to handle state equations of arbitrary order.

Transfer Function Methods

Many computer-aided design and analysis packages use the state variable notation in their user's manuals. Since their capabilities often include transfer function analysis, it is helpful to see how the transfer function concepts are handled with vector–matrix notation.

A transfer function is a relation between an input and an output. When there are two inputs and two outputs, there are four possible input–output pairs, and therefore four possible transfer functions for the system. These can be found by either reducing the system diagram, or algebraic manipulation of the transformed equations. If the diagram is in state variable form, usually the easiest method is to obtain the state equations and then use the algebraic method. These equations are obtained by choosing the state variables as the outputs of the integrators. The right-hand side of each state equation is given by the input to the integrator.

The transfer function describes the effect of a given input on a given output. It does not describe the effects of other inputs or initial conditions on that output. Therefore, the other inputs and all initial conditions are set to zero when deriving a transfer function. That is, the transfer function between the output y_i and the input u_j is

$$T_{ij}(s) = \left. \frac{Y_i(s)}{U_j(s)} \right|_{\substack{U_k(s)=0,\, k \neq j \\ x_k(0)=0,\, \text{for all } k}} \tag{5.7-20}$$

Note that the derivative (or integral) of a vector or matrix is the vector or matrix of the derivatives (or integrals) of the elements. Therefore, since the Laplace transform is an integration operator, the transform of a vector is the vector consisting of the transforms of the elements; namely,

$$\mathbf{X}(s) = \begin{bmatrix} X_1(s) \\ X_2(s) \end{bmatrix} \tag{5.7-21}$$

Therefore,

$$\mathscr{L}(\dot{\mathbf{x}}) = \begin{bmatrix} sX_1(s) - x_1(0) \\ sX_2(s) - x_2(0) \end{bmatrix} = s\mathbf{X}(s) - \mathbf{x}(0) \tag{5.7-22}$$

Transforming (5.7-8) and (5.7-9) and using this result, we obtain

$$s\mathbf{X}(s) - \mathbf{x}(0) = \mathbf{AX}(s) + \mathbf{BU}(s) \tag{5.7-23}$$

$$\mathbf{Y}(s) = \mathbf{CX}(s) + \mathbf{DU}(s) \tag{5.7-24}$$

The first of these can be arranged as

$$(s\mathbf{I} - \mathbf{A})\mathbf{X}(s) = \mathbf{x}(0) + \mathbf{BU}(s) \tag{5.7-25}$$

where the identity matrix $\mathbf{I}$ has been used to allow $\mathbf{X}(s)$ to be factored out of the expression. The matrix

$$\mathbf{R}(s) = s\mathbf{I} - \mathbf{A} \tag{5.7-26}$$

is the *resolvent matrix*. Its properties will tell us much about the system's behavior.

The transfer function is defined for zero initial conditions. With $\mathbf{x}(0) = \mathbf{0}$, (5.7-25) gives

$$\mathbf{X}(s) = (s\mathbf{I} - \mathbf{A})^{-1}\mathbf{B}\mathbf{U}(s) \tag{5.7-27}$$

where the resolvent matrix inverse $(s\mathbf{I} - \mathbf{A})^{-1}$ is assumed to exist. Substitute this into (5.7-24), and factor out $\mathbf{U}(s)$ to obtain

$$\mathbf{Y}(s) = [\mathbf{C}(s\mathbf{I} - \mathbf{A})^{-1}\mathbf{B} + \mathbf{D}]\mathbf{U}(s) \tag{5.7-28}$$

The matrix multiplying $\mathbf{U}(s)$ is the *transfer function matrix* $\mathbf{T}(s)$.

$$\mathbf{T}(s) = \mathbf{C}(s\mathbf{I} - \mathbf{A})^{-1}\mathbf{B} + \mathbf{D} \tag{5.7-29}$$

The elements of $\mathbf{T}(s)$ are the transfer functions as defined by (5.7-20); (5.7-29) provides an alternative way of computing these elements instead of by diagram reduction or algebraic manipulation.

Example 5.8

Use (5.7-29) to find $T_{12}(s)$ for the model given in Example 5.7.

For this model,

$$\mathbf{A} = \begin{bmatrix} -4 & 2 \\ 0 & -3 \end{bmatrix} \qquad \mathbf{B} = \begin{bmatrix} 4 & 0 \\ 0 & 5 \end{bmatrix}$$

$$\mathbf{C} = \begin{bmatrix} 1 & 1 \\ 0 & 1 \end{bmatrix} \qquad \mathbf{D} = \begin{bmatrix} 3 & 0 \\ 0 & 0 \end{bmatrix}$$

Thus,

$$s\mathbf{I} - \mathbf{A} = s\begin{bmatrix} 1 & 0 \\ 0 & 1 \end{bmatrix} - \begin{bmatrix} -4 & 2 \\ 0 & -3 \end{bmatrix} = \begin{bmatrix} (s+4) & -2 \\ 0 & (s+3) \end{bmatrix}$$

As shown in Appendix A, the inverse of a (2×2) matrix $\mathbf{M}$ is

$$\mathbf{M}^{-1} = \frac{1}{m_{11}m_{22} - m_{12}m_{21}}\begin{bmatrix} m_{22} & -m_{22} \\ -m_{21} & m_{11} \end{bmatrix} \tag{5.7-30}$$

Letting $\mathbf{M} = s\mathbf{I} - \mathbf{A}$, we see that

$$(s\mathbf{I} - \mathbf{A})^{-1} = \frac{1}{(s+4)(s+3)}\begin{bmatrix} (s+3) & 2 \\ 0 & (s+4) \end{bmatrix} = \begin{bmatrix} \dfrac{1}{s+4} & \dfrac{2}{(s+4)(s+3)} \\ 0 & \dfrac{1}{s+3} \end{bmatrix}$$

and

$$\mathbf{T}(s) = \begin{bmatrix} 1 & 1 \\ 0 & 1 \end{bmatrix}\begin{bmatrix} \dfrac{1}{s+4} & \dfrac{2}{(s+4)(s+3)} \\ 0 & \dfrac{1}{s+3} \end{bmatrix}\begin{bmatrix} 4 & 0 \\ 0 & 5 \end{bmatrix} + \begin{bmatrix} 3 & 0 \\ 0 & 0 \end{bmatrix}$$

The transfer functions are obtained by carrying out the indicated multiplications.

They are

$$T_{11}(s) = \frac{4}{s+4} + 3 = \frac{3s+16}{s+4}$$

$$T_{12}(s) = \frac{10}{(s+4)(s+3)} + \frac{5}{s+3} = \frac{5(s+6)}{(s+4)(s+3)}$$

$$T_{21}(s) = 0$$

$$T_{22}(s) = \frac{5}{s+3}$$

All of the transfer functions are obtained at once with the matrix method. It is cumbersome when the inverse of the resolvent matrix is difficult to form. If only some of the transfer functions are required, the algebraic method or diagram reduction is sometimes easier to apply.

The matrix form of the state equations is valid for a model of any order, although we have used only second-order examples. An nth-order model will have n state variables, and the state vector $\mathbf{x}$ will have n elements. Equations (5.7-8) and (5.7-9) represent this general case. The power of this notation is demonstrated by the fact that the general transfer function relation (5.7-29) was obtained for a system of any order by using the matrix form of the state equations. A second advantage of (5.7-8) is that it has the appearance of a first-order model, and therefore the first-order techniques in Chapters 3 and 4 can be mimicked at least formally, keeping in mind the requirements of the matrix algebra. For example, the Laplace transformation of (5.7-8) to obtain (5.7-27) follows very closely the transfer function development in Chapter 4 for the first-order model.

The state vector notation also allows a compact description of the system in graphical form, called the state vector diagram. This can be either a block diagram or a signal flow graph. Both are shown in Figure 5.14 for the system described by (5.7-8) and (5.7-9). The reduced transfer function diagram is shown in Figure 5.14*c*. In the vector diagram, the signals (now multiple or vector quantities) are shown with a double-lined arrow.

The Transition Matrix

We have obtained the free response of the second-order model from its reduced form in Chapter 3. When we encounter higher order models, it will not be convenient to use the reduced form. For this reason, we now introduce a matrix method for obtaining the free response.

Earlier, we used the Laplace transformation of the state equation $\dot{\mathbf{x}} = \mathbf{A}\mathbf{x} + \mathbf{B}\mathbf{u}$ to obtain

$$(s\mathbf{I} - \mathbf{A})\mathbf{X}(s) = \mathbf{x}(0) + \mathbf{B}\mathbf{U}(s) \tag{5.7-31}$$

For the free-response calculation we can take $\mathbf{u}(t)$ to be zero. This gives

$$(s\mathbf{I} - \mathbf{A})\mathbf{X}(s) = \mathbf{x}(0) \tag{5.7-32}$$

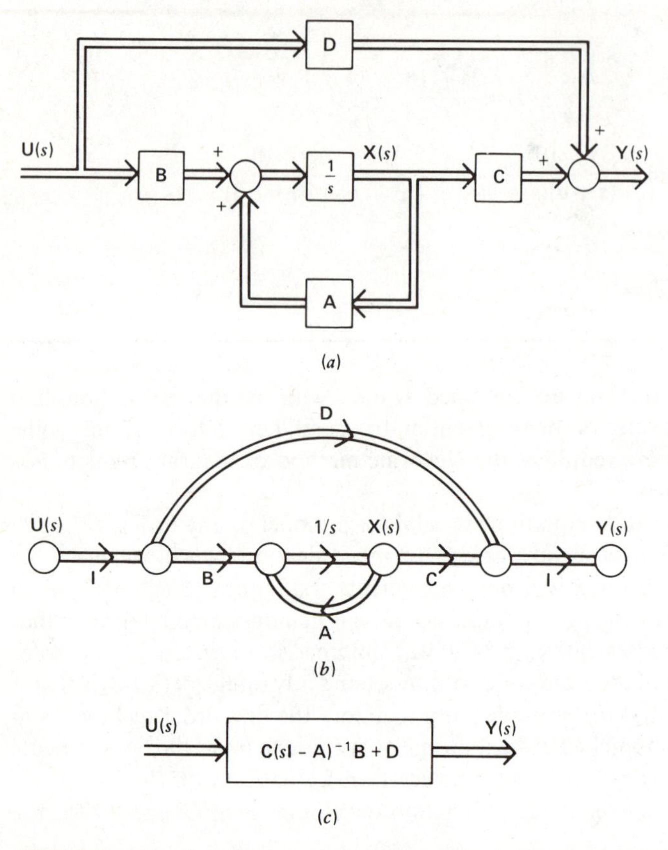

FIGURE 5.14 Graphical representations of the state vector model: $\dot{\mathbf{x}} = \mathbf{A}\mathbf{x} + \mathbf{B}\mathbf{u}$, $\mathbf{y} = \mathbf{C}\mathbf{x} + \mathbf{D}\mathbf{u}$. (*a*) Block diagram for the Laplace transform variables. (*b*) Signal flow graph. (*c*) Reduced diagram showing the transfer function matrix between y and u.

or

$$\mathbf{X}(s) = (s\mathbf{I} - \mathbf{A})^{-1}\mathbf{x}(0) \tag{5.7-33}$$

This says that the inverse of the resolvent matrix can be considered as the transfer function matrix relating the free response $\mathbf{x}(t)$ to the initial condition $\mathbf{x}(0)$. Since $\mathbf{x}(0)$ is a vector of constants, (5.7-33) shows the free response to have the form

$$\mathbf{x}(t) = \boldsymbol{\phi}(t)\mathbf{x}(0) \tag{5.7-34}$$

where the matrix $\boldsymbol{\phi}(t)$ is given by

$$\boldsymbol{\phi}(t) = \mathscr{L}^{-1}[(s\mathbf{I} - \mathbf{A})^{-1}] \tag{5.7-35}$$

The matrix $\boldsymbol{\phi}(t)$ is the *transition matrix.*

One advantage of expressing the free response in the form (5.7-33) is that the effects of the initial conditions are clearly distinguished from the intrinsic or natural behavior of the system as expressed by $\boldsymbol{\phi}(t)$. This is especially useful when we wish to evaluate $\mathbf{x}(t)$ numerically at one or more values of t for a variety of initial conditions.

Equation (5.7-33) represents only one of several ways by which the transition matrix can be computed. Other methods, some of which are suitable for implementation on a digital computer, are presented later in this section. Now, we illustrate the use of (5.7-35) with an example.

Example 5.9

Compute the transition matrix given the state equations

$$\dot{x}_1 = x_2$$
$$\dot{x}_2 = -4x_1 - 5x_2$$

The resolvent matrix is

$$s\mathbf{I} - \mathbf{A} = \begin{bmatrix} s & -1 \\ 4 & (s+5) \end{bmatrix}$$

and

$$\begin{aligned} \mathbf{\Phi}(s) = (sI - A)^{-1} &= \frac{1}{s(s+5)+4}\begin{bmatrix} (s+5) & 1 \\ -4 & s \end{bmatrix} \\ &= \frac{1}{(s+1)(s+4)}\begin{bmatrix} (s+5) & 1 \\ -4 & s \end{bmatrix} \end{aligned} \tag{5.7-36}$$

Thus, with a partial fraction expansion,

$$\Phi_{11}(s) = \frac{(s+5)}{(s+1)(s+4)} = \frac{4}{3}\frac{1}{s+1} - \frac{1}{3}\frac{1}{s+4}$$

or

$$\phi_{11}(t) = \tfrac{4}{3}e^{-t} - \tfrac{1}{3}e^{-4t}$$

Performing the same procedure with the other three entries in the matrix, we obtain

$$\phi_{12}(t) = \tfrac{1}{3}e^{-t} - \tfrac{1}{3}e^{-4t}$$
$$\phi_{21}(t) = \tfrac{4}{3}e^{-4t} - \tfrac{4}{3}e^{-t}$$
$$\phi_{22}(t) = \tfrac{4}{3}e^{-4t} - \tfrac{1}{3}e^{-t}$$

Example 5.10

Compute the transition matrix for the system of the previous example, given the state diagram shown in Figure 5.15*a*.

This approach considers the elements of $\mathbf{\Phi}(s)$ to be the transfer functions between the two outputs $X_1(s)$ and $X_2(s)$, and the two inputs $x_1(0)$ and $x_2(0)$. Set $x_2(0) = 0$ temporarily, and reduce the diagram to find the transfer functions between $X_1(s)$, $X_2(s)$ and $x_1(0)$. From Figure 5.15*b*, the result is

$$\Phi_{11}(s) = \frac{X_1(s)}{x_1(0)} = \frac{s+5}{s(s+5)+4}$$

$$\Phi_{21}(s) = \frac{X_2(s)}{x_1(0)} = \frac{-4}{s(s+5)+4}$$

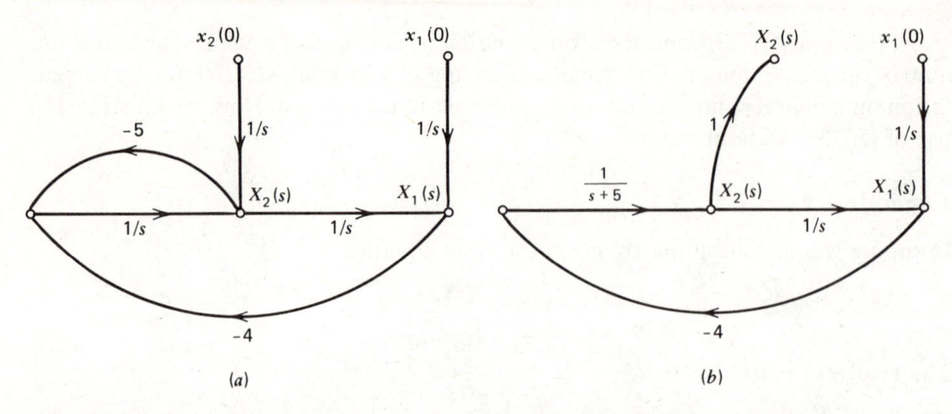

FIGURE 5.15 (*a*) State diagram for Example 5.9. (*b*) Partially reduced diagram with $x_2(0) = 0$.

These are the entries in the first column of $\mathbf{\Phi}(s)$, as given by (5.7-36). Similarly, with $x_1(0) = 0$, we can reduce the diagram to obtain $\Phi_{12}(s)$ and $\Phi_{22}(s)$. The rest of the procedure follows as in Example 5.9 in order to obtain $\phi(t)$.

Calculating the Forced Response From $\phi(t)$

If $\mathbf{u}(t) \neq 0$, the resulting response can be found from (5.7-25) as follows.

$$\mathbf{X}(s) = (s\mathbf{I} - \mathbf{A})^{-1}\mathbf{x}(0) + (s\mathbf{I} - \mathbf{A})^{-1}\mathbf{B}\mathbf{U}(s) \tag{5.7-37}$$

The scalar form of the convolution theorem (Table B.2, entry 9) can be extended to the vector case as follows. If $\mathscr{L}[\mathbf{v}(t)] = \mathbf{V}(s)$ and $\mathscr{L}[\boldsymbol{\phi}(t)] = \mathbf{\Phi}(s)$, the theorem states that

$$\mathscr{L}^{-1}[\mathbf{\Phi}(s)\mathbf{V}(s)] = \int_0^t \boldsymbol{\phi}(t-\tau)\mathbf{v}(\tau)d\tau \tag{5.7-38}$$

If we take $\mathbf{v}(t) = \mathbf{B}\mathbf{u}(t)$, then

$$\mathscr{L}^{-1}[\mathbf{\Phi}(s)\mathbf{B}\mathbf{U}(s)] = \int_0^t \boldsymbol{\phi}(t-\tau)\mathbf{B}\mathbf{u}(\tau)d\tau \tag{5.7-39}$$

Since $\boldsymbol{\phi}(t) = \mathscr{L}^{-1}[(s\mathbf{I} - \mathbf{A})^{-1}]$, (5.7-39) must be the forced response. Therefore, the complete response from (5.7-37) is

$$\mathbf{x}(t) = \boldsymbol{\phi}(t)\mathbf{x}(0) + \int_0^t \boldsymbol{\phi}(t-\tau)\mathbf{B}\mathbf{u}(\tau)d\tau \tag{5.7-40}$$

Equation (5.7-40) provides the basis for several useful computational procedures as we will see.

Evaluating the integrals of the $\boldsymbol{\phi}(t)$ elements is tedious for high-order systems. However, with the use of computer subroutines for numerical evaluation of the integrals and for matrix multiplication, we can develop a powerful program for analyzing the response of linear systems of high order. More detailed discussion of these methods and results for other input types are now given.

Properties of the Transition Matrix

Some useful properties of the state transition matrix can be derived from (5.7-34). Set $t = 0$ in (5.7-34) to show that

$$\boldsymbol{\phi}(0) = \mathbf{I} \tag{5.7-41}$$

For $\mathbf{u} = \mathbf{0}$, $\dot{\mathbf{x}} = \mathbf{A}\mathbf{x}$. Use this with the derivative of (5.7-34) to obtain

$$\begin{aligned}\dot{\mathbf{x}} = \dot{\boldsymbol{\phi}}(t)\mathbf{x}(0) &= \mathbf{A}\mathbf{x}(t)\\ &= \mathbf{A}\boldsymbol{\phi}(t)\mathbf{x}(0)\end{aligned}$$

or

$$\dot{\boldsymbol{\phi}}(t) = \mathbf{A}\boldsymbol{\phi}(t) \tag{5.7-42}$$

Thus, $\boldsymbol{\phi}(t)$ is the solution of the differential equation (5.7-42) with the initial condition (5.7-41).

Equation (5.7-34) relates $x(t)$ to the initial state at time zero. However, because we are dealing with the free response of a constant coefficient equation, we could also write the solution in terms of the state at any time, say, t_1, as

$$\mathbf{x}(t + t_1) = \boldsymbol{\phi}(t)\mathbf{x}(t_1) \tag{5.7-43}$$

But $\mathbf{x}(t_1) = \boldsymbol{\phi}(t_1)\mathbf{x}(0)$, and thus

$$\mathbf{x}(t + t_1) = \boldsymbol{\phi}(t)\boldsymbol{\phi}(t_1)\mathbf{x}(0)$$

If we had used (5.7-34) immediately, we would have obtained

$$\mathbf{x}(t + t_1) = \boldsymbol{\phi}(t + t_1)\mathbf{x}(0)$$

Comparing the last two equations shows that

$$\boldsymbol{\phi}(t + t_1) = \boldsymbol{\phi}(t)\boldsymbol{\phi}(t_1) \tag{5.7-44}$$

for any t and t_1. Also, since $\boldsymbol{\phi}(t + t_1) = \boldsymbol{\phi}(t_1 + t)$, the commutative property holds.

$$\boldsymbol{\phi}(t)\boldsymbol{\phi}(t_1) = \boldsymbol{\phi}(t_1)\boldsymbol{\phi}(t) \tag{5.7-45}$$

Finally, if we let $t = t_1 = \Delta$ in (5.7-44), we see that

$$\boldsymbol{\phi}(2\Delta) = \boldsymbol{\phi}(\Delta)\boldsymbol{\phi}(\Delta)$$

or in general,

$$\boldsymbol{\phi}(n\Delta) = \boldsymbol{\phi}^n(\Delta) \tag{5.7-46}$$

This property will be especially useful in numerical solutions.

Series Solution for $\boldsymbol{\phi}(t)$

Sometimes we do not require the solution for $x(t)$ in functional form, but only at discrete points in time. If we wish to investigate a variety of initial conditions, the transition matrix is particularly well suited. In such cases, we can use a numerical method for computing $\boldsymbol{\phi}(t)$.

The development of the series method to be presented now can be motivated by noting that for the first-order model $\dot{x} = ax$,

$$\phi(t) = e^{at} = 1 + at + \frac{(at)^2}{2!} + \cdots + \frac{(at)^k}{k!} + \cdots$$

This suggests a solution for the vector case of the form

$$\phi(t) = \mathbf{I} + \mathbf{A}t + \frac{\mathbf{A}^2 t^2}{2!} + \cdots + \frac{\mathbf{A}^k t^k}{k!} + \cdots \tag{5.7-47}$$

That this is the solution can be verified by noticing that it satisfies property (5.7-41), $\phi(0) = \mathbf{I}$, and that it satisfies the differential equation (5.7-42), $\dot{\phi}(t) = \mathbf{A}\phi(t)$. To show this, differentiate (5.7-47).

$$\begin{aligned}\dot{\phi}(t) &= \mathbf{A} + \mathbf{A}^2 t + \cdots + \frac{A^k t^{k-1}}{(k-1)!} + \cdots \\ &= \mathbf{A}\left[\mathbf{I} + \mathbf{A}t + \cdots + \frac{\mathbf{A}^{k-1} t^{k-1}}{(k-1)!} + \cdots\right] \\ &= \mathbf{A}\phi(t)\end{aligned}$$

This completes the proof of (5.7-47). Because of the similarity of this series expression with the series for e^{at}, the transition matrix $\phi(t)$ is sometimes termed the *matrix exponential* and written as $e^{\mathbf{A}t}$.

The series expression for $\phi(t)$ is particularly well suited for computer application, because it merely requires repeated multiplication of the matrix **A**. Suppose we desire the free response at a series of instants a time Δ apart. First, compute $\phi(\Delta)$ from (5.7-47) with $t = \Delta$. The number of terms in the series obviously determines its accuracy. H. M. Paynter of MIT has suggested the following approach (Reference 5).

1. Define p to be the element of the matrix $\mathbf{A}\Delta$ that has the maximum absolute value; that is, $p = \max|a_{ij}\Delta|$, where a_{ij} represents the elements of the $(n \times n)$ system matrix **A**.
2. Determine a number q such that the following relation is satisfied:

$$\frac{1}{q!}(np)^q e^{np} \leqslant 0.001 \tag{5.7-48}$$

3. Compute $\phi(\Delta)$ by truncating the series (5.7-47) after the term involving $\mathbf{A}^q$.

The criterion (0.001) in step 2 was determined from simulations of somewhat typical cases and therefore is only an approximate guide.

With $\phi(\Delta)$ thus computed, properties (5.7-43) and (5.7-46) can be used to determine $\mathbf{x}(t)$ at the times $\Delta, 2\Delta, 3\Delta, \ldots$. For a given system matrix **A**, $\phi(\Delta)$ need be computed only once. If the response is desired for several sets of initial conditions $\mathbf{x}(0)$, this method has advantages over the methods discussed earlier in this chapter. Note that the larger Δ is chosen, the more terms must be carried in the series for comparable accuracy.

Total Response

Because the total response is the sum of the free and the forced responses, we have

$$\mathbf{x}(t) = \boldsymbol{\phi}(t)\mathbf{x}(0) + \int_0^t \boldsymbol{\phi}(t-\tau)\mathbf{B}\mathbf{u}(\tau)d\tau \tag{5.7-49}$$

Example 5.11

Determine the response of the time-invariant system $\dot{\mathbf{x}} = \mathbf{A}\mathbf{x} + \mathbf{B}\mathbf{u}$ to an input vector $\mathbf{u}$ consisting of (a) step functions starting at $t = 0$ and (b) ramp functions starting at $t = 0$.

(a) Let the vector $\mathbf{u}$ be $(m \times 1)$. Then $\mathbf{B}$ is $(n \times m)$. Assume the step inputs have different magnitudes and denote the $(m \times 1)$ vector of the magnitudes by $\mathbf{p}$. The solution from (5.7-49) is

$$\mathbf{x}(t) = \boldsymbol{\phi}(t)\mathbf{x}(0) + \int_0^t \boldsymbol{\phi}(t-\tau)\mathbf{B}\mathbf{p}\, d\tau$$

From property (5.7-44), $\boldsymbol{\phi}(t-\tau) = \boldsymbol{\phi}(t)\boldsymbol{\phi}(-\tau)$. Thus,

$$\mathbf{x}(t) = \boldsymbol{\phi}(t)\mathbf{x}(0) + \boldsymbol{\phi}(t)\int_0^t \boldsymbol{\phi}(-\tau)d\tau\, \mathbf{B}\mathbf{p}$$

From the series expression for $\boldsymbol{\phi}(t)$,

$$\boldsymbol{\phi}(-\tau) = \mathbf{I} - \mathbf{A}\tau + \frac{\mathbf{A}^2\tau^2}{2!} - \cdots$$

and

$$\begin{aligned}\int_0^t \boldsymbol{\phi}(-\tau)d\tau &= \mathbf{I}t - \frac{\mathbf{A}t^2}{2!} + \frac{\mathbf{A}^2 t^3}{3!} - \cdots \\ &= \mathbf{A}^{-1}[\mathbf{I} - \boldsymbol{\phi}(-t)] \\ &= \mathbf{A}^{-1}[\mathbf{I} - \boldsymbol{\phi}^{-1}(t)]\end{aligned}$$

if $\mathbf{A}^{-1}$ exists. Since $\mathbf{A}\boldsymbol{\phi} = \boldsymbol{\phi}\mathbf{A}$, $\boldsymbol{\phi}\mathbf{A}^{-1} = \mathbf{A}^{-1}\boldsymbol{\phi}$, and we obtain

$$\mathbf{x}(t) = \boldsymbol{\phi}(t)\mathbf{x}(0) + \mathbf{A}^{-1}[\boldsymbol{\phi}(t) - \mathbf{I}]\mathbf{B}\mathbf{p} \tag{5.7-50}$$

(b) A vector of ramp functions starting at $t = 0$ can be written as

$$\mathbf{u}(t) = t\mathbf{q}$$

where $\mathbf{q}$ is the $(m \times 1)$ vector of the slopes of the ramps. The response is

$$\begin{aligned}\mathbf{x}(t) &= \boldsymbol{\phi}(t)\mathbf{x}(0) + \int_0^t \boldsymbol{\phi}(t-\tau)\mathbf{B}\tau\mathbf{q}\, d\tau \\ &= \boldsymbol{\phi}(t)\mathbf{x}(0) + \boldsymbol{\phi}(t)\int_0^t \boldsymbol{\phi}(-\tau)\tau\, d\tau\, \mathbf{B}\mathbf{q} \\ &= \boldsymbol{\phi}(t)\mathbf{x}(0) + \boldsymbol{\phi}(t)\left(\frac{\mathbf{I}t^2}{2} - \frac{2\mathbf{A}t^3}{3!} + \frac{3\mathbf{A}^2}{4!}t^4 - \cdots\right)\mathbf{B}\mathbf{q}\end{aligned}$$

Inserting the series expression for $\phi(t)$ and multiplying the two series, we can show that

$$\mathbf{x}(t) = \phi(t)\mathbf{x}(0) + \mathbf{A}^{-2}[\phi(t) - \mathbf{I} - \mathbf{A}t]\mathbf{B}\mathbf{q} \tag{5.7-51}$$

The preceding results are especially convenient to use if the transition matrix is already available from an analysis of the free response.

Other methods are available for computing the transition matrix. These are based on eigen-analysis techniques and the Cayley–Hamilton theorem. The state–space representation is also useful for deriving results dealing with the controllability and observability (measurability) of dynamic systems. These are advanced topics beyond the scope of this text. The interested reader should consult chapter 9 of Reference 4 for more details.

5.8 COMPUTER-AIDED ANALYSIS AND DESIGN

Computer methods play an important role in analyzing dynamic systems and designing controllers for them. Here, we give an overview of the available computer techniques and discuss their proper role in assisting the engineer.

The categories of computer methods needed by the control system engineer are:

1. Transfer function synthesis.
2. Time response calculation.
3. Characteristic-root solution.
4. Frequency response calculation.

Many commercial software packages are available from a variety of sources to do the preceding tasks, but a listing of these packages would soon be out of date. Major computer manufacturers, such as Hewlett Packard, distribute packages for their machines. The best sources for current information are the newsletters of the Dynamic Systems and Control Division of the ASME and the Control Systems Society of the IEEE.

Transfer Function Synthesis

Transfer function synthesis is the process of determining the overall transfer function of a system given the transfer functions of the individual subsystems and the manner in which they are connected. For example, given the general block diagram shown in Figure 5.16, a computer routine can be written to perform the block diagram algebra required to find the overall transfer functions: $T_1(s) = C(s)/R(s)$ and $T_2(s) = C(s)/D(s)$, given the subsystem transfer functions $G_i(s)$, $i = 1, 2, 3, 4$. Of course, the arrangement shown in Figure 5.15 is not the only one possible, and computer routines can be developed for other arrangements.

A process closely related to transfer function synthesis is the automatic generation of the system's state equations and transfer functions from a state diagram description such as the one shown in Figure 5.17 (see Section 5.7, Figure 5.14*a* for a

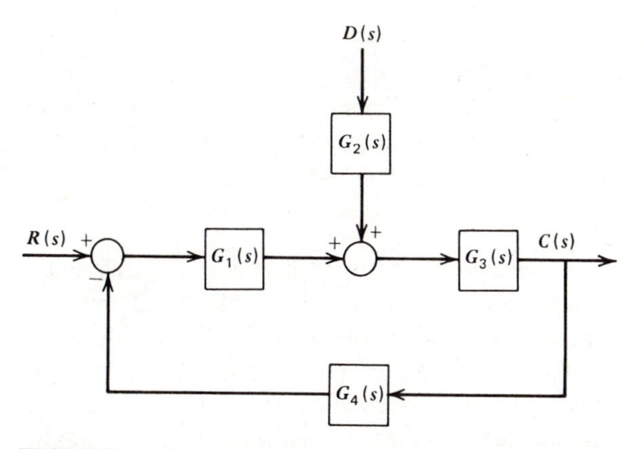

FIGURE 5.16 General block diagram for a single-loop system.

simpler form). This represents the structure of many common control systems, as we will see in later chapters. For this system, the vector equations are

$$\dot{\mathbf{x}} = \mathbf{Ax} + \mathbf{Bu} \tag{5.8-1}$$

$$\mathbf{y} = \mathbf{Cx} \tag{5.8-2}$$

$$\mathbf{u} = \mathbf{K}(\mathbf{r} - \mathbf{Fy}) \tag{5.8-3}$$

Following the approach in Section 5.7, the overall transfer function between **y** and **r** can be found using Laplace transforms. From (5.8-1) and (5.8-3), we obtain

$$\begin{aligned} \mathbf{X}(s) &= (s\mathbf{I} - \mathbf{A})^{-1}\mathbf{Bu} \\ &= (s\mathbf{I} - \mathbf{A})^{-1}\mathbf{BK}[\mathbf{R}(s) - \mathbf{FY}(s)] \end{aligned}$$

From (5.8-2),

$$\begin{aligned} \mathbf{Y}(s) &= \mathbf{CX}(s) \\ &= \mathbf{C}(s\mathbf{I} - \mathbf{A})^{-1}\mathbf{BK}[\mathbf{R}(s) - \mathbf{FY}(s)] \end{aligned}$$

If the inverse matrix exists, the solution for **Y**(*s*) is

$$\mathbf{Y}(s) = [\mathbf{I} + \mathbf{C}(s\mathbf{I} - \mathbf{A})^{-1}\mathbf{BKF}]^{-1}\mathbf{C}(s\mathbf{I} - \mathbf{A})^{-1}\mathbf{BKR}(s) \tag{5.8-4}$$

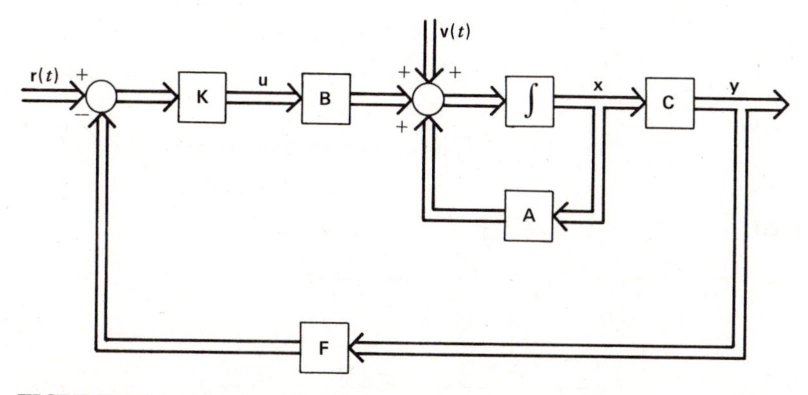

FIGURE 5.17 State diagram representation of a control system.

For higher order systems, the algebra required to perform the matrix operations needed by (5.8-4) is tedious, and computer routines can be written to perform them automatically. The engineer need only enter the coefficient values of the matrices **A**, **B**, **C**, **K**, and **F**. An example of such a program is given in Reference 6.

Time Response Calculation

Time response calculation can be done in two ways. In the first way, the differential equation model of the system is solved using a numerical method like those presented in Sections 5.1–5.3. The time response is then available as a listing or plot of the output versus time. There are many software packages available for solving differential equations numerically, but it is not difficult to develop your own (see Appendix C for a listing). References 6–12 provide listings and discussion of some available packages. Some of them can accept a system description based on either the state diagram, the transfer function, or the differential equations.

One such package is TUTSIM, which is available for the IBM PC and Apple Computers[1]. It and CSMP (Refs. 7, 8, 9) are examples of block diagram based simulation languages. To use such a language, the user first constructs a block diagram of the system to be simulated. This diagram is then converted to a simulation diagram, which the program's user interface is designed to accept as input. For example, to use TUTSIM to simulate the unit ramp response of the system shown in Figure 5.18*a*, we construct the simulation diagram shown in Figure 5.18*b*. Block 1 represents the time variable t, a unit ramp. Block 2 represents the first-order system block, whose parameters are K, T, and the initial condition IC of its response y. Block 3 is a gain block with a multiplier C. TUTSIM's input consists of first the model structure, followed by the model's parameters. The model structure consists of the following information for each block: the block number, its type, and which blocks produce its inputs (with the appropriate sign). The model's parameters are specified by the block number and its appropriate parameter values, for each block. If $K=10$, $T=5$, $IC=3$, and $C=7$ for the example shown, the input data lines would be as follows:

Model Structure

1, TIM	(Block 1 has no input.)
2, FIO, 1, −3	(The input to Block 2 from Block 3 has its sign changed.)
3, GAI, 2	(The output from Block 2 is multiplied by C.)

Model Parameters

2, 10, 5, 3	($K=10$, $T=5$, $y(0)=3$)
3, 7	($C=7$)

---

[1]Available from APPLIED i, 200 California Avenue, Palo Alto CA 94306.

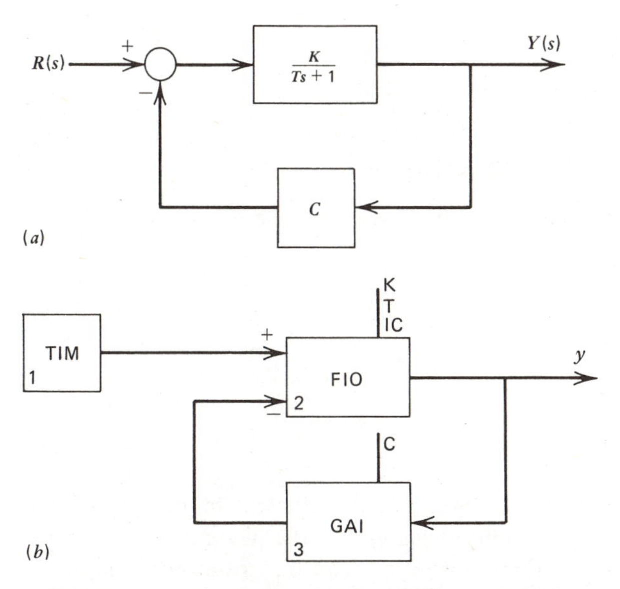

FIGURE 5.18 Simulation with TUTSIM. (*a*) Block diagram of the system to be simulated. (*b*) TUTSIM simulation diagram.

The program then produces a graphic display of the response. It is easy to see how a simulation of a complex system can be easily built up from its component subsystems. In addition, these diagram languages usually contain provisions for incorporating nonlinearities, special input types, and logical decision functions.

The second way to calculate the time response is to obtain the response in closed form. The most common way this is done by computer is to use the transfer function. Once the input function and the transfer function are specified, the program can generate the partial fraction expansion for the response, from which it is easy to recover the time response in closed form (see Appendix B). Another approach is to specify the coefficients of the matrices **A** and **B** and have the computer determine the transition matrix $\phi(t)$, and the free and forced responses, following the development of Example 5.11.

Characteristic-Root Solution

In order for the partial fraction expansion to be generated, the program must have a routine for computing the roots of a polynomial. There are many available packages for doing this, but the most useful ones will also have a routine for automatically finding the coefficients of the polynomial, given the system structure in terms of a block diagram, transfer function, or state description. Another useful feature is the ability to vary a system parameter and have the computer determine the resulting variation in the characteristic roots. A general approach to this problem is the *root locus* method, to be presented in Chapter 8. Reference 6 contains a listing of such a program.

Frequency Response Calculation

Transfer functions can usually be factored into the following form:

$$T(s) = K \frac{N_1(s)N_2(s) \ldots}{D_1(s)D_2(s) \ldots}$$

where the $N_i(s)$ and $D_i(s)$ terms usually take the forms (see Section 4.5):

1. A constant
2. s^n
3. $\tau s + 1$
4. $s^2 + 2\zeta\omega_n s + \omega_n^2, \quad \zeta < 1$
5. e^{-Ds}

This fact allows a modular computer program to be written for computing the system's frequency response, the log magnitude $m(\omega)$, and the phase angle $\phi(\omega)$ for a range of values of the frequency ω. Five subroutines can be written for computing $m(\omega)$ and $\phi(\omega)$ for each of the preceding terms with the frequency ω as an input. A main program that varies ω and calls these subroutines can then be written. The outputs $m(\omega)$ and $\phi(\omega)$ from each subroutine can then be combined using the additive property given by (4.5-22) and (4.5-23). The listing in Reference 6 has these capabilities.

Plotting Requirements

The results of most of the preceding computer methods should be displayed graphically for maximum effectiveness. The plotting requirements differ according to the type of analysis. For time response calculation, the response must be plotted versus time, with time increasing from left to right on a linear scale. The frequency response plots are made, however, with the frequency ω increasing on a logarithmic scale. However, the root locus plots and the polar form of the frequency response plots require extra attention. Both of these plots have linear scales on each axis, but the independent variable is not plotted on either axis. This requires the plotting device to be able to move back and forth as well as up and down. If this is not possible, as with a line printer, then the program must sort the two-dimensional data in ascending or descending order. Reference 6 contains a program listing for this purpose.

The current generation of personal, or desktop, computers is well equipped to handle these kinds of programs. Most have a screen graphics capability that is easily programmed. In addition, hard copies of plots can now be obtained with relatively low-cost plotters or dot matrix printers by doing a "screen dump."

Computer Methods Versus Analytical Methods

The advent of low-cost, widely available computing capability has diminished the importance of traditional analytical methods, such as the Laplace transform or the transition matrix, for finding the time response. Such methods are helpful for presenting general concepts and useful for first- and second-order linear, time-invariant models; and we have introduced them earlier for these reasons. However, for higher order models and models that are time-invariant or nonlinear, the numerical integration methods in Sections 5.1–5.3 are preferred *if* the engineer will need them often enough to justify buying or programming the software.

These points can be illustrated by considering two approaches to finding the time response of a linear, second-order, time-invariant model. Figure 5.19*a* shows the diagram of the procedure necessary to find the closed-form solution; while Figure 5.19*b* shows the numerical integration approach. We assume that each method is given the numerical values for the model's coefficients, the initial conditions, and the initial and final times t_0 and t_f. It is obvious that more coding is required for the analytical method. If the engineer cannot estimate the size of the required time step Δt for the numerical integration method, then a subroutine can be included to find the characteristic roots and the dominant time constant, from which Δt can be computed automatically. If this is necessary, then the differences between the code lengths for the two methods is not so great. However, the following points should be noted.

1. If a plot of the time response is desired for numerous cases, then some form of computer implementation is desired.
2. The numerical integration program can easily be extended to include higher order models. If the model is expressed in state variable form [see (5.7-8)], then only the dimension statements for the arrays need to be changed. On the other hand, a separate version of the program for the analytical method must be written for each model type (second order, third order, etc.)

Our conclusion is that if we frequently need time response plots for a variety of model types, then the numerical integration method is the one to program. Similar conclusions apply to the other types of computer methods. If we frequently need to calculate transfer functions, frequency response plots, and characteristic roots for a variety of models, then the time or expense required to develop or buy computer packages is justified.

Computer Methods and System Design

In the first five chapters, we have not devoted much attention to system design, but have rather concentrated on modeling and analysis. However, the remaining chapters deal exclusively with control systems design, which relies heavily on modeling and analysis. It is therefore appropriate at this point to put the role of modeling, analysis, and computer simulation into perspective, as they relate to system design. Figure 5.20 is adapted from Reference 13. It shows how the design process is iterative and makes use of both analysis and computer simulation, in general. First, the system of interest is defined along with the required performance specifications. Then a model of the system is formulated using the techniques in Chapter 2. In order to use the analysis methods in Chapter 3 or the transfer function techniques in Chapter 4, the model must be lumped-parameter, linear, and time-invariant. If this is not the case, then the model must be linearized or otherwise simplified. If this cannot be done without

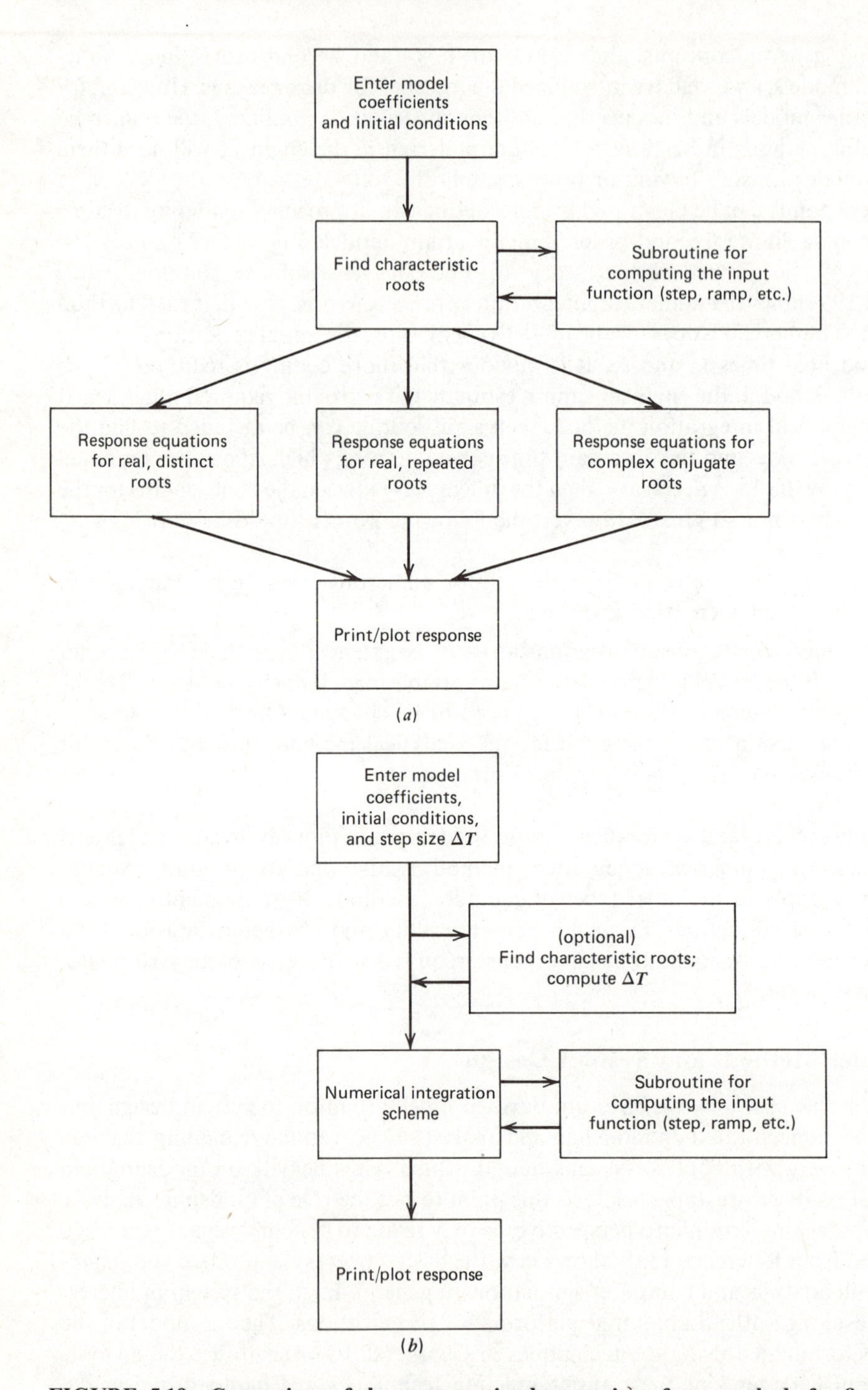

FIGURE 5.19 Comparison of the computer implementation of two methods for finding the response of a second-order linear system. (*a*) Use of the closed-form solution. (*b*) Use of numerical integration.

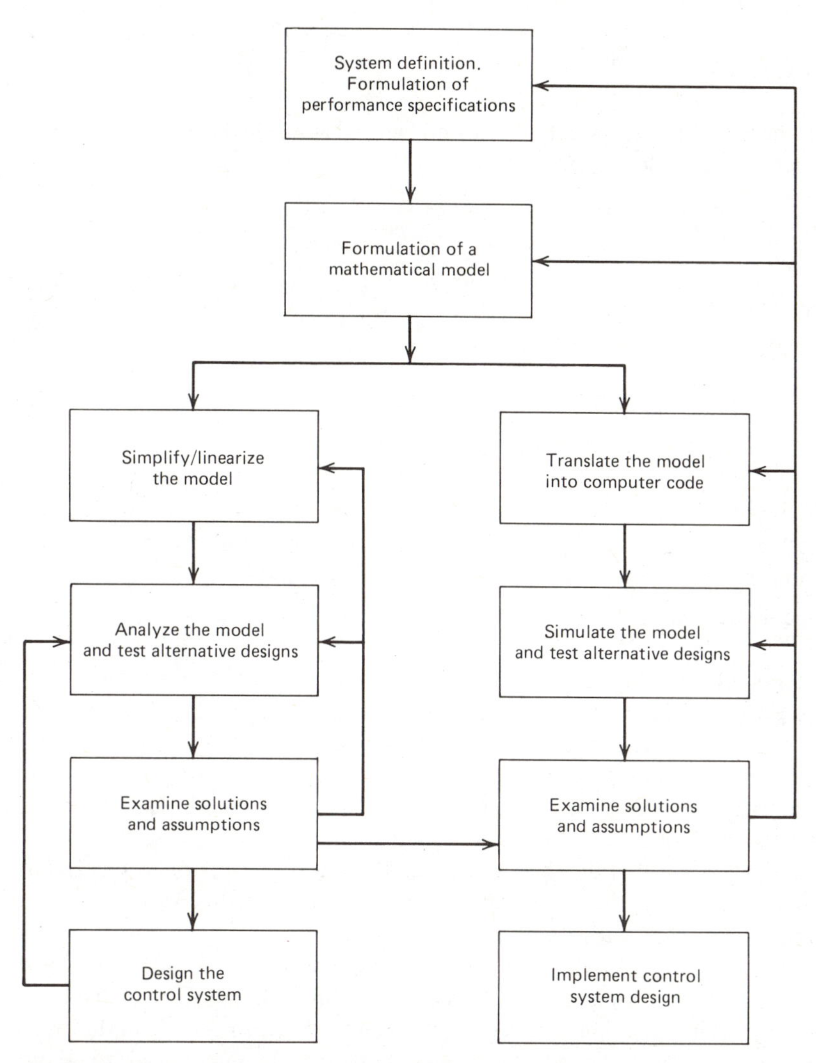

FIGURE 5.20 The roles of mathematical analysis and computer simulation in an iterative approach to control systems design.

sacrificing realism, then the model must be expressed in computer code and its performance studied by means of computer simulation. If the model can be simplified, it can be analyzed using such linear methods as transfer functions to see if the system design meets the performance specifications. If it does not, then an iterative process is used to see if the model and the performance specifications are reasonable. If they are, then the design must be modified. If they are not reasonable, then they must be changed. Often, if the design performs well according to the analysis of the simplified model, its performance is checked by simulating the computer model, which is presumably more realistic, since it contains more details of the physical system. Depending on these results, the design is accepted or modified.

5.9 SUMMARY

In order for the system response to be found with the mathematical methods in Chapters 3 and 4, the model must:

1. be linear;
2. be time-invariant;
3. be of relatively low order; and
4. have a relatively simple input function.

If any of the preceding conditions are not true, then a numerical method is useful. In this chapter, we have introduced three such methods:

1. the Euler method;
2. the predictor–corrector method (specifically, the modified Euler method, which uses an Euler predictor with a trapezoidal corrector); and
3. the Runge–Kutta method.

The Runge–Kutta method is more difficult to program but the most powerful of the three. It is probably the most commonly used method for applications of the type covered in this text. It has several versions, depending on the order of the approximation and on how the coefficients are selected. We have presented the second- and fourth-order methods and have given coefficients for the three common fourth-order versions: Gill's, Ralston's, and the classical versions.

The state variable form is the most convenient one to use for simulation, and in Section 5.4 we presented methods for converting a model into this form. They are especially useful for models with numerator dynamics.

Selecting the proper step size Δt is important, and in Sections 5.5 and 5.6, we have attempted to show how this is accomplished. The dominant root of a linear or linearized model provides a guide for selecting Δt. The time rate of change of the input provides another clue, and the practical step function was introduced to avoid numerical approximation difficulties caused by the pure step function input.

The chapter concludes with a discussion of general methods for computer-aided analysis and design in control system engineering. Vector–matrix representation provides the basis for most of these applications, and this was introduced in Section 5.7. Section 5.8 then provided an introduction to the applications.

REFERENCES

1. A. Ralston, *A First Course in Numerical Analysis*, McGraw-Hill, New York, 1965.
2. R. W. Hornbeck, *Numerical Methods*, Quantum Publishers, New York, 1975.
3. J. H. Ginsberg and J. Genin, *Dynamics*, 2nd ed., John Wiley & Sons, New York, 1984.

4. W. J. Palm III, *Modeling, Analysis, and Control of Dynamic Systems*, John Wiley & Sons, New York, 1983.
5. Y. Takahashi, M. J. Rabins, and D. Auslander, *Control*, Addison-Wesley, Reading, Mass., 1970.
6. J. L. Melsa and S. K. Jones, *Computer Programs for Computational Assistance in the Study of Linear Control Theory*, 2nd ed., McGraw-Hill, New York, 1973.
7. Anon., *Continuous System Simulation Language*, Version 3, User's Guide, Control Data Corp., Sunnyvale, Calif., 1971.
8. E. O. Doebelin, *System Modeling and Response*, John Wiley & Sons, New York, 1980.
9. F. H. Speckart and W. L. Green, *A Guide to Using CSMP*, Prentice-Hall, Englewood Cliffs, N.J., 1976.
10. G. A. Korn and J. V. Wait, *Digital Continuous System Simulation*, Prentice-Hall, Englewood Cliffs, N.J., 1978.
11. A. Pugh, III, *DYNAMO User's Manual*, MIT Press, Cambridge, Mass., 1973.
12. F. O. Smetana and A. O. Smetana, *FORTRAN Codes for Classical Methods in Linear Dynamics*, McGraw-Hill, New York, 1982.
13. K. P. White, "Mathematical Models of Dynamic Physical Systems," in *The Handbook of Mechanical Engineering*, M. Kutz, ed., John Wiley, New York, in press.

PROBLEMS

5.1 Use the Euler method to solve the equation $\dot{y} = y^2 + 1$, with $y(0) = 0$ for $0 \leqslant t \leqslant 1$ using (a) $\Delta t = 0.1$ and (b) $\Delta t = 0.01$. (c) Compare the results with the exact solution $y(t) = \tan t$.

5.2 Repeat Problem 5.1 using the modified Euler method.

5.3 Repeat Problem 5.1 using a version of the Runge–Kutta algorithm. Appendix C contains a program implementing Gill's version.

5.4 A clue to selecting a proper step size for a numerical solution can be obtained from the approximate speed of the system's response if this is known and from the rate of change of any input function. Consider the tank model

$$C\dot{h} = q_i - 0.0589\sqrt{h}$$

where $C = \pi$ ft^2 and time is in seconds. The model's solution for $q_i = 0$ can be found by making the substitution: $h = z^2$. This gives a linear model in terms of z. The result for $q_i = 0$, is,

$$h(t) = (\sqrt{h(0)} - 0.009375t)^2$$

Suppose that the inflow rate is $q_i = 0.01t$ and $h(0) = 9$ ft. Choose a numerical algorithm and a step size and solve for $h(t)$ over the interval $0 \leqslant t \leqslant 100$. Solve the equation for at least two different step sizes to check the accuracy.

5.5 The following model form results from a dilution problem like that developed in Problem 2.22. The variable y represents the amount of salt in a tank of brine as a function of time.

$$\dot{y} + \frac{2}{10+2t}y = 4$$

Use a numerical method to find the response for $0 \leqslant t \leqslant 10$, if $y(0) = 0$.

5.6 Use a numerical method to find the response of the model

$$\dot{y} + 2ty = t$$

for $0 \leqslant t \leqslant 2$ if $y(0) = 0$.

5.7 Suppose the two-tank system shown in Figure 2.21 has the parameter values: $A_1 = 3\ \text{ft}^2$, $A_2 = 10\ \text{ft}^2$, and $g/R_1 = g/R_2 = 0.004\ \text{ft}^2/\text{sec}$. Even though the model, which was derived in Example 2.9, is linear, numerical solution methods can be useful if the input flow rate $q(t)$ is not given in a convenient analytical form. For example, suppose that $q(t)$ is given in tabular form, as shown below. Use a numerical method to compute the response of h_1 and h_2 for $0 \leqslant t \leqslant 5000$ sec. The initial heights are $h_1(0) = 0$, $h_2(0) = 5$ ft.

t (sec)	q (ft^3/sec)
0	0
1000	0.004
2000	0.003
3000	0.001
>4000	0

5.8 Van der Pol's equation is a nonlinear model for some oscillatory processes. It is

$$\ddot{y} - b(1 - y^2)\dot{y} + ay = 0$$

(a) Put the model in state variable form and linearize it.
(b) Let $a = 1$ and use a numerical technique to find $y(t)$ for
 (1) $b = 0.1,\quad y(0) = \dot{y}(0) = 1,\quad 0 \leqslant t \leqslant 25$
 (2) $b = 0.1,\quad y(0) = \dot{y}(0) = 3,\quad 0 \leqslant t \leqslant 25$
 (3) $b = 3,\quad y(0) = \dot{y}(0) = 1,\quad 0 \leqslant t \leqslant 25$
(c) Compare the results of (b) with the linearized behavior.

5.9 Given the following transfer functions, find a state variable model for each. Also, obtain the expressions relating the initial values $x_i(0)$ of the state variables to the initial values of y and its derivatives.

(a) $$\frac{Y(s)}{V(s)} = \frac{s+2}{s^2+4s+3}$$

(b) $$\frac{Y(s)}{V(s)} = \frac{1}{s^3+6s^2+4s+6}$$

(c) $$\frac{Y(s)}{V(s)} = \frac{s^2 + 3s + 1}{s^2 + 4s + 3}$$

(d) $$\frac{Y(s)}{V(s)} = \frac{s + 10}{s^3 + 6s^2 + 4s + 6}$$

5.10 Use a numerical method to compute the response $y(t)$ of the state model found in Problem 5.9(*d*) for $v(t) = (2 - t)e^{-t}$, $y(0) = 10$, $\dot{y}(0) = 3$, $\ddot{y}(0) = 1$.

5.11 Use a numerical method to compute the time it will take to fill a tank with liquid to a height of 10 feet, if the input flow rate is $q = 8$ ft^3/minute. The initial liquid height is $h(0) = 1$ ft. The tank's model is

$$\dot{h} = q - 2\sqrt{h}$$

5.12 A certain vibratory system has a nonlinear spring relation, and is modeled by

$$\ddot{x} = 0.2(u - x) + 10(u - x)^3 + 2(\dot{u} - \dot{x})$$

Use a numerical method with the practical step function $u(t) = 1 - e^{-at}$ to find the model's unit step response. Use a linearized approximation to the model for $u \approx x \approx 0$ to estimate the required step size and an approximate value of a to use. The initial conditions are $x(0) = \dot{x}(0) = 0$.

5.13 Repeat Problem 5.12 for the model

$$\ddot{x} = 10(u - x) + 5(u - x)^3 + 2(\dot{u} - \dot{x})$$

5.14 Consider the gantry robot shown in Figure 2.18. Its equations of motion were derived in Example 2.6, and are given by (2.4-36) and (2.4-37). Suppose that $M = 2$ slugs, $m = 1$ slug, $a = 2$ ft, $T = 0$, $g = 32.2$ ft/sec^2, and that f is step input of magnitude 3 lbs. The initial conditions are $x_1(0) = \theta(0) = \dot{x}_1(0) = \dot{\theta}(0) = 0$.

(a) Put the model into state variable form.

(b) Use a numerical solution method to compute the maximum angular displacement of θ.

5.15 Given the following model:

$$\dot{x}_1 = -5x_1 + 3x_2 + 2u_1$$
$$\dot{x}_2 = -4x_2 + 6u_2$$

and the output equations

$$y_1 = x_1 + 3x_2 + 2u_1$$
$$y_2 = x_2$$

(a) Draw the state diagram.

(b) Find the characteristic roots.

5.16 With the state diagram shown in Figure P5.16, find the following.

(a) The state and output equations.

(b) The transfer function matrix by diagram reduction and by formula (5.7-29).

(c) The reduced model forms for each output.

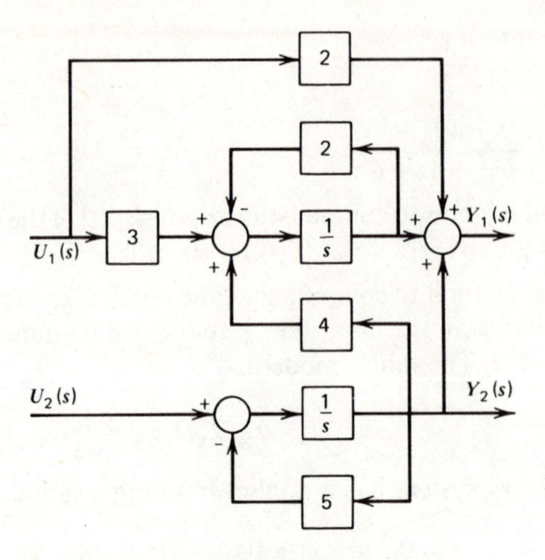

FIGURE P5.16

5.17 Find the characteristic roots for the following model:

$$\dot{x}_1 = x_2$$
$$\dot{x}_2 = x_3$$
$$\dot{x}_3 = x_2 - 2x_3 + u$$

5.18 Given the system matrix

$$\mathbf{A} = \begin{bmatrix} 0 & 1 \\ -k & -c \end{bmatrix}$$

where $k > 0$ and $c > 0$, use the Laplace transform method to compute the entries of the transition matrix for the following:

(a) The underdamped case.
(b) The overdamped case.

5.19 Given the system matrix

$$\mathbf{A} = \begin{bmatrix} -\alpha & \beta \\ -\beta & -\alpha \end{bmatrix}$$

where $\beta > 0$, use the Laplace transform method to compute the entries of the transition matrix.

CHAPTER SIX
Feedback Control Systems

We now come to the third and final topic of this work: the control of dynamic systems. The modeling and analysis techniques developed in the previous chapters can be readily applied to the design of control systems. Our emphasis on linear and linearized models is justified by the fact that feedback improves linearity and acts to keep the linearization accurate by preventing large deviations from the reference operating condition.

We begin our study of feedback control with an introduction to its history, some examples, and its terminology (Sections 6.1 and 6.2). Sensors are required to provide the measurements necessary for feedback, and actuators are needed to supply the control forces. We present an overview of the devices available for these purposes (Sections 6.3 and 6.4). The heart of the controller is the element used to implement the control logic, called the *control law*, and in Sections 6.5–6.7, we introduce and analyze the most common laws in use today. Finally, we discuss the design of devices used to implement the control law physically. These can be categorized as analog electronic (Section 6.8), pneumatic (Section 6.9), hydraulic (Section 6.10), and digital. The latter case is treated in Chapters 10 and 11, where we develop additional algorithms that take advantage of the features of digital computers used as controllers.

6.1 FEEDBACK CONTROL: CONCEPTS, HISTORY AND APPLICATIONS

Some of the basic concepts of feedback and its uses were introduced in Chapter 1, as applied to static systems. It was shown that feedback can improve a system's linearity and sensitivity to parameter variations. The associated cost of this improvement is a decrease in amplification or gain. In this and the remaining chapters, we demonstrate that these properties of feedback also apply to the dynamic case, and we show how to use these properties to design systems for controlling variables commonly found in machines and processes.

A control system may be defined as a system whose purpose is to regulate or adjust the flow of energy in some desired manner. A *closed-loop* or *feedback control system* uses measurements of the output to modify the system's actions in order to achieve the intended goal. An *open-loop* control system does not use feedback, but adjusts the flow of energy according to a prescribed schedule. Examples of open-loop systems are cam-driven elements and timers. The term *automatic control system* is sometimes used to distinguish a system that acts without outside intervention from a manually controlled system that requires regular supervision by an operator. Thus, an automatic control system can be either open or closed loop.

Open-loop controllers are relatively inexpensive but limited to situations where events are quite predictable. For example, a timer can be used to switch a building's lights on at night and off in the morning. However, if a particular day is very dark, the system will not give adequate performance. If a light level sensor is added to the controller, it becomes a closed-loop system. The challenging design problems lie in the area of feedback controllers, to which we will henceforth limit our attention.

In addition to improving linearity and parameter sensitivity, feedback in a control system enables the controller to respond to changes in commands and disturbances acting on the system. Perhaps the most familiar control system is the thermostat for controlling the temperature inside a house. The command is the desired temperature setting. The thermostat measures the actual temperature, compares it with the command temperature, and turns the furnace on or off depending on the difference between the actual and command temperatures. A disturbance in the system is anything that changes the actual inside temperature other than the heating system. A change in outside temperature or opening doors are examples of disturbances. By sensing the change in inside temperature, the thermostat takes action to keep the temperature near its desired value.

It is helpful to consider some simple historical examples of control systems at this point.

History of Feedback Control

Examples of feedback control systems have been identified dating back to the third century B.C. Ktesibios, a Greek living in Alexandria, is credited with building a self-regulating flow device for use in water clocks (Reference 1). Such a clock uses a float indicator to mark the passage of time (Figure 6.1). Its accuracy depends on how constant the inflow rate of water is to the tank. This flow rate can change if the supply pressure changes. Ktesibios introduced a secondary tank with a float and valve tapered to match the opening of the supply line. As long as the water height in this secondary tank remains constant, the flow rate into the main water clock tank will not change. Suppose that the supply flow rate increases. This raises the level in the secondary tank, and the float tends to close off the supply line and decrease the flow. The level then drops, and the supply flow increases. The sensitivity of the system to changes in supply pressure depends in part on the area of the secondary tank. Rapid fluctuations in supply pressure will have less effect on the water clock if this area is made larger.

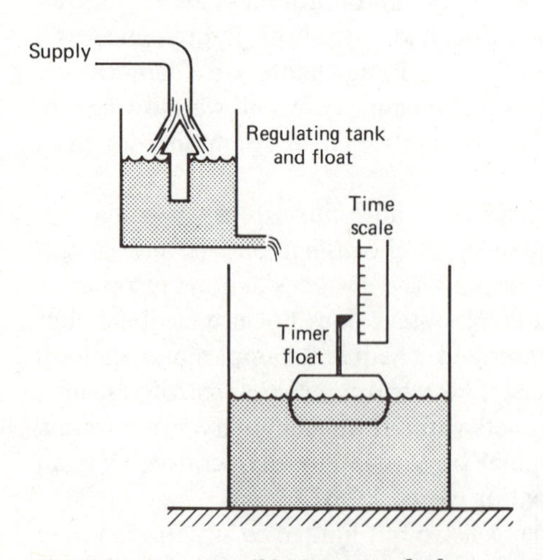

FIGURE 6.1 Ktesibios's water clock.

Little progress was made in control systems during the next 2000 years. Several examples of such systems appear in windmill designs used by the Dutch in the fifteenth century. The windmills were used to pump water and grind grain more efficiently by means of an auxiliary propeller at a right angle to the main blades.

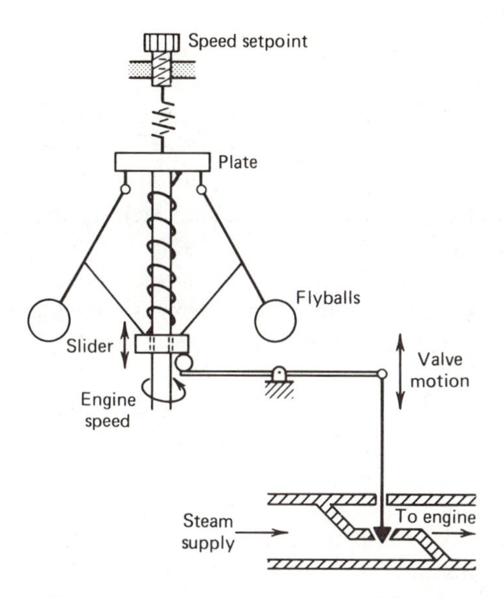

FIGURE 6.2 James Watt's flyball governor for speed control of a steam engine.

This was connected to a gear train that turned the main blades so they always faced into the wind.

The arrival of the machine age was accompanied by a large increase in the number of control system designs. The development of the steam engine led to the need for a speed control device to maintain constant speed in the presence of changes in load torque or steam pressure. In 1788, James Watt of Glasgow developed his now famous flyball governor for this purpose (Figure 6.2). Watt took the principle of sensing speed with the centrifugal pendulum of Thomas Mead and used it in a feedback loop on a steam engine (Reference 2). As the motor speed increases, the flyballs move outward and pull the slider upward. The upward motion of the slider closes the steam valve, thus causing the engine to slow down. If the engine's speed is too slow, the spring force overcomes that due to the flyballs, and the slider moves down to open the steam valve. The desired speed can be set by moving the plate to change the compression in the spring.

The principle of the flyball governor is still used in speed control applications. Typically, a hydraulic servomotor is connected to the slider to provide the high forces required to move large supply valves.

Analytical tools for control problems were first developed by J. C. Maxwell in 1868 for application to telescope position control, but a general understanding of the principle of feedback control was not developed until the twentieth century.

From 1900 to 1940, significant developments occurred in large-scale power generation, aeronautics, chemical industries, and electronics. The development of the vacuum tube amplifier in particular resulted in many analytical techniques for the design of feedback systems. Black, Bode, and Nyquist of the Bell Telephone Laboratories are responsible for many of these principles, especially those involving frequency response.

The advent of World War II gave additional impetus to the design of new types of control systems. The rapid increase in aircraft speeds made the manually controlled antiaircraft gun obsolete. The perfection of the radar-controlled gun was a major engineering advancement of the period, because it represented the first application to a mechanical system of the mathematical methods developed for electronic amplifiers. We might say that control theory as a science first appeared at this time, because a set of general principles were formulated that could be applied to any type of system.

Norbert Wiener, who helped develop the radar-controlled gun, contributed greatly to the emerging science with analytical techniques and by enlarging its purview to include applications other than to machines. He coined the term *cybernetics* to describe the study of control and communication in humans, animals, and machines (Reference 3). The term is derived from the Greek word for the person who steers a ship (the ship's controller). Although we have described a control system as one that regulates the flow of energy, the concepts can be applied to systems involving the flow of information, money, or other quantities. Control systems are found in abundance in biological and ecological systems. Also, feedback system analysis and design techniques have been applied to economic and social systems.

Following World War II, system design techniques were consolidated primarily around transfer function methods based on the Laplace transform. Significant advances in applications were made in the areas of aircraft and missile guidance systems and nuclear power.

By the late 1950s, the digital computer had begun to be applied to the design of control systems and used as a control element itself. Rediscovery of the state–space point of view, originally developed in classical mechanics, was made at this time. The 1960s were dominated by theoretical developments based on this approach. Initially, difficulty was encountered in applying digital computers as controllers for many reasons, and the typical hardware during this decade consisted of pneumatic and hydraulic devices and operational amplifiers. The most significant applications were probably in the space program.

At the start of the 1970s, it was realized that the so-called modern control theory based on the state–space approach could not entirely replace the "classical" theory

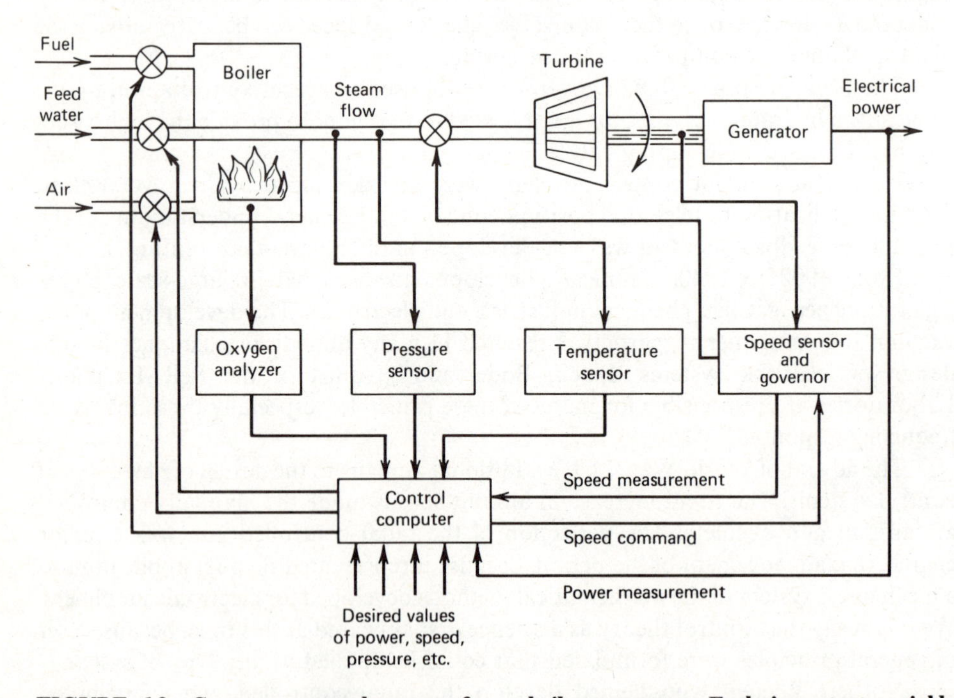

FIGURE 6.3 Computer control system for a boiler generator. Each important variable requires its own controller. The interaction between variables calls for coordinated control of all loops.

based on frequency response and the Laplace transform. The modern control engineer should have a working knowledge of both approaches. In terms of hardware, the decade witnessed the widespread use of digital controllers, first as large-scale machines, then as minicomputers, and finally as microcomputers.

The 1980s should see the microcomputer continue to replace other devices as control elements, although there will always be a need for operational amplifiers, pneumatics, and hydraulics because of simplicity, reliability, and force requirements. Increasing demands for fuel economy will require improved control systems for aircraft and automobile engines as well as other types of power plants. Classical control theory, with its emphasis on single-input, single-output systems, is not capable of dealing with some of these applications, because several variables must be controlled, as shown in Figure 6.3.

In the future, we can expect significant applications of automatic control systems in the space program because of the expense of sending people into space. The use of robots in this area as well as industrial situations will become more widespread. Robotics has much to contribute in areas where the task is hazardous or tedious. High-speed transportation systems on land, sea, and in the air will require better control systems for improved passenger safety and comfort. Also, lighter weight structures can be used if improved controllers are available to deal with the problems caused by the flexibility of such structures. With more stringent requirements for safety and efficiency in all areas, there will be much to do in both the theory and the practice of feedback control systems.

The Role of Modeling and Analysis

The modeling techniques in the previous chapters are required for the successful design of control systems. Transients due to disturbances and changes in the command input make it essential to develop a dynamic model for the system to be controlled. The structure of the resulting control system will be heavily dependent on the form of this model. Thus, the model should contain the most important behavior of the system, but no more than is necessary to achieve a satisfactory design. Quite often, a first- or second-order model in terms of the system's dominant roots is satisfactory.

The analytical techniques developed thus far are extremely useful in predicting the performance of a control system. The quantifiers of transient performance—such as the time constant, damping ratio, and natural frequency—enable the designer to adjust the parameters of the controller to suit the particular system to be controlled. For example, consider the problem of balancing a cylindrical object on end in the palm of your hand. Such an object could be a pencil or broom handle. The form of the mathematical model is the same for both, but the moments of inertia have different values. The broom handle is easier to balance because of its higher inertia. Our hand, eyes, and brain constitute the control system in this example, and the rapidity with which we respond to the object's motion must be faster for the pencil than for the broom handle. If we were to design a controller to replace the hand–eye–brain system, the parameters of the controller (which we will call *gains*) must be different for each object.

Analytical techniques for determining the stability of a control system are important, because the introduction of a feedback loop around a stable system can result in an unstable system. The parameters of the controller must be selected to avoid this, and they depend on the system to be controlled. If we respond too quickly in balancing the pencil, our hand will overshoot too far and cause the pencil to topple too far in the opposite direction. The system will be unstable. On the other hand, the broom handle is more forgiving of fast (or slow) response because of its higher inertia. In general, we will see that instability is more of a problem in control systems with higher performance specifications.

The previous example shows that stability and speed of response can be conflicting objectives. In addition, control system accuracy can be another competing requirement. Simple control systems with loose performance specifications can sometimes be designed by trial and observation because of the self-correcting nature of feedback. However, the performance requirements and expense of modern systems can be quite high, and an analytical approach is usually necessary to design a control system with desirable accuracy, speed of response, and stability characteristics.

6.2 CONTROL SYSTEM STRUCTURE

The electromechanical position control system shown in Figure 6.4 illustrates the structure of a typical control system. A load with an inertia I is to be positioned at some desired angle θ_r. A dc motor is provided for this purpose. The system contains viscous damping, and a disturbance torque T_d acts on the load, in addition to the motor torque T. The origin of the disturbance torque depends on the particular application. If the load to be positioned is a radar antenna, for example, wind gusts produce a torque whose magnitude and time of occurrence are unknown to some extent. If the motor is to control the position of a valve in a flow line (such as a damper in a forced-hot-air heating system), the disturbance torque would result from a change in fluid forces caused by a change in supply pressure in the line. In both examples, the effects of Coulomb friction could also be modeled as a disturbance, in the sense that the magnitude of such a torque is difficult to predict (coefficients of friction are not easily computed and not usually constants).

Because of the disturbance, the angular position θ of the load will not necessarily equal the desired value θ_r. For this reason, a potentiometer is used to measure the

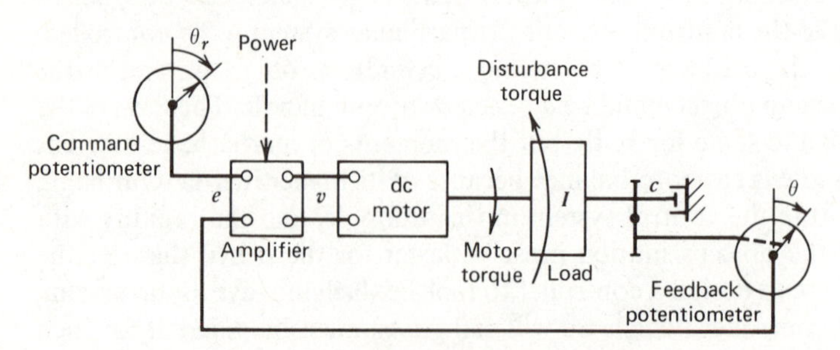

FIGURE 6.4 Position control system using a dc motor.

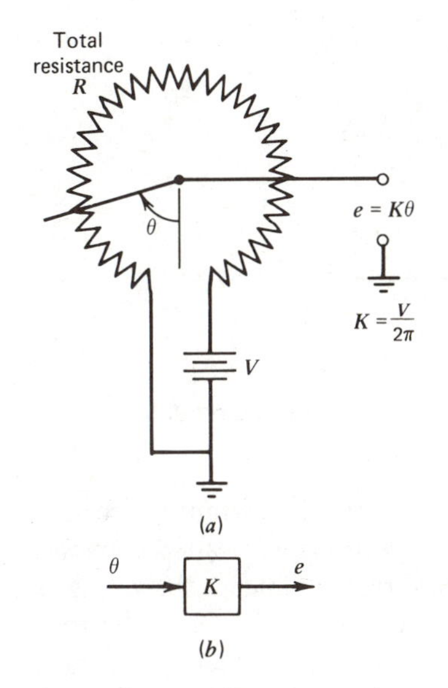

FIGURE 6.5 Rotary potentiometer and its block diagram.

displacement θ. A potentiometer consists of a wire-wound resistor with a sliding electrical contact, or wiper. As shown in Figure 6.5, when the wiper rotates, the resistance between the wiper contact and ground changes. This produces a voltage at the wiper that is a function of its angular displacement. If the resistance of the winding is uniform, and if the wiper circuit draws negligible current, the wiper voltage is proportional to the angular displacement, so that $e = K\theta$. A simple circuit analysis can be used to show that $K = V/2\pi$, where V is the constant voltage supplied to the potentiometer. Potentiometers are also available for applications in which the displacement is a translation, not a rotation.

The potentiometer voltage representing the controlled position θ is compared to the voltage generated by the command potentiometer. This device enables the operator to dial in the desired angle θ_r. The amplifier sees the difference e between the two potentiometer voltages. The basic function of the amplifier is to increase the small error voltage e to the voltage level required by the motor and to supply the current required by the motor to drive the load. In addition, the amplifier may shape the voltage signal in certain ways to improve the performance of the system. We will return to this aspect later.

The system operates as follows. If both potentiometers have the same constant K, the error voltage is

$$e = K(\theta_r - 0)$$

When θ does not equal θ_r, a nonzero voltage e appears at the input terminals of the amplifier. Assume for now that the amplifier produces a voltage v proportional to e and with the same sign; that is, $v = K_a e$, where K_a is the amplifier gain. Then if $\theta_r > \theta$, e and v are positive, and the motor produces a torque T that increases θ. This continues until $e = 0$ and therefore $\theta = \theta_r$. Similar events occur if $\theta_r < \theta$. When the action of a disturbance torque causes θ to deviate from θ_r, the same process acts to restore θ to its desired value. Thus, the control system is seen to provide two basic functions: (1) to respond to a command input that specifies a new desired value for the controlled variable and (2) to keep the controlled variable near the desired value in spite of disturbances. The presence of the feedback loop, with which the measurement of the controlled variable is used to alter the operation of the motor, is seen to be vital to both functions.

A block diagram of this system will help analyze its performance. This is shown in Figure 6.6, where the transfer functions for the amplifier and motor have been left in

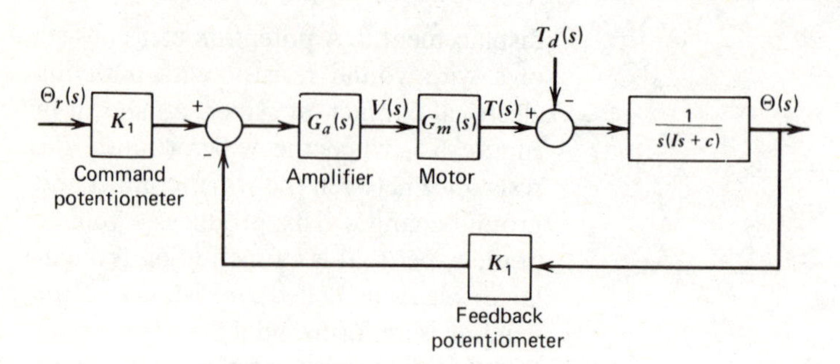

FIGURE 6.6 Block diagram of the dc position control system shown in Figure 6.4.

general form so that we can consider various models later. It is important to recall at this time some assumptions of block diagram representations. Cascaded elements must be "nonloading." For example, the motor cannot directly influence the voltage v coming from the amplifier (except by way of the feedback loop). Thus, the motor must have a high input impedance. Similarly, the potentiometer derivation assumes that the wiper current is negligible; that is, the amplifier must have a high input impedance. If these assumptions are not valid, the entire subsystem must be represented by a single transfer function derived from basic principles rather than from diagram reduction. Refer to Chapter 4 for more discussion of this point.

The power supplies required for the potentiometers and the amplifier are not shown in block diagrams of control system logic, because they do not contribute to the control logic. However, their existence cannot be ignored. The motor's torque-current relation is $T = K_t i$. Therefore, as the load on the motor is increased, it calls for more current from the amplifier. If the available current is limited (which is true for all real systems), we must take this limitation into account before accepting any controller designs as final.

Some additional modeling approximations might also be contained in the diagram. We have taken the motor, damping and disturbance torques to be acting on the inertia I of the load. This implies that the inertia of the motor has been lumped into that of the load. In reality, the damping torque, for example, might result from electromagnetic effects within the motor itself, and the motor torque acts directly on the rotor. But the approximation is usually a good one, especially when the shaft connecting the motor and load is stiff.

A Standard Diagram

The electromechanical positioning system fits the general structure of a control system (Figure 6.7). This figure also gives some standard terminology. Not all systems can be forced into this format, but it serves as a reference for discussion.

The controller is generally thought of as a logic element that compares the command signal with the measurement of the output and decides what should be done. The input and feedback elements are transducers for converting one type of signal into another type. The potentiometer converts displacement into voltage. This allows the error detector directly to compare two signals of the same type (e.g., two

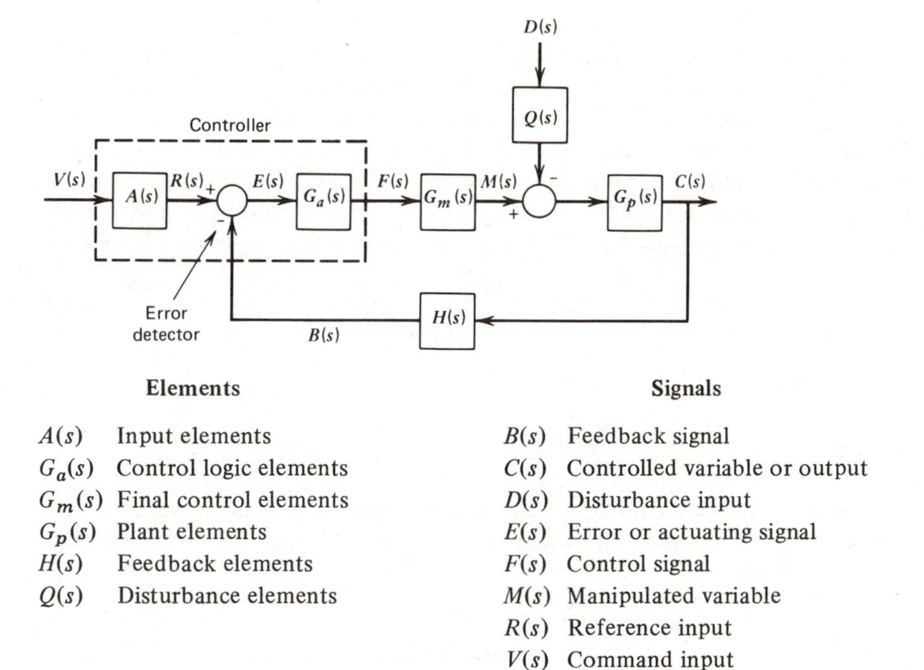

FIGURE 6.7 Terminology and basic structure of a feedback control system.

voltages). Not all functions show up as separate physical elements. The error detector in Figure 6.4 is simply the input terminals of the amplifier. The control logic elements produce the control signal, which is sent to the *final control elements.* These are the devices that develop enough torque, pressure, heat, etc., to influence the elements under control. Thus, the final control elements are the "muscle" of the system, while the control logic elements are the "brain." In Chapter 6, we are primarily concerned with the design of the logic to be used by this brain.

The object to be controlled is the *plant.* The *manipulated variable* is generated by the final control elements for this purpose. The disturbance input also acts on the plant. This is an input over which the designer has no control, and perhaps for which little information is available about the magnitude, functional form, or time of occurrence. The disturbance can be a random input, such as wind gusts on a radar antenna, or deterministic, such as coulomb friction effects. In the latter case, we can include the friction force in the system model by using a nominal value for the coefficient of friction. The disturbance input would then be the deviation of the friction force from this estimated value and would represent the uncertainty in our estimate.

Several control system classifications can be made with reference to Figure 6.7. A *regulator* is a control system in which the controlled variable is to be kept constant in spite of disturbances. An example is the temperature control system for a house. Once the desired temperature is set, say, at 68°F, the controller is to keep the room temperature near this value. The command input for a regulator is its *set point.* On the other hand, a *follow-up system* is supposed to keep the controlled variable near the command value, which is changing with time. An example of a follow-up system is a machine tool in which a cutting head must trace a specific path in order to shape the product properly. This is also an example of a *servomechanism*, which is a control system whose controlled variable is a mechanical position, velocity, or acceleration. The thermostat system is not a servomechanism but a *process control system*, where the controlled variable describes a thermodynamic process. Typically, such variables are temperature, pressure, flow rate, liquid level, chemical concentration, and so forth.

The *primary*, or *reference* transfer function for the system shown in Figure 6.7 is

$$\frac{C(s)}{V(s)} = A(s)\frac{G_a(s)G_m(s)G_p(s)}{1 + G_a(s)G_m(s)G_p(s)H(s)} \tag{6.2-1}$$

The *disturbance* transfer function is

$$\frac{C(s)}{V(s)} = -Q(s)\frac{G_p(s)}{1 + G_a(s)G_m(s)G_p(s)H(s)} \tag{6.2-2}$$

An important measure of a control system's performance is the error signal $E(s)$. Block diagram algebra can be used to show that for the system in Figure 6.7, $E(s)$ is related to $V(s)$ and $D(s)$ as follows:

$$E(s) = \frac{A(s)V(s) + G_p(s)H(s)Q(s)D(s)}{1 + G_a(s)G_m(s)G_p(s)H(s)} \tag{6.2-3}$$

The configuration shown in Figure 6.7 is that of a basic control system, but it is not the only configuration used. As we will see, improvements in system performance can sometimes be made by using the configurations shown in Figure 6.8. We will study the applications for these arrangements in this chapter; we will see even more configurations in Chapter 7. For future reference, we now give the transfer functions and the error signal relationships. As a simplified special case, we take $A(s) = Q(s) = 1$. Note that this implies that $V(s) = R(s)$.

The system shown in Figure 6.8*a* uses *feedforward compensation*; its usefulness will be seen in Section 6.5. For visual clarity, we now suppress the Laplace variable s in the subsystem transfer functions. From the block diagram, we can write the following relations:

$$F(s) = G_a E(s) + G_f R(s)$$

$$C(s) = [G_m F(s) - D(s)]G_p$$

$$E(s) = R(s) - HC(s)$$

Set $D(s) = 0$, and eliminate all the other variables except $C(s)$ and $R(s)$ to obtain the primary transfer function

$$\frac{C(s)}{R(s)} = \frac{G_p G_m (G_a + G_f)}{1 + G_a G_m G_p H} \tag{6.2-4}$$

Now replace $D(s)$, and set $R(s) = 0$. Then eliminate all variables except $C(s)$ and $D(s)$ to obtain the disturbance transfer function

$$\frac{C(s)}{D(s)} = \frac{-G_p}{1 + G_a G_m G_p H} \tag{6.2-5}$$

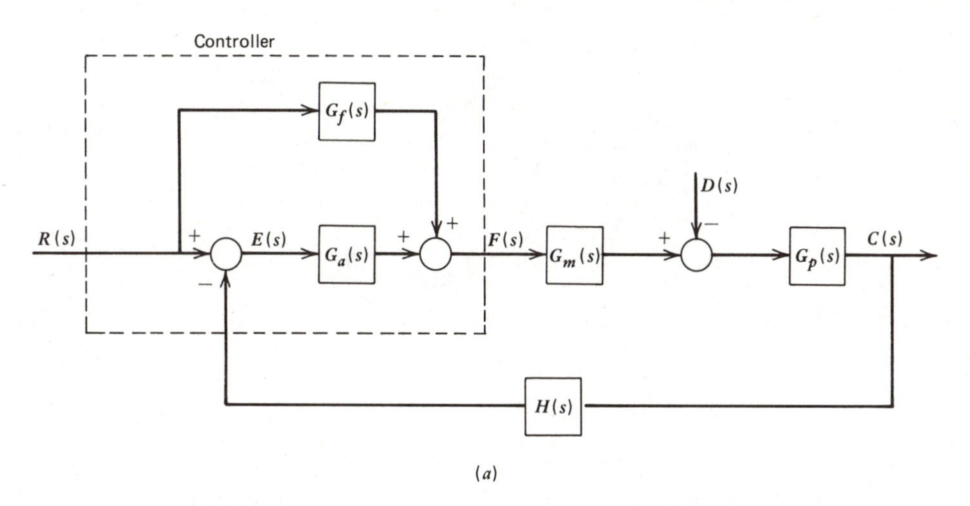

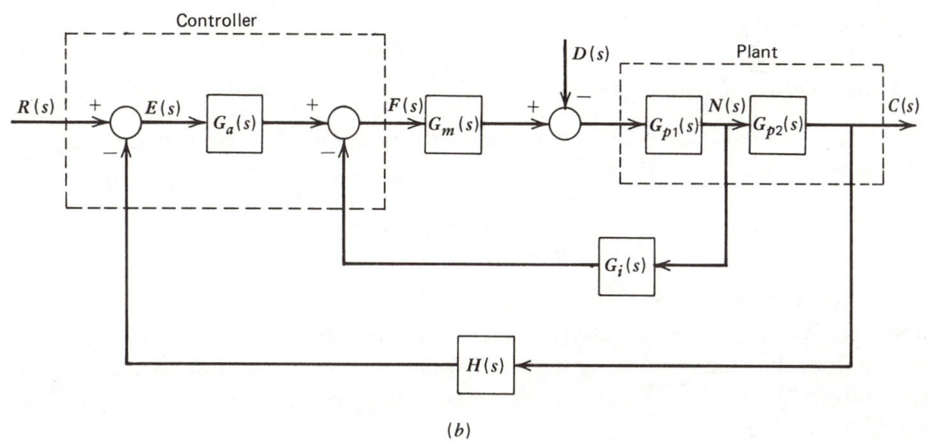

FIGURE 6.8 Alternate control system configurations. (*a*) Feedforward compensation. (*b*) Internal feedback compensation.

Finally, the error relation is

$$E(s) = \frac{(1 - G_f G_m G_p H)R(s) + G_p H D(s)}{1 + G_a G_m G_p H} \tag{6.2-6}$$

The system shown in Figure 6.8*b* has *internal feedback compensation.* The plant is modeled with two transfer functions $G_{p1}(s)$ and $G_{p2}(s)$, and the intermediate plant variable $N(s)$ is fed back for use by the controller. We will see the usefulness of this in Section 6.7. The system relations are:

$$F(s) = G_a E(s) - G_i N(s)$$
$$E(s) = R(s) - HC(s)$$
$$C(s) = G_{p2} N(s)$$
$$N(s) = G_{p1}[G_m F(s) - D(s)]$$

By alternately setting $D(s) = 0$ and then $R(s) = 0$ and eliminating variables as before, we obtain

$$\frac{C(s)}{R(s)} = \frac{K}{P} \quad \textbf{(6.2-7)}$$

$$\frac{C(s)}{D(s)} = \frac{L}{P} \quad \textbf{(6.2-8)}$$

$$E(s) = \frac{(1 + G_{p1}G_mG_i)}{P}R(s) - \frac{LH}{P}D(s) \quad \textbf{(6.2-9)}$$

where

$$K = G_{p1}G_{p2}G_mG_a \quad \textbf{(6.2-10)}$$

$$L = -G_{p1}G_{p2} \quad \textbf{(6.2-11)}$$

$$P = 1 + G_{p1}G_{p2}G_aG_mH + G_{p1}G_mG_i \quad \textbf{(6.2-12)}$$

The following examples illustrate these concepts and provide practice in describing the control system.

A Temperature Controller

The oven shown in Figure P2.37 is now considered a candidate for control. If we neglect the thermal capacitance of the heater, then $C_1 = 0$, and the model is the following.

$$C_2\frac{dT_2}{dt} = q - \frac{1}{R_2}(T_2 - T_o) \quad \textbf{(6.2-13)}$$

A control system for this process is shown in Figure 6.9. A power amplifier is used to supply current to the resistance-type heater. The input to the amplifier consists of the voltage from the command potentiometer, which represents the desired value of the temperature T_2, and the voltage from the temperature sensor. These voltages are K_1T_r and K_2T_2, respectively.

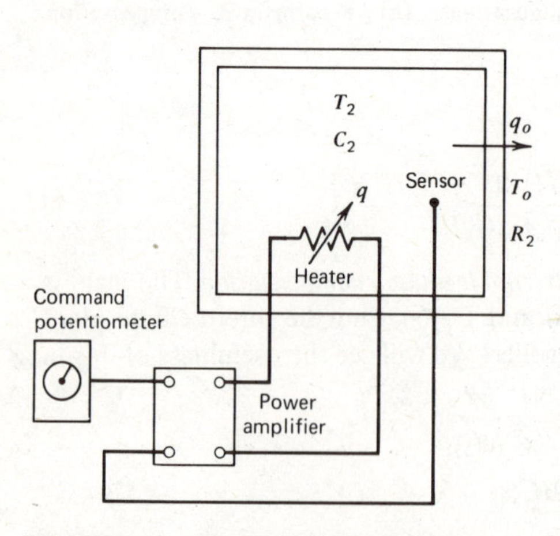

FIGURE 6.9 Electrical temperature control system.

Suppose the ambient temperature T_o is reasonably constant at 70°F. In order for the oven temperature T_2 to be held constant at some desired value, say, 200°F, the heat input rate from the heater must equal that lost through the resistance R_2; that is, for $T_2 = T_r = 200$°F, the steady-state heater output must be

$$q_{ss} = \frac{1}{R_2}(200° - 70°) = \frac{130°}{R_2}$$

The amplifier circuit must be constructed so that the potentiometer setting will produce this heat flow when $T_r = T_2 = 200$°F.

The second function of the amplifier circuit is to compare the voltage from the potentiometer with that from the temperature sensor and to modify the heater output by an amount m. We can thus express the heater output as

$$q = q_{ss} + m \tag{6.2-14}$$

If the rate m can be made proportional to the difference between the potentiometer voltage and the sensor voltage, then

$$m = K_a e \tag{6.2-15}$$

$$e = K_1 T_r - K_2 T_2 = K_2(T_r - T_2) \tag{6.2-16}$$

where K_a is a gain associated with the amplifier and the scale factor of the potentiometer has been chosen so that $K_1 = K_2$.

It is convenient and standard practice to choose the block diagram variables as deviations from reference values corresponding to a zero-error equilibrium operating condition. Let us choose the temperatures 200° and 70° as our references for the oven and ambient temperatures, and define the following deviations:

$$x = T_2 - 200°$$

$$v = T_o - 70°$$

The deviation $m = q - q_{ss}$ is the corresponding variable for heat rate, and the governing differential equation becomes

$$C_2 \frac{dx}{dt} = m - \frac{1}{R_2}(x - v) \tag{6.2-17}$$

A change in ambient temperature from 70° is represented by a nonzero value for the disturbance variable v. On the other hand, to allow the possibility of selecting a new oven temperature, define a set point deviation u as

$$u = T_r - 200°$$

In this case, $e = K_2(u - x)$; the diagram is shown in Figure 6.10. Note that the diagram does not show the equilibrium reference values but only the deviations from these values. If the new oven temperature is desired to be 250°, the input u is a step function with a magnitude of 250° − 200° = 50°.

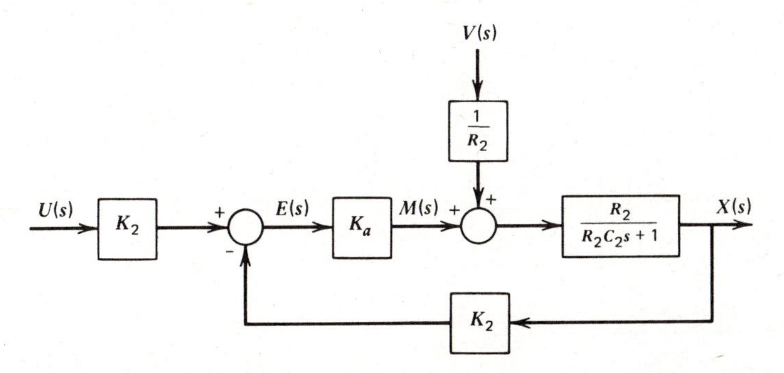

FIGURE 6.10 Block diagram of the temperature control system shown in Figure 6.9.

The selection of deviation variables was not explicitly made in the previous position control system. This is because any angular position of the load is an equilibrium operating condition when the motor is turned off and no disturbance torque is acting. In this case, the reference motor torque and voltage are zero, and the variables T and v can be considered to be deviations from these zero reference values. If the nominal disturbance torque were not zero, the reference motor torque and voltage would be nonzero as well, and deviations would be defined from these values.

The primary reason for employing deviation variables is that they readily allow a linearized model to be used to describe the system. Physical systems are inherently nonlinear, but most controllers are the result of a successful application of linear systems analysis. This is because the principal function of the controller is to eliminate deviations from some desired operating value. The next example demonstrates this point.

A Liquid Level Controller

A simplified illustration of a liquid level controller commonly found in industry is shown in Figure 6.11. The liquid is supplied under pressure p to the control valve. The valve motion is produced by a diaphragm with pneumatic pressure on one side and a resisting spring on the other. The air pressure is produced by a pneumatic amplifier

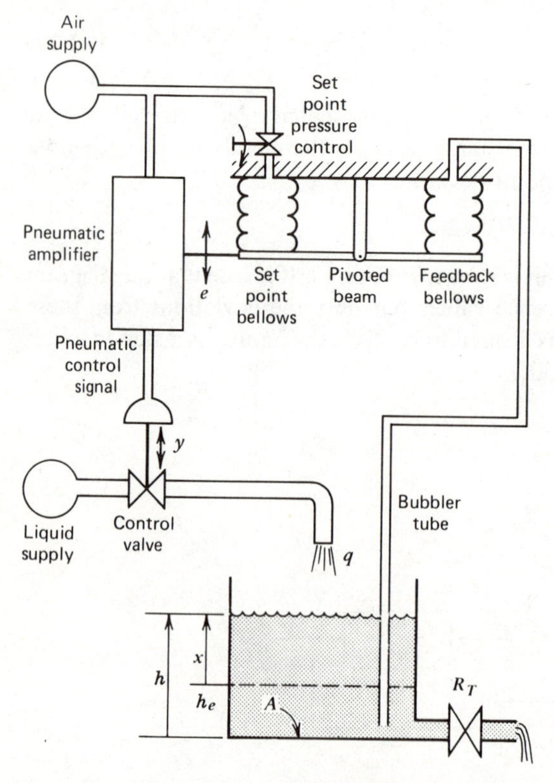

FIGURE 6.11 Pneumatic liquid level control system.

whose output is affected by the motion of a beam with a bellows at each end (this amplifier is discussed in Section 6.4). A bubbler tube filled with air indicates the liquid level by sensing the hydrostatic pressure in the tank. The air pressure in the tube is connected to the feedback bellows. The set point of the controller indicates the desired liquid level. This is set by changing the pressure in the set point bellows via an adjustable restriction in the supply line. The system is balanced when the level is at the desired value and the feedback pressure equals the set point pressure. If the level is below the desired value, the beam pivots counterclockwise. The amplifier and valve are designed to respond to this motion by increasing the flow rate q. The flow q is a function of p and the valve motion y.

Two possible disturbances to this system are changes in the supply pressures of the air and the liquid. Here, we consider a change in the liquid supply pressure p. Assume that the outlet restriction is nonlinear. Then the governing equations are

$$\rho A \frac{dh}{dt} = q - \sqrt{\frac{\rho g h}{R_T}} \tag{6.2-18}$$

$$q = f(p, y) \tag{6.2-19}$$

Let the level h_e denote the nominal equilibrium level, q_e the flow rate required to maintain this level, p_e the nominal supply pressure, and y_e the nominal valve position. Define the deviations

$$x = h - h_e$$

$$z = y - y_e$$

$$v = p - p_e$$

$$w = q - q_e$$

Measure the beam displacement e from its position at this equilibrium condition. This displacement is an indication of the system error, and the beam acts like an error detector. The desired level h_r is not always h_e. Therefore, define the set point deviation $u = h_r - h_e$.

Linearization of the tank and valve equation gives

$$\rho A \frac{dx}{dt} = w - \frac{1}{R} x \tag{6.2-20}$$

$$\frac{1}{R} = \frac{1}{2} \sqrt{\frac{\rho g}{R_T h_e}} \tag{6.2-21}$$

$$w = K_1 z + K_2 v \tag{6.2-22}$$

$$K_1 = \left(\frac{\partial f}{\partial y} \right)_e, \quad K_2 = \left(\frac{\partial f}{\partial p} \right)_e \tag{6.2-23}$$

The linearized resistance is R and positive. The constant K_1 is positive because of the assumed operation of the valve; K_2 is positive, because an increase in supply pressure causes an increase in flow rate.

The beam motion e depends on the set point pressure and feedback pressure, which in turn depend on h_r and h, respectively. Thus, $e = g(h_r, h)$, and in linearized

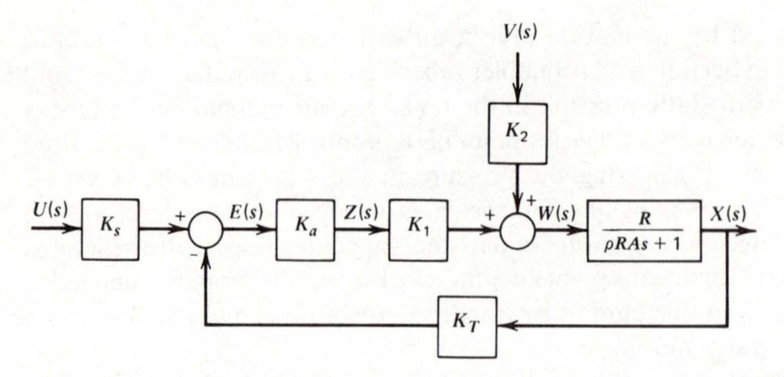

FIGURE 6.12 Block diagram of the liquid level control system of Figure 6.11.

form,

$$e = K_s u - K_T x \tag{6.2-24}$$

$$K_s = \left(\frac{\partial g}{\partial h_r}\right)_e, \qquad K_T = -\left(\frac{\partial g}{\partial h}\right)_e \tag{6.2-25}$$

since e is already a deviation. The set point and bubbler tube constants K_s and K_T depend on the beam geometry, bellows area and stiffness, and air supply pressure. The latter quantity can be absorbed into K_s, because we have assumed it is constant.

The valve displacement y is a function of the beam motion via the pneumatic amplifier. In linearized form, this gives

$$z = K_a e \tag{6.2-26}$$

where K_a is the linearization constant for the amplifier valve combination.

The linearized description of the system is now complete, and the block diagram can be drawn (Figure 6.12). Caution must be observed when using this model. For a large enough change in set point or supply pressure away from their nominal values h_e and p_e, the linearized model is inaccurate, and the linearization constants should be recomputed at the new operating condition. However, in many cases the original linearized description is quite adequate.

6.3 TRANSDUCERS AND ERROR DETECTORS

The control system structure shown in Figure 6.7 indicates a need for physical devices to perform several types of functions. Here, we present a brief overview of some available transducers and error detectors. Actuators and devices used to implement the control logic are discussed in Sections 6.4 and 6.8–6.10. Space limitations prevent a detailed discussion of these devices. The technology is rapidly changing, and the most up-to-date information is available only in manufacturers' literature. Some of these devices have been analyzed in previous examples; models for the others can be obtained by applying the techniques in Chapter 2 or by consulting appropriate references (for example, see References 4–7).

Displacement and Velocity Transducers

A *transducer* is a device that converts one type of signal into another type. An example is the potentiometer, which converts displacement into voltage. In addition to this conversion, the transducer can be used to make measurements. In such applications, the term *sensor* is more appropriate. For our purposes, we will assume that the sensor measurements are accurate. A detailed discussion of sensor errors is given in Reference 4. The design of control systems that compensates for imperfect measurements relies on filtering theory and is an advanced topic (Reference 8).

In addition to the potentiometer, displacement can also be measured electrically with a *linear variable differential transformer* (LVDT) or a *synchro*. An LVDT measures the linear displacement of a movable magnetic core through a primary winding and two secondary windings (Figure 6.13). An ac voltage is applied to the primary. The secondaries are connected together and also to a detector that measures the voltage and phase difference. A phase difference of 0° corresponds to a positive core displacement, while 180° indicates a negative displacement. The amount of displacement is indicated by the amplitude of the ac voltage in the secondary. The detector converts this information into a dc voltage e_o such that $e_o = Kx$. The LVDT is sensitive to small displacements. Two of them can be wired together like the potentiometers in Section 6.2 to form an error detector.

A synchro is a rotary differential transformer with angular displacement as either the input or output. A *transmitter* and a *receiver* are often used in pairs where a remote indication of angular displacement is needed. When a transmitter is used with a synchro *control transformer*, two angular displacements can be measured and compared (Figure 6.14). The output voltage e_o is approximately linear with angular difference within $\pm 70°$, so that $e_o = K(\theta_1 - \theta_2)$.

Displacement measurements can be used to obtain forces and accelerations. For example, the displacement of a calibrated spring indicates the applied force. The accelerometer described in Section 4.6 is another example. Still another is the *strain gage* used for force measurement. It is based on the fact that the resistance of a fine wire changes as it is stretched. The change in resistance is detected by a circuit that can be calibrated to indicate the applied force.

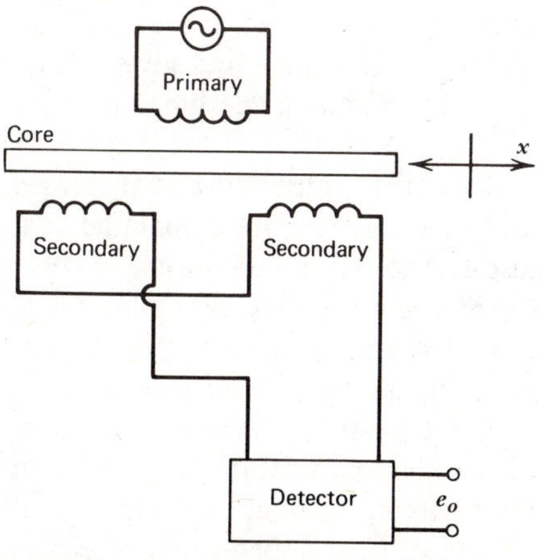

Primary
Core
x
Secondary
Secondary
Detector
e_o

FIGURE 6.13 Linear variable differential transformer (LVDT).

Velocity measurements in control systems are most commonly obtained with a *tachometer*. This is essentially a dc generator (the reverse of a dc motor). The input is mechanical (a velocity). The output is a generated voltage proportional to the velocity. Translational velocity is usually measured by converting it into angular velocity with gears, for example. Tachometers using ac signals are also available.

Other velocity transducers include a

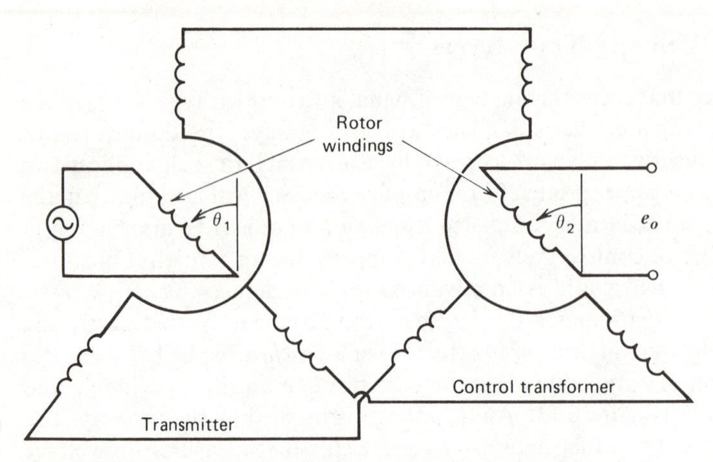

FIGURE 6.14 Synchro transmitter control transformer.

magnetic pickup that generates a pulse every time a gear tooth passes. If the number of gear teeth is known, a pulse counter and timer can be used to compute the angular velocity. A similar principle is employed by an *optical encoder*. A light beam is broken by a rotating slotted disk, and a photoelectric cell converts this information into pulses that can be analyzed as before. The outputs of these devices are especially suitable for digital control purposes.

Temperature Transducers

When two wires of dissimilar metals are joined together as in Figure 6.15, a voltage is generated if the junctions are at different temperatures. If the reference junction is kept at a fixed, known temperature, the thermocouple can be calibrated to indicate the temperature at the other junction in terms of the voltage v. An ice water bath can be used for the reference temperature, but many modern systems provide an electronic equivalent.

Electrical resistance changes with temperature. Platinum gives a linear relation between resistance and temperature, while nickel is less expensive and gives a large resistance change for a given temperature change. Semiconductors designed with this property are called *thermistors*.

Different metals expand at different rates when the temperature is increased. This fact is used in the bimetallic strip transducer found in most home thermostats (Figure 6.16). Two dissimilar metals are bonded together to form the strip. In most thermostats, the strip is actually coiled, but the figure illustrates the principle. As the temperature rises, the strip curls, breaking contact and shutting off the furnace. The temperature gap can be adjusted by changing the distance between the contacts. The motion also moves a

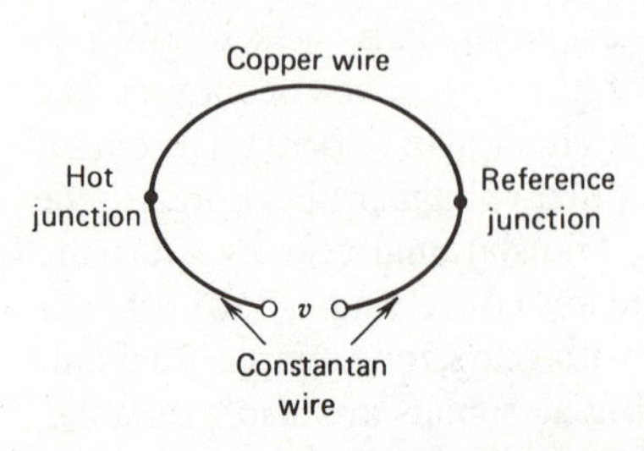

FIGURE 6.15 Type-*T* thermocouple.

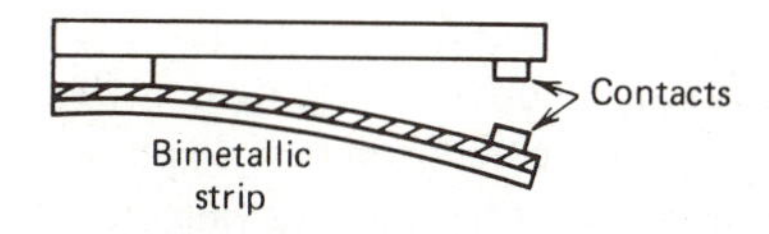

FIGURE 6.16 Electrical temperature sensor using a bimetallic strip.

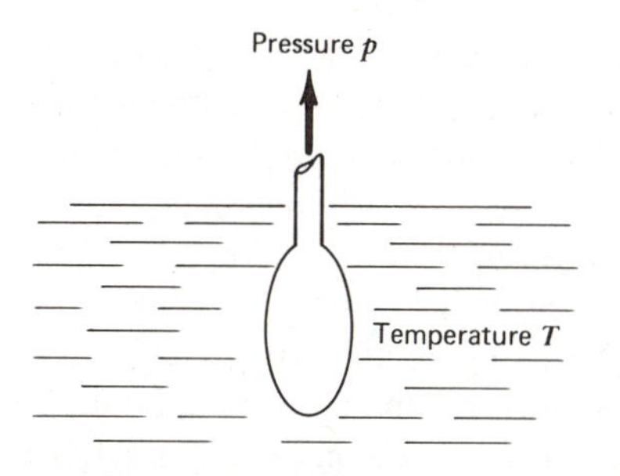

FIGURE 6.17 Pneumatic temperature sensor.

pointer on the temperature scale of the thermostat.

The pressure of a fluid inside a bulb will change as its temperature changes (Figure 6.17). If the bulb fluid is air, the device is suitable for use in pneumatic temperature controllers. If the air mass in the bulb is constant and if the thermal resistance and capacitance of the bulb walls are negligible, the perfect gas law can be applied to obtain

$$p = KT \tag{6.3-1}$$

where the constant K depends on the gas and the bulb volume.

Flow Transducers

Flow rates can be measured by introducing a flow restriction, such as an orifice plate, and measuring the pressure drop across the restriction. The flow–rate–pressure relation is $\Delta p = Rq^2$, where R can be found from the device calibration. The pressure drop can be sensed by converting it into the motion of a diaphragm. Figure 6.18 illustrates a related technique. The venturi-type flowmeter measures the static pressures in the constricted and unconstricted flow regions. Bernoulli's principle relates the pressure difference to the flow rate. This pressure difference is converted into displacement by the diaphragm.

Error Detectors

The error detector is simply a device for finding the difference between two signals. This function is sometimes an integral feature of sensors, such as in the synchro transmitter–transformer combination. A beam on a pivot provides a way of comparing displacements, forces, or pressures, as was done with the pneumatic level controller (Figure 6.11). A similar concept is used with the diaphragm element shown in Figure 6.18. A detector for voltage difference can be obtained with the position control system shown in Figure 6.4. An amplifier intended for this purpose is a *differential amplifier*. Its output is proportional to the difference between the two inputs.

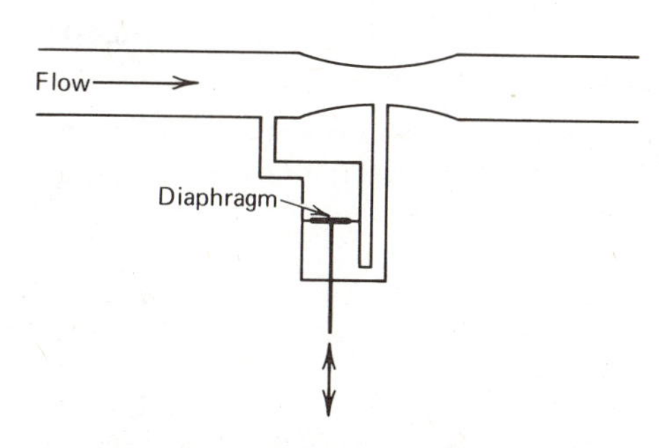

FIGURE 6.18 Venturi-type flowmeter. The diaphragm displacement indicates the flow rate.

Relative motion can be detected with sufficient accuracy for some applications by the simple device shown in Figure 6.19.

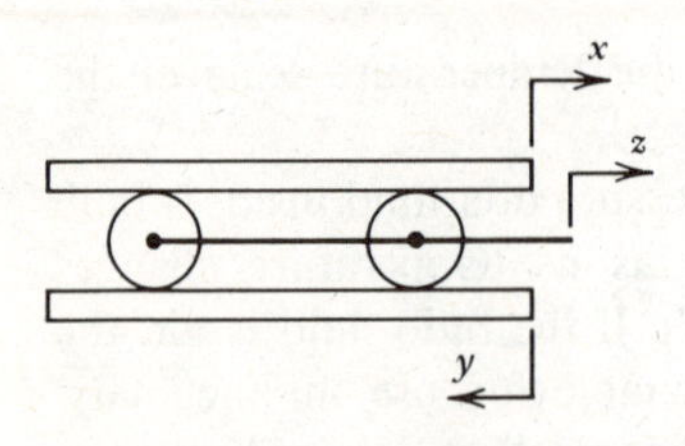

FIGURE 6.19 Rack-and-gear differential for detecting displacement differences $z = \frac{1}{2}(x - y)$.

In order to detect differences in other types of signals, such as temperature, the signals are usually converted to a displacement or pressure. One of the detectors mentioned previously can then be used.

Dynamic Response

The quantitative transducer and detector models presented earlier are static models and as such imply that the components respond instantaneously to the variable being sensed. Of course, any real component has a dynamic response of some sort, and this response time must be considered in relation to the controlled process when a sensor is selected. For example, the time constant of a thermocouple in air typically is between 10 and 100 sec. If the thermocouple is in a liquid, the thermal resistance is different, and so will be the time constant (it will be smaller). If the controlled process has a time constant at least 10 times greater, we probably would be justified in using a static sensor model. The effect of this assumption will be considered in a later example.

We will assume in general that the sensors have been selected properly in relation to the plant dynamics, and we will thus use a static model in most cases.

6.4 ACTUATORS

An *actuator* is the final control element that operates on the low-level control signal to produce a signal containing enough power to drive the plant for the intended purpose. The armature-controlled dc motor, the hydraulic servomotor, and the pneumatic bellows are common examples of actuators. Dynamic models of their behavior were developed earlier.

Electromechanical Actuators

From Example 4.6, the transfer function for the armature-controlled dc motor is

$$\frac{\Omega(s)}{V(s)} = \frac{K_T}{LIs^2 + (RI + cL)s + cR + K_e K_T} \tag{6.4-1}$$

where $\Omega(s)$ and $V(s)$ are the angular velocity and input voltage transforms (see Figure 6.20*a*). The armature inductance is often negligible. In that case, the transfer function becomes a first-order one.

$$\frac{\Omega(s)}{V(s)} = \frac{K_T}{RIs + cR + K_e K_T} \tag{6.4-2a}$$

$$= \frac{K}{\tau s + 1} \tag{6.4-2b}$$

Another motor configuration is the field-controlled dc motor, whose model was developed in Chapter 2 (problem 2.19). See Figure 6.20*b*. In this case, the armature

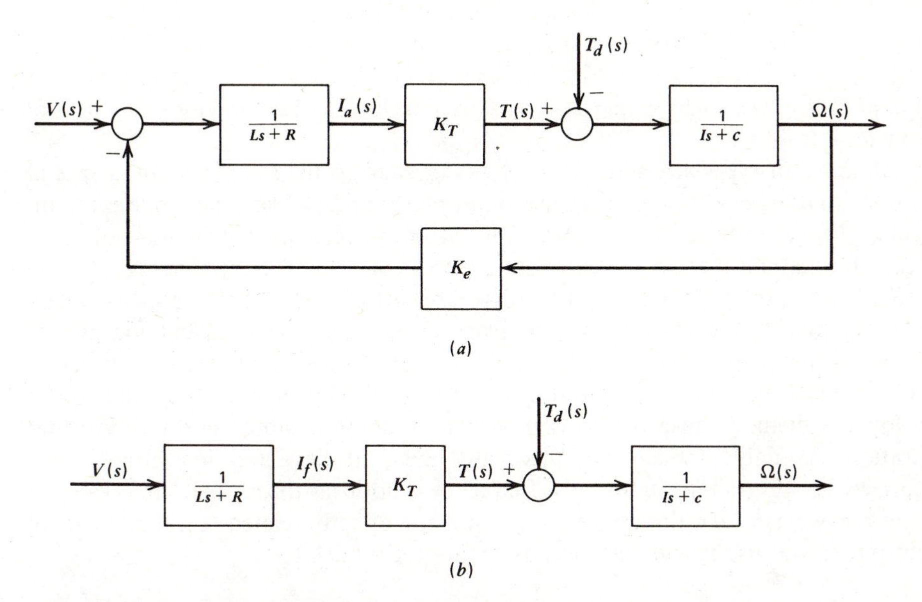

FIGURE 6.20 Block diagrams for two motor systems with voltage inputs and speed outputs. (*a*) Armature-controlled motor. (*b*) Field-controlled motor.

current is kept constant, and the field voltage v is used to control the motor. The transfer function is

$$\frac{\Omega(s)}{V(s)} = \frac{K_T}{(R+Ls)(Is+c)} = \frac{K_T/Rc}{\left(\dfrac{L}{R}s+1\right)\left(\dfrac{I}{c}s+1\right)} \tag{6.4-3}$$

where R and L are the resistance and inductance of the field circuit and K_T is the torque constant. No back emf exists in this motor to act as a self-braking mechanism. The field control requires less power than the armature-controlled motor. However, a constant current is more difficult to provide than the constant voltage needed for a constant field (this field can also be provided by a permanent magnet).

From (6.4-3), we see that the time constant of the electrical subsystem is L/R. This is usually small compared to that of the mechanical subsystem I/c, and the motor is frequently approximated by the first-order model

$$\frac{\Omega(s)}{V(s)} = \frac{K_T/Rc}{\dfrac{I}{c}s+1} = \frac{K_T/R}{Is+c}, \qquad \text{if}\ \frac{L}{R} \ll \frac{I}{c} \tag{6.4-4}$$

where the motor torque is now given approximately by

$$T(s) = \frac{K_T}{R}V(s) \tag{6.4-5}$$

Two-phase ac motors can be used to provide a low-power, variable-speed actuator. This motor type can accept the ac signals directly from LVDTs and synchros without demodulation. However, it is difficult to design ac amplifier circuitry to do other than proportional action (see Section 6.5). For this reason, the ac motor is not found in control systems so often as dc motors. The transfer function for this type is of the form of (6.4-2*b*).

An actuator especially suitable for digital systems is the *stepper motor*, a special dc motor that takes a train of electrical input pulses and converts each pulse into an angular displacement of a fixed amount. Motors are available with resolutions ranging from about four steps per revolution to as many as 800 steps per revolution. For 36 steps per revolution, the motor will rotate by 10° for each pulse received. When not being pulsed, the motors lock in place. Thus, they are excellent for precise positioning applications, such as required with printers and computer tape drives. A disadvantage is that they are low-torque devices. Mathematical models are difficult to develop for them because of the complexity of stepping-motor construction and operation. Available models are piecewise linear at best and inaccurate when linearized. However, if the input pulse frequency is not near the resonant frequency of the motor, we can take the output rotation to be directly related to the number of input pulses and use that description as the motor model.

Hydraulic Actuators

The hydraulic servomotor with a load was analyzed in Example 2.10. Its transfer function from (2.7-15) is,

$$\frac{X(s)}{Z(s)} = \frac{C_1 A}{C_2 m s^2 + (cC_2 + A^2)s + C_2 k} \tag{6.4-6}$$

where z is the input displacement and x is the output displacement. For no damping and no stiffness, and a negligible load, it was shown that (6.4-6) reduces to the first-order model

$$\frac{X(s)}{Z(s)} = \frac{C_1}{As} \tag{6.4-7}$$

Rotational motion can be obtained with a *hydraulic motor*, which is, in principle, a pump acting in reverse (fluid input and mechanical rotation output). Such motors can achieve higher torque levels than electric motors. A hydraulic pump driving a hydraulic motor constitutes a *hydraulic transmission.*

A popular actuator choice is the *electrohydraulic* system, which uses an electric actuator to control a hydraulic servomotor or transmission by moving the pilot valve or the swash plate angle of the pump. Such systems combine the power of hydraulics with the advantages of electrical systems. For example, modern aircraft control systems use hydraulic actuators to move the aerodynamic control surfaces. These motions are commanded by the pilot's stick motions, which are converted into electrical signals to be sent to the actuators. This avoids the need to run long, damage prone hydraulic lines from the cockpit to the actuators. The small electrical lines can be run in multiple sets for redundancy.

Figure 6.21 shows a hydraulic motor whose pilot valve motion is caused by an armature-controlled dc motor. The transfer function between the motor voltage and

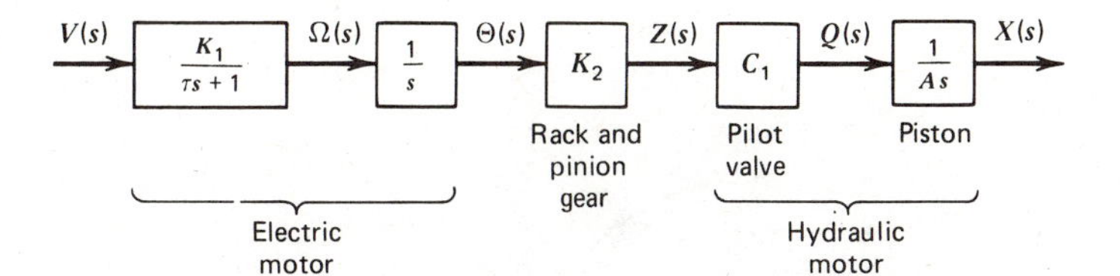

FIGURE 6.21 Electrohydraulic system for translation.

the piston displacement is

$$\frac{X(s)}{V(s)} = \frac{K_1 K_2 C_1}{As^2(\tau s + 1)} \tag{6.4-8}$$

If the rotational inertia of the electric motor is small, then $\tau \approx 0$.

Pneumatic Actuators

Pneumatic actuators are commonly used, because they are simple to maintain and use a readily available working medium. Compressed air supplies with the pressures required are commonly available in factories and laboratories. No flammable fluids or electrical sparks are present, so these devices are considered the safest to use with chemical processes. Their power output is less than that of hydraulic systems but greater than that of electric motors.

A device for converting pneumatic pressure into displacement is the bellows considered in Example 2.12 and shown in Figure 6.22. The transfer function for a linearized model is of the form

$$\frac{X(s)}{P(s)} = \frac{K}{\tau s + 1} \tag{6.4-9}$$

where x and p are deviations of the bellows displacement and input pressure from nominal values. The restriction in the input line acts as a resistance and the bellows volume as a capacitance. The time constant τ is the product of the resistance and capacitance.

In many control applications, a device is needed to convert small displacements into relatively large pressure changes. The *nozzle flapper* serves this purpose (Figure 6.23*a*). The input displacement y moves the flapper with little effort required. This changes the opening at the nozzle orifice. For a large enough opening, the nozzle's back pressure is approximately the same as atmospheric pressure p_a. At the other extreme position with the flapper completely blocking the orifice, the back pressure equals the supply pressure p_s. This variation is shown in Figure 6.23*b*. Typical supply pressures are between 30 and 100 psi (absolute). The orifice diameter is approximately 0.01 in. Flapper displacement is usually less than one orifice diameter.

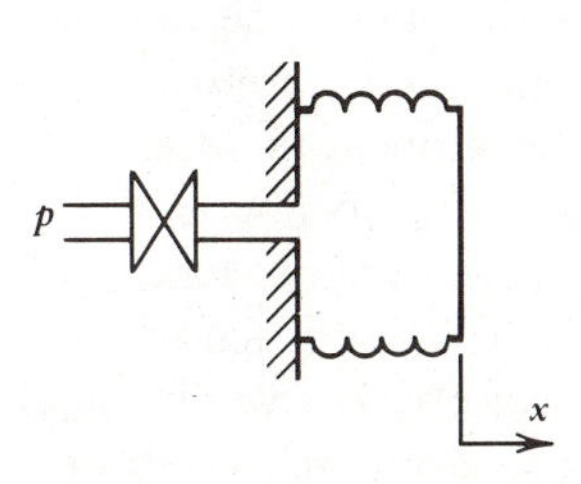

FIGURE 6.22 Pneumatic bellows.

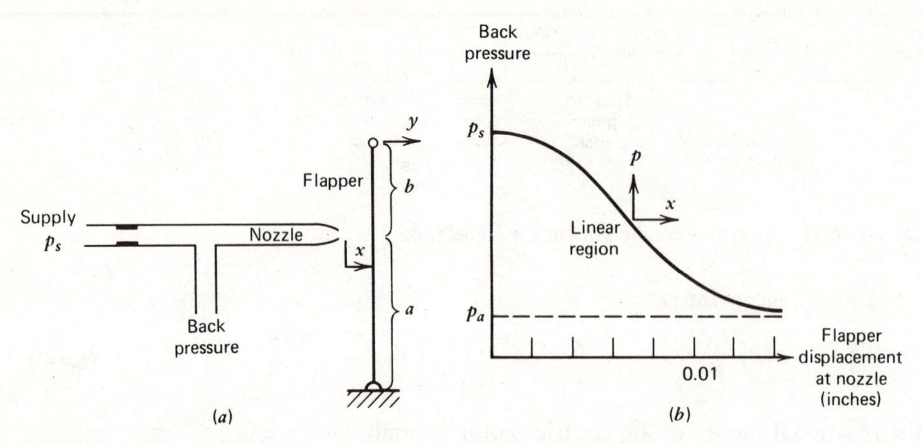

FIGURE 6.23 Pneumatic nozzle flapper amplifier and its characteristic curve.

The nozzle flapper is operated in the linear portion of the back pressure curve. Let p and x denote deviations in back pressure and flapper displacement from the nominal values. The input displacement y is zero at the nominal condition. Then the linearized back pressure relation is

$$p = -K_f x \tag{6.4-10}$$

where $-K_f$ is the slope of the curve and a very large number. From the geometry of similar triangles, we have

$$x = \frac{a}{a+b} y \tag{6.4-11}$$

Thus,

$$p = -\frac{aK_f}{a+b} y \tag{6.4-12}$$

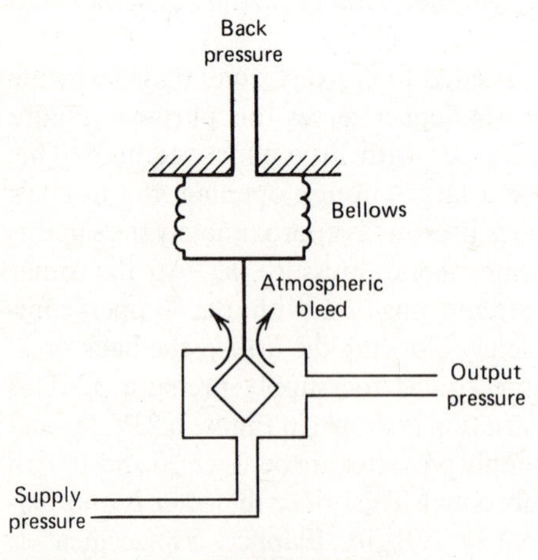

FIGURE 6.24 Pneumatic relay.

In its operating region, the nozzle flapper's back pressure is well below the supply pressure. If more output pressure is required, a pneumatic relay or amplifier can be used. Figure 6.24 illustrates this concept. As the back pressure increases, the relay closes off the supply line, and the output pressure approaches atmospheric pressure. As the back pressure decreases, the valve shuts off the atmospheric bleed, and the output pressure approaches the supply pressure. The relay is said to be reverse acting, because an increase in back pressure produces a decrease in output.

The output pressure from the relay

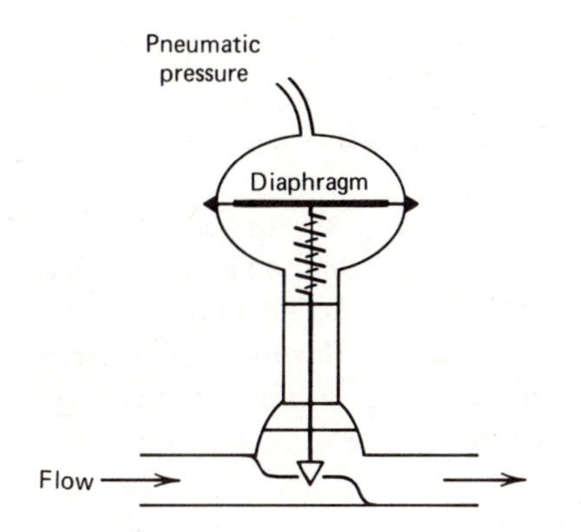

FIGURE 6.25 Pneumatic flow control valve.

can be used to drive a final control element like the pneumatic actuating valve shown in Figure 6.25. The pneumatic pressure acts on the upper side of the diaphragm and is opposed by the return spring.

The nozzle flapper is a *force distance* type of actuator, because its input is a displacement and its output a force (pressure). The other type of pneumatic actuator is the *force balance* type, in which both the input and output are pressures. The pressures are made to act across diaphragms whose resulting motion activates a pneumatic relay. Details can be found in the specialized literature (e.g., Reference 7).

6.5 CONTROL LAWS

The control logic elements are designed to act on the actuating (error) signal to produce the control signal. The algorithm that is physically implemented for this purpose is the *control law* or *control action.* A non zero error signal results from either a change in command or a disturbance. The general function of the controller is to keep the controlled variable near its desired value when these occur. More specifically, the control objectives might be stated as follows:

1. Minimize the steady-state error.
2. Minimize the settling time.
3. Achieve other transient specifications, such as minimizing the maximum overshoot.

In practice, the design specifications for a controller are more detailed. For example, the bandwidth might also be specified along with a safety margin for stability. We never know the numerical values of the system's parameters with true certainty, and some controller designs can be more sensitive to such parameter uncertainties than other designs. So a parameter sensitivity specification might also be included. We will return to this topic later, but for now, a general understanding of control objectives is sufficient for our purpose.

The following two control laws form the basis of many control systems.

Two-Position Control

Two-position control is the most familiar type perhaps because of its use in home thermostats. The control output takes on one of two values. With the *on–off* controller, the controller output is either on or off (fully open or fully closed). Such is the case with the thermostat furnace system. The controller output is determined by

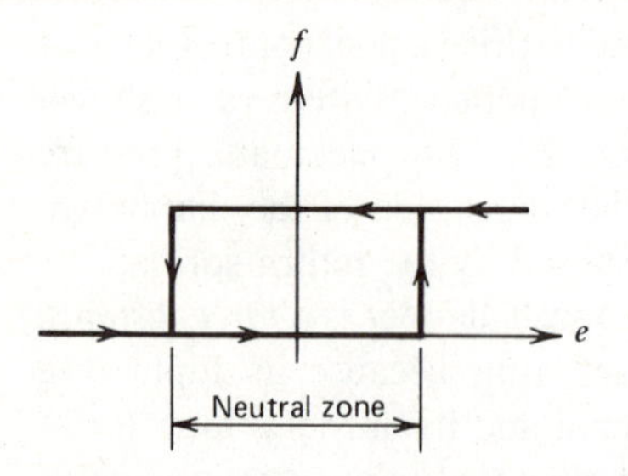

FIGURE 6.26 Transfer characteristics of the on–off controller. The actuating error is $e = r - c$, where r = set point, c = controlled variable, and f = control signal.

the magnitude of the error signal. The switching diagram for the on–off controller with hysteresis is shown in Figure 6.26.

An example of an application of an on–off controller to a liquid level system is shown in Figure 6.27*a*. The time response shown in Figure 6.27*b* with a solid line is for an ideal system in which the control valve acts instantaneously. The controlled variable cycles with an amplitude that depends on the width of the neutral zone or gap. This zone is provided to prevent frequent on–off switching, or *chattering*, which can shorten the life of the device. The cycling frequency also depends on the time constant of the controlled process and the magnitude of the control signal.

In a real system, as opposed to ideal, the sensor and control valve will not respond instantaneously, but have their own time constants. The valve will not close at the instant the height reaches the desired level. There will be some delay during which flow continues into the tank. The result is shown by the dotted line in Figure 6.27*b*. The opposite occurs when the valve is turned on. This unwanted effect can be reduced by decreasing the neutral zone, but the cycling frequency increases if this is done.

The overshoot and undershoot in on–off control will be acceptable only when the time constant of the process is large compared to the time lag of the control elements. This lag is related to the time constants of the elements as well as to their distance from the plant. If the control valve in Figure 6.27*a* is far upstream from the tank, a significant lag can exist between the time of control action and its effect on the plant. Another source of time lag is the capacitance of the controller itself. For example, if the heater capacitance in the temperature controller of Section 6.2 is appreciable, the heater will continue to deliver energy to the oven even after it has been turned off.

An example close to home demonstrates how the capacitance of the plant affects

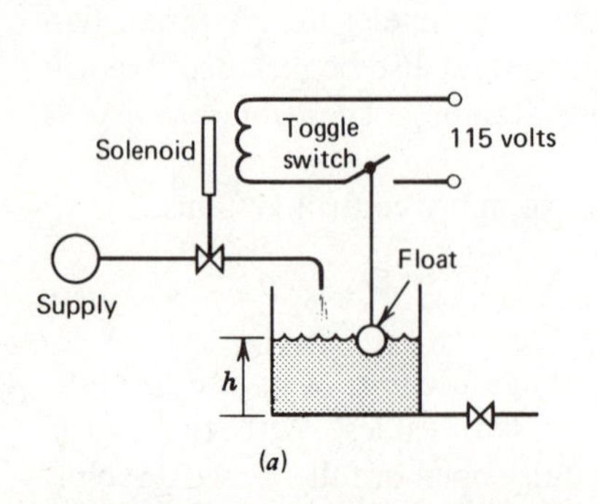

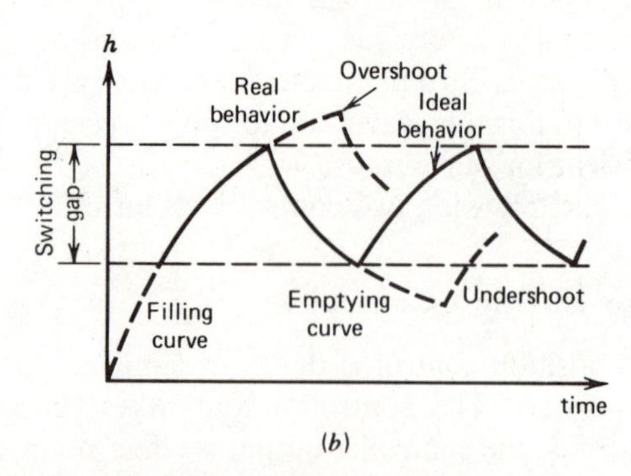

FIGURE 6.27 (*a*) Liquid level control with on–off action. (*b*) Time response.

the suitability of on–off control. On–off control of the hot water valve in a shower will obviously be unsuitable, but it is acceptable for a bath, because the thermal capacitance is greater.

Another type of two-position control is the *bang–bang* controller whose switching diagram is shown in Figure 6.28*a*. This controller is distinguished from on–off control by the fact that the direction or sign of the control signal can have two values. A motor with constant torque that can reverse quickly might be modeled as a bang–bang device. Because such perfect switching is impossible, a more accurate model would include a dead zone (Figure 6.28*b*). When the error is within the zone, the controller output is zero.

Proportional Control

Two-position control is acceptable for many applications in which the requirements are not too severe. In the home heating application, the typical 2°F temperature gap is hardly detectable by the occupants. Thus, the system is acceptable. However, many situations require finer control.

Consider the tank system shown in Figure 6.27*a*. To replace the two-position controller, we might try setting the control valve manually to achieve a flow rate that balances the system at the desired level. We might then add a controller that adjusts this setting in proportion to the deviation of the level from the desired value. This is *proportional control*, the algorithm in which the change in the control signal is proportional to the error. Recall the convention that block diagrams for controllers are drawn in terms of the deviations from a zero-error equilibrium condition. Applying this convention to the general terminology in Figure 6.7, we see that proportional control is described by

$$F(s) = K_p E(s) \tag{6.5-1}$$

where $F(s)$ is the deviation in the control signal and K_p is the *proportional gain.* If the total valve displacement is $y(t)$ and the manually created displacement is x, then

$$y(t) = K_p e(t) + x$$

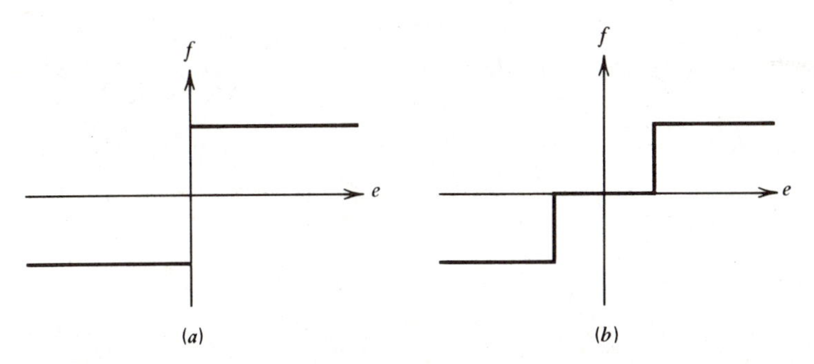

FIGURE 6.28 Transfer characteristics. (*a*) Ideal bang-bang control. (*b*) Bang-bang control with a dead zone. The control signal is *f*; the error signal is *e*.

The percent change in error needed to move the valve full scale is the *proportional band.* It is related to the gain as follows:

$$K_p = \frac{100}{\text{band } \%} \tag{6.5-2}$$

The zero-error valve displacement x is the *manual reset.*

Proportional Control of a First-Order System

To investigate the behavior of proportional control, consider the speed control system shown in Figure 6.29; it is identical to the position controller shown in Figure 6.4 except that a tachometer replaces the feedback potentiometer. A linear differential amplifier produces an output proportional to the difference between the input voltages. If the power amplifier is also linear, we can combine their gains into one, denoted K_p. The system is thus seen to have proportional control in which the motor voltage is proportional to the difference between the command voltage and the feedback voltage from the tachometer.

Assume that the motor is field-controlled with a negligible electrical time constant, so that the models (6.4-4) and (6.4-5) apply. The disturbance is a torque T_d, for example, resulting from friction. Choose the reference equilibrium condition to be $T_d = T = 0$ and $\omega_r = \omega = 0$. The block diagram is shown in Figure 6.30. For a meaningful error signal to be generated, K_1 and K_2 should be chosen to be equal. With this simplification, the diagram becomes that shown in Figure 6.31, where $K = K_1 K_p K_T / R$.

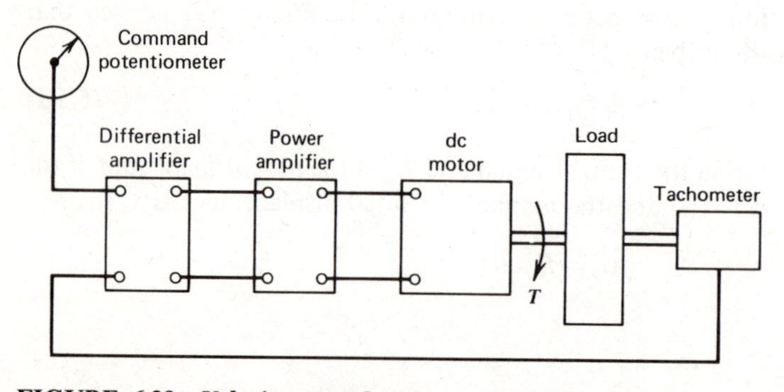

FIGURE 6.29 Velocity control system using a dc motor.

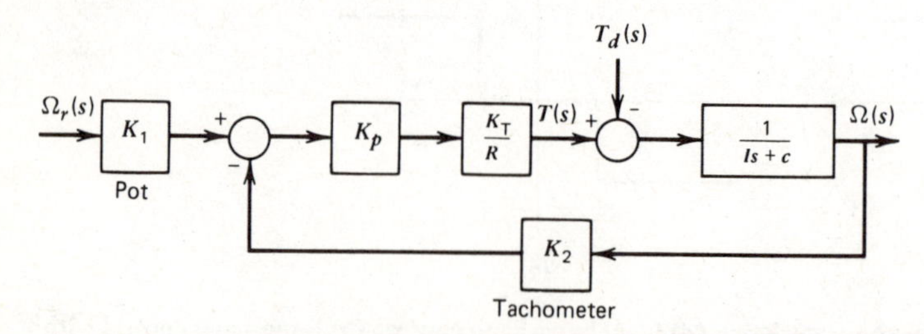

FIGURE 6.30 Block diagram of the velocity control system in Figure 6.29.

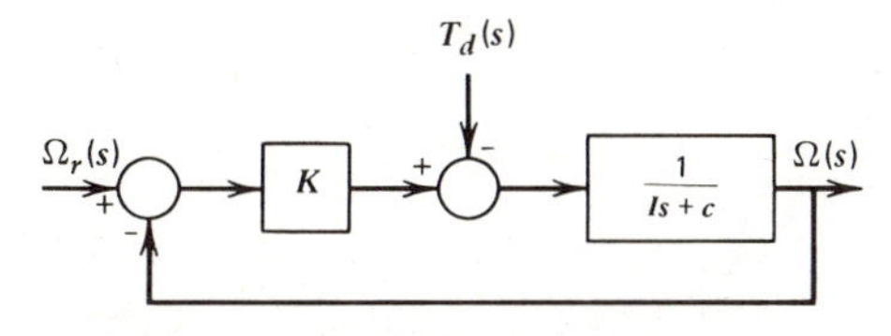

FIGURE 6.31 Simplified form of Figure 6.30 for the case $K_1 = K_2$; $K = K_1 K_p K_T / R$.

The transfer functions are

$$\frac{\Omega(s)}{\Omega_r(s)} = \frac{K}{Is + c + K} \tag{6.5-3}$$

$$\frac{\Omega(s)}{T_d(s)} = \frac{-1}{Is + c + K} \tag{6.5-4}$$

A change in desired speed can be simulated by a step input for ω_r. Linearity allows us to use a unit step and scale the results accordingly. For $\Omega_r(s) = 1/s$,

$$\Omega(s) = \frac{K}{Is + c + K} \frac{1}{s}$$

The response can be computed by partial fraction expansion as before. The velocity approaches the steady-state value

$$\omega_{ss} = \lim_{s \to 0} s \frac{K}{Is + c + K} \frac{1}{s} = \frac{K}{c + K} < 1$$

Thus, the final value is less than the desired value of 1, but it might be close enough if the damping c is small. The time required to reach this value is approximately four time constants, or $4\tau = 4I/(c + K)$.

A sudden change in load torque can also be modeled by a unit-step function $T_d(s) = 1/s$. The response due solely to the disturbance is found from (6.5-4).

$$\Omega(s) = \frac{-1}{Is + c + K} \frac{1}{s} \tag{6.5-5}$$

The steady-state effect of the disturbance is found with the final value theorem to be $-1/(c + K)$. If $(c + K)$ is large, this error will be small.

The performance of the proportional control law thus far can be summarized as follows. For a first-order system whose inputs are step functions,

1. The output never reaches its desired value even in the absence of a disturbance if resistance is present ($c \neq 0$), although it can be made arbitrarily close by choosing the gain K large enough. This is *offset* error.
2. The output approaches its final value without oscillation. The time to reach this value is inversely proportional to K.
3. The output error due to the disturbance at steady state is inversely proportional to the gain K. This error is present even in the absence of resistance ($c = 0$).

Figure 6.32 shows the two types of steady-state error that can exist in this system. In the example shown, we wish the speed to be $\omega = 1$. Thus, $\Omega_r(s) = 1/s$. In Figure 6.32*a*, no disturbance is acting, and the offset error is as shown. In 6.32*b*, we temporarily set $\Omega_r(s) = 0$ to see the effect of the disturbance. The speed shown is for a unit-step disturbance $T_d(s) = 1/s$. The steady-state error due to the disturbance is

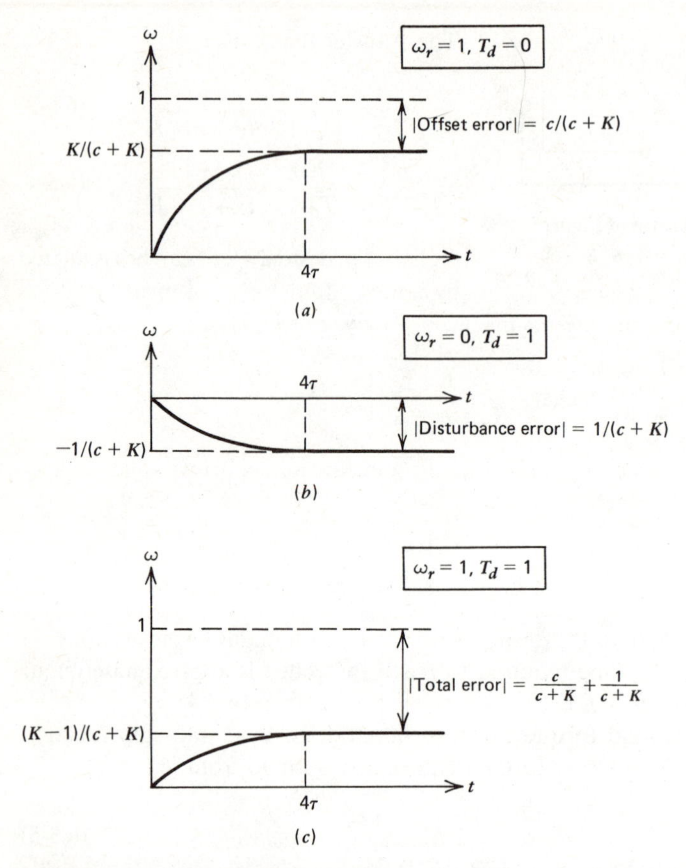

FIGURE 6.32 Comparison of two sources of error in a first-order system with proportional control. (*a*) System response to a unit step reference input, with no disturbance. (*b*) System response to a unit step disturbance with zero reference input. (*c*) Total response when both the reference input and the disturbance are unit step functions.

$-1/(c+K)$. If *both* inputs act simultaneously on the system, the actual speed behaves as shown in 6.32*c*; it is the sum of the speed curves shown in 6.32*a* and 6.32*b*. The total error is the sum of the offset error and the error due to the disturbance.

As the gain K is increased, the time constant becomes smaller, and the response faster. Thus, the chief disadvantage of proportional control is that it results in steady-state errors and can be used only when the gain can be selected large enough to reduce the effect of the largest expected disturbance. Since proportional control gives zero error for only one load condition (the reference equilibrium), the operator must change the manual reset by hand (hence the name).

An advantage to proportional control is that the control signal responds to the error instantaneously (in theory at least). It is used in applications requiring rapid action. Processes with time constants too small for the use of two-position control are likely candidates for proportional control.

The results of this analysis can be applied to any first-order system of the form in

Figure 6.31. The temperature controller and the liquid level controller in Section 6.2 are proportional controllers, and their block diagrams can easily be put into this form.

Example 6.1

Suppose the plant shown in Figure 6.31 has $I = 2$ and $c = 3$. Find the gain K required to give a maximum steady-state offset error of 0.2 if ω_r is a unit-step input. Evaluate the resulting time constant and steady-state error due to the disturbance if T_d is also a unit step.

The steady-state offset error is

$$\text{offset error} = \omega_{ss} - \omega_r = \frac{K}{3+K} - 1 = -0.2$$

Therefore, $K = 12$. The time constant is

$$\tau = \frac{I}{c+K} = \frac{2}{3+K} = \frac{2}{15}$$

and the steady-state error due to the disturbance is

$$\text{disturbance error} = \frac{-1}{c+K} = \frac{-1}{15} = -0.067$$

If *both* inputs are applied, the actual steady-state speed is

$$\begin{aligned}\omega_{ss} &= \omega_r + \text{offset error} + \text{disturbance error}\\ &= 1 - 0.2 - 0.067 = 0.733\end{aligned}$$

Feedforward Compensation

From Figure 6.32*a*, we see that it is possible to "fool" the controller into giving a zero steady-state offset error by giving the controller a reference input ω_r that is higher than the desired speed value. In particular, if we give the controller a command input that is $(c + K)/K$ times higher than the desired speed, the steady-state offset error will be zero (see Figure 6.33*a*). This introduces the principle of *input compensation.* We can design the controller to automatically adjust itself to eliminate the offset error if we use the arrangement shown in Figure 6.33*b*, where the gain K_f is chosen to eliminate the offset error. This form of input compensation is called feedforward compensation (see Figure 6.8*a*). It is equivalent to multiplying a step input by $(c + K)/K$, but we will see later that it has the additional advantage of being able to reduce errors due to ramp inputs as well. The primary transfer function can be found from (6.2-4). It is

$$\frac{\Omega(s)}{\Omega_r(s)} = \frac{K_f + K}{Is + c + K} \tag{6.5-6}$$

For a unit-step input, $\omega_{ss} = (K_f + K)/(c + K)$. Thus, if we choose $K_f = c$, then $\omega_{ss} = 1$, and the offset error is zero. Note that the feedforward compensation does not affect the disturbance response [the disturbance transfer function is the same as that given by (6.5-4)].

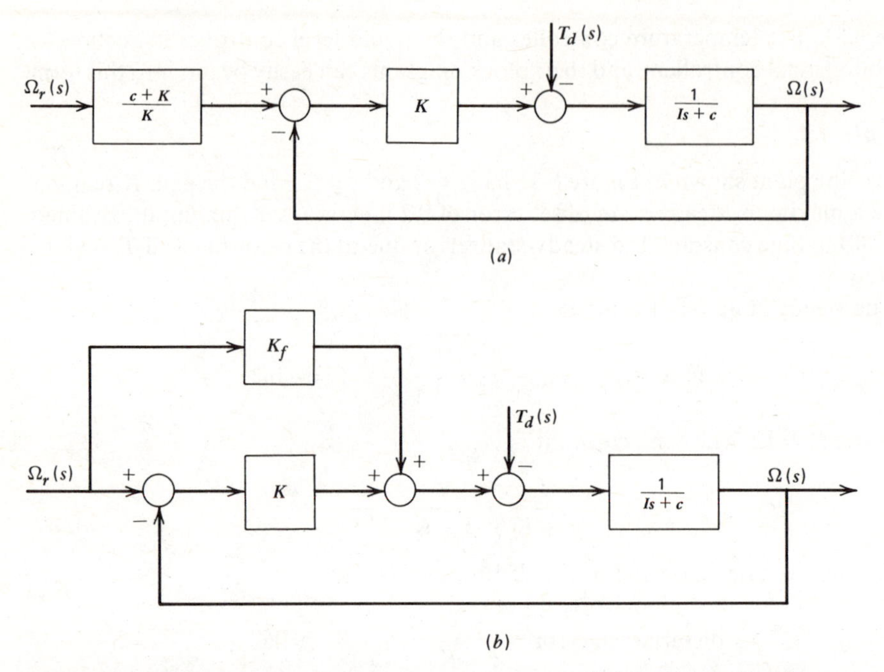

FIGURE 6.33 Input compensation to augment proportional control. (*a*) This arrangement eliminates error with a step input. (*b*) Feedforward compensation eliminates the errors due to both step and ramp inputs.

The effectiveness of feedforward compensation depends on how accurately we know the value of c, because we must set $K_f = c$. Usually, the damping coefficient c is difficult to determine precisely, and therefore feedforward compensation is not widely used by itself. We will see in later sections that there are better methods for eliminating the offset error. Sometimes feedforward compensation is used in combination with these methods.

Proportional Control with Ramp Inputs

A command input that is a ramp function is a good test of a control system's ability to follow a constantly changing command. Although the ramp function continues to increase without limit, in practice we are interested in the system's response only until the transient response disappears. The reason for this is illustrated by Figure 6.34, which is a typical rotational-speed profile that represents the desired speed of a computer tape reel, for example. In order to reach a certain part of the tape as quickly as possible, the reel should move at its maximum speed ω_m. But the reel can accelerate and decelerate no faster than some rate α due to motor torque and tape stress limitations.

In order to follow such a command input, the controller must be able to deal with both step and ramp commands (the step command corresponds to the constant speed ω_m). The performance specifications might be given as follows: The speed ω must be within $\omega_m \pm a$ for $t_1 \leqslant t \leqslant t_2$, and ω must satisfy $0 \leqslant \omega(t_3) \leqslant b$, where t_1, t_2, t_3, a, and b are given numbers (see Figure 6.34).

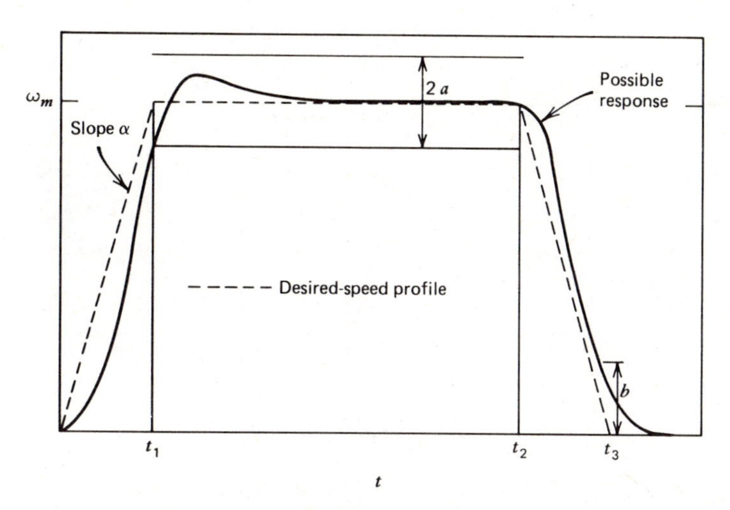

FIGURE 6.34 A typical desired speed profile and accuracy specifications.

Let us see whether proportional control could satisfy these specifications. With the feedforward arrangement shown in Figure 6.33*b*, the error relation can be found from (6.2-6) with $T_d(s) = 0$. It is

$$E(s) = \frac{Is + c - K_f}{Is + c + K}\Omega_r(s) \tag{6.5-7}$$

For $\Omega_r(s) = \alpha/s^2$, the steady-state error is

$$e_{ss} = \begin{cases} \infty, & K_f \neq c \\ \dfrac{\alpha I}{c + K}, & K_f = c \end{cases}$$

Although the predicted e_{ss} is ∞ for $K_f \neq c$, the positive-slope ramp input in this application does not exist for $t > t_1$. We could make $|e_{ss}(t_1)| \leqslant a$ as required, with feedforward compensation, by proper choice of K. However, this requires a trial-and-error solution method to find K (find the ramp response in closed-form to see this). If we know c accurately enough, the simplest way is to set $K_f = c$ and choose K such that

$$e_{ss} = \frac{\alpha I}{c + K} = a$$

Thus, $K = (\alpha I - ac)/a$ will give $|e(t_1)| \leqslant a$ as required. It remains to be seen whether the rest of the specifications will be satisfied. The response to the speed profile can be found analytically or by simulation with the computed values of K and K_f.

As an illustration, suppose that $I = c = 1$, $\alpha = 100$, $t_1 = 0.1$, $t_2 = 0.5$, and $t_3 = 0.6$. If we require that $a = b = 1$ and $\tau = 0.01$, then the necessary gains are $K = 99$ and $K_f = 1$. The desired speed profile and the resulting response are shown in Figure 6.35, from which it can be seen that the specifications have been satisfied.

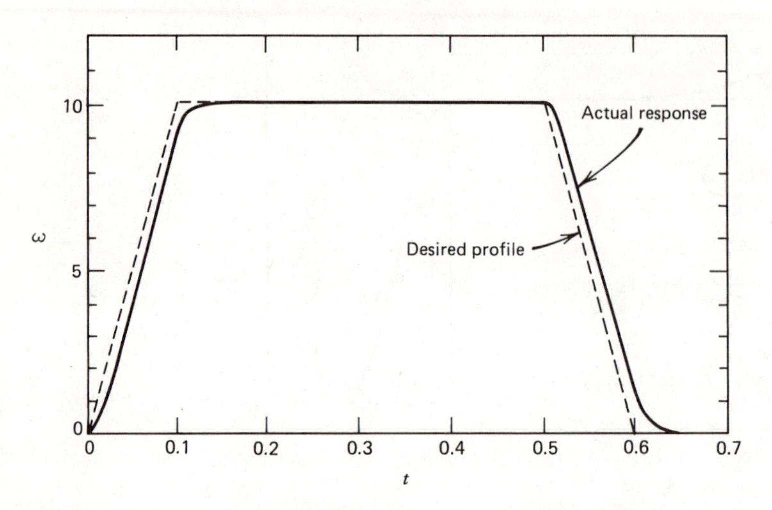

FIGURE 6.35 Comparison of the desired speed profile with the controller's actual response.

Proportional Control of a Second-Order System

Proportional control of a neutrally stable second-order plant is represented by the position controller in Figure 6.6 if the amplifier transfer function is a constant $G_a(s) = K_a$. Let the motor transfer function be $G_m(s) = K_T/R$ as before. The modified block diagram is given in Figure 6.36, with $K = K_1 K_a K_T/R$. The transfer functions are

$$\frac{\Theta(s)}{\Theta_r(s)} = \frac{K}{Is^2 + cs + K} \tag{6.5-8}$$

$$\frac{\Theta(s)}{T_d(s)} = \frac{-1}{Is^2 + cs + K} \tag{6.5-9}$$

The closed-loop system is stable if I, c, and K are positive. For no damping ($c = 0$), the closed-loop system is neutrally stable.

With no disturbance and a unit-step command, $\Theta_r(s) = 1/s$, the steady-state output is

$$\theta_{ss} = \frac{K}{K} = 1$$

The offset error is thus zero if the system is stable ($c > 0$, $K > 0$). The steady-state output deviation due to a unit-step disturbance is $-1/K$. This deviation can be reduced by choosing K large.

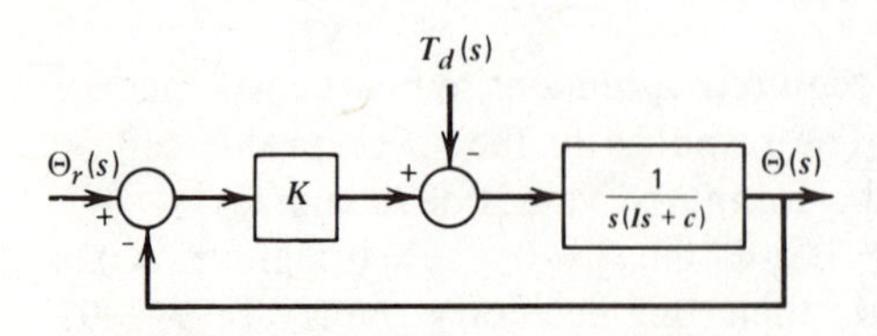

FIGURE 6.36 Position servo with proportional control.

Note that the offset error for proportional control of a second-order system is not always zero. If the plant's transfer function were instead $1/(Is^2 + cs + k)$, then the offset error would be $k/(k + K)$. The zero offset error in our particular example occurs because $k = 0$, which indicates that the plant transfer function contains a pure integrator (this integrator is shown in factored form in Figure 6.36).

The transient behavior is indicated by the damping ratio.

$$\zeta = \frac{c}{2\sqrt{IK}}$$

For slight damping, the response to a step input will be very oscillatory and the overshoot large. The situation is aggravated if the gain K is made large to reduce the deviation due to the disturbance.

The steady-state error of this system for a unit-ramp command input is $e_{ss} = c/K$. Thus, if c is large, the system is not very oscillatory, but the ramp error is large. For a ramp disturbance, $e_{ss} = \infty$. We conclude therefore that proportional control of this type of second-order plant is not a good choice. We will see shortly how to improve the design. The reader is encouraged to investigate the performance of feedforward compensation with this system.

6.6 INTEGRAL CONTROL

The offset error that occurs with proportional control is a result of the system reaching an equilibrium in which the control signal no longer changes. This allows a constant error to exist. If the controller is modified to produce an increasing signal as long as the error is nonzero, the offset might be eliminated. This is the principle of *integral control.* In this mode, the change in the control signal is proportional to the *integral* of the error. In the terminology of Figure 6.7, this gives

$$F(s) = \frac{K_I}{s} E(s) \tag{6.6-1}$$

where $F(s)$ is the deviation in the control signal and K_I is the *integral gain.* In the time domain, the relation is

$$f(t) = K_I \int_0^t e(t)\, dt \tag{6.6-2}$$

if $f(0) = 0$. In this form, it can be seen that the integration cannot continue indefinitely, because it would theoretically produce an infinite value of $f(t)$. This implies that special care must be taken to reinitialize the controller. This requirement is considered later.

Integral Control of a First-Order System

Integral control of the velocity in the system in Figure 6.29 has the block diagram shown in Figure 6.37, where $K = K_1 K_I K_T / R$. The integrating action of the amplifier is physically obtained by the techniques to be presented in Section 6.8. The closed-

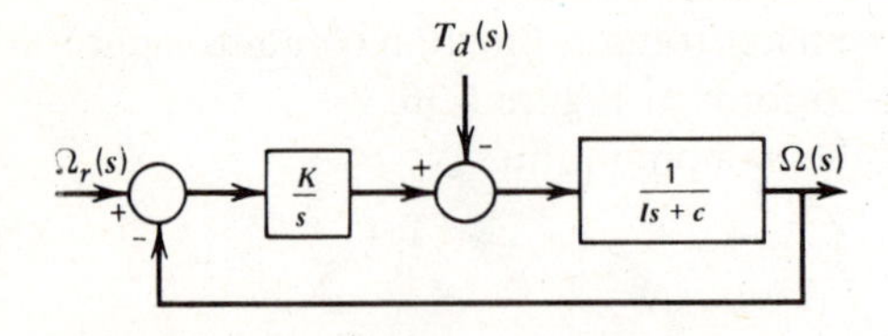

FIGURE 6.37 Velocity servo with integral control.

loop transfer functions are

$$\frac{\Omega(s)}{\Omega_r(s)} = \frac{K}{Is^2 + cs + K} \tag{6.6-3}$$

$$\frac{\Omega(s)}{T_d(s)} = \frac{-s}{Is^2 + cs + K} \tag{6.6-4}$$

The control system is stable for I, c, and K positive. For a unit-step command input, $\omega_{ss} = K/K = 1$; so the offset error is zero. For a unit-step disturbance, the steady-state deviation is zero if the system is stable. The steady-state performance using integral control is thus excellent for this plant with step inputs.

The damping ratio is

$$\zeta = \frac{c}{2\sqrt{IK}}$$

For slight damping, the response will be oscillatory rather than exponential as with proportional control. Improved steady-state performance has thus been obtained at the expense of degraded transient performance. The conflict between steady-state and transient specifications is a common theme in control system design. As long as the system is underdamped, the time constant is $\tau = 2I/c$ and not affected by the gain K, which only influences the oscillation frequency in this case. It might be physically possible to make K small enough so that $\zeta \geqslant 1$, and the nonoscillatory feature of proportional control recovered, but the response would tend to be sluggish. Transient specifications for fast response often require $\zeta < 1$. The difficulty with $\zeta < 1$ is that τ is fixed by c and I. If c and I are such that $\zeta < 1$, then τ is large if $I \gg c$.

Integral Control of a Second-Order System

Proportional control of the position servomechanism in Figure 6.36 gives a nonzero steady-state deviation due to the disturbance. Integral control applied to this system results in the block diagram in Figure 6.38 and the transfer functions

$$\frac{\Theta(s)}{\Theta_r(s)} = \frac{K}{Is^3 + cs^2 + K} \tag{6.6-5}$$

$$\frac{\Theta(s)}{T_d(s)} = \frac{-s}{Is^3 + cs^2 + K} \tag{6.6-6}$$

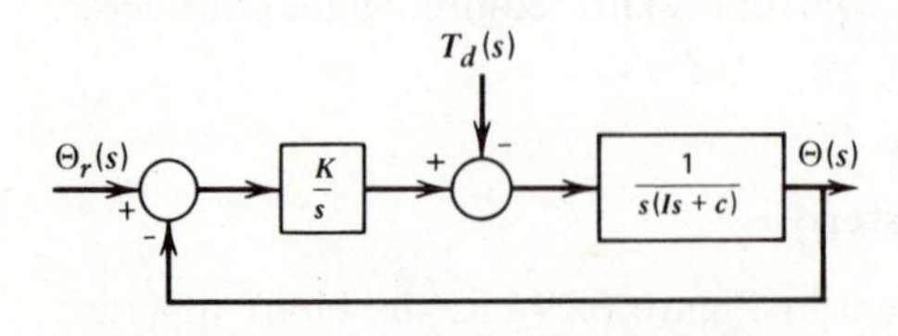

FIGURE 6.38 Position servo with integral control.

With the Routh criterion, we immediately see that the system is not stable because of the missing s term. Thus, the final value theorem cannot be applied.

Integral control is useful in improving steady-state performance, but in general it does not improve and may even degrade transient performance. Improperly applied, it can produce an unstable control system. It is best used in conjunction with other control modes.

Proportional-plus-Integral Control

Integral control raised the order of the system by one in the preceding examples but did not give a characteristic equation with enough flexibility to achieve acceptable transient behavior. The instantaneous response of proportional-control action might introduce enough variability into the coefficients of the characteristic equation to allow both steady-state and transient specifications to be satisfied. This is the basis for using *proportional-plus-integral control* (PI control). The algorithm for this two-mode control is

$$F(s) = K_p E(s) + \frac{K_I}{s} E(s) \tag{6.6-7}$$

The integral action provides an automatic, not manual, reset of the controller in the presence of a disturbance. For this reason, it is often called *reset action.*

The algorithm is sometimes expressed as

$$F(s) = K_p\left(1 + \frac{1}{T_I s}\right) E(s) \tag{6.6-8}$$

where T_I is the *reset time.* The reset time is the time required for the integral action signal to equal that of the proportional term if a constant error exists (a hypothetical situation). The reciprocal of reset time is expressed as repeats per minute and is the frequency with which the integral action repeats the proportional-correction signal.

The proportional-control gain must be reduced when used with integral action. The integral term does not react instantaneously to a zero-error signal but continues to correct, which tends to cause oscillations if the designer does not take this effect into account.

PI Control of a First-Order System

Proportional-plus-integral-control action applied to the speed controller in Figure 6.29 gives the diagram shown in Figure 6.39. The transfer functions are

$$\frac{\Omega(s)}{\Omega_r(s)} = \frac{K_p s + K_I}{Is^2 + (c + K_p)s + K_I} \tag{6.6-9}$$

$$\frac{\Omega(s)}{T_d(s)} = \frac{-s}{Is^2 + (c + K_p)s + K_I} \tag{6.6-10}$$

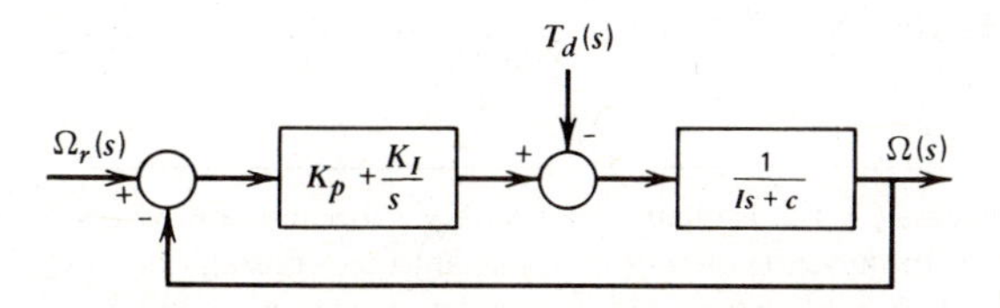

FIGURE 6.39 Velocity servo with PI control.

where the gains K_p and K_I are related to the component gains as before. The system is stable for positive values of K_p and K_I. For $\Omega_r(s) = 1/s$, $\omega_{ss} = K_I/K_I = 1$, and the offset error is zero, as with only integral action. Similarly, the deviation due to a unit-step disturbance is zero at steady state. The damping ratio is

$$\zeta = \frac{c + K_p}{2\sqrt{IK_I}} \tag{6.6-11}$$

The presence of K_p allows the damping ratio to be selected without fixing the value of the dominant time constant. For example, if the system is underdamped, the time constant is†

$$\tau = \frac{2I}{c + K_p}, \qquad (\zeta < 1) \tag{6.6-12}$$

The gain K_p can be picked to obtain the desired time constant, while K_I is used to set the damping ratio. A similar flexibility exists if $\zeta = 1$. Complete description of the transient response requires the numerator dynamics present in the transfer functions to be accounted for (see Section 4.9).

Example 6.2

Suppose that $I = c = 1$ and the performance specifications require that $\tau = 0.2$. (a) Find the gain values for PI-control such that:

1. $\zeta = 0.707$

2. $\zeta = 1$.

(b) Discuss the design for $\zeta > 1$ and the unit-step response for all three cases.

(a) For both cases (1) and (2), (6.6-12) gives

$$\tau = \frac{2}{1 + K_p} = 0.2$$

Therefore, $K_p = 9$. For case (1), (6.6-11) gives

$$\zeta = 0.707 = \frac{10}{2\sqrt{K_I}}$$

Thus, $K_I = 50$. For case (2), the same procedure gives $K_I = 25$.

(b) Since $\zeta > 1$ is required, we cannot use (6.6-12) to find τ. There will be two real roots, and the dominant root must be $s = -1/\tau = -5$. For this to be the dominant root, the second root should be far to the left. Choosing a separation factor of 10, we

---

†Although (6.6-12) applies strictly to only the case $\zeta < 1$, it is often used when $\zeta = 1$, because it provides a quick but rough estimate of the time constant. In this case, there is no dominant time constant, because both roots are equal, and predictions such as the 2% settling time that are based on τ are not so accurate as they are for $\zeta < 1$ or for first-order systems.

place the secondary root at $s = -50$. The characteristic polynomial must therefore be

$$(s+5)(s+50) = s^2 + 55s + 250 = 0$$

Comparing this with the denominator in (6.6-9) gives equations for the required gain values.

$$1 + K_p = 55$$
$$K_I = 250$$

Thus, $K_p = 54$.

Figure 6.40 shows the unit-step response for the three designs. The following table summarizes the important response characteristics. The overshoot decreases as ζ increases, as expected. But the 10–90% rise time defined in Figure 3.13 is the smallest for $\zeta = 1.74$. A common misconception is that response is sluggish for $\zeta > 1$, but this is clearly not the case here. The gains $K_p = 54$, $K_I = 250$ give a good response with a very small rise time and a small overshoot. However, high gain values are required, and this might make the physical system expensive or difficult to build.

In general, the larger the root separation factor, the larger ζ is (see Table 3.4). But the large root separation factor is not the cause of the fast response. Rather, it is due to the numerator dynamics, whose effect is increased by the high gain values used in the third case. This effect can be seen by comparing the actual overshoot and rise time with those predicted by the second-order model without numerator dynamics (see Section 3.5). These values are given in the following table.

						Without Numerator Dynamics	
Case	K_p	K_I	ζ	**Overshoot (%)**	**Rise Time (10–90%)**	**Overshoot (%)**	**Rise Time (10–90%)**
1	9	50	0.707	16	0.14	4	0.36
2	9	25	1	8	0.18	0	0.66
3	54	250	1.74	5	0.03	0	0.44

In summary, Example 6.2 shows that the numerator dynamics introduced by PI control produce larger overshoots but faster rise times, as compared to a system without numerator dynamics. The *I* action is responsible for the numerator dynamics [set $K_I = 0$ in (6.6-9) and (6.6-10), and cancel the s term to see this]. But the I action is desirable, because it gives zero steady-state error for step inputs. In Chapter 7, we will see other control laws that have this error response but produce no overshoot for $\zeta \geq 1$. However, the ramp response characteristics are not so good as those with PI control. The choice must always be made on the basis of the particular performance specifications given.

If a unit-ramp command input is applied, the steady-state error is $e_{ss} = c/K_I$, which is zero only if no damping is present. For a unit-ramp disturbance, $e_{ss} = 1/K_I$

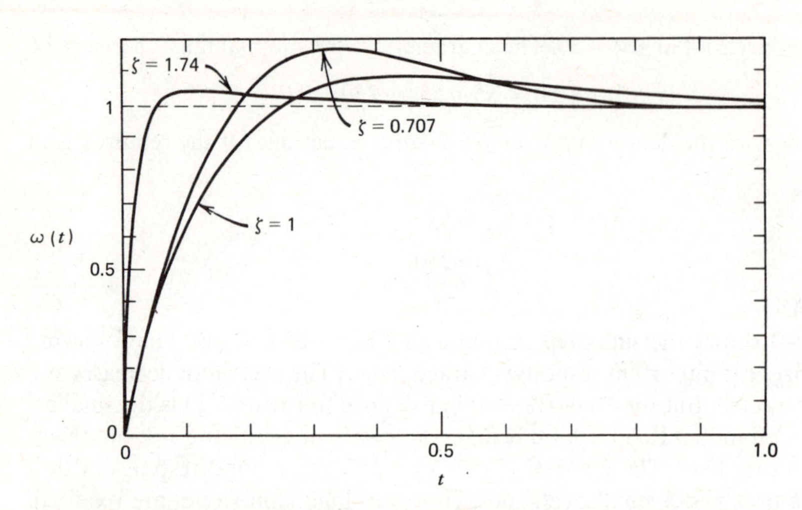

FIGURE 6.40 Unit-step responses for the designs obtained in Example 6.2.

even if damping is absent. The reader should verify these statements and investigate the performance of feedforward compensation when used to augment PI control.

PI Control of a Second-Order System

Integral control for the position servomechanism in Figure 6.38 resulted in a third-order system that is unstable. With a proportional term, the diagram becomes that in Figure 6.41, with the transfer functions

$$\frac{\Theta(s)}{\Theta_r(s)} = \frac{K_p s + K_I}{Is^3 + cs^2 + K_p s + K_I} \tag{6.6-13}$$

$$\frac{\Theta(s)}{T_d(s)} = \frac{-s}{Is^2 + cs^2 + K_p s + K_I} \tag{6.6-14}$$

The steady-state performance is acceptable as before if the system is assumed to be stable. This is true if the Routh criterion is satisfied; that is, if I, c, K_p, and K_I are positive and $cK_p - IK_I > 0$. The difficulty here occurs when the damping is slight. For small c, the gain K_p must be large in order to satisfy the last condition, and this can be difficult to implement physically. Such a condition can also result in an unsatisfactory time constant, but a detailed analysis of the cubic equation is beyond our scope now. The root locus method of Chapter 8 provides the tools for analyzing this design further.

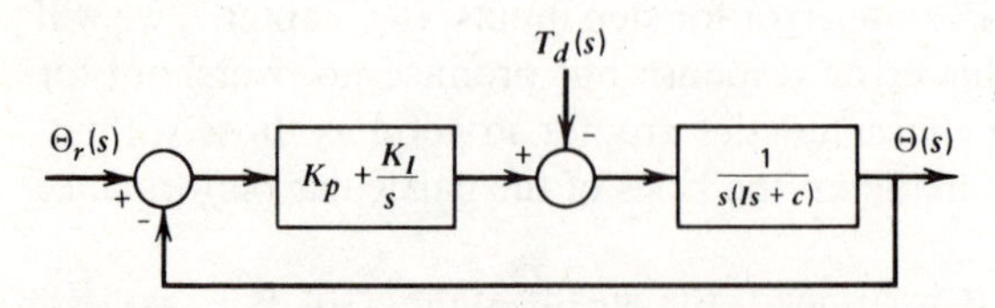

FIGURE 6.41 Position servo with PI control.

6.7 DERIVATIVE CONTROL

Integral action tends to produce a control signal even after the error has vanished, which suggests that the controller should be made aware that the error is approaching zero. One way to accomplish this is to design the controller to react to the derivative of the error with the *derivative control* law.

$$F(s) = K_D s E(s) \tag{6.7-1}$$

where K_D is the *derivative gain.* This algorithm is also called *rate action.* It is used to damp out oscillations.

Since it depends on only the error rate, derivative control should never be used alone. When used with proportional action, the following PD-control algorithm results.

$$\begin{aligned} F(s) &= (K_p + K_D s)E(s) \\ &= K_p(1 + T_D s)E(s) \end{aligned} \tag{6.7-2}$$

where T_D is the *rate time* or *derivative time.* With integral action included, the proportional-plus-integral-plus-derivative (PID) control law is obtained.

$$F(s) = \left(K_p + \frac{K_I}{s} + K_D s\right)E(s) \tag{6.7-3}$$

This is a *three-mode* controller.

PD Control of a Second-Order System

Designing a controller with all three modes increases the cost of the system (except perhaps for digital systems, where the only change is a software modification). There are applications of the position servomechanism in which a nonzero deviation resulting from the disturbance can be tolerated, but an improvement in transient response over the proportional control result is desired. Integral action would not be required, and rate action can be substituted to improve the transient response. Application of PD control to this system gives the block diagram in Figure 6.42 and the following transfer functions:

$$\frac{\Theta(s)}{\Theta_r(s)} = \frac{K_p + K_D s}{Is^2 + (c + K_D)s + K_p} \tag{6.7-4}$$

$$\frac{\Theta(s)}{T_d(s)} = \frac{-1}{Is^2 + (c + K_D)s + K_p} \tag{6.7-5}$$

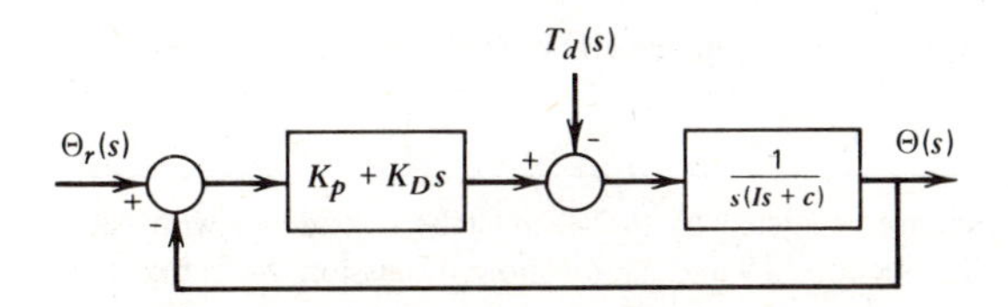

FIGURE 6.42 Position servo with PD control.

The system is stable for positive values of K_D and K_p. The presence of rate action does not affect the steady-state response for step inputs, and the steady-state results are identical to those with P control; namely, zero offset error and a deviation of $-1/K_p$ due to the disturbance. The damping ratio is

$$\zeta = \frac{c + K_D}{2\sqrt{IK_p}}$$

For P control, $\zeta = c/2\sqrt{IK_p}$. Introducing rate action allows the proportional gain K_p to be selected large to reduce the steady-state deviation, while K_D can be used to achieve an acceptable damping ratio. The rate action also helps to stabilize the system by adding damping (if $c = 0$, the system with P control is not stable).

The feasibility of constructing a differentiating device is contradicted by the principle of integral causality (Chapter 2)†. However, in the next section, techniques are presented for obtaining an approximation of such a device. For now, we note that in the present example, the equivalent of derivative action can be obtained by using a tachometer to measure the angular velocity of the load. The block diagram is shown in Figure 6.43. The gain of the amplifier–motor–potentiometer combination is K_1, and K_2 is the tachometer gain. The transfer functions are

$$\frac{\Theta(s)}{\Theta_r(s)} = \frac{K_1}{Is^2 + (c + K_1K_2)s + K_1} \tag{6.7-6}$$

$$\frac{\Theta(s)}{T_d(s)} = \frac{-1}{Is^2 + (c + K_1K_2)s + K_1} \tag{6.7-7}$$

Comparison with (6.7-4) and (6.7-5) shows that the system with tachometer feedback does not possess numerator dynamics. This system will therefore be somewhat more sluggish than the system with pure PD control. Otherwise, the tachometer feedback arrangement gives a similar characteristic equation. The gains K_1 and K_2 can be chosen to yield the desired damping ratio and steady-state deviation as was done with

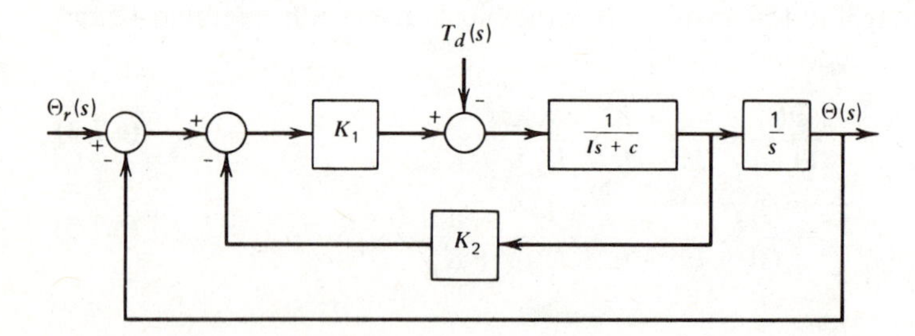

FIGURE 6.43 Tachometer feedback arrangement to replace PD control for the position servo.

†Therefore, the interpretation of the step response predicted by (6.7-4) must be considered with care, especially for the initial part of the response. See Sections 4.9 and 5.6 for more discussion. Note that this caution is not necessary in interpreting the step response of a controller having I action even though it has numerator dynamics, because a step input is not applied to a physical differentiator.

K_p and K_D. The tachometer—compensated system has a larger steady-state error in response to a ramp command, and both systems have an infinite steady-state error for a ramp disturbance.

PID Control

The position servomechanism design with PI control is not completely satisfactory because of the difficulties encountered when the damping c is small. This problem can be solved by using the full PID-control law. From Figure 6.44, the following transfer functions are derived:

$$\frac{\Theta(s)}{\Theta_r(s)} = \frac{K_D s^2 + K_p s + K_I}{Is^3 + (c + K_D)s^2 + K_p s + K_I} \tag{6.7-8}$$

$$\frac{\Theta(s)}{T_d(s)} = \frac{-s}{Is^3 + (c + K_D)s^2 + K_p s + K_I} \tag{6.7-9}$$

A stable system results if all gains are positive and if

$$(c + K_D)K_p - IK_I > 0 \tag{6.7-10}$$

The presence of K_D relaxes somewhat the requirement that K_p be large to achieve stability. The steady-state errors are zero, and the transient response can be improved, because three of the coefficients of the characteristic equation can be selected. To make further statements requires the analysis techniques in later chapters.

Proportional, integral, and derivative actions and their various combinations are not the only control laws possible, but they are the most common. It has been estimated that 90% of all controllers are of the PI type. This percentage will probably decrease as digital control with its great flexibility becomes more widely used. But the PI and PID controllers will remain for some time the standard against which any new designs must compete.

The conclusions reached concerning the performance of the various control laws are strictly true for only the plant model forms considered. These are the first-order model without numerator dynamics and the second-order model with a root at $s = 0$ and no numerator zeros.

The analysis of a control law for any other linear system follows the preceding pattern. The overall system transfer functions are obtained, and all of the linear system analysis techniques can be applied to predict the system's performance. If the performance is unsatisfactory, a new control law is tried and the process repeated. When this process fails to achieve an acceptable design, more systematic methods of altering the system's structure are needed; they appear in Chapters 7–9. Also, the

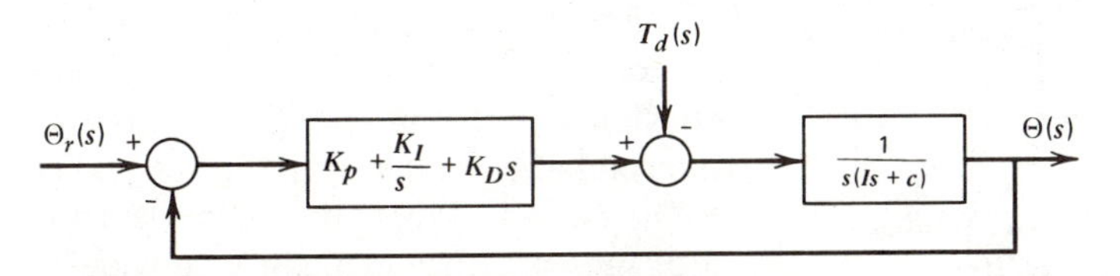

FIGURE 6.44 Position servo with PID control.

methods in Chapters 8 and 9, are needed if the design requires a detailed analysis of a model of third order or higher.

We have used step functions as the principal test signals, because they are the most common and perhaps represent the severest test of system performance. Impulse, ramp, and sinusoidal test signals are also employed. The type to use should be made clear in the design specifications.

6.8 ELECTRONIC CONTROLLERS

The control law must be implemented by a physical device before the control engineer's task is complete. The earliest devices were purely kinematic and mechanical elements, such as gears, levers, and diaphragms, that usually obtained their power from the controlled variable. Most controllers are now analog electronic, hydraulic, pneumatic, or digital electronic devices. We now consider the analog electronic type. Digital control is taken up in Chapters 10 and 11.

Feedback Compensation and Controller Design

Most controllers that implement versions of the PID algorithm are based on the following feedback principle. Consider the single-loop system shown in Figure 6.45. If the open-loop transfer function is large enough so that $|G(s)H(s)| \gg 1$, the closed-loop transfer function is approximately given by

$$T(s) = \frac{G(s)}{1 + G(s)H(s)} \cong \frac{G(s)}{G(s)H(s)} = \frac{1}{H(s)}$$

The principle states that a power unit $G(s)$ can be used with a feedback element $H(s)$ to create any desired transfer function $T(s)$. The power unit must have a gain high enough so that $|G(s)H(s)| \gg 1$, and the feedback elements must be selected so that $H(s) = 1/T(s)$. The latter task can sometimes require ingenious design.

This principle was used in Chapter 1 to explain the design of a feedback amplifier. The power unit is the uncompensated amplifier with a high gain of K. The feedback element is a resistance network (Figure 6.46). Thus, $G(s) = K$ and $H(s) = R_2/R_1$. If $KR_2/R_1 \gg 1$, then $T(s) \cong R_1/R_2$. This design eliminates the difficulty associated with the age and temperature variability of the gain K. The feedback amplifier gain R_1/R_2 is more stable but smaller than the original gain. Although the original problem was not a control problem, the design can be used as a proportional controller. The gain R_1/R_2 can be adjusted for a particular application if one resistance is a potentiometer.

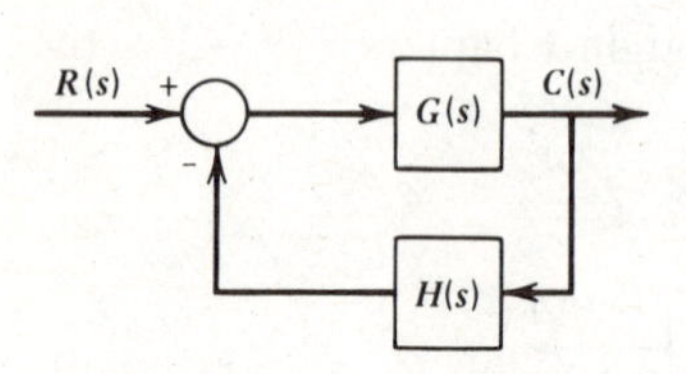

FIGURE 6.45 Principle of feedback compensation. $T(s) \cong 1/H(s)$ if $|G(s)H(s)| \gg 1$.

The feedback amplifier model is a static one in which $G(s)$ and $H(s)$ are constants. For this case, it is easy to interpret the meaning of the inequality $|G(s)H(s)| \gg 1$. When dynamic elements are present, the inequality obviously depends on the value of the Laplace

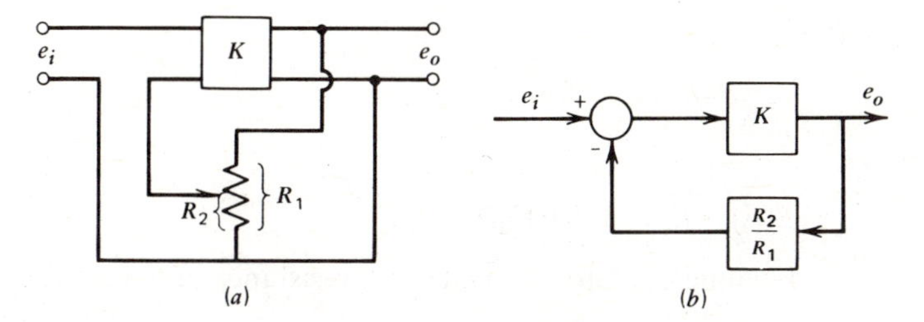

FIGURE 6.46 Feedback compensation of an amplifier.

operator s. In practice, the inequality is considered to be satisfied if it is true for the range of frequencies ω over which the device will be operated, where $s = i\omega$.

Op Amp Circuits

The op amp is a high-gain amplifier with a high-input impedance. In Chapter 1, a multiplying circuit was designed by making use of these properties. The procedure is generalized here for the purpose of designing PID controllers. The uncompensated amplifier is shown in Figure 6.47*a*, in which the sign reversal property is displayed. This will require using an inverter to maintain proper voltage signs.

A circuit diagram of the op amp with general feedback and input elements is shown in Figure 6.47*b*. A similar but simplified form is given in Figure 6.47*c*. The

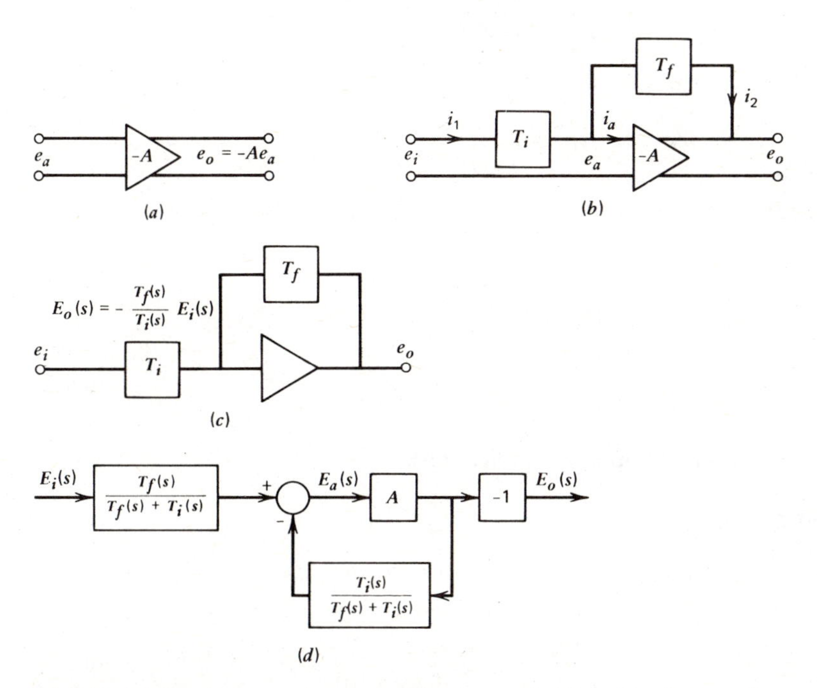

FIGURE 6.47 (*a*) Operational amplifier (op amp). (*b*) and (*c*) Its impedance representations. (*d*) Its block diagram.

impedance $T_i(s)$ of the input elements is defined such that

$$E_i(s) - E_a(s) = T_i(s)I_1(s)$$

For the feedback elements,

$$E_a(s) - E_o(s) = T_f(s)I_2(s)$$

(Recall that impedance is the dynamic operator equivalent to resistance in the static case.)

The high internal impedance of the op amp implies that $i_a \cong 0$, and thus $i_1 \cong i_2$. The final relation we need is the amplifier relation

$$e_o = -Ae_a$$

or

$$E_o(s) = -AE_a(s) \tag{6.8-1}$$

When the preceding relations are used to eliminate $I_1(s)$ and $I_2(s)$, the result is

$$E_a(s) = \frac{T_f(s)}{T_f(s) + T_i(s)} E_i(s) + \frac{T_i(s)}{T_f(s) + T_i(s)} E_o(s)$$

With this and (6.8-1), the block diagram in Figure 6.47*d* can be constructed. The transfer function between $E_i(s)$ and $E_o(s)$ is

$$T(s) = \frac{E_o(s)}{E_i(s)} = -\frac{T_f(s)}{T_f(s) + T_i(s)} \frac{A}{1 + AH(s)}$$

where

$$H(s) = \frac{T_i(s)}{T_f(s) + T_i(s)}$$

Since A is large (of the order of 10^5 to 10^8), $|AH(s)| \gg 1$, and we obtain

$$T(s) \cong -\frac{T_f(s)}{T_i(s)}$$

and

$$\frac{E_o(s)}{E_i(s)} = -\frac{T_f(s)}{T_i(s)} \tag{6.8-2}$$

This is the basic relation for op-amp applications.

Proportional Control

A proportional controller can be obtained with two resistors, as shown in Figure 6.48*a*. For this circuit, $T_f(s) = R_f$, $T_i(s) = R_i$, and

$$\frac{E_o(s)}{E_i(s)} = -\frac{R_f}{R_i} \tag{6.8-3}$$

The gain can be made adjustable by using a potentiometer for one of the resistances.

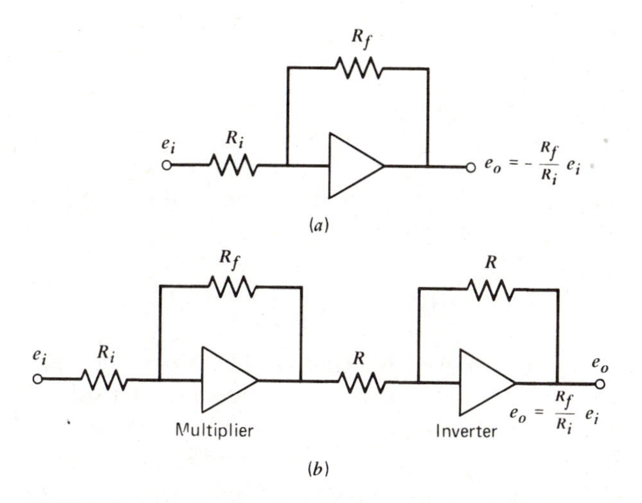

FIGURE 6.48 (*a*) Op amp implementation of proportional control. (*b*) Use of an inverter to maintain voltage polarity.

An *inverter* is such a circuit with $R_f = R_i$. It simply inverts the sign of the input voltage without changing its magnitude. It can be used in cascade with other elements to maintain proper sign relations, as shown in Figure 6.48*b*.

This multiplier circuit can be modified to act as an adder (Figure 6.49). The output relation is

$$e_o = -\frac{R_3}{R_1}e_1 - \frac{R_3}{R_2}e_2 \tag{6.8-4}$$

When used with an inverter, this circuit implements the proportional control law plus the manual reset term. The voltage e_1 corresponds to the error, R_3/R_1 is the proportional gain, and R_3e_2/R_2 is the manual reset. An application requiring such a circuit is the temperature controller in Figure 6.9.

PI Controllers

The impedance of a capacitor is found from its voltage–current relation in the Laplace domain

$$E(s) = \frac{1}{Cs}I(s)$$

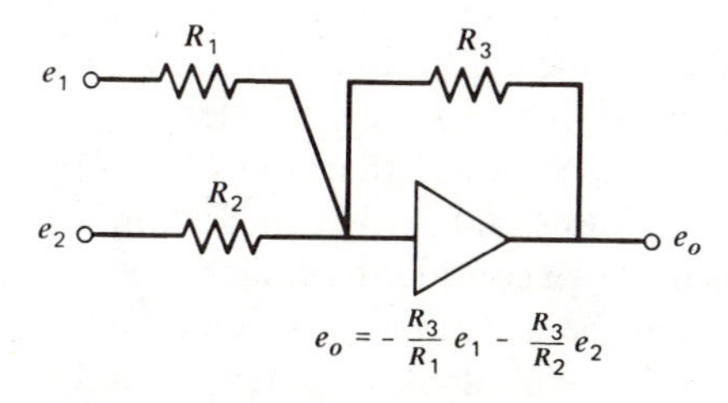

FIGURE 6.49 Op-amp adder circuit.

The impedance is $1/Cs$. An integral controller is obtained by using this element in the feedback circuit, with a resistance as the input element (Figure 6.50). For this circuit,

$$\frac{E_o(s)}{E_i(s)} = -\frac{1}{RCs} \tag{6.8-5}$$

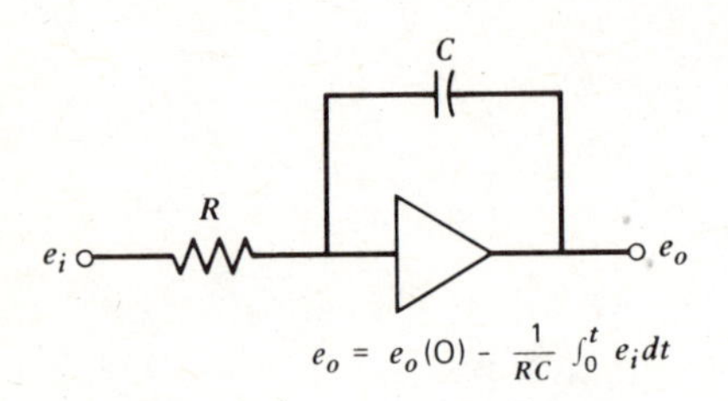

FIGURE 6.50 Op amp integrator circuit.

The integral gain is $K_I = 1/RC$. If a larger gain is required, a multiplier can be cascaded with this circuit.

Equation (6.8-5) is true if the initial voltage on the capacitor is zero. Any nonzero initial voltage $e_o(0)$ is added to the output to give

$$e_o(t) = -\frac{1}{RC}\int_0^t e_i(t)\,dt + e_o(0) \tag{6.8-6}$$

Many industrial controllers provide the operator with a choice of control modes, and the operator can switch from one mode to another when the process characteristics or control objectives change. When a switch occurs, it is necessary to provide any integrators with the proper initial voltages, or else undesirable transients will occur when the integrator is switched into the system. Commercially available controllers usually have built-in circuits for this purpose. A diagram of one such design for PI control is given in Appendix C of Reference 9.

Proportional-plus-Integral control can be implemented with the circuit in Figure 6.51. The impedance of the series resistance and capacitance is $T_f(s) = R_f + 1/Cs$. If the capacitance is initially discharged,

$$\frac{E_o(s)}{E_i(s)} = -\frac{R_f Cs + 1}{R_i Cs} = -\frac{R_f}{R_i} - \frac{1}{R_i Cs} \tag{6.8-7}$$

The corresponding gains are

$$K_p = \frac{R_f}{R_i}$$

$$K_I = \frac{1}{R_i C}$$

Output Limitation

In practice, the final control elements are always incapable of delivering energy to the controlled system above a certain rate. Also, we will see that integral control can suffer from a nonlinear effect called *reset windup*, which is in part caused by the finite capacity of the final control elements. This problem can be eliminated by limiting the output of the controller so that it cannot command the final control elements to deliver more power than they can. This topic is discussed further in Section 7.3. For now, we consider how such a limitation can be accomplished. (Some power amplifiers will already contain such circuitry; consult the manufacturer's data.)

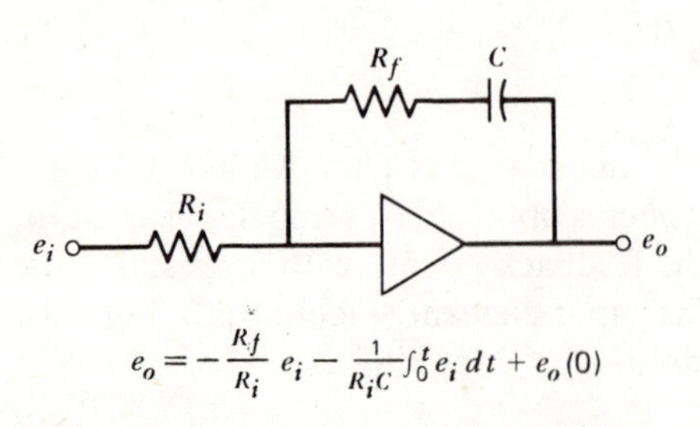

FIGURE 6.51 Op amp implementation of PI control.

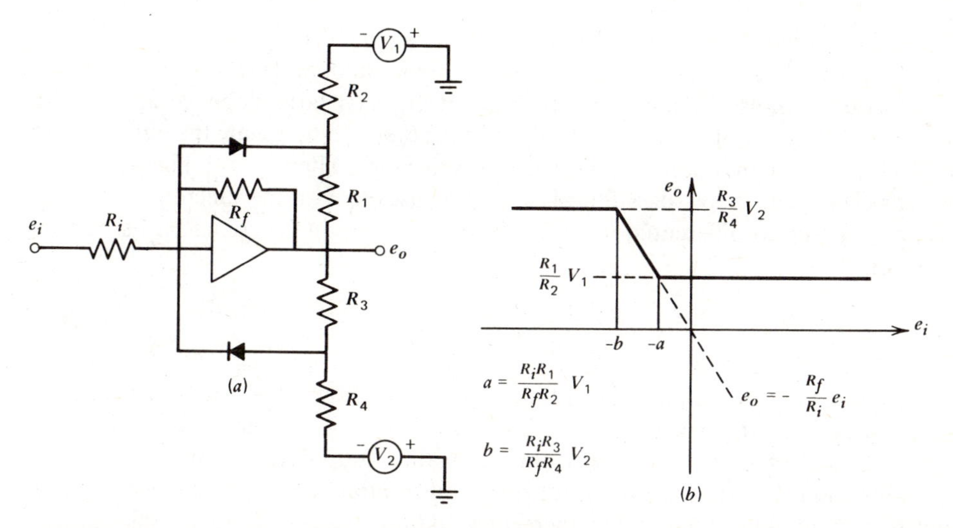

FIGURE 6.52 **(*a*) Diode circuit for limiting the output of an op amp multiplier. (*b*) Transfer characteristic.**

One way of limiting the output of a multiplier is by using diodes and bias voltages V_1 and V_2, as shown in Figure 6.52*a*. The diodes short circuit the feedback circuit whenever the output e_o lies outside of the desired range. The input–output relation is shown in Figure 6.52*b* for $V_1 < V_2 (V_1 > 0, V_2 > 0)$. Other transfer characteristics can be achieved with slightly different circuits (see, for example, chapter 3 of Reference 9 and chapter 10 of Reference 10). For PI control, the feedback resistor R_f is placed in series with a capacitor, and the same limiting circuit design can be used. The concept can be extended to other control circuits.

PD Control

In theory, a differentiator can be created by interchanging the resistance and capacitance in the integrator. The result is shown in Figure 6.53*a*. The input–output relation for this ideal differentiator is

$$\frac{E_o(s)}{E_i(s)} = -RCs \qquad (6.8\text{-}8)$$

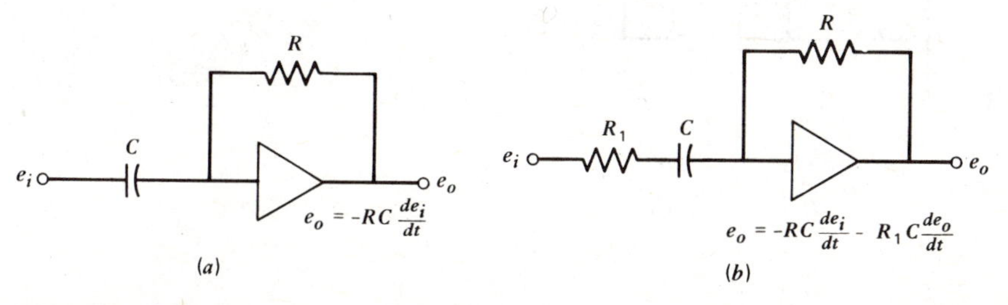

FIGURE 6.53 **Op amp implementations of a differentiator. (*a*) Ideal differentiator. (*b*) Practical differentiator.**

The difficulty with this design is that no electrical signal is "pure." Contamination always exists as a result of voltage spikes, ripple, and other transients generally categorized as "noise." These high-frequency signals have large slopes compared with the more slowly varying primary signal, and thus they will dominate the output of the differentiator. In practice, this problem is solved by filtering out high-frequency signals either with a low-pass filter inserted in cascade with the differentiator or by using a redesigned differentiator, such as the one shown in Figure 6.53*b*. Its transfer function is

$$\frac{E_o(s)}{E_i(s)} = -\frac{RCs}{R_1Cs+1} \tag{6.8-9}$$

The frequency response plot of this transfer function shows that it acts like the ideal differentiator for frequencies up to about $\omega = 1/R_1C$. For higher frequencies, the attenuation curve has zero slope rather than the 20 db/decade slope required for differentiation, and thus the circuit does not differentiate the high-frequency signals, but merely amplifies them. Since their amplitudes are generally small, this effect is negligible. For $\omega < 1/R_1C$, the derivative gain of (6.8-9) is $K_d = RC$.

A practical PD controller can be constructed in a similar manner (Figure 6.54). The impedance of the input elements is

$$T_i(s) = \frac{R_1 + R_2 + R_1R_2Cs}{R_2Cs+1}$$

The transfer function reduces to

$$\frac{E_o(s)}{E_i(s)} = -\frac{R(R_2Cs+1)}{R_1+R_2+R_1R_2Cs} \tag{6.8-10}$$

For the ideal PD controller, $R_1 = 0$ and

$$\frac{E_o(s)}{E_i(s)} = -\frac{R(R_2Cs+1)}{R_2} = -\frac{R}{R_2}(R_2Cs+1) \tag{6.8-11}$$

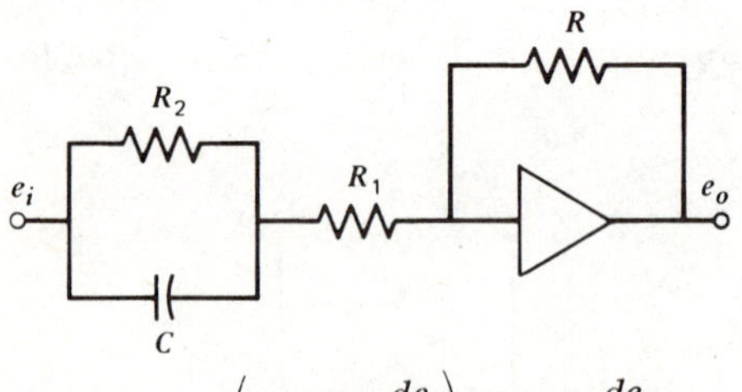

$$e_o = -K_p\left(e_i + T_D\frac{de_i}{dt}\right) - \alpha T_D\frac{de_o}{dt}$$

$$K_p = \frac{R}{R_1+R_2} \qquad T_D = R_2C \qquad \alpha = \frac{R_1}{R_1+R_2}$$

FIGURE 6.54 Practical op amp implementation of PD control.

The attenuation curve for the ideal controller breaks upward at $\omega = 1/R_2C$ with a slope of 20 db/decade. The curve for the practical controller does the same but then becomes flat for $\omega > (R_1 + R_2)/R_1R_2C$. This provides the same limiting effect at high frequencies as that of the practical differentiator (6.8-9).

The transfer function (6.8-10) can be written as

$$\frac{E_o(s)}{E_i(s)} = -\frac{K_p(1 + T_Ds)}{(1 + \alpha T_Ds)} \tag{6.8-12}$$

where

$$K_p = \frac{R}{R_1 + R_2}$$

$$T_D = R_2C$$

$$\alpha = \frac{R_1}{R_1 + R_2}$$

The circuit limits frequencies above $\omega = 1/\alpha T_D$. For lower frequencies, the transfer function is that of PD control (6.7-2). The proportional gain is K_p, and the rate constant is T_D.

PID Control

Proportional-plus-Integral-plus-Derivative control can be implemented by joining the PI and PD controllers in parallel, but this is expensive because of the number of op amps and power supplies required. Instead, the usual implementation is that shown in Figure 6.55. The impedances for the series and parallel RC elements were given previously. From (6.8-2), the transfer function is

$$\frac{E_o(s)}{E_i(s)} = -\frac{(RCs + 1)(R_2C_1s + 1)}{Cs(R_1 + R_2 + R_1R_2C_1s)} \tag{6.8-13a}$$

$$= -\left(\frac{RC + R_2C_1}{R_2C} + \frac{1}{R_2Cs} + RC_1s\right)\frac{\beta}{(\beta R_1C_1s + 1)}$$

$$= -\left(K_p + \frac{K_I}{s} + K_Ds\right)\frac{1}{(\beta R_1C_1s + 1)} \tag{6.8-13b}$$

Where

$$\beta = \frac{R_2}{R_1 + R_2}$$

$$K_p = \beta\frac{RC + R_2C_1}{R_2C}$$

$$K_I = \frac{\beta}{R_2C}$$

$$K_D = \beta RC_1$$

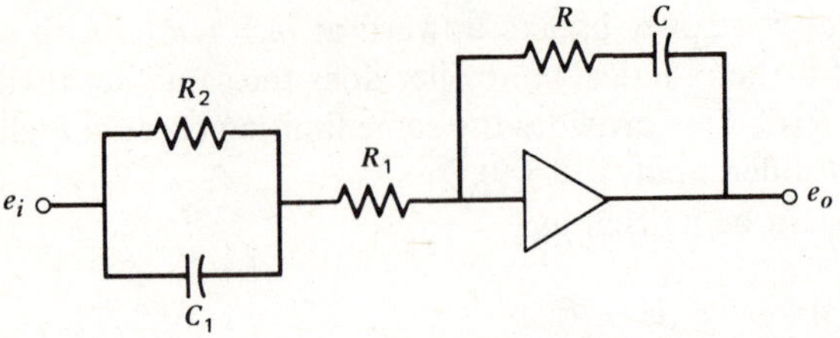

$$e_o = -\left(K_P e_i + K_I \int_0^t e_i dt + K_D \frac{de_i}{dt}\right) - \beta R_1 C_1 \frac{de_o}{dt}$$

$$\beta = \frac{R_2}{R_1 + R_2} \qquad K_P = \beta \frac{RC + R_2 C_1}{R_2 C}$$

$$K_I = \frac{\beta}{R_2 C} \qquad K_D = \beta R C_1$$

FIGURE 6.55 Practical op amp implementation of PID control.

The denominator term $\beta R_1 C_1 s + 1$ limits the effect of frequencies above $\omega = 1/\beta R_1 C_1$. When $R_1 = 0$, ideal PID control results. This is sometimes called the *noninteractive* algorithm, because the effect of each of the three modes is additive, and they do not interfere with one another. The form given by (6.8-13*a*) for $R_1 \neq 0$ is the *real* or *interactive* algorithm. This name results from the fact that historically it was difficult to implement noninteractive PID control with mechanical or pneumatic devices.

Preliminary calculation of the control gains is made as in Sections 6.5–6.7, using the transient and steady-state performance specifications. The resistance and capacitance values can then be selected in terms of the gains. Usually, there is sufficient freedom to choose feasible values for these electrical components, because there are more such elements than control gains. In the PID controller, there are three resistors and two capacitors, but only three gains. If the limiting upper frequency is specified, the problem involves four constraints and five values to be chosen. Four of the values can be related to the fifth one, which can be selected to satisfy another criterion (cost, size, etc.).

System Implementation

Figure 6.56 shows how the complete control system is implemented, using PI control as an example. The input voltage v_r represents the command $r(t)$. The inverter changes the sign of the measured output voltage v_c; this enables the adder to function as a comparator if R_1, R_2, and R_3 are chosen to be equal. For this case, the voltage v_m, which is applied to the power amplifier and is the manipulated variable, is given by

$$v_m = \frac{R_f}{R_i}(v_r - v_c) + \frac{1}{R_i C}\int_0^t (v_r - v_c)\, dt$$

Thus, PI control is obtained, with $K_p = R_f/R_i$ and $K_I = 1/R_i C$.

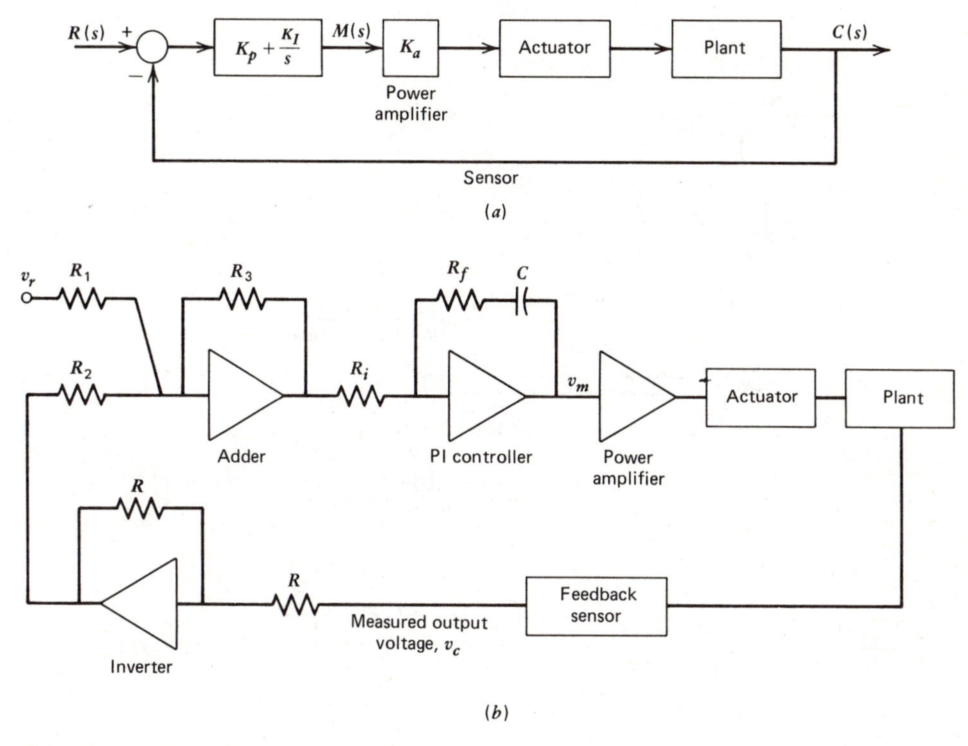

FIGURE 6.56 Implementation of a PI-controller using op amps. (*a*) Diagram of the system. (*b*) Diagram showing how the op amps are connected.

6.9 PNEUMATIC CONTROLLERS

The nozzle flapper introduced in Section 6.4 is a high-gain device that is difficult to use without modification. The gain K_f, known only imprecisely, is sensitive to changes induced by temperature and other environmental factors. Also, the linear region over which (6.4-10) applies is very small. However, the device can be made useful by compensating it with feedback elements, using the general principle of Figure 6.45 as a guide. The overall gain is reduced, but the result is a controller with an increased linear range and decreased sensitivity to parameter changes and uncertainties.

Proportional Control

In order to obtain proportional action, the feedback compensation principle states that the feedback element around the nozzle flapper should have a constant transfer function. Let the input to the controller be the flapper displacement y and the output be a controlled pressure (see Figures 6.23*a* and 6.25). Then we can use the flapper itself as an error-detecting beam. The feedback element must then be a transducer capable of converting the output pressure into a displacement of the lower end of the flapper. The result is shown in Figure 6.57*a*. The controlled output pressure p_o represents a deviation from the nominal output pressure corresponding to the reference

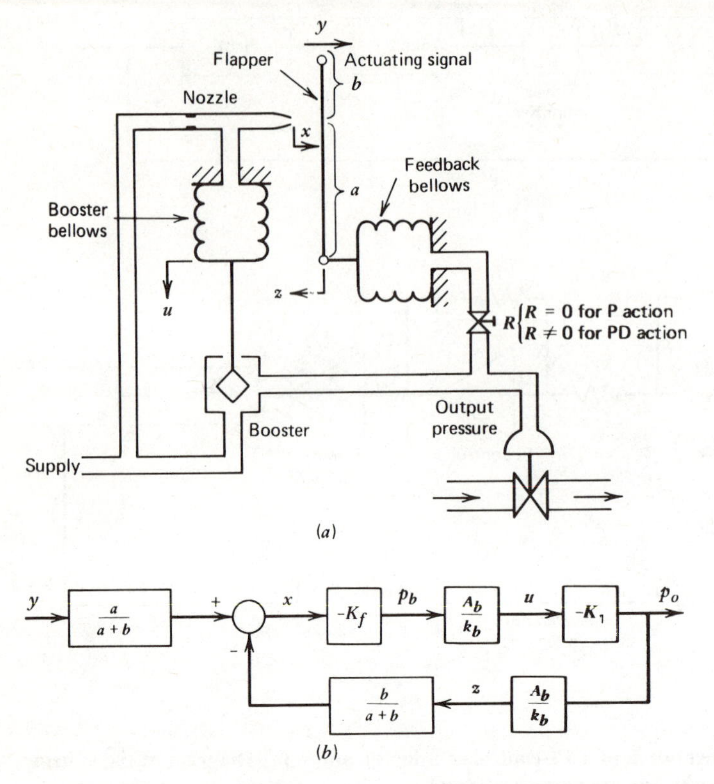

FIGURE 6.57 ***(a)*** **Pneumatic implementation of P and PD control via feedback compensation of the nozzle-flapper element.** ***(b)*** **Block diagram for the P-control case ($R = 0$).**

equilibrium operating condition. The total pressure (p_o plus nominal) can be used to operate a pneumatic flow valve, for example.

A pneumatic *relay* or *booster* allows a large airflow and amplifies the back pressure, as in Figure 6.24. As the actuating displacement y moves to the right, the feedback bellows acts as a pivot, and the back pressure decreases. This causes the booster bellows to retract and open the booster; the output pressure then increases. This increased pressure causes the feedback bellows to expand to the left and reduce the nozzle flapper distance (the upper end of the flapper is held constant at its new position during this motion).

The system is modeled as follows in terms of linearized variables. The geometric relation for the flapper is

$$x = \frac{a}{a+b}y - \frac{b}{a+b}z$$

Assume both bellows are identical with areas A_b and equivalent spring constants k_b. Newton's law applied to a negligible mass gives for the feedback bellows

$$p_o A_b = k_b z$$

and

$$p_b A_b = k_b u$$

for the booster bellows. The time constants of the bellows are negligible if their inlet resistance is small ($R = 0$). Finally, the linearized relation for the booster operation is

$$p_o = -K_1 u$$

where $K_1 > 0$. The resulting block diagram is shown in Figure 6.57*b*.

The transfer function for the controller is approximately

$$\frac{p_o}{y} = \frac{ak_b}{bA_b} \tag{6.9-1}$$

if $K_f \gg (a+b)k_b^2/bA_b^2K_1$. The action is proportional. Note that we need not be concerned with the precise value of K_f as long as we know that the latter inequality is satisfied. The gain in (6.9-1) is smaller than the uncompensated gain K_f. This is the price paid for reduced sensitivity to uncertainties in K_f.

To demonstrate how the feedback improves the linearity, derive the relation between x and y. From the block diagram,

$$x = \frac{a}{a+b} \frac{y}{1 + \dfrac{b}{a+b}\left(\dfrac{A_b}{k_b}\right)^2 K_1 K_f}$$

Since K_f is large, x remains small and thus within the linear region for a wider range of y values than with the uncompensated system.

The proportional gain can be adjusted by using a linkage that changes the ratio a/b. An example of such a design is shown in Figure 6.58. This can be used with the pivoted-beam, double-bellows arrangement shown in Figure 6.11. The force on the left side of the beam results from the set point pressure. The force on the right results from the feedback bellows. If the feedback pressure is below the set point pressure, the flapper moves away from the nozzle, and the output pressure is increased. The sensitivity of x to the beam motion is affected by the pivot location of the sector link. Moving the pivot clockwise decreases the sensitivity and thus decreases the proportional gain.

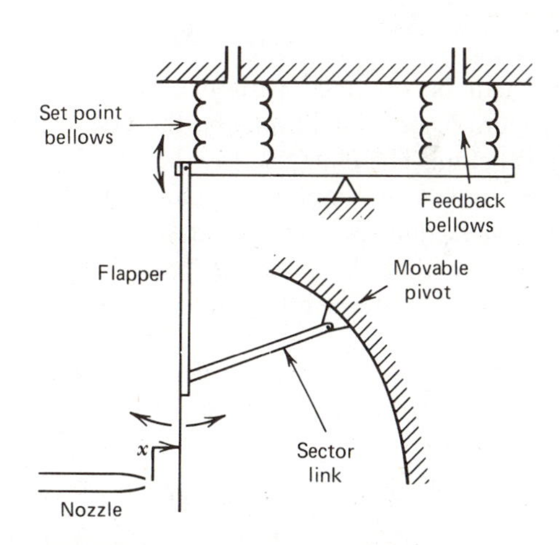

FIGURE 6.58 Typical mechanism for adjusting the gain of a nozzle-flapper controller.

PD Control

Derivative action can be synthesized pneumatically with an adjustable restriction in the input line of a bellows, as shown in Figure 6.22. The transfer function is given by (6.4-9). Proportional-plus-Derivative

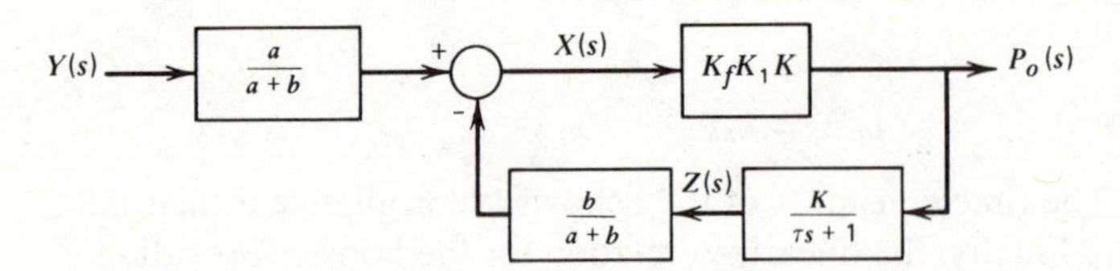

FIGURE 6.59 Block diagram of the system shown in Figure 6.57 for PD control ($R \neq 0$).

control is obtained from the system in Figure 6.57 if the inlet resistance R of the feedback bellows is nonzero. In this case, the time constant τ in (6.4-9) is not zero, and the bellows' pressure lags behind changes in the output pressure p_o. The resulting model is given by Figure 6.59. The bellows constant K is A_b/k_b. The system's transfer function is

$$\frac{P_o(s)}{Y(s)} = \frac{a}{a+b} \frac{K_f K_1 K}{1 + \dfrac{bK_f K_1 K^2}{(a+b)(\tau s+1)}}$$

If $K_f > |(a+b)(\tau s+1)/bK_1 K^2|$,

$$\frac{P_o(s)}{Y(s)} = \frac{a}{bK}(\tau s+1) \tag{6.9-2}$$

The proportional gain is $K_p = a/bK$; the derivative gain is $K_D = a\tau/bK$. Note that this result reduces to (6.9-1) when the bellows time constant τ is zero. This occurs when the adjustable resistance R is fully open ($R = 0$). When the resistance is fully closed ($R = \infty$), the feedback loop does not operate, and the system becomes a two-position controller, because the nozzle back pressure will take either one of the two extreme values (p_a or p_s), depending on the direction of motion of y.

PID Control

Proportional-plus-Integral and PID control can be obtained with two opposing bellows in the feedback path, as shown in Figure 6.60*a*. For simplicity, the booster subsystem is not shown but can be included as before. The plate used as the interface between the bellows is connected to the lower end of the flapper. Assume that the bellows have the same area A_b, spring constant k_b, and pneumatic capacitance C. Usually, the resistances are such that $R_I \gg R_D$, so that the time constant $R_I C$ of bellows 2 is larger than that for bellows 1. A force balance at the interface for negligible system mass gives,

$$k_b Z(s) = \frac{A}{R_D Cs+1} P_o(s) - \frac{A}{R_I Cs+1} P_o(s)$$

This gives the feedback loop shown in Figure 6.60*b*, with $K = A_b/k_b$. The transfer function of the feedback loop reduces to

$$H(s) = \frac{bK}{a+b} \frac{(R_I - R_D)Cs}{(R_I Cs+1)(R_D Cs+1)} \tag{6.9-3}$$

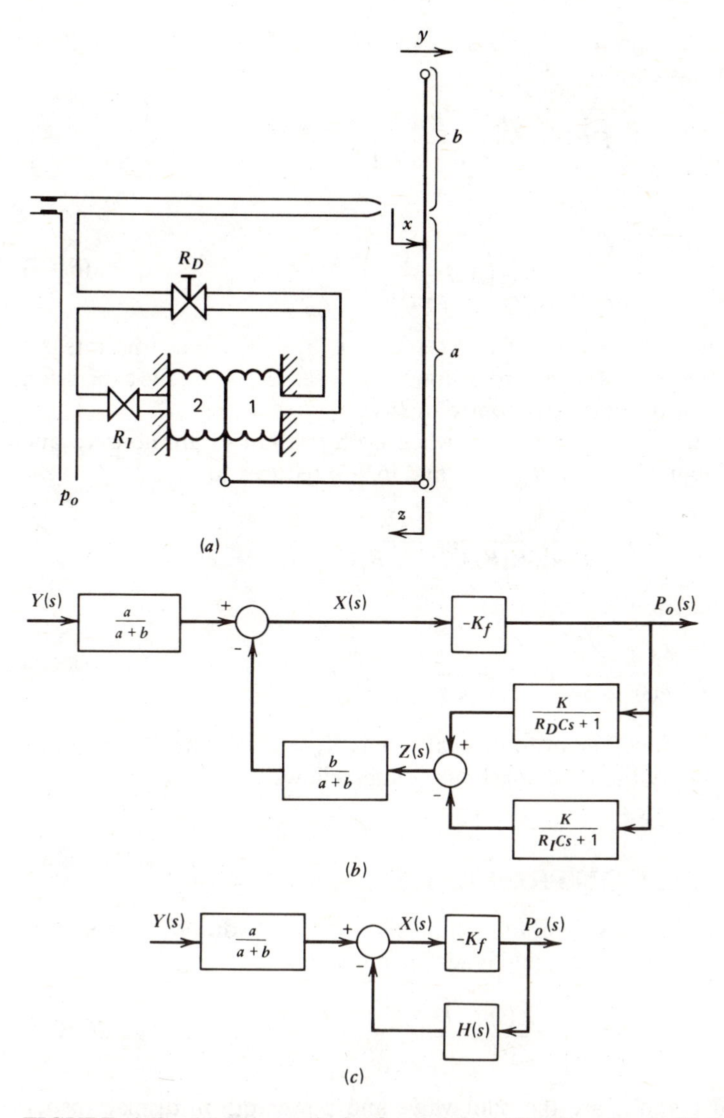

FIGURE 6.60 (*a*) **Pneumatic implementation of PID control.** (*b*) **Block diagram showing multiple feedback paths.** (*c*) **Reduced diagram.**

The simplified diagram is given in Figure 6.60*c*. The closed-loop transfer function is

$$\frac{P_o(s)}{Y(s)} = \frac{a}{a+b}\,\frac{-K_f}{1-K_fH(s)}$$

If $K_f \gg |1/H(s)|$,

$$\frac{P_o(s)}{Y(s)} = \frac{a}{(a+b)H(s)} \tag{6.9-4}$$

Proportional-plus-Integrator control is obtained if the inlet resistance to bellows 1 is zero ($R_D = 0$). In this case,

$$H(s) = \frac{bK}{a+b}\frac{R_I Cs}{R_I Cs + 1}$$

and

$$\frac{P_o(s)}{Y(s)} = \frac{a}{bK}\left(1 + \frac{1}{R_I Cs}\right) \tag{6.9-5}$$

The proportional gain is $K_p = a/bK$. The reset time is $T_I = R_I C$, and the integral gain is $K_I = K_p/T_I$. Note that if the inlet to bellows 2 is closed ($R_I = \infty$), the expression (6.9-5) reduces to that of proportional control (6.9-1).

The full PID-control law is obtainable when both resistances are nonzero and $R_I \gg R_D$. With this inequality, $R_I \pm R_D \cong R_I$, and (6.9-3) reduces to

$$H(s) = \frac{bK}{a+b}\frac{R_I Cs}{R_I R_D C^2 s^2 + R_I Cs + 1}$$

The system transfer function is

$$\frac{P_o(s)}{Y(s)} = \frac{a}{bK}\left(1 + \frac{1}{R_I Cs} + R_D Cs\right) \tag{6.9-6}$$

which is the PID-control law. The proportional gain is $K_p = a/bK$. The reset and rate times T_I and T_D are simply the time constants of the bellows.

6.10 HYDRAULIC CONTROLLERS

The basic unit for synthesizing hydraulic controllers is the hydraulic servomotor analyzed in Section 2.7. For a negligible load, the model is

$$\frac{X(s)}{Z(s)} = \frac{K}{s} \tag{6.10-1}$$

where $K = C_1/A$ and z and x are the pilot valve and power piston displacements. Using this model for design purposes presumes that other, more powerful final control elements are used if the load is appreciable. We note in passing that the nozzle flapper concept is also used in hydraulic controllers. Reference 11 discusses these devices and others in some detail.

In contrast to the high-gain amplifier used as the basis for electronic and pneumatic controllers, the hydraulic servomotor is an integrator. Nevertheless, the principle of feedback compensation can still be applied. With a servomotor in the forward path, the structure of a hydraulic controller is as shown in Figure 6.61. The closed-loop transfer function is

$$\frac{X(s)}{Y(s)} = \frac{K/s}{1 + (K/s)H(s)}$$

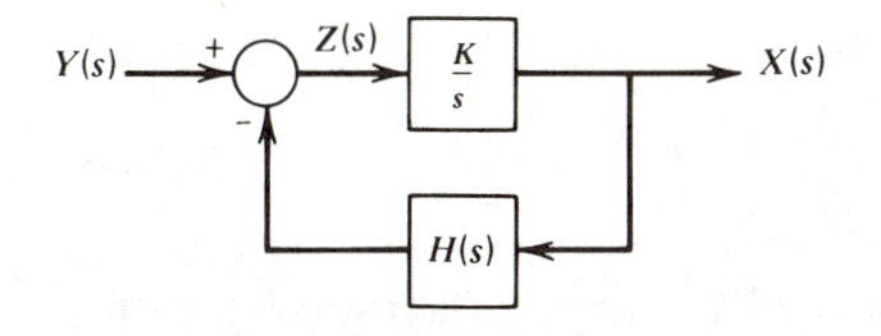

FIGURE 6.61 Feedback compensation of a hydraulic integrator.

where y is a command displacement similar to that for the pneumatic flapper. If $K \gg |s/H(s)|$, the transfer function becomes

$$\frac{X(s)}{Y(s)} = \frac{1}{H(s)} \qquad \textbf{(6.10-2)}$$

The preceding inequality is satisfied if the servovalve constant C_1 is large in relation to the power piston area A. Physically, this means that the valve flow rate must be large compared to the volume rate swept out by the piston; that is, the unit's response must be rapid.

Proportional Control

Equation (6.10-2) shows that $H(s)$ must be constant in order for proportional control to be obtained. Since $H(s)$ relates two displacements, this suggests a pivoted beam. The design is shown in Figure 6.62*a*. The beam is called a *walking beam* because of its motion during the unit's operation. Its function is similar to that of the pneumatic flapper. Assume that the displacements x, y, and z are measured from reference positions corresponding to the line-on-line configuration of the pilot valve and some rest position of the load. For small deviations from this reference equilibrium, the geometry of the beam is such that

$$z = \frac{b}{a+b}y - \frac{a}{a+b}x \qquad \textbf{(6.10-3)}$$

This equation and (6.10-1) complete the model. The block diagram is shown in Figure 6.62*b*. The feedback transfer function is

$$H(s) = \frac{a}{a+b}$$

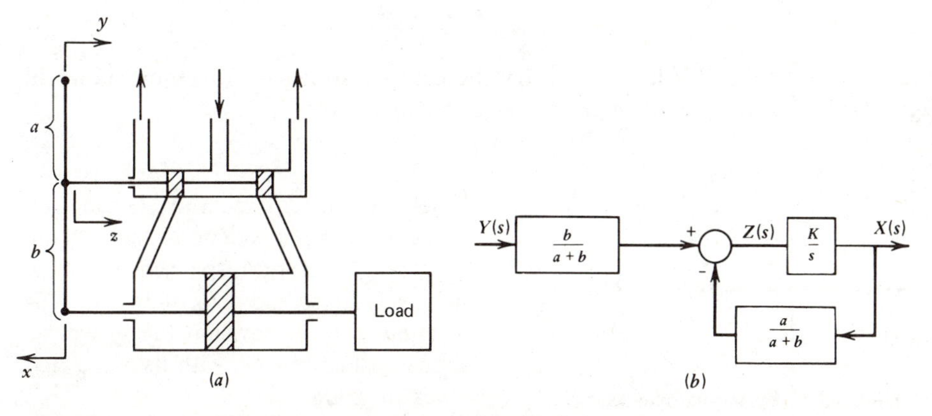

FIGURE 6.62 Walking-beam servomotor and its block diagram.

If $K \gg |(a+b)s/a|$, the closed-loop transfer function is

$$\frac{X(s)}{Y(s)} = \frac{b}{a+b}\frac{a+b}{a} = \frac{b}{a} \tag{6.10-4}$$

The proportional gain is $K_p = b/a$. It can be adjusted by varying the ratio b/a with a linkage.

In addition to a proportional controller, the servomotor with a walking beam can also be used as a servomechanism. If we assume that the gain K is not high enough for the preceding approximation to be valid, the following transfer function applies:

$$\frac{X(s)}{Y(s)} = \frac{b/a}{\tau s + 1} \tag{6.10-5}$$

where $\tau = (a+b)/aK$. The operation is as follows. For an actuating displacement y to the right, the lower end of the beam acts as a pivot, because the inertia and friction of the power piston are greater than those of the pilot valve. This motion moves the pilot valve to the right and causes the power piston to move to the left. The beam now pivots about the upper end, which is held fixed by the mechanism that produces the actuating displacement. This motion moves the pilot valve to the left until the flow is shut off and the load comes to rest at a new position. If y is a step input, the load reaches its new position in approximately 4τ time units. If the gain K is high, τ is small and the motion approximates the instantaneous action of proportional control (6.10-4).

PI Control

To obtain PI control, (6.10-2) indicates that the feedback transfer function $H(s)$ must be of the form

$$H(s) = \frac{a_1 s}{a_2 + a_3 s} = \frac{W(s)}{X(s)}$$

where w is a displacement created by the feedback element in response to x. The displacement x and w must be related by

$$a_3 \dot{w} + a_2 w = a_1 \dot{x}$$

With some thought, this leads us to try the damper–spring combination shown in Figure 6.63. For a negligible mass, the force balance gives

$$c\dot{w} + kw = c\dot{x}$$

which is in the required form. The motion w occurs at the lower end of the walking beam and affects the displacement z in the same way that x did in (6.10-3). With this in mind, we can envision the system as shown in Figure 6.64, with its associated block diagram.

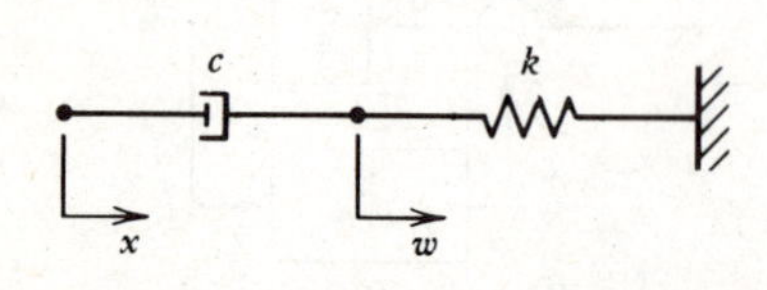

FIGURE 6.63 Spring-damper subsystem used to create PI control.

The transfer function for the closed-

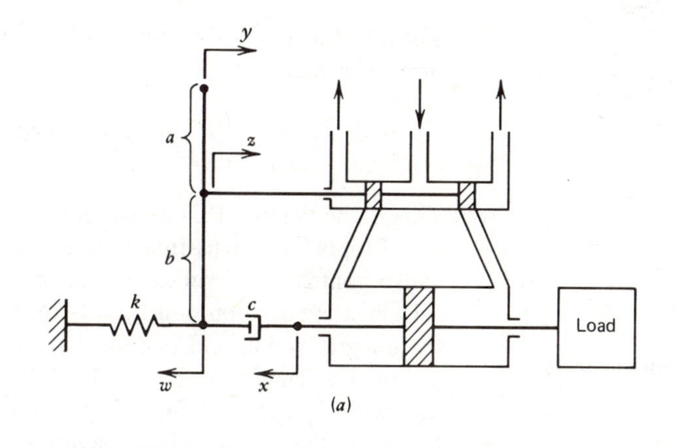

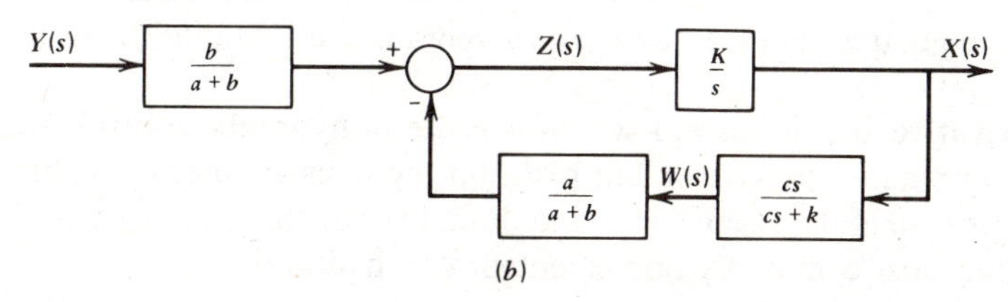

FIGURE 6.64 Hydraulic implementation of PI control and its block diagram.

loop system is

$$\frac{X(s)}{Y(s)} = \frac{b}{a}\left(1 + \frac{k}{cs}\right) \tag{6.10-6}$$

if $K \gg |k(a+b)(cs+k)/ac|$. The proportional gain b/a is adjustable with a linkage, while the integral gain is adjusted by changing the damping constant c. This cannot be done with the typical dashpot we have seen thus far (for example, see Figure 2.22), but the design shown in Figure 6.65 is suitable for this purpose. The damping is varied by changing the resistance R. It can be shown with the approach of Section 2.7 that the damping constant is approximately given by $c = RA^2$, where R relates the volume flow rate to the pressure drop, and A is the area of the damper's piston.

Derivative Action

In order to obtain PD control, $H(s)$ must be of the form

$$H(s) = \frac{b_1}{b_2 + b_3 s} = \frac{W(s)}{X(s)}$$

Thus,

$$b_3\dot{w} + b_2 w = b_1 x$$

This action can be synthesized by interchanging the locations of the spring and damper elements in Figure 6.64 so that

$$c\dot{w} + kw = kx$$

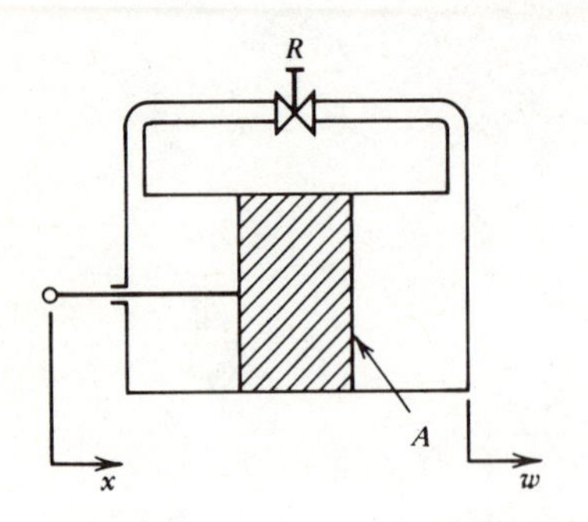

FIGURE 6.65 Damper with adjustable resistance.

The resulting overall system transfer function for a high-gain K is

$$\frac{X(s)}{Y(s)} = \frac{b}{a}\left(\frac{c}{k}s + 1\right) \qquad \textbf{(6.10-7)}$$

This is the desired PD-control law.

Proportional-plus-Integral-plus-Derivative control was synthesized pneumatically by using two opposing bellows, one for the integral action and one for derivative action. The same principle can be applied here with the spring–damper pairs playing the role of the bellows, where the time constant of one pair is much greater than the other. The algebra is similar to the previous cases.

We note that derivative action has not seen much use in hydraulic controllers. This action supplies damping to the system, but hydraulic systems are usually highly damped intrinsically because of the viscous working fluid. Proportional-plus-Integral control is the algorithm most commonly implemented with hydraulics.

6.11 SUMMARY

Chapter 6 introduced the basic concepts, devices, and algorithms of feedback control. We have emphasized the PID family of control laws, since they are satisfactory for many applications. A proposed design is analyzed on paper in light of specifications relating to stability, accuracy, and speed of response. If the control law appears satisfactory, the engineer chooses the control medium: analog electronic, pneumatic, hydraulic, or digital. This choice involves many factors, such as the type of plant being controlled, weight, power, reliability, or cost, and it is impossible to make general recommendations in this regard.

Often, however, simple adjustment of the gains in a PID algorithm will not yield satisfactory performance. In this case, a new control law must be found or more controllers added to the system. This brings us to the topic of the next chapter, which treats alternative control system configurations.

REFERENCES

1. O. Mayr, *The Origins of Feedback Control*, MIT Press, Cambridge, Mass., 1970.
2. A. N. Burstall, *A History of Mechanical Engineering*, MIT Press, Cambridge, Mass., 1969.
3. N. Wiener, *Cybernetics*, John Wiley & Sons, New York, 1948.
4. E. Doebelin, *Measurement Systems*, McGraw-Hill, New York, 1975.
5. *Transducer Compendium*, Instrument Society of America, Pittsburg.

6. L. S. Marks, *Mechanical Engineer's Handbook*, McGraw-Hill, New York, 1967.

7. J. Truxal, ed., *Control Engineer's Handbook*, McGraw-Hill, New York, 1958.

8. Y. Takahashi, M. Rabins, and D. Auslander, *Control*, Addison-Wesley, Reading, Mass., 1970.

9. R. Phelan, *Automatic Control Systems*, Cornell Univ. Press, Ithaca, New York, 1977.

10. J. Truxal, *Introductory Systems Engineering*, McGraw-Hill, New York, 1971.

11. D. McCloy and H. Martin, *The Control of Fluid Power*, Longman, London, 1973.

PROBLEMS

6.1 Consider the dc motor position controller shown in Figure 6.4. Assume that the amplifier has a gain K_a. Modify the block diagram in Figure 6.6, and find the transfer function between θ and θ_r, and that between θ and T_d, for the following cases:

(a) An armature-controlled dc motor. Do not neglect the motor's inductance.
(b) A field-controlled dc motor. Do not neglect the motor's inductance.

6.2 Consider the temperature control system shown in Figure 6.9. Modify the block diagram shown in Figure 6.10 to include the effect of the heater's thermal capacitance C_1 and thermal resistance R_1 (see Problem 2.37). Find the transfer function between x and u and that between x and v.

6.3 A common error detector is the Wheatstone bridge, shown in Figure P6.3. A fixed voltage V is applied, while R_3 and R_4 are potentiometers and represent the reference input and the controlled variable, respectively. Find the relation between the error voltage e and the resistances R_3 and R_4.

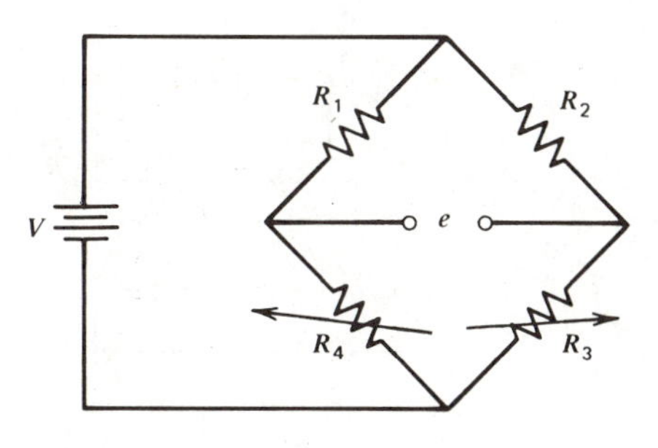

FIGURE P6.3

6.4 The model for a one-degree-of-freedom gyroscope is

$$I\ddot{\theta} + c\dot{\theta} + k\theta = H\omega_i$$

where θ is the angular displacement of the gyro's spin axis and is the unit's output

(Figure P6.4). The inertia about the output axis is I, and H is the angular momentum of the wheel, which spins at a constant velocity.

The gyro can be used as a sensor to measure either the angular displacement (an *integrating* gyro) or angular velocity (a *rate* gyro).

(a) Show that the gyro measures displacement if $k = 0$.
(b) Show that the gyro measures velocity if $k \neq 0$.

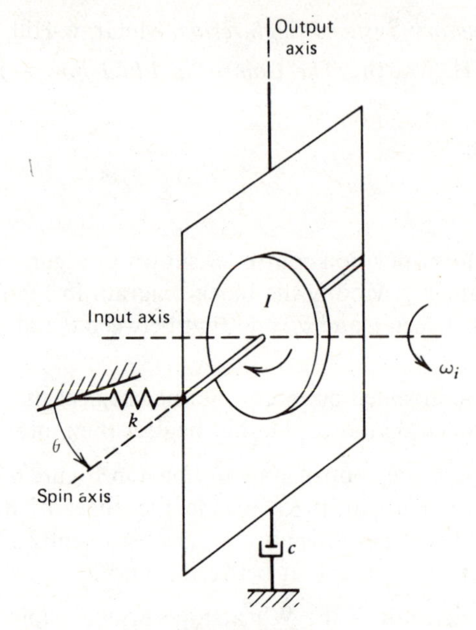

FIGURE P6.4

6.5 Hydraulic controllers are used to drive cutting tools in machining applications. (see Figure P6.5). The command input x_d is a voltage representing the desired

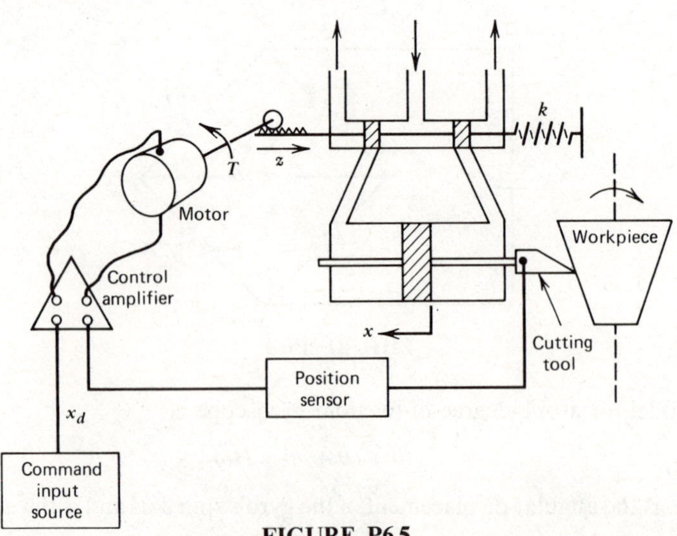

FIGURE P6.5

position of the cutting tool as a function of time. It can be generated by a cam drive, a punched paper tape, or a computer program. Develop a block diagram for the system, and find the transfer function between the tool displacement x and x_d. Neglect the inertia, inductance, and shaft elasticity of the motor. Treat the reactive force F of the cutting tool as a disturbance.

6.6 Figure P6.6 shows a hydraulic system for controlling the flow rate q. The adjustment screw is used to select the desired flow rate q_d. It adjusts the opposing force on the pilot valve by adjusting the compression in the spring. Explain the operation of the system, and draw the block diagram with q_d as the input and q as the output. Find the transfer function. Name two possible disturbances to the system.

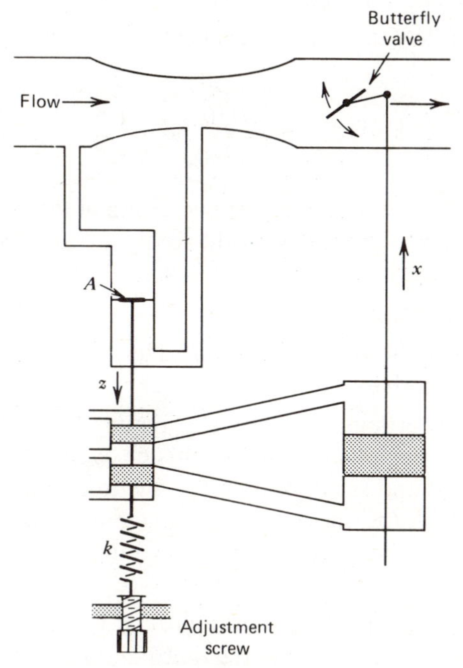

FIGURE P6.6

6.7 The motion of an oscillatory system is to be controlled by a relay controller. The relay has a maximum available force magnitude of $f_{max} = 50$. Plot the phase plane trajectories of the system for the following:

(a) No dead zone in the relay.

(b) A dead zone of total width 2 (see Figure P6.7).

6.8 For the liquid level system shown in Figure 6.27, the outlet resistance is $R = 5/6$ min/ft^2, the tank area is $A = 12$ ft^2, and the inlet flow rate when switched on is a constant 12 ft^3/min. The desired height is 5 ft, and the switching gap is 2 ft, so that $h = 5 \pm 1$ ft in the ideal case.

(a) Assume that $h(0) = 4$ ft (the start of the filling curve). Plot the height as a function of time for one cycle. What is the cycle time? Assume that the switch and the float act instantaneously.

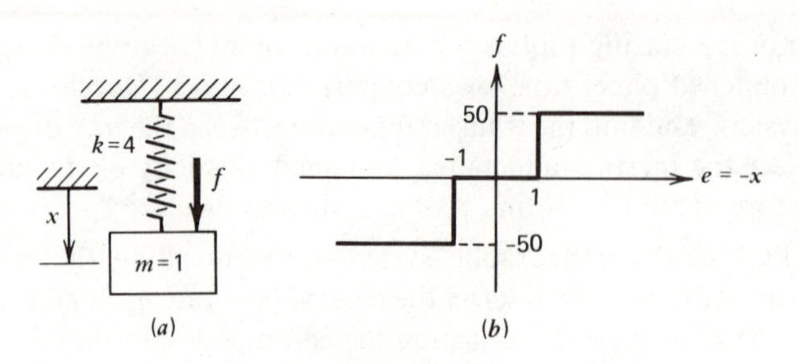

FIGURE P6.7

(b) Now assume that the float is a first order system with a time constant τ, so that the transfer function between the height measurement h_m and the actual height h is

$$\frac{H_m(s)}{H(s)} = \frac{1}{\tau s + 1}$$

Assume that $h(0) = 4$ ft and plot two cycles of the response $h(t)$ for the two following cases. Compare with the ideal case analyzed in part a.

(1) $\tau = 2$ min.
(2) $\tau = 0.5$ min.

6.9 Suppose the plant shown in Figure 6.31 has $I = 10$ and $c = 5$, with unit-step command and disturbance inputs.

(a) Find the proportional gain value K required to give a maximum steady-state error of magnitude 0.1 for a unit-step command. Evaluate the resulting time constant and steady-state error due to a unit-step disturbance.
(b) Repeat (a) with the maximum error magnitude specified to be 0.01.
(c) Find the gain value K required to give a time constant of 0.1. Evaluate the steady-state errors due to each input.

6.10 Suppose that the plant shown in Figure 6.31 has $I = 10$ and $c = 5$. The performance specifications require the time constant to be $\tau = 1$ and the steady-state error to be zero for a step command. Use P control with feedforward compensation.

(a) Find the necessary gain values K and K_f. Evaluate the resulting steady-state error due to a step disturbance.
(b) Suppose we are certain only that the value of c lies in the range $4.5 \leqslant c \leqslant 5.5$. For the values of K and K_f found in (a), evaluate the range of uncertainty in τ and in the steady-state errors.

6.11 Repeat Problem 6.10 for $\tau = 0.1$. Comment on the design's sensitivity to uncertainties in the value of c in light of the specification on τ.

6.12 For the system shown in Figure 6.36, $I = 10$ and $c = 5$. For unit-step inputs, use P control, and find the value of the gain K required to satisfy the following specifications:

(a) The steady-state error due to the command should be zero, and that due to the disturbance should be not greater than 0.1 in magnitude. Find the resulting values of ζ and τ.

(b) The steady-state error due to the command should be zero and $\zeta = 1$. Find the resulting values of τ and the disturbance error.

(c) The steady-state error due to the command should be zero, and that due to the disturbance should be not greater than 0.1 in magnitude. The damping ratio should be $\zeta = 1$.

6.13 Modify the system diagram shown in Figure 6.36 to include feedforward compensation. Investigate its ability to improve the steady-state error for step and ramp commands.

6.14 For the integral control system shown in Figure 6.37, $I = 10$, and the performance specifications require the steady-state errors due to step command and disturbance inputs to be zero. Find the required gain value K so that $\zeta = 1$. Evaluate the resulting time constant. Do this for each of the following values of c:

(a) $c = 5$

(b) $c = 0.1$.

6.15 For the PI controller shown in Figure 6.39, suppose that $I = 10$ and $c = 5$. The performance specifications require that $\tau = 0.4$.

(a) Find the gain values required to achieve $\zeta = 0.707$.

(b) Find the gain values required to achieve $\zeta = 1$.

(c) Find the gain values required to achieve a root separation factor of 10.

(d) Compare the unit-step command responses in (a–c). Evaluate the overshoot and the 10–90% rise time.

6.16 For the designs found in (a–c) of Problem 6.15, evaluate the steady-state error due to a unit-ramp command and due to a unit-ramp disturbance.

6.17 Modify the system diagram shown in Figure 6.39 to include feedforward compensation. Investigate its ability to eliminate the steady-state error for step and ramp commands.

6.18 Consider the P control, PD control, and tachometer feedback compensation systems shown in Figures 6.36, 6.42, and 6.43. Suppose that $I = 10$ and $c = 5$. The specifications require the steady-state error due to a unit step command to be zero and that due to a unit step disturbance to be no greater than 0.1 in magnitude. Also, we require that $\zeta = 0.707$.

(a) Find the values of the required gains for each design that follows, and evaluate the resulting time constant.

(1) P control (Figure 6.36).

(2) PD control (Figure 6.42).

(3) P control with tachometer feedback (Figure 6.43).

(b) Compare the performance of the three designs for this problem.

6.19 Repeat Problem 6.18, requiring the time constant to be 0.1 instead of requiring $\zeta = 0.707$. Also find the resulting value of ζ.

6.20 Derive expressions for the steady-state error due to unit-ramp commands and disturbances for the systems shown in Figures 6.42 and 6.43.

6.21 Figure P6.21 shows a position control system with PD control. Find the values of the gains K_p and K_D to meet all of the following specifications:

(1) No steady-state error with a step input.
(2) A damping ratio of 0.9.
(3) A dominant time constant of 1.

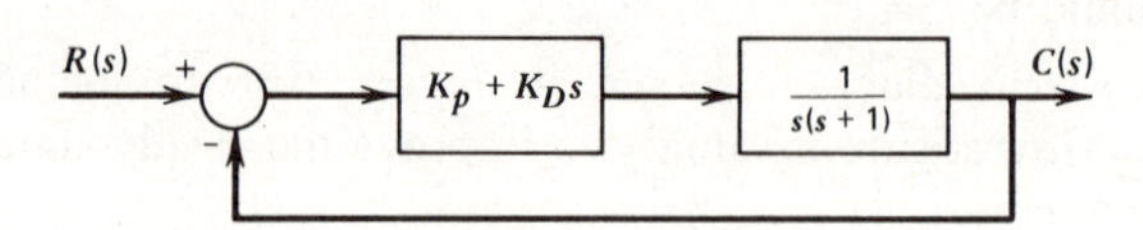

FIGURE P6.21

6.22 Figure P6.22 shows a PID-control system with the same plant as in Problem 6.21. For the values of K_p and K_D found previously,

(a) Find the maximum value K_I can have without causing the system to be unstable.

(b) If we relax the requirement that $\zeta = 0.9$ but still require that $\tau = 1$, then $s = -1$ is one characteristic root. Is it possible to select K_I so that $s = -1$ is the dominant root? If so, find the value of K_I required to do this.

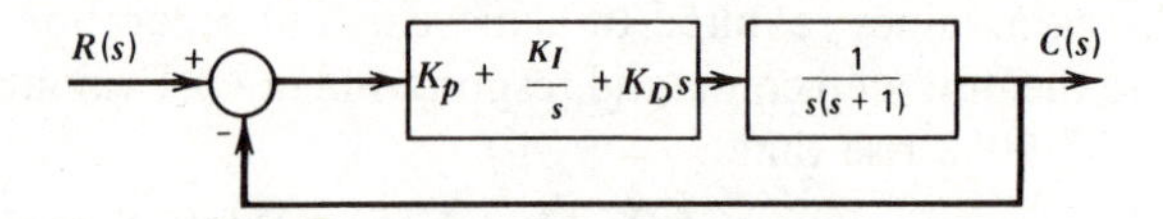

FIGURE 6.22

6.23 In the absence of a disturbance, the original system shown in Figure P6.23*a* gave a satisfactory steady-state output of $c_{ss} = 1$ when the input was a unit step, and it had a satisfactory time response (the output took 4/5 sec to reach 98% of its final value). However, with a step disturbance, the performance is not satisfactory, so the controller $G(s)$ and the feedback loop have been added to compensate for the disturbance without changing the satisfactory aspects of the original system. This new system is shown in Figure P6.23*b*.

Design the controller. The choice is limited to *P*, *I*, or PI control. With unit-step command and disturbance inputs, the specifications are

(1) No steady-state offset error in the absence of a disturbance.
(2) No steady-state deviation resulting from the disturbance.
(3) The output must respond as quickly as possible to the command input but without oscillation.
(4) The output must take no longer than 4/5 sec to reach and stay within approximately 98% of the final value in the absence of a disturbance. (The dominant-root approximation is acceptable here.)

6.24 A speed control system with proportional control is shown in Figure P6.24. Find the minimum gain K_p such that the speed deviation due to a 500 lb-in. step

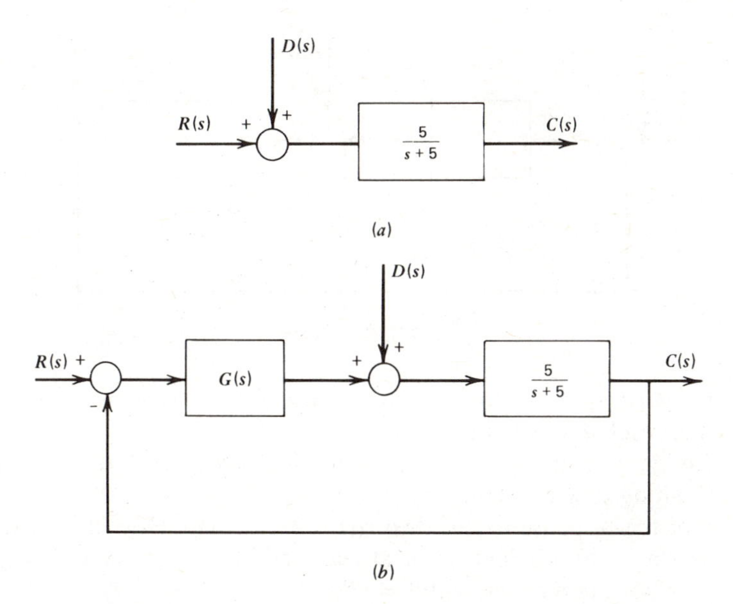

FIGURE P6.23 (*a*) Original system. (*b*) New system.

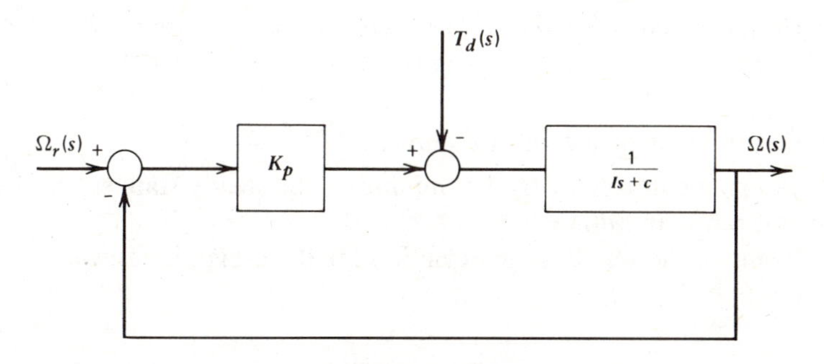

FIGURE P6.24

disturbance will not exceed 2 rad/sec. The plant's time constant is 2 sec and $I = 200$ lb-in.-sec^2. The gain K_p must be positive for physical reasons.

6.25 Integral control of the plant

$$G_p(s) = \frac{K}{\tau s + 1}$$

results in a system that is too oscillatory. Will D action improve this situation?

6.26 Consider the first-order system shown in Figure P6.26 and suppose that $I = 10$ and $c = 0.1$. Consider the following performance specifications:

(1) The magnitude of the steady-state error must be no more than 0.01 when $r(t)$ is a unit ramp and $d(t) = 0$.

(2) The damping ratio must be unity.

(3) The dominant time constant must be no greater than 0.1.

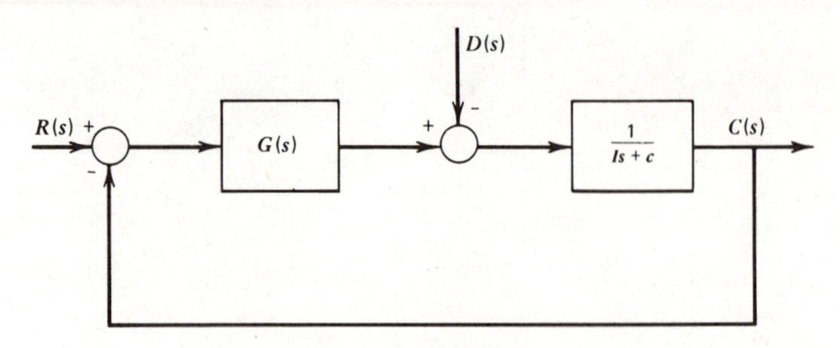

FIGURE P6.26

(a) Design a controller to meet the first specification. What is the damping ratio that results? What is the time constant?

(b) Design a controller $G(s)$ to meet the first two specifications. Evaluate the resulting time constant.

(c) Plot the response in (b) when $r(t)$ is a unit step. How can we modify the design in (b) to eliminate or at least reduce the overshoot? Why does an overshoot occur even with $\zeta = 1$?

(d) For the design obtained in (b), what is the deviation caused by a unit-step disturbance? By a unit-ramp disturbance?

(e) Design a controller $G(s)$ to meet all three specifications. Plot the response for $r(t)$ a unit step and $d(t) = 0$. Discuss the results in light of the second and third specifications.

6.27 Consider the system shown in Figure P6.27.

(a) Design a controller $G(s)$ that minimizes the steady-state error when $r(t)$ is a unit ramp and $d(t) = 0$.

(b) Evaluate the steady-state deviation for the design in (a) when $d(t)$ is a unit ramp.

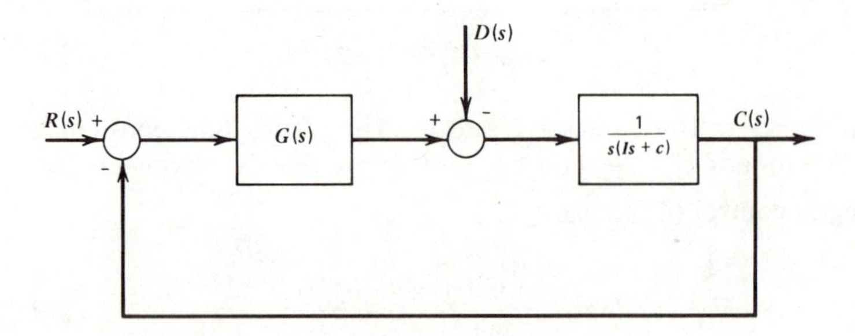

FIGURE P6.27

6.28 Consider the system shown in Figure P6.28. Suppose that $I = 10$, $c = 0.1$, and $a = 1$.

(a) Design a controller $G(s)$ to give a minimum steady-state error when $r(t)$ is a unit step and $d(t) = 0$. The dominant time constant should be no greater than 0.1 sec, and the damping ratio ζ should be unity.

(b) Compute the step response of the design in (a). What is the effect of the numerator dynamics $(s + a)$ on the design and its performance?

(c) Compute the steady-state error of the design in (a) if $r(t)$ is a unit ramp and $d(t) = 0$.

(d) Evaluate the steady-state deviation resulting from a unit-step disturbance.

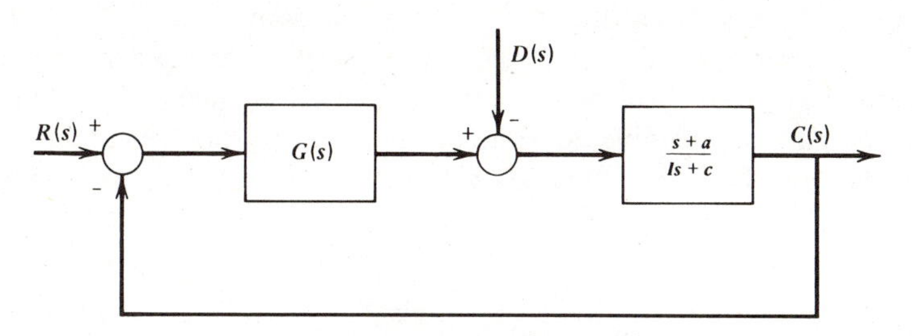

FIGURE P6.28

6.29 The plant shown in Figure P6.29 is representative of a vibratory system. Suppose that $m = 1$, $c = 1$, and $k = 4$. Suppose also that $r(t)$ is a unit step and $d(t) = 0$.

(a) Design a controller $G(s)$ so that the magnitude of the steady-state error e_{ss} is no greater than 0.1. Evaluate the resulting time constant and damping ratio, and compare with those of the plant.

(b) Suppose we wish the dominant time constant to be no greater than 1 sec, and that $|e_{ss}| \leqslant 0.1$. Design the controller. Evaluate the damping ratio.

(c) Now suppose that instead of specifying the time constant, we specify that the damping ratio should be no less than 0.707. In addition, we require that $|e_{ss}| \leqslant 0.1$. Design the controller, and evaluate the time constant.

(d) What type of controller might be used to satisfy all three specifications (e_{ss}, τ and ζ)? Do not compute the gains.

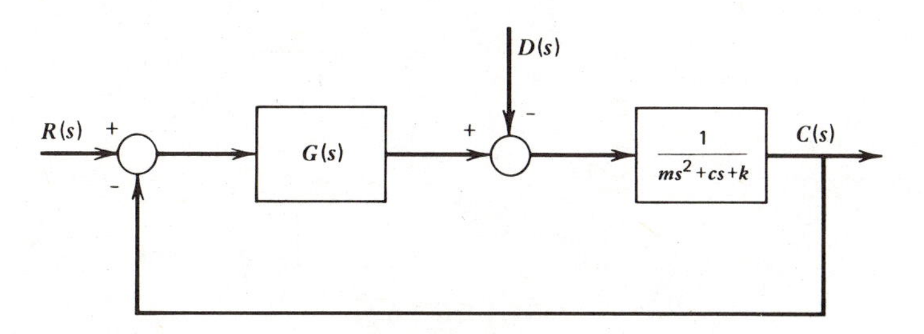

FIGURE P6.29

6.30 The attitude of a vehicle moving through a fluid represents an inherently unstable plant, because the fluid forces do not act through the vehicle's center of mass, but through a point called the center of pressure. They thus create a net torque that acts to rotate the vehicle. The attitude can be controlled by control surfaces, such as elevators and ailerons on missiles and aircraft, by thrusters on rockets, and by diving planes and fins on submarines and other underwater vehicles.

An example of such a system is shown in Figure P6.30. A linearized model of the missile's dynamics is

$$I\ddot{\theta} = T + C_n L\theta$$

where θ is the missile's attitude angle, I is the missile's inertia about the indicated axis of rotation, C_n is the normal-force coefficient, L is the distance between the center of mass and the center of pressure, and T is the applied control torque.

The plant is unstable, and the primary control objective is to stabilize the missile's attitude so that $\theta \cong 0$. Develop a control law for this purpose.

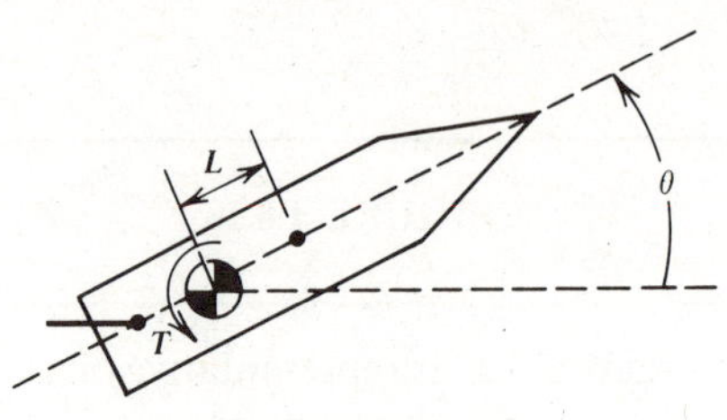

FIGURE P6.30

6.31 The control variable for the tank system shown in Figure P6.31 is q_1, the flow rate into tank 1. It is desired to control the height h_2 with a measurement of h_2.

Develop a control law that will give a zero steady-state error for a step command h_{2d} and a zero steady-state deviation for a step disturbance q_d.

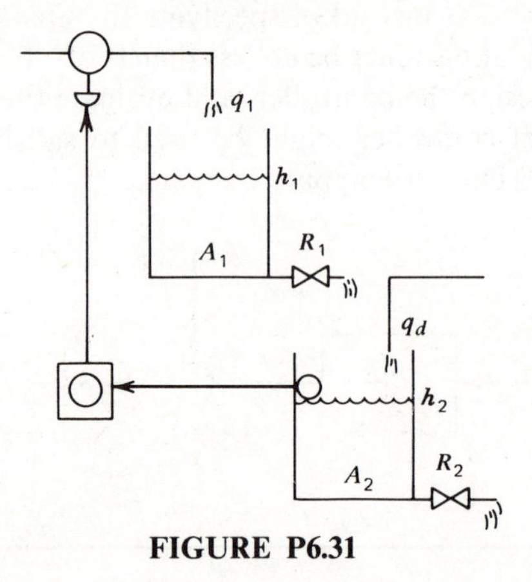

FIGURE P6.31

6.32 The system shown in Figure P6.32 can be thought of as representing the problem of stabilizing the attitude of a rocket during takeoff. The applied force f represents that from the side thrusters of the rocket. The linearized form of Newton's law for the system reduces to

$$ML\ddot{\theta} - (M+m)g\theta = f$$

where f is the control variable.

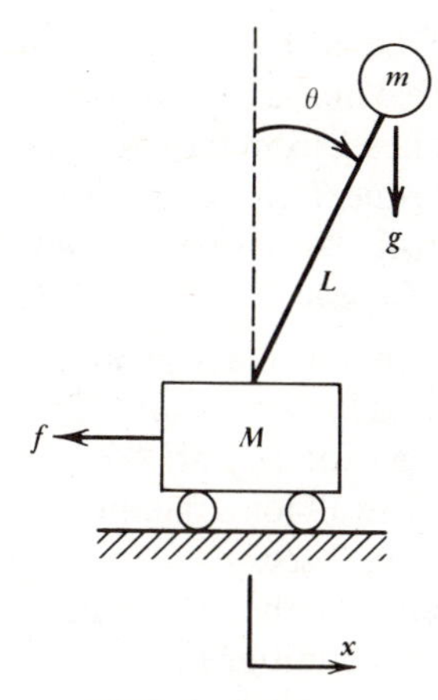

FIGURE P6.32

Design a control law to maintain θ near zero. The specifications are $\zeta = 0.707$ and a 2% settling time of 10 sec. The parameter values are $M = 50$ slugs, $m = 10$ slugs, $L = 30$ ft, and $g = 32.2$ ft/sec^2.

6.33 In many applications, such as tape drives or cutting tools, the controlled object must be accelerated to a certain speed, held at that speed, and then decelerated to zero speed at prescribed times. One such speed profile is shown in Figure P6.33. Suppose that the plant is $1/(s+1)$ and the motor amplifier to be used has a gain of $K_m = 10$.

(a) Design a controller such that the speed $c(t)$ at $t = 0.4$ is within 2 rad/sec of the desired value of 100 rad/sec at that time, assuming that $c(0) = \dot{c}(0) = 0$.

(b) Plot the response for one cycle, and compare it to the desired profile. What is the maximum deviation from the desired speed on each part of the profile?

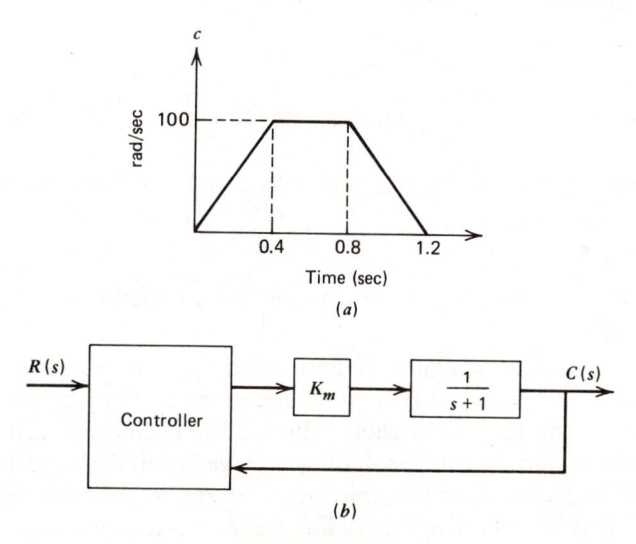

FIGURE P6.33

6.34 The problem is to design a controller to position an unbalanced load of mass m. We can represent the unbalance as a pendulum with a point mass m located at the center of mass, a distance L from the center of rotation (see Figure P6.34). The control variable is the torque T to be supplied by a motor whose dynamics we will ignore for now. We will also assume for now that the motor can produce as much torque as needed. Take $mg = 1$ lb, $g = 32.2$ ft/sec^2, and $L = 0.25$ ft.

(a) We desire to position the mass m at $\theta = \pi$ rad. Obtain the governing dynamic equations and linearize them. Use the linearized model to design a controller that will position m at $\theta = \pi$ rad exactly at steady state. The transient specifications call for a damping ratio $\zeta \geqslant 0.707$ and a 2% settling time of no greater than 2 sec.

(b) Compute the response for the design obtained in (a), using the linearized model with the initial conditions

(1) $\theta(0) = 2.9$ rad, $\dot{\theta}(0) = 0$.
(2) $\theta(0) = 1.6$ rad, $\dot{\theta}(0) = 0$.

Does the response satisfy that demanded by the specifications? (i.e., does the response settle quickly without much overshoot?).

(c) Simulate the response of the design obtained in part a using the *nonlinear* dynamic model with the initial conditions given in (b). Evaluate the performance of the controller.

(d) What is the maximum static torque that the motor must be capable of supplying in order for the load to be kept at $\theta = \pi$ rad?

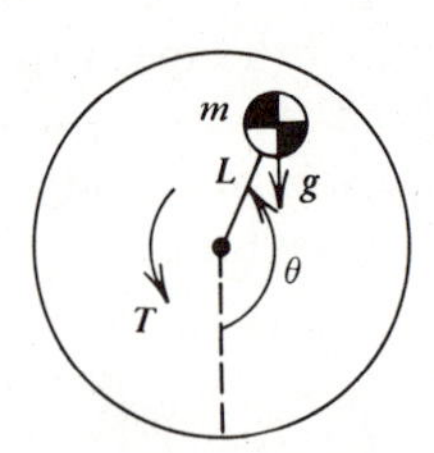

FIGURE P6.34

6.35 Repeat Problem 6.34 except position the load at $\theta = 0$. For the simulations, use

(1) $\theta(0) = 0.2$, $\dot{\theta}(0) = 0$
(2) $\theta(0) = 1.5$, $\dot{\theta}(0) = 0$

Discuss how the performance of this design compares with that found in Problem 6.34.

6.36 The following problem is typical of those found in the textile and wire industries. A reel containing yarn, wire, or cable, for example, must be unwound while maintaining a constant linear cable velocity. The tachometer voltage indicates this velocity v and is compared to a voltage from the potentiometer that represents the desired velocity v_d (see Figure P6.36). As the cable unwinds due to its own weight, the effective reel radius decreases, and thus the reel's angular velocity $\dot{\theta}$ must increase.

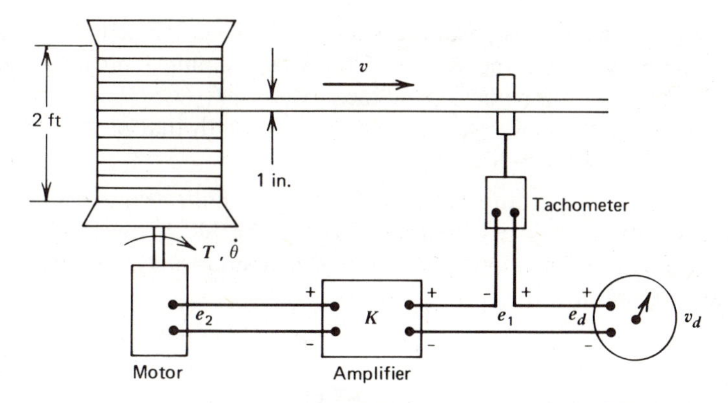

FIGURE P6.36

Suppose that the reel's effective radius is 2 ft when empty and 3 ft when full. The reel's shape resembles a hollow cylinder, and its inertia is taken to be

$$I = 16.5R^4 - 150 \text{ ft-lb-sec}^2$$

where R is the effective radius of the reel, which will be a function of time. The reel width is 2 ft, and the cable diameter is 1 in. Assume that the motor's inertia is negligible, and take its transfer function to be

$$\frac{T(s)}{E_2(s)} = \frac{5}{2s+1}$$

where $T(s)$ and $E_2(s)$ are the motor torque and voltage. Let the tachometer time constant be 1 sec so that

$$\frac{E_1(s)}{V(s)} = \frac{2}{s+1}$$

where $E_1(s)$ is the output voltage of the tachometer and $V(s)$ is the cable velocity. The potentiometer constant is $K_1 = 2$ V/ft/sec. The amplifier gain K implements proportional control.

(a) For a given layer of cable, we assume that R and I are constant to obtain an invariant model. Draw the block diagram and obtain the transfer function $V(s)/V_d(s)$ in terms of R and I.

(b) Use the Routh–Hurwitz criterion to find the range of K over which the system will be stable. Compare the ranges at the two limits represented by an empty and a full reel.

6.37 Determine the resistance values required to obtain an electronic PI controller with $K_p = 2$ and $K_I = 0.04$. Use a 1-μf capacitor.

6.38 (a) Determine the resistance values to obtain an electronic PD controller with $K_p = 1$, $T_D = 1$ sec. The circuit should limit frequencies above 10 rad/sec. Use a 1 μf capacitor.

(b) Plot the frequency response of the circuit. Compare with the plot of the PD controller without frequency limiting.

6.39 **(a)** Determine the resistance values to obtain an electronic PID controller with $K_p = 5$, $K_I = 0.7$, and $K_D = 2$. The circuit should limit frequencies above 250 rad/sec. Take one capacitance to be 1 μf.

(b) Plot the frequency response and compare with that of the PID controller without frequency limiting.

6.40 A pneumatic temperature controller using a bulb transducer is shown in Figure P6.40. Explain its operation. Draw the block diagram, and find the transfer function. The input is the desired temperature T_d, and the output is the actual temperature T.

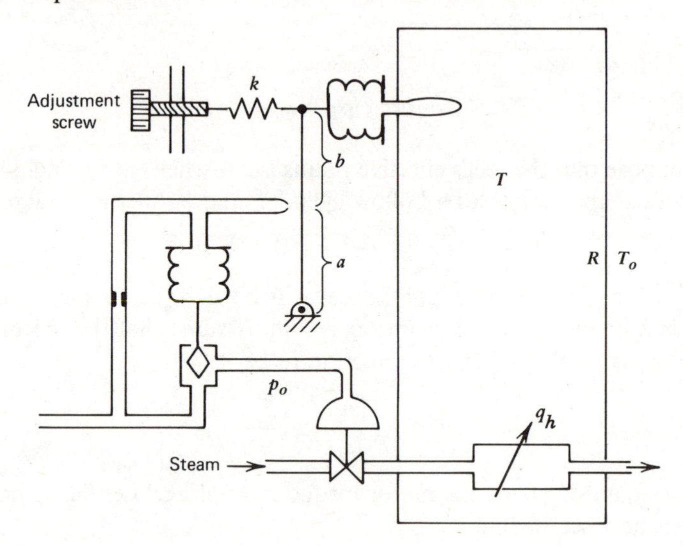

FIGURE P6.40

6.41 The pneumatic stack controller shown in Figure P6.41 can be operated with the upper outlet restriction either fully closed or partly open. Obtain the transfer function between p_o and $(p_r - p_c)$ for each mode of operation. Identify the control action in each case.

6.42 Final control elements often have their own feedback loops to improve their performance. An example is the pneumatic valve positioner shown in Figure P6.42. If the position of the actuating valve does not correspond to that demanded by the control pressure, the positioner will act until the valve position is correct. Explain the operation of the system and draw a schematic block diagram.

6.43 The hydraulic jet pipe controller shown in Figure P6.43*a* is an alternative to the spool valve design. Instead of moving the spool valve, the input displacement z moves the jet pipe, or nozzle, which ejects fluid at high pressure. When the pipe is in the neutral position in the center, no fluid enters either side of the piston chamber.

(a) Derive the transfer function between the output displacement x and the input displacement z. What type of control action results?

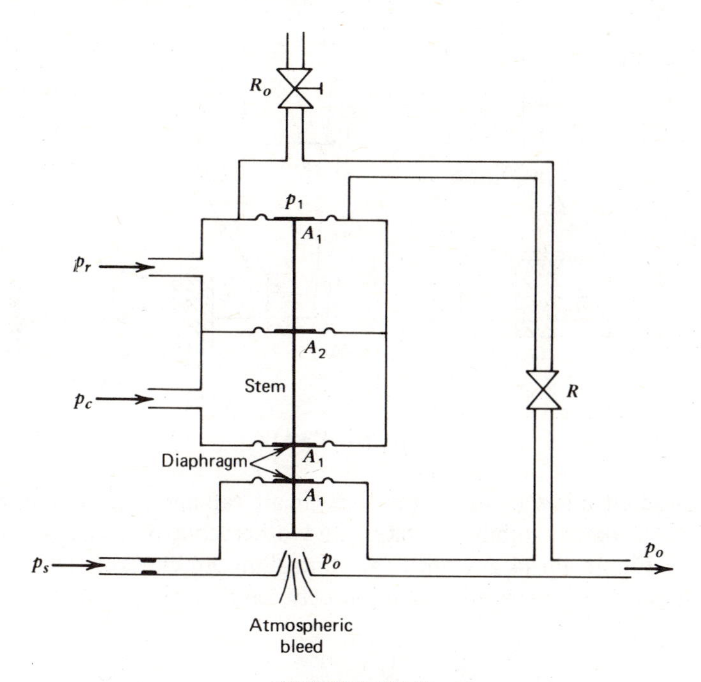

FIGURE P6.41

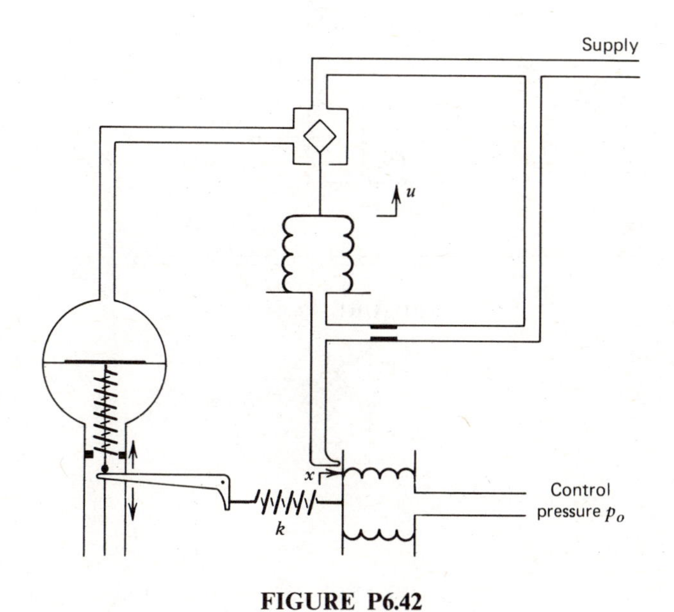

FIGURE P6.42

(b) The system in (a) has been modified with a linkage. Derive the transfer function between y and x. What type of control action results?

(c) How could the jet pipe unit be used to produce PI action?

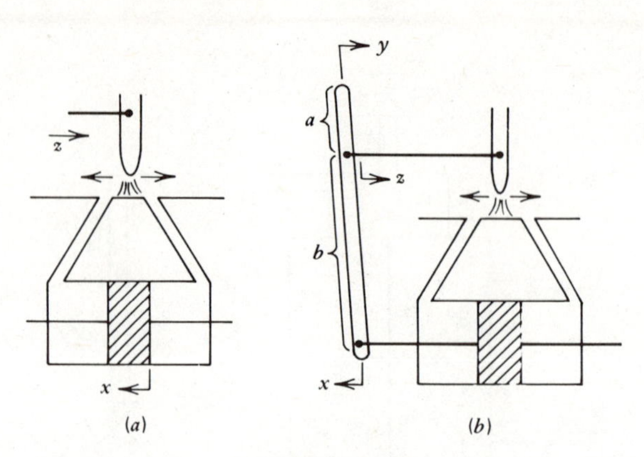

FIGURE P6.43

6.44 Hydraulic controllers are used extensively in aircraft and ships for controlling the motion of rudders, stabilizers, flaps, and other control surfaces. The system shown in Figure P6.44 is a unit for controlling an elevator. Draw the block diagram and find the transfer function between θ and y.

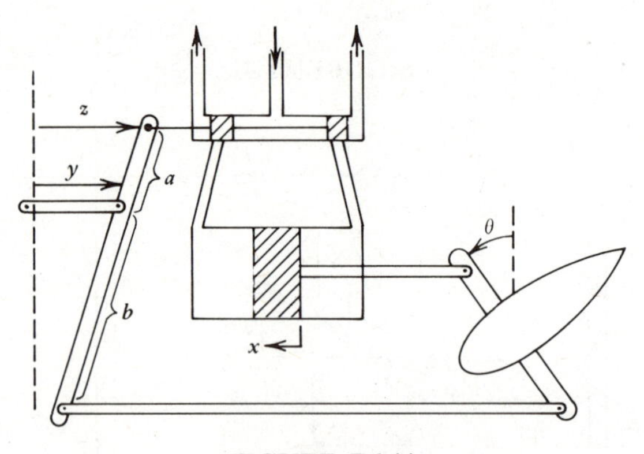

FIGURE P6.44

CHAPTER SEVEN

Control System Design: Modeling Considerations and Alternative Control Structures

The models used to describe the plant, sensors, and actuators of a control system greatly influence the design and performance of a control law. Once the form of the law is selected, its gains must be computed from the system model and the given performance specifications. Often a transfer function model of the plant is not available, and the gains must be computed from experimentally determined open-loop response data. Also, systematic procedures for calculating the control gains to achieve the desired stability, accuracy, and speed of response characteristics can be difficult to develop, because such specifications are often loosely stated. For example, the results of choosing the gains to give a dominant root with a damping ratio of $\zeta = 0.7$ can be unclear, because the other roots might have a significant effect on the response. Thus, more direct procedures for computing the gains are sometimes required. Methods for treating such cases are developed in Section 7.1.

Assumptions made to reduce the order of the model and to obtain a linear model can have a significant effect on the controller's performance. In particular, when performance specifications are stringent and large control gains must be used, behavior not predicted by a reduced-order model can become significant. In addition, selecting high gain values tends to drive the control elements to such an extent that they overload or "saturate" and thus exhibit nonlinear behavior. Design procedures that avoid these unwanted effects are covered in Sections 7.2 and 7.3.

When a PID-type control law fails to yield the desired performance, it must be either modified or replaced by an entirely different control scheme. Inserting additional control elements to modify the control law is called *compensation*, and Section 7.4 discusses several types of compensation schemes. Compensation often leads to a satisfactory design if the PID law is capable of satisfying most of the performance specifications.

When compensating the PID law still does not produce an acceptable design, a different control scheme must be developed. Multiple feedback loops have been effective, by using either all the state variables to form the actuating signal (Section 7.5), or an inner loop to modify the controller's output. Section 7.6 presents one such scheme that has recently been developed, called *pseudo derivative feedback*. It has advantages over the traditional PID algorithms and is often selected first instead of trying to achieve satisfactory performance by compensating the PID law.

Unacceptable performance often results when two different control loops affect each other's behavior. If this interaction is significant, the design must be approached from the viewpoint of the total system, using a multiple-input, multiple-output control scheme (Section 7.7).

7.1 SELECTING CONTROLLER GAINS

Once the form of the control law has been selected, the gains must be computed in light of the performance specifications. In the examples of the PID family of control laws in Sections 6.5–6.7, the damping ratio, dominant time constant, and steady-state error were taken as the primary indicators of system performance in the interests of simplicity. In practice, the criteria are usually more detailed. For example, the rise time and maximum overshoot as well as the other transient response specifications in Section 3.5 may be encountered. Requirements can also be stated in terms of frequency response characteristics, such as bandwidth, resonant frequency, and peak amplitude. Whatever specific form they take, a complete set of specifications for control system performance should generally include the following considerations, for given forms of the command and disturbance inputs:

1. Equilibrium specifications.
   - **(a)** Stability.
   - **(b)** Steady-state error.
2. Transient specifications.
   - **(a)** Speed of response.
   - **(b)** Form of response (degree of damping).
3. Sensitivity specifications.
   - **(a)** Sensitivity to parameter variations.
   - **(b)** Sensitivity to model inaccuracies.
   - **(c)** Noise rejection (bandwidth, etc.).
4. Nonlinear effects.
   - **(a)** Stability.
   - **(b)** Final control element capabilities.

In addition to these performance stipulations, the usual engineering considerations of initial cost, weight, maintainability, and so forth must be taken into account. The considerations are highly specific to the chosen hardware, and it is difficult to deal with such issues in a general way. Here, we limit ourselves to the performance considerations listed.

Two approaches exist for designing the controller. The proper one depends on the quality of the analytical description of the plant to be controlled. If an accurate model of the plant is easily developed, the approach is to design a specialized controller for the particular application. The range of adjustment of controller gains in this case can usually be made small, because the accurate plant model allows the gains to be precomputed with confidence. This technique reduces the cost of the controller and can usually be applied to electromechanical systems.

The second approach is used when the plant is relatively difficult to model, which is often the case in process control. A standard controller with several control modes and wide ranges of gains is used, and the proper mode and gain settings are obtained by testing the controller on the process in the field. Some guidelines for such tuning are given in this section. This approach should be considered when the cost of developing an accurate plant model might exceed the cost of controller tuning in the field. Of course, the plant must be available for testing for this approach to be feasible.

Performance Indices

The performance criteria encountered thus far require a set of conditions to be specified—for example, one for steady-state error, one for damping ratio, and one for the dominant time constant. If there are many such conditions and if the system is of high order with several gains to be selected, the design process can become quite complicated, because transient and steady-state criteria tend to drive the design in different directions. An alternative approach is specifying the system's desired performance by means of one analytical expression called a *performance index*. Powerful analytical and numerical methods are available that allow the gains to be systematically computed by minimizing (or maximizing) this index. This approach forms the basis for many advanced design methods. Here, we introduce the concept.

To be useful, a performance index must be selective. The index must have a sharply defined extremum in the vicinity of the gain values that give the desired performance. If the numerical value of the index does not change very much for large changes in the gains from their optimal values, the index will not be selective. We will discuss only minimizing the index. Applications requiring maximization of an index can be converted into minimization problems by changing the sign of the index.

Any practical choice of a performance index must be easily computable, either analytically, numerically, or experimentally. Four common choices for an index are the following:

$$J = \int_0^\infty |e(t)|\, dt \qquad \text{(IAE)} \qquad \textbf{(7.1-1)}$$

$$J = \int_0^\infty t|e(t)|\, dt \qquad \text{(ITAE)} \qquad \textbf{(7.1-2)}$$

$$J = \int_0^\infty e^2(t)\, dt \qquad \text{(ISE)} \qquad \textbf{(7.1-3)}$$

$$J = \int_0^\infty te^2(t)\, dt \qquad \text{(ITSE)} \qquad \textbf{(7.1-4)}$$

where $e(t)$ is the system error. This error is usually the difference between the desired and the actual values of the output. However, if $e(t)$ does not approach zero as $t \to \infty$, the preceding indices will not have finite values. In this case, $e(t)$ can be defined as $e(t) = c(\infty) - c(t)$, where $c(t)$ is the output variable. If the index is to be computed numerically or experimentally, the infinite upper limit can be replaced by the limit t_f, where t_f is large enough so that $e(t)$ is negligible for $t > t_f$.

The *integral absolute-error* (IAE) criterion (7.1-1) expresses mathematically that the designer is not concerned with the sign of the error, only its magnitude. In some applications, the IAE criterion describes the fuel consumption of the system. The index says nothing about the relative importance of an error occurring late in the response versus an error occurring early. Because of this, the index is not so selective as the *integral-of-time-multiplied absolute-error* criterion (ITAE) (7.1-2). Since the multiplier t is small in the early stages of the response, this index weights early errors less heavily than later errors, which makes sense physically. No system can respond instantaneously, and the index is lenient accordingly, while punishing any system that allows a nonzero error to remain for a long time. Neither criterion allows highly underdamped or highly overdamped systems to be optimum. The ITAE criterion usually results in a system whose step response has a slight overshoot and well-damped oscillations.

The *integral squared-error* (ISE) and *integral-of-time-multiplied squared-error* (ITSE) criteria are analogous to the IAE and ITAE criteria except that the square of the error is employed, for three reasons: (1) in some applications, the squared error represents the system's power consumption; (2) squaring the error weights large errors much more heavily than small errors; (3) the squared error is much easier to handle analytically. The derivative of a squared term is easier to compute than that of an absolute value and does not have a discontinuity at $e = 0$. These differences are important when the system is of high order with multiple error terms. This case is treated at an advanced level (see, for example, chapter 10 of Reference 1 and chapter 10 of Reference 2).

The ISE and ITSE criteria are similar to the least-squares technique for curve fitting. We can think of the process as one of finding the response curve that minimizes the total squared error. The ITSE index is analogous to weighted least-squares, with the weighting factor being proportional to time. Systems designed with the ISE criterion tend to be fast and oscillatory. The ITSE index provides better selectivity than the ISE. The following example illustrates the procedure.

Example 7.1

We want to use proportional control to keep the state x and the rate $\dot{x}$ near zero for some arbitrary initial condition $x(0)$. The state equation is $\dot{x} + x = u$, where the input u is given by the proportional law $u = -Kx$. (Here, the desired output is $x = 0$; thus, the error is $0 - x = -x$.) Find the gain K.

Since we want to keep both x and $\dot{x}$ near zero, the performance index should reflect this. Choosing a weighted sum of the squares of the deviations, we obtain

$$J = \int_0^\infty [w_1 x^2(t) + w_2 \dot{x}^2(t)]\, dt \qquad \textbf{(7.1-5)}$$

where w_1 and w_2 are the constant weighting factors. With the control law $u = -Kx$, the system equation is $\dot{x} + x = -Kx$. Its solution is $x(t) = x(0) \exp[-(1+K)t]$. Substituting this expression into (7.1-5) and integrating gives

$$J = x^2(0)[w_1 + w_2(1+K)^2]\frac{1}{2(1+K)} \qquad \textbf{(7.1-6)}$$

To minimize J with respect to K, we require that

$$\frac{\partial J}{\partial K} = 0, \qquad \frac{\partial^2 J}{\partial K^2} \geqslant 0$$

For a stable system, $1 + K > 0$, and the minimizing value of K is

$$K = -1 + \sqrt{\frac{w_1}{w_2}} \tag{7.1-7}$$

Some insight is gained from examining the extreme values of the weights. For $w_1 = 0$, no weight is given to minimizing x, and the solution $K = -1$ results in $x(t) = x(0)$. The velocity is thus minimized. For $w_2 = 0$, $K \to \infty$, which indicates that the controller is trying to drive the state to zero in an infinitesimal amount of time. The general solution for K lies between these two extremes. If we weight x more heavily than $\dot{x}$, $w_1 > w_2$ and K is large. We might argue that the problem has not been solved, but merely shifted from one of finding K to finding w_1 and w_2. In a sense, this is true, and we rarely use the performance index concept for such simple systems. However, when many gains must be computed, this approach offers a systematic way of doing so.

The general second-order system with zero steady-state error for a unit-step input has the form

$$\frac{C(s)}{R(s)} = \frac{\omega_n^2}{s^2 + 2\zeta\omega_n s + \omega_n^2}$$

If we introduce a new time scale T such that $t = \omega_n T$, the transfer function becomes

$$\frac{C(s)}{R(s)} = \frac{1}{s^2 + 2\zeta s + 1} \tag{7.1-8}$$

The step response can easily be found in terms of the parameter ζ and the various performance indices computed, but the algebra required is enough to make us think twice about attempting an analytical approach with a third-order system. For $e(t) = 1 - c(t)$, the results show that the ITAE index and the index

$$J = \int_0^\infty t(|e| + |\dot{e}|)\, dt \tag{7.1-9}$$

have the most sharply defined minima as functions of ζ. For these two indices, the optimum value of ζ occurs at approximately $\zeta = 0.7$. The other indices might be more selective when applied to a different system and should not be discarded on the basis of this example.

In conclusion, we note that the closed-form solution for the response is not required to evaluate a performance index. For a given set of parameter values, the response and resulting index value can be computed numerically—for example, with the Runge–Kutta method in Chapter 5 and Appendix C. Using systematic search procedures, such as those given in Reference 3, the optimum solution can be obtained; this makes the procedure suitable for use with nonlinear systems.

The Ziegler–Nichols Rules

The difficulty of obtaining accurate transfer function models for some processes has led to the development of empirically based rules of thumb for computing the optimum gain values for a controller. Commonly used guidelines are the Ziegler–Nichols rules, which have proved so helpful that they are still in use 40 years after their development (References 4 and 5). The rules actually consist of two separate methods. The first method requires the open-loop step response of the plant, while the second uses the results of experiments performed with the controller already installed. While primarily intended for use with systems for which no analytical model is available, the rules are also helpful even when a model can be developed.

Ziegler and Nichols developed their rules from experiments and by analyzing various industrial processes. Using the IAE criterion with a unit-step response, they found that controllers adjusted according to the following rules usually had a step response that was oscillatory but with enough damping so that the second overshoot was less than 25% of the first (peak) overshoot. This is the *quarter-decay* criterion, and it is sometimes used as a specification.

The first method, the *process reaction method* relies on the fact that many processes have an open-loop step response like that shown in Figure 7.1. This is the process *signature* characterized by two parameters, R and L. The R is the slope of a line tangent to the steepest part of the response curve, and L is the time at which this line interests the time axis. First- and second-order linear systems do not yield positive values for L, and so the method cannot be applied to such systems. However, third-order and higher linear systems with sufficient damping do yield such a response. If so, the Ziegler–Nichols rules recommend the controller settings given in Table 7.1.

The *ultimate-cycle method* uses experiments with the controller in place. All control modes except proportional are turned off, and the process is started with the proportional gain K_p set at a low value. The gain is slowly increased until the process begins to exhibit sustained oscillations. Denote the period of this oscillation by P_u and the corresponding ultimate gain by K_{pu}. The Ziegler–Nichols recommendations are given in Table 7.1 in terms of these parameters. The proportional gain is lower for PI control than for P control and higher for PID control, because I action increases the order of the system and thus tends to destabilize it; thus, a lower gain is needed. On the other hand D action tends to stabilize the system; hence, the proportional gain can be increased without degrading the stability characteristics. Because rules were developed for a typical case out of many types of processes, final tuning of the gains in the field is usually necessary. Example 7.2 shows how the ultimate-cycle method can be applied to an analytical model.

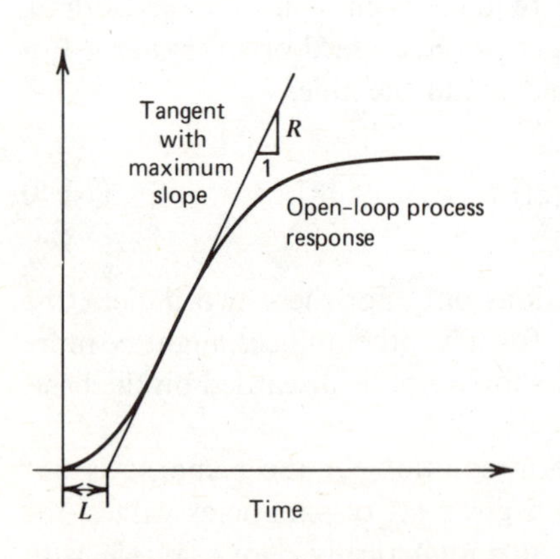

FIGURE 7.1 Process signature for a unit-step input. The parameters R and L are determined for use with the process-reaction method of Ziegler and Nichols.

TABLE 7.1 The Ziegler–Nichols Rules

Controller transfer function $G(s) = K_p\left(1 + \frac{1}{T_I s} + T_D s\right) = K_p + \frac{K_I}{s} + K_D s$

Control Mode	Process Reaction Method	Ultimate Cycle Method
P control	$K_p = 1/RL$	$K_p = 0.5\ K_{pu}$
PI control	$K_p = 0.9/RL$	$K_p = 0.45\ K_{pu}$
	$T_I = 3.3L$	$T_I = 0.83\ P_u$
PID control	$K_p = 1.2/RL$	$K_p = 0.6\ K_{pu}$
	$T_I = 2L$	$T_I = 0.5\ P_u$
	$T_D = 0.5L$	$T_D = 0.125\ P_u$

Example 7.2

Proportional control of a third-order plant without numerator dynamics leads to the following characteristic equation:

$$a_3 s^3 + a_2 s^2 + a_1 s + a_0 + K_p = 0 \tag{7.1-10}$$

where K_p is the proportional gain. Find the ultimate period P_u and the associated gain K_{pu}.

Sustained oscillation occurs when a pair of roots is purely imaginary and the rest of the roots have negative real parts. To find when this occurs, set $s = i\omega_u$, where ω_u is the as-yet-unknown frequency of oscillation. Then $s^2 = -\omega_u^2$ and $s^3 = -i\omega_u^3$. Substituting these into (7.1-10) and collecting real and imaginary parts, we obtain

$$(a_0 + K_{pu} - a_2\omega_u^2) + i(a_1\omega_u - a_3\omega_u^3) = 0$$

Both the real and imaginary terms must be zero, so we obtain two equations to solve for the two unknowns ω_u and K_{pu}.

$$a_0 + K_{pu} - a_2\omega_u^2 = 0$$
$$\omega_u(a_1 - a_3\omega_u^2) = 0$$

The latter equation yields the nonoscillatory solution $\omega_u = 0$ as well as the one of interest—namely,

$$\omega_u = \pm\sqrt{\frac{a_1}{a_3}} \tag{7.1-11}$$

Substituting this into the first equation, we see that the ultimate gain is given by

$$K_{pu} = a_2\omega_u^2 - a_0 = \frac{a_2 a_1}{a_3} - a_0 \tag{7.1-12}$$

The ultimate period is

$$P_u = \frac{2\pi}{\omega_u} \tag{7.1-13}$$

In order for sustained oscillations to occur, the third root must be negative. The first two roots are $s_1 = i\omega_u$ and $s_2 = -i\omega_u$. The third root can be determined by multiplying the three factors and comparing with the polynomial (7.1-10) after the cubic coefficient has been normalized to unity. This gives

$$(s - i\omega_u)(s + i\omega_u)(s - s_3) = 0$$

or

$$s^3 - s_3 s^2 + \omega_u^2 s - s_3 \omega_u^2 = 0$$

From (7.1-10), we immediately see that $s_3 = -a_2/a_3$. If $s_3 < 0$, the values given by (7.1-12) and (7.1-13) can be used with Table 7.1 to compute the optimum gains. Note that this analysis uses only proportional control, but the gains for other control laws can be computed with these results. The power of the method is shown by the fact that we did not have to solve for the roots of the cubic (7.1-10). Note also that I action applied to this system would result in a fourth-order characteristic equation to solve were it not for the Ziegler–Nichols rules.

For most systems, the value of K_{pu} represents the largest gain that can be used without producing an unstable system. This is a convenient result, as is the result for the third root s_3. In general, for the polynomial

$$s^n + a_{n-1}s^{n-1} + \cdots + a_1 s + a_0 = 0 \qquad \textbf{(7.1-14)}$$

the sum of the roots equals $-a_{n-1}$. This fact is frequently very useful.

Example 7.3

Use the ultimate-cycle method to design P, PI, and PID controllers for the following plant. Compare the step responses for each design.

$$G_p(s) = \frac{1}{(s+1)(s+2)(s+3)}$$

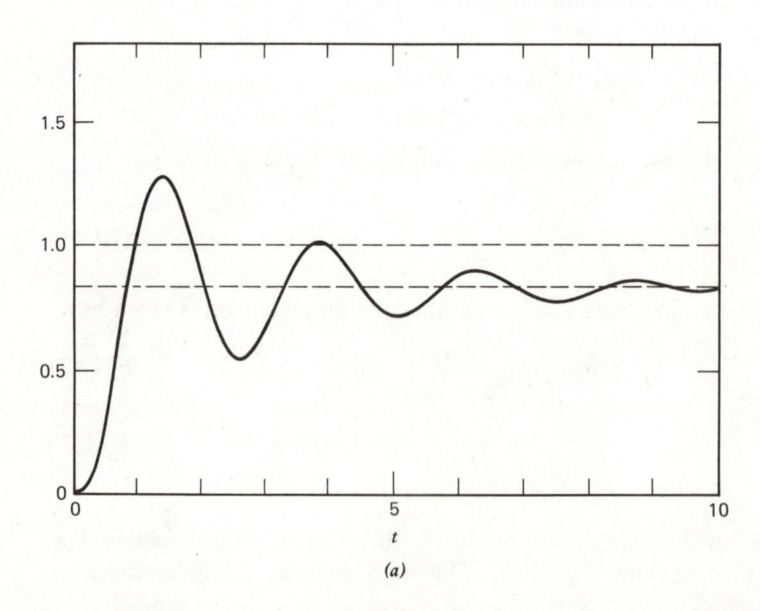

(a)

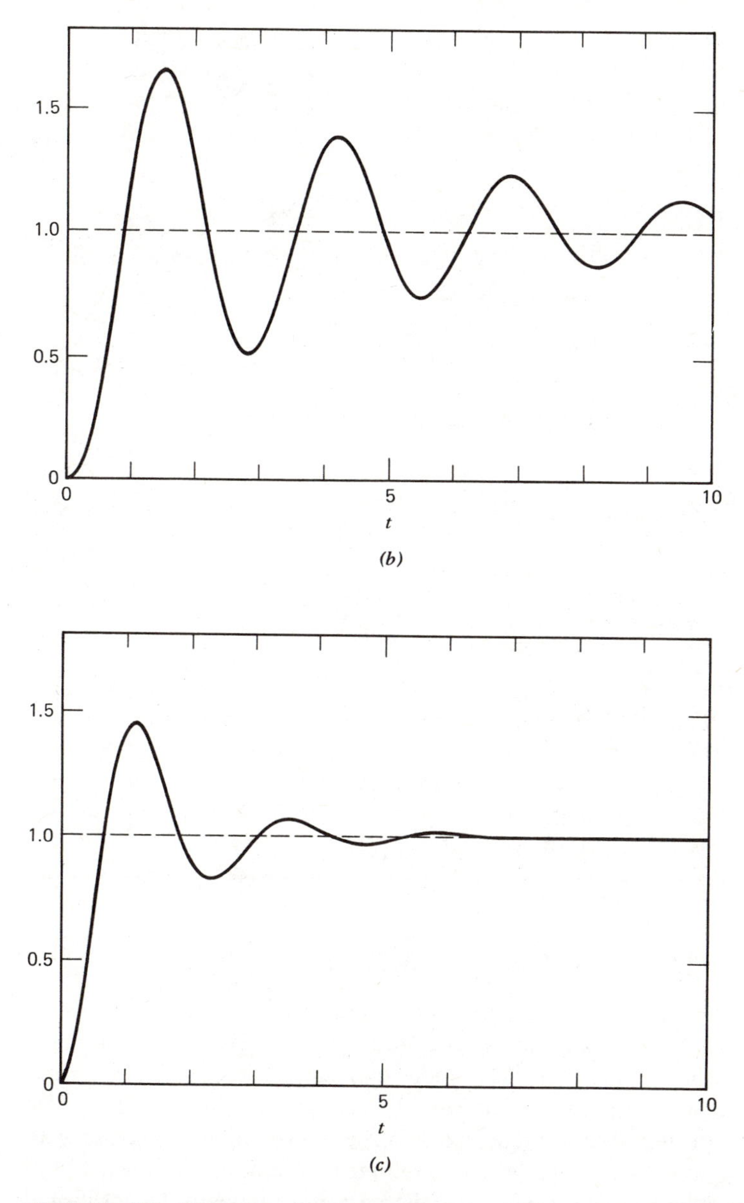

FIGURE 7.2 Unit-step responses of the control systems designed in Example 7.3. (*a*) P-control. (*b*) PI-control. (*c*) PID-control.

Proportional control applied to this plant gives a characteristic equation of the form in (7.1-10), where $a_3 = 1$, $a_2 = 6$, $a_1 = 11$, and $a_0 = 6$. From the results in Example 7.2, we have

$$K_{pu} = 60 \qquad \omega_u = \sqrt{11} = 3.317 \qquad P_u = \frac{2\pi}{3.317} = 1.89$$

For P control, Table 7.1 gives $K_p = 0.5(60) = 30$. For PI control,

$$K_p = 0.45(60) = 27 \qquad T_I = 0.83(1.89) = 1.57 \qquad K_I = \frac{K_p}{T_I} = 17.18$$

Similarly, for PID control,

$$K_p = 0.6(60) = 36 \qquad T_I = 0.5(1.89) = 0.947$$
$$T_D = 0.125(1.89) = 0.236 \qquad K_I = 36/0.947 = 38.1$$
$$K_D = K_p T_D = 36(0.236) = 8.52$$

The closed-loop transfer function for P control is

$$T(s) = \frac{30}{s^3 + 6s^2 + 11s + 36}$$

For PI control,

$$T(s) = \frac{27s + 17.18}{s^4 + 6s^3 + 11s^2 + 33s + 17.18}$$

For PID control

$$T(s) = \frac{8.52s^2 + 36s + 38.1}{s^4 + 6s^3 + 19.52s^2 + 42s + 38.1}$$

The unit-step responses are shown in Figure 7.2. The steady-state error is 0.1667 for P control and zero for PI and PID. Note the improvement as the control law is changed from P to PI to PID. First, the steady-state error is eliminated, and then the oscillations and settling time are reduced. This does not imply that PID should always be used, because the cost of implementation will be greater for PID in general. But Figure 7.2 shows the trend that is often found.

7.2 DESIGN WITH LOW-ORDER MODELS

The development of any model is influenced by our limited knowledge of the system under study as well as by our desire to keep the model simple enough to be useful. A real system always has more energy storage modes than can be described by a model, given the time and cost constraints that inevitably exist for the analysis. Thus, the designer is always working with a model of an order lower than is necessary to describe the system dynamics completely. The system responses that are omitted from the model are termed the *parasitic modes*, and the parasitic roots are the additional roots that these modes would add to the characteristic equation.

The art of engineering includes the ability to decide which modes are important and estimate the effects of any parasitic modes on the performance of the system. Here, we investigate this process as it pertains to designing feedback control system.

Design of a DC Servo System

Consider PI control of velocity using an armature-controlled dc motor. The block diagram of this control system is given in Figure 7.3. The salient points of the example

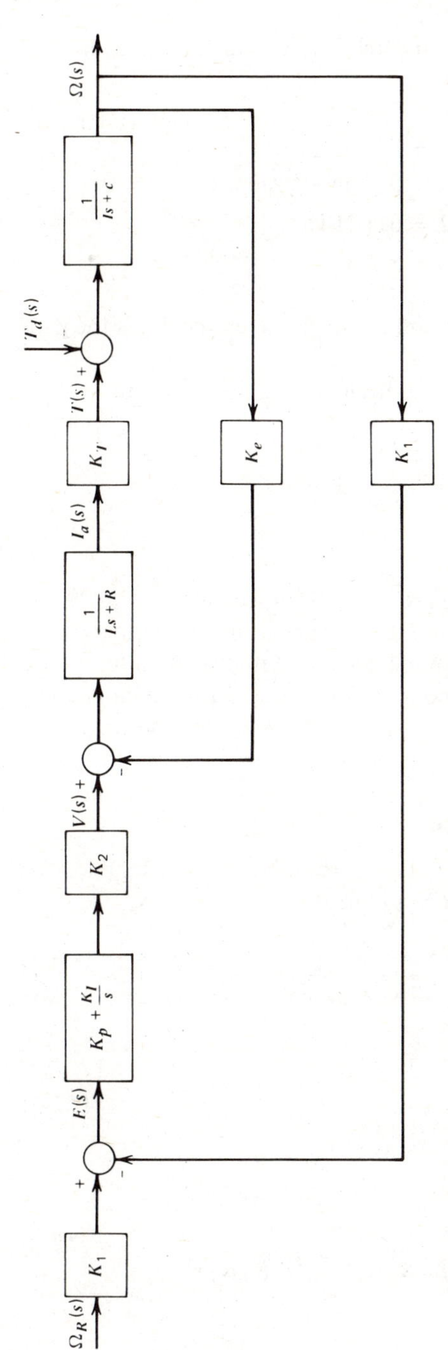

FIGURE 7.3 PI control of velocity with an armature-controlled dc motor.

are more easily understood if we use specific and realistic numerical values for the parameters. We use the following values.

$K_e = 0.199$ V/rad/sec	$R = 0.43\ \Omega$
$K_T = 26.9$ oz-in./A	$c = 0.07$ in.-oz/rad/sec
$I = 0.1$ oz-in.-sec^2	$L = 0.0021$ H
$K_1 = 0.027$ V/rad/sec	$K_2 = 5$ V/V

where the gain K_1 is that of the command pot and the tachometer and K_2 is the gain of the power amplifier.

The time constants of the armature and the mechanical subsystems are

$$\tau_a = \frac{L}{R}$$

$$\tau_m = \frac{I}{c}$$

With the values described, the time constants are $\tau_a = 0.0049$ sec and $\tau_m = 1.429$ sec. Since τ_m is 290 times larger than τ_a, we conjecture that the armature inductance might be neglected to reduce the order of the system model by one. With this assumption, we can attempt the PI controller design with a second-order model. With the inductance retained in the model, a third-order system must be analyzed, and the analytical expressions for the roots in terms of K_p and K_I are difficult to use to select the gains.

Design Assuming Negligible Inductance

Neglect the inductance temporarily, and set $L = 0$. The resulting transfer function for the reference input velocity ω_R is obtained from Figure 7.3.

$$T_1(s) = \frac{\Omega(s)}{\Omega_R(s)} = \frac{K_1 K_2 K_T (K_p s + K_I)}{RIs^2 + (cR + K_e K_T + K_1 K_2 K_T K_p)s + K_1 K_2 K_T K_I} \tag{7.2-1}$$

For the given values,

$$T_1(s) = \frac{3.6315(K_p s + K_I)}{0.043s^2 + (5.3832 + 3.6315K_p)s + 3.6315K_I} \tag{7.2-2}$$

The disturbance transfer function is

$$T_2(s) = \frac{\Omega(s)}{T_d(s)} = \frac{-Rs}{RIs^2 + (cR + K_T K_e + K_T K_1 K_2 K_p)s + K_T K_1 K_2 K_I} \tag{7.2-3}$$

or

$$T_2(s) = \frac{-0.43s}{0.043s^2 + (5.3832 + 3.6315K_p)s + 3.6315K_I} \tag{7.2-4}$$

For now, assume that the performance specifications require the system to deal with step changes in the reference velocity and the disturbance torque. For P action alone ($K_I = 0$), (7.2-2) shows that the system is first order and the steady-state response to a set point change has the value

$$\omega_{ss} = \frac{3.6315K_p}{5.3832 + 3.6315K_p}$$

For ω_{ss} to be close to unity as desired, K_p must be large. Equation (7.2-4) with $K_I = 0$ shows that P action alone cannot give a zero steady-state offset due to the disturbance. Thus, the usual disadvantages of proportional control are encountered here.

If we attempt to use I action alone ($K_p = 0$), (7.2-4) shows that the step disturbance effects will be eliminated but the damping coefficient in the characteristic equation is now beyond our control. The equation is

$$0.043s^2 + 5.3832s + 3.6315K_I = 0$$

For stability, K_I must be positive. We can select K_I to achieve the degree of damping desired, but the time constant is then beyond our control. For example, for $\zeta \leqslant 1$, the time constant is 0.016 sec regardless of the K_I value as long as $K_I \geqslant 46.6$. Note that when $K_p = 0$, most of the damping is caused by the back emf. If this damping is not great enough to achieve the desired time constant, proportional control must be added.

The PI control will give a zero offset at steady state for both the set point and disturbance inputs for any stable combination of K_p and K_I. The gains can then be used to select the damping ratio and dominant time constant. Let us assume that we can accept a 5% overshoot in the response to a step set point change. From Figure 3.14*a*, a 5% overshoot specification is seen to correspond to $\zeta = 0.707$ approximately.

From (7.2-2), the roots of the denominator can be seen to give the following time constant if $\zeta \leqslant 1$:

$$\tau = \frac{0.086}{5.3832 + 3.6315K_p} \tag{7.2-5}$$

Suppose that a time constant no larger than 0.01 sec is desired. For $\tau = 0.01$, the proportional gain must be $K_p = 0.8856$, and therefore the integral gain must be $K_I = 237$ for $\zeta = 0.707$. The characteristic roots will be $s = -100 \pm 100i$.

Now, suppose that we want $\tau \leqslant 0.001$. Then (7.2-5) gives $K_p = 22.2$. For $\zeta = 0.707$, $K_I = 23{,}700$. The roots are $s = -1000 \pm 1000i$. These two cases are tabulated as cases 1 and 2 in the first column of Table 7.2.

The Effects of Armature Inductance

Let us check the performance of each design by predicting its behavior more accurately with the armature inductance included in the model. From Figure 7.3, the two transfer functions are

$$T_1(s) = \frac{\Omega(s)}{\Omega_R(s)} = \frac{K_T K_1 K_2(K_p s + K_1)}{\Delta(s)} \tag{7.2-6}$$

$$T_2(s) = \frac{\Omega(s)}{T_d(s)} = \frac{-s(Ls + R)}{\Delta(s)} \tag{7.2-7}$$

TABLE 7.2 The Effects of Parasitic Armature Inductance L

Armature time constant = 0.0049 sec	
Second-Order Design ($L=0$)	**Third-Order Behavior ($L=0.0021$)**
1. $\tau \leqslant 0.01$, $\zeta = 0.707$ $K_p = 0.8856$ $K_I = 237$ $s = -100 \pm 100i$	$\tau = 0.027$, $\zeta = 0.21$ $s = -131$, $-37 \pm 172i$ Dominant time constant is too large. Damping ratio is too small.
2. $\tau \leqslant 0.001$, $\zeta = 0.707$ $K_p = 22.2$ $K_I = 23{,}708$ $s = -1000 \pm 1000i$	$s = -167$, $206 \pm 788i$ An unstable system.
3. $\tau_1 = 0.05$, $\tau_2 = 0.005$ $\zeta > 1$ $K_p = 1.123$ $K_I = 47.37$ $s = -20, -200$	$\tau = 0.0505$, $\zeta > 1$ $s = -19.8$, $-92.8 \pm 181i$ Dominant root is close to desired value.

where the characteristic polynomial is

$$\Delta(s) = LIs^3 + (RI + cL)s^2 + (cR + K_eK_T + K_TK_1K_2K_p)s + K_TK_1K_2K_I \tag{7.2-8}$$

For $L = 0.0021$ and the other parameter values as given before, we obtain

$$\Delta(s) = 0.00021s^3 + 0.04315s^2 + (5.3832 + 3.6315K_p)s + 3.6315K_t \tag{7.2-9}$$

Examining the transfer functions shows that the presence of a nonzero inductance L does not change the steady-state performance predicted by the second-order model. Let us see if the same is true for the transient performance. For case 1, substitute the values $K_p = 0.8856$ and $K_I = 237$ into (7.2-9), and solve for the roots. They are $s = -131$, $-37 \pm 172i$. The dominant behavior is given by the conjugate pair. The dominant time constant is $\tau = 1/37 = 0.027$, which is almost three times larger than can be tolerated according to the given specifications. The damping ratio for the dominant pair is $\zeta = 0.21$ and indicates an overshoot of approximately 50%, ten times larger than desired. Obviously, the design is poor, because the roots of the quadratic model do not approximate the dominant roots of the cubic model.

For case 2, with $K_p = 22.2$ and $K_I = 23{,}708$, the roots of (7.2-9) are $s = -167$, $206 \pm 788i$. The third-order model is thus unstable, whereas the second-order model predicted the dominant roots would be $s = -1000 \pm 1000i$. These results are summarized in Table 7.2.

Case 2 shows that the quadratic approximation is degraded as the desired time constant is made smaller. This gives a clue as to what the underlying difficulty is. The armature time constant τ_a is neglected in the quadratic approximation. For case 1 the desired time constant τ is about twice the value of τ_a (0.01 versus 0.0049), whereas in case 2, τ is about five times *smaller* than τ_a (0.001 vs. 0.0049). Thus, we have attempted to obtain a system whose time constant is barely dominant relative to the neglected time constant τ_a in case 1 and not dominant in case 2.

The solution to the difficulty lies in specifying a time constant that is much larger than τ_a. For $\zeta < 1$, (7.2-5) shows that the largest time constant obtainable in the quadratic model is $\tau = 0.016$ if K_p must be positive for physical reasons. This value is only three times larger than τ_a, so we must look at the overdamped case. For a dominant time constant $\tau = 0.05 = 10\tau_a$ and a secondary time constant much smaller, say, 0.005, the required gains from the quadratic model are $K_p = 1.123$ and $K_I = 47.37$. The predicted roots are $s = -20, -200$. With these gain values substituted into (7.2-9), the cubic roots are $s = -19.8, -92.8 \pm 181i$. Thus, the dominant root of the third-order model is very close to that of the second-order model. The response of the third-order model will show oscillations that are not predicted by the second-order one, but these will decay five times more quickly than the transients produced by the dominant root. This is case 3 in Table 7.2.

The imaginary designer in this example sought to avoid using a third-order model to select the gains K_p and K_I. In fact, such a model is often not too difficult to work with if the techniques of the following chapters are employed. However, the example was constructed to illustrate the effect of parasitic roots. In practice, for this particular application, the designer has the third-order model available but might not have an accurate value for the armature inductance.

In other applications, however, it might be difficult or expensive to develop the higher order model necessary to predict the effects of parasitic modes. Such modes commonly result from the neglected dynamics of sensors and final control elements. For example, the interaction of the water surface and the float in Figure 6.27*a* can produce a significant parasitic mode if proportional control is used with a high gain (small time constant). Another example is the neglected elasticity of the shaft connecting a load to a motor. If the system is a high-performance one, the motor torques will be large and easily reversible, and the shaft twist might be significant. In general, parasitic modes will be important in systems whose specifications call for a time constant close to that of any neglected mode. The model to be used in such a design should therefore include such modes.

It is difficult to state specific guidelines for when parasitic roots will become important. In addition to the preceding observations, we will also note that the parasitic roots will make a greater contribution to the system's response when the system's inputs are rapidly changing. Proper treatment of parasitic roots requires experience and sometimes experiments.

7.3 NONLINEARITIES AND CONTROLLER PERFORMANCE

All physical systems have nonlinear characteristics of some sort, although they can often be modeled as linear systems provided the deviations from the linearization reference condition are not too great. Under certain conditions, however, the nonlinearities have significant effects on the system's performance. Here, we consider some of these effects and how they can be avoided or accounted for in control system design.

One such situation can occur during the start-up of a controller if the initial conditions are much different from the reference condition for linearization. The linearized model is then not accurate, and nonlinearities govern the behavior. If the nonlinearities are mild, there might not be much of a problem. Where the nonlinearities are severe, such as in process control, special consideration must be given to start-up. Usually, in such cases, the control signal sent to the final control elements is manually adjusted until the system variables are within the linear range of the controller. Then the system is switched into automatic mode. Digital computers are often used to replace the manual adjustment process, because they can be readily coded to produce complicated functions for the start-up signals. Care must also be taken when switching from manual to automatic. For example, the integrators in electronic controllers must be provided with the proper initial conditions. An example of a circuit with this feature is given in Reference 6 (pp. 237–239).

Limit Cycles

A *limit cycle* is a self-sustaining oscillation whose characteristics are determined by the *system itself* and not by the initial conditions or the inputs. Limit cycles are found in only nonlinear systems. The sustained oscillations found in neutrally stable linear systems, such as an undamped mass spring system, are not limit cycles, because their amplitude depends on the initial conditions. The oscillation of a linear system produced at steady state by periodic forcing function is also not a limit cycle, because the frequency of oscillation is determined by the input frequency.

The existence of a limit cycle is characterized by a closed curve on the phase plane plot. If all the state trajectories approach the limit cycle curve, it is said to be a stable limit cycle. An example is given by the liquid level controller shown in Figure 6.27 with on–off control. The familiar system model is

$$\dot{h} = -\frac{1}{RC}h + \frac{1}{C}q_m \qquad (7.3\text{-}1)$$

for the filling curve and

$$\dot{h} = -\frac{1}{RC}h \qquad (7.3\text{-}2)$$

for the emptying curve. The flow rate q_m is the maximum flow rate through the supply valve. If we wish to keep the height at h_d, a constant flow rate of $q_d = h_d/R$ is required. If the controller allows a fluctuation Δ in the level above and below h_d, the controller switches from the filling to the

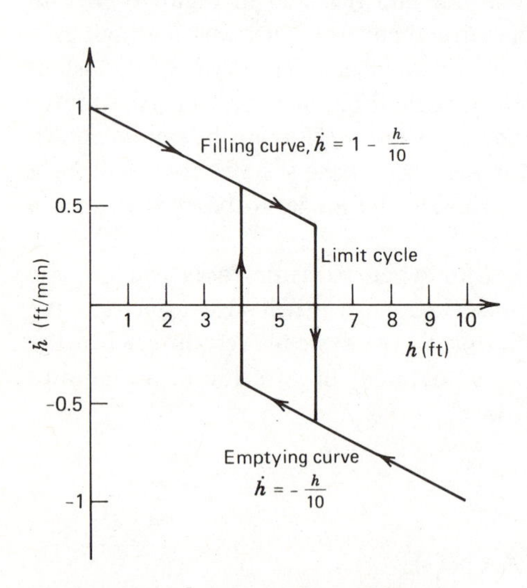

FIGURE 7.4 Limit cycle produced by on–off control of the liquid-level system of Figure 6.27. The tank's time constant is $RC = 10$ min, and the switching gap is 2 ft.

emptying curve when $h = h_d + \Delta$. When $h = h_d - \Delta$, the controller switches back to the filling curve.

These characteristics can be easily summarized with a phase plot of $\dot{h}$ versus h. On this plot, (7.3-1) is a straight line with a slope of $-1/RC$ and an intercept of q_m/C. The emptying curve (7.3-2) has the same slope but a zero intercept. Figure 7.4 shows the limit cycle for a specific case with the following parameter values:

$R = 5/6 \text{ min/ft}^2$	$C = 12 \text{ ft}^2$
$h_d = 5 \text{ ft}$	$\Delta = 1 \text{ ft}$
$q_d = 6 \text{ ft}^3/\text{min}$	$q_m = 12 \text{ ft}^3/\text{min}$

The only trajectories possible are those shown in the figure. For any initial height greater than 4 ft, the system follows the emptying curve until the switching at $h = 4$ ft. The switching is assumed to be instantaneous; thus, h is discontinuous at that point. Similarly, for $h(0) < 6$ ft, the filling curve is followed until $h = 6$ ft. Once the height is within the range $4 \leqslant h \leqslant 6$, the system follows the limit cycle, which is the closed curve. The solutions of (7.3-1) and (7.3-2) can be used to show that 8.1 min are required for one cycle. Because all the trajectories lead to the limit cycle it is stable.

When the system model is of first or second order and piecewise linear, it is not too demanding to use the phase plot to determine the existence of a limit cycle. Otherwise, it is generally difficult to do so, and more advanced methods are required (see, for example, chapter 12 of Reference 2). Digital simulation can be used to detect such cycles, but the step size must be chosen carefully. If not, round-off and truncation errors can either mask the presence of a cycle or induce oscillations that might be falsely interpreted as a cycle.

Control systems designed to be stable on the basis of a linearized model can be unstable if any nonlinearities present are strong enough to cause a limit cycle. This can occur in servomechanisms in which Coulomb friction or gear backlash is present. Another cause of this behavior is saturation of the final control elements when overdriven by the controller.

Reset Wind-up

The saturation nonlinearity shown in Figure 7.5 represents the practical behavior of many actuators and final control elements. For example, a motor–amplifier combination can produce a torque proportional to the input voltage over a limited range. However, no amplifier can supply an infinite current; there is a maximum current and thus a maximum torque that the system can produce in either direction (clockwise or counterclockwise). Another example is a flow control valve. Once it is fully open, the flow rate cannot be made greater without

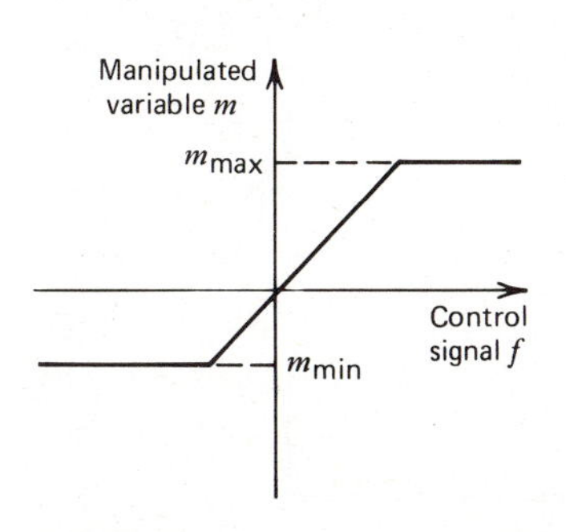

FIGURE 7.5 Saturation nonlinearity.

increasing the supply pressure. This shows that the saturation nonlinearity is not always symmetric; if m is flow rate, f is valve position, and $m_{\min} = 0$.

The final control elements are said to be overdriven when they are commanded by the controller to do something they cannot do. Since the limitations of the final control elements are ultimately due to the limited rate at which they can supply energy, it is important for all system performance specifications and controller designs to be consistent with the energy delivery capabilities of the elements to be used.

Any controller with integral action can exhibit the phenomenon called *reset windup* or *integrator buildup* when overdriven, if it is not properly designed. Consider the PI-control law.

$$m(t) = K_p e(t) + K_I \int_0^t e(t)\,dt \tag{7.3-3}$$

For a step change in set point, the proportional term responds instantly and saturates immediately if the set point change is large enough. On the other hand, the integral term does not respond so fast. It integrates the error signal and saturates some time later if the error remains large for a long enough time. As the error decreases, the proportional term is no longer saturated. However, the integral term continues to increase as long as the error has not changed sign, and thus the manipulated variable remains saturated. Even though the output is very near its desired value, the manipulated variable remains saturated until after the error has reversed sign. The result can be a large overshoot in the response of the controlled variable.

Consider PI control of a plant that is a pure integrator, such as a unit inertia with torque input, velocity output, and no damping (Figure 7.6). Assume that the torque is limited to a maximum magnitude M and the slope of the linear region of the saturation function is unity, for simplicity. For system operation within the linear region, we can use the transfer function

$$\frac{\Omega(s)}{\Omega_R(s)} = \frac{K_p s + K_I}{s^2 + K_p s + K_I} \tag{7.3-4}$$

where $T(s) = V(s)$ for $|v(t)| < M$. The system will be critically damped if $K_p^2 = 4K_I$. In this case, the time constant will be $\tau = 2/K_p$.

As an example, take the case where $M = 1$ and the input is a unit step. For the specifications $\zeta = 1$ and $\tau = 0.4$, the gains must be $K_p = 5$, $K_I = 6.25$. For zero initial

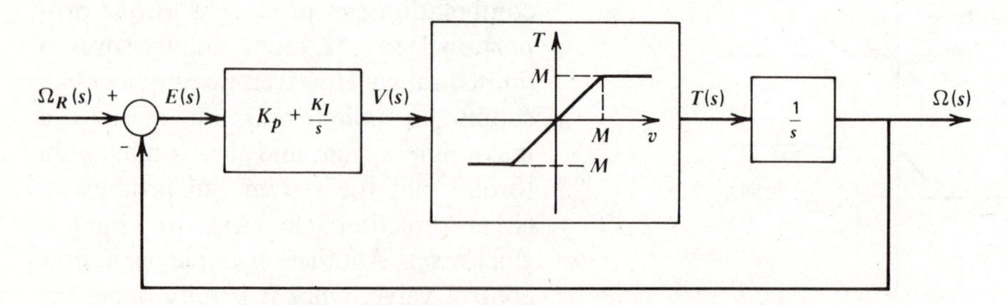

FIGURE 7.6 PI control of an inertia load with a saturating-torque motor.

velocity, the error is $e(0) = 1 - 0 = 1$. The system immediately saturates, because the proportional term is $K_p e = 5$. In the saturated mode, $T(t) = 1$ and $\dot{\omega} = 1$. Thus, at start-up, $\omega(t) = t$ and $e(t) = 1 - t$, until $v = 1$. The voltage function during start-up is found from (7.3-3), with $m = v$, to be

$$v(t) = 5 + 1.25t - 3.12t^2$$

The system enters the linear region when $v(t) = 1$ at $t = 1.35$.

For $t > 1.35$, the response can be obtained from (7.3-4) with the two solutions matched at $t = 1.35$. The governing equation is

$$\ddot{\omega} + 5\dot{\omega} + 6.25\omega = 5\dot{\omega}_R + 6.25\omega_R$$

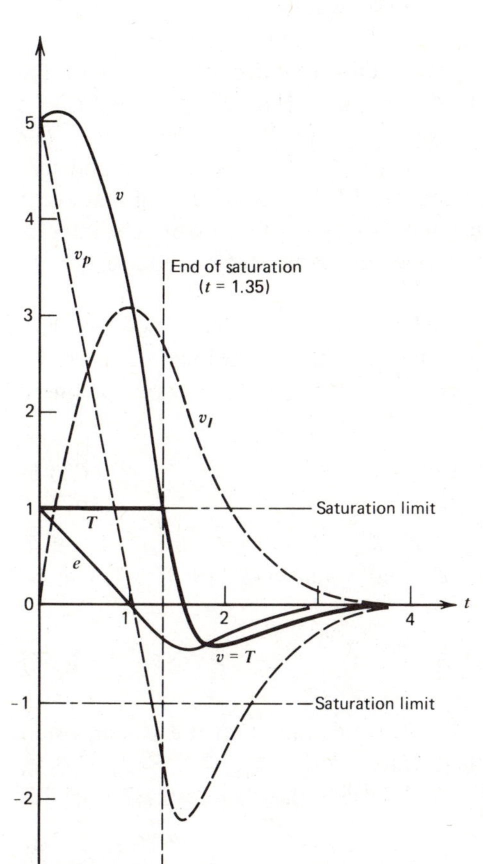

FIGURE 7.7 Example of the reset wind-up phenomenon in a PI controller. The integral term v_I does not begin to decrease until the error e changes sign. This extends the time during which the controller is saturated ($|v| \geq 1$). The lower limit is not exceeded here since $v = v_p + v_I > -1$.

where $\omega_R = 1$ and $\dot{\omega}_R = 0$ for $t > 1.35$. The solution is

$$\omega(t) = 54.58te^{-2.5t} - 63.46e^{-2.5t} + 1$$

with the matching conditions $\omega(1.35) = 1.35$, $\dot{\omega}(1.35) = 1$. The remaining variables are obtained from $\omega(t)$ through their defining relations. The results are plotted in Figure 7.7. The proportional action's contribution to the voltage $v(t)$ is denoted $v_p(t)$, while that due to the integral action is $v_I(t)$. Note that the error changes sign at $t = 1$, but the integral term v_I keeps the system saturated until $t = 1.35$. The velocity output is plotted in Figure 7.8 and displays the overshoot that is often caused by the reset wind-up phenomenon.

Design with Power-Limited Elements

It has been stated that control system performance can be improved if the limited energy-delivering capacity of the final control elements is taken into account. The previous example of the reset wind-up phenomenon provides a convenient illustration of how this improvement can be accomplished.

Reset wind-up can be avoided if we make sure the controller never allows the final control elements to remain saturated. However, the limiters for electronic controllers shown in Figure 6.52 will not

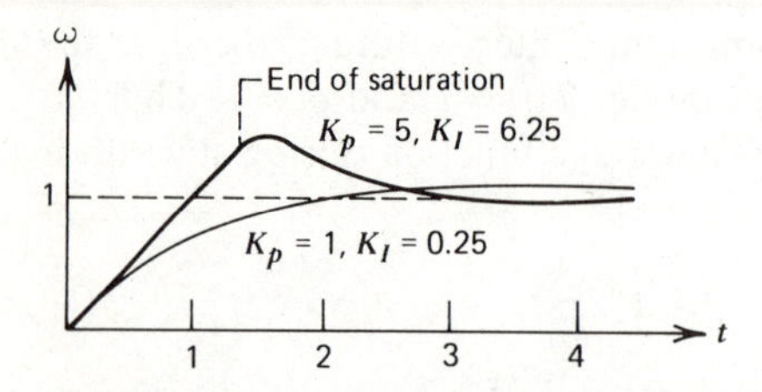

FIGURE 7.8 Unit step response of a PI controller with a saturation nonlinearity. The overshoot for the case $K_p = 5$, $K_I = 6.25$ results from reset wind-up. The overshoot is decreased when the gains are lowered to prevent saturation.

prevent reset wind-up; they merely prevent the actuators from receiving high voltages that might damage them. For the PI controller in the previous example, the voltages $v_p(t)$ and $v_I(t)$ would not be allowed to exceed the value that causes the torque to saturate, but the overshoot in the velocity response would still occur. The limiters protect equipment in applications where the input magnitudes are not known in advance.

Reset wind-up is prevented by selecting the gains so that saturation will never occur. This requires knowledge of the maximum input magnitude that the system will encounter. For example, suppose the error signal is $e(t) = r(t) - c(t)$, where r and c are the input and output signals. The maximum error in a well-designed system will occur at $t = 0$ for a step input and zero initial output. Thus, in a PI controller described by (7.3-3), saturation will first occur because of the proportional term, since it responds instantly to the error. The integral term might cause saturation also, but at a later time, because its output requires time to accumulate.

Denote the maximum expected value of $r(t)$ by r_{max} and the value of the controller output at saturation by m_{max}. Then under the stated conditions, the maximum error is r_{max}. Since the integral contribution is zero at $t = 0$, the maximum gain for linear operation is

$$K_p = \frac{m_{max}}{r_{max}} \tag{7.3-5}$$

The time constant for the model (7.3-4) is $\tau = 2/K_p$ if $\zeta \leq 1$. Thus, K_p is usually set to the value given above to minimize τ. The required value of K_I for (7.3-4) can be expressed in terms of ζ as

$$K_I = \frac{K_p^2}{4\zeta^2} \tag{7.3-6}$$

For the previous example, $m_{max} = M = 1$, and we assume that the system's step inputs will have a magnitude no greater than unity, so that $r_{max} = 1$. From (7.3-5), $K_p = 1$. With $\zeta = 1$ as before, (7.3-6) gives $K_I = 0.25$. Using these values in (7.3-4) with a unit-step input, we obtain the response

$$\omega(t) = 0.5te^{-0.5t} - e^{-0.5t} + 1$$

The other variables are found from $\omega(t)$.

$$e(t) = 1 - \omega(t)$$
$$v_p(t) = e(t)$$
$$v_I(t) = 0.25te^{-0.5t}$$
$$v(t) = v_p(t) + v_I(t)$$

From these expressions, it can be seen that v_p, v_I, and v do not exceed the saturation limit of 1. The system is always within the linear region. Its response is shown in Figure 7.8 along with that of the previous example. With the present choice of gains, no reset wind-up occurs, and the overshoot is much less than before. However, the response now takes longer to settle down to its steady-state value. The new time constant is 2, as opposed to 0.4 for the previous design.

In order to decide which response is optimal, the designer must look more closely at the requirements for the particular application. In many systems, the occurrence of saturation and response overshoot indicates that the system is using more energy than is necessary to accomplish the task. If so, the preceding design method is preferred. The technique was applied to PI control of a pure integrator, but it can be generalized as follows.

1. For a model whose only nonlinearity results from saturation, assume the system is operating in the linear region, and compute the response for the most severe inputs the system is expected to deal with.
2. Compute the maximum gain values that can be allowed without inducing saturation. Steps 1 and 2 can be done analytically for first- and second-order systems. For higher order systems, computer simulation can be used.
3. Use other criteria to decide whether the gains should be set to these maximum values or lower values. Such criteria might be the desired values of time constants or damping ratio or the performance indices (IAE, ISE, etc.).

7.4 COMPENSATION

It is sometimes impossible to adjust the gains in a PID-type control law in order to satisfy all of the performance specifications. If a control law can be found that satisfies most of the specifications, a common design technique is to insert a *compensator* into the system. This is a device that alters the response of the controller so that the overall system will have satisfactory performance.

Compensation alters the structure of the control system, and compensation techniques are categorized accordingly. The three categories generally recognized are *series compensation*, *parallel* (or *feedback*) *compensation*, and *feed-forward disturbance compensation*.† The three structures are loosely illustrated in Figure 7.9, where we assume the final control elements have a unity transfer function. The transfer function of the controller is $G_1(s)$. The feedback elements are represented by $H(s)$ and the compensator by $G_c(s)$. We assume that the plant is unalterable, as is usually the case in control system design.

The choice of compensation structure depends on what type of specifications must be satisfied. The following examples are intended to provide some insight into this selection. More examples of compensation are given in the following chapters.

---

†Another type of compensation is feed-forward compensation of the *command* input. This was analyzed in Chapter 6.

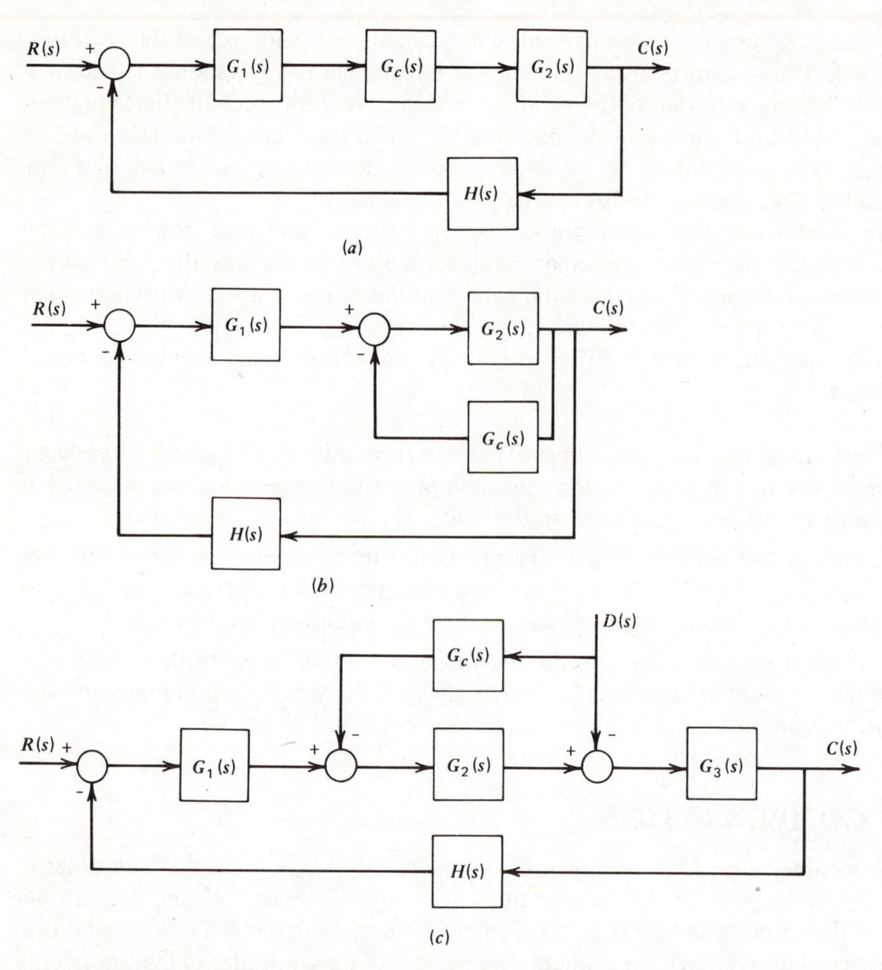

FIGURE 7.9 General structures of the three compensation types. (*a*) Series. (*b*) Parallel (or feedback). (*c*) Feed-forward. The compensator transfer function is $G_c(s)$.

The physical devices used as compensators are similar to the pneumatic, hydraulic, and electrical devices treated previously. Once the transfer function is determined, a device can be selected with criteria similar to those used to select controller elements.

Feed-Forward Disturbance Compensation

The control algorithms considered thus far have counteracted disturbances by using measurements of the output. One difficulty with this approach is that the effects of the disturbance must show up in the output of the plant before the controller can begin to take action. On the other hand, if we can measure the disturbance, the response of the controller can be improved by using the measurement to augment the control signal sent from the controller to the final control elements. This is the essence of feed-forward disturbance compensation.

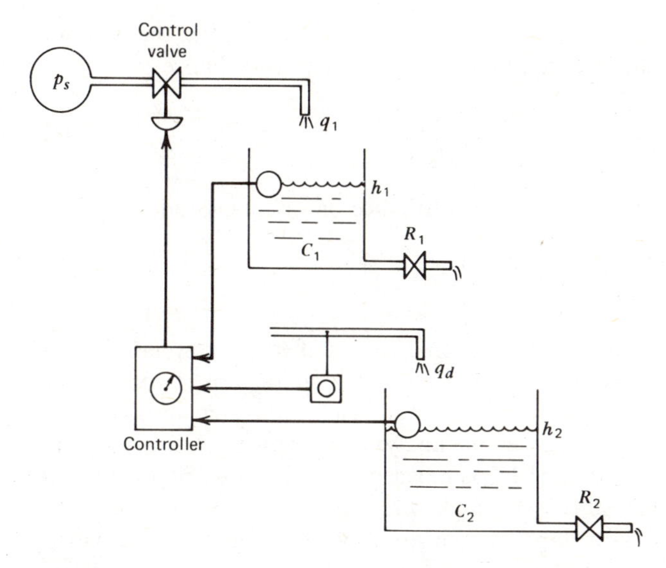

FIGURE 7.10 Liquid-level system used for feed-forward compensation and cascade control examples.

Consider the two-tank system shown in Figure 7.10. The height h_2 in the second tank is to be controlled by adjusting the flow rate q_1 into the first tank. A measurement of the height h_2 is used to do this. The system disturbance is the flow rate q_d that enters the second tank at unpredictable times and might result from catalysts being recycled into the process, for example. The block diagram is given in Figure 7.11. The feed-forward compensator is a simple-gain K_f in this case; dynamic elements can be used if necessary. The PI action could be used for the controller transfer function $G_1(s)$ to cancel the effects of the disturbance, but the integral term increases the order of the system. This makes it more oscillatory, less stable, and more difficult to analyze.

The performance of the feed-forward disturbance compensator is revealed by the disturbance transfer function.

$$\frac{H_2(s)}{Q_d(s)} = \frac{R_2(R_1C_1s + 1 - K_fK_v)}{(R_1C_1s + 1)(R_2C_2s + 1) + R_2K_vK_p} \tag{7.4-1}$$

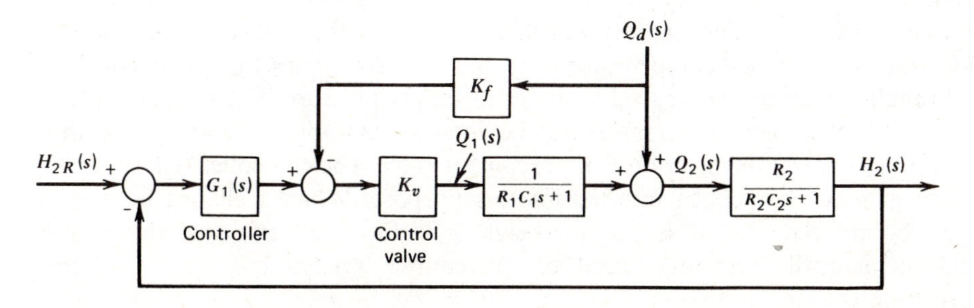

FIGURE 7.11 Feed-forward compensation applied to the liquid-level system of Figure 7.10.

where P action is used for the controller $[G_1(s) = K_p]$. If K_p is positive, the system is stable. The steady-state deviation resulting from a unit-step disturbance is

$$\Delta h_{2ss} = \frac{R_2(1 - K_f K_v)}{1 + R_2 K_v K_p} \tag{7.4-2}$$

The deviation can be made zero for a step disturbance of any magnitude if

$$K_f = \frac{1}{K_v} \tag{7.4-3}$$

The feed-forward disturbance compensation does not affect the response of h_2 to the command h_{2R}. In particular, K_f does not affect the characteristic roots of the system. Its effect on the disturbance response is shown by the step response of a second-order system with numerator dynamics analyzed in Section 4.9, Example 4.15. The numerator zero of (7.4-1) is $s=0$ if (7.4-3) is satisfied; and it is $s=-1/R_1C_1$ if $K_f=0$ (no compensation). From the results in Example 4.15, we can conclude that the feed-forward disturbance compensator improves the speed of response to a disturbance.

A difficulty with feed-forward compensation is that it is an open-loop technique in that it contains no self-correcting action. If the value of K_v is not accurately known, the gain K_f selected from (7.4-3) will not cancel the disturbance completely. The usefulness of the method must be determined in light of the specific performance requirements and the uncertainty in K_v. If the uncertainty is large, self-correcting action can be obtained by using a PI control law for $G_1(s)$ in conjunction with the feed-forward compensation.

Feedback Compensation and Cascade Control

We have already encountered several examples of feedback compensation. Using a tachometer to obtain velocity feedback, as in Figure 6.43 is such a case. The feedback-compensation principle in Figure 6.45 is another. Here, we consider another form, called *cascade control*, in which another controller is inserted within the loop of the original control system. The new controller can be used to achieve better control of variables within the forward path of the system. Its set point is manipulated by the first controller.

Cascade control can be applied to the system shown in Figure 7.10 to reduce the effects of a change in the supply pressure p_s. The block diagram is shown in Figure 7.12 with the cascade controller inserted. For now, we neglect the disturbance q_d and the feed-forward disturbance compensator. The constant K_s relates the change in flow rate through the control valve to the change in supply pressure. The idea of cascade control as applied here is to measure the internal variable h_1 and to use this information to adjust the position y of the control valve. The controller $G_2(s)$ and the feedback gain K_1 have been inserted for this purpose. When a change in supply pressure occurs, the effect is seen in the height h_1 before h_2. Using cascade control should increase the response speed of the control system in reacting to the disturbance.

The effect of the cascade controller is more easily seen with specific values for the

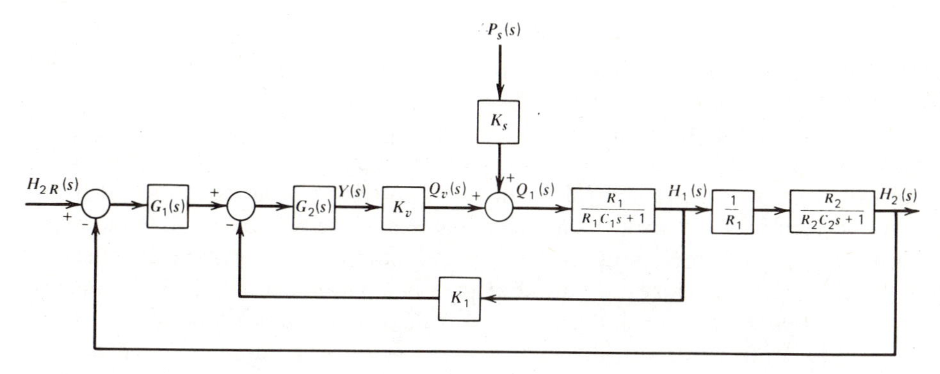

FIGURE 7.12 Cascade control applied to the liquid-level system of Figure 7.10.

parameters. To this end, let

$$R_1 = R_2 = C_1 = 1, \qquad C_2 = 2$$
$$K_1 = K_v = K_s = 1$$

and assume that P action is used for both controllers, so that $G_1(s) = K_{p1}$, $G_2(s) = K_{p2}$. The resulting transfer function for the disturbance is

$$\frac{H_2(s)}{\Delta P_s(s)} = \frac{1}{2s^2 + (3 + 2K_{p2})s + 1 + K_{p1}K_{p2} + K_{p2}} \tag{7.4-4}$$

If no compensation is used $[K_1 = 0,\ G_2(s) = 1]$,

$$\frac{H_2(s)}{\Delta P_s(s)} = \frac{1}{2s^2 + 3s + 1 + K_{p1}} \tag{7.4-5}$$

For a unit-step disturbance, the compensated system gives a steady-state deviation of

$$\Delta h_{2ss} = \frac{1}{K_{p2}(K_{p1} + 1) + 1} \tag{7.4-6}$$

With no compensation,

$$\Delta h_{2ss} = \frac{1}{1 + K_{p1}} \tag{7.4-7}$$

We immediately see that the damping coefficient of the uncompensated system is not affected by the control gain K_{p1}. This gain can be selected to achieve a desired time constant, damping ratio, or steady-state deviation, but it is highly unlikely that a single value will satisfy all three criteria. On the other hand, the compensated system has the damping ratio

$$\zeta = \frac{3 + 2K_{p2}}{2\sqrt{2(1 + K_{p1}K_{p2} + K_{p2})}} \tag{7.4-8}$$

The inner-loop gain K_{p2} can be adjusted to increase the speed of response, while the outer-loop gain K_{p1} can be used to minimize the steady-state deviation.

To see this, assume that the maximum gain value available is 9, due to physical constraints. The minimum deviation for the uncompensated system is $\Delta h_{2ss} = 0.1$ with $K_{p1} = 9$; the time constant is 1.33, and the damping ratio is 0.34; this indicates a large overshoot. For the compensated system with $K_{p1} = K_{p2} = 9$, the results are $\Delta h_{2ss} = 0.01$, $\tau = 0.19$, and $\zeta = 0.78$. The improvement is substantial but must be weighed against the cost of an extra controller and an extra sensor to measure the height h_1.

If the feed-forward disturbance compensator is reintroduced to handle the disturbance q_d, the gain K_f must be recomputed, because the cascade controller has changed the gain in the forward path. For previous parameter values, K_f should be unity without cascade control, but $K_f = 1/9$ is the proper choice otherwise.

Cascade control exists in disguise in many systems, because the final control elements usually include feedback loops to improve their performance. For example, a power amplifier has internal circuits to regulate voltage level while supplying current to the load. We usually assume that such subsystems act as perfect open-loop devices. If not, cascade control must be used to regulate their performance. For example, the control valve in Figure 7.10 will have improved performance in the presence of a supply pressure disturbance if the flow rate q_1 is measured and used with a controller to adjust the valve position y.

Cascade control is frequently used when the plant cannot be satisfactorily approximated with a model of second order or lower. This is because the difficulty of analysis and control increases rapidly with system order. The characteristic roots of a second-order system can easily be expressed in analytical form. This is not so for third order or higher, and fewer general design rules are available. When faced with the problem of controlling a high-order system, the designer should first see if the performance requirements can be relaxed so that the system can be approximated with a low-order model. The discussion of parasitic roots in Section 7.2 should serve as a guide in this regard. If this is not possible, the designer should attempt to divide the plant into subsystems, each of which is second order or lower. A controller is then designed for each subsystem. Finally, if the expense of extra hardware is prohibitive or if it is impossible to insert controllers at internal points in the system, the design techniques of the following chapters can be applied.

7.5 STATE VARIABLE FEEDBACK

There are techniques for improving system performance that do not fall entirely into one of the three compensation categories considered in the previous section. In some forms, these techniques can be viewed as a type of feedback compensation, while in other forms, they constitute a modification of the control law. We now consider two such techniques in this and the following section.

State variable feedback (SVFB) is a technique that uses information about all the system's state variables to modify either the control signal or the actuating signal. These two forms are illustrated in Figure 7.13. The former is a version of internal feedback compensation relative to the outer-control loop. An example of this approach is the cascade control scheme shown in Figure 7.12, where the state variables are the heights h_1 and h_2. The tachometer feedback scheme shown in Figure 6.43

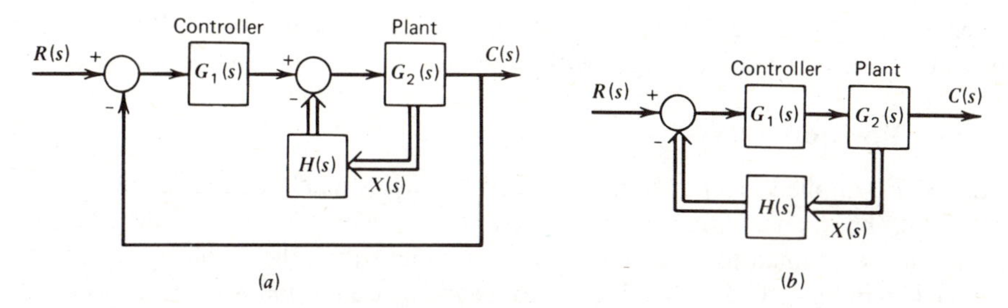

FIGURE 7.13 Two forms of state-variable feedback. (*a*) Internal compensation of the control signal. (*b*) Modification of the actuating signal.

illustrates the latter form. Both forms require all the state variables to be measurable or at least derivable from other information. For example, a position signal can theoretically be passed through a differentiator to obtain the velocity if a tachometer cannot be used. Devices or algorithms used to obtain state variable information other than directly from measurements are variously termed *state reconstructors, estimators, observers,* or *filters* in the literature.

The advantages of SVFB can be summarized as follows. In some applications, instrumentation costs might be less than the cost of implementing a complex control algorithm or adding extra actuators. If so, it would be more economical to measure all the state variables and use this information to reduce the controller's complexity. For example, P control can sometimes be substituted for PD control if the state variable corresponding to rate is measured. In addition to cost consideration, SVFB offers significant improvement in performance, because the state variables contain a complete description of the plant's dynamics. With SVFB, we thus have a better chance of placing the characteristic roots of the closed-loop system in locations that will give desirable performance. This concept is explored in more detail for high-order systems in Reference 1, chapter 10.

Avoiding D Action in the Controller

We have seen that D action can be used to improve the system's speed of response. However, it is advantageous to avoid using D action in the controller if it is possible to obtain an equivalent effect with a compensator elsewhere in the system. The reason for this is simple. When a change in command input occurs, the error signal and therefore the controller are instantaneously affected. If the change is sudden, as with a step input, the physical limitations of the differentiating device mean that an accurate derivative is not computed, and the actual performance of the system will be degraded relative to the ideal performance predicted by the mathematical model. If D action is required, a good design practice is to place the differentiator at a point in the loop where the signals are more slowly varying. Then the physical behavior of the differentiator will more closely correspond to its behavior as predicted by the mathematical model. A frequent choice for a differentiator location is in the outer feedback loop of the system. The output is usually the result of several integrations and therefore will exhibit behavior that is smoothed (filtered) and slowly varying with respect to the other signals in the system. Compare the time history of an inertia's displacement with the time history of its driving torque to see this point. With SVFB compensation, we can often dispense with D action in the controller.

SVFB in a Position Servo

Consider SVFB applied to a position control system using a field-controlled dc motor (Figure 7.14). The state variables are θ, ω, and the field current i_f. The transfer function $G_1(s)$ is assumed to contain the controller algorithm, the amplifier gain, and the feedback gain for θ. The feedback gains for ω and i_f are K_2 and K_3. We neglect the system damping c.

A similar system was treated in Section 6.7, where field inductance was neglected. It was found that D action could be avoided in the controller by using tachometer feedback, but the resulting P-control system is not so fast as the one with PD control. Here, we indicate how SVFB can be used with P control to improve the response.

Consider the PD-control case first. Let $G_1(s) = K_p + K_D s$ and $K_2 = K_3 = 0$ in Figure 7.14. This represents PD control with only position feedback. The transfer function is

$$\frac{\Theta(s)}{\Theta_R(s)} = \frac{K_T(K_p + K_D s)}{LIs^3 + RIs^2 + K_T K_D s + K_T K_p} \tag{7.5-1}$$

The system is stable if $K_p/K_D < R/L$. The numerator dynamics will speed up the response, but the coefficient of s^2 in the characteristic polynomial is unaffected by the control gains. Therefore, it might not be possible to select these gains to achieve the desired transient response.

Now, consider P control with SVFB. The gains K_2 and K_3 are nonzero, and $G_1(s) = K_1$. The transfer function is

$$\frac{\Theta(s)}{\Theta_R(s)} = \frac{K_1 K_T}{LIs^3 + (RI + IK_1 K_3)s^2 + K_1 K_2 K_T s + K_1 K_T} \tag{7.5-2}$$

The system is stable if $(R + K_1 K_3)K_2 > L$. The coefficient of s^2 is now affected by the gains, as are the coefficient of s, and the constant term. We therefore have more influence over the transient behavior with this design—achieved by feeding back all the state variables. If the field current is not fed back, $K_3 = 0$, and the s^2 coefficient is again beyond our influence. The system is unstable if no velocity feedback is used ($K_2 = 0$).

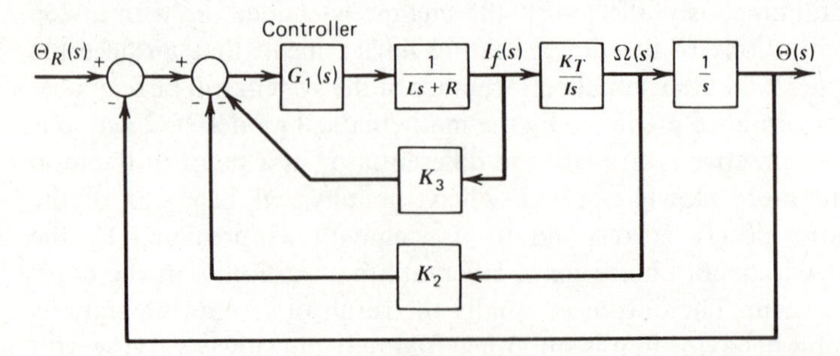

FIGURE 7.14 Position servo with a field-controlled dc motor and state-variable feedback.

7.6 PSEUDO-DERIVATIVE FEEDBACK

By now, it should be clear that a great variety of control system structures and algorithms can be used to solve a given problem. The final choice is intimately connected to the accuracy of the plant model and to the performance requirements. The field has seen many developments within the last 20 years, not only in terms of new hardware, such as microprocessors, but also in terms of new algorithms for control schemes as well as for analyzing such schemes. The state variable methods are an example of this. Another more recent development is *pseudo-derivative feedback* (PDF), a new control structure that captures the advantages of D action without the attendant difficulties caused by a differentiator located in the forward path of the controller.

The concept of PDF was developed by R. M. Phelan of Cornell University. His monograph (Reference 6) is a thorough presentation of the properties of PDF controllers and a comparison of their performance with that of traditional control systems. Here, we present an introduction to PDF and some examples to illustrate its advantages.

The Principle of One Master

Criticism can be leveled at the PID family of control laws, because they ask the controller to respond simultaneously to signals that can be conflicting. If $f(t)$ is the control signal, the actuating signal $e(t)$ produces $f(t)$ according to the differential equation

$$\frac{df}{dt} = K_I e + K_P \frac{de}{dt} + K_D \frac{d^2 e}{dt^2}$$

For example, if $e(t)$ is sinusoidal, the terms $\dot{e}$ and $\ddot{e}$ are 90° and 180° out of phase with $e(t)$, and the controller is forced to reconcile these differing signals to produce $f(t)$.

The result of this observation is the *principle of one master*, in the terminology of Reference 6. It states that the control algorithm should contain no more than one operation in its forward path. From the discussion in Section 7.5, it is obvious that this single operation cannot be D action. Under some circumstances where the performance requirements are not severe, the action can be on–off control or proportional control. The latter can be used with PDF when the plant is first order with negligible damping (a pure integrator), no disturbances occur, and fast response is not required. However, a criticism of P action in the forward path is that it unrealistically says that the control element can respond instantaneously to a step input. Thus, Phelan recommends I action for the control algorithm because of its physical realizability and ability to counteract disturbances.

PI versus PDF Control of a Second-Order Plant

To illustrate the concept of PDF control, consider the control of the displacement of a pure inertia plant, as shown in Figure 7.15. For PID control, $H(s) = 1$ and

$$G(s) = \left(K_p + \frac{K_I}{s} + K_D s \right)$$

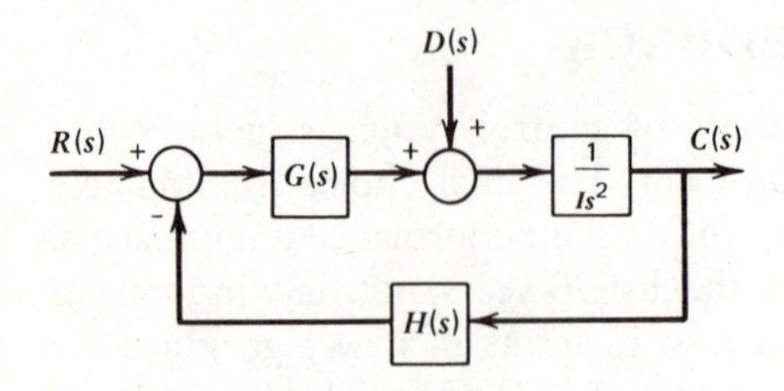

FIGURE 7.15 Control of a second-order plant.

The transfer function for the command input is

$$\frac{C(s)}{R(s)} = \frac{K_D s^2 + K_p s + K_I}{Is^3 + K_D s^2 + K_p s + K_I} \tag{7.6-1}$$

Each control gain corresponds to one coefficient in the characteristic polynomial. This eases the task of selecting the gains to achieve the desired transient performance. The disturbance transfer function has an s in the numerator, and the steady-state deviation for a step disturbance will be zero. The design has promise except for the physically unrealistic use of a differentiator in the forward path.

Suppose the D action is removed from the forward path and used in the feedback loop. The resulting design is a PI controller with proportional-plus-derivative feedback. In terms of Figure 7.15,

$$G(s) = K_p + \frac{K_I}{s}$$

$$H(s) = 1 + K_D s$$

The disturbance rejection properties are unchanged, but the command transfer function becomes

$$\frac{C(s)}{R(s)} = \frac{K_p s + K_I}{Is^3 + K_p K_D s^2 + (K_p + K_I K_D)s + K_I} \tag{7.6-2}$$

The controller might be more difficult to tune, because the coefficients of s and s^2 in the characteristic polynomial depend on more than one gain. Adjusting one control gain will change two coefficients in the polynomial. One approach is to take a cue from the Ziegler–Nichols ultimate-cycle method and set $K_I = 0$. Select the gains K_p and K_D to provide critical damping with a large value for K_p. It can be shown with the Routh criterion that this procedure gives a system that cannot be made unstable when K_I is reintroduced, in theory at least. In practice, Phelan notes, the fact that a physical differentiator cannot produce a pure derivative means that the system will be unstable if K_I is large enough. Therefore, K_I should be made just large enough to provide acceptable response.

The prediction of the response of (7.6-2) is made more difficult by the numerator dynamics. For such systems, the response is not entirely described by the characteristic polynomial, and both numerators and denominators must be analyzed when comparing systems. The occurrence of numerator dynamics is a consequence of the violation of the principle of one master. If the principle is followed, only a constant will appear in the numerator of the command transfer function. Thus, only the command variable, and not its integrals or derivatives, will appear as a forcing function in the system's differential equations—except for disturbances, of course.

The difficulties with tuning and numerator dynamics disappear if PDF control is used. Its logic can be derived as follows. The numerator dynamics in (7.6-2) are caused

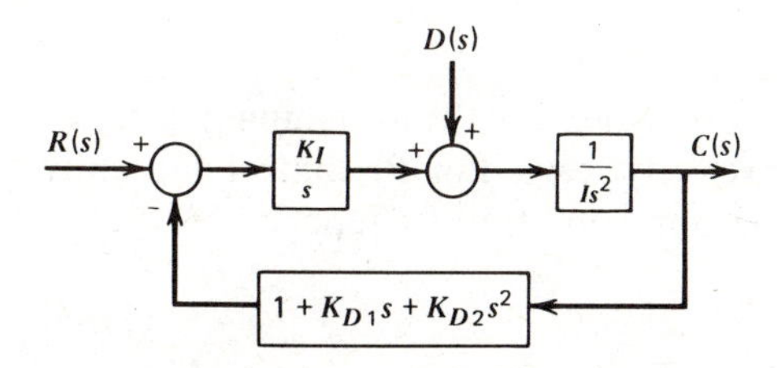

FIGURE 7.16 Hypothetical system used to motivate the development of PDF control logic.

by the K_p term. Therefore, we reserve P action for the feedback loop. With only I action in the forward path, the control signal will depend on the output, its derivative, and its integral only if the feedback path has the transfer function

$$H(s) = 1 + K_{D1}s + K_{D2}s^2$$

This is shown in Figure 7.16. The command transfer function is

$$\frac{C(s)}{R(s)} = \frac{K_I}{Is^3 + K_I K_{D2}s^2 + K_I K_{D1}s + K_I} \tag{7.6-3}$$

The numerator dynamics have disappeared, and the polynomial coefficients are under the designer's influence, but the physical system will require devices to compute not only the first but also the second derivative of the output. Reliable second-order derivatives of signals are impossible to obtain except in a few applications, and we therefore seek a better way.

To avoid this difficulty, we note that the system in Figure 7.16 integrates the derivatives of the output, an operation that makes little sense physically. Removing one s operator from the K_{D1} and K_{D2} terms is equivalent to integration. If this is done and the result inserted directly into the control signal as it leaves the integrator, the same effect will be achieved except for multiplication by K_I. The result is the PDF controller shown in Figure 7.17. Its command transfer function is

$$\frac{C(s)}{R(s)} = \frac{K_I}{Is^3 + K_{D2}s^2 + K_{D1}s + K_I} \tag{7.6-4}$$

The disturbance transfer function has a simple s term in the numerator. A single gain is associated with each polynomial coefficient, thus simplifying tuning of the controller. Another advantage is that the system can be treated as one with an inner and an outer loop, which simplifies the procedure for selecting the control gain. We will return to this after developing some insights from PDF control applied to a first-order plant.

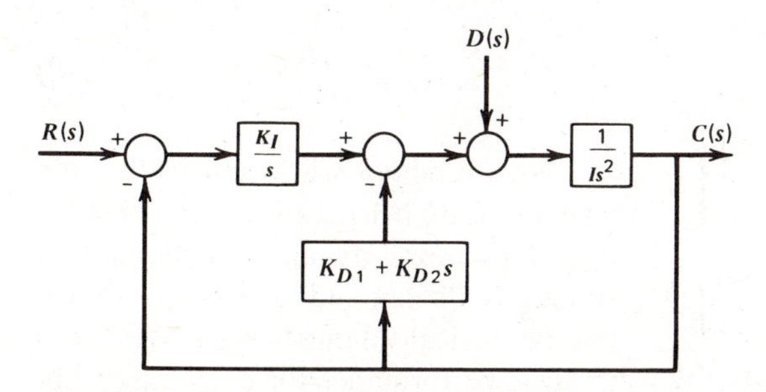

FIGURE 7.17 PDF control of a second-order plant.

PDF Control of a First-Order Plant

Consider the problem of controlling a first-order plant with no damping term, as shown in Figure 7.18. One application is a velocity servo. For PI control and unity feedback, $G(s) = K_p + K_I/s$ and $H(s) = 1$. The command transfer function is

$$\frac{C(s)}{R(s)} = \frac{K_p s + K_I}{Is^2 + K_p s + K_I} \tag{7.6-5}$$

In contrast to the second-order plant, no derivative action is required for stability, and thus no differentiator is required. Changing K_p or K_I only changes one coefficient in the characteristic polynomial. Finally, the disturbance transfer function has only an s in the numerator, so steady-state deviations are canceled. Except for violating the one-master principle, an admittedly somewhat vague concept at this point, the design seems to be satisfactory.

Following the one-master principle, we remove the proportional term from the controller. (The integral term must remain to cancel the disturbance.) However, with $K_p = 0$, the system is not stable, as can be seen from the denominator of (7.6-5). Therefore, we must change the structure to retain influence over the s term in the characteristic polynomial. This can be done with I control and proportional-plus-derivative feedback, but, as before, it makes no sense physically to differentiate a signal only to integrate it immediately afterward. Therefore, we insert a term proportional to the output immediately after the I action, which gives the PDF controller shown in Figure 7.19. The command transfer function is

$$\frac{C(s)}{R(s)} = \frac{K_I}{Is^2 + K_{D1}s + K_I} \tag{7.6-6}$$

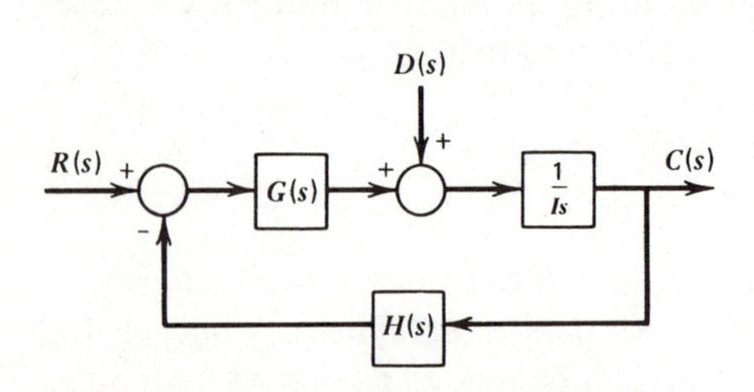

FIGURE 7.18 Traditional control system structure for a first-order plant with no damping.

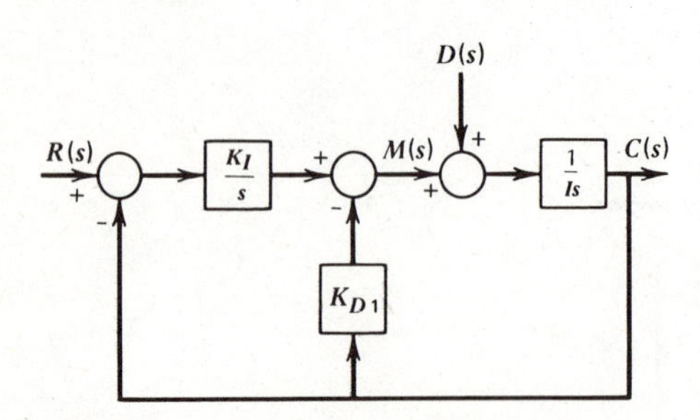

FIGURE 7.19 PDF control of a first-order plant with no damping.

Each polynomial coefficient depends on only one gain, while the numerator dynamics have been eliminated. The step disturbance is rejected at steady state. The system damping ratio is

$$\zeta = \frac{K_{D1}}{2\sqrt{IK_I}} \tag{7.6-7}$$

and the undamped natural frequency is

$$\omega_n = \sqrt{\frac{K_I}{I}} \tag{7.6-8}$$

No overshoot will occur in the step response for $\zeta \geqslant 1$ because of the absence of numerator dynamics. Simulation studies by Phelan and others have shown that the critically damped system provides the best performance for most cases, but these results are for specific values of the parameter I and the controller saturation limits. Therefore, values of ζ other than unity should not necessarily be ignored in the design.

The step response curve can be normalized with the parameters ζ and ω_n. This is shown in Figure 3.9 where the nondimensional response is plotted versus the nondimensional time $\omega_n t$. Such general plots are useful but must be carefully interpreted. Figure 3.9 shows the overshoot greatly increases as ζ decreases, but this conclusion is valid only if we are comparing systems that have the same value of ω_n. The same is true for the measures of response time (rise time, settling time, etc.). When the comparison is for different values of ω_n, the response should be presented in dimensional form as a function of true time, without any normalizing constants. See Figures 3.10–3.12 for examples.

When this was done for the specific parameter values used in Reference 6, it was seen that the step response curves were close to one another for values of ζ from about 0.6 to 1.0. A traditional design precept advises that ζ should be less than unity to achieve a response that reaches the vicinity of its final value rather quickly; a commonly recommended value is $\zeta = 0.7$. However, if the response curves for various ζ values are bunched together, there is little to be gained by designing ζ to be less than 1.0, especially if the existence of overshoot indicates excessive energy consumption.

For the step response in (7.6-6), the IAE and ITAE performance indices are minimized for $\zeta \approx 1$ and show relatively slight increase for $\zeta > 1$ for the parameter values considered. By plotting the response versus true time, the system in these cases was not observed to be very sluggish even for large values of ζ. In fact, the response was always faster than that achieved with P control with proportional feedback. Thus, highly damped systems should not be ruled out as possible designs. Example 6.2 demonstrates this. As will be seen shortly, a large damping ratio will allow the use of a large integral gain K_I to counteract disturbances more effectively.

Selecting the PDF Gains

The presence of integration in the PDF algorithm means that reset windup can occur if the gains are not chosen properly. This can be accomplished in a manner similar to that used in Section 7.3 for PI control. For a step input, it was easy to see from (7.3-3) that the maximum error, and therefore the maximum control signal, occurs at $t = 0$ for PI control. However, this is not true for most control algorithms in general and for PDF control in particular. The maximum control signal will usually occur at some time after the application of the input. This time must be determined by an extremum-seeking technique. For relatively simple systems, this can be done analytically. Here, we illustrate the method for the PDF controller shown in Figure 7.19 for the critically damped case.

As before, let r_{max} be the largest expected magnitude of the step input $r(t)$ and m_{max} the value of the manipulated variable at which saturation occurs. From Figure 7.19,

$$M(s) = \frac{K_I}{s} E(s) - K_{D1} C(s) \tag{7.6-9}$$

or

$$M(s) = IsC(s) \tag{7.6-10}$$

For $\zeta = 1$, the step response with zero initial conditions is found from (7.6-6).

$$c(t) = r_{\max}\left[1 - \left(1 + \frac{K_{D1}}{2I}t\right)e^{-(K_{D1}/2I)t}\right] \tag{7.6-11}$$

From (7.6-10),

$$m(t) = I\frac{dc}{dt} = r_{\max}\frac{K_{D1}^2}{4I}te^{-(K_{D1}/2I)t} \tag{7.6-12}$$

The maximum value of $m(t)$ is attained when dm/dt is zero. Applying this condition to (7.6-12), we find the maximum occurs at

$$t_{\max} = \frac{2I}{K_{D1}} \tag{7.6-13}$$

and has the value

$$m(t_{\max}) = r_{\max}\frac{K_{D1}}{5.44} \tag{7.6-14}$$

Note that the latter result is independent of I.

In order for the system response to remain within the linear region,

$$m(t_{\max}) \leqslant m_{\max} \tag{7.6-15}$$

From (7.6-14), we obtain the maximum value allowed for K_{D1}.

$$K_{D1,\max} = 5.44\frac{m_{\max}}{r_{\max}} \tag{7.6-16}$$

Since $\zeta = 1$, (7.6-7) gives the corresponding value of K_I.

$$K_{I,\max} = \frac{K_{D1,\max}^2}{4I} \tag{7.6-17}$$

PDF versus PI Control

The analysis of PI control following (7.6-5) indicated that the design might be acceptable. Let us compare its behavior with that of PDF control (7.6-6). A valid comparison can be made between competing designs only when the performance of each is optimized for identical constraints. In particular, the test inputs and the available energy-delivery capability of the final control elements should be the same for each system, and each system should have its gain adjusted to use this maximum capability. Therefore, we assume that the gains have been chosen with (7.6-16) and (7.6-17) for PDF control and with (7.3-5) and (7.3-6) for PI control.

Assume that oscillations are not desirable, and let $\zeta = 1$ for the PI case. Then the integral gain K_I for PDF is 29.6 times the value of K_I for the PI case. We would therefore expect the effects of a disturbance to be reduced faster with the PDF design. This will be shown to be true.

Consider the unit-step response first, and assume that $m_{\max} > 1$, so that no saturation occurs. For PDF control, the response is given by (7.6-11) with $t_{\max} = 1$.

For PI control, the critically damped response is

$$c(t) = 1 - \left(1 - \frac{K_p}{2I}t\right)e^{-(K_p/2I)t} \tag{7.6-18}$$

This relation can be used to show that an overshoot of 13.5% occurs at $t = 4I/K_p$. No overshoot exists for the PDF system.

Our immediate reaction might be to increase the damping ratio of the PI design to eliminate the overshoot. For $\zeta > 1$, the analytical solution for the unit-step response in (7.6-5) shows that the overshoot occurs at

$$t_p = \frac{I\zeta}{K_p\sqrt{\zeta^2 - 1}} \ln\left(\frac{x - y}{x + y}\right) \tag{7.6-19}$$

where $x = 1 - 2\zeta^2$ and $y = 2\zeta\sqrt{\zeta^2 - 1}$. An overshoot always exists for $\zeta > 1$; its magnitude becomes smaller and the time of occurrence is later as ζ is increased. For example, if $\zeta = 2$, the overshoot is 4.8% at $t = 6.08\ I/K_p$. The overshoot cannot be eliminated, because the numerator coefficients of (7.6-5) are coupled to those of the denominator, and the latter are partly constrained by the need to avoid saturation of the final control elements.

A similar effect occurs when a step disturbance is applied to the system. Both disturbance transfer functions have the form

$$\frac{C(s)}{D(s)} = \frac{s}{as^2 + bs + c}$$

where the denominator is given by that of either (7.6-5) or (7.6-6). Thus, both systems eliminate the disturbance effects at steady state, but the transient behavior is quite different. For $\zeta = 1$ and a unit-step input disturbance, the response of the PI design is

$$c(t) = \frac{t}{I}e^{-(K_p/2I)t} \tag{7.6-20}$$

The peak is

$$c_{\max} = \frac{0.735}{K_p} \tag{7.6-21}$$

and occurs at

$$t_p = \frac{2I}{K_p} \tag{7.6-22}$$

For the critically damped PDF design, the results are identical to the preceding equations, with K_p replaced by K_{D1}. Comparing (7.3-5) with (7.6-16), we see that $K_{D1} = 5.44K_p$. Thus, $c_{\max}$ and the time to recover, t_p, are 5.44 times greater for the PI controller. In addition, the IAE index can be shown to be 29.6 times larger for the PI controller.

Increasing the damping ratio of the PI controller does not improve the situation. For $\zeta > 1$, the unit-step disturbance response is

$$c(t) = \frac{I\zeta}{K_p\sqrt{\zeta^2 - 1}}(e^{s_1 t} - e^{s_2 t}) \tag{7.6-23}$$

where the characteristic roots are

$$s_{1,2} = \frac{K_p}{2I}\left(-1 \pm \frac{\sqrt{\zeta^2 - 1}}{\zeta}\right) \tag{7.6-24}$$

The peak is

$$c_{\max} = \frac{-1}{s_2}\left(\frac{s_2}{s_1}\right)^{s_1/(s_1 - s_2)} \tag{7.6-25}$$

and occurs at

$$t_p = \frac{1}{s_1 - s_2} \ln \frac{s_2}{s_1} \tag{7.6-26}$$

As ζ is increased, $c_{\max}$ and t_p increase. For example, with $\zeta = 2$, $c_{\max} = 0.874/K_p$ and $t_p = 3.04\ I/K_p$.

To see why this counterintuitive behavior occurs, look at the characteristic equation of the PI design.

$$s^2 + as + b = 0$$

where $a = K_p/I$ and $b = K_p^2/4I^2\zeta^2$. With K_p fixed from (7.3-5) in terms of the saturation constraints, we can think of the system as a mass–spring–damper combination with a fixed mass and a fixed damping constant. As the damping ratio is increased, the equivalent spring constant b is decreased. Thus, the restoring force of the system is smaller, and the peak caused by a disturbance is larger. This is another example where nondimensional response plots in terms of ω_n and ζ can be misleading. As we change the damping ratio of this system, the undamped natural frequency ω_n is also changed.

The relative performance of the PI controller is not entirely negative, however. The step input is the most severe input to be encountered because of its suddenness; therefore, we will continue to use the gains computed previously for a step input while investigating the response to a unit-ramp command input. The PI design for $\zeta = 1$ gives

$$c(t) = t(1 - e^{-(K_p/2I)t}) \tag{7.6-27}$$

while for PDF control,

$$c(t) = t - A + (t + A)e^{-2t/A} \tag{7.6-28}$$

where $A = 4I/K_{D1}$. The steady-state error is zero for PI control. For PDF control, $e_{ss} = A$. If the design specifications place heavy weight on the ramp response, the previous disadvantages of the PI system might be ignored.

The main points of this comparison can be summarized as follows. PDF control generally gives performance superior to that obtained by PI control, at least for the first-order plant with no damping term. Its performance with other types of plants and inputs must be investigated on an individual basis. When making comparisons, use nondimensional results with care, and allow all competing systems to have final control elements with the same rate of energy delivery. Select the gains to use this delivery capability for a fair comparison.

Selecting PDF Gains for a Second-Order Plant

In theory, the gains of the PDF controller shown in Figure 7.17 can be selected in the same way as for the first-order plant. However, the analytical expressions would be extremely cumbersome. For given plant parameters and saturation limits, numerical simulation might be used, but three gains must be selected, and thus the number of possible combinations can be large. Here, we present some guidelines developed by Phelan. The approach is also useful for cascade control problems that involve inner-loop and outer-loop controllers.

As shown in Figure 7.20, the PDF controller of Figure 7.17 can be broken down into two loops. With no disturbance, the inner loop has the transfer function

$$\frac{C(s)}{R_i(s)} = \frac{1}{Is^2 + K_{D2}s + K_{D1}} \tag{7.6-29}$$

with the damping ratio

$$\zeta_i = \frac{K_{D2}}{2\sqrt{IK_{D1}}} \tag{7.6-30}$$

Certainly, no overshoot will occur if $\zeta_i \geqslant 1$, but values of $\zeta_i < 1$ can also be used sometimes, since the input r_i can never be so severe as a step function. This is because r_i is the output of an integrator. In particular, the results of many numerical simulations for a wide range of gain values have shown that the smoothest, quickest step response without overshoot for the total system occurs when $\zeta_i \cong 0.7$ and

$$K_{D1} \cong 8\frac{m_{\max}}{r_{\max}} \tag{7.6-31}$$

For the specific case, $I = 1$, $m_{\max} = 0.25$, and $r_{\max} = 1$, it has been shown that for a step input, the ITAE index is minimized for $\zeta \cong 0.7$ and the index increases only slightly for $\zeta \gg 1$ (References 6 and 7).

A design procedure is as follows. For $\zeta_i = 0.7$, (7.6-30) and (7.6-31) yield values for K_{D1} and K_{D2}. With these values, analyze the command input-step response for increasingly larger values of K_I until saturation or overshoot occurs. A larger value of K_I is desired to compensate for disturbances. This procedure should result in most cases in a ballpark estimate of the required gains. Further simulations would be used to fine tune the results.

The controller output is

$$m(t) = K_I \int e\,dt - K_{D1}c - K_{D2}\dot{c} \tag{7.6-32}$$

The only increase in m comes from the integral term, and both the speed of response and the possibility of overshoot increase with increasing values of K_I. With the response relatively insensitive to

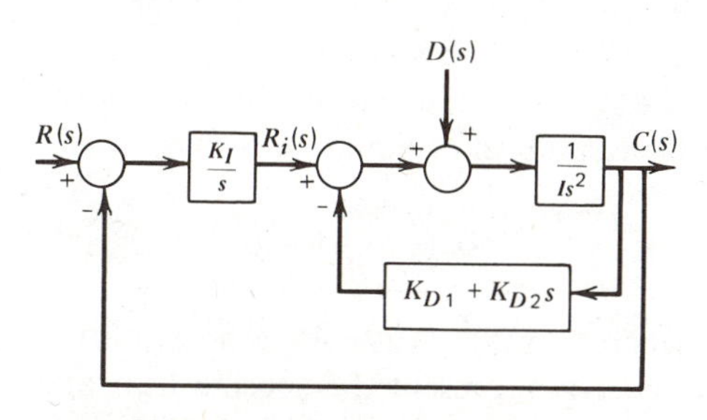

FIGURE 7.20 Inner-outer loop configuration of the PDF controller for a second-order plant.

ζ_i, we see that the choice of K_I is the most critical. The $K_{D1}c$ term is most important near steady state when c is large. Thus, K_{D1} can be adjusted to eliminate overshoot. In the early stages of the response, c and the error integral are small, and the controller output is heavily influenced by the $K_{D2}\dot{c}$ term. Increasing K_{D2} will therefore tend to decrease the maximum value of m, prevent saturation, and reduce oscillations.

Modified versions of PDF control are also useful. These are explored in the problems at the end of the chapter.

7.7 INTERACTING CONTROL LOOPS

We have seen that it is important for the control system designer to be aware of the effects of the system's parasitic modes even though they are not explicitly modeled. A related effect occurs when simplified models are used to design controllers for complex systems in which several variables are to be controlled. If these variables affect each other, the control loops are said to be interacting. In some cases, the interaction is so slight that we can neglect it and design the controller for each loop as if it were a separate system. The stringency of the performance specifications will also determine whether or not this approach can be used.

As the demands on technology become more severe for safety, environmental, and efficiency considerations, many traditional systems with separate-loop control designs must now be redesigned to include the interactive effects. For example, a steam turbine plant for generating electricity requires control of the turbine speed, steam flow rate, steam temperature, and steam pressure, among others. Historically, separate controllers were designed for each of these variables, but the system's performance could not be improved beyond a certain point without taking the interactions into account. Modern control theory has led to improvements in this respect, as it has also done with internal combustion engine and jet engine control.

As a simple example, consider the liquid level system shown in Figure 7.21. A chemical reaction requires the concentration as well as the temperature of the reactants to be closely controlled. This requires the level h and the liquid temperature T to be controlled. We do this here by changing the flow rate and the hot–cold ratio of the incoming liquid streams. A simple model can be derived as follows. Conservation of liquid mass gives

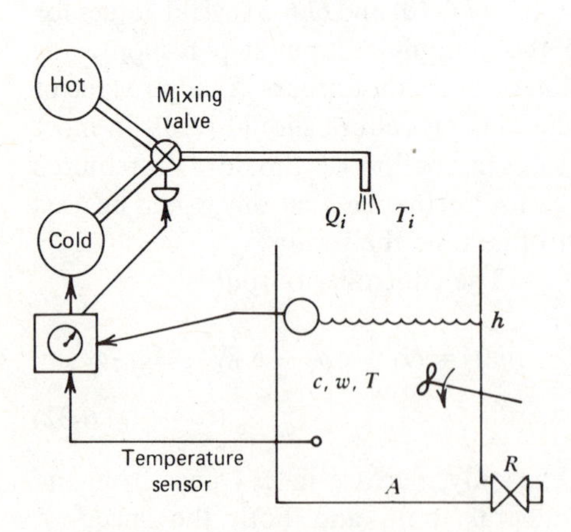

FIGURE 7.21 System in which the level-control and temperature-control loops interact.

$$A\dot{h} = -\frac{g}{R}h + Q_i \qquad \textbf{(7.7-1)}$$

where Q_i is the input-volume flow rate. Let c and w represent the specific heat and density of the liquid, and assume these are constants. The heat energy in the tank liquid is $cwAhT$, where T is the temperature of the tank liquid and is assumed to be uniform from stirring. The rate of change of this energy equals the incoming rate minus the outflow rate, or

$$\frac{d}{dt}(cwAhT) = cwQ_iT_i - \frac{g}{R}hcwT \tag{7.7-2}$$

Carrying out the differentiation gives

$$AT\dot{h} + Ah\dot{T} = Q_iT_i - \frac{g}{R}hT$$

Substituting (7.7-1) for $\dot{h}$ results in

$$Ah\dot{T} = Q_i(T_i - T) \tag{7.7-3}$$

Equations (7.7-1) and (7.7-3) form the model of the system. We note that (7.7-1) is uncoupled from the temperature model, but the derivative coefficient in (7.7-3) is a function of h. Therefore, (7.7-3) is nonlinear unless the height h is constant. Also, the input variable Q_i for the height subsystem appears in the temperature equation. Thus, the temperature subsystem is coupled to the height subsystem in two ways.

Possible Design Methods

The system is to be controlled by changing the manipulated variables Q_i and T_i. Three approaches to designing a control system to do this are the following.

1. Treat the coupling variables (h and Q_i here) as disturbances to the temperature control loop, and take the temperature and height systems to be separate. Design a controller for each. This requires the temperature model to be linearized about nominal values for h and Q_i.
2. Linearize the coupled model set (7.7-1) and (7.7-3), and treat the problem as a linear multivariable system. The two outputs would be h and T, and the two command inputs would be h_R and T_R, the desired height and temperature. This results in a higher order model than the first approach, but it provides a better description of the interaction between the subsystems. Linear control theory can be used with both approaches.
3. Treat the problem as a coupled, nonlinear system. No general control theory is available to support this approach, although many ad hoc procedures exist. Their success depends on the nature of the nonlinearity.

Equations (7.7-1) and (7.7-3) have the following special form:

$$\dot{y}_1 = f_1(y_1, v_1) \tag{7.7-4}$$

$$\dot{y}_2 = f_2(y_1, y_2, v_1, v_2) \tag{7.7-5}$$

where v_1 and v_2 are the manipulated variables corresponding to Q_i and T_i; y_1 and y_2 correspond to h and T. Figure 7.22 is a diagram of this special coupled configuration. More complex coupling configurations are possible, but we will use this relatively

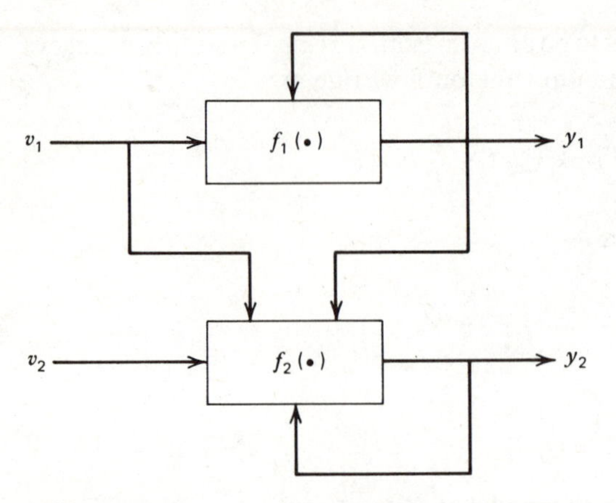

FIGURE 7.22 **A second-order system in which the coupling variables are the input v_1 and the state variable y_1.**

simple form as the basis for our discussion. The linearized system model is

$$\dot{x}_1 = a_{11}x_1 + b_{11}u_1 \tag{7.7-6}$$

$$\dot{x}_2 = a_{21}x_1 + a_{22}x_2 + b_{21}u_1 + b_{22}u_2 \tag{7.7-7}$$

The first design approach would use a model like that shown in Figure 7.23. The control laws $G_1(s)$ and $G_2(s)$ are designed separately. The latter's design would treat u_1 and x_1 as disturbances, modeled perhaps as step functions. If the response of the resulting control system in Figure 7.23*b* is fast relative to the changes in u_1 and x_1, the design might be satisfactory.

The second approach can be described by a block diagram like that in Figure

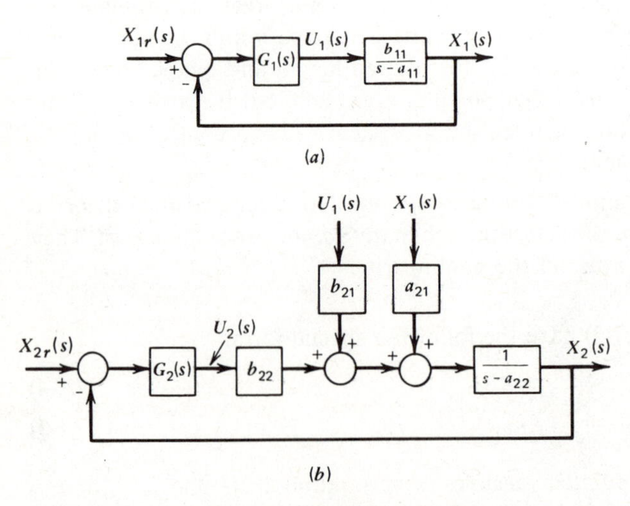

FIGURE 7.23 **Separate control loops for the system (7.7-6) and (7.7-7). The coupling variables are treated as disturbances to the second loop (*b*).**

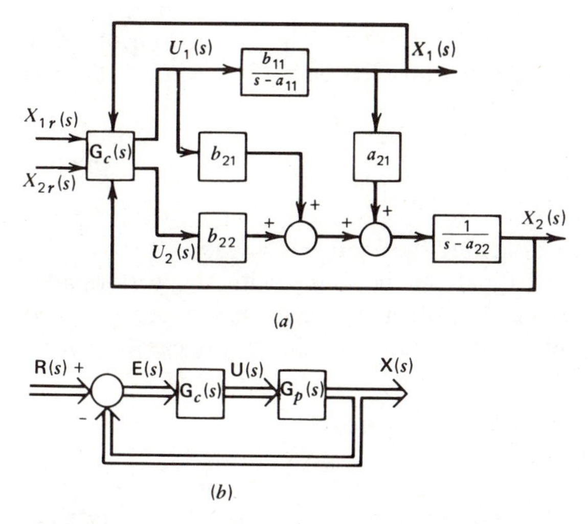

FIGURE 7.24 Control system with interacting loops. (*a*) Specific diagram for (7.7-6) and (7.7-7). (*b*) General scheme for the vector case.

7.24*a*. In this case, the control would be expressed as a transfer function matrix $\mathbf{G}_c(s)$.† The general form is shown in Figure 7.24*b*, where $\mathbf{G}_p(s)$ is the transfer function matrix for the plant. For now, we assume that interaction between the two subsystems is undesirable, and we consider how a controller might be designed to reduce the interaction.

Noninteraction in Control Systems

The absence of interaction between the x_1 and x_2 subsystems is indicated by a diagonal transfer function matrix $\mathbf{T}(s)$ between $\mathbf{X}(s)$ and $\mathbf{R}(s)$. The transfer function matrix for the system in Figure 7.24*b* is obtained by writing the equations for each component and solving for $\mathbf{X}(s)$ in terms of $\mathbf{R}(s)$.

$$\mathbf{X}(s) = \mathbf{G}_p(s)\mathbf{U}(s)$$
$$\mathbf{U}(s) = \mathbf{G}_c(s)\mathbf{E}(s)$$
$$\mathbf{E}(s) = \mathbf{R}(s) - \mathbf{X}(s)$$

Thus,

$$\mathbf{X}(s) = \mathbf{G}_p(s)\mathbf{G}_c(s)[\mathbf{R}(s) - \mathbf{X}(s)]$$
$$[\mathbf{I} + \mathbf{G}_p(s)\mathbf{G}_c(s)]\mathbf{X}(s) = \mathbf{G}_p(s)\mathbf{G}_c(s)\mathbf{R}(s)$$

or

$$\mathbf{X}(s) = \mathbf{T}(s)\mathbf{R}(s) \tag{7.7-8}$$

where

$$\mathbf{T}(s) = [\mathbf{I} + \mathbf{G}_p(s)\mathbf{G}_c(s)]^{-1}\mathbf{G}_p(s)\mathbf{G}_c(s) \tag{7.7-9}$$

The matrix **I** is the identity matrix.

†The transfer function matrix is defined in Section 5.7.

If $\mathbf{x}$ and $\mathbf{r}$ are two-dimensional, $\mathbf{T}(s)$ is a (2×2) matrix. The system is uncoupled if the command r_1 has no effect on x_2 and if r_2 has no effect on x_1. This is true if $\mathbf{T}(s)$ has the form

$$\mathbf{T}(s) = \begin{bmatrix} T_{11}(s) & 0 \\ 0 & T_{22}(s) \end{bmatrix} \tag{7.7-10}$$

This condition provides the basis for designing the control law $\mathbf{G}_c(s)$.

The desired forms of $T_{11}(s)$ and $T_{22}(s)$ can be found with the performance specifications for the design. Thus, $\mathbf{T}(s)$ can be taken as a given function, and (7.7-7) can be solved for $\mathbf{G}_c(s)$. To do this, multiply both sides from the left to clear the inverse matrix. This gives

$$[\mathbf{I} + \mathbf{G}_p(s)\mathbf{G}_c(s)]\mathbf{T}(s) = \mathbf{G}_p(s)\mathbf{G}_c(s)$$

Now, solve for $\mathbf{G}_c(s)$

$$\mathbf{G}_c(s) = \mathbf{G}_p^{-1}(s)\mathbf{T}(s)[\mathbf{I} - \mathbf{T}(s)]^{-1} \tag{7.7-11}$$

We have assumed that the inverse of the plant matrix $\mathbf{G}_p(s)$ exists.

The plant transfer function matrix $\mathbf{G}_p(s)$ relates $\mathbf{X}(s)$ to $\mathbf{U}(s)$ and can be obtained by either reducing the block diagram shown in Figure 7.24*a* or using the plant's resolvent matrix and the result (5.7-29), repeated here.

$$\mathbf{G}_p(s) = [s\mathbf{I} - \mathbf{A}]^{-1}\mathbf{B} \tag{7.7-12}$$

The matrices $\mathbf{A}$ and $\mathbf{B}$ are found from the linear state equations, of which (7.7-6) and (7.7-7) are examples. Here, these matrices have the form

$$A = \begin{bmatrix} a_{11} & 0 \\ a_{21} & a_{22} \end{bmatrix}, \quad B = \begin{bmatrix} b_{11} & 0 \\ b_{21} & b_{22} \end{bmatrix} \tag{7.7-13}$$

The design procedure is best illustrated with a specific example.

Example 7.4

Compute the controller transfer function required to achieve noninteraction in the system described by (7.7-13). Assume that

$$\mathbf{A} = \begin{bmatrix} -3 & 0 \\ 0 & -4 \end{bmatrix}, \quad \mathbf{B} = \begin{bmatrix} 1 & 0 \\ -1 & 2 \end{bmatrix}$$

The desired closed-loop time constants for the control loops are $\tau_1 = 1$, $\tau_2 = 2$.

The last condition indicates that the desired overall transfer function is

$$\mathbf{T}(s) = \begin{bmatrix} \dfrac{1}{s+1} & 0 \\ 0 & \dfrac{1}{2s+1} \end{bmatrix} \tag{7.7-14}$$

and the term in (7.7-11) becomes

$$\mathbf{T}(s)[\mathbf{I}-\mathbf{T}(s)]^{-1} = \begin{bmatrix} \frac{1}{s} & 0 \\ 0 & \frac{1}{2s} \end{bmatrix} \tag{7.7-15}$$

Using (7.7-12) to compute $G_p(s)$, we obtain

$$\mathbf{G}_p(s) = \begin{bmatrix} \frac{1}{s+3} & 0 \\ \frac{-1}{s+4} & \frac{2}{s+4} \end{bmatrix} \tag{7.7-16}$$

From (7.7-11), the required controller matrix is

$$\mathbf{G}_c(s) = \begin{bmatrix} s+3 & 0 \\ \frac{s+3}{2} & \frac{s+4}{2} \end{bmatrix} \begin{bmatrix} \frac{1}{s} & 0 \\ 0 & \frac{1}{2s} \end{bmatrix}$$

$$= \begin{bmatrix} \frac{s+3}{s} & 0 \\ \frac{s+3}{2s} & \frac{s+4}{4s} \end{bmatrix}$$

The controller elements are seen to be PI algorithms. For example, the manipulated variable u_2 is produced by

$$U_2(s) = \left(\frac{1}{2} + \frac{3}{2s}\right)E_1(s) + \left(\frac{1}{4} + \frac{1}{s}\right)E_2(s)$$

where e_1 and e_2 are the errors in x_1 and x_2; that is,

$$e_1 = r_1 - x_1$$

$$e_2 = r_2 - x_2$$

The control system's block diagram is shown in Figure 7.25.

If the diagram in Figure 7.25 is reduced to find the transfer functions relating x_1 and x_2 to r_1 and r_2, the matrix given by (7.7-14) would be obtained. In so doing, some numerator terms would be canceled by denominator terms. These canceled terms represent patterns of the system response that have been isolated from the inputs r_1 and r_2 by the design of the controller. A problem occurs when the system is acted on by disturbances that have not been accounted for in the model. These disturbances could excite these hidden response patterns and produce uncontrolled behavior.

Obviously, these more advanced control schemes are getting us into deeper water, and to analyze them properly, we need additional methods. In particular, the concepts of pole-zero cancellation are useful here. They are covered in Chapter 9.

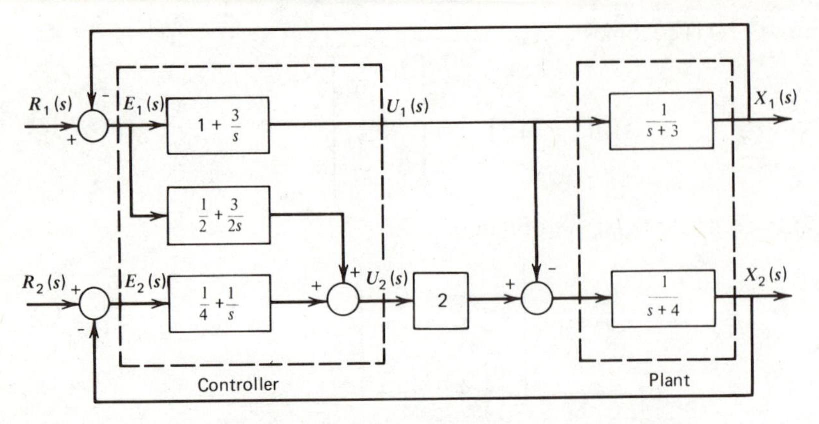

FIGURE 7.25 Noninteracting control system design of Example 7.4.

7.8 SUMMARY

It is apparent that some of the issues and control schemes discussed in this chapter will often require analytical techniques more sophisticated than those presented thus far. The full PID law has three gains to be selected. This selection can be difficult when the plant model must be of high order to include important parasitic modes. The inclusion of compensation devices can also raise the order of the system, and state variable feedback requires several feedback gains to be computed. Thus, practical design problems can easily involve high-order models and multiple gains.

Fortunately, several techniques exist to aid the designer in this matter. Chapter 8 presents some useful graphical techniques, and Chapter 9 applies them to control system design. These are helpful for high-order systems and allow the designer to take advantage of the computational power of the digital computer.

REFERENCES

1. W. Palm, *Modeling, Analysis, and Control of Dynamic Systems*, John Wiley & Sons, New York, 1983.
2. Y. Takahashi, M. Rabins, and D. Auslander, *Control*, Addison-Wesley, Reading, Mass., 1970.
3. L. Hasdorff, *Gradient Optimization and Nonlinear Control*, John Wiley & Sons, New York, 1976.
4. J. G. Ziegler and N. B. Nichols, "Optimum Settings for Automatic Controls," *ASME Transactions* vol. 64, no. 8, 1942, 759.
5. J. G. Ziegler and N. B. Nichols, "Process Lags in Automatic Control Circuits," *ASME Transactions* vol. 65, no. 5, 1943, 433.
6. R. Phelan, *Automatic Control Systems*, Cornell Univ. Press, Ithaca, New York, 1977.

7. A. G. Ulsoy, "Optimal Pseudo-derivative Feedback Control," M. S. thesis, Cornell Univ., Ithaca, New York, 1975.

PROBLEMS

7.1 (a) The following table gives the measured open-loop response of a system to a unit-step input. Use the process reaction method to find the controller gains for P, PI, and PID control.

Time (min)	Response
0	0
0.5	0.02
1.0	0.05
1.5	0.16
2.0	0.28
2.5	0.42
3.0	0.58
3.5	0.70
4.0	0.80
4.5	0.86
5.0	0.92
5.5	0.95
6.0	0.97
7.0	0.98

(b) A certain plant has the open-loop transfer function

$$G_p(s) = \frac{0.5}{(s^2 + s + 1)(s + 0.5)}$$

Use the ultimate-cycle method to compute the controller gains for P, PI, and PID control.

(c) The response in (a) was actually generated from the plant in (b) for the purpose of this example. Compare the controller gains computed with each method.

7.2 (a) Use the ultimate-cycle method to design P, PI, and PID controllers for the following plant:

$$G_p(s) = \frac{1}{s^3 + 5s^2 + s + 2}$$

(b) Plot and compare the unit-step responses for the three designs obtained in (a).

7.3 **(a)** Consider (b) in Problem 6.8, where the float dynamics are significant. Plot the phase plane trajectories. Does this model have a limit cycle? Compare with the results for the ideal case shown in Figure 7.4.

(b) Consider (b) in Problem 5.8. Plot the results on the phase plot. Does a limit cycle exist?

7.4 Use analytical methods or numerical simulation to determine the value of ζ that minimizes the IAE index for the step response of the following system:

$$\frac{C(s)}{R(s)} = \frac{1}{s^2 + 2\zeta s + 1}$$

Plot the unit-step response for the optimal value of ζ, and find the maximum percent overshoot.

7.5 Consider the PI-control system shown in Figure 7.6 except now the plant is such that

$$\Omega(s) = \frac{1}{s+2} T(s)$$

It is desired to obtain a closed-loop system having $\zeta = 1$ and $\tau = 0.1$.

(a) Find the required values of K_p and K_I, neglecting any saturation of the control elements.

(b) Let $m_{max} = 1 = r_{max}$. Find K_p and K_I. Compare the unit-step response of the two designs.

7.6 Many of the causes of damping are nonlinear in nature, such as Coulomb friction. For this reason, we try to model their effects as disturbances to the system, so that a linear system model can be used. Consider a mass with a control force f applied to it along with the Coulomb friction force. The equations of motion are

$$m\ddot{x} = f - \mu mg \qquad \dot{x} > 0$$
$$m\ddot{x} = f + \mu mg \qquad \dot{x} < 0$$

(a) Draw the block diagram of a system to control the position x with the friction modeled as a disturbance.

(b) What type of disturbance input could the designer use to test the response of the controller?

(c) Design a controller to position the mass at $x = 0$.

7.7 Given the plant

$$G_p(s) = \frac{1}{s(\tau_p s + 1)}$$

where $\tau_p \ll 1$.

(a) Neglect τ_p and use proportional control to design a system whose closed-loop time constant is 1.

(b) Same as (a) but obtain a time constant of 0.1.

(c) Suppose that $\tau_p = 0.06$. Evaluate the performance of the designs in (a) and (b) in terms of meeting their specifications.

7.8 The system shown in Figure P7.8*a* is a representation of a common industrial situation in which the displacement or velocity of a mass m_2 must be controlled. The force (or torque) that must do the controlling is constrained to act indirectly on m_2 through the mass m_1 and the elastic element k. For example, m_1 might represent the mass of the motor's armature and the gear reducer, f the applied motor torque, and k the elasticity of the coupling shaft.

(a) Find the transfer function $G_p(s)$ between the displacement x_2 and the force f as input.

(b) What difficulties will be encountered if we attempt to design a single-loop controller, as shown in Figure P7.8*b*, to control the displacement x_2?

(c) Formulate a cascade control system to control x_2. Make the inner-loop control the force $k(x_1 - x_2)$ acting on the mass m_2. Assume that $k(x_1 - x_2)$ can be measured by a force transducer, and use this measurement in an inner loop. Use the outer loop to control the set point r_1 of the inner loop. The set point r_1 is the desired value of the force $k(x_1 - x_2)$. Find $G_{p1}(s)$ and suggest forms for the control laws $G_{c1}(s)$ and $G_{c2}(s)$, as shown in Figure P7.8*c*.

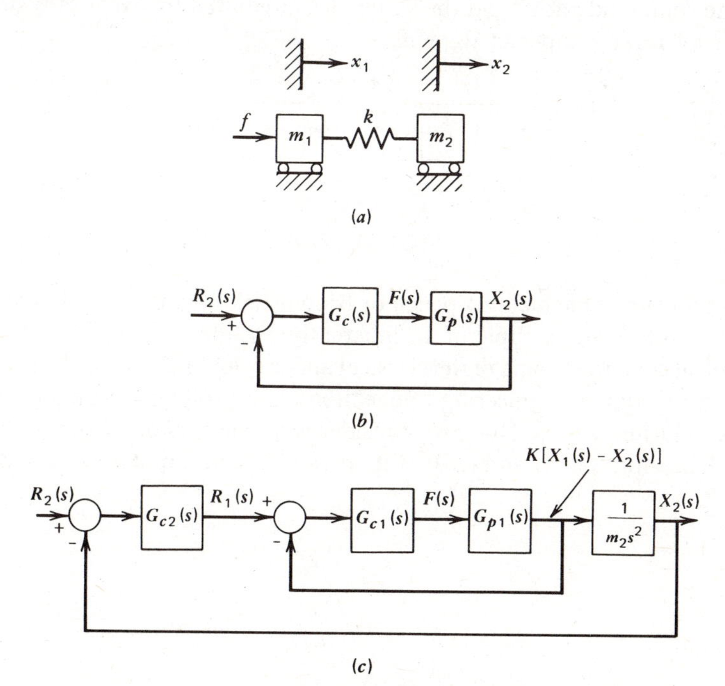

FIGURE P7.8

7.9 A system with feed-forward compensation is shown in Figure P7.9.

(a) With no compensation ($K_f = 0$), find the value of the proportional gain K_p to give $\zeta = 1$.

(b) Find the value of K_f required to give zero steady-state error to a unit-step disturbance. Use the value of K_p from part a.

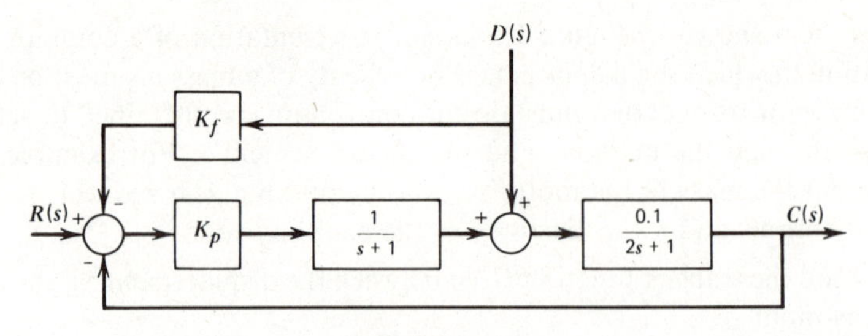

FIGURE P7.9

(c) Compare the steady-state and transient responses of the designs for the following:

(1) A unit-step disturbance.
(2) A unit-impulse disturbance.
(3) A unit-ramp disturbance.

7.10 The control system shown in Figure P7.10 has a PD-type compensator inserted in the command path. Find the value of K_D required to give a zero steady-state error for a ramp input of slope m.

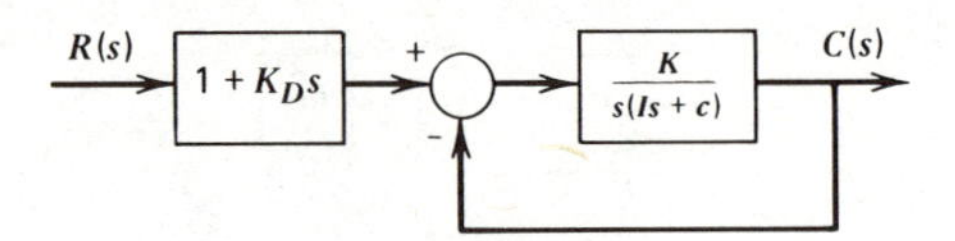

FIGURE P7.10

7.11 The problem of balancing a pencil or broomstick in your hand was discussed in Section 6.1 as an example of an inherently unstable system. Variations of this problem constitute some of the classical and most interesting problems in control theory. A similar engineering application is controlling the angular pitch of a rocket during takeoff. Here, we consider a simple version of the problem.

The situation is shown in Figure P7.11. The input u is an acceleration

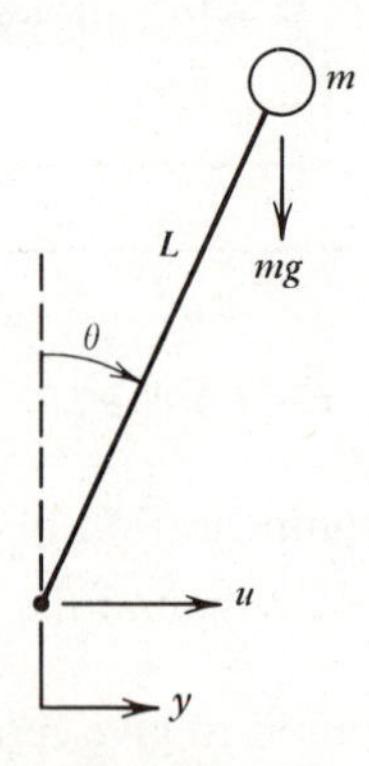

FIGURE P7.11

provided by the control system and applied in the horizontal direction to the lower end of the rod. The horizontal displacement of the lower end is y. The linearized form of Newton's law for small angles gives

$$mL\ddot{\theta} = mg\theta - mu$$

Put this into state variable form by letting $x_1 = \theta$ and $x_2 = \dot{\theta}$. Construct a state variable feedback controller by letting $u = k_1 x_1 + k_2 x_2$. Over what ranges of values of k_1 and k_2 will the controller stabilize the system (keep θ near 0)? What does this formulation imply about the displacement y?

7.12 We desire to stabilize the plant $1/(s^2 - 2)$ with a feedback controller. The closed-loop system should also have a damping ratio of $\zeta = 0.707$ and a dominant time constant of $\tau = 0.1$.

(a) Use the arrangement shown in Figure P7.12*a*. Find the values of K_p and K_D.
(b) Use the arrangement shown in Figure P7.12*b*. Find the values of K_p and K_1.
(c) Compare the two designs in light of their unit-step response.

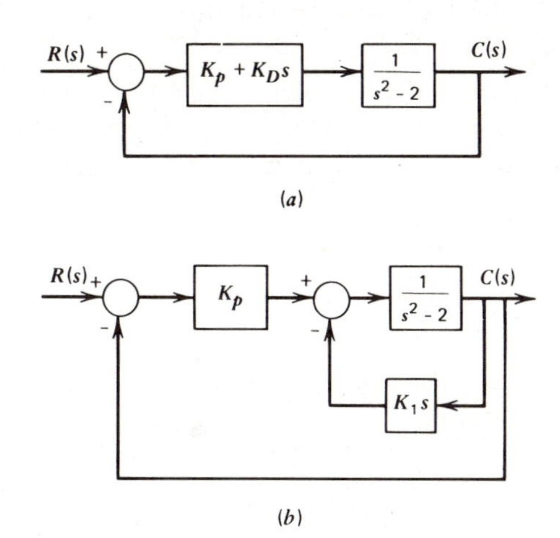

FIGURE P7.12

7.13 Compare P control with PDF control (Figure 7.19) for the case $I = 2$, $m_{\max} = 4$, and $r_{\max} = 8$ with step inputs of the following magnitudes:

(a) 4.
(b) 8.
(c) 12.

7.14 Compare the performance of the critically damped PI and PDF controllers for the plant $1/Is$ with the following:

(a) A unit-ramp disturbance.
(b) A sinusoidal disturbance.
(c) A sinusoidal-command input.

7.15 Consider PDF control as in Figure 7.19 except that the plant's transfer function is $1/(Is + c)$.

(a) Find the transfer functions $C(s)/R(s)$ and $C(s)/D(s)$.
(b) Discuss the effect of c on the choice of K_{D1}.
(c) Discuss the effect of c on the choice of K_I.
(d) Obtain $m(t)$ for a unit-step reference input when $\zeta = 1$. Find t_{max}, the time when $m(t)$ reaches its maximum.
(e) Let $I = c = 1$ and $r_{max} = m_{max} = 1$. Use numerical simulation to design a PDF controller, and compare its performance with that of a PI controller. (*Hint*. Start with the gain values for the case $c = 0$, and use the results from (b) and (c) to choose values for K_I and K_{D1} for the simulation.)

7.16 Consider the PDF controller shown in Figure 7.20 with $I = 1$, $r_{max} = 1$ and $m_{max} = 0.25$. Select the gains K_{D1}, K_{D2}, and K_I to eliminate overshoot.

7.17 Consider again the speed control problem for a tape drive, treated in Section 6.5 and Problem 6.33. The desired speed profile is shown again in Figure P7.17*a*. As in Problem 6.33, the plant's transfer function is $1/(s + 1)$, and the gain of the motor amplifier is K_m (See Figure P7.17*b*). We desire to design a controller such that the speed $c(t)$ at $t = 0.4$ is within 2 rad/sec of the desired value of 100 rad/sec, assuming that $c(0) = \dot{c}(0) = 0$.

(a) Find the gains for a PI controller to meet the specifications.
(b) Find the gains for a PDF controller to meet the specifications.
(c) Compare the designs obtained in (a) and (b) with the design found in Problem 6.33.

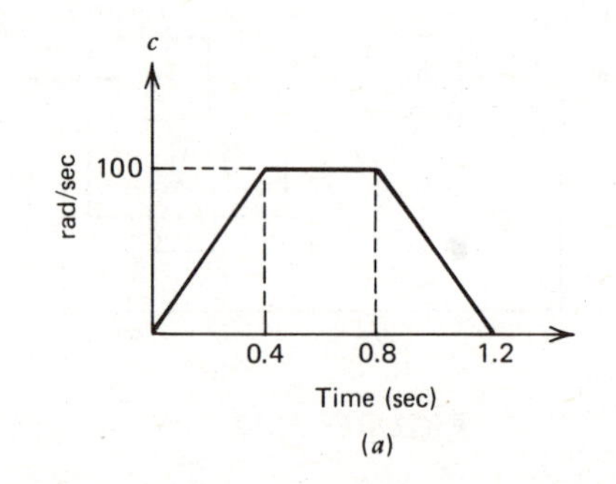

(*a*)

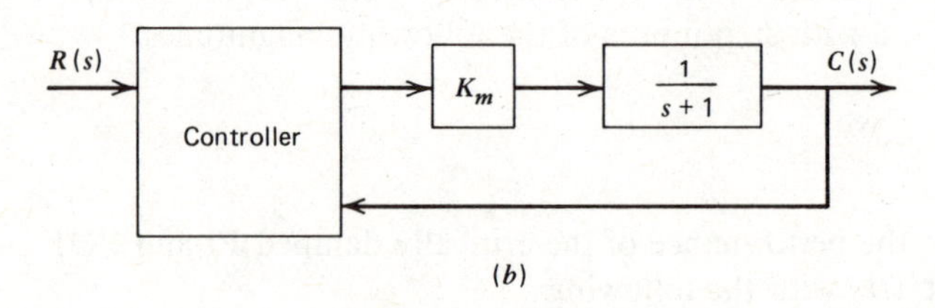

(*b*)

FIGURE P7.17

7.18 A modification of PDF control called *PDF plus* is obtained by including a derivative operation in the inner feedback loop. Figure P7.18 shows PDF-plus control applied to a first-order plant.

(a) Obtain the transfer function $C(s)/R(s)$ and $C(s)/D(s)$.

(b) Show that if K_{D2} can be made arbitrarily large, the response to the disturbance $D(s)$ can be made arbitrarily small without changing the speed of response of the basic PDF controller (with $K_{D2} = 0$).

(c) What limits the magnitude of K_{D2} in practice?

(d) How can the concepts of PDF-plus control be extended to handle a second-order plant $1/Is^2$?

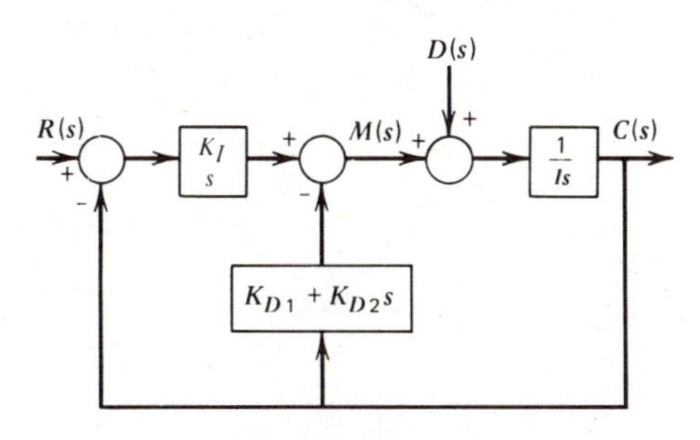

FIGURE P7.18

7.19 When the coupled nonlinear equations of the temperature liquid level system (7.7-1) and (7.7-3) are linearized about the equilibrium solution with T_i and Q_i constant, the linearized equations are uncoupled. Explain this effect. (*Hint.* Consider the linearization of the equation $\dot{y} = y^2$.)

7.20 In Example 7.4 suppose that because of aging effects, the system matrix changes to

$$\mathbf{A} = \begin{bmatrix} -3.3 & 0 \\ 0 & -4 \end{bmatrix}$$

Analyze the effect of this change on the performance of the controller designed in that example.

7.21 Compute the controller transfer function required to achieve noninteraction for a system: $\dot{\mathbf{x}} = \mathbf{Ax} + \mathbf{Bu}$, where

$$\mathbf{A} = \begin{bmatrix} -5 & 0 \\ 0 & -8 \end{bmatrix} \qquad \mathbf{B} = \begin{bmatrix} 1 & 0 \\ -2 & 5 \end{bmatrix}$$

The desired closed-loop time constants for the control loops are $\tau_1 = 1$, $\tau_2 = 3$. Draw the block diagram of the resulting system.

CHAPTER EIGHT
The Root Locus Plot

It is not difficult to analyze a second-order linear system, because its characteristic equation can be solved for the roots in closed form. Rooting formulas exist for third- and fourth-order polynomials, but their complexity is such that they do not generate much insight into the system's behavior. The Routh–Hurwitz criterion is useful for stability analysis, but it cannot give a complete picture of the transient response. Thus, we need design methods for systems of third order and higher.

Here, we develop graphical methods for representing the system's behavior. The *root locus plot* is a plot of the location of the characteristic roots in terms of some system parameter, such as the proportional gain. The theory of polynomial equations is sufficiently well developed to allow us to sketch the general behavior of the roots without actually solving for them in many cases. This information is often sufficient to make design decisions.

Although the methods in this chapter were developed before the widespread use of digital computers and calculators for computation, the methods' usefulness has been enhanced by such machines. The plots can be sketched by hand or generated with a computer-based plotter if available.

8.1 THE ROOT LOCUS CONCEPT

The root locus is a plot of the locations of the roots of an equation as some real-valued parameter varies, possibly from $-\infty$ to $+\infty$. To see this more clearly, consider the polynomial equation

$$ms^2 + cs + k = 0 \tag{8.1-1}$$

Suppose that $m = c = 1$ and we wish to display the root locations as k varies. For now, let $k \geqslant 0$. The roots are

$$s = \frac{-1 \pm \sqrt{1-4k}}{2} \tag{8.1-2}$$

It is easily seen that if $k < 0.25$, the roots are real and distinct. They are repeated if $k = 0.25$ and complex conjugates if $k > 0.25$. In the latter case, the time constant is always 2. The root locations for $0 \leqslant k < \infty$ can thus be plotted in the s plane as in Figure 8.1*a*. The root locations when $k = 0$ are denoted by X and are at $s = 0, -1$. The arrows on the plot indicate the direction of root movement as k increases. As k approaches ∞, the roots approach $s = -0.5 \pm i\infty$.

Equation (8.1-1) is the characteristic equation of the commonly found transfer function

$$T(s) = \frac{1}{ms^2 + cs + k} \tag{8.1-3}$$

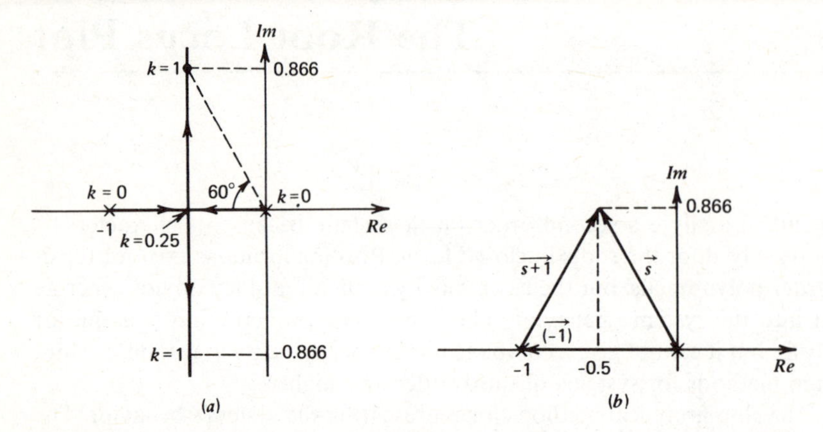

FIGURE 8.1 (*a*) Root locus plot for $s^2 + s + k = 0$, $k \geqslant 0$. (*b*) Vector diagram showing the calculation of *k* at the point $s = -0.5 + i0.866$.

Suppose the desired damping ratio is $\zeta = 0.5$, with $m = c = 1$. The corresponding characteristic roots are $s = -0.5 \pm i0.866$ and are marked in Figure 8.1*a*. These roots can be obtained from the plot by marking off the line corresponding to $\cos \beta = \zeta = 0.5$, or $\beta = 60°$. The intersection of this line with the root–locus plot gives the upper root of the conjugate pair. At this point, the reader might wish to review the graphical interpretations of the parameters τ, ω_n, and ζ in terms of the root locations. See Section 3.4 (Figures 3.7 and 3.8).

The root locus plot can thus be used to determine characteristic-root locations that give specified behavior, such as $\zeta = 0.5$, because the plot represents the solution of the characteristic equation (8.1-1). Once the root locations have been determined, the plot can be used to compute the value of *k* required to give these characteristic roots. To see how this is done, rewrite (8.1-1) with $m = c = 1$.

$$k = -s(s+1) \tag{8.1-4}$$

Since *k* is a real number, the right-hand side must also be real. At the desired root locations $s = -0.5 \pm i0.866$, (8.1-4) gives

$$\begin{aligned} k &= -(-0.5 + i0.866)(-0.5 + 1 + i0.866) \\ &= -(-\tfrac{1}{4} - \tfrac{3}{4}) = 1 \end{aligned} \tag{8.1-5}$$

Thus, if $k = 1$, the damping ratio will be 0.5. This result could have been obtained directly from (8.1-1), but we are preparing the way for more difficult problems.

The vector properties of complex numbers can be used to simplify calculating *k*. (A review of these properties is given in Appendix A, Section A.5.) With reference to (8.1-4) and Figure 8.1*b*, we note that the vector $\vec{s} = -0.5 + i0.866$ has its head at this point and its tail at the origin. From the figure, we see that the property of vector addition gives

$$\vec{s} = (\overrightarrow{-1}) + (\overrightarrow{s+1}) \tag{8.1-6}$$

From this, we deduce that $(\overrightarrow{s+1})$ has its head at $s = -0.5 + i0.866$ and its tail at -1.

In general, the number $(s + a)$ is represented by a vector with its head at the point s and its tail at $-a$.

The magnitude of a product of complex numbers is the product of the magnitudes. Therefore, for $k \geqslant 0$, (8.1-4) can be expressed as

$$|k| = k = |-s(s+1)| = |-s||s+1| = |s||s+1| \tag{8.1-7}$$

Since the magnitude of a complex number is the length of its corresponding vector, we can evaluate k by measuring on the plot the lengths of the vectors $\vec{s}$ and $\overrightarrow{(s+1)}$. This gives

$$k = 1(1) = 1$$

When many such factors are involved, it is easier to measure vector lengths on the plot rather than multiply several complex numbers, as was required in (8.1-5). The resulting accuracy depends on the size of the graph, but acceptable accuracy is usually obtained with a graph of moderate size.

Sensitivity to Parameter Variation

With the design selected to be $k = 1$, we now wish to see what is the effect of the parameter c not having its nominal value of $c = 1$. This is a common problem, because we never know parameter values exactly. To this end, we write (8.1-1) with $m = k = 1$ and $c = 1 + \Delta$.

$$s^2 + (1 + \Delta)s + 1 = 0 \tag{8.1-8}$$

Here, Δ represents the deviation of c from its nominal value. To simplify the discussion, we consider only $\Delta \geqslant 0$. The roots of (8.1-8) are

$$s = \frac{-(1+\Delta) \pm \sqrt{2\Delta + \Delta^2 - 3}}{2} \tag{8.1-9}$$

We can plot the roots as shown in Figure 8.2. Even though we would normally

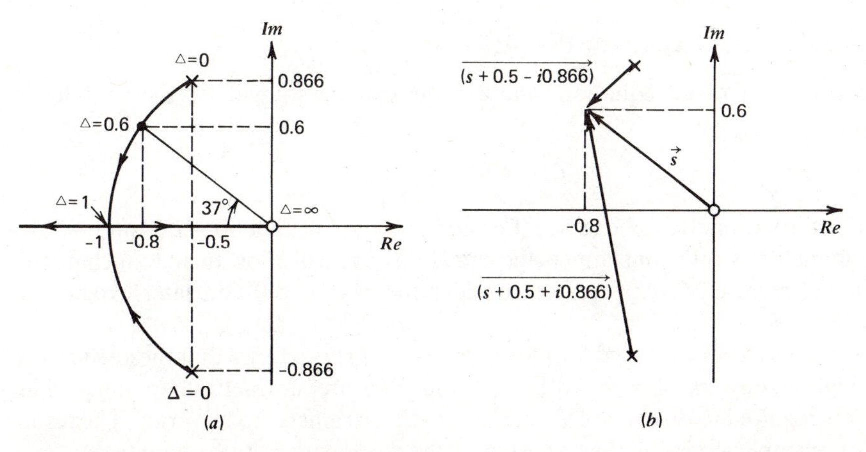

FIGURE 8.2 **(*a*) Root-locus plot for $s^2 + (1 + \Delta)s + 1 = 0$, $\Delta \geqslant 0$. (*b*) Vector diagram showing the calculation of Δ at $s = -0.8 + i0.6$.**

consider only small deviations from the nominal, we show the roots for $0 \leqslant \Delta < \infty$ for general interest. The root locations when $\Delta = 0$ are marked by X and coincide with the roots when $c = k = 1$. As $\Delta \to \infty$, one root approaches $s = -\infty$ while the other approaches $s = 0$. A finite termination point is denoted by a circle (○). The arrows here represent the motion of the roots as Δ increases.

This plot can be used to determine how much of a change in c is required to change the damping ratio by a certain amount from its desired value of $\zeta = 0.5$. The method is similar to that used earlier. Rewrite (8.1-8) as

$$\Delta = \frac{-(s^2 + s + 1)}{s} = \frac{-(s + 0.5 - i0.866)(s + 0.5 + i0.866)}{s} \tag{8.1-10}$$

To see what value of Δ will increase ζ from 0.5 to 0.8, we use a protractor to draw the line corresponding to $\cos\beta = 0.8$, or $\beta = 37°$. It intersects the locus at $s = -0.8 + i0.6$, as shown in Figure 8.2*a*. To evaluate Δ at this point, we note that (8.1-10) involves vectors whose heads are at this point and whose tails are at the points denoted by ○ and X. We measure the lengths of these vectors to be

$$|s| = 1$$
$$|(s + 0.5 - i0.866)| = 0.4$$
$$|(s + 0.5 + i0.866)| = 1.5$$

Equation (8.1-10) gives

$$\Delta = |\Delta| = \frac{|s + 0.5 - i0.866||s + 0.5 + i0.866|}{|s|}$$
$$= \frac{0.4(1.5)}{1} = 0.6$$

If $\Delta = 0.6$ (or $c = 1.6$), the damping ratio will be 0.8.

Formulation of the General Problem

The general form of an equation whose roots can be studied by the root locus method is

$$D(s) + KN(s) = 0 \tag{8.1-11}$$

where K is the parameter to be varied. For now, we take the functions $D(s)$ and $N(s)$ to be polynomials in s with constant coefficients. Later, we will allow them to include the dead-time element e^{-Ds}. At first, we consider the case $K \geqslant 0$ and later extend the results to $K < 0$.

The guides to be developed for plotting the root locus require that the coefficients of the highest powers of s in both $N(s)$ and $D(s)$ are normalized to unity. The multipliers required to do this are absorbed into the parameter to be varied. The result is K. For example, consider the variation of the parameter b in the equation

$$3s^2 + 5bs + 2 = 0$$

This can be written as

$$s^2 + \frac{2}{3} + \frac{5b}{3}s = 0$$

From (8.1-11), we see that $D(s) = s^2 + 2/3$, $N(s) = s$, and $K = 5b/3$. The root locus plot would be made in terms of the parameter K and the values for b recovered from K.

Comparing the general form (8.1-11) with the two previous examples shows that for the first example with $m = c = 1$ and k the parameter, $K = k$, $N(s) = 1$, and $D(s) = s^2 + s$. For the second example, (8.1-8) shows that $K = \Delta$, $N(s) = s$, and $D(s) = s^2 + s + 1$.

Another standard form of the problem is obtained by rewriting (8.1-11) as

$$1 + KP(s) = 0 \tag{8.1-12}$$

where

$$P(s) = \frac{N(s)}{D(s)} \tag{8.1-13}$$

Terminology

The standard terminology is as follows. The roots of $N(s) = 0$ are called the *zeros* of the problem. The name refers to the fact that they are the finite values of s that make $P(s)$ zero. The roots of $D(s) = 0$ are the *poles* of the problem. They are the finite values of s that make $P(s)$ become infinite.

Referring to (8.1-11), we see that when $K = 0$, the roots of the equation are the poles. As $K \to \infty$, the roots of (8.1-11) approach the zeros. These facts will be useful.

For most of our applications, the root locus method will be applied to the characteristic equation of a feedback system. Therefore, it is helpful to relate the preceding terminology to such a system. A general single-loop system is shown in Figure 8.3 without a disturbance input. The transfer function is

$$T(s) = \frac{C(s)}{R(s)} = \frac{G(s)}{1 + G(s)H(s)} \tag{8.1-14}$$

In general, $G(s)$ and $H(s)$ will consist of numerator and denominator polynomials, denoted by

$$G(s) = \frac{N_G(s)}{D_G(s)} \tag{8.1-15}$$

$$H(s) = \frac{N_H(s)}{D_H(s)} \tag{8.1-16}$$

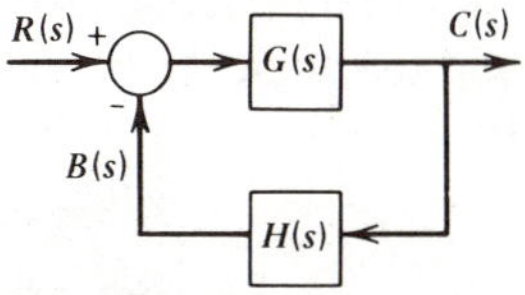

The transfer function can be written as

$$T(s) = \frac{N_G(s)D_H(s)}{D_G(s)D_H(s) + N_G(s)N_H(s)} \tag{8.1-17}$$

The characteristic equation is

$$D_G(s)D_H(s) + N_G(s)N_H(s) = 0 \tag{8.1-18}$$

FIGURE 8.3 Terminology for the single-loop system.

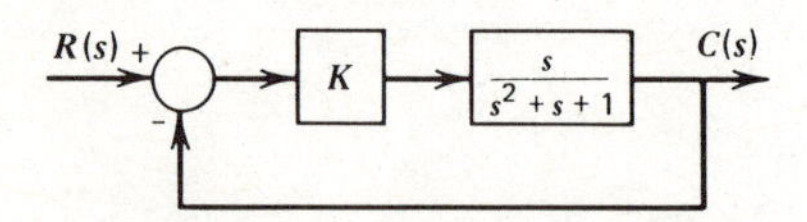

FIGURE 8.4 Feedback system with open-loop poles at $s = -0.5 \pm i0.866$ and an open loop zero at $s = 0$. The root locus for this system for $K \geqslant 0$ is given by Figure 8.2a, with $\Delta = K$.

The *open-loop transfer function* of this system is $G(s)H(s)$ and so named because it is the transfer function relating the feedback signal $B(s)$ to the input $R(s)$ if the loop is opened or broken at $B(s)$. That is,

$$B(s) = G(s)H(s)R(s) \tag{8.1-19}$$

The *closed-loop transfer function* is $T(s)$ given by (8.1-14). Comparing (8.1-18) with (8.1-11), we see that the poles of the characteristic equation are also the poles of $G(s)H(s)$ if the parameter K is a multiplicative factor in $N_G(s)N_H(s)$, as is often the case. Similarly, the zeros of the problem are also the zeros of $G(s)H(s)$. Thus, the terms *open-loop poles* and *open-loop zeros* are often used for the poles and zeros on the root locus plot. The *closed-loop poles* are the characteristic roots, since they are the finite values of s that make the closed-loop transfer function become infinite.

For example, consider proportional control of a second-order plant, as shown in Figure 8.4. The open- and closed-loop transfer functions are

$$G(s)H(s) = \frac{Ks}{s^2 + s + 1} \tag{8.1-20}$$

$$T(s) = \frac{Ks}{s^2 + (1+K)s + 1} \tag{8.1-21}$$

The open-loop poles are found from $s^2 + s + 1 = 0$ to be $s = -0.5 \pm i0.866$. The open-loop zero is $s = 0$. With $K = \Delta$, the characteristic equation for this system is the same as that given by (8.1-8). Therefore, the root locus plot is given by Figure 8.2*a*. Note that the open-loop poles are the characteristic root locations when $K = 0$ and represent the starting points of the locus. The open-loop zero is one of the characteristic roots when $K = \infty$ and represents the termination point of one of the root locus paths. Finally, we observe that the open-loop poles and zeros can be found without having a value for K. The same is not true for the closed-loop poles.

Angle and Magnitude Criteria

The general problem in the form (8.1-12) can be written as

$$KP(s) = -1 \tag{8.1-22}$$

From this, we see that two requirements must be met if s is to be a root of this equation. Equation (8.1-22) is an equation between two complex numbers. As such, the equality requires the magnitudes of each side to be equal and the angles to be the same. These are the magnitude and angle criteria, respectively.

$$|KP(s)| = 1 \tag{8.1-23}$$

$$\measuredangle KP(s) = \measuredangle(-1) = n180^\circ \qquad n = \pm 1, \pm 3, \pm 5, \ldots \tag{8.1-24}$$

For now, we consider only $K \geqslant 0$. Thus, $|K| = K$ and $\measuredangle K = 0^\circ$, so that (8.1-23) becomes

$$K = \frac{1}{|P(s)|} \tag{8.1-25}$$

Also, since the angle of a product is the sum of the angles (see Section A.5 in Appendix A),

$$\measuredangle KP(s) = \measuredangle K + \measuredangle P(s) = \measuredangle P(s)$$

Equation (8.1-24) implies that

$$\measuredangle P(s) = n180^\circ \qquad n = \pm 1, \pm 3, \pm 5, \ldots \tag{8.1-26}$$

Equations (8.1-25) and (8.1-26) are the only two conditions that every point on the root locus must satisfy. It is interesting to note that the angle criterion (8.1-26) does not contain K. The *shape* of the root locus plot is determined *entirely* by the angle criterion. All of the plotting guides to follow, except two, are the result of this single condition. The magnitude criterion (8.1-25) is used only to assign the associated value of K to a designated point s on the root locus. This process is referred to as *scaling* the locus.

8.2 PLOTTING GUIDES

Let us collect the insights generated thus far and formalize them as guides for plotting the root locus. Our primary reference will be (8.1-12) and its associated magnitude and angle criteria, (8.1-25) and (8.1-26), for $K \geqslant 0$. The poles are the finite s values that make $P(s)$ infinite. The zeros are the finite values of s that make $P(s)$ zero. We assume that the order of the numerator of $P(s)$ is less than or equal to the order of the denominator. We require the real and imaginary axes to have the same scale. Each path on the plot corresponds to one root of (8.1-12). These paths are referred to as the loci or branches.

Guide 1

The root locus plot is symmetric about the real axis. This is because complex roots occur in conjugate pairs. Thus, we need deal with only the upper half-plane of the plot.

Guide 2

The number of loci equals the number of poles of P(s). This guide tells us how many paths we must account for. The proof is taken from (8.1-11). If the order of $D(s)$ is greater than that of $N(s)$, then the order of the equation, and thus the number of roots, is determined by $D(s)$. But the order of $D(s)$ is the number of poles of $P(s)$.

Guide 3

The loci start at the poles of P(s) *with* $K = 0$ *and terminate with* $K = \infty$ *either at the zeros of* P(s) *or at infinity.* When $K = 0$, (8.1-11) shows that the roots are given by $D(s) = 0$; in other words, by the poles. When $K \to \infty$, there are only two ways that (8.1-11) can be satisfied. The first way requires that $N(s) \to 0$; that is, that s approach a zero of $P(s)$. If $N(s)$ does not approach zero as $K \to \infty$, then $D(s)$ must approach infinity, which can occur only if $s \to \infty$. This is the basis for the guide.

The assumption concerning the relative order of $D(s)$ and $N(s)$ is true in control system application as well as in most other types of problems. When the assumption is not satisfied, it is often because of an improper formulation of the problem. For example, consider again (8.1-1), but this time with $c = k = 1$ and m as the parameter K. Rearranging in terms (8.1-12), we obtain

$$1 + \frac{Ks^2}{s+1} = 0$$

Thoughtless application of Guides 2 and 3 results in a contradiction. Guides 2 and 3 state that there is one path and it starts at $s = -1$. Guide 3 says that two termination points exist at $s = 0$, because there are two zeros at that point. The problem is improperly formulated, because when $m = K = 0$, the order of the equation drops from two to one. When $K \neq 0$, there are two roots; when $K = 0$, there is only one.

The variation of the leading coefficient can be studied if the order of the equation remains the same throughout the variation. For example, if $c = k = 1$ in (8.1-1) and the nominal value of m is 1, let $m = 1 + K$, and write (8.1-1) as

$$1 + \frac{Ks^2}{s^2 + s + 1} = 0$$

The guides for $K \geqslant 0$ can now be applied without any difficulty. Both paths terminate at $s = 0$.

Behavior of the Loci on the Real Axis

In order to develop more plotting guides, we now apply the angle criterion to some typical problems.

Consider proportional control of a first-order plant $1/(s + a)$ with a proportional gain K. The characteristic equation is

$$K = -(s + a)$$

If we express the number $(s + a)$ in complex exponential form (Section A.5, Appendix A), we have

$$K = -r_a e^{i\phi_a}$$

See Figure 8.5*a*. By definition, $r_a > 0$. Thus, the preceding equation implies that $K = r_a$ and $\exp(i\phi_a) = -1$. This requires that $\phi_a = n180°$, $n = \pm 1, \pm 3, \ldots$. It says that the root locus lies entirely on the real axis and entirely on the left of the pole at $s = -a$, as shown in Figure 8.5*b*. Of course, we could have deduced this without the angle criterion, but the approach is useful.

Now, suppose the proportional control is applied to a second-order plant $1/(s + a)(s + b)$. The characteristic equation is

$$K = -(s + a)(s + b)$$

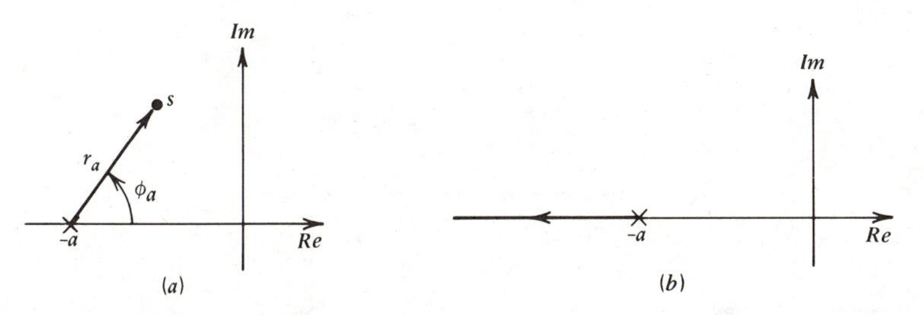

FIGURE 8.5 Single-pole system $s + a + K = 0$. (*a*) Complex exponential representation. (*b*) Root-locus plot for $K \geqslant 0$.

or

$$K = -r_a r_b e^{i(\phi_a + \phi_b)}$$

See Figure 8.6*a*. The requirements are $K = r_a r_b$ and

$$\phi_a + \phi_b = n180°, \qquad n = \pm 1, \pm 3, \ldots \tag{8.2-1}$$

The latter condition implies that the root locus *must* exist on the real axis between the poles and *cannot* exist anywhere else on the real axis. This can be seen by noting that the tips of the vectors must touch and trying different values of ϕ_a and ϕ_b that satisfy this constraint and the relation (8.2-1). The only values that satisfy both on the real axis are: $\phi_a = n180°$, $n = \pm 1, \pm 3, \ldots$; and $\phi_b = m360°$, $m = 0, \pm 1, \pm 2. \ldots$. The root locus is shown in Figure 8.6*b*, but discussion of how the behavior off the real axis was determined is deferred for now.

A third-order plant $1/(s + a)(s + b)(s + c)$ with proportional control gives the characteristic equation

$$K = -(s + a)(s + b)(s + c)$$

or

$$K = -r_a r_b r_c e^{i(\phi_a + \phi_b + \phi_c)}$$

The two conditions for the root locus are $K = r_a r_b r_c$ and

$$\phi_a + \phi_b + \phi_c = n180° \qquad n = \pm 1, \pm 3, \ldots \tag{8.2-2}$$

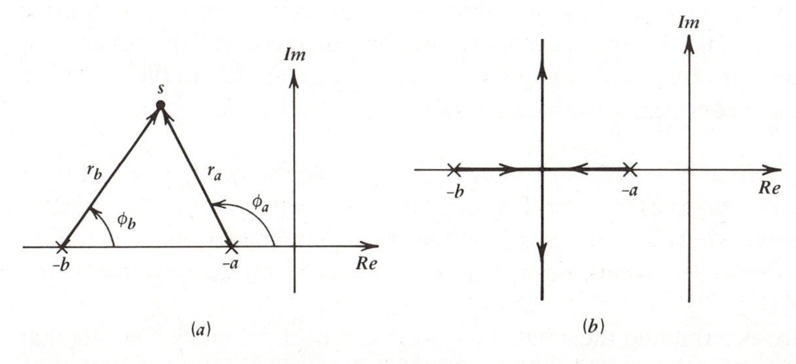

FIGURE 8.6 System with two real poles $(s + a)(s + b) + K = 0$. (*a*) Complex representation. (*b*) Root-locus plot for $K \geqslant 0$.

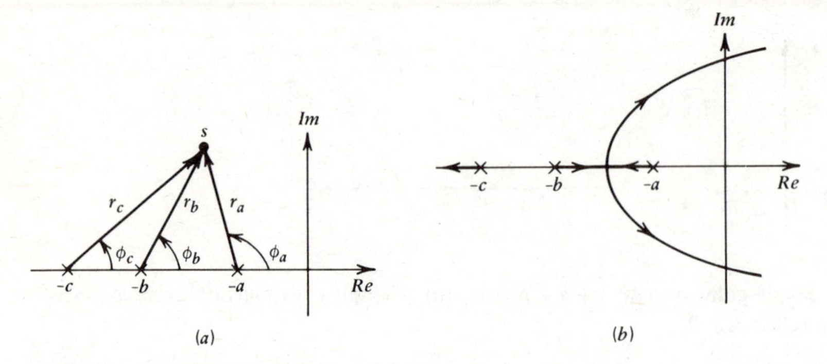

FIGURE 8.7 System with three real poles $(s+a)(s+b)(s+c)+K=0$. (*a*) Complex representation. (*b*) Root-locus plot for $K \geqslant 0$.

Analyzing Figure 8.7*a* in the same manner as the previous case shows that the only possible root locations on the real axis correspond to

$$\phi_a = n180°, \qquad n = \pm 1, \pm 3, \ldots$$
$$\phi_b = \phi_c = m360°, \qquad m = 0, \pm 1, \pm 2, \ldots$$

or

$$\phi_a = \phi_b = n180°, \qquad n = \pm 1, \pm 3, \ldots$$
$$\phi_c = m360°, \qquad m = 0, \pm 1, \pm 2, \ldots$$

In other words, the root locus exists on the real axis to the left of the pole $s = -c$ and between the poles at $s = -a$ and $s = -b$. The root locus is shown in Figure 8.7*b*.

We can generalize these results to obtain Guide 4. For a rigorous proof of this and the other guides, see Reference 1.

Guide 4

The root locus can exist on the real axis only to the left of an odd number of real poles and/or zeros; furthermore, it must exist there. Our numbering system is as follows. The real pole or zero that lies the farthest to the right is number 1; the next real pole or zero is number 2; etc. If a pole or zero falls on top of another, number each one. The phrase *only to the left* means that the locus *cannot* exist on the real axis to the *right* of an odd-numbered real pole or zero. Some examples are shown in Figure 8.8. In this figure, no implication is made concerning the location of the locus off the real axis.

If we refer to either Figure 8.1*a* or 8.6*b*, we can see that the two loci depart from the poles and head toward each other. It is impossible for two root paths to coincide over any finite length (but they can cross). So the two loci must leave the real axis at some point (the *breakaway point*). From Guide 1, they must do this in a symmetric way.

A method for determining the breakaway point can be developed by noting that the loci approach each other as K increases. Thus, at the breakaway point, K attains the *maximum* value it has on the real axis in the vicinity of the breakaway point. The

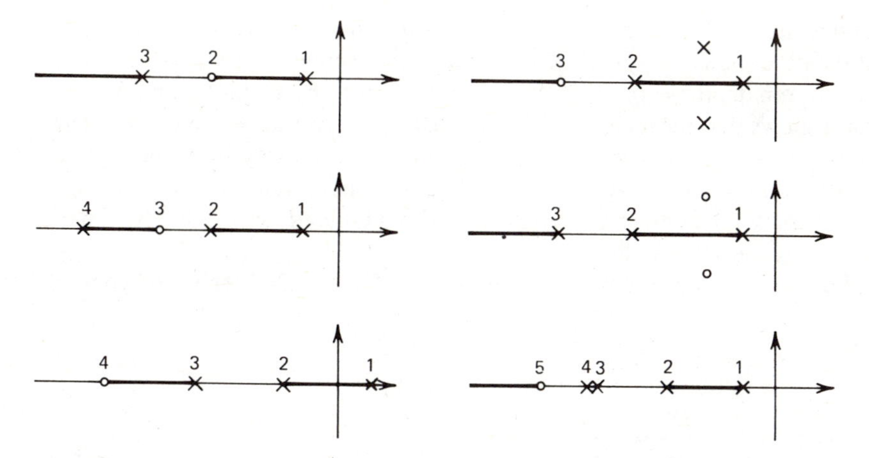

FIGURE 8.8 Location of the root locus on the real axis for various pole-zero configurations. The locus off the real axis is not shown.

value of s corresponding to the breakaway point can be found by computing dK/ds from (8.1-25), setting $dK/ds = 0$, and solving for the value of s. In general, multiple solutions will occur. The extraneous ones can be discarded with a knowledge of the location of the locus on the real axis from Guide 4. Usually, the second derivative need not be computed to distinguish between a minimum and a maximum.

Figure 8.2*a* shows that the locus can also enter the real axis. The point at which this occurs is the *break-in* point. After the locus has entered the real axis, K continues to increase. Thus, at a break-in point, K attains the *minimum* value it has on the real axis in the vicinity of the break-in point. The location of the break-in point is determined from $dK/ds = 0$ in exactly the same manner as for a breakaway point.

Although the same method is used to find the breakaway and break-in points, no difficulty is encountered in identifying the type of point, since this is usually obvious once the first four guides have been applied. It is possible that there are multiple breakaway or break-in points. Hence, we speak of K attaining a maximum or minimum value only in the vicinity of each such point.

Guide 5

The locations of breakaway and break-in points are found by determining where the parameter K *attains a local maximum or minimum on the real axis.*

You should use this guide to convince yourself of the locations of the breakaway and break-in points shown in Figures 8.1*a* and 8.2*a*. We now illustrate Guide 5 with an example.

Example 8.1

Determine the root locus plot for the equation

$$1 + \frac{K(s+6)}{s(s+4)} = 0 \tag{8.2-3}$$

The poles are $s = 0, -4$; the zero is $s = -6$. Thus, there are two loci. One starts at $s = 0$ and the other at $s = -4$, with $K = 0$. One path terminates at $s = -6$, while the other must terminate at $s = \infty$. This information is given by Guides 2 and 3.

Guide 4 shows that the locus exists on the real axis between $s = 0$ and $s = -4$ and to the left of $s = -6$. Therefore, the two paths must break away from the real axis between $s = 0$ and $s = -4$. From the location of the termination point at $s = -6$, we know that the locus must return to the real axis. From Guide 1, we see that both paths must break in at the same point.

These points are found from Guide 5 as follows. Solve (8.2-3) for K and compute dK/ds.

$$K = -\frac{s(s+4)}{s+6}$$

$$\frac{dK}{ds} = -\frac{(s+6)(2s+4) - s(s+4)}{(s+6)^2} = 0$$

This is satisfied for finite s if the numerator is zero.

$$s^2 + 12s + 24 = 0$$

The candidates are

$$s = -6 \pm 2\sqrt{3} = -2.54, \ -9.46$$

There is no need to check for a minimum or a maximum, because we know from

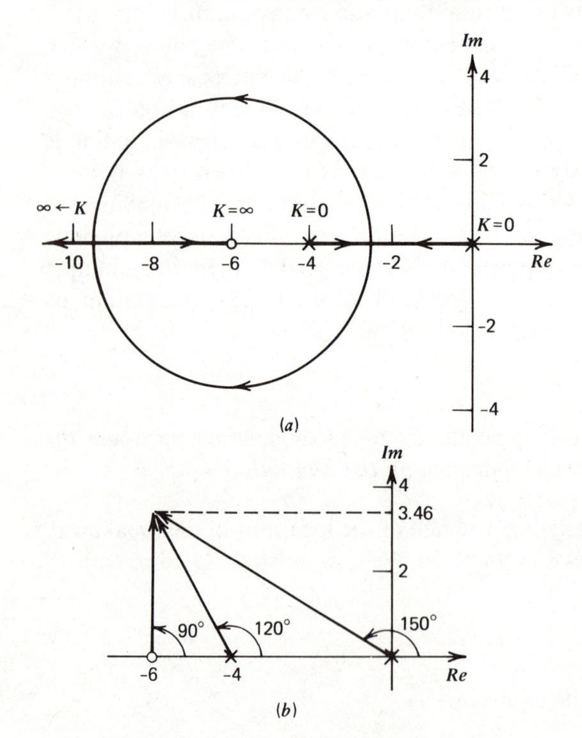

FIGURE 8.9 **(*a*) Root-locus plot for $s(s+4) + K(s+6) = 0$, for $K \geqslant 0$. (*b*) Angle criterion applied to the test point $s = -6 + i3.46$.**

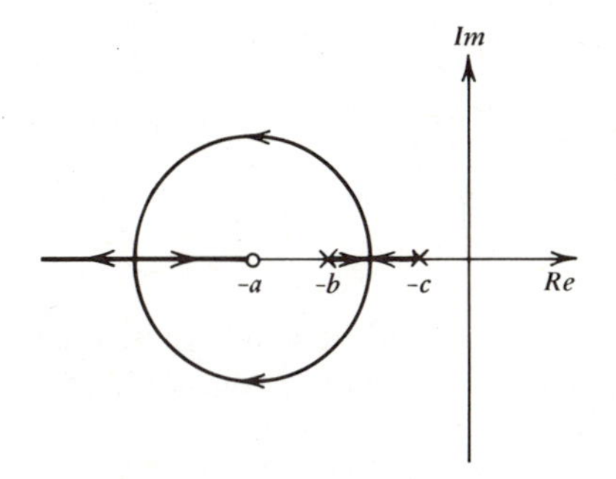

FIGURE 8.10 **Root-locus plot for $(s+b)(s+c)+K(s+a)=0$ for $K \geqslant 0$ and $a>b>c>0$.**

Guide 4 that the breakaway point must be $s=-2.54$ and the break-in point $s=-9.46$. This leaves only the shape of the locus off the real axis to be determined. We note that the breakaway and break-in points are symmetrically placed with a distance of 3.46 from the zero at $s=-6$. The simplest way for this to occur is with a circle of radius 3.46 centered at the zero. We draw this circle as shown in Figure 8.9*a*. To confirm this, we can select trial points along the circle and apply the angle criterion. For example, the point $s=-6+3.46i$ lies on the locus, because

$$\measuredangle \frac{s(s+4)}{s+6} = \measuredangle s + \measuredangle (s+4) - \measuredangle (s+6)$$
$$= 150^\circ + 120^\circ - 90^\circ$$
$$= 180^\circ$$

Several points can be quickly checked with a protractor. This would show that the locus is indeed a circle.

Elementary geometry can be used to show that the root locus of the equation

$$(s+b)(s+c)+K(s+a)=0 \tag{8.2-4}$$

is a circle centered on the zero at $s=-a$ if $a, b, c>0$ and $a>b>c$. This case is shown in Figure 8.10. The radius of the circle can be determined once the breakaway and break-in points are found.

Behavior of the Locus for Large K

As $K \to \infty$, the loci approach either a zero or $s=\infty$. We now investigate how to determine the shape of the loci in the latter case. The vectors to a point on the locus at $s=\infty$ from the poles and zeros are parallel and make an angle θ with the real axis. This is shown in Figure 8.11 for a particular pole–zero pattern. If we let P and Z denote the number of poles and zeros, respectively, the angle criterion states that

$$\measuredangle P(s) = Z\theta - P\theta = n180^\circ$$
$$n = \pm 1, \pm 3, \ldots$$

Solving for θ, we obtain

$$\theta = \frac{n180^\circ}{Z-P} \qquad n = \pm 1, \pm 3, \ldots \tag{8.2-5}$$

This leads to Guide 6.

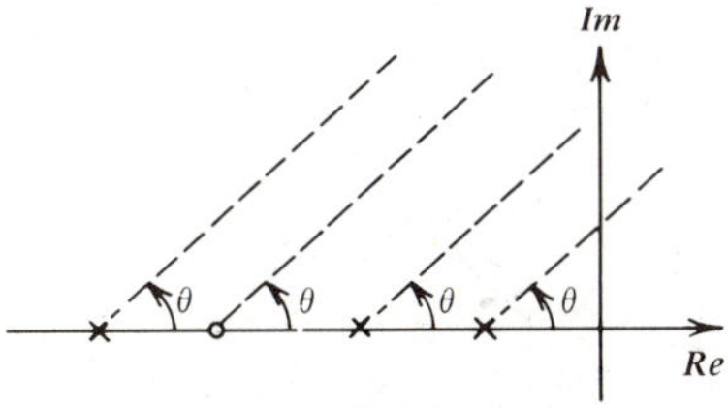

FIGURE 8.11 **Asymptotic angle θ for a particular pole-zero configuration.**

Guide 6

The loci that do not terminate at a zero approach infinity along asymptotes. The angles that the asymptotes make with the real axis are found from (8.2-5), where n *is chosen successively as* $n = +1, -1, +3, -3, \ldots,$ *until enough angles have been found.* Figure 8.12 shows the most commonly found patterns.

Unless θ is 0° or 180°, the asymptotes cannot be drawn until we know where they intersect the real axis. This is given by Guide 7. The proof follows from the angle criterion, but is detailed (see Reference 1).

Guide 7

The asymptotes intersect the real axis at the common point

$$\sigma = \frac{\Sigma s_p - \Sigma s_z}{P - Z} \tag{8.2-6}$$

where Σs_p *and* Σs_z *are the algebraic sums of the values of the poles and zeros.*

Note that this guide states that the asymptotes all intersect at the same point.

Referring to (8.1-1) with $m = c = 1$ and $K = k$, the poles are $s = 0, -1$, and there are no zeros. Thus, $P = 2$, $Z = 0$, and

$$\Sigma s_p = 0 + (-1) = -1$$

$$\Sigma s_z = 0$$

We need two asymptotes. Therefore, from (8.2-5)

$$\theta_1 = \frac{180^\circ}{0 - 2} = -90^\circ$$

$$\theta_2 = \frac{-180^\circ}{0 - 2} = +90^\circ$$

From (8.2-6),

$$\sigma = \frac{-1 - 0}{2 - 0} = -\frac{1}{2}$$

As shown in Figure 8.1*a*, the locus lies on the asymptotes. The asymptote in the upper half-plane has the angle $+90^\circ$. Both asymptotes intersect at $s = -1/2$.

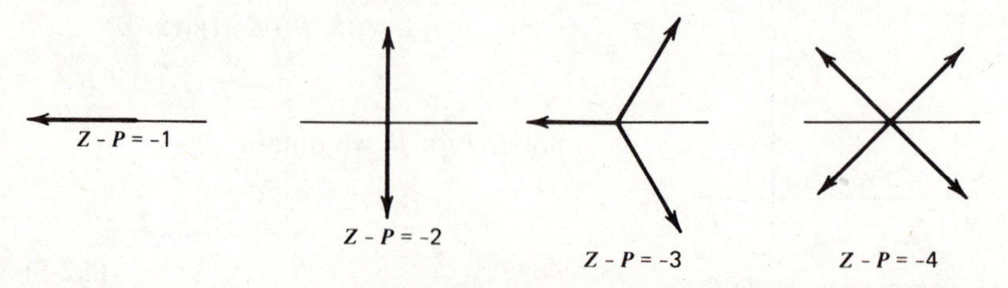

FIGURE 8.12 Asymptotic angles for commonly occurring cases.

An example with complex poles is given by (8.1-8) with $K = \Delta$. The poles are $s = -0.5 \pm i0.866$. The zero is $s = 0$. Thus, $P = 2$, $Z = 1$, and

$$\Sigma s_p = (-0.5 + i0.866) + (-0.5 - i0.866) = -1$$
$$\Sigma s_z = 0$$

Therefore, the one asymptote required has the angle

$$\theta = \frac{180°}{1-2} = -180°$$

This is shown in Figure 8.2*a*. The asymptote lies on the real axis, and we do not have to calculate an intersection point.

Some systems become unstable for certain values of the parameter K. This occurs if any path crosses the imaginary axis into the right half-plane. The value of K at which this happens can be determined from the Routh–Hurwitz criterion (Section 3.8 and Appendix F). Often, the substitution $s = i\omega$ into the polynomial equation of interest is quicker and gives the crossing location as well as the value of K. This procedure was applied to the cubic polynomial in Section 7.1, Example 7.2. This gives Guide 8.

Guide 8

The points at which the loci cross the imaginary axis and the associated values of K *can be found by the Routh–Hurwitz criterion or by substituting* $s = i\omega$ *into the equation of interest. The frequency* ω *is the* crossover frequency.

Example 8.2

Plot the root locus for the equation

$$s^3 + 3s^2 + 2s + K = 0 \qquad \text{for } K \geq 0 \tag{8.2-7}$$

We can factor the equation as follows:

$$1 + \frac{K}{s(s+1)(s+2)} = 0$$

The poles are $s = 0, -1, -2$, and there are no zeros. All three paths approach asymptotes as $K \to \infty$. The locus exists on the real axis between $s = 0$ and -1 and to the left of $s = -2$.

To find the breakaway point, compute dK/ds.

$$K = -(s^3 + 3s^2 + 2s)$$
$$\frac{dK}{ds} = -(3s^2 + 6s + 2) = 0$$

The candidates are $s = -0.423, -1.58$. The breakaway point must obviously be $s = -0.423$, because the locus cannot exist between $s = -1$ and $s = -2$. With this value of s, (8.2-7) gives $K = 0.385$. The significance of the second solution will be discussed later.

The three asymptotic angles are found from (8.2-5).

$$\theta = \frac{n180^\circ}{0-3} = -n60^\circ \qquad n = \pm 1, \pm 3, \ldots$$

Thus,

$$\theta = -60^\circ, +60^\circ, \pm 180^\circ$$

Note that for the last angle it does not matter whether we use $n = +3$ or $n = -3$. The intersection is found from (8.2-6).

$$\sigma = \frac{0 + (-1) + (-2) - 0}{3 - 0} = -1$$

The two paths that start at $s = 0$ and -1 approach the $\pm 60^\circ$ asymptotes. The path starting at $s = -2$ follows the 180° asymptote and thus lies entirely on the real axis.

The 60° asymptotes indicate that two paths will cross the imaginary axis and generate two unstable roots. Substituting $s = i\omega$ into (8.2-7) gives

$$-i\omega^3 - 3\omega^2 + 2i\omega + K = 0$$

or

$$i\omega(2 - \omega^2) = 0$$
$$K = 3\omega^2$$

The solution $\omega = 0$, $K = 0$ corresponds to the pole at $s = 0$. The solution of interest is $\omega = \pm\sqrt{2}$, $K = 6$. The locus can now be sketched. This is shown in Figure 8.13*a*. The system has three roots. All are real and negative for $0 \leqslant K \leqslant 0.385$. For $0.385 < K < 6$, the system is stable with two complex roots and one real root. For $K > 6$, the system is unstable due to two complex roots with positive real parts.

In many applications, it is not necessary to determine the precise location of the locus off the real axis, and the first eight guides are often quite sufficient for obtaining ample information about the system's behavior. When more accuracy is required, the angle criterion can be employed as follows (see Figure 8.13*b*). We know the position of the locus at the breakaway and crossover points. We select a series of test points along a line parallel to the real axis, starting to the left of where the locus could be, for example. At the leftmost point, we draw the vectors from the poles and zeros and apply the angle criterion. Unless we are extremely lucky, $\measuredangle P(s)$ will not equal $n180^\circ$, $n = \pm 1, \pm 3, \ldots$. In the previous example,

$$\measuredangle P(s) = \measuredangle \frac{1}{s(s+1)(s+2)} = -\measuredangle s - \measuredangle (s+1) - \measuredangle (s+2)$$

Suppose the first test point gives $\measuredangle P(s) = 200^\circ$. We select another point to the right of the first. Suppose this gives 175°. We have passed the point that would give 180° and therefore know that the locus lies between the two test points. This iteration procedure is repeated until the desired accuracy is reached. Then a new line of test points is selected above the first, and the plot is extended in this way.

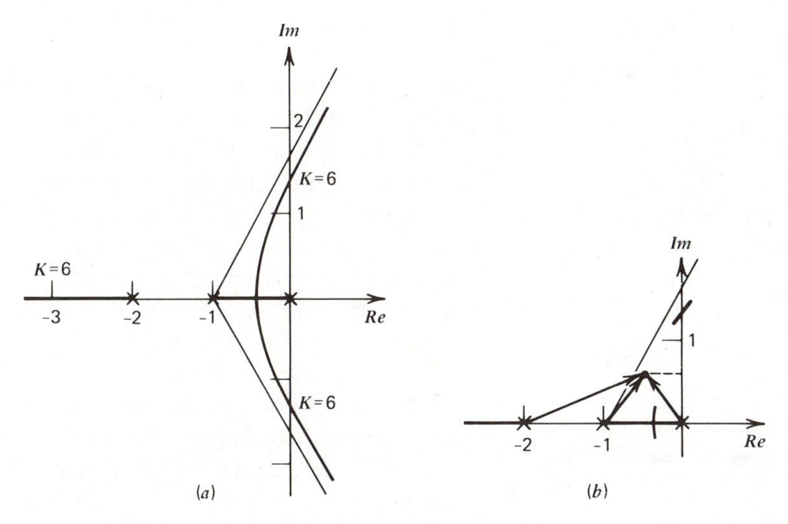

FIGURE 8.13 **(*a*) Root-locus plot for $s(s+1)(s+2)+K=0$, for $K \geqslant 0$. (*b*) Use of test points with the angle criterion to plot the locus off axis.**

The root locus method is very useful in designing aircraft control systems. Thirty years ago, a significant part of the time of many engineers employed in that industry was taken up with constructing root locus plots with this trial-and-error method, which is still useful for sketching the locus. A device that helps the process is the Spirule.†

Today, we are fortunate to have computer programs to generate root locus plots. The author has found the program listed in Reference 2 to be useful. You must have a working knowledge of the plotting guides when using such a program in order to interpret the plots and apply them. For those not wishing to support a general-purpose root locus program, note that the analytical solutions are available for the roots of polynomials up to order four. Since the majority of applications give models of this order or less, you can develop a very useful root locus program by coding these solutions.

Locus Behavior near Complex Poles and Zeros

In order to apply the preceding iterative procedure, it is helpful to know the angle at which the locus leaves a complex pole (the *angle of departure*) and the angle at which it terminates at a complex zero (the *angle of arrival*). To determine these angles, we again call on the angle criterion.

Guide 9

Angles of departure and angles of arrival are determined by choosing an arbitrary point infinitesimally close to the pole or zero in question and applying the angle criterion.

---

†The Spirule is available from The Spirule Co., 9728 El Venado Ave., Whittier, Calif. 90603.

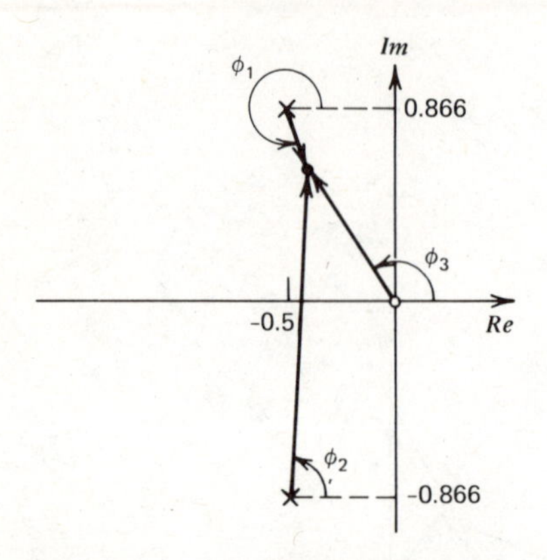

FIGURE 8.14 Application of the angle criterion to determine the angle of departure ϕ_1.

The guide is easily explained with an example. Equation (8.1-8) with $K = \Delta$ gives the root locus shown in Figure 8.2*a*. Assume that we do not know the location of the locus. The pole–zero pattern is shown in Figure 8.14. To determine the departure angle for the upper pole, we select an arbitrary test point and draw the vectors to this point from the poles and zeros. The test point is infinitesimally close to the upper pole, but the figure exaggerates the distance from the pole to the test point so that we may see the vector properly. The equation in question can be written as (8.1-10), repeated here as

$$K = -\frac{(s + 0.5 - i0.866)(s + 0.5 + i0.866)}{s}$$

The angle criterion requires that

$$\measuredangle K = 0° = \measuredangle(-1) + \phi_1 + \phi_2 - \phi_3$$

or

$$\phi_1 + \phi_2 - \phi_3 = n180° \qquad n = \pm 1, \pm 3, \ldots \tag{8.2-8}$$

The key step is to assume that the point s is so close to the upper pole that the angles ϕ_2 and ϕ_3 are given approximately by the angles to the upper pole. These are $\phi_2 = 90°$, $\phi_3 = 90° + \tan^{-1}(0.5/0.866) = 120°$. From (8.2-8), ϕ_1 must be

$$\phi_1 = 120° - 90° + n180°$$

With $n = 1$, $\phi_1 = 210°$. Any other choice of odd-numbered n would give an equivalent result. A glance at Figure 8.2*a* will show that this departure angle is correct. From symmetry, we can easily obtain the departure angle for the lower pole.

Additional Guides

Two more guides can be obtained from the theory of equations and the magnitude criterion.

Guide 10

For the polynomial equation

$$s^n + a_{n-1}s^{n-1} + \cdots + a_1 s + a_0 = 0 \tag{8.2-9}$$

the sum of the roots $r_1, r_2, \ldots, r_n$ *is*

$$r_1 + r_2 \cdots + r_n = -a_{n-1} \tag{8.2-10}$$

Note the unity coefficient of s^n. This guide is useful for determining the location of the remaining roots once some roots have been found. Consider equation (8.2-7) in Example 8.2. When $K = 6$, two of the roots were found to be $s = \pm i1.41$. The third root can be found with $r_1 = +i1.41$, $r_2 = -i1.41$ in (8.2-10).

$$i1.41 - i1.41 + r_3 = -3$$

Thus, $r_3 = -3$ when $K = 6$.

Also, given a root or root pair that is suspected of being dominant, Guide 10 sometimes provides a quick way of telling whether or not the remaining roots are close to the candidate for dominance.

The final guide formally establishes the usefulness of the magnitude criterion.

Guide 11

Once the root locus is drawn, it is scaled with the magnitude criterion (8.1-25).

For example, the circular locus of Example 8.1 corresponds to (8.2-3). The magnitude criterion states that

$$|K| = K = \frac{|s||s+4|}{|s+6|}$$

Given any point s on the locus, K can be computed from this equation by either direct substitution of values of s or measuring the lengths of the vectors s, $(s+4)$, $(s+6)$, and combining the lengths according to the preceding equation.

Figure 8.15 shows some other common root locus plots for third-order systems. They can be used to test your knowledge of the plotting guides. Table 8.1 summarizes the guides for $K \geqslant 0$.

8.3 SOME NUMERICAL AIDS

Sometimes the root locus guides require solving for the roots of a higher order equation. For example, the condition $dK/ds = 0$ used to solve for breakaway or break-in points can easily result in a cubic equation. While an analytical result is known, it is often easier to use *Newton iteration* (Appendix A, Section A.2), because from Guide 4, we know the general location of the breakaway and break-in points and thus have a good starting point for the iteration. In addition, we need only to find as many roots as there are breakaway and break-in points.

The second case requiring numerical methods is the solution for the crossover frequency and ultimate gain (Guide 8). When $s = i\omega$ is substituted into the equation of interest, two equations result: one for the real part and one for the imaginary part. The equation for ω is of lower order than the original. When the original equation is of seventh order, a cubic equation must be solved for ω. In this case and those of higher order, Newton iteration can be used to advantage. Again, a good starting point is usually obtainable from the value of s where the asymptote crosses the imaginary axis.

As shown in Section A.4, Newton iteration is based on the truncated Taylor series expansion of a function $f(x)$ about a point x_o (see Appendix A, Section A.1 for a

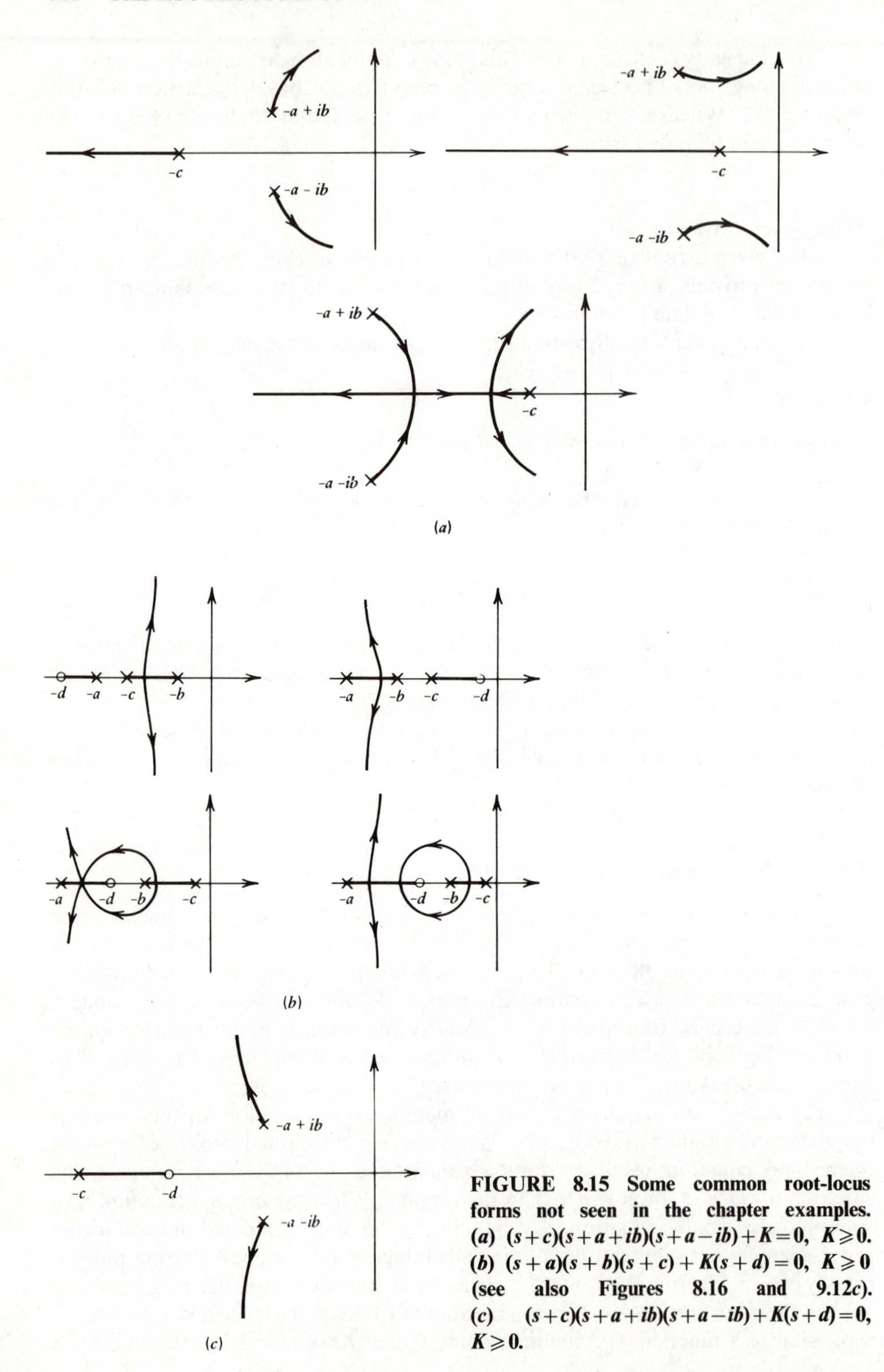

FIGURE 8.15 Some common root-locus forms not seen in the chapter examples. (*a*) $(s+c)(s+a+ib)(s+a-ib)+K=0,\ K\geqslant 0$. (*b*) $(s+a)(s+b)(s+c)+K(s+d)=0,\ K\geqslant 0$ (see also Figures 8.16 and 9.12*c*). (*c*) $(s+c)(s+a+ib)(s+a-ib)+K(s+d)=0,\ K\geqslant 0$.

TABLE 8.1 Plotting Guides for the Primary Root Locus

Standard Form:

$$1 + KP(s) = 0, \qquad K \geq 0$$
$$P(s) = N(s)/D(s)$$
$$N(s) = s^m + b_{m-1}s^{m-1} + \cdots + b_1 s + b_0$$
$$D(s) = s^n + a_{n-1}s^{n-1} + \cdots + a_1 s + a_0$$
$$m \leq n$$

Terminology:

The *zeros* of $P(s)$ are the roots of $N(s) = 0$.

The *poles* of $P(s)$ are the roots of $D(s) = 0$.

Angle Criterion: $\measuredangle P(s) = n180°$, $n = \pm 1, \pm 3, \ldots$

Magnitude Criterion: $K = 1/|P(s)|$

Plotting Guides:

Guide 1

The root-locus plot is symmetric about the real axis.

Guide 2

The number of loci equals the number of poles of P(s).

Guide 3

The loci start at the poles of P(s) *with* K = 0, *and terminate with* K = ∞ *either at the zeros of* P(s) *or at infinity.*

Guide 4

The root locus can exist on the real axis only to the left of an odd number of real poles and/or zeros; furthermore, it must exist there.

Guide 5

The locations of breakaway and break-in points are found by determining where the parameter K *attains a local maximum or minimum on the real axis.*

Guide 6

The loci that do not terminate at a zero approach infinity along asymptotes. The angles that the asymptotes make with the real axis are found from the following equation, where n *is chosen successively as* n = +1, −1, +3, −3, ..., *until enough angles have been found.*

$$\theta = \frac{n180°}{Z - P}, \qquad n = \pm 1, \pm 3, \ldots$$

Guide 7

The asymptotes intersect the real axis at the common point

$$\sigma = \frac{\Sigma s_p - \Sigma s_z}{P - Z}$$

where Σs_p *and* Σs_z *are the algebraic sums of the values of the poles and zeros.*

Guide 8

The points at which the loci cross the imaginary axis and the associated values of K *can be found with the Routh–Hurwitz criterion or by substituting* $s = i\omega$ *into the equation of interest. The frequency* ω *is the crossover frequency.*

Guide 9

Angles of departure and angles of arrival are determined by choosing an arbitrary point infinitesimally close to the pole or zero in question and applying the angle criterion.

Guide 10

For the polynomial equation

$$s^n + a_{n-1}s^{n-1} + \cdots + a_1 s + a_0 = 0$$

the sum of the roots $r_1, r_2, \ldots, r_n$ *is*

$$r_1 + r_2 + \cdots + r_n = -a_{n-1}$$

Guide 11

Once the root locus is drawn, it is scaled with the magnitude criterion.

discussion of the Taylor series). When terms of second order and higher are neglected, the series becomes

$$f(x) \cong f(x_o) + \left(\frac{df}{dx}\right)_{x=x_o} (x - x_o) \tag{8.3-1}$$

If x is the root of the equation $f(x) = 0$ and x_o is an initial guess for x, we can set (8.3-1) equal to zero and solve for the estimate of the root x. This gives

$$x \cong x_o - \frac{f(x_o)}{f'(x_o)} \tag{8.3-2}$$

where

$$f'(x_o) = \left(\frac{df}{dx}\right)_{x=x_o} \tag{8.3-3}$$

We can repeat the process until it converges to an answer with the desired accuracy. Thus, the algorithm can be expressed as

$$x_k \cong x_{k-1} - \frac{f(x_{k-1})}{f'(x_{k-1})} \qquad k = 1, 2, \ldots \tag{8.3-4}$$

Example 8.3

Sketch the root locus for the equation

$$1 + \frac{K(s+1.5)}{s(s+1)(s+2)} \tag{8.3-5}$$

The poles are $s = 0, -1, -2$, and the zero is $s = -1.5$ (see Figure 8.16). A breakaway point exists between $s = 0$ and $s = -1$. It is found as follows. From (8.3-5),

$$K = -\frac{s(s+1)(s+2)}{s+1.5} = -\frac{s^3 + 3s^2 + 2s}{s+1.5}$$

$$\frac{dK}{ds} = -\frac{(s+1.5)(3s^2 + 6s + 2) - (s^3 + 3s^2 + 2s)}{(s+1.5)^2} = 0$$

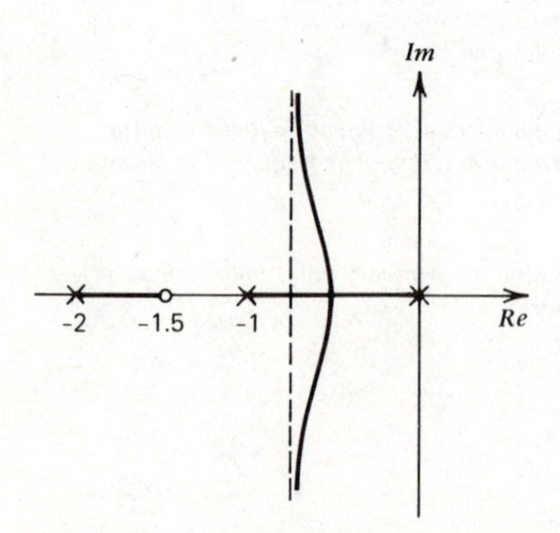

FIGURE 8.16 Root-locus plot for $s(s+1)(s+2) + K(s+1.5) = 0$, $K \geq 0$.

Setting the numerator to zero gives

$$f(s) = 2s^3 + 7.5s^2 + 9s + 3 = 0$$

from which

$$f'(s) = 6s^2 + 15s + 9$$

The Newton iteration formula is

$$s_k \cong s_{k-1} - \frac{f(s_{k-1})}{f'(s_{k-1})} \tag{8.3-6}$$

The logical trial point is the center of the region $(-1, 0)$, or $s_0 = -0.5$. Instead, to show the details of the Newton iteration, we choose to start at $s_0 = 0$. The steps are

$$s_1 = s_0 - \frac{f(s_0)}{f'(s_0)} = 0 - \frac{f(0)}{f'(0)} = -0.333$$

$$s_2 = -0.333 - \frac{f(-0.333)}{f'(-0.333)} = -0.333 - 0.163$$

$$= -0.496$$

$$s_3 = -0.496 - 0.045 = -0.541$$

$$s_4 = -0.541 - 0.003 = -0.544$$

The process has stabilized in the second decimal place, so we stop, because no more decimal places can be shown on our plot. The breakaway point is approximately at $s = -0.54$.

If we had chosen the logical trial point $s_0 = -0.5$, the iteration would have been

$$s_1 = -0.5 - 0.042 = -0.542$$

The answer is obtained in one step. The information provided by Guide 4 concerning the location of the locus on the real axis is invaluable in choosing a starting point for the Newton iteration. If a little common sense or insight is included in making the choice, the iteration's convergence can be greatly improved.

The rest of the locus can be sketched with the remaining guides. The asymptotic angles are

$$\theta = \frac{n180^\circ}{1-3} = \pm 90^\circ$$

Their intersection occurs at

$$\sigma = \frac{0-1-2-(-1.5)}{3-1} = \frac{-1.5}{2} = -0.75$$

The results of this example can be used to compare the performance of two control systems (Figure 8.17). The plant is $1/(s+1)(s+2)$. Figure 8.17a shows an integral controller with this plant. The root locus of this system is shown in

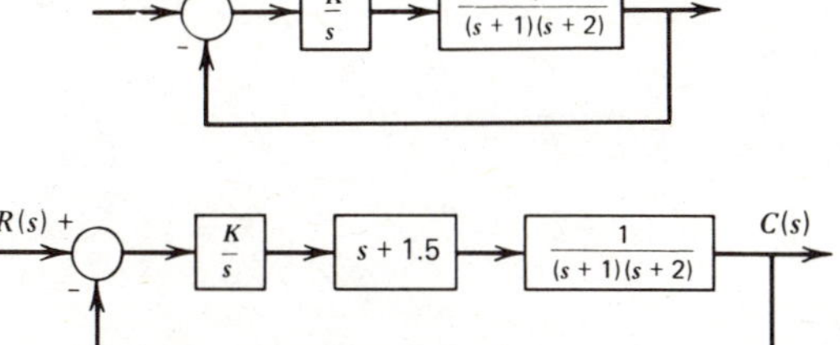

FIGURE 8.17 Effect of an open-loop zero on system performance. (*a*) Integral control of a second-order plant. (*b*) Series compensation of the controller in part *a*. The corresponding root loci are shown in Figures 8.13*a* and 8.16.

Figure 8.13*a*. From this, it was seen that the system is unstable for $K \geqslant 6$. Figure 8.17*b* shows the same system with a series compensator inserted after the controller. The compensator's transfer function $(s + 1.5)$ produces the zero in the root locus plot shown in Figure 8.16. The effect of the zero is to shift the locus away from the right half-plane. The new system is always stable, and the good response characteristics of the original controller for $\zeta = 1$ are preserved. We will return to the subject of series compensators later in this chapter and in Chapter 9.

8.4 RESPONSE CALCULATIONS FROM THE ROOT LOCUS

Once the system's characteristic roots have been determined with the root locus plot, the plot can be used to simplify the calculations required to obtain the transient and frequency responses. For a given input function, the transform of the system's output is expressed in a partial fraction expansion (see Appendix B). The coefficients of the expansion are easily calculated with the vector interpretation of complex numbers, and the transient response is obtained from this expansion. Frequency response is computed similarly, with the factors in the transfer function taken to be vector functions of the frequency ω. Two examples are sufficient to demonstrate the techniques.

Example 8.4

Proportional control of a second-order system is shown in Figure 8.18. This system has the circular locus shown in Figure 8.9*a*. Find the gain K required to give a damping ratio of $\zeta = 0.9$. With this gain, compute the response to a unit-step disturbance.

The desired damping ratio corresponds to a line making an angle of $\cos^{-1} 0.9 = 26°$ with the negative real axis. This line is shown in Figure 8.19*a*. It intersects the locus at two places. We choose the intersection at $s = -6.95 + i3.4$, because this gives the smallest time constant. The required gain is found from the locus to be

$$K = \frac{|s||s+4|}{|s+6|} = \frac{7.74(4.50)}{3.53} = 9.87$$

See Figure 8.19*a*.

The disturbance transfer function is

$$\frac{C(s)}{D(s)} = \frac{s+6}{s(s+4) + K(s+6)}$$

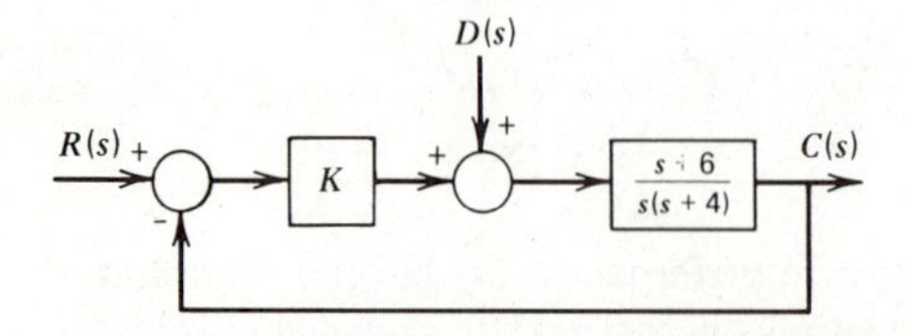

FIGURE 8.18 Proportional control of a second-order system.

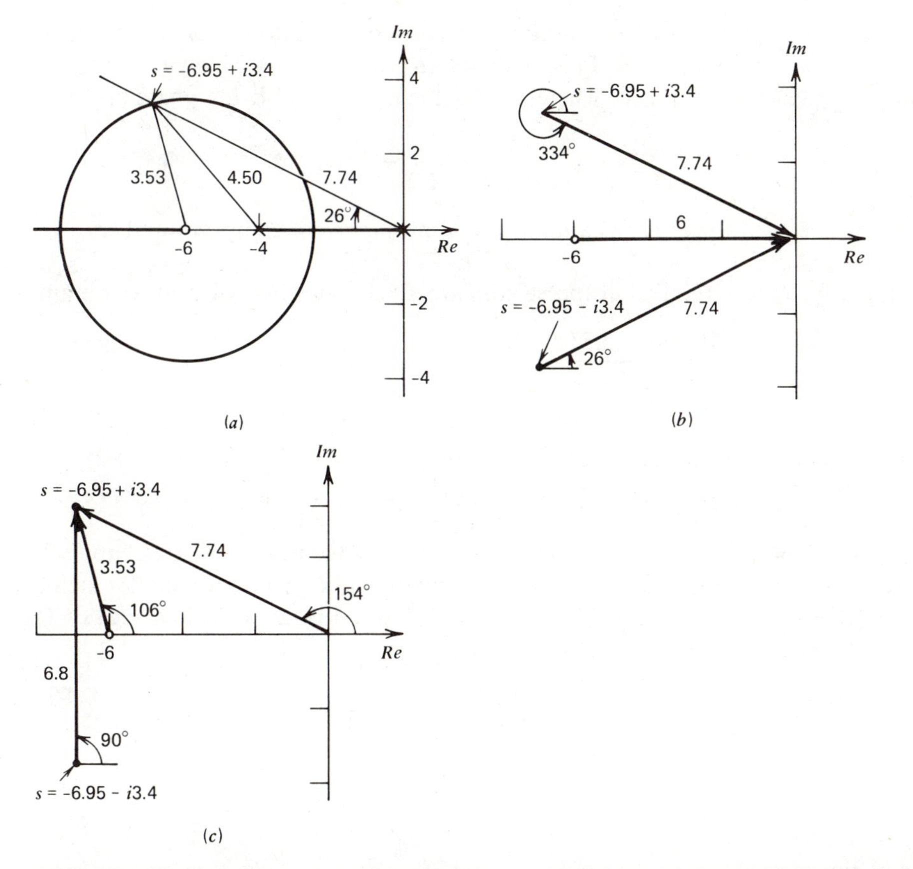

FIGURE 8.19 Response calculations with the locus of the system in Figure 8.18. (*a*) Calculation of the gain at the point $s = -6.95 + i3.4$. (*b*) Calculation of the constant C_1 in (8.4-3). (*c*) Calculation of the constant C_2 in (8.4-4).

With $K = 9.87$ and $D(s) = 1/s$, we obtain

$$C(s) = \frac{s+6}{s(s+6.95-i3.4)(s+6.95+i3.4)} \tag{8.4-1}$$

The partial fraction expansion is

$$C(s) = \frac{C_1}{s} + \frac{C_2}{s+6.95-i3.4} + \frac{C_3}{s+6.95+i3.4} \tag{8.4-2}$$

Where

$$C_1 = \lim_{s\to 0}\left[\frac{s+6}{(s+6.95-i3.4)(s+6.95+i3.4)}\right] \tag{8.4-3}$$

$$C_2 = \lim_{s\to -6.95+i3.4}\left[\frac{s+6}{s(s+6.95+i3.4)}\right] \tag{8.4-4}$$

$$C_3 = \text{complex conjugate of } C_2 \tag{8.4-5}$$

The constant C_1 can be evaluated using the vector representation of each factor with the tips of the vectors at $s=0$. This is shown in Figure 8.19*b*. Thus, if a complex number is represented by $r \measuredangle \phi$, where r is the magnitude and ϕ the phase angle,

$$C_1 = \frac{6 \measuredangle 0^\circ}{(7.74 \measuredangle 334^\circ)(7.74 \measuredangle 26^\circ)} = 0.100 \measuredangle 0^\circ$$

where we have used the multiplication/division properties of complex numbers [see (A.5-15) and (A.5-16)].

Similarly, for C_2, the tips of the vectors are at $s = -6.95 + i3.4$, and we obtain

$$C_2 = \frac{3.53 \measuredangle 106^\circ}{(7.74 \measuredangle 154^\circ)(6.8 \measuredangle 90^\circ)} = 0.067 \measuredangle 222^\circ$$

See Figure 8.19*c*. Therefore, $C_3 = 0.067 \measuredangle 138^\circ$.

With these values, the response is given by the inverse transform of (8.4-2).

$$c(t) = 0.1 + 0.067e^{i222^\circ}e^{(-6.95+i3.4)t} + 0.067e^{i138^\circ}e^{(-6.95-i3.4t)} \qquad \textbf{(8.4-6)}$$

This can be reduced to the form of sines and cosines by the techniques in Chapter 3. However, a simpler method is to replace the calculation of C_2 with the complex factor $R(-a+ib)$ developed in Section 4.9. See (4.9-1), (4.9-6), and (4.9-7). Comparing (8.4-1) with (4.9-1), we see that

$$C(s) = \frac{s+6}{s[(s+6.95)^2 + (3.4)^2]} \qquad \textbf{(8.4-7)}$$

and

$$P(s) = \frac{s+6}{s}$$

From (4.9-6),

$$R(-6.95+i3.4) = \lim_{s \to -6.95+i3.4} \left(\frac{s+6}{s}\right)$$

$$= \frac{3.53 \measuredangle 106^\circ}{7.74 \measuredangle 154^\circ} = 0.456 \measuredangle 312^\circ$$

where $a = 6.95$ and $b = 3.4$. From (5.7-29), the contribution to the response from the complex roots is

$$0.134e^{-6.95t} \sin(3.4t + 312^\circ)$$

This is added to the response terms due to the other roots. Here, this is C_1/s, and the response is

$$c(t) = 0.1 + 0.134e^{-6.95t} \sin(3.4t + 312^\circ)$$

Example 8.5

Demonstrate how the closed-loop frequency response to the command $R(s)$ can be determined from the root locus plot. Use the system shown in Figure 8.18 with the gain $K = 9.87$ as computed in the previous example.

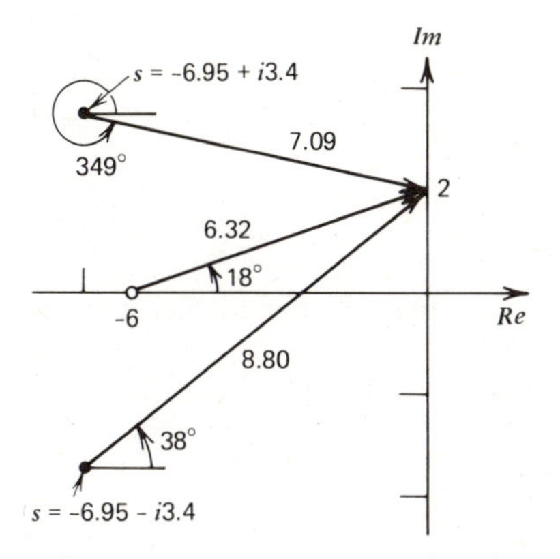

FIGURE 8.20 Calculation of the frequency response from the root locus.

The desired transfer function with $s = i\omega$ is

$$T(i\omega) = \frac{C(i\omega)}{R(i\omega)} = \frac{9.87(i\omega + 6)}{(i\omega + 6.95 - i3.4)(i\omega + 6.95 + i3.4)} \tag{8.4-8}$$

The vector for each factor has its tip at $s = i\omega$. We select a value for ω and compute the magnitude and phase angle for $T(i\omega)$ at this frequency. For example, at $\omega = 0$, the vectors are the same as those shown in Figure 8.19*b*. Thus,

$$T(0) = \frac{9.87(6 \measuredangle 0^\circ)}{(7.74 \measuredangle 334^\circ)(7.74 \measuredangle 26^\circ)} = 0.989 \measuredangle 0^\circ$$

For $\omega = 2$, Figure 8.20 shows that

$$T(i2) = \frac{9.87(6.32 \measuredangle 18^\circ)}{(7.09 \measuredangle 349^\circ)(8.80 \measuredangle 38^\circ)} = 0.999 \measuredangle -9^\circ$$

The process is repeated until the desired range of frequencies is covered. The magnitude ratio and phase angle curves can then be plotted.

8.5 THE COMPLEMENTARY ROOT LOCUS

The development of the plotting guides was based on the angle and magnitude criteria, (8.1-23) and (8.1-24), with the assumption that the parameter K is positive or zero. We now consider the construction of the locus when $K \leqslant 0$ for the equation

$$1 + KP(s) = 0 \tag{8.5-1}$$

This is the *complementary root locus.* Its plotting guides are found from the same criteria, repeated here

$$|KP(s)| = 1 \tag{8.5-2}$$

$$\measuredangle KP(s) = \measuredangle -1 \tag{8.5-3}$$

If $K \leqslant 0$, (8.5-2) becomes

$$-K|P(s)| = 1$$

or

$$K = \frac{-1}{|P(s)|} \tag{8.5-4}$$

Thus, the same scaling procedure is used, with a sign reversal included. Equation (8.5-3) becomes

$$\measuredangle K + \measuredangle P(s) = \measuredangle -1$$

or

$$180^\circ + \measuredangle P(s) = 180^\circ$$

The angle criterion reduces to

$$\measuredangle P(s) = n360^\circ \qquad n = 0, \pm 1, \pm 2, \ldots \tag{8.5-5}$$

For this reason, the complementary locus is sometimes referred to as the 0° locus, while the primary locus ($K \geqslant 0$) is called the 180° locus.

The guides for plotting the complementary locus can be obtained from (8.5-4) and (8.5-5) in the same way as those for the primary locus. We limit ourselves to their statement and denote them as 1a, 2a, etc., to distinguish them from the guides for the primary locus. Any differences are noted in the list of guides. We continue to assume that $P(s)$ is a ratio of two finite polynomials with constant coefficients.

Guide 1a

The locus is symmetric about the real axis.

Guide 2a

The number of loci equals the number of poles of P(s).

Guide 3a

The loci start at the poles of P(s) *with* K = *0 and terminate with* K = $-\infty$ *either at the zeros of* P(s) *or at infinity.*

Guide 4a

The locus exists in a section on the real axis only if the number of real poles and/or zeros to the right of the section is even; furthermore, it must exist there. Thus, the complementary locus exists on the real axis wherever the primary locus is absent. For the purpose of this guide, the number zero is even.

Guide 5a

The locations of breakaway and break-in points are found by determining where the parameter K *attains a local minimum or maximum on the real axis.* This is the same as Guide 5 for the primary locus. The extraneous roots that were discarded previously

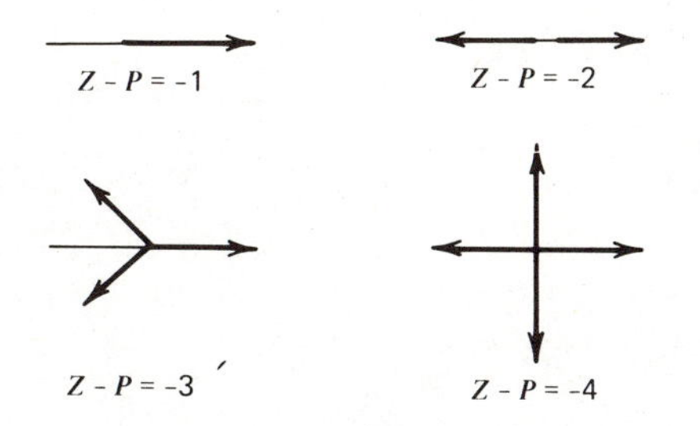

FIGURE 8.21 Common asymptote configurations for $K \leqslant 0$.

when applying Guide 5 are now seen to be the breakaway or break-in points of the complementary locus. This occurred in Example 8.2 with the root $s = -1.58$.

Guide 6a

The loci that do not terminate at a zero approach infinity along asymptotes. The asymptotic angles relative to the real axis are found from

$$\theta = \frac{n360^\circ}{Z - P} \qquad n = 0, \pm 1, \pm 2, \ldots \tag{8.5-6}$$

where Z *and* P *are the number of zeros and poles and* n *is increased until enough angles have been found.* Some of these possibilities are shown in Figure 8.21.

Guide 7a

All the asymptotes for both the primary and complementary loci intersect at a common point on the real axis. This point is given by (*8.2-6*), *repeated here.*

$$\sigma = \frac{\Sigma s_p - \Sigma s_z}{P - Z} \tag{8.5-7}$$

where the summations are for the values of the poles and zeros.

Guide 8a

The points at which the loci cross the imaginary axis and the associated values of K *can be found as with Guide 8, namely, by the Routh–Hurwtz criterion or substituting* $s = i\omega$ *into* (*8.5-1*).

Guide 9a

The angles of departure from complex poles, and the angles of arrival at complex zeros, can be found by applying the angle criterion (*8.5-5*) *to a trial point infinitesimally close to the pole or zero in question.* Note that the angle criterion for the complementary locus must be used in this case.

Guide 10a

This is the same as Guide 10. See (*8.2-9*) *and* (*8.2-10*). This guide does not result from the angle and magnitude criteria.

Guide 11a

Once the complementary locus is drawn, it is scaled with the magnitude criterion (*8.5-4*). Table 8.2 summarizes the guides for $K \leqslant 0$.

TABLE 8.2 Plotting Guides for the Complementary Root Locus

Standard Form:

$$1 + KP(s) = 0, \qquad K \leqslant 0$$

$$P(s) = N(s)/D(s)$$

$$N(s) = s^m + b_{m-1}s^{m-1} + \cdots + b_1 s + b_0$$

$$D(s) = s^n + a_{n-1}s^{n-1} + \cdots + a_1 s + a_0$$

$$m \leqslant n$$

Terminology:

The *zeros* of $P(s)$ are the roots of $N(s) = 0$.

The *poles* of $P(s)$ are the roots of $D(s) = 0$.

Angle Criterion: $\measuredangle P(s) = n360°$, $n = 0, \pm 1, \pm 2, \pm 3, \ldots$

Magnitude Criterion: $|K| = 1/|P(s)|$

Plotting Guides:

Guide 1a

The locus is symmetric about the real axis.

Guide 2a

The number of loci equals the number of poles of P(s).

Guide 3a

The loci start at the poles of P(s) *with* K = 0, *and terminate with* K = −∞ *either at the zeros of* P(s) *or at infinity.*

Guide 4a

The locus exists in a section on the real axis only if the number of real poles and/or zeros to the right of the section is even; furthermore, it must exist there.

Guide 5a

The locations of breakaway and break-in points are found by determining where the parameter K *attains a local minimum or maximum on the real axis.*

Guide 6a

The loci that do not terminate at a zero approach infinity along asymptotes. The asymptotic angles relative to the real axis are found from

$$\theta = \frac{n360°}{Z - P}, \qquad n = 0, \pm 1, \pm 2, \ldots$$

where Z *and* P *are the number of zeros and poles, and* n *is increased until enough angles have been found.*

Guide 7a

All the asymptotes for both the primary and complementary loci intersect at a common point on the real axis. This point is given by (8.2-6), repeated here.

$$\sigma = \frac{\Sigma s_p - \Sigma s_z}{P - Z}$$

where the summations are for the values of the poles and zeros.

Guide 8a

The points at which the loci cross the imaginary axis and the associated values of K *can be found as with Guide 8, namely by the Routh–Hurwitz criterion or by substituting* s = iω *into the equation of interest.*

Guide 9a

The angles of departure from complex poles, and the angles of arrival at complex zeros, can be found by applying the angle criterion to a trial point infinitesimally close to the pole or zero in question.

Guide 10a

This is the same as Guide 10. See Table 8.1.

Guide 11a

Once the complementary locus is drawn, it is scaled with the magnitude criterion.

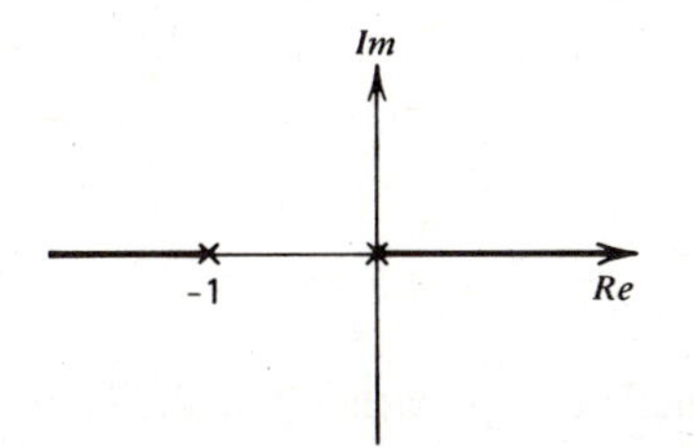

FIGURE 8.22 Root locus for $s(s+1)+K=0$, $K \leqslant 0$.

Examples

The primary root locus of the equation

$$1+\frac{K}{s(s+1)} \tag{8.5-8}$$

is shown in Figure 8.1*a*. Its complementary locus is easily obtained from the complementary guides and is shown in Figure 8.22.

For the equation

$$1+\frac{Ks}{s^2+s+1}=0 \tag{8.5-9}$$

the primary locus is given by Figure 8.2*a*. The solutions for the breakaway or break-in points are $s=\pm 1$. The break-in point at $s=-1$ is for the primary locus, while that at $s=+1$ is for the complementary locus. At $s=1$, $K=-3$. For $K<0$, Guide 9a gives the departure angle as $\phi = 120° - 90° - n360°$, or $\phi_1 = 30°$. From Guide 8a, the crossover frequency is $\omega = 1$ at $K=-1$. The entire locus off the real axis is seen to be a unit circle centered at $s=0$, as shown in Figure 8.23.

The latter example illustrates the usual situation where the complementary locus is useful. Control system gains are usually positive for physical reasons, and the primary locus can be used to determine the proper gain values. Then a sensitivity study is often performed to ascertain the effects of the variation or uncertainty in another parameter. For example, the primary locus in Figure 8.1*a* was used to set the gain k to obtain the desired response. With k set equal to 1, the coefficient c in (8.1-1) was allowed to vary about its nominal value of 1. Figure 8.2*a* is the locus with the variation Δ as the parameter. It is valid only for $\Delta \geqslant 0$. The complementary locus given by Figure 8.23 for $\Delta \leqslant 0$ is required to obtain the effects of variations in c below the nominal value of 1. Normally, we are interested in only relatively small variations about the nominal, so the plot need not be constructed for the entire range $-\infty \leqslant \Delta \leqslant +\infty$.

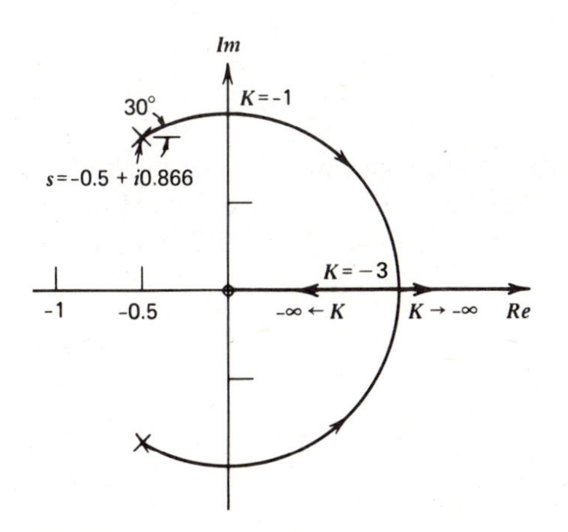

FIGURE 8.23 Root locus for $s(s+1)+Ks=0$, $K \leqslant 0$.

Another application of the complementary root locus is in the analysis of *nonminimum phase* systems, which possess poles and zeros in the right half-plane. The name is descriptive of only the frequency response characteristics (see Reference 3). A positive-feedback system can be nonminimum phase. Another example is the *reverse-reaction process*. The response of this process to a positive-step input exhibits a negative initial slope for zero initial conditions. This is a model of some boiling processes in which the liquid level temporarily drops when bubbles collapse as the result of an input of cold liquid.

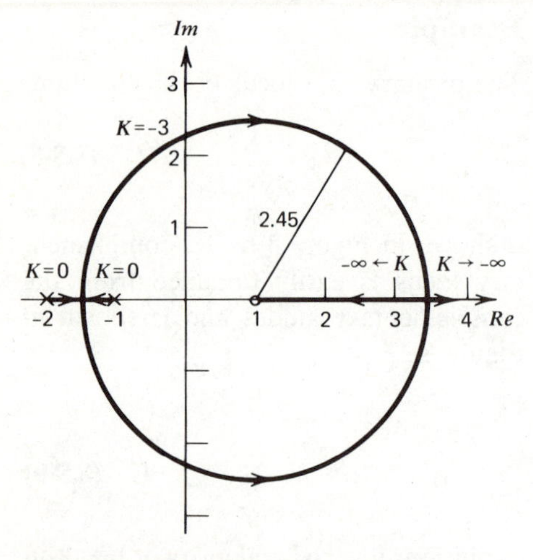

FIGURE 8.24 Root locus for the reverse reaction process $(s+1)(s+2)+K(s-1)=0$, $K \geqslant 0$.

Example 8.6

Proportional control with a gain K_p is to be applied to a reverse-reaction process described by

$$\frac{1-s}{(s+1)(s+2)}$$

Plot the root locus for $K_p \geqslant 0$.

The characteristic equation for the system with P control is

$$1+\frac{-K_p(s-1)}{(s+1)(s+2)}=0 \quad \textbf{(8.5-10)}$$

The root locus parameter is $K=-K_p$. Since $K \leqslant 0$, we use the plotting guides for the complementary root locus. The breakaway point is at $s=-1.45$, and the break-in point is at $s=3.45$. The locus is shown in Figure 8.24. The control system is unstable if $K<-3$; that is, if $K_p>3$.

8.6 APPLICATIONS

The root locus is useful for determining how the control law can be modified to improve performance and for setting the required gain values. It is also useful for determining the effects of parameter variations and parasitic modes. The following examples illustrate these applications.

PI Control of a Second-Order Plant

PI control of a second-order plant leads to a third-order model. The methods in previous chapters are not always sufficient for dealing with such a problem. However, the root locus and Bode plots provide the necessary design tools. For example, let the plant transfer function be

$$G_p(s)=\frac{1}{s^2+a_2s+a_1} \quad \textbf{(8.6-1)}$$

where $a_2>0$ and $a_1>0$. PI control applied to this plant gives the closed-loop command transfer function

$$T_1(s)=\frac{K_ps+K_I}{s^3+a_2s^2+a_1s+K_ps+K_I} \quad \textbf{(8.6-2)}$$

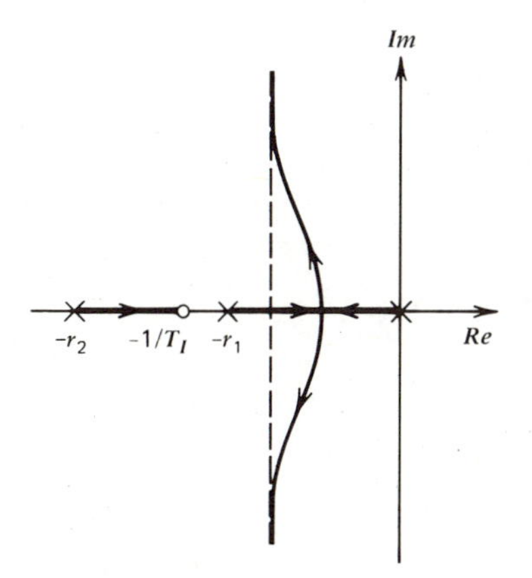

FIGURE 8.25 Root-locus plot for PI control of a second-order plant. The characteristic equation is $s(s+r_1)(s+r_2)+K_p(s+1/T_I)=0$.

This is the same form as that of the dc motor problem in Section 7.2, with the armature inductance included in the model [see (7.2-6)].

Note that the Ziegler–Nichols rules cannot be used to compute the gains K_p and K_I. The second-order plant (8.6-1) does not have the S-shaped signature of Figure 7.1, so the process reaction method does not apply. The ultimate-cycle method requires K_I to be set to zero and the ultimate gain K_{pu} determined. With $K_I=0$ in (8.6-2), the resulting system is stable for all $K_p>0$, and thus a positive ultimate gain does not exist.

Take the PI control law, and assume that the characteristic roots of the plant (8.6-1) are real values $-r_1$ and $-r_2$ such that $-r_2 < -r_1$. In this case, the open-loop transfer function of the control system is

$$G(s)H(s)=\frac{K_p(s+1/T_I)}{s(s+r_1)(s+r_2)} \tag{8.6-3}$$

One design approach is to select T_I and plot the locus with K_p as the parameter. If the zero at $s=-1/T_I$ is located to the right of $s=-r_1$, the dominant time constant cannot be made as small as is possible with the zero located between the poles at $s=-r_1$ and $s=-r_2$ (Figure 8.25). A large integral gain (small T_I and large K_p) is desirable for reducing the overshoot due to a disturbance, but the zero should not be placed to the left of $s=-r_2$, because the dominant time constant will be larger than that obtainable with the placement shown in Figure 8.25 for large values of K_p. Sketch the root locus plots to see this. A similar situation exists if the poles of the plant are complex.

The steady-state error for a step input will be zero. The values of K_p and T_I can then be selected from the root locus to achieve specifications on the transient performance.

Example 8.7

We wish to control the following plant so that there will be zero steady-state error with a step command input.

$$G_p(s)=\frac{1}{(s+1)(s+2)} \tag{8.6-4}$$

Using integral control gives the desired error response, but the system can become unstable, and we cannot obtain a dominant time constant of less than $\tau=1/0.423=2.36$. This can be seen from the root locus in Figure 8.26. (a) Use the root locus method to

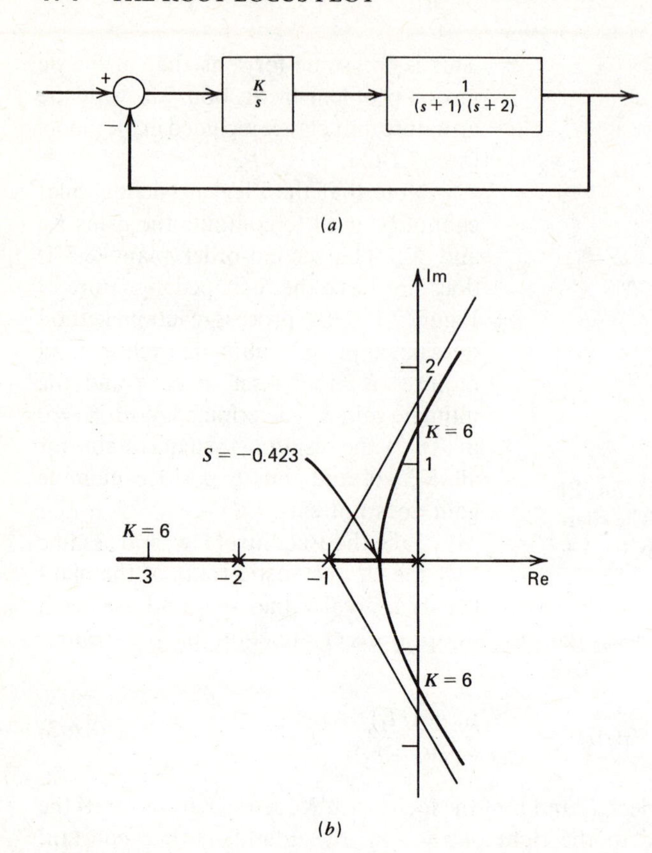

FIGURE 8.26 Integral control of a second-order plant. (*a*) Block diagram. (*b*) Root-locus plot.

modify the integral controller so that the system will always be stable. What is the smallest dominant time constant that can be achieved with this design? (b) Set the gains of the new controller to give the smallest possible dominant time constant without producing an oscillatory step response.

(a) If we modify the controller so that it introduces a zero between $s = -1$ and $s = -2$, the root locus will be pulled to the left, and the system cannot become unstable. This is shown in Figure 8.27*a* and *b*, where the zero is set at $s = -b$, $1 \leqslant b \leqslant 2$. With this restriction, the smallest dominant time constant that can be achieved lies near the asymptote at $s = \sigma$. From Guide 7, Equation (8.2-6), this asymptote is at

$$\sigma = \frac{-1-2+b}{3-1} = \frac{b-3}{2} \tag{8.6-5}$$

Thus, $-1 \leqslant \sigma \leqslant -0.5$. The smallest σ is given by $b = 1$. Thus, the smallest dominant time constant achievable with this design is $\tau = 1$, but this requires a high gain value to place the roots near the asymptote. The root locus for $b = 1$ is shown in Figure 8.27*c*.

(b) For $\zeta \geqslant 1$, the smallest dominant time constant occurs at the breakaway point in Figure 8.27*c*, which is $s = -0.93$. Thus, $\tau = 1/0.93 = 1.08$. The gain at this point is

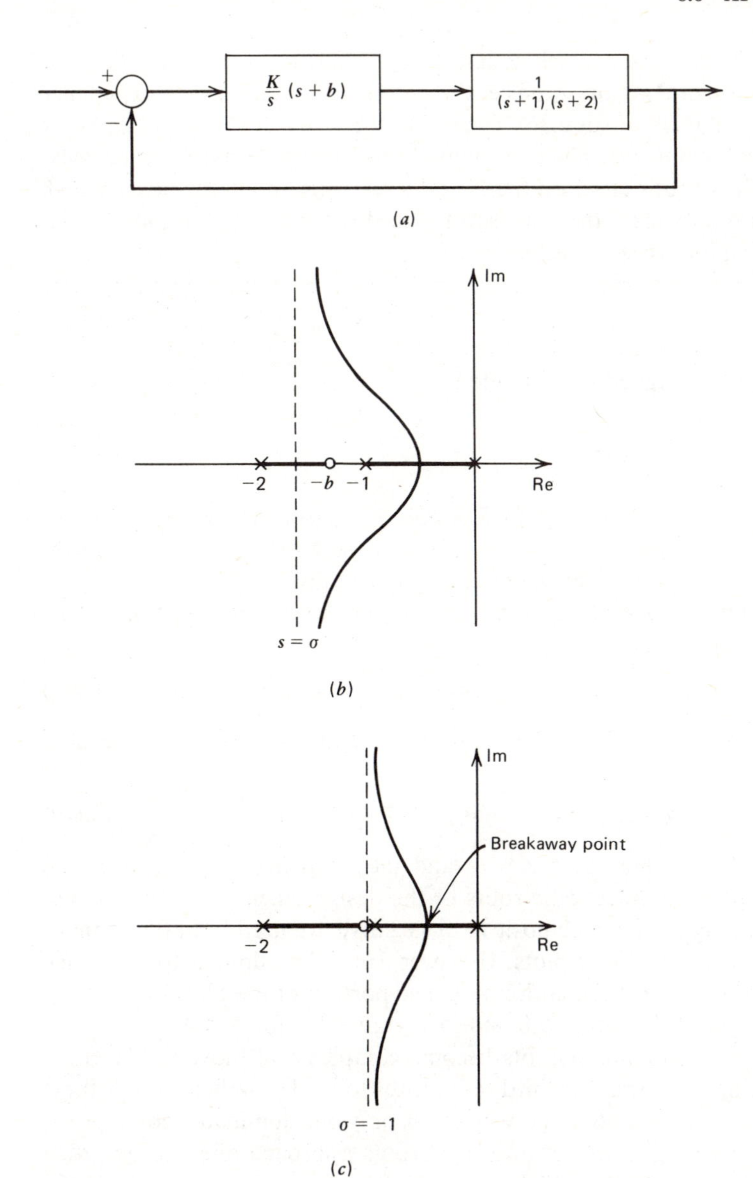

FIGURE 8.27 The effect of including a zero in the controller transfer function. (*a*) Block diagram. (*b*) Resulting root locus for $1 \leqslant b \leqslant 2$. (*c*) Root locus for $b = 1$.

$K = 0.995$. The controller transfer function is

$$\frac{K}{s}(s+b) = \frac{0.995}{s}(s+1)$$

$$= 0.995 + \frac{0.995}{s} \tag{8.6-6}$$

which is PI control. Thus, we see that P action is responsible for introducing the zero at $s = -b$. The addition of P action stabilizes the system and makes possible a smaller dominant time constant (1.08 versus 2.36). The third root is now at $s = -1.14$ and now much closer to the dominant root. Since the dominant-root approximation is now less accurate, the secondary root's effect on the system's response should be investigated.

Other results are possible if the zero is not placed between $s = -1$ and $s = -2$. The reader should explore these possibilities.

Example 8.8

Consider the controller designed in Example 8.7. The plant transfer function is

$$G_p(s) = \frac{1}{(s+c)(s+2)} \quad \textbf{(8.6-7)}$$

The nominal value of $c = 1$ was used in Example 8.7. Now, assume that we are uncertain of the value of c, but we know it is in the range $0.5 \leqslant c \leqslant 1.5$. Use the root locus to investigate the sensitivity of the design to uncertainty in c.

For the gain values found in Example 8.7, the characteristic equation of the system with PI control is

$$s^3 + (c+2)s^2 + (2c + 0.995)s + 0.995 = 0 \quad \textbf{(8.6-8)}$$

Define Δ to be the deviation in c from its nominal value. Thus, $\Delta = c - 1$. In terms of Δ, (8.6-8) can be written as

$$s^3 + 3s^2 + 2.995s + 0.995 + \Delta(s^2 + 2s) = 0 \quad \textbf{(8.6-9)}$$

Thus, the root locus parameter K is $K = \Delta$, and the zeros are at $s = 0, -2$. The poles are identical to the characteristic roots of the design when $c = 1$; that is, $s = -0.93, -0.93, -1.14$. Since Δ can be positive or negative, we need both the primary and the complementary root locus plots. However, since Δ is limited to the range $-0.5 \leqslant \Delta \leqslant 0.5$, we do not need the entire plot. The portion of the plot for Δ in this range is shown in Figure 8.28*a* for $\Delta \geqslant 0$ and in Figure 8.28*b* for $\Delta \leqslant 0$.

For $c < 1 (\Delta < 0)$, the dominant roots become complex and move to the right. This gives a larger time constant. The third root moves to the left. When $c = 0.5$, these roots are $s = -0.42 \pm 0.65i, -1.66$. For $c > 1 (\Delta > 0)$, only one dominant root appears, and it moves to the right. The second and third roots approach one another, then become complex and move to the left. For $c = 1.5$, the roots are $s = -0.34, -1.58 \pm 0.65i$. Thus, the dominant time constant is sensitive to the indicated uncertainty in the value of c. The nominal value of τ is 1.08, but it can be as large as $1/0.34 = 2.94$ if $c = 0.5$.

Note that by choosing Δ to be the deviation from the nominal value of $c = 1$, we need not solve the cubic polynomial in (8.6-9) to find the poles. They are already known, since they are the characteristic roots of the system when $c = 1$. If Δ were instead chosen to be $\Delta = c - 0.5$, then Δ would always be positive, and only the primary root locus guides would be needed. But then a cubic would have to be solved to find the poles.

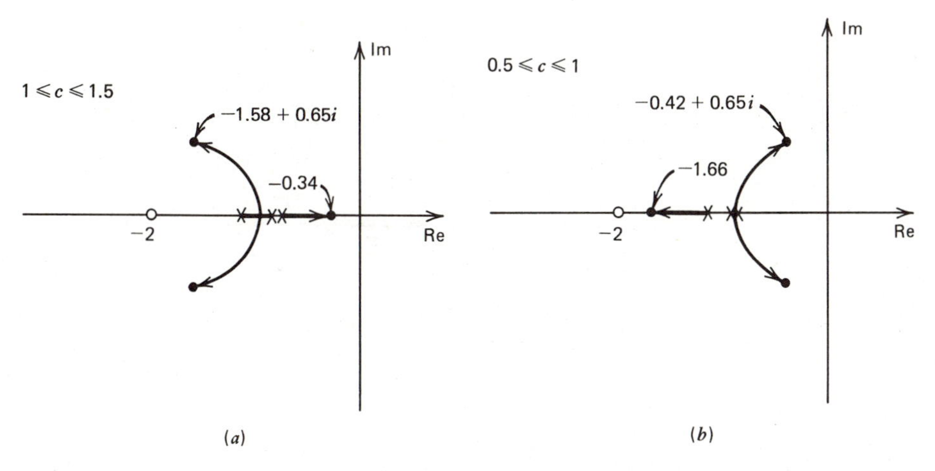

FIGURE 8.28 Root locus plots showing the parameter sensitivity of the PI controller of Example 8.8. (*a*) Plot for $\Delta \geqslant 0$. (*b*) Plot for $\Delta \leqslant 0$.

Example 8.9

The PI controller designed in Example 6.2 was for the speed control system shown in Figure 6.30. The gain values for $\zeta = 1.74$ and $\tau = 0.2$ were $K_p = 54$ and $K_I = 250$. The model assumed that the time constant of the field-controlled motor is negligible. Use the root locus plot to examine the sensitivity of the design to this assumption.

The modified block diagram showing the motor's transfer function is given in Figure 8.29*a*. The system's characteristic equation is

$$\tau_m s^3 + (1 + \tau_m)s^2 + 55s + 250 = 0 \tag{8.6-10}$$

Since the variable parameter τ_m in (8.6-10) is the coefficient of the highest power, we must be careful in applying the root locus guides. When $\tau_m = 0$, (8.6-10) changes from a cubic to a quadratic equation, and we must set up our analysis to prevent this from happening as we vary τ_m. Recall that the guides are valid only if the number of poles is greater than or equal to the number of zeros. This restriction prevents the equation's order from changing as the root locus parameter K is varied.

One way of avoiding this difficulty here is to choose Δ to be the deviation in τ_m from a small but positive value, say, 0.01. Thus, we define $\Delta = \tau_m - 0.01$. In terms of Δ, (8.6-10) becomes

$$s^3 + 101s^2 + 5500s + 25000 + 100\Delta s^2(s + 1) = 0 \tag{8.6-11}$$

Thus, the root locus parameter K must be $K = 100\Delta$. The poles are not known beforehand and must be found from the cubic (8.6-11) with $\Delta = 0$. They are $s = -4.98$, $-48.01 \pm 52.1i$. The zeros are $s = 0, 0, -1$. The plot is shown in Figure 8.29*b*. The crossover point occurs when $K = 27.205$ or $\tau_m = 0.282$. Thus, our controller is unstable if $\tau_m \geqslant 0.282$.

Our analysis of parasitic modes in Section 7.2 indicated that we should design for a closed-loop time constant that is large compared to the time constants of any

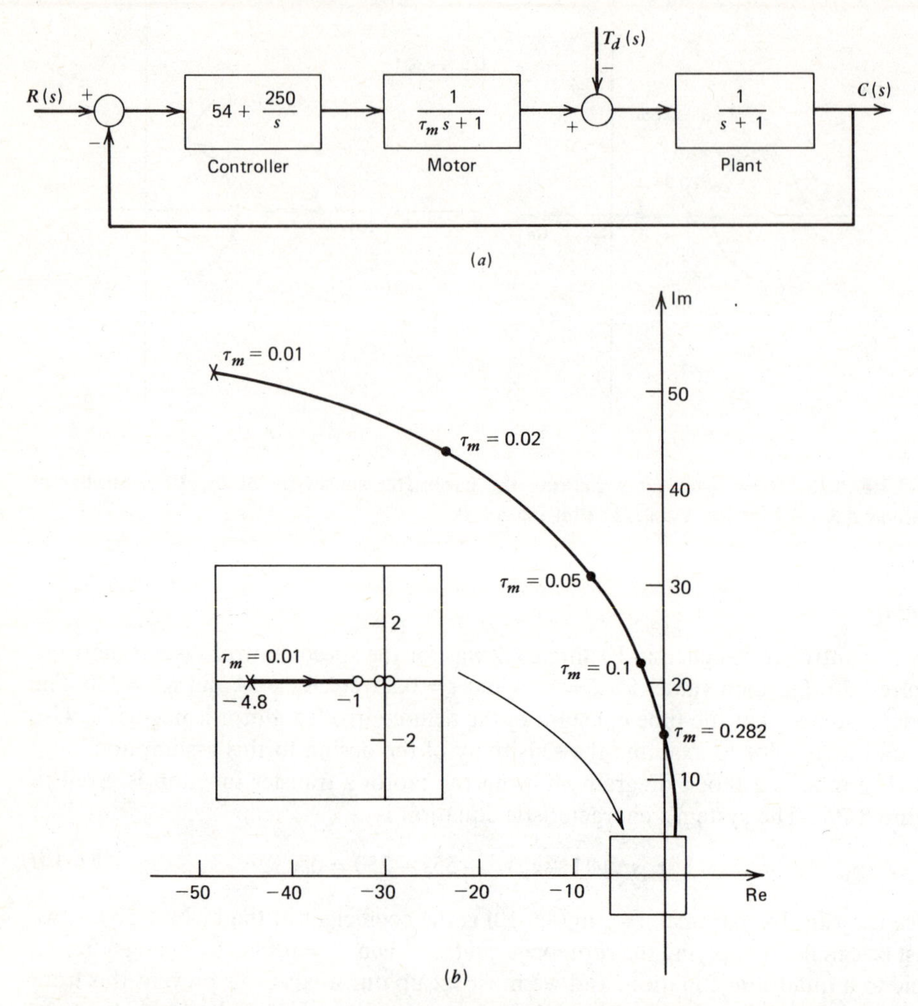

FIGURE 8.29 The effect of motor time constant on controller performance. (*a*) Block diagram. (*b*) Root locus plot.

neglected parasitic modes. Recall that here we designed for a closed-loop time constant $\tau = 0.2$, so this example confirms our previous observation. The root locus plot also shows that when τ_m becomes larger than 0.07, the dominant time constant becomes larger than the desired value of 0.2, and the dominant response becomes oscillatory, which might not be desirable.

8.7 SUMMARY

The root locus plot is perhaps the most useful tool in control systems design, because it allows the designer to obtain a global view of the behavior of the system's characteristic roots. It is used to set the control gains and to see how the control law can be modified to improve the system's performance. Once the nominal design is found, the effects of parameter uncertainties and parasitic modes can be analyzed with the root locus. In the next chapter, we will see more applications of the method.

REFERENCES

1. R. H. Cannon, *Dynamics of Physical Systems*, McGraw-Hill, New York, 1967.
2. J. L. Melsa and S. K. Jones, *Computer Programs for Computational Assistance in the Study of Linear Control Theory*, McGraw-Hill, New York, 1973.
3. Y. Takahashi, M. Rabins, and D. Auslander, *Control*, Addison-Wesley, Reading, Mass., 1970.

PROBLEMS

8.1 In the following equations, identify the root locus plotting parameter K and its range in terms of the parameter b and its indicated range.

(a) $s^2 + 4s + 3b = 0, \quad b \geqslant 0$
(b) $s^2 + (2 + b)s + 3 + 2b = 0, \quad b \geqslant 0$
(c) $s^2 + (2 + b)s + 2 + 2b = 0, \quad b \leqslant 0$
(d) $s^3 + 4bs^2 + 3s + b = 0, \quad b \geqslant 0$

8.2 In (a)–(g), sketch the root locus plot for the given characteristic equation for $K \geqslant 0$. Use all the guides that are applicable.

(a) $s(s + 4) + K = 0$
(b) $s(s + 4)(s + 6) + K = 0$
(c) $s^2 + 2s + 2 + K(s + 2) = 0$
(d) $s(s + 2) + K(s + 3) = 0$
(e) $s(s^2 + 2s + 2) + K = 0$
(f) $s(s + 1)(s + 8) + K(s + 2) = 0$
(g) $s(s + 2) + K(3 - s) = 0$

8.3 In (a)–(e), sketch the root locus plot for $K \leqslant 0$ (the complementary locus) for the given characteristic equation. Use all applicable guides.

(a) $s(s + 4) + K = 0$
(b) $s^2 + 2s + 2 + K(s + 2) = 0$
(c) $s(s^2 + 2s + 2) + K = 0$
(d) $s(s + 4)(s + 6) + K = 0$
(e) $s(s + 2) + K(s + 3) = 0$

8.4 The root locus plot obtained in Problem 8.2d is for a unity-feedback system with the open-loop transfer function

$$G(s) = \frac{K(s + 3)}{s(s + 2)}$$

(a) Set the gain K so that the damping ratio is $\zeta = 0.95$ with the smallest possible dominant time constant.

(b) For this value of K, compute the response to a unit-step input. Use the root locus plot to evaluate the coefficients in the partial fraction expansion.

8.5 A common situation in position control systems is that the inertia of the object to be positioned changes during the control process. The control of the angular position of a satellite (to aim a telescope, for example) must account for the change in the satellite's inertia due to the fuel consumption of the control jets. The inertia of a tape reel or a paper roll changes as it unwinds, and the controller must be designed to handle this variation.

Figure P8.5 shows P control of a second-order plant with an inertia $I = 20$. Use the root locus to investigate the change in the characteristic roots of the closed-loop system if I can vary by $\pm 10\%$.

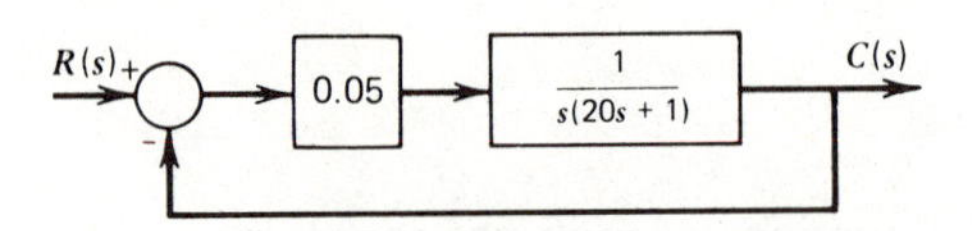

FIGURE P8.5

8.6 Consider Problem 6.32. Use the root locus to evaluate the performance of the resulting controller if the smaller mass m has a range of uncertainty of $5 \leqslant m \leqslant 15$. The relevant specifications are $\zeta = 0.707$ and $t_s = 10$ sec.

8.7 Consider Problem 6.36. Plot the root locus for the system in terms of the amplifier gain K for the two limits represented by (a) an empty reel and (b) a full reel.

8.8 Consider Problem 6.26e. Use the root locus to evaluate the performance of the resulting control system in light of the given specifications.

(a) When the final control elements have a transfer function

$$G_c(s) = \frac{1}{\tau s + 1} \qquad 0 \leqslant \tau \leqslant 1$$

(b) When the feedback sensor has a transfer function of the form in (a) and the dynamics of the final control elements are negligible.

8.9 Consider Problem 6.32. Use the root locus to evaluate the performance of the resulting controller in light of the given specifications.

(a) When the final control elements have a transfer function

$$G_c(s) = \frac{1}{\tau s + 1} \qquad 0 \leqslant \tau \leqslant 1$$

(b) When the feedback sensor has a transfer function of the form in (a) and the dynamics of the final control elements are negligible.

8.10 Consider Problem 6.28. Use the root locus to evaluate the performance of the resulting controller relative to the specifications $\tau = 0.1, \zeta = 1$ when the following parameters vary about their nominal values:

(a) $0.05 \leqslant c \leqslant 0.15$
(b) $0.5 \leqslant a \leqslant 1.5$

8.11 Consider Problem 7.12a. Use the root locus to evaluate the performance of the resulting controller in light of the specifications $\zeta = 0.707$, $\tau = 0.1$ if the plant transfer function $G_p(s)$ has an uncertainty Δ, where

$$G_p(s) = \frac{1}{s^2 - 2 - \Delta} \qquad -1 \leqslant \Delta \leqslant 1$$

8.12 A certain plant has the transfer function

$$G_p(s) = \frac{4p}{(s^2 + 4\zeta s + 4)(s + p)}$$

where the nominal values of ζ and p are $\zeta = 0.5$, $p = 1$.

(a) Use Ziegler–Nichols tuning to compute the PID gains. Find the resulting closed-loop characteristic roots.
(b) Use the root locus to determine the effect of a variation in the parameter ζ over the range $0.4 \leqslant \zeta \leqslant 0.6$.
(c) Use the root locus to determine the effect of a variation in the parameter p over the range $0.5 \leqslant p \leqslant 1.5$.

8.13 The open-loop transfer function of a certain plant is

$$G_p(s) = \frac{1}{s(s + 1)(s + 8)}$$

It is not possible with proportional control to achieve a dominant time constant of less than 2.08 sec for this plant. Use PD control to improve the response, so that the dominant time constant is 0.5 sec or less and the damping ratio is 0.5 or greater. To do this, first select a suitable value for T_D, then plot the locus with K_p as the variable.

8.14 Proportional control of a liquid level system gives the open-loop transfer function

$$G(s) = \frac{K}{\tau s + 1}$$

where unity feedback is used $[H(s) = 1]$ and τ is the time constant of the plant. This transfer function assumes that the float used to measure the liquid height responds instantaneously to changes in the liquid level. However, the float possesses dynamics of its own that resemble those of a mechanical oscillator. These dynamics introduce parasitic roots into the system.

Assume that $\tau = 1$ and the float dynamics are such that the feedback transfer function is

$$H(s) = \frac{8}{s^2 + 4s + 8}$$

Sketch the root locus plots for the unity feedback and nonunity feedback cases.

For both cases, select K to give the smallest dominant time constant possible. Discuss how the parasitic roots affect this choice. What is the effect of using a large value of K in each case?

8.15 Consider the velocity control of a load driven by an armature-controlled dc motor, as discussed in Section 7.2 (see Figure 7.3).

(a) If we neglect the armature inductance L, the plant is first order. With the parameter values given in Section 7.2, the open-loop transfer function for PI control is

$$G(s)H(s) = \frac{3.6315K_p(s + 1/T_I)}{s(0.043s + 5.3832)}$$

We desire $\zeta = 0.707$ and the dominant time constant to be less than 0.01 sec. Two roots will satisfy these conditions if the zero at $s = -1/T_I$ is selected to the left of the plant's pole. Choose T_I and plot the locus for K_p. Find the values of K_p that correspond to both root locations. Which one gives the smallest dominant time constant?

(b) If the armature inductance is included, the open-loop transfer function with PI control is

$$G(s)H(s) = \frac{3.6315K_p(s + 1/T_I)}{s[0.1Ls^2 + (0.07L + 0.043)s + 5.3832]}$$

For each of the two designs generated in (a), plot the root locus in terms of L as the locus variable for $0 \leqslant L \leqslant 0.003$. Which design is more sensitive to a nonzero inductance L?

8.16 Design a PI controller for the system described in Problem 8.15b. Take $L = 0.0021$. The specifications state that the dominant roots must have $\zeta \geqslant 0.5$ and a time constant of no more than 0.02 sec. Choose a trial value for T_I, then plot the locus for K_p. If a solution is not obtained, try a new T_I and repeat the process.

CHAPTER NINE

Applications of Graphical Methods to System Design

This chapter contains a further discussion of how the root locus and frequency response plots can be applied to designing a control system. We introduce the *Nyquist stability theorem* and some additional performance criteria: the *phase* and *gain margins*, and the *static error coefficients*.

The frequency response plots of the system's open-loop transfer function contain much information about the behavior of the closed-loop system. These plots are easily generated even for high-order systems, and they allow the proper control gain to be selected simply by adjusting the scale factor on the plot. The technique is especially useful, because it does not require the values of the characteristic roots. This is helpful for analyzing high-order systems and systems with dead time. The latter are especially difficult to treat with rooting methods, since they possess an infinite number of roots.

When a gain adjustment must be supplemented with a series compensator, the open-loop frequency response plots enable the designer to select the appropriate compensator parameters. In this context, we introduce two widely used compensators, the *series lead* and the *series lag* and develop graphical design methods for them.

The root locus and frequency response plots in this chapter can be sketched by hand with the aid of a calculator. If these methods are to be used regularly for design applications, they can easily be adapted for computer use.

9.1 SYSTEM DESIGN WITH OPEN-LOOP FREQUENCY RESPONSE PLOTS

Although we already have some powerful methods for designing linear control systems, there are several advantages to performing the design with the system's open-loop frequency response characteristics. First, frequency response data are often easier to obtain experimentally. This is especially useful when it is difficult to develop a transfer function model for the plant and actuators from basic principles. Second, the method to be developed is easier to use for higher order systems with dead-time elements than the root locus method. Third, a compensator to improve the system's response is sometimes more easily designed with this approach. Finally, this technique is sometimes useful for determining the existence of limit cycles and instability in nonlinear systems. We will pursue the first three topics; the last consists of a group of specialized methods whose applications are limited to specific problem types (see chapter 12 of Reference 1).

The Nyquist Stability Theorem

Recall that a plot of the frequency transfer function $T(i\omega)$ in vector form is the polar plot. On it, we plot the location of the tip of the vector as ω varies from 0 to ∞. The axes of the plot are the real and imaginary parts of $T(i\omega)$. Thus, the angle and magnitude of $T(i\omega)$ are represented on the same plot.

The Nyquist stability theorem is a powerful tool for linear system analysis. Its proof requires considerable development and will not be attempted here (see Reference 2). Instead, we will concentrate on the aspect of the theorem that is useful for control systems design. In many control applications, the plant has no poles with positive real parts, and typical controllers do not introduce poles of this type. If the open-loop system has no poles with positive real parts, we can concentrate our attention on the region around the point $-1 + i0$ on the polar plot of the *open-loop* transfer function. Because of its relationship to the theorem, the polar plot is also known as the *Nyquist plot*.

Figure 9.1 shows the polar plot of the open-loop transfer function of two arbitrary systems, both of which are assumed to be open-loop stable. The Nyquist stability theorem is stated as follows:

> The system is closed-loop stable if and only if the point $-1 + i0$ lies to the left of the open-loop Nyquist plot relative to an observer traveling along the plot in the direction of increasing frequency ω.

Therefore, the system described by Figure 9.1*a* is closed-loop stable. A system exhibits sustained oscillations (neutral stability) if the plot passes through the $-1 + i0$ point. It is unstable if the point lies to the right of the plot, as in Figure 9.1*b*.

The basis for the Nyquist theorem is easily understood. Let $e(t)$ and $b(t)$ be the actuating and feedback signals, respectively (see Figure 9.2). Suppose that the input $r(t)$ is sinusoidal with frequency ω_f, and assume that the phase angle of the open-loop transfer function is $-180°$. This means that at steady state, the signals $e(t)$ and $b(t)$ will

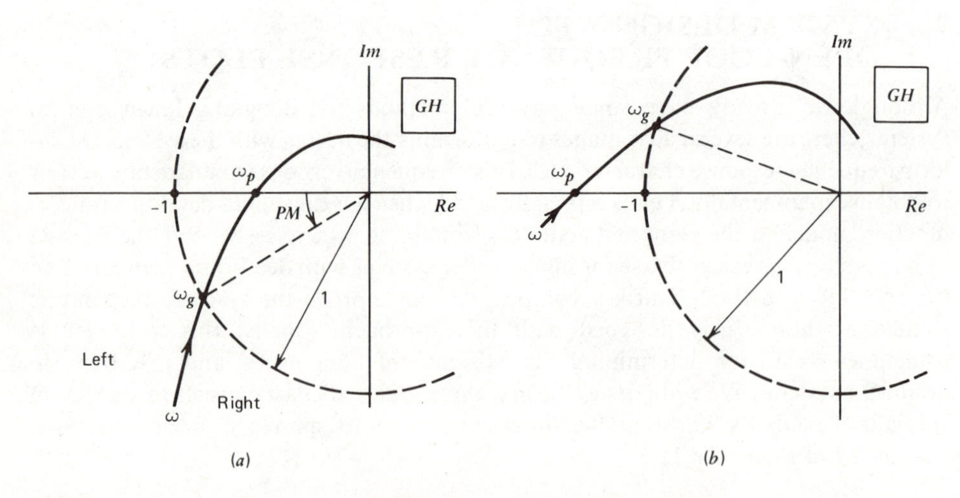

FIGURE 9.1 Nyquist plots. (*a*) Stable system. (*b*) Unstable system.

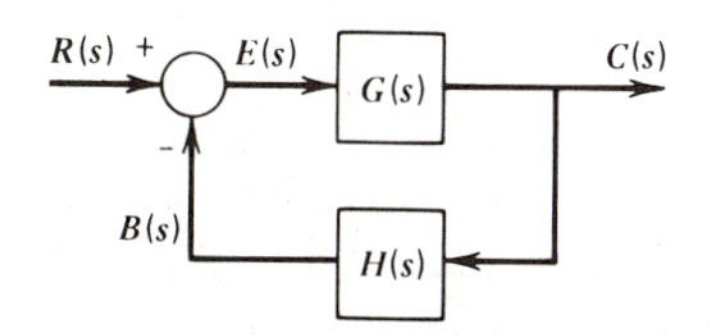

FIGURE 9.2 Single-loop system used in the development of the Nyquist stability theorem.

be sinusoidal with the frequency ω_f and 180° out of phase with each other [when $e(t)$ reaches a peak, $b(t)$ is at a minimum, and vice versa]. When the signal $b(t)$ has its sign changed at the comparator, the result, $-b(t)$, will be in phase with the signal $e(t)$.

Imagine that the input is now switched off. When the open-loop plot passes through the $-1 + i0$ point at the frequency ω_f, the gain relating $b(t)$ to $e(t)$ is unity at this frequency. Therefore, $e(t) = -b(t)$ and the original oscillation will be sustained at a constant amplitude. If the $-1 + i0$ point lies to the left of the plot, the gain between $e(t)$ and $b(t)$ will be less than unity. In this case, the amplitude of $b(t)$ will be less than that of $e(t)$, and the oscillations will gradually diminish. The system is then stable. Finally, if the $-1 + i0$ point lies to the right of the plot, the amplitude of $b(t)$ will be larger than that of $e(t)$, and the oscillations will grow in magnitude as the signal travels around the loop. The system is then unstable.

Phase and Gain Margins

The Nyquist theorem provides a convenient measure of the relative stability of a system. A measure of the proximity of the plot to the $-1 + i0$ point is given by the angle between the negative real axis and a line from the origin to the point where the plot crosses the unit circle (see Figure 9.1*a*). The frequency corresponding to this intersection is denoted as ω_g. This angle is the *phase margin* (PM), and it is positive when measured down from the negative real axis. The absence of a positive- or zero-phase margin thus indicates an unstable system (Figure 9.1*b*). The phase margin is the phase at the frequency ω_g where the magnitude ratio or "gain" of $G(i\omega)H(i\omega)$ is unity (0 db).

The frequency ω_p, the phase crossover frequency, is the frequency at which the phase angle is $-180°$. The *gain margin* (GM) is the difference in decibels between the unity gain condition (0 db) and the value of $|GH|$ db at the phase crossover frequency. Thus,

$$\text{gain margin} = -|G(i\omega_p)H(i\omega_p)| \text{ db} \tag{9.1-1}$$

A system is stable only if the phase and gain margins are both positive.

H. W. Bode contributed greatly to the development of design methods based on the open-loop frequency response. For this reason, the rectangular frequency response plots are sometimes called *Bode plots*. The phase and gain margins can be illustrated on the Bode plots shown in Figure 9.3*a*. The situation shown in Figure 9.3*b* represents an unstable system that lacks positive gain and phase margins.

The phase and gain margins can be stated as safety margins in the design specifications. A typical set of such specifications follows:

$$\text{gain margin} \geqslant 8 \text{ db} \quad \text{and} \quad \text{phase margin} \geqslant 30° \tag{9.1-2}$$

In common design situations, only one of these equalities can be met, and the other margin is allowed to be greater than its minimum value. It is not desirable to make the

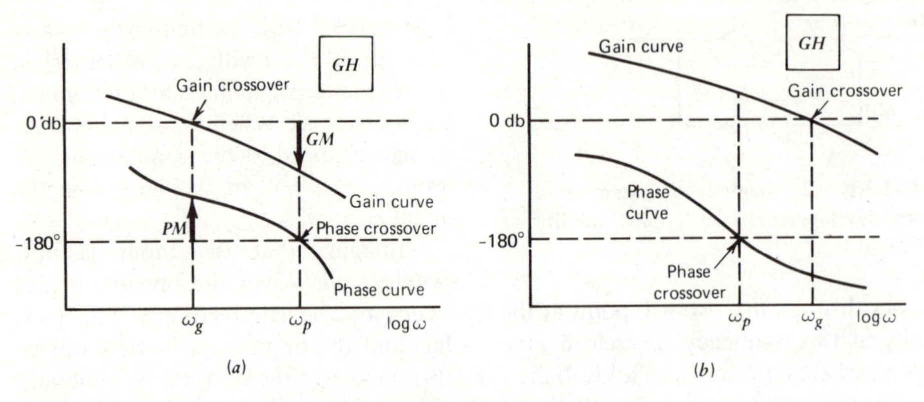

FIGURE 9.3 Bode plots. (*a*) Stable system. (*b*) Unstable system.

margins too large, since this results in a low gain. The design will be sluggish and might have a large steady-state error. We note that another common set of specifications is

$$\text{gain margin} \geqslant 6 \text{ db} \quad \text{and} \quad \text{phase margin} \geqslant 40^\circ \tag{9.1-3}$$

The 6-db limit corresponds to the quarter amplitude decay response obtained with the gain settings given by the Ziegler–Nichols ultimate-cycle method (Table 7.1).

Open-Loop Design for PID Control

Some general comments can be made about the effects of proportional, integral, and derivative control actions on the phase and gain margins. The P action does not affect the phase curve at all and thus can be used to raise or lower the open-loop gain curve until the specifications for the gain and phase margins are satisfied. If I action or D action is included, the proportional gain is selected last. Therefore, when using this approach to the design, it is best to write the PID algorithm with the proportional gain factored out, as

$$F(s) = K_p\left(1 + \frac{1}{T_I s} + T_D s\right)E(s) \tag{9.1-4}$$

The D action affects both the phase and gain curves. Therefore, selecting the derivative gain is more difficult than the proportional gain. The increase in phase margin due to the positive phase angle introduced by D action is partly negated by the derivative gain, which reduces the gain margin. Increasing the derivative gain increases the speed of response, makes the system more stable, and allows a larger proportional gain to be used to improve the system's accuracy. However, if the phase curve is too steep near $-180°$, it is difficult to use D action to improve the performance.

The I action also affects both the gain and phase curves. It can be used to increase the open-loop gain at low frequencies. However, it lowers the phase crossover frequency ω_p and thus reduces some of the benefits provided by D action. If required, the D-action term is usually designed first, followed by I action and P action, respectively.

The classical design methods based on the Bode plots obviously have a large component of trial and error because usually both the phase and gain curves must be manipulated to achieve an acceptable design. Given the same set of specifications, two designers can use these methods and arrive at substantially different designs. Many rules of thumb and ad hoc procedures have been developed, but a general foolproof procedure does not exist. Many such procedures and examples can be found in older references and the technical literature (see References 3–10). An experienced designer can often obtain a good design quickly with these techniques. Using a computer plotting routine greatly speeds up the design process.

Closed-Loop Response and the Nichols Chart

We have seen how the open-loop Bode diagrams can be used to set the parameters of the controller. Once this is done, computing the closed-loop response can be done from the same diagrams. The system shown in Figure 9.2 has the closed-loop frequency transfer function

$$T_c(i\omega) = \frac{G(i\omega)}{1 + G(i\omega)H(i\omega)} \tag{9.1-5}$$

Define the magnitude and phase of this transfer function as

$$m_c = |T_c(i\omega)| \text{ db} \tag{9.1-6}$$

$$\phi_c = \measuredangle\, T_c(i\omega) \tag{9.1-7}$$

From (9.1-5),

$$T_c(i\omega) = \frac{G(i\omega)H(i\omega)}{1 + G(i\omega)H(i\omega)} \frac{1}{H(i\omega)} \tag{9.1-8}$$

and thus

$$m_c = \frac{G(i\omega)H(i\omega)}{1 + G(i\omega)H(i\omega)} \text{ db} - |H(i\omega)| \text{ db} \tag{9.1-9}$$

$$\phi_c = \measuredangle \frac{G(i\omega)H(i\omega)}{1 + G(i\omega)H(i\omega)} - \measuredangle\, H(i\omega) \tag{9.1-10}$$

If a computer is used to calculate the complex numbers $G(i\omega)$ and $H(i\omega)$ to generate the open-loop plots, these numbers can also be used in (9.1-9) and (9.1-10) to generate the closed-loop plots simultaneously. If the system has unity feedback [$H(s) = 1$], (9.1-9) and (9.1-10) are simplified considerably.

The Nichols chart provides a manual technique for obtaining closed-loop response information from the open-loop data (Reference 11). The chart is usually stated in terms of a unity-feedback system. If nonunity feedback exists, the block diagram can be rearranged to give an equivalent unity-feedback system. With $H(s) = 1$, define m_o and ϕ_o to be the magnitude and phase of the open-loop transfer function.

$$m_o = |G(i\omega)| \text{ db} \tag{9.1-11}$$

$$\phi_o = \measuredangle\, G(i\omega) \tag{9.1-12}$$

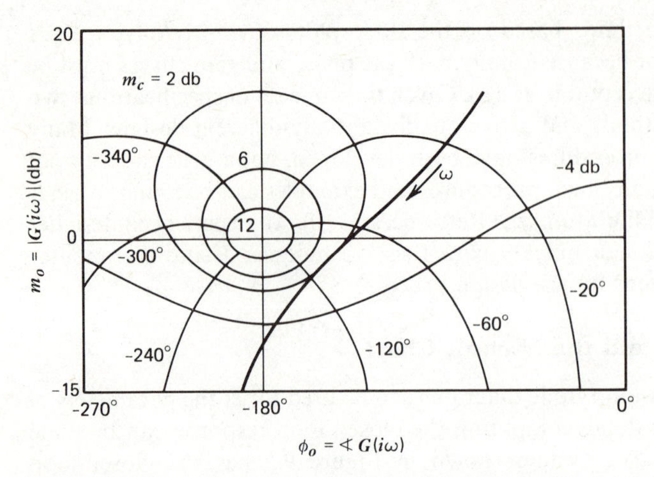

FIGURE 9.4 Nichols chart. The heavy line is a plot of the open-loop response of a particular system.

The Nichols chart is a plot of m_o versus ϕ_o. Superimposed on the chart are contours of constant m_c and ϕ_c values. Figure 9.4 shows a Nichols chart with a few contours displayed. Most design work requires many contours. These can be obtained with commercially available templates (for example, Berol's Rapidesign #331) or by computer-generated plots.

The procedure for obtaining the closed-loop response from the Nichols chart is as follows.

1. From the open-loop Bode plots, the values of m_o and ϕ_o can be read off for a particular frequency. These are used to plot a point on the Nichols chart. This step is repeated for a number of frequency values. The result is a single curve on the Nichols chart, such as shown by the heavy line in Figure 9.4.
2. From the intersection of this curve with the contours at particular frequencies, values of m_c and ϕ_c can be read off the plot. Use these values to construct the closed-loop Bode plots.

This procedure can be modified to handle systems with nonunity feedback. Equations (9.1-9) and (9.1-10) provide the basis for this modification.

The intersections of the closed-loop response curve with the magnitude and phase axes are given by the phase and gain margins, respectively (see Figure 9.5). Thus, the curve for a stable system must pass below the origin of the magnitude–phase axes. Interpretations for unstable systems are meaningless, because the frequency response plots are not defined for such systems, since a steady state does not exist.

Constructing the closed-loop response plots from the Nichols chart can be tedious. However, the chart is useful for determining the gain required to achieve a specified maximum- (resonance) peak value m_{cp} in the closed-loop response. It can be shown with Fourier series analysis that the frequency components of an input signal near the resonance peak dominate the system's response (Reference 1, section 9-4).

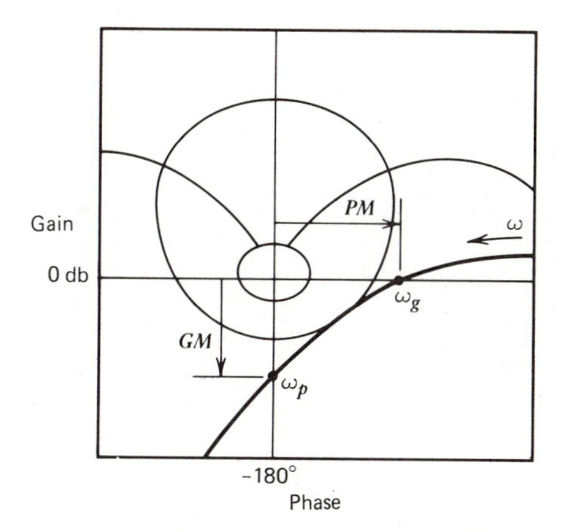

FIGURE 9.5 Interpretation of phase and gain margins on the Nichols chart.

Thus, specifying this peak value is an important design criterion. The control gain required to achieve the desired m_{cp} is found by plotting the closed-loop response line on the chart for a unity gain. The amount of gain that must be added (in decibels) is then found from the number of decibels by which the line must be translated to become tangent to the m_{cp} contour. A specification in common use is that M_{cp} should be less than 2 (m_{cp} less than 6 db). A typical value is $M_{cp} = 1.3$; that is, $m_{cp} = 2.28$ db. For many systems, this specification is approximately equivalent to the gain and phase margins specified by (9.1-2)—namely, $GM \geqslant 8$ db and $PM \geqslant 30°$.

9.2 SERIES COMPENSATION AND PID CONTROL

Of the types of compensation discussed in Section 7.4, series (or cascade) compensation is particularly well suited for design with the open-loop Bode plots. Usually, the system is analyzed with proportional control action included (often with a unity gain). The series compensator is then included to meet the performance specifications. The effect of the compensator is readily seen on the Bode plot, because the compensator's gain and phase have an additive effect on the plot of the uncompensated open-loop system.

PD Control as Series Compensation

The P control with either I or D action, or both, may be considered a form of series compensation. The PD law is

$$F(s) = K_p(1 + T_D s)E(s) \tag{9.2-1}$$

The term $(1 + T_D s)$ can be considered as a series compensator to the proportional controller. The D action adds an open-loop zero at $s = -1/T_D$. Its Bode plots are shown in Figure 9.6*a*. From these, it can be seen that the usefulness of D action is that it adds phase shift at higher frequencies. It is thus said to give phase "lead." However, it also increases the gain at these frequencies, since the derivative term gives more response for rapidly changing signals.

PI Control as Series Compensation

The PI control is described by

$$F(s) = K_p\left(1 + \frac{1}{T_I s}\right)E(s) = \frac{K_p}{s}\left(s + \frac{1}{T_I}\right)E(s) \tag{9.2-2}$$

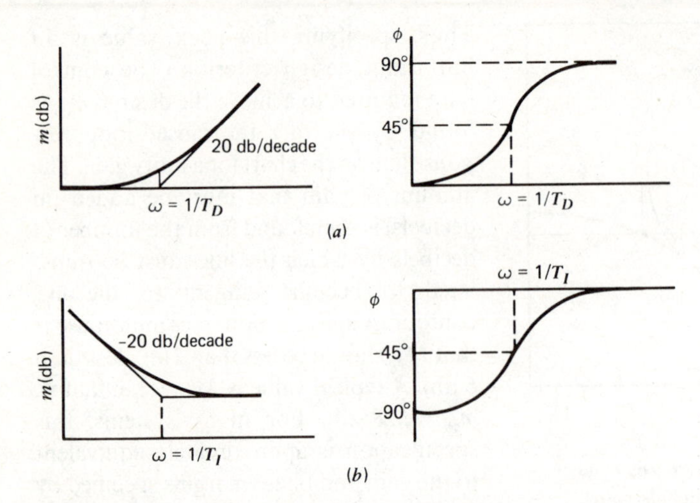

FIGURE 9.6 Bode plots for series compensators. (*a*) PD action $1+T_D s$. (*b*) PI action $(s+1/T_I)/s$.

The integral action can thus be considered to add an open-loop pole at $s=0$ and a zero at $s=-1/T_I$. The Bode plots for the term $(T_I s+1)/T_I s$ are shown in Figure 9.6*b*. The compensation adds gain at the lower frequencies but decreases the phase. It can be considered as a phase lag compensator.

Design Considerations with the Bode Plot

It is difficult to state a rigid set of rules to follow for designing a series compensator because of the variety of specifications and plant types that occur. However, the following considerations should be kept in mind.

1. In order to minimize the steady-state error, the open-loop gain should be kept as high as possible in the low-frequency range of the Bode plot.
2. At intermediate frequencies (near the gain crossover frequency), a slope of -20 db/decade in the gain curve will help to provide an adequate phase margin.
3. At high frequencies, small gain is desirable in order to attenuate high-frequency disturbances, such as electronic noise, or mechanical vibrations induced by gear teeth, shaft elasticity, or hydraulic and pneumatic pressure fluctuations, and so forth.

In addition to shaping the gain curve, the phase curve must often be shaped by means of a series compensator in order to achieve phase and gain margin specifications, for example. The PID-type compensators do not always allow sufficient flexibility to do this. In Sections 9.4–9.7 we will consider more general types of compensators, called lead and lag compensators, that provide this flexibility. Specifications used to design these compensators include the types seen thus far as well as the static error coefficients, to be presented in Section 9.3.

9.3 STATIC ERROR COEFFICIENTS

Consider the single-loop system shown in Figure 9.7. The error signal is related to the input as follows:

$$E(s) = \frac{1}{1 + G(s)H(s)} R(s) \tag{9.3-1}$$

Assuming that the final value theorem can be applied, it gives

$$e_{ss} = \lim_{s \to 0} \frac{sR(s)}{1 + G(s)H(s)} \tag{9.3-2}$$

The *static error coefficient* C_i is defined as

$$C_i = \lim_{s \to 0} s^i G(s)H(s) \qquad i = 0, 1, 2, \ldots \tag{9.3-3}$$

A system is of *type n* if $G(s)H(s)$ can be written as $s^n F(s)$. Table 9.1 relates the steady-state error to the system-type number for three command inputs and can be used to design a system for minimum error.

If the input is a unit step, we obtain the steady-state error

$$e_{ss} = \lim_{s \to 0} \frac{1}{1 + G(s)H(s)} = \frac{1}{1 + C_p} \tag{9.3-4}$$

where

$$C_p = \lim_{s \to 0} G(s)H(s) = C_0 \tag{9.3-5}$$

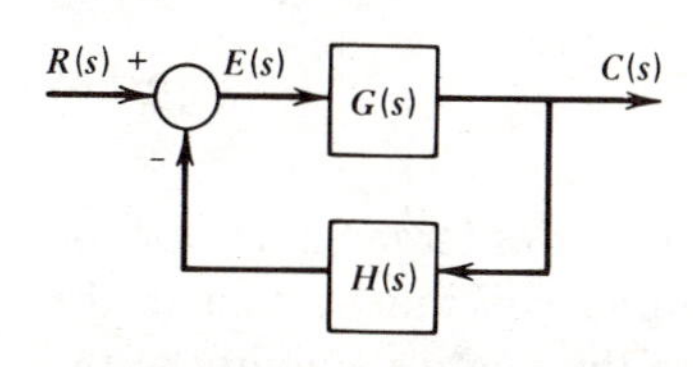

FIGURE 9.7 System configuration for definition of static error coefficients.

The constant C_p is the *static position error coefficient*. The name derives from servomechanism applications in which the output is a position. The error between the desired and the actual positions at steady state is given by (9.3-4). The larger C_p is, the smaller the error. C_p is finite and e_{ss} nonzero when P control is applied to a

TABLE 9.1 Steady-State Error versus System-Type Number

$R(s)$ \ e_{ss}	System-Type Number 0	1	2	3
Step $1/s$	$1/(1 + C_0)$	0	0	0
Ramp $1/s^2$	∞	$1/C_1$	0	0
Parabola $1/s^3$	∞	∞	$1/C_2$	0

first-order plant. When I control is applied to such a plant, the resulting C_p will be infinite, and the error e_{ss} will be zero. In general, it can be shown that a unity feedback system will have a nonzero steady-state error if no integration occurs in the forward path.

The *static velocity error coefficient* is obtained for a unit-ramp input as follows:

$$e_{ss} = \lim_{s \to 0} \frac{s}{1 + G(s)H(s)} \frac{1}{s^2}$$

$$= \frac{1}{C_v} \tag{9.3-6}$$

where the velocity coefficient is

$$C_v = \lim_{s \to 0} sG(s)H(s) = C_1 \tag{9.3-7}$$

Note that the error here is not an error in velocity, but a position error that results when a unit-ramp input is applied. For type 0 systems, $C_v = 0$; for type 1 systems, C_v is finite but nonzero; for type 2 and higher, $C_v = \infty$. Thus, to eliminate the steady-state error in a unity feedback system with a ramp input, at least two integrations are required in the forward loop. For a type 1 system with unity feedback, the output velocity at steady state equals that of the input (the slope of the ramp), but an error exists between the desired and the actual positions.

Other error coefficients can be interpreted in a similar manner. For example, the *static acceleration error coefficient* C_a is defined for a unit-acceleration input $R(s) = 1/s^3$ as

$$C_a = \lim_{s \to 0} s^2 G(s)H(s) = C_2 \tag{9.3-8}$$

Thus,

$$e_{ss} = \frac{1}{C_a} \tag{9.3-9}$$

Reduced steady-state error can thus be seen to result from large values for the error coefficients (high gain) and from increasing the number of integrations in the open-loop transfer function. Both remedies can decrease the relative stability of the system, however. A good design must achieve a balance between these objectives.

9.4 LEAD AND LAG COMPENSATORS

We now consider three commonly used compensators, the lead, lag, and lag–lead compensators. These can be easily realized electrically with *RC* networks and perhaps an amplifier. Mechanical equivalents are developed in the chapter problems.

Lead and Lag Compensator Circuits

A variety of compensator circuits have been developed, but the lead and lag compensators shown in Figures 9.8 and 9.10 are the simplest ways of adjusting gains and phases in a variety of frequency ranges. In addition to the PID compensators, they are the most commonly used.

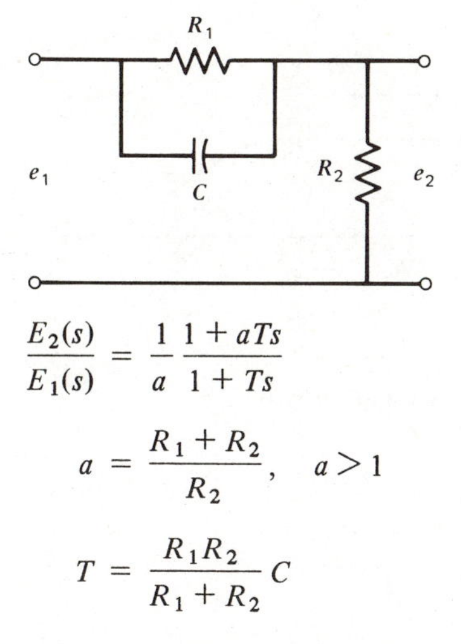

FIGURE 9.8 Passive-lead compensator.

When used as series compensators, the circuits shown must see a small impedance at the source (e_1) and a large impedance at the load (e_2). Sometimes an isolating amplifier is inserted to ensure the validity of this assumption. Taking these impedances to be zero and infinity, respectively, we can derive the circuits' transfer functions with the methods in Chapter 2. For the lead compensator (Figure 9.8).

$$\frac{E_2(s)}{E_1(s)} = \frac{R_2 + R_1 R_2 Cs}{R_1 + R_2 + R_1 R_2 Cs} = \frac{1}{a}\frac{1 + aTs}{1 + Ts}$$

$$= \frac{s + \dfrac{1}{aT}}{s + \dfrac{1}{T}} \tag{9.4-1}$$

where

$$a = \frac{R_1 + R_2}{R_2} \qquad a > 1 \tag{9.4-2}$$

$$T = \frac{R_1 R_2}{R_1 + R_2} C \tag{9.4-3}$$

Note that the circuit would be useless as a series compensator without the resistance R_1, since it would not pass dc signals.

For the lag compensator (Figure 9.10),

$$\frac{E_2(s)}{E_1(s)} = \frac{1 + R_2 Cs}{1 + (R_1 + R_2)Cs} = \frac{1 + aTs}{1 + Ts}$$

$$= a\frac{s + \dfrac{1}{aT}}{s + \dfrac{1}{T}} \tag{9.4-4}$$

where

$$a = \frac{R_2}{R_1 + R_2} \qquad a < 1 \tag{9.4-5}$$

$$T = C(R_1 + R_2) \tag{9.4-6}$$

The Bode plots of the lead and lag compensators are shown in Figures 9.9 and 9.11. The compensators' usefulness can be briefly described as follows. When used in series with a proportional gain K_p, the lead compensator increases the gain margin. This allows the gain K_p/a to be made larger than is possible without the compensator. The result is a decrease in the closed-loop bandwidth and an increase in the speed of response.

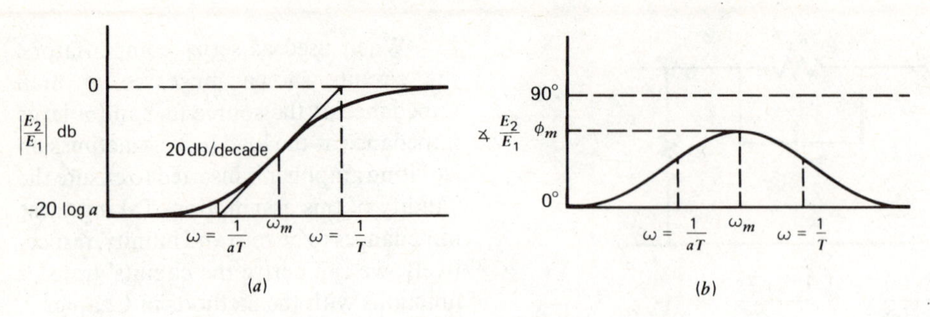

FIGURE 9.9 Bode plots for the lead compensator shown in Figure 9.8.

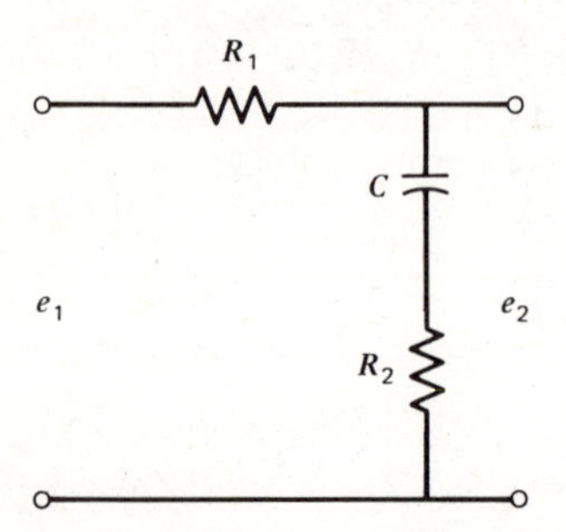

FIGURE 9.10 Passive-lag compensator.

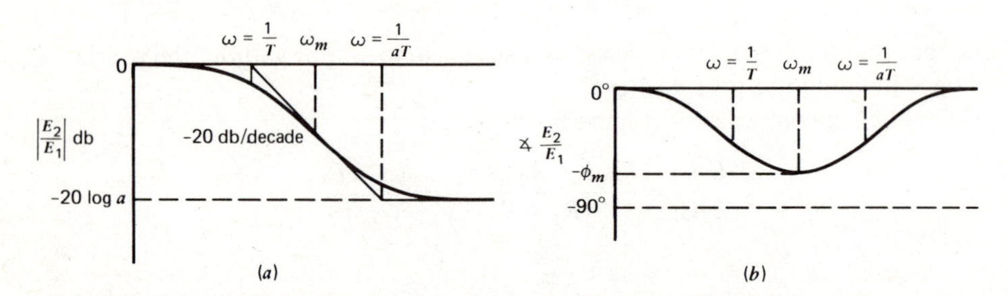

FIGURE 9.11 Bode plots for the lag compensator shown in Figure 9.10.

On the other hand, the lag compensator is used when the speed of response and damping of the closed-loop system are satisfactory, but the steady-state error is too large. The lag compensator allows the gain to be increased without substantially changing the resonance frequency ω_r and the resonance peak m_p of the closed-loop system.

9.5 ROOT LOCUS DESIGN OF COMPENSATORS

When the performance specifications, such as required time constant, damping ratio, and so forth are given in terms of root locations, the root locus method is the preferred way of designing the compensator.

Lead Compensators

The effects of the compensator in terms of time domain specifications (characteristic roots) can be shown with the root locus plot. Consider the second-order plant with the real distinct roots $s = -\alpha, -\beta$. The root locus for this system with proportional control is shown in Figure 9.12*a*. The smallest dominant time constant obtainable is τ_1, marked in the figure. With lead compensation, the root locus becomes that shown in Figure 9.12*b*. The pole and zero introduced by the compensator reshape the locus so that a smaller dominant time constant can be obtained. This is done by choosing the proportional gain high enough to place the roots close to the asymptotes.

Designing a lead compensator with the root locus is done as follows. We assume that the specifications do not include required values for the static error coefficients. If they do, the design should be done with the Bode plot. The same is true if phase and gain margins are specified.

1. From the time domain specifications (time constant, damping ratio, etc.), the required locations of the dominant closed-loop poles are determined.
2. From the root locus plot of the uncompensated system, determine whether or not the desired closed-loop poles can be obtained by adjusting the open-loop gain. If not, determine the net angle associated with the desired closed-loop pole by drawing vectors to this pole from the open-loop poles and zeros. The difference between this angle and $-180°$ is the *angle deficiency.*
3. Locate the pole and zero of the compensator so that they will contribute the angle required to eliminate the deficiency. A method for doing this is presented in Example 9.1.

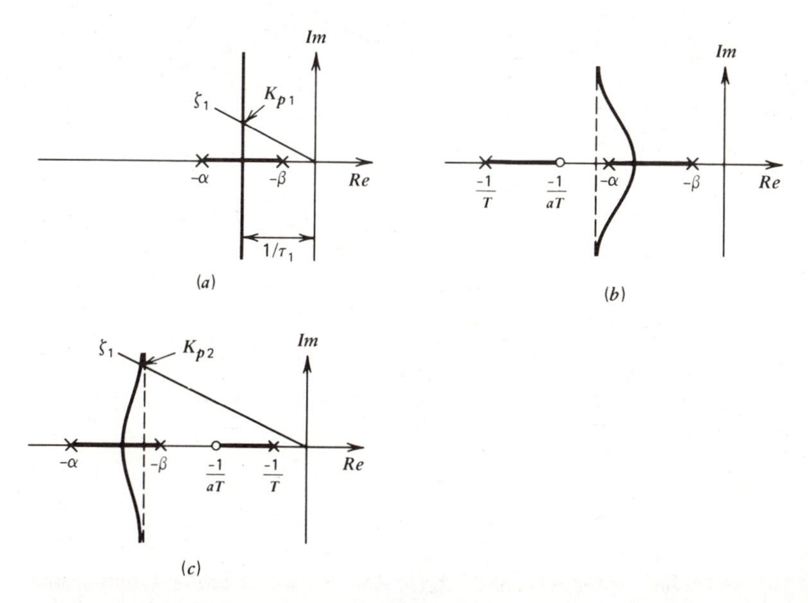

FIGURE 9.12 Effects of series lead and lag compensators. (*a*) Uncompensated system's root locus. (*b*) Root locus with lead compensation. (*c*) Root locus with lag compensation.

4. Compute the required value of the open-loop gain from the root locus plot.
5. Check the design to see if the specifications are met. If not, adjust the locations of the compensator's pole and zero.

Example 9.1

Consider the system shown in Figure 9.13*a*. Its root locus plot is given in Figure 9.13*b*. In Section 8.1 (Figure 8.1), it was shown that a value of $K = 1$ gives $\zeta = 0.5$ and closed-loop poles at $s = -\frac{1}{2} \pm i\frac{1}{2}\sqrt{3}$. The resulting time constant is 2. Assume that the transient specifications require that $\zeta = 0.5$ with a time constant of $\tau = 1$ and no steady-state error specifications are given. This information implies that the closed-loop poles should be at $s = -1 \pm i\sqrt{3}$. This performance cannot be obtained with a gain change in the present system. Design a compensator to meet the specifications.

Since the specifications are given in terms of time domain characteristics, we use the root locus method. From Figure 9.13*c*, the angle of the uncompensated system at the desired root location is

$$\measuredangle \left. \frac{1}{s(s+1)} \right|_{s=-1+i\sqrt{3}} = -90° - 120° = -210°$$

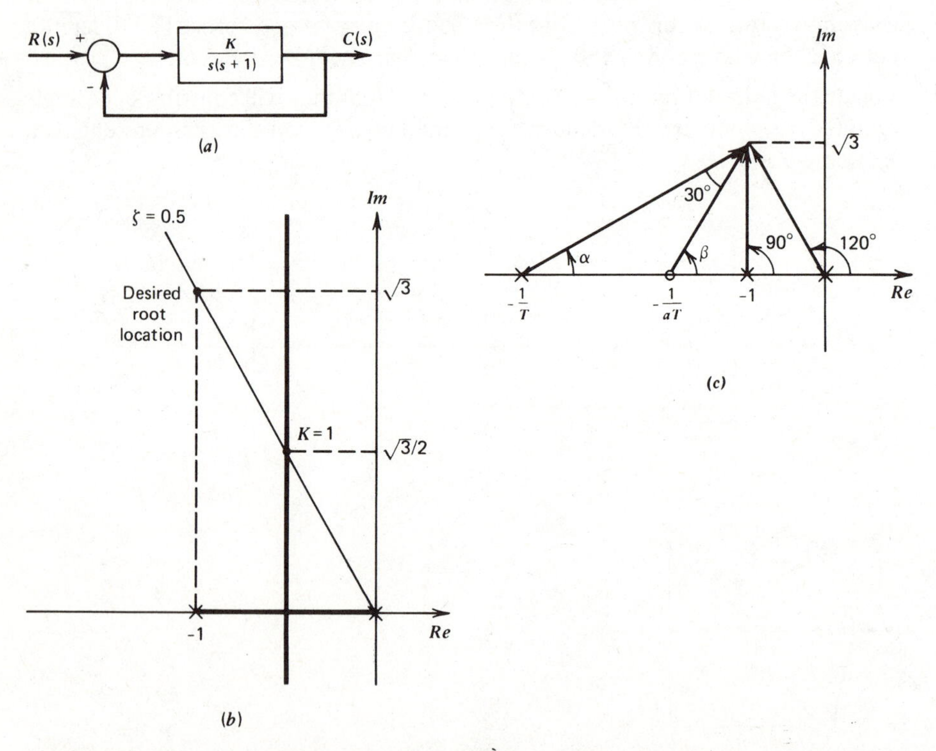

FIGURE 9.13 (*a*) System for Examples 9.1 and 9.2. (*b*) Root locus plot for the system shown in (*a*) showing the actual and desired root locations. (*c*) Vector diagram showing the contributions of the poles and zeros at the desired-root location.

Thus, the angle deficiency is $-210° + 180° = -30°$. We therefore need a lead compensator to increase the phase angle by 30°.

The choice of the poles and zeros of the compensator is somewhat arbitrary. We must pull the original locus to the left, so we place the pole and zero to the left of the pole at $s = -1$. The difference between the angle contributions of the pole and zero of the compensator must be 30°. This specifies the size of the apex angle of the triangle shown in Figure 9.13*c*. Since the inverse of a appears in the numerator of the transfer function (9.4-1), we should select its value to be as small as possible in order to minimize the additional amount of gain required to cancel its effect. Keeping physical realizability in mind, we choose $a = 10$. Since $\beta = \alpha + 30°$, simple trigonometry applied to Figure 9.13*c* gives the required value for T. The tangents of α and β are

$$\tan \alpha = \frac{T\sqrt{3}}{1 - T}$$

$$\tan \beta = \frac{aT\sqrt{3}}{1 - aT}$$

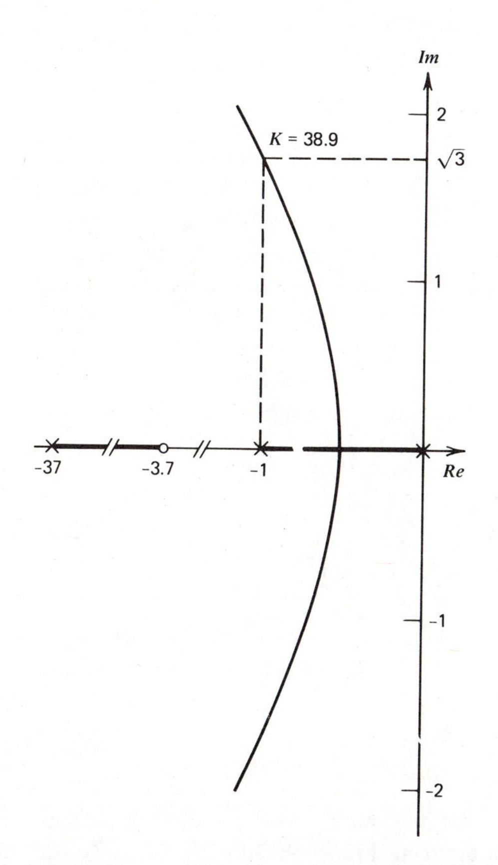

FIGURE 9.14 Root locus plot for the compensated system in Example 9.1.

Also,

$$\tan \beta = \tan (\alpha + 30^\circ) = \frac{\tan \alpha + \tan 30^\circ}{1 - \tan \alpha \tan 30^\circ}$$

Eliminating α and β yields

$$4aT^2 + 2(1 - 2a)T + 1 = 0$$

For $a = 10$, $T = 0.027$, 0.923. The second solution results in the pole lying to the left of $s = -1$ and the zero lying to the right of $s = -1$, so we use $T = 0.027$. From (9.4-2) and (9.4-3), the resistances required with a 1-μf capacitance are $R_1 = 270\ k\Omega$ and $R_2 = 30\ k\Omega$. The compensator pole is at $s = -37.0$, and the zero is at $s = -3.7$.

The open-loop transfer function of the compensated system is

$$G_c(s)G(s)H(s) = \frac{K(s + 3.7)}{s(s + 37)(s + 1)}$$

Its root locus is shown in Figure 9.14. From this, the gain K required to place the roots at $s = -1 \pm i\sqrt{3}$ is $K = 38.9$. The third root is far to the left at $s = -36$ for this value of K. Its effect on the transient behavior is probably slight. This can be checked by analysis or simulation before the design is made final.

Lag Compensators

With reference to the system shown in Figure 9.12*a*, suppose that the desired damping ratio ζ_1 and desired time constant τ_1 are obtainable with a proportional gain of K_{p1}, but the resulting steady-state error $\alpha\beta/(\alpha\beta + K_{p1})$ for a step input is too large. We need to increase the gain while preserving the desired damping ratio and time constant. With the lag compensator, the root locus is as shown in Figure 9.12*c*. By considering specific numerical values, we can show that for the compensated system, roots with a damping ratio ζ_1 correspond to a high value of the proportional gain. Call this value K_{p2}. Thus, $K_{p2} > K_{p1}$, and the steady-state error will be reduced.

The effect of the lag compensator on the time constant can be seen as follows. The open-loop transfer function is

$$G(s)H(s) = \frac{aK_p\left(s + \dfrac{1}{aT}\right)}{(s + \alpha)(s + \beta)\left(s + \dfrac{1}{T}\right)} \tag{9.5-1}$$

If the value of T is chosen large enough, the pole at $s = -1/T$ in (9.5-1) is approximately canceled by the zero at $s = -1/aT$, and the open-loop transfer function is given approximately by

$$G(s)H(s) = \frac{aK_p}{(s + \alpha)(s + \beta)} \tag{9.5-2}$$

Thus, the system's response is governed approximately by the complex roots corresponding to the gain value K_{p2}. By comparing Figure 9.12*a* with 9.12*c*, we see that the compensation leaves the time constant relatively unchanged.

From (9.5-2), it can be seen that since $a < 1$, K_p can be selected as the larger value K_{p2}. The ratio of K_{p1} to K_{p2} is approximately given by the parameter a.

Design by pole-zero cancellation can be difficult to accomplish, because a response pattern of the system is essentially ignored. The pattern corresponds to the behavior generated by the canceled pole and zero, and this response can be shown to be beyond the influence of the controller. In this example, the canceled pole gives a stable response, because it lies in the left-hand plane. However, another input not modeled here, such as a disturbance, might excite the response and cause unexpected behavior. The designer should therefore proceed with caution. None of the physical parameters of the system are known exactly, so exact pole zero cancellation is not possible. A root locus study of the effects of parameter uncertainty and a simulation study of the response are often advised before the design is accepted as final.

Based on this example, we can outline an approach to the design of a lag compensator with the root locus as follows:

1. From the root locus of the uncompensated system, the gain K_{p1} is found that will place the roots at the locations required to give the desired relative stability and transient response.
2. Let K_{p2} denote the value of the gain required to achieve the desired steady-state performance. The parameter a is the ratio of these two gain values $a = K_{p1}/K_{p2} < 1$.
3. The value of T is then chosen large so that the compensator's pole and zero are close to the imaginary axis. This placement should be made so that the compensated locus is relatively unchanged in the vicinity of the desired closed-loop poles. This will be true if the angle contribution of the lag compensator is close to zero.
4. Locate the desired closed-loop poles on the compensated locus and set the open-loop gain so that the dominant roots are at this location (neglecting the existence of the compensator's pole and zero).
5. Check the design to see if the specifications are met. If not, adjust the locations of the compensator's pole and zero.

Example 9.2

Consider the system shown in Figure 9.13a. If $K = 1$, then $\zeta = 0.5$ and $s = -\frac{1}{2} \pm i\sqrt{3}/2$. The static velocity error coefficient is $C_v = 1$. Suppose that the transient response thus obtained is satisfactory but the error must be decreased by increasing C_v to 10. Design a compensator to do this.

A lag compensator is indicated, since the steady-state error is too large. The gain K_{p1} required to achieve the desired transient performance has already been established as $K_{p1} = K = 1$. The second step is to determine the value of the parameter a. For this system, the coefficient C_v is

$$C_v = \lim_{s \to 0} s \frac{K_p}{s(s+1)} = K_p$$

and K_{p2} is the value of K_p that gives $C_v = 10$. Thus, $K_{p2} = 10$, and the parameter $a = K_{p1}/K_{p2} = 1/10$. The compensator's pole and zero must be placed close to the imaginary axis, with the ratio of their distances being 1/10. Noting that the plant has a pole at $s = -1$, we select locations well to the right of this pole, say, at $s = -0.01$ and $s = -0.1$ for the pole and zero, respectively. This gives $T = 100$.

The open-loop transfer function of the compensated system is thus

$$G_c(s)G(s)H(s) = \frac{0.1K_c(s+0.1)}{s(s+1)(s+0.01)}$$

The root locus is shown in Figure 9.15. For the desired damping ratio of $\zeta = 0.5$, the locus shows that the dominant roots are at $s = -0.449 \pm i0.778$ with $K_c = 9$. The third root is at $s = -0.111$.

The error coefficient is thus $C_v = 9$ and less than the desired value of 10. Also, the dominant roots differ somewhat from the desired locations at $s = -0.5 \pm i0.866$. If these differences are too large, the compensator's pole and zero can be placed closer to the imaginary axis, say, at $s = -0.01$ and $s = -0.001$, respectively, with $T = 1000$. This will decrease the compensator's influence on the locus near the desired root locations.

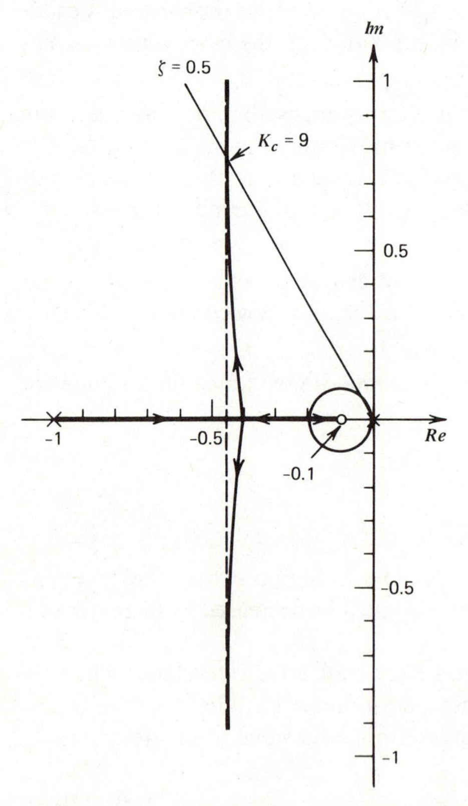

FIGURE 9.15 Root locus plot for $G(s)H(s) = 1/s(s+1)$ with the lag compensator $G_c(s) = 0.1K_c(s+0.1)/(s+0.01)$.

With $T = 100$ and the capacitance $C = 1\ \mu f$, $R_1 = 90\ M\Omega$, and $R_2 = 10\ M\Omega$. The improved performance obtained with $T = 1000$ requires that $R_1 = 900\ M\Omega$ and $R_2 = 100\ M\Omega$.

The Root Contour Method

The root contour method can be used with the root locus plots to select the values of more than one parameter. For lead or lag compensation, the pertinent parameters are a and T. To use the method, one parameter is fixed at a reasonable value, and the other is varied to generate a root locus plot. From this plot, reasonable values of the second parameter are chosen, and the locus is constructed for the variation in the first parameter. For example, lag compensation requires that $0 < a < 1$, so we can choose $a = 0.5$, say, and plot the locus for $0 < T < \infty$. Each point on this T locus is a potential pole for a locus in terms of the deviation Δ, where $a = 0.5 + \Delta$. The Δ loci can be constructed for $-0.5 \leqslant \Delta \leqslant 0.5$ at strategic points along the T locus. The values of a and T are then selected with the aid of these plots.

For lead compensation, $a > 1$; thus, a starting value of $a = 1$ might be used. In this case, $a = 1 + \Delta$ and $0 \leqslant \Delta \leqslant \infty$. The number of loci generated with this approach can be quite large, and a computer-plotting routine is advised.

9.6 BODE DESIGN OF COMPENSATORS

When the performance specifications are given in terms of the frequency response, such as phase or gain margins, then the compensator should be designed with the Bode plot.

Lead Compensators

For any particular value of the parameter a, the lead compensator can provide a maximum phase lead ϕ_m (see Figure 9.9). This value, and the frequency ω_m at which it occurs, can be found as a function of a and T from the Bode plot. The frequency ω_m is the geometric mean of the two corner frequencies of the compensator. Thus,

$$\omega_m = \frac{1}{T\sqrt{a}} \tag{9.6-1}$$

By evaluating the phase angle of the compensator at this frequency, we can show that

$$\sin \phi_m = \frac{a-1}{a+1} \tag{9.6-2}$$

or

$$a = \frac{1 + \sin \phi_m}{1 - \sin \phi_m} \tag{9.6-3}$$

These relations are useful in designing a lead compensator with the Bode plot.

System design in the frequency domain is usually done with phase and gain margin requirements. We assume that the specifications include these requirements as well as a specification for the allowable steady-state error. The purpose of the lead compensator is to use the maximum phase lead of the compensator to increase the phase of the open-loop system near the gain crossover frequency while not changing the gain curve near that frequency. This is usually not entirely possible, because the gain crossover frequency is increased in the process, and a compromise must be sought between the resulting increase in bandwidth and the desired values of the phase and gain margins.

A suggested design method is:

1. Set the gain K_p of the uncompensated system to meet the steady-state error requirement.
2. Determine the phase and gain margins of the uncompensated system from the Bode plot, and estimate the amount of phase lead ϕ required to achieve the margin specifications. The extra phase lead required can be used to estimate the value for ϕ_m to be provided by the compensator. Thus, a can be found from (9.6-3).
3. Choose T so that ω_m from (9.6-1) is located at the gain crossover frequency of the compensated system. One way of doing this is to find the frequency at which the gain of the uncompensated system equals $-20 \log \sqrt{a}$. Choose this frequency to be the new gain crossover frequency. This frequency corresponds to the frequency ω_m at which ϕ_m occurs.
4. Construct the Bode plot of the compensated system to see if the specifications have been met. If not, the choice for ϕ_m needs to be evaluated and the process repeated. It is possible that a solution does not exist.

The procedure can be quickly accomplished in the hands of an experienced designer, because intuition based on past designs can play a large role in the process. It is difficult to state rules that are foolproof.

The attempt to design a lead compensator can be unsuccessful if the required value of a is too large. This can occur with plants that are not stable or have a low relative stability with a rapidly decreasing phase curve near the gain crossover frequency. In the former case, the extra phase lead required can be too large. In the latter case, the phase angle at the compensated gain crossover frequency is much less than at the uncompensated gain crossover. Thus, the extra phase required can be excessive. The difficulty with a large value of a is that the resistance and capacitance values that result might be incompatible or impossible to obtain physically. The usual range for a is $1 < a < 20$. Additional phase lead can sometimes be obtained by cascading more than one lead compensator.

In spite of these potential difficulties, the lead compensator has a record of many successful applications.

Example 9.3

The system shown in Figure 9.13a can yield a static velocity error coefficient of $C_v = 10/\text{sec}$ if K is set to 10. However, if this is done, the transient performance is unsatisfactory. Design a compensator to give a gain margin of at least 6 db and a phase margin of at least 40°.

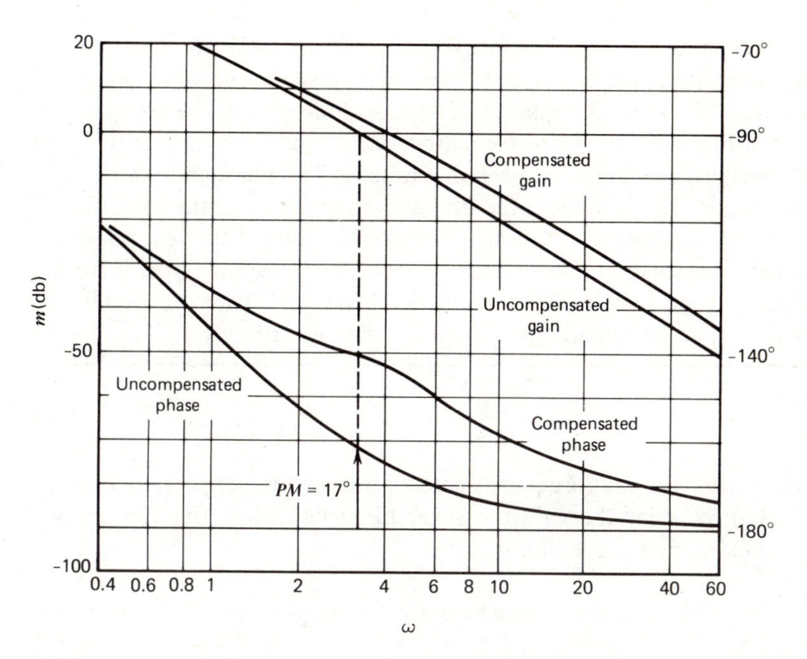

FIGURE 9.16 Bode plots for the uncompensated and compensated systems in Example 9.3.

The Bode plot of the open-loop uncompensated system is shown in Figure 9.16. The phase margin is 17°, and the gain margin is infinite, because the phase curve never falls below $-180°$. Thus, $40° - 17° = 23°$ must be added to the phase curve in order to meet the specifications, so we choose to try a lead compensator with $\phi_m = 23°$. From (8.9-19), we have

$$a = \frac{1 + \sin 23°}{1 - \sin 23°} = 2.283$$

From step 3, we compute

$$-20 \log \sqrt{a} = -3.585 \text{ db}$$

The open-loop gain of the uncompensated system is -3.585 db at approximately $\omega = 4$ rad/sec. Take this frequency to be ω_m, and solve for T from (9.6-1).

$$T = \frac{1}{4\sqrt{2.283}} = 0.1655$$

From (9.4-1), these values of a and T give a compensator with the transfer function

$$G_c(s) = \frac{1}{2.283} \frac{0.3778s + 1}{0.1655s + 1}$$

The open-loop transfer function of the compensated system is

$$G_c(s)G(s)H(s) = \frac{K(0.3778s + 1)}{2.283s(s + 1)(0.1655s + 1)} \qquad \textbf{(9.6-4)}$$

The original choice of $K = 10$ no longer gives $C_v = 10$, because the compensator introduces the attenuation factor 1/2.283. Thus, we must choose $K = 10(2.283) = 22.83$ to achieve $C_v = 10$. With K equal to 22.83, the Bode plot of the compensated system is shown in Figure 9.16. The phase margin is 37°, and the gain margin is still infinite, so the specifications have not been met exactly. The phase margin is 3° less than the required value. This illustrates the trial-and-error nature of this method. The guides listed in the design steps are based on approximations of the real phase and gain curves of the compensator. The difficulty could have been avoided if we had aimed for a phase margin higher than that specified, say, 45°. Then ϕ_m would have been used as $45° - 17° = 28°$, and a and T would differ accordingly.

Lag Compensators

Lag compensation uses the high-frequency attenuation of the network to keep the phase curve unchanged near the gain crossover frequency while this frequency is lowered. A suggested design procedure is:

1. Set the open-loop gain K_p of the uncompensated system to meet the steady-state error requirements.
2. Construct the Bode plots for the uncompensated system, and determine the frequency at which the phase curve has the desired phase margin. Determine the number of decibels required at this frequency to lower the gain curve to 0 db. Let this amount be $m' > 0$ db and this frequency be ω'_g. Then a is found from

$$a = 10^{-m'/20} \tag{9.6-5}$$

3. The first two steps alter the phase curve. However, this curve will not be changed appreciably near ω'_g if T is chosen so that $\omega'_g \gg 1/aT$. A good choice is to place the frequency $1/aT$ one decade below ω'_g. Any larger separation might result in a system with a slow response.
4. Construct the Bode plots of the compensated system to see if all the specifications are met. If not, choose another value for T and repeat the process.

As with lead compensation, this procedure is one of trial and error and might not work. The physical elements must be realizable, so a common range for a is $0.05 < a < 1$. The compensator introduces a lag angle that is not accounted for in the preceding procedure. To account for this effect, the designer might add 5° to 10° to the specified phase margin before starting the design process.

Example 9.4

Consider the uncompensated system in Example 9.3 shown in Figure 9.13*a*. The characteristic roots when $K = 1$ are $s = -0.5 \pm i0.866$. Suppose that the transient response given by these roots is acceptable, but we wish $C_v = 10$. Setting $K = 10$ will accomplish this but will alter the transient performance. Design a compensator to improve the system so that $C_v = 10$, $PM \geqslant 40°$, and $GM \geqslant 6$ db.

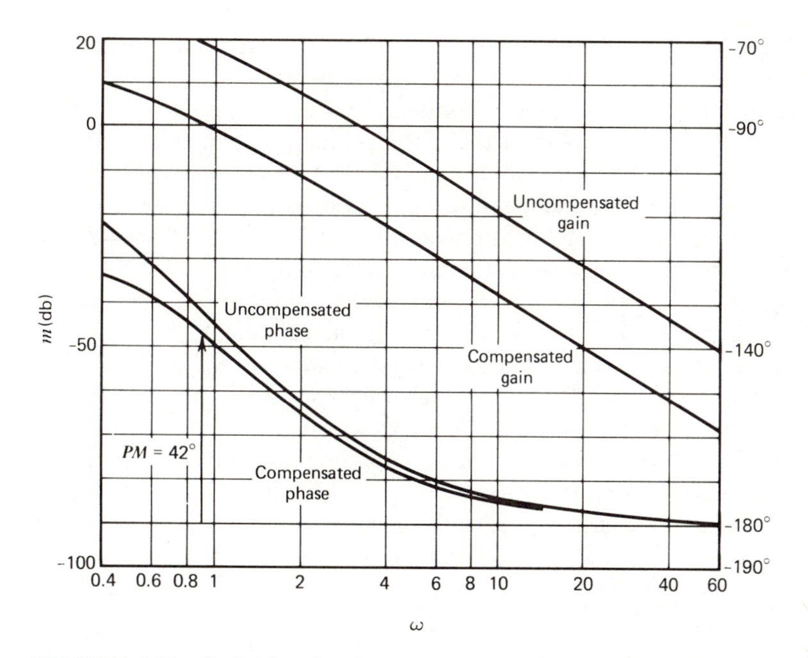

FIGURE 9.17 Bode plots for the uncompensated and compensated systems in Example 9.4.

Since the transient response is satisfactory, a lag compensator is indicated. Set $K = 10$ to achieve the desired value of C_v, and construct the Bode plots for the uncompensated open-loop system. These are the same as those shown in Figure 9.16 and repeated in Figure 9.17. Keeping in mind the lag introduced by the compensator, we attempt to achieve a phase margin of $40° + 5° = 45°$. The uncompensated system would have a phase margin of 45° if the gain crossover occurs at $\omega = 1$. Thus, $\omega_g' = 1$. The gain at this frequency is 18 db, so $m' = 18$. From (9.6-5),

$$a = 10^{-18/20} = 0.126$$

For step 3, we choose T to place $\omega = 1/aT$ one decade below $\omega = 1$. Thus, $1/aT = 0.1$, and $T = 79.4$. The lag compensator that results is

$$G_c(s) = \frac{10s + 1}{79.4s + 1}$$

The open-loop transfer function of the compensated system is

$$G_c(s)G(s)H(s) = \frac{K(10s + 1)}{s(s + 1)(79.4s + 1)}$$

With $K = 10$, $C_v = 10$, and the Bode plots are given in Figure 9.17. The phase margin is 42°, and the gain margin is infinite, so the specifications have been met.

On the surface it appears that Examples 9.3 and 9.4 are concerned with solving the same problem; namely, compensate the system in Figure 9.13*a* so that $C_v = 10$, $PM \geqslant 40°$, and $GM \geqslant 6$ db. However, this is not the case. The difference lies in what is taken as satisfactory in the performance of the uncompensated system and what must be improved. In Example 9.3, the transient response that results when $K = 10$ was judged to be poor, and the lead compensator was used to improve the transient performance. In Example 9.4, acceptable transient response was obtained when $K = 1$, but the steady-state error was then unacceptable, and a lag compensator was therefore used. The compensation in Example 9.3 was obtained by shifting the phase curve, whereas that in Example 9.4 was obtained by shifting the gain curve.

When $K = 1$, the roots of the uncompensated system are $s = -0.5 \pm i0.866$. With the lag compensator in Example 9.4, the new characteristic roots are $s = -0.107$, $-0.453 \pm i0.985$ and thus very close to the desired roots if the effect of the root at $s = -0.107$ is negligible because of the pole zero cancellation of the compensator. This is the same process that occurred in Example 9.2.

On the other hand, the characteristic roots of the lead-compensated system in Example 9.4 are $s = -3.66$, $-1.68 \pm i3.69$. These represent an improvement in the transient response since the uncompensated roots lie at $s = -0.5 \pm i3.12$ when $K = 10$.

9.7 LAG—LEAD COMPENSATION

The lead and lag compensators are complementary to each other in that one improves the transient performance while the other improves steady-state performance. In situations where one of these fails to produce a satisfactory design, a cascade combination of the two can be used. This is lag–lead compensation. A lag and a lead cascaded together give the transfer function

$$\frac{E_2(s)}{E_1(s)} = \frac{1}{a} \frac{1 + aT_1 s}{1 + T_1 s} \frac{1 + bT_2 s}{1 + T_2 s} \tag{9.7-1}$$

A simpler approach is to use the single network shown in Figure 9.18. With the usual impedance assumptions, we obtain

$$\frac{E_2(s)}{E_1(s)} = \frac{1 + (R_1C_1 + R_2C_2)s + R_1C_1R_2C_2s^2}{1 + (R_1C_1 + R_1C_2 + R_2C_2)s + R_1C_1R_2C_2s^2}$$

$$= \frac{1 + aT_1 s}{1 + T_1 s} \frac{1 + bT_2 s}{1 + T_2 s} \tag{9.7-2}$$

where

$$aT_1 = R_1C_1 \qquad a > 1 \tag{9.7-3}$$

$$bT_2 = R_2C_2 \tag{9.7-4}$$

$$T_1 + T_2 = R_1C_1 + R_1C_2 + R_2C_2 \tag{9.7-5}$$

$$b = \frac{1}{a} \tag{9.7-6}$$

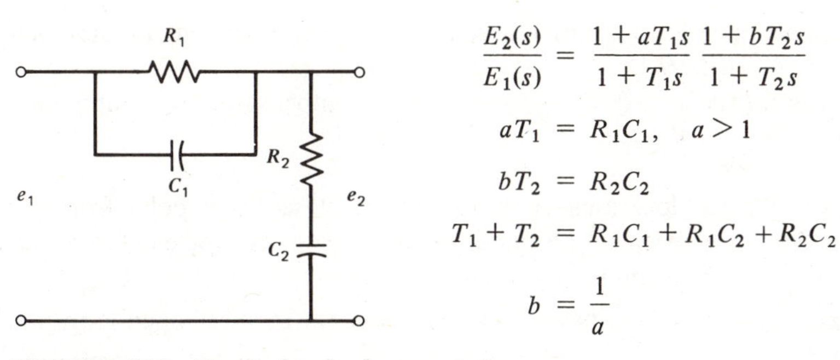

$$\frac{E_2(s)}{E_1(s)} = \frac{1 + aT_1 s}{1 + T_1 s}\frac{1 + bT_2 s}{1 + T_2 s}$$

$$aT_1 = R_1C_1, \quad a > 1$$

$$bT_2 = R_2C_2$$

$$T_1 + T_2 = R_1C_1 + R_1C_2 + R_2C_2$$

$$b = \frac{1}{a}$$

FIGURE 9.18 Passive lag–lead compensator.

The Bode plots for this compensator are shown in Figure 9.19 for $T_2 > bT_2 > aT_1 > T_1$. The maximum phase shift ϕ_m occurs at $\omega_{m1} = 1/T_1\sqrt{a}$, and the attenuation at this frequency is equal in magnitude but opposite in sign to that of the lead compensator alone. Thus, the lag–lead compensator can be more effective than the lead compensator. The plots also show that the compensator affects the gain and phase in only the intermediate frequency range from $\omega = 1/T_2$ to $1/T_1$.

Let the compensator transfer function in (9.7-2) be written as

$$G_c(s) = G_1(s)G_2(s) \tag{9.7-7}$$

where $G_1(s)$ represents the lead compensation and $G_2(s)$ the lag compensation.

$$G_1(s) = \frac{1 + aT_1 s}{1 + T_1 s} \tag{9.7-8}$$

$$G_2(s) = \frac{1 + T_2 s/a}{1 + T_2 s} \tag{9.7-9}$$

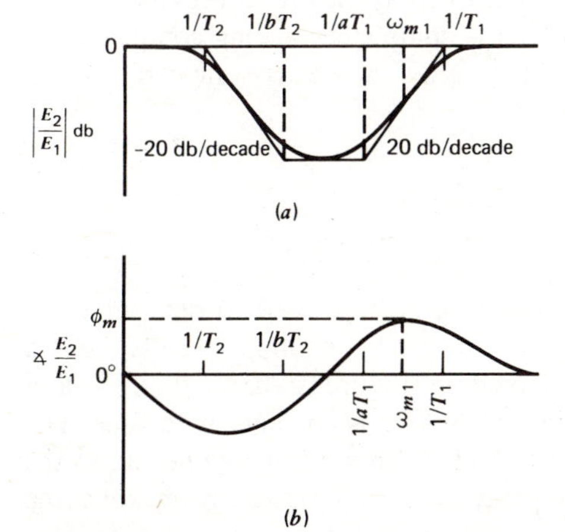

FIGURE 9.19 Bode plots for the lag–lead compensator shown in Figure 9.18. The frequency ω_{m1} is the geometric mean frequency for the lead compensator alone: $\omega_{m1} = 1/T_1\sqrt{a}$.

The uncompensated open-loop transfer function is $G(s)H(s)$, so that of the compensated system is $G_c(s)G(s)H(s)$.

The root locus approach to designing the lag–lead compensator is a combination of the approaches used for the lead and lag compensators.

1. Determine the desired locations of the dominant closed-loop poles from the transient performance specifications, and calculate the phase angle deficiency ϕ. This deficiency must be supplied by $G_1(s)$.
2. Use the steady-state error specifications to determine the required open-loop gain for $G_c(s)G(s)H(s)$.
3. Let s_d be the desired location of the dominant closed-loop poles. Assume the lag coefficient T_2 is large enough to achieve pole zero cancellation when s is near s_d; that is,

$$|G_2(s_d)| \cong 1 \tag{9.7-10}$$

With (9.7-10) satisfied, T_1 and a can be found from the magnitude criterion for the locus and the phase lead requirement from step 1; that is,

$$|G_1(s_d)||G(s_d)H(s_d)| = 1 \tag{9.7-11}$$

$$\measuredangle G_1(s_d) = \phi \tag{9.7-12}$$

4. With the parameters a and T_1 now selected, choose T_2 so that condition (9.7-10) is satisfied. As before, check the design to see if the specifications are satisfied and the circuit elements are realizable.

Designing a lag–lead compensator with the Bode plot follows the procedures for the lead and lag compensators discussed previously. Presumably the lag-lead is to be designed because of a deficiency in both the transient and steady-state performance of the uncompensated system. A designer who thinks that the transient response constitutes the most serious deficiency can choose to apply the procedures for the lead compensator first. When this part of the design is completed, the lag compensation can be designed. The opposite procedure could be used if the steady-state performance were worse than the transient.

9.8 SYSTEMS WITH DEAD-TIME ELEMENTS

The shifting property of the Laplace transform can also be used to determine the response of a system with *dead time.* This occurs when an input results from a transporting fluid, for example. This is shown in Figure 9.20. A fluid with a temperature y flows through a pipe. The fluid velocity u is constant with time. The pipe length is L, so it takes a time $D = L/u$ for the fluid to move from one end to the other. Let $y_1(t)$ denote the incoming fluid temperature and $y_2(t)$ the temperature of the fluid leaving the pipe.

Now, suppose that the temperature of the incoming fluid suddenly increases. If this is modeled as a step function, the result is shown in Figure 9.20. If no heat energy

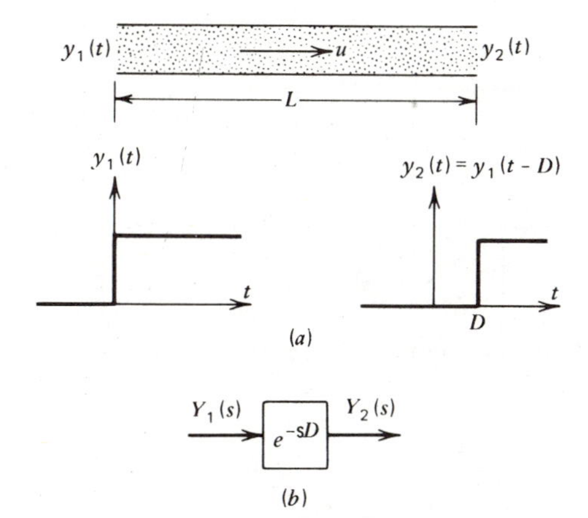

FIGURE 9.20 Process with dead time $D = L/u$. (a) Time plots of the input and the response. (b) Block diagram.

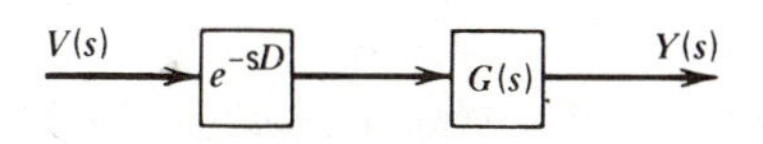

FIGURE 9.21 System diagram with dead time.

is lost, then $y_2(t)$, the temperature at the output, is $y_1(t-D)$, where $y_1(t)$ is the temperature at the input. Thus, a time D later, the output temperature suddenly increases.

A similar effect occurs for any change in $y_1(t)$; in general, we may write

$$y_2(t) = y_1(t-D)$$

From the shifting theorem,

$$Y_2(s) = e^{-Ds}Y_1(s) \tag{9.8-1}$$

This result is shown in block diagram form in Figure 9.20. Figure 9.21 shows the general case of an element with dead time. In this case,

$$Y(s) = e^{-Ds}G(s)V(s) \tag{9.8-2}$$

The presence of dead time means that the system does not have a characteristic equation of finite order. In fact, there are an infinite number of characteristic roots for a system with dead time. This can be seen by noting that the term e^{-Ds} can be expanded in an infinite series as

$$e^{-Ds} = \frac{1}{e^{Ds}} = \frac{1}{1 + Ds + (D^2s^2/2!) + \cdots} \tag{9.8-3}$$

Methods for dealing with such models are now presented.

Dead-Time in Control Systems

Some systems have an unavoidable time delay in the signal flow between components. The delay usually results from the physical separation of the components and typically occurs as a delay between a change in the manipulated variable and its effect on the plant or as a delay in the measurement of the output. Examples of each are shown in Figures 9.22 and 9.23.

Figure 9.22*a* shows a temperature control system in which hot air is delivered to the space to be controlled. The heating element is located a distance L from the plant. This distance can be long, as in large buildings heated by a central source. The velocity of the air in the duct is v, so that a delay $D = L/v$ exists between a change in the temperature of the air leaving the heating element and its effect on the temperature of the space. If the plant is modeled as a first-order system, a proportional controller would have the block diagram in Figure 9.22*b*. The heat flow rate from the heater is $q_1(t)$ and that affecting the plant is $q_2(t)$. Thus, $q_2(t) = q_1(t-D)$, so that $Q_2(s) = e^{-sD}Q_1(s)$.

The second example involves a measurement delay (Figure 9.23). Ribbons of hot

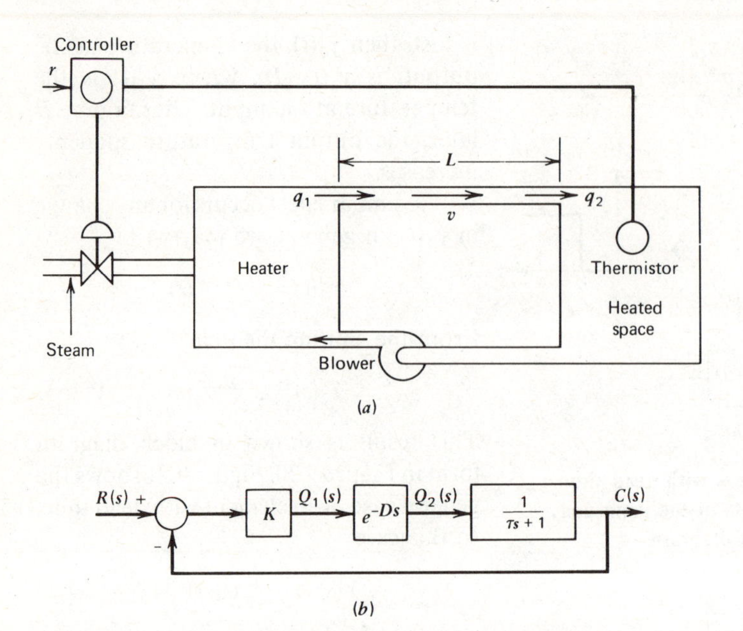

FIGURE 9.22 **(*a*) Temperature controller with dead time in the forward path. (*b*) Block diagram for proportional control and a first-order plant.**

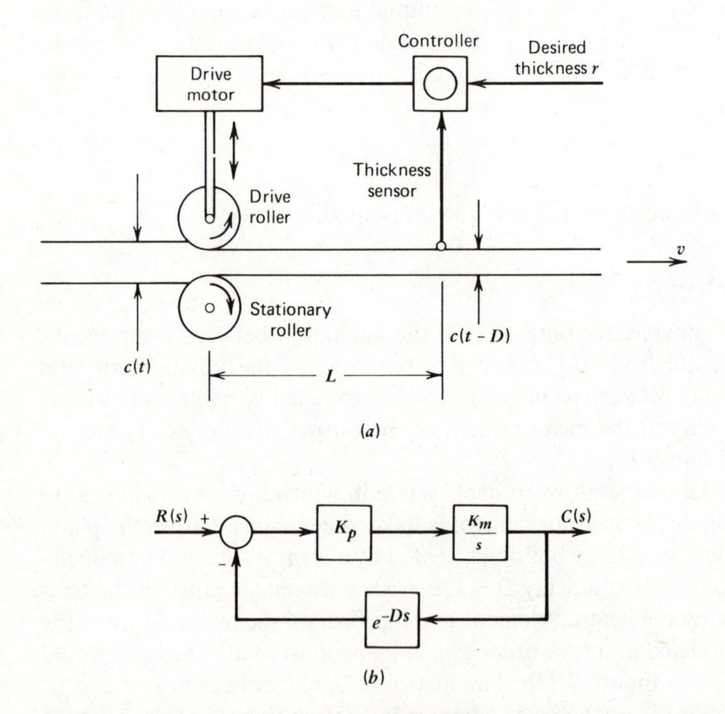

FIGURE 9.23 **(*a*) System for metal rolling. (*b*) Thickness control system with proportional action and dead time in the feedback loop.**

metal are drawn through rollers at high speed to produce sheet metal of a desired thickness. The upper roller's position is controlled to obtain this thickness. For obvious reasons, the thickness sensor cannot be placed at the rollers; therefore, it is placed downstream. The delay between the measured thickness $c(t-D)$ and the thickness $c(t)$ at the rollers is $D = L/v$, and it can be significant, because the high ribbon velocity requires a fast control system. The drive motor and roller displacement can be modeled as an integrator. The system diagram for proportional control is given in Figure 9.23*b*. The delay element now appears in the feedback path.

The time delay described in the examples mentioned is variously referred to as *dead time*, *transportation lag*, or *pure-time delay* to distinguish it from the delay associated with the time constant of a first-order element. We will see that the presence of such delays tends to cause oscillations and make the system less stable. They should be avoided or reduced if at all possible. This cannot always be done, so we must know how to analyze systems with dead time. The root locus provides one means of doing so.

Root Locus for Dead-Time Elements

The basic equation studied previously with the root locus is (8.1-12). Here, we generalize its form to include the effect of dead time. Our basic form is

$$1 + KP(s)e^{-Ds} = 0 \tag{9.8-4}$$

where $P(s)$ is the ratio of two polynomials with constant coefficients. Write s as $s = \sigma + i\omega$ to show that

$$|e^{-Ds}| = |e^{-D\sigma}e^{-i\omega D}| = e^{-D\sigma} \tag{9.8-5}$$

since

$$|e^{-i\omega D}| = |\cos \omega D - i \sin \omega D| = 1 \tag{9.8-6}$$

Therefore, the magnitude criterion for (9.8-4) is

$$|K| = \frac{1}{e^{-D\sigma}|P(s)|} \tag{9.8-7}$$

The angle criterion of (9.8-4) becomes

$$\measuredangle K + \measuredangle P(s) + \measuredangle e^{-Ds} = \measuredangle(-1) \tag{9.8-8}$$

Note that

$$\begin{aligned} \measuredangle e^{-Ds} &= \measuredangle e^{-D\sigma} + \measuredangle e^{-i\omega D} \\ &= 0^\circ + \measuredangle(\cos \omega D - i \sin \omega D) \\ &= -\omega D \end{aligned} \tag{9.8-9}$$

Thus, for $K \geqslant 0$, (9.8-8) gives

$$\measuredangle P(s) = n180^\circ + \omega D \qquad n = \pm 1, \pm 3, \ldots \tag{9.8-10}$$

For $K \leqslant 0$, we obtain

$$\measuredangle P(s) = n360^\circ + \omega D \qquad n = 0, \pm 1, \pm 2, \ldots \tag{9.8-11}$$

The presence of the imaginary part ω means that the angle criterion depends on the root location. If $D=0$, there are n points in the s plane that satisfy the angle criterion, where n is the order of the denominator of $P(s)$. However, if $D \neq 0$, the presence of ω in the angle criterion means that there will be an infinite number of roots for any given K value.

The plotting guides can be developed as before from the magnitude and angle criteria. We do not state all of the guides explicitly here, because some of them are identical to the case where $D=0$. For example, the symmetry property of Guide 1 is unchanged. The guides for finding the locus on the real axis (Guide 4) are unchanged, because $\omega=0$ on this axis, and thus the angle criterion is identical to the case with no dead time. Similarly, the procedure for determining the locations of breakaway and break-in points remains the same (Guide 5). Also, the crossover points are found with the substitution $s=i\omega$, but the Routh–Hurwitz criterion does not apply (Guide 8). Finally, Guides 9 and 11 apply without changes.

The remaining guides require some modifications. To distinguish them, we denote them as Guide 2b, 3b, and so forth, and simultaneously consider the case for $K \geqslant 0$ and $K \leqslant 0$.

Guide 2b

The number of loci is infinite for a dead-time system. This follows from the discussion of the angle criterion.

Guide 3b

The loci start at the poles of P(s) *and at* $\sigma=-\infty$, *with* $K=0$, *and terminate at the zeros of* P(s) *or at* $\sigma=+\infty$, *with* $K=\pm\infty$. The proof of this guide is similar to that of Guide 3, except that the term $e^{-D\sigma}$ now appears in the magnitude criterion (9.8-7). From this equation, we see that $K=0$ when $\sigma=-\infty$. Similarly, when $\sigma=+\infty$, $K=\pm\infty$. Note that we must determine the asymptotes that describe the starting location at $\sigma=-\infty$. These are found with Guide 6b.

Guide 6b

The loci that do not terminate at a zero approach infinity along asymptotes. There is an infinite number of asymptotes, and they are all parallel to the real axis. The intersections of these asymptotes and the $K=0$ *asymptotes with the imaginary axis are given by*

$$\omega=\frac{N\pi}{D} \tag{9.8-12}$$

Let P *be the number of finite poles and* Z *the number of finite zeros. Then choose* N *as follows. Let* n *take the values* $n=\pm1, \pm3, \ldots$, *and* m *the values* $m=0, \pm2, \pm4, \ldots$. *For the termination asymptotes* ($K=\pm\infty$),

1. $K \geqslant 0$

$$N=n \tag{9.8-13}$$

2. $K \leqslant 0$

$$N=m \tag{9.8-14}$$

For the starting asymptotes (K = *0*),

3. $K = 0+$

$$N = \begin{cases} n & \text{if} \quad P - Z \text{ even} \\ m & \text{if} \quad P - Z \text{ odd} \end{cases} \tag{9.8-15}$$

4. $K = 0-$

$$N = \begin{cases} m & \text{if} \quad P - Z \text{ even} \\ n & \text{if} \quad P - Z \text{ odd} \end{cases} \tag{9.8-16}$$

Note that Guide 7 does not apply, because the asymptotes do not intersect the real axis. With respect to Guide 8, we note that there will be an infinite number of intersections of the locus with the imaginary axis. In general, we need not find all of these, because the dominant roots of the system usually lie on the first path to cross the imaginary axis. This is shown in Example 9.5. Guide 10 applies only to polynomials with a finite number of roots and cannot be applied to a transcendental equation like (9.8-4).

Guide 11b

Once the root locus is drawn, it is scaled with the magnitude criterion (*9.8-7*).

These guides are summarized in Table 9.2.

Example 9.5

Plot the root locus for the equation

$$1 + \frac{Ke^{-Ds}}{s} = 0 \tag{9.8-17}$$

The finite pole is $s = 0$, and there are no finite zeros. The locus for $K > 0$ exists on the real axis to the left of $s = 0$, and that for $K < 0$ exists to the right. On the real axis, with $K \geqslant 0$, the locus starts at $s = 0$ *and also* at $s = -\infty$. These two paths move toward each other, so there must be a breakaway point. This is found from

$$\frac{dK}{ds} = \frac{d}{ds}(-se^{Ds}) = 0$$

or

$$-Dse^{Ds} - e^{Ds} = 0$$

This gives

$$s = -\frac{1}{D}$$

This path will approach one of the asymptotes. To find these, we use Guide 6b. For this problem, $P - Z = 1$. Thus, the termination asymptotes intersect the imaginary

TABLE 9.2 Root Locus Plotting Guides for Systems with Dead Time

Standard Form:

$$1 + KP(s)e^{-Ds} = 0, \qquad -\infty \leqslant K \leqslant \infty$$
$$P(s) = N(s)/D(s)$$
$$N(s) = s^m + b_{m-1}s^{m-1} + \cdots + b_1 s + b_0$$
$$D(s) = s^n + a_{n-1}s^{n-1} + \cdots + a_1 s + a_0$$
$$m \leqslant n$$

Terminology:

The *zeros* of $P(s)$ are the roots of $N(s) = 0$.

The *poles* of $P(s)$ are the roots of $D(s) = 0$.

Angle Criteria: $\measuredangle P(s) = n180° + \omega D,\ n = \pm 1, \pm 3, \ldots$ for $K \geqslant 0$

$\measuredangle P(s) = n360° + \omega D,\ n = 0, \pm 1, \pm 2, \ldots$ for $K \leqslant 0$

$s = \sigma + i\omega$

Magnitude Criterion: $|K| = 1/e^{-D\sigma}|P(s)|$

Plotting Guides:

Guide 1b

The root-locus plot is symmetric about the real axis.

Guide 2b

The number of loci is infinite for a dead-time system.

Guide 3b

The loci start at the poles of P(s) *and at* $\sigma = -\infty$, *with* K = 0, *and terminate at the zeros of* P(s) *or at* $\sigma = +\infty$, *with* K = $\pm\infty$.

Guide 4b

For K $\geqslant$ 0

The root locus can exist on the real axis only to the left of an odd number of real poles and/or zeros; furthermore, it must exist there.

For K $\leqslant$ 0

The locus exists in a section on the real axis only if the number of real poles and/or zeros to the right of the section is even; furthermore, it must exist there.

Guide 5b

For K $\geqslant$ 0

The locations of breakaway and break-in points are found by determining where the parameter K *attains a local maximum or minimum on the real axis.*

For K $\geqslant$ 0

The locations of breakaway and break-in points are found by determining where the parameter K *attains a local minimum or maximum on the real axis.*

Guide 6b

The loci that do not terminate at a zero approach infinity along asymptotes. There is an infinite number of asymptotes, and they are all parallel to the real axis. The intersections of these asymptotes and the K = 0 *asymptotes with the imaginary axis are given by*

$$\omega = \frac{\mathrm{N}\pi}{\mathrm{D}}$$

Let P *be the number of finite poles and* Z *the number of finite zeros. Then choose* N *as follows. Let* n *take the values* n = $\pm 1, \pm 3, \ldots$, *and* m *the values* m = $0, \pm 2, \pm 4, \ldots$. *For the termination asymptotes* (K = $\pm\infty$),

1. $K \geqslant 0$ $\qquad N = n$
2. $K \leqslant 0$ $\qquad N = m$

TABLE 9.2 –*continued*

For the starting asymptotes (K = 0),

3. $K = 0+$ $$N = \begin{cases} n & \text{if} \quad P - Z \text{ even} \\ m & \text{if} \quad P - Z \text{ odd} \end{cases}$$

4. $K = 0-$ $$N = \begin{cases} m & \text{if} \quad P - Z \text{ even} \\ n & \text{if} \quad P - Z \text{ odd} \end{cases}$$

Guide 7b
The asymptotes are parallel to the real axis.

Guide 8b
There will be an infinite number of intersections of the locus with the imaginary axis.

Guide 9b
The angles of departure from complex poles, and the angles of arrival at complex zeros, can be found by applying the angle criterion to a trial point infinitesimally close to the pole or zero in question.

Guide 10b
For dead-time systems, there is no guide corresponding to Guides 10 and 10a.

Guide 11b
Once the root locus is drawn, it is scaled with the magnitude criterion.

---

axis at $\omega = N\pi/D$ where

$$N = 0, \pm 2, \pm 4, \ldots \qquad \text{for } K \leqslant 0$$
$$N = \pm 1, \pm 3, \ldots \qquad \text{for } K \geqslant 0$$

The starting asymptotes intersect the axis at the points given by

$$N = \pm 1, \pm 3, \ldots \qquad \text{for } K \leqslant 0$$
$$N = 0, \pm 2, \pm 4, \ldots \qquad \text{for } K \geqslant 0$$

These are shown by dotted lines in Figure 9.24.

To find the crossover point, let $s = i\omega$ to obtain from (9.8-17),

$$i\omega + Ke^{-i\omega D} = 0$$

or

$$i(\omega - K \sin \omega D) + K \cos \omega D = 0$$

The crossover points are infinite in number and are given by

$$\cos \omega D = 0$$

$$K = \frac{\omega}{\sin \omega D}$$

Thus, if $K > 0$, the crossover points for $\omega > 0$ are

$$\omega D = \frac{\pi}{2}, \frac{5\pi}{2}, \ldots$$

$$K = \omega$$

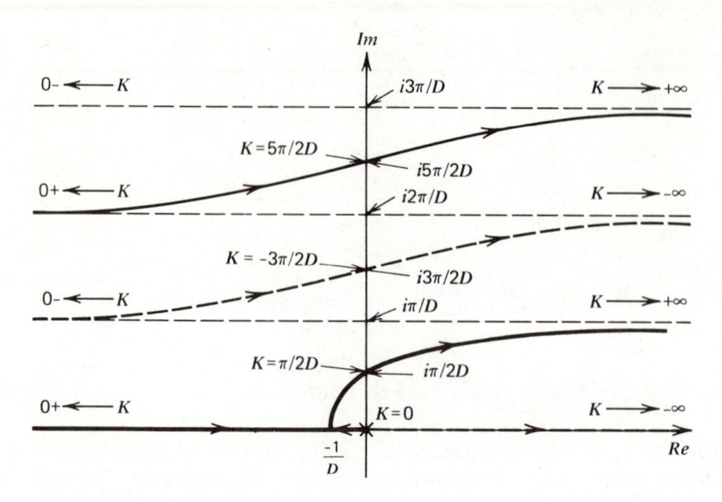

FIGURE 9.24 Root locus plot for $s + Ke^{-sD} = 0$, $-\infty \leqslant K \leqslant \infty$.

For $K < 0$, they are

$$\omega D = \frac{3\pi}{2}, \frac{7\pi}{2}, \ldots$$

$$K = -\omega$$

for $\omega > 0$. For $\omega < 0$, the crossing points are the negative of the preceding values.

The complete root locus can now be sketched. It is shown in Figure 9.24, where the heavy dotted lines are the loci for $K < 0$. The system is stable if $0 < K < \pi/2D$, the value for the gain at the first crossover point. The plot shows that the roots on the branches that lie outside the region $-\pi \leqslant \omega D \leqslant \pi$ are far to the left of the roots on the primary branch for stable values of K. Thus, we usually consider only the primary branch when choosing a value for K. The selection of K is made in the same way as before.

The root locus shows that the system will always exhibit oscillatory behavior because of the infinite number of paths with nonzero imaginary parts. However, if the roots on the primary branch are selected to be real, the oscillations will not dominate the response.

Frequency Response of Dead-Time Elements

The Nyquist theorem is particularly useful for systems with dead-time elements, especially when the plant is of an order high enough to make the root locus method cumbersome. A delay in either the manipulated variable or the measurement will result in an open-loop transfer function of the form

$$G(s)H(s) = e^{-Ds}P(s) \tag{9.8-18}$$

For this case, (9.8-6) and (9.8-9) show that

$$|G(i\omega)H(i\omega)| = |P(i\omega)||e^{-i\omega D}| = |P(i\omega)| \tag{9.8-19}$$

and

$$\begin{aligned} \measuredangle G(i\omega)H(i\omega) &= \measuredangle P(i\omega) - \measuredangle e^{-i\omega D} \\ &= \measuredangle P(i\omega) - \omega D \end{aligned} \tag{9.8-20}$$

Thus, the dead time decreases the phase proportionally to the frequency ω, but it does not change the gain curve. This makes the analysis of its effects easier to accomplish with the open-loop frequency response plot.

Example 9.6 shows how these methods can be applied to a system whose transfer function cannot be obtained analytically.

Example 9.6

The frequency response for a particular plant was determined experimentally, and the results are shown in Figure 9.25*a*. It is intended to use proportional control for this plant with a unity feedback loop, as shown in Figure 9.25*b*. It is known that there will be dead time between the controller action and its effect on the plant.

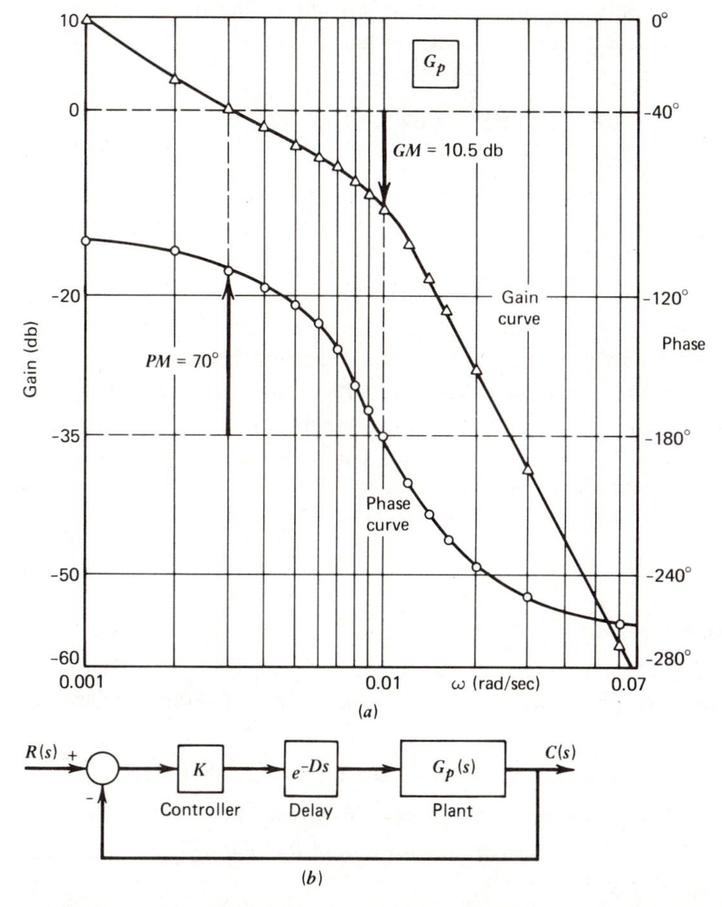

FIGURE 9.25 Design information for Example 9.6. (*a*) Experimentally determined open-loop frequency response for the plant. (*b*) Proportional controller with unity feedback and dead time.

(a) Design the controller to achieve the specifications given by (9.1-2), namely, $GM \geqslant 8$ db and $PM \geqslant 30°$.

(b) How large can the dead time be before the system becomes unstable?

(a) Neglect the dead-time for now and design the controller. The plot in Figure 9.25*a* shows that if the gain K is unity, the phase crossover frequency ω_p is 0.01 rad/sec. and the gain crossover occurs at $\omega_g = 0.003$ rad/sec. The phase margin will be 70° and the gain margin 10.5 db.

Increasing the value of K above unity affects only the gain curve. This curve can be raised by $10.5 - 8 = 2.5$ db without violating the specifications given for PM and GM. Thus, for $GM = 8$ db, K must be such that $20 \log K = 2.5$, or $K = 1.334$. With this value of K, the phase margin becomes 60°, and the new gain crossover frequency is $\omega_g = 0.0045$ rad/sec. This can be seen by translating the gain curve upward by 2.5 db.

(b) From (9.8-9) and (9.8-10), we see that the dead time affects only the phase margin. The system is stable only if the phase margin is positive. This occurs only if the dead time's contribution to the phase curve at the *new* gain crossover frequency is greater than $-60°$. The stability requirement thus is

$$0.0045D \leqslant 60° = 1.047 \text{ rad}$$

or

$$D \leqslant 232.7 \text{ sec}$$

9.9 SUMMARY

Control system design is greatly facilitated by using the root locus and frequency response plots. The performance of different control laws and compensation techniques can be quickly compared especially if the plots are generated with an interactive computer program. The designer needs some intuition acquired through experience in order to make the most of these classical design methods, but, nevertheless, they have proved to be the most successful methods to date.

REFERENCES

1. Y. Takahashi, M. Rabins, and D. Auslander, *Control*, Addison-Wesley, Reading, Mass., 1970.
2. S. H. Lehnigk, *Stability Theorems for Linear Motions*, Prentice-Hall, Englewood Cliffs, N.J., 1966.
3. R. Oldenburger (ed.), *Frequency Response*, Macmillan, New York, 1956.
4. J. G. Truxal (ed.), *Control Engineers Handbook*, McGraw-Hill, New York, 1958.
5. H. Chestnut and R. W. Mayer, *Servomechanisms and Regulating System Design*, vol. I, 2d ed., John Wiley, New York, 1960.
6. *ASME Quarterly Transactions on Dynamic Systems, Measurement, and Control* (formerly the *Journal of Basic Engineering*), ASME, New York.

7. *IEEE Monthly Transactions on Automatic Control*, IEEE, New York.
8. *Control Engineering* (monthly), Penton Press, Cleveland.
9. *Instruments and Control Systems* (monthly), Chilton Publishing Co., Philadelphia.
10. *Automatica* (monthly), Pergamon, Elmsford, N.Y.
11. H. M. James, N. B. Nichols, and R. S. Philips, *Theory of Servomechanisms*, McGraw-Hill, New York, 1947.

PROBLEMS

9.1 Two mechanical compensators are shown in Figure P9.1. Derive the transfer functions for each. The input is the displacement x_i, and the output is the displacement x_o. Show that the system in (a) is a lead compensator and the one in (b) is a lag compensator.

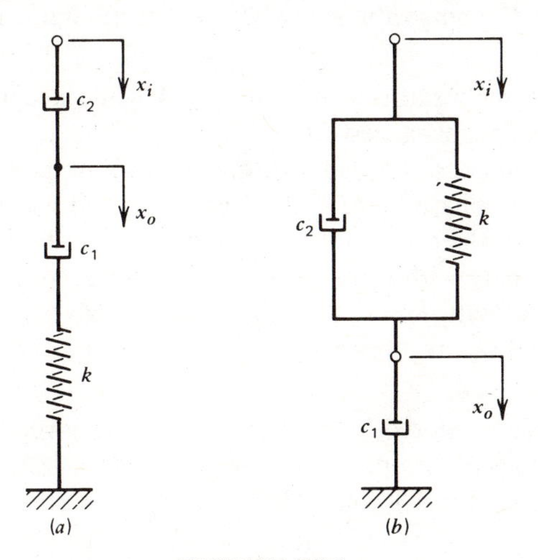

FIGURE P9.1

9.2 Consider a plant whose open-loop transfer function is

$$G(s)H(s) = \frac{1}{s(s^2 + 4s + 13)}$$

The complex poles near the origin give only slightly damped oscillations that are considered undesirable. Insert a gain K_c and a compensator $G_c(s)$ in series to speed up the closed-loop response of the system. Consider the following for $G_c(s)$:

(a) The lead compensator (9.4-1).
(b) The lag compensator (9.4-4).
(c) The so-called reverse-action compensator

$$G_c(s) = \frac{1 - T_1 s}{T_2 s + 1}$$

Sketch the root locus plots for the compensated system using each compensator. Use K_c as the locus parameter. Without computing specific values for the compensator parameters, determine which compensator gives the best response.

9.3 The transfer functions for the most basic lead and lag compensators are

(a) $T(s) = K(\tau s + 1)$ (lead)

(b) $T(s) = \dfrac{K}{\tau s + 1}$ (lag)

Plot the Bode diagrams for these and compare with those of the compensator circuits shown in Figures 9.8–9.11. Discuss the relative performance of each.

9.4 A particular control system with unity feedback has the open-loop transfer function

$$G(s) = \frac{K}{s(s+2)}$$

With $K = 4$, the damping ratio is $\zeta = 0.5$, the natural frequency is $\omega_n = 2$ rad/sec, and $C_v = 2$/sec.

(a) Design a compensator to obtain $\omega_n = 4$ while keeping $\zeta = 0.5$. Find the compensator's resistances if $C = 1\ \mu f$.

(b) Design a compensator that will give a static velocity error coefficient of $C_v = 20$/sec, a phase margin of at least 40°, and a gain margin of at least 6 db. How do the open-loop poles and zeros of the resulting compensated system compare with those of the system designed in (a)?

(c) Suppose that with $K = 4$, the original system gives a satisfactory transient response, but C_v must be increased to $C_v = 20$/sec. Design a compensator to do this.

9.5 The bandwidth of a closed-loop system is usually given roughly by the gain crossover frequency of the open-loop system. This is a useful design guide. Design a lead compensator for a system whose open-loop transfer function is

$$G(s)H(s) = \frac{72}{(s+1)(s+3)^2}$$

The compensated system should have a phase margin of 40° or more and a closed-loop bandwidth approximately the same as the gain crossover frequency of the open-loop uncompensated system.

9.6 Consider the system whose open-loop transfer function is given in the previous problem. Design a lag–lead compensator to give a phase margin of no less than 40°, $C_p \geqslant 10$, and the same gain crossover frequency as with the uncompensated system.

9.7 A unity feedback system has the open-loop transfer function

$$G(s) = \frac{0.625}{s(0.5s+1)(0.125s+1)}$$

Design a compensator to give $s = -2 \pm i2\sqrt{3}$ and $C_v = 80$/sec.

9.8 A feedback system with dead time is shown in Figure P9.8.

(a) Find the transfer function.

(b) The characteristic equation for this system has an infinite number of roots. Show this by writing the root as a complex number $s = a + ib$ and obtaining the equations that must be solved to find a and b.

(c) For $D = 0.2$, set $b = 0$, and use Newton iteration to find the only real characteristic root (see Appendix A for the Newton iteration procedure for finding the roots of an equation). Estimate the system's time constant.

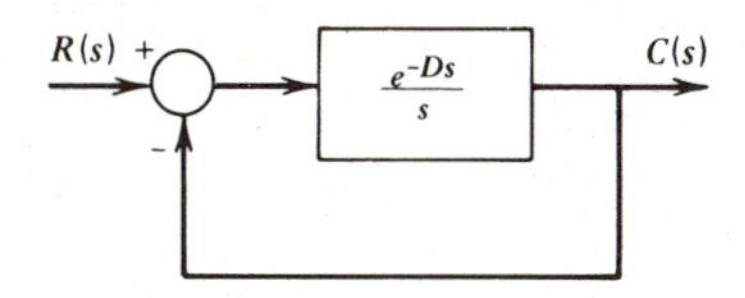

FIGURE P9.8

9.9 Sketch the root locus plot for the given characteristic equation for $K \geqslant 0$. Use all the guides that are applicable.

(a) $s + 1 + Ke^{-Ds} = 0 \qquad D = 1$

(b) $s(s + 1) + Ke^{-Ds} = 0 \qquad D = 1$

9.10 Sketch the root locus plot for $K \leqslant 0$ (the complementary locus) for the given characteristic equation. Use all applicable guides.

(a) $s + 1 + Ke^{-Ds} = 0 \qquad D = 1$

(b) $s(s + 1) + Ke^{-Ds} = 0 \qquad D = 1$

9.11 A control system with a dead time of D sec has the open-loop transfer function

$$G(s)H(s) = \frac{10e^{-sD}}{0.1s + 1}$$

How large can the dead time be if the phase margin must be at least 40°?

9.12 The Pade approximation to the dead-time element is

$$e^{-sD} = \frac{1}{[1 + (Ds/n)]^n}$$

The approximation improves as n becomes larger. For relatively high-order systems with dead time, the techniques presented in this chapter can be difficult to apply, and the Pade approximation for low values of n is often used to model the effects of dead time.

Use the Pade approximation to construct the root locus for the system

$$G(s)H(s) = \frac{Ke^{-s}}{s}$$

for $n = 1$ and $n = 2$. Compare with the exact plot given in Figure 9.24.

CHAPTER TEN

Discrete-Time Models and Sampled-Data Systems

Using a digital computer as a controller is widespread and continues to increase. As we will see in Chapter 11, digital control offers several advantages over the analog methods in Chapters 6–9. These include increased flexibility of the control algorithm, because it exists only in software and can be readily changed as the requirements or the applications change. Also, a hardware analog of the algorithm need not be designed. In many cases, the digital controller is also less expensive.

Designing digital controllers requires some additional considerations not found in analog designs. The need for *analog-to-digital* (A/D) and *digital-to-analog* (D/A) conversion of the measurement and control signals involves the processes of *sampling* and *quantization.* The presence of sampling requires us to look at the system as a *sampled-data* system, and this means that continuous-time models of the system are not so convenient to use as a *discrete-time* model. Therefore, Chapter 10 begins with the study of discrete-time models: their origin and the analysis methods needed to use them to design digital controllers. We will use the same approach in developing analytical methods for discrete-time models as was used for the continuous-time case. The emphasis is initially on first-order models, so that the fundamental concepts are not lost in a jungle of algebraic manipulations. Then the results are extended to higher order cases. Engineering examples of sampled-data systems are used throughout the chapter to illustrate the analysis techniques and to prepare for the treatment of digital control systems in Chapter 11.

10.1 ORIGIN OF DISCRETE-TIME MODELS

Discrete-time models can arise from the description of systems that are inherently of a discrete-time nature, in the analysis of digital measurement and control systems, and in numerical solution methods for differential equations.

Intrinsically Discrete Systems

In the models we have encountered in previous chapters, the independent variable (time) was continuous. This is the natural situation in physical systems. However, in some circumstances, particularly with nonphysical systems, it is more convenient to conceive of events as occurring at discrete instants. An imposed sequencing or scheduling of events naturally leads to a discrete formulation of the time variable. For example, a bank might compute interest for a savings account once per year on the amount in the account at the end of the year. If $y(k)$ represents the amount of money in the account at the beginning of the year k, then $y(1)$ is the amount initially deposited, $y(2)$ is the amount after the first interest payment is added, and so forth. If interest is computed every three months (quarterly), then $y(2)$ would still represent the amount after the first interest payment. That is, in discrete-time models we normally use integer values to represent the instants at which events occur. We will see shortly that such a formulation allows us to compute easily the total amount in the account after any specified number of interest periods.

As another example, a student's progress through a college curriculum is indicated by the number of semesters completed. Suppose that $y(k)$ represents the number of mechanical engineering majors who have just finished semester k, where $k = 1$ represents the first semester of freshman year. Then $y(8)$ represents the number of graduating seniors. Note that the time index k is quite distinct from physical time, because the length of time between semesters is variable (due mainly to the summer recess). Such a model type is said to be event-oriented, not time-oriented. Although this representation ignores changes that occur between the discrete time points, such as a student who leaves at midsemester, the formulation would be useful to a registrar who needs to allocate classroom space for the following semester. A continuous-time model of student enrollment would in fact be too cumbersome for the registrar's purposes.

Digital Systems

The widespread use of digital computers has had a profound impact on engineering practice and education. With the computational burden eased by computers, it is now possible to use many numerical analysis techniques whose potential was previously unrealized. Also, with the recent breakthroughs in circuit miniaturization, digital devices are now commonly used in experiments and industrial processes for data collection and system control. Using computers and other digital devices in control systems is discussed in Chapter 11. We now introduce some basic digital devices and the mathematical methods needed to analyze their behavior.

Digital devices can handle mathematical relations and operations only when expressed as a finite set of numbers rather than as infinite-valued functions. Thus, any continuous measurement signal must be converted into a set of pulses by *sampling* — the process by which a continuous-time variable is measured at distinct, separated instants of time. The sequence of measurements replaces the smooth curve of the measured variable versus time, and the infinite set of numbers represented by the smooth curve is replaced by a finite set of numbers. This is done because a digital device cannot store a continuous signal. Each pulse amplitude is then rounded off to one of a finite number of levels depending on the characteristics of the machine. This is called *quantization*. Consequently, a digital device is one in which the signals are quantized in both time and amplitude. In an analog device, the signals are analog; that is, they are continuous in time and are not quantized in amplitude.

Figures 10.1*a*–10.1*c* show an example of a analog signal, the sampling pulses, and the pulses after quantization for a simplified hypothetical machine with only four quantization levels (0, 1, 2, and 3). The sampling is done at times $t_0, t_1, t_2, \ldots$, and the index k denotes these sampling times by $k = 0, 1, 2, \ldots$. The quantized signal is shown in Figure 10.1*c*, where we have assumed that any value in the interval $[0.5, 1.5)$ is rounded off to 1, and so forth. After quantization, the signal is *coded* by converting the

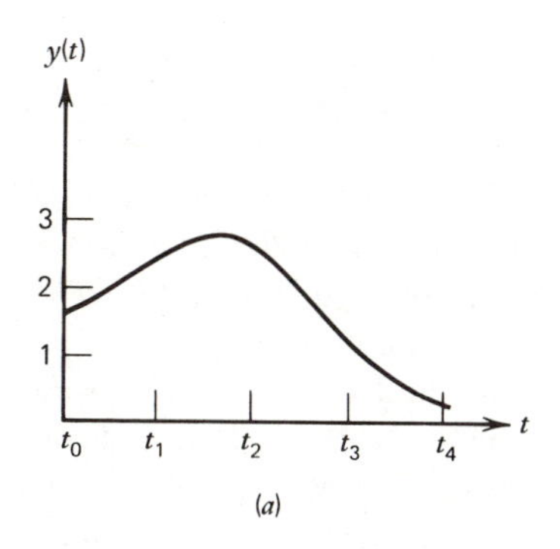

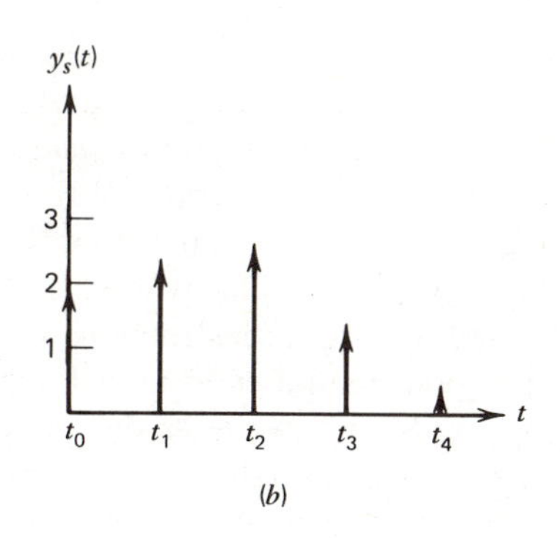

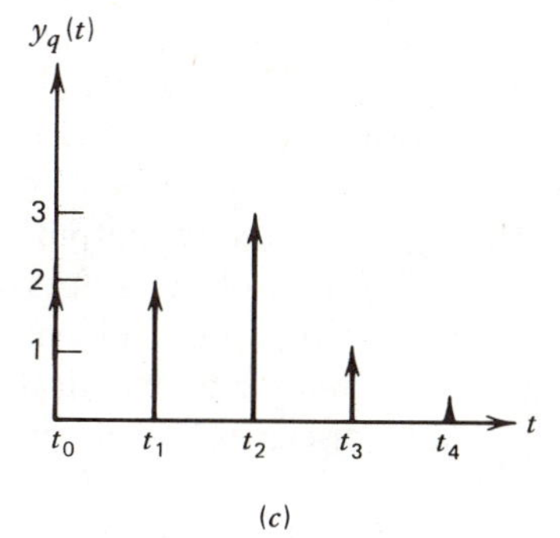

	$y_q(t_k)$	
k	Decimal	Binary
0	2	10
1	2	10
2	3	11
3	1	01
4	0	00

(d)

FIGURE 10.1 Sampling, quantization, and coding of an analog signal. (*a*) Original analog signal as a function of time. (*b*) Sampled signal. (*c*) Quantized signal for a hypothetical device with two bits. (*d*) Decimal-coded and binary-coded representations of the quantized signal.

quantization level of each pulse (0, 1, 2, or 3 here) into an equivalent binary number that the digital device can accept and store. Figure 10.1*d* shows the resulting binary number sequence for our example at the discrete times $k = 0, 1, 2, 3, 4, \ldots$. This is only one example of a binary coding scheme; several are in common use (Reference 1). The device that performs the sampling, quantization, and coding is an *analog-to-digital (A/D) converter*.

The number of binary digits carried by the machine is its *word length*, and this is obviously an important characteristic related to the device's resolution—the smallest change in the input signal that will produce a change in the output signal. The hypothetical machine in Figure 10.1 has two binary digits and thus four quantization levels. Any change therefore in the input over the interval $[0.5, 1.5)$ produces no change in the output. With three binary digits, 2^3, or 8 quantization levels can be obtained, and the resolution of the device would improve.

The A/D converters in common use have a word length of 10 bits (a bit is one binary digit). This means that an input signal can be resolved to one part in 2^{10}, or 1 in 1024. If the input signal has a range of 10 V, the resolution is 10/1024, or approximately 0.01 V. Thus the input must change by at least 0.01 V in order to produce a change in the output. High-resolution devices are available with more bits, but their cost is higher and their speed of conversion is slower.

A concept pertinent to A/D conversion is *filtering*, a collection of techniques for removing the effects of noise and measurement error from a signal. For example, the series *RC* circuit seen in Chapter 2 is used to filter out noise from voltage signals and it is an analog filter. In Chapter 11, we will learn how to use an algorithm called a *digital filter*, which is stored in the software of the computer. Thus, the difference between analog and digital filtering is that the latter is performed by a physical device (a circuit) inserted into the system specifically for the purpose of filtering, while digital filtering can be performed by a set of instructions stored in the computer.

When a digital computer is used for measurement and control, the natural question to ask is: Should we filter the measurement signals with an analog device before they are passed through an A/D converter, or should the A/D conversion be done on the noisy measurements with the filtering done afterward by a digital algorithm? Digital signals are more easily detected than analog signals and more resistant to noise degradation in transmission and computing. Thus, since A/D conversion must eventually take place if a digital computer is involved, the trend is toward digital filtering in such applications.

In many digital system applications, it is necessary to convert binary outputs from the digital device into a usable form for the equipment being controlled (motors, heaters, etc.). This usually means a continuous or piecewise-continuous voltage. A *digital-to-analog (D/A) converter* performs this function. A common mathematical algorithm used in a D/A converter is the *zero-order data hold*, discussed in Section 10.6.

The preceding discussion deals with hardware considerations. Another implication for digital systems concerns the proper form of models to be used in light of the discrete nature of digital operations versus the continuous nature of the physical devices and processes being measured and controlled by the computer. One view is that the engineer from the start should model the system to be measured or controlled by a set of discrete-time equations called difference equations. This model form is immediately compatible with the digital operations of the computer. The difficulty with this approach is that physical laws are not often stated in discrete-time form. For example, displacement and velocity are continuous variables, and Newton's laws of motion are naturally stated as differential equations. Thus, the other view is that the initial system model should be in the form in which the appropriate physical laws are most conveniently and clearly stated. If this form is a set of differential equations, the numerical analysis techniques introduced in Chapter 5 can be used to convert them into a discrete-time formulation. This procedure mimics what actually happens in the physical system where sampling converts the continuous physical variables to discrete pulses. Because of this correspondence, here we will adopt the second viewpoint.

Numerical Solution of Continuous-Time Models

Just as sampling is used to convert a continuous-time variable into a discrete-time variable, a similar process is used to convert a continuous-time model (a differential equation) into a discrete-time model. This form allows a solution to be obtained with a digital computer or calculator (inherently discrete-time devices). This technique is useful when a closed-form solution is not obtainable or assumptions required for the linearization of a nonlinear model are not satisfied.

Several algorithms for obtaining difference equations from differential equations were presented in Chapter 5. Recall, for example, the Euler method applied to the model

$$\frac{dy}{dt} = ry \qquad r = \text{constant} \tag{10.1-1}$$

If the time increment Δt is chosen small enough, the derivative can be replaced by the approximate expression

$$\frac{dy}{dt} \cong \frac{y(t + \Delta t) - y(t)}{\Delta t} \tag{10.1-2}$$

Replacing the left-hand side of (10.1-1) with this expression, we obtain

$$y(t + \Delta t) = y(t) + ry(t)\,\Delta t \tag{10.1-3}$$

This can be written in several forms. If (10.1-2) is applied successively at the instants t_k, where $t_{k+1} = t_k + \Delta t$, we can write the model as

$$y(t_{k+1}) = (1 + r\,\Delta t)\,y(t_k) \tag{10.1-4}$$

or

$$y(k + 1) = (1 + r\,\Delta t)\,y(k) \tag{10.1-5}$$

where $y(k)$ represents $y(t)$ evaluated at $t = t_k$, $k = 0, 1, 2, \ldots$.

While (10.1-5) can be solved recursively in a straightforward manner, it is tedious to do this if we wish to compute y for a large value of k. Therefore, more sophisticated solution techniques are needed, and they are developed in this chapter.

10.2 FREE RESPONSE

Suppose that in our savings account example the bank compounds interest annually at 5%. If we have invested one dollar initially, $y(0) = \$1.00$, after one year we would have accumulated an amount

$$y(1) = \$1.00 + 0.05(\$1.00) = \$1.05$$

At the end of the second year the amount would be

$$y(2) = \$1.05 + 0.05(\$1.05) = \$1.1025$$

In general, an initial amount $y(0)$ invested at 5% will yield at the end of the first year an amount

$$y(1) = y(0) + 0.05y(0) = 1.05y(0)$$

and for the second year

$$y(2) = y(1) + 0.05y(1) = 1.05y(1)$$

From this, it is easy to obtain the general formula

$$y(k) = y(k - 1) + 0.05y(k - 1)$$

or

$$y(k) = 1.05y(k-1) \qquad k = 1, 2, 3, \ldots \tag{10.2-1}$$

We can calculate the accumulated savings at any year k either by repetitive multiplication by the factor 1.05 or by realizing that the process can be described by the formula

$$y(k) = (1.05)^k y(0) \tag{10.2-2}$$

This apparently is the solution of (10.2-1) with the initial condition $y(0)$. The form is convenient because it allows the amount in any year k to be computed in one step. With this insight, we can hypothesize that the solution of the difference equation

$$y(k) = ay(k-1) \qquad k = 1, 2, 3, \ldots \tag{10.2-3}$$

is

$$y(k) = a^k y(0) \qquad k = 0, 1, 2, 3, \ldots \tag{10.2-4}$$

when a is a constant. This is affirmed by noting that (10.2-4) implies

$$y(k-1) = a^{k-1}y(0) = a^{-1}[a^k y(0)] = a^{-1}y(k)$$

which is equivalent to (10.2-3). Equation (10.2-4) also satisfies the initial condition at $k = 0$, since $a^0 = 1$.

Equation (10.2-3) represents a *linear first-order difference equation.* The term *linear* has the same meaning for difference equations as it does for differential equations. Similarly, a linear difference equation can be recognized as a linear relation between the state variables and inputs (if any). The *order* of a difference equation is the maximum difference between the time indices in the equation. Thus, the equation

$$y(k) = 5y(k-1)y(k-2)$$

is *second order* and nonlinear as a result of the cross-product term between $y(k-1)$ and $y(k-2)$.

The solution behavior of the difference equation (10.2-3) differs greatly from that of the corresponding differential equation (10.1-1). Seven types of behavior can arise from (10.2-4) depending on the value of a.

1. $a > 1$. The solution's magnitude grows with time, and the solution keeps the sign of $y(0)$.
2. $a = 1$. The solution remains constant at $y(0)$.
3. $0 < a < 1$. The solution decays in magnitude and keeps the sign of $y(0)$.
4. $a = 0$. The solution jumps from $y(0)$ to zero at $k = 1$ and remains there.
5. $-1 < a < 0$. The magnitude decays, but the solution alternates sign at each time step (an oscillation with a period of two time units).
6. $a = -1$. The magnitude remains constant at $y(0)$, but the solution alternates sign at each step.
7. $a < -1$. The magnitude grows, and the sign of the solution alternates at each time step.

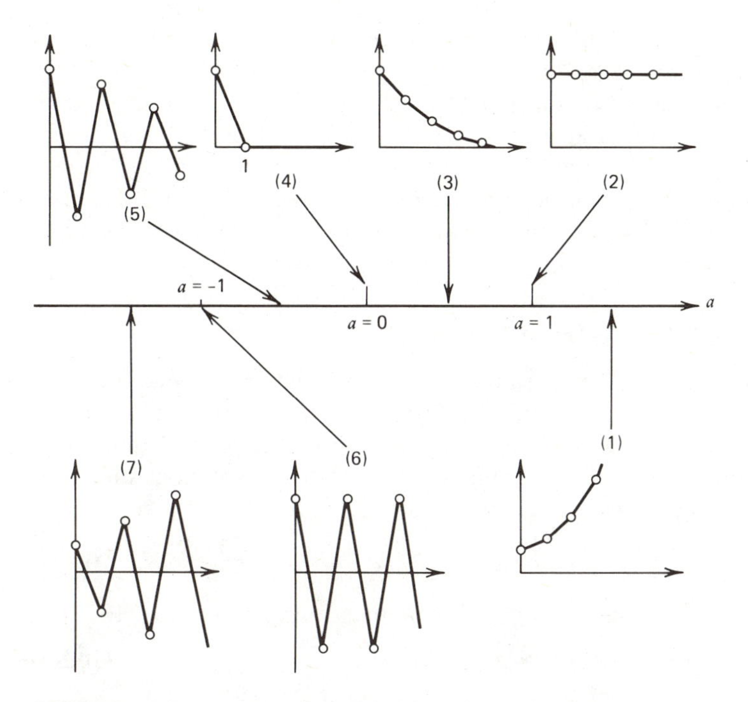

FIGURE 10.2 Free response of the linear first-order equation (10.2-4) for various values of the coefficient *a*. Numbered cases refer to the discussion in the text.

This behavior is summarized in Figure 10.2 for a positive $y(0)$. The smaller plots represent $y(k)$ versus k for each of the seven numbered cases. The small circles represent the solution. The lines connecting the circles are visual aids only. The difference equation makes no statement about what happens between integer values of k.

Stability

The *concepts* of point equilibrium and stability can be immediately transferred from the differential equation case (Section 3.1). However, *determining* a point equilibrium and its stability properties is done differently. Since a point equilibrium is a condition of no change, it is found for a differential equation by setting the time derivatives to zero. For a difference equation, it is found by dropping all time indices; that is, by setting $y(k) = y(k-1) = y_e$ for all $k \geqslant 0$, where y_e is the equilibrium value. With (10.2-3), this implies that

$$y_e = ay_e$$

which is satisfied if either

$$1 = a$$

or

$$y_e = 0$$

Thus, if $a \neq 1$, the only point equilibrium possible for (10.2-3) is $y_e = 0$. If $a = 1$, any value of y is an equilibrium. This can be seen in Figure 10.2, case (2).

The criterion for stability of the point equilibrium can be ascertained from Figure 10.2. For an equilibrium at $y = 0$, the only cases in which the solution approaches and remains at equilibrium are the cases where $-1 < a < 1$. In other words, the model is *stable* if $|a| < 1$, *unstable* if $|a| > 1$, and *neutrally stable* if $|a| = 1$. Contrast this with the criterion for (10.1-1). For that equation, instability occurs if $r > 0$, stability if $r < 0$, and neutral stability if $r = 0$.

Relation to the Continuous-Time Model

It is apparent from Figure 10.2 that simple difference equations produce types of behavior not found in simple differential equations. In (10.1-1) no oscillations are possible. But since it is possible to derive a difference equation from a differential equation—with the Euler method, for example—then their two solutions should be related somehow. For comparison, the solution of (10.1-1) is repeated here.

$$y(t) = y(0)e^{rt} \tag{10.2-5}$$

If time t_k corresponds to the discrete time index k, then from (10.2-4) and (10.2-5)

$$y(t_k) = y(0)e^{rt_k} \tag{10.2-6}$$

and

$$y(t_k) = y(k) = a^k y(0)$$

Comparing these two expressions gives

$$e^{rt_k} = a^k$$

or

$$rt_k = \ln a^k = k \ln a \tag{10.2-7}$$

This indicates that a comparison between the two models can be made only if $a > 0$, since $\ln a$ is undefined for $a \leqslant 0$. This is not unexpected because oscillatory behavior occurs in the discrete-time model only if $a < 0$. Equation (10.2-7) can be used to find the value of r required to make a continuous-time process equivalent to one in discrete time, and vice versa.

Example 10.1

One bank offers a savings plan with interest compounded *continuously*, while another offers a plan with 5% interest compounded *annually*. What rate must be used for continuous compounding to make it equivalent to 5% annual compounding? For annual compounding at 5%, how much is an investment worth after 10 years?

For annual compounding at 5%, $a = 1.05$ and

$$y(k) = 1.05y(k-1)$$

where k measures the number of years of investment. For continuous compounding, the growth of the investment is described by (10.1-1), where r is the instantaneous rate of growth (the interest rate). Let t_k measure the number of years of investment. Then $k = 1$ corresponds to $t_k = 1$, and from (10.2-7),

$$r = \ln 1.05 = 0.04879 = 4.879\%$$

Thus, continuous compounding at 4.879% is equivalent to 5% compounded annually. After $k = 10$ years, the initial investment $y(0)$ has grown to $y(10) = (1.05)^{10} y(0) = 1.62889 y(0)$. The investment has increased by approximately 163%.

It is now clear that the form z^k plays the same fundamental solution role in linear difference equations that the form e^{st} plays in linear differential equations. We will see that this also holds true for higher order equations, and thus a firm grasp of the first-order model's behavior will greatly promote understanding behavior in higher order models. The differences and similarities of the two first-order models are summarized in Table 10.1.

The form of the first-order equation given by (10.1-5) is equivalent to (10.2-3) with $a = 1 + r\Delta t$ and the index k starting from 1 instead of 0. In general, the linear model given by (10.2-3) is equivalent to

$$y(k+1) = ay(k) \qquad k = 0, 1, 2, \ldots$$

Both equations have the same solution: (10.2-4).

TABLE 10.1 Comparison of Differential and Difference Equations

Characteristic	Differential Equation $\dot{y} = ry$	Difference Equation $y(k) = ay(k-1)$
1. Solution	$y(t) = y(0)e^{rt}$	$y(k) = y(0)a^k$
2. Solution behavior	No oscillation	Oscillations of period two if $a < 0$
3. Stability		
Stable if	$r < 0$	$\lvert a\rvert < 1$
Neutrally stable if	$r = 0$	$\lvert a\rvert = 1$
Unstable if	$r > 0$	$\lvert a\rvert > 1$
4. Time constant (time to decay by 63%)	$t = \tau = -1/r,\ r < 0$	$k = -1/\ln\lvert a\rvert,\ \lvert a\rvert < 1$
5. Relation between the two models	(a) time t_k corresponds to time k (b) $rt_k = k \ln a$ if $a > 0$. (c) no relation if $a \leqslant 0$.	

10.3 SAMPLING

Sampling extracts a discrete-time signal from a continuous-time signal. If the sampling frequency is not selected properly, the resulting sample sequence will not accurately represent the original signal. Fortunately the proper frequency is readily determined in many cases by means of the sampling theorem, which we present shortly.

Impulse Representation of Sampling

Sampling can be represented by the opening and closing of a switch, as shown in Figure 10.3. In digital systems this is accomplished electronically, not mechanically. Let $y(t)$ be the continuous-time signal and $y^*(t)$ the discrete-time signal resulting from sampling $y(t)$ briefly every T seconds. *Uniform sampling* occurs when the sampling period T is constant. Only this case will be treated here. Figure 10.3*a* shows the three processes associated with analog-to-digital conversion. These processes correspond to the plots in Figure 10.1. The digital signal that results from the coding process is $y_d(t)$. Since the quantization and coding algorithms used to obtain $y_d(t)$ are not germane to the purposes of block diagram representation, the simple representation of Figure 10.3*b* is often used, and $y^*(t)$ is taken to be equivalent to $y_d(t)$.

Figure 10.4 illustrates the sampling process. Figure 10.4*a* shows the signal to be sampled. In Figure 10.4*b*, the duration Δ represents the length of time the sampling switch remains closed. The circuitry in the sampler is designed to take this duration into account, and thus the sampler output corresponds to an average value of y over the interval Δ (Reference 1). One representation of the sampling process considers the sampler outputs to be impulses with a strength equal to the sampled value of y at the sampling times. This is shown in Figure 10.4*c*.

Let $\delta(t)$ be the unit-impulse function. Then $\delta(t - kT)$ is a unit impulse occurring at time kT. The sampled sequence, denoted $y^*(t)$, can be represented by a train of impulses occurring at the sampling times kT, each having a strength equal to the

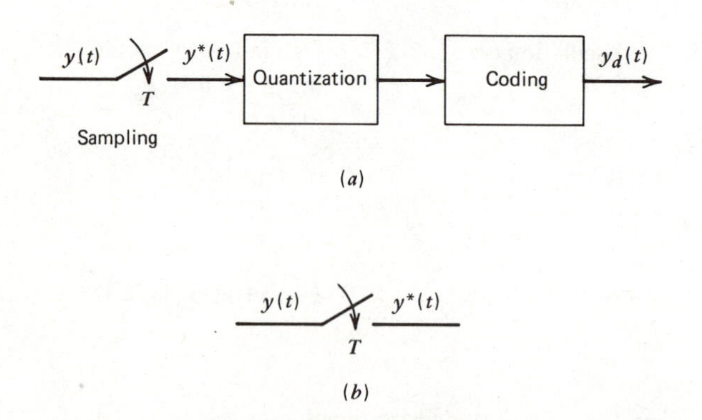

FIGURE 10.3 Block diagram representations of analog-to-digital conversion. (*a*) Full representation showing the sampling, quantization, and coding processes. (*b*) The sampler representation. This symbol is normally used instead of that shown in (*b*).

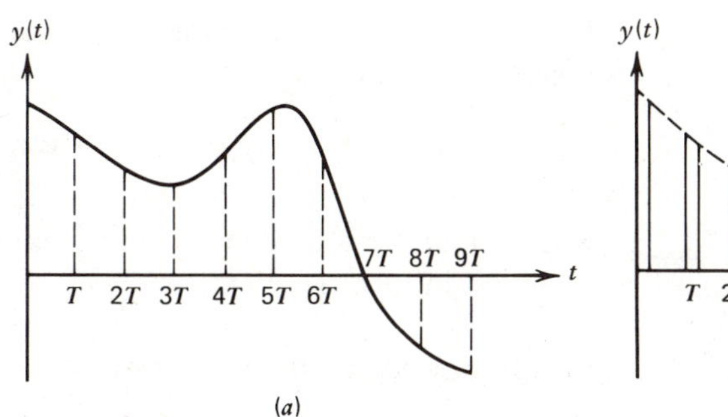

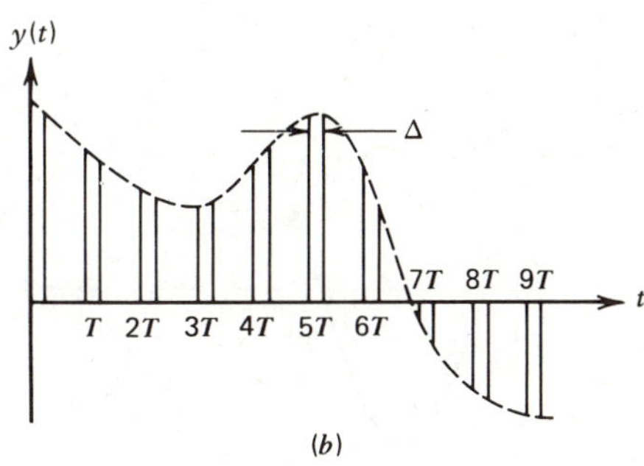

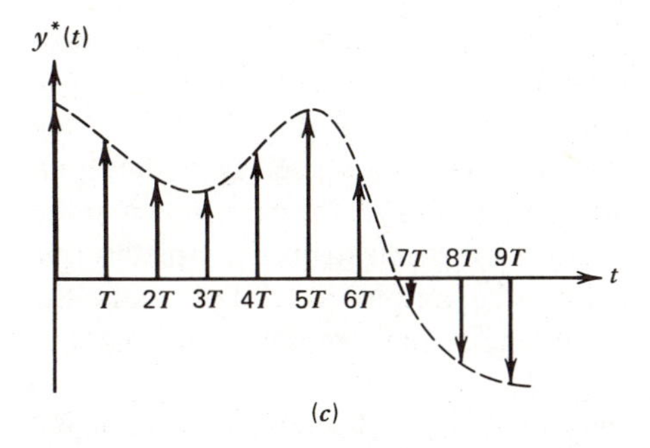

FIGURE 10.4 Sampling process. (*a*) The original signal. (*b*) Illustration of the sampling duration Δ. (*c*) Impulse representation of the sampled signal.

sample value at that time; that is,

$$y^*(t) = y(0)\delta(t) + y(T)\delta(t-T) + y(2T)\delta(t-2T) + \cdots$$
$$= \sum_{i=0}^{N} y(iT)\delta(t-iT) \qquad \textbf{(10.3-1)}$$

where N is the number of samples.

Only the values of $y(t)$ at the sampling instants are retained by the digital memory. Thus, the continuous time variable t is no longer needed and can be replaced with the discrete index k, where $t = kT$. The times $t = 0$, T, 2T, ... correspond to $k = 0, 1, 2, \ldots$, and (10.3-1) can be written as

$$y^*(kT) = \sum_{i=0}^{N} y(iT)\delta(kT - iT) \qquad \textbf{(10.3-2)}$$

In this form, the sampling period T is obviously superfluous once it is specified, and we can simply use the index k, as was done with the Euler method (10.1-4) and (10.1-5). In this notation,

$$y_k^* = y^*(kT)$$
$$\delta_{k-i} = \delta(kT - iT)$$

and we have

$$y_k^* = \sum_{i=0}^{N} y_i \delta_{k-i}$$

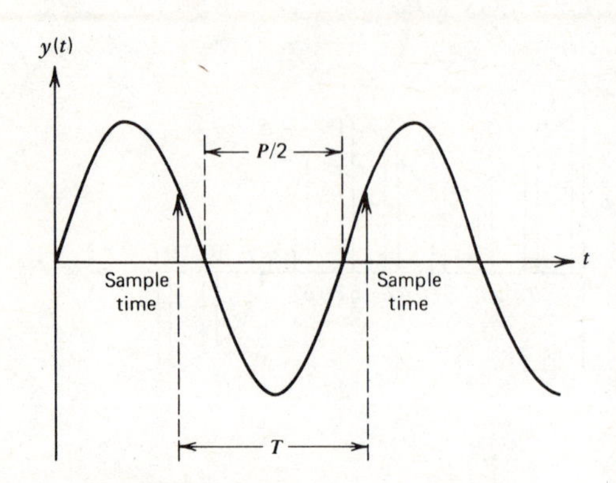

FIGURE 10.5 Sampling a sinusoid. The adjacent samples are too far apart to detect the oscillation in $y(t)$.

If the subscripts are cumbersome, the preceding is replaced by

$$y^*(k) = \sum_{i=0}^{N} y(i)\delta(k-i) \quad \textbf{(10.3-3)}$$

where it is understood that $y(k)$ represents y_k, which in turn represents $y(kT)$.

Frequency Content of Signals

The proper value of the sampling period depends on the nature of the signal being sampled. This is easily shown with a sinusoid of period P (Figure 10.5). If the sampling period T is slightly greater than the half-period $P/2$, the case shown in the figure, it is possible to miss completely one lobe of the sinusoid. If $T < P/2$, each lobe will always be sampled at least once, and the oscillation will be detected. Stated in terms of frequencies, the sampling frequency $1/T$ must be at least twice the sinusoidal frequency $1/P$.

The issue is not so clear when the signal does not consist of sinusoids. If the signal $y(t)$ is periodic with period P, it can be represented by a Fourier series (Appendix A) as

$$y(t) = \sum_{n=0}^{\infty} A_n \sin\left(\frac{n\pi t}{P} + \phi_n\right) \quad \textbf{(10.3-4)}$$

where $A_n \geqslant 0$ is the amplitude of the nth harmonic and ϕ_n is its phase angle. The plot of A_n versus the frequency $n\pi/P$ is called the *spectrum* of the signal, in analogy with the light spectrum used in physical optics. For example, the half-sine function shown in Figure 4.28 has the Fourier series

$$\begin{aligned} y(t) &= \frac{1}{\pi} + \frac{1}{2}\sin 2\pi t - \frac{2}{3\pi}\cos 4\pi t - \frac{2}{15\pi}\cos 8\pi t - \frac{2}{35\pi}\cos 12\pi t + \cdots \\ &= \frac{1}{\pi} + \frac{1}{2}\sin 2\pi t + \frac{2}{3\pi}\sin\left(4\pi t + \frac{3\pi}{2}\right) \\ &\quad + \frac{2}{15\pi}\sin\left(8\pi t + \frac{3\pi}{2}\right) + \frac{2}{35\pi}\sin\left(12\pi t + \frac{3\pi}{2}\right) + \cdots \end{aligned}$$

The spectrum of this signal is shown in Figure 10.6. Notice how the amplitudes decrease with frequency for the higher frequencies. This characteristic is typical of physical signals because of "inertia" effects.

The spectrum of a periodic signal provides a convenient representation of the relative importance of each frequency component. Thus, it would be useful to have a similar representation for signals that are not periodic (aperiodic). This is provided by the *Fourier transform* of the signal and is defined to be

$$\mathscr{F}[y(t)] = Y(\omega) = \int_{-\infty}^{\infty} y(t)e^{-i\omega t}\,dt \quad \textbf{(10.3-5)}$$

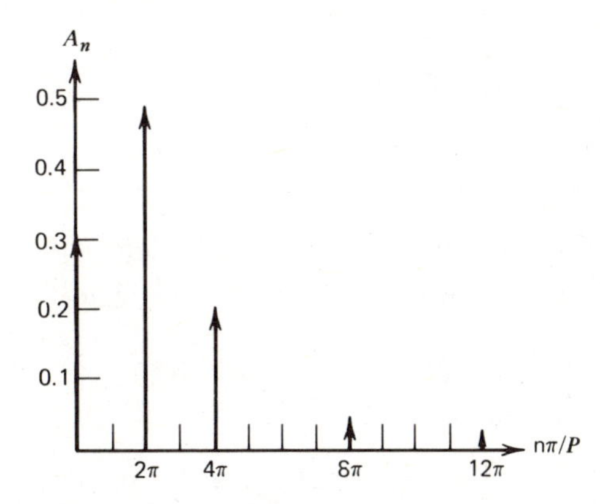

FIGURE 10.6 Spectrum of the half-sine function.

The inverse transform is

$$\mathscr{F}^{-1}[Y(\omega)] = y(t) = \frac{1}{2\pi}\int_{-\infty}^{\infty} Y(\omega)e^{i\omega t}\,dt \tag{10.3-6}$$

Note the similarity to the Laplace transform. Here, the integral is taken over all time rather than only over positive time. The transform variable ω is a real number and is taken to be circular frequency. If $y(t) = 0$ for $t \leqslant 0$ and if the region of convergence for the Laplace transform includes the imaginary axis, then

$$Y(\omega) = [Y(s)]_{s=i\omega} \tag{10.3-7}$$

where $Y(s)$ is the Laplace transform of $y(t)$.

The Euler formula relating the complex exponential in (10.3-5) to $\sin \omega t$ and $\cos \omega t$ can be used to present a heuristic derivation of the Fourier transform pair from the Fourier series formulas for A_n (Appendix A). The essence of the argument is to allow the period P to approach infinity while the separation in signal frequencies is allowed to approach zero. In the limit, the summation in (10.3-4) approaches the integral in (10.3-6). The details are given in Reference 2, a good source of information dealing with sampling and digital filtering. The important point here is that the magnitude $|Y(\omega)|$ of the Fourier transform can also be used as a measure of the signal's frequency content for aperiodic signals, just as A_n is used for periodic signals. The plot of $|Y(\omega)|$ versus ω is the spectrum of $y(t)$.

The energy content of a sinusoid is proportional to the square of its amplitude. The energy contained in a general aperiodic signal is found by integrating the square of the amplitude $|Y(\omega)|$ over all frequencies and is

$$\int_0^{\infty} |Y(\omega)|^2\, d\omega$$

Example 10.2

Compute the Fourier transform of the signal

$$y(t) = \begin{cases} 0 & t < 0 \\ e^{-t/\tau} & t \geqslant 0 \end{cases}$$

Find its spectrum and the frequency ω_u for which 99% of the signal's energy lies in the range $0 \leqslant \omega \leqslant \omega_u$.

From (10.3-6),

$$Y(\omega) = \int_{-\infty}^{0} 0\, dt + \int_0^{\infty} e^{-t/\tau}e^{-i\omega t}\, dt$$

$$= \int_0^\infty e^{-[(1/\tau)+i\omega]t}\,dt = \left.\frac{-e^{-[(1/\tau)+i\omega]t}}{(1/\tau)+i\omega}\right|_0^\infty$$

$$= \frac{\tau}{1+i\omega\tau}$$

This is the same form as the frequency transfer function of the first-order system with a time constant τ. The result could also have been obtained from (10.3-7) with $Y(s) = \tau/(\tau s + 1)$. The magnitude is

$$|Y(\omega)| = \frac{\tau}{\sqrt{1+\tau^2\omega^2}}$$

This is plotted in Figure 4.11 in logarithmic form. The spectrum is flatter for smaller values of τ and falls off with ω more rapidly for larger values of τ. Thus, a rapidly decaying exponential has a higher frequency content than one that decays slowly.

The upper frequency ω_u is found from

$$\int_0^{\omega_u} |Y(\omega)|^2\,d\omega = 0.99\int_0^\infty |Y(\omega)|^2\,d\omega$$

With the given expression for $Y(\omega)$, this becomes

$$\int_0^{\omega_u} \frac{\tau^2}{1+\tau^2\omega^2}\,d\omega = 0.99\int_0^\infty \frac{\tau^2}{1+\tau^2\omega^2}\,d\omega$$

$$\tan^{-1}(\tau\omega_u) - \tan^{-1}(0) = 0.99[\tan^{-1}(\infty) - \tan^{-1}(0)]$$

or

$$\omega_u = \frac{1}{\tau}\tan\left(\frac{0.99\pi}{2}\right) = \frac{63.657}{\tau}\,\frac{\text{rad}}{\text{unit time}}$$

If $\tau = 1$ sec, 99% of the exponential signal's energy lies in a frequency band of $0 < \omega < 63.657$ rad/sec.

Aliasing

Now that we are able to determine the frequency content of a signal, we need to know what effects the higher frequencies have on the sampling process. Uniform sampling of the sinusoid

$$y(t) = \sin(\omega_1 t + \phi)$$

produces the sequence

$$y(kT) = \sin(k\omega_1 T + \phi)$$

No loss of information is incurred in sampling if T is selected so that $0 \leqslant \omega_1 \leqslant \pi/T$, as shown earlier. If we sample the following signals,

$$x(t) = \sin\left[\left(\omega_1 + \frac{2\pi n}{T}\right)t + \phi\right]$$

$$z(t) = \sin\left[\left(\frac{2\pi n}{T} - \omega_1\right)t + \pi - \phi\right]$$

the resulting sequences can be seen to be identical to $y(kT)$ after applying several trigonometric identities. These are

$$\begin{aligned} x(kT) &= \sin\left[\left(\omega_1 + \frac{2\pi n}{T}\right)kT + \phi\right] \\ &= \sin(k\omega_1 T + \phi) \\ z(kT) &= \sin\left[\left(\frac{2\pi n}{T} - \omega_1\right)kT - \pi + \phi\right] \\ &= \sin(k\omega_1 T + \phi) \end{aligned}$$

for *any* positive integer n. The signals $y(t)$ and $x(t)$ *differ* in circular frequency by an integral multiple of $2\pi/T$. Similarly, the *sum* of the frequencies of $y(t)$ and $z(t)$ is an integral multiple of $2\pi/T$.

We conclude that uniform sampling cannot distinguish between two sinusoidal signals when their circular frequencies have a sum or difference equal to $2\pi n/T$, where n is any positive integer. This means that the only effective frequency range for uniform sampling is $0 \leqslant \omega \leqslant \pi/T$. That is, the signal being sampled must have no circular frequency content above π/T if uniform sampling is not to distort the signal's information. The frequency π/T (radians per unit time) is called the *Nyquist frequency* or the *folding frequency*, because all frequencies in the signal are folded into the interval $0 \leqslant \omega \leqslant \pi/T$ by uniform sampling (Reference 1). This phenomenon is called *aliasing*. It is typically seen in motion pictures of a rotating spoked wheel or aircraft propeller. As the object rotates faster, it appears to slow down and then stop or even rotate backward. The sampling process produced by the picture frames "aliases" the high rotation speed into the lower frequency interval defined by the Nyquist frequency.

The Sampling Theorem

The concept of aliasing leads directly to the sampling theorem.

Sampling Theorem

A continuous-time signal $y(t)$ can be reconstructed from its uniformly sampled values $y(kT)$ if the sampling period T satisfies

$$T \leqslant \frac{\pi}{\omega_u}$$

where ω_u is the highest frequency contained in the signal; that is,

$$Y(\omega) = 0 \qquad \omega > \omega_u$$

where $Y(\omega)$ is the Fourier transform of $y(t)$.

With the principal exception of a pure sinusoid, most physical signals have no finite upper frequency ω_u. Their spectra $|Y(\omega)|$ approach zero only as $\omega \to \infty$. In such cases, we estimate ω_u by finding the frequency range containing most of the signal's energy, as in Example 10.2. With a conservative engineering design philosophy, a safety factor between two and ten is applied to determine the sampling rate. This factor is also necessary because we do not have in practice an infinite sequence of impulses as required by the sampling theorem. Because of aliasing, a low-pass filter, called a *guard* filter, is inserted before the sampler to eliminate frequencies above the Nyquist frequency π/T. By definition, these frequencies do not contribute significantly to the signal's energy. Thus, they are equivalent to noise in the system and should be filtered out anyway.

Example 10.3

Compute the uniform sampling period required for the signal $e^{-t/\tau}$.

From Example 10.2, the upper frequency for the 99% energy criterion was $\omega_u = 63.657/\tau$. From the sampling theorem,

$$T \leqslant \frac{\pi}{\omega_u} = \frac{\pi\tau}{63.657} = 0.049\tau$$

With a safety factor of ten, the sampling period is $T = 0.0049\tau$. The guard filter would be designed to filter out frequencies above π/T.

In many applications, the Fourier transform integral in (10.3-5) is impossible to evaluate in closed form. Also, the signal might be given in tabular or graphical form. In such cases, we may compute the integral numerically. With standard integration algorithms like the trapezoidal rule, the number of arithmetic operations goes as N^2, where $2N$ is the number of data points or subintervals. Most computer installations support a version of the Fast Fourier Transform (FFT) algorithm, which is a numerical method requiring $N \log N$ arithmetic operations. Thus, for large N, the savings in computation time and accuracy is quite large (Reference 2).

10.4 THE z TRANSFORM

The impulse representation of sampling provides a convenient way of introducing the z transform. The first three terms of (10.3-2) are

$$y^*(t) = y(0)\delta(t) + y(T)\delta(t - T) + y(2T)\delta(t - 2T) + \cdots \qquad \textbf{(10.4-1)}$$

If both sides are transformed with the Laplace transform, we obtain

$$Y^*(s) = y(0) + y(T)e^{-Ts} + y(2T)e^{-2Ts} + \cdots$$

where the *shifting property* of the Laplace transform has been used. From Table B.1 in Appendix B, this property states that for two functions $g(t)$ and $f(t)$, where

$$g(t) = \begin{cases} f(t-a) & t \geqslant a \\ 0 & t < a \end{cases}$$

the transform of $g(t)$ is

$$\mathscr{L}[g(t)] = e^{-as}\mathscr{L}[f(t)]$$

Here, $f(t) = \delta(t)$ and $\mathscr{L}[\delta(t)] = 1$.

Define a new variable z as follows:

$$z = e^{Ts} \tag{10.4-2}$$

and write $Y^*(s)$ as a function of z.

$$Y(z) = y(0) + y(T)\frac{1}{z} + y(2T)\frac{1}{z^2} + \cdots \tag{10.4-3}$$

The transformed variable $Y(z)$ is the z *transform* of the function $y^*(t)$. The transformation is written as $\mathscr{Z}[y(t)]$, and it is a power series in $1/z$. Comparing (10.4-1) and (10.4-3) shows that $1/z$ is a *delay* operator representing a time delay T. Also, $1/z^2$ represents a delay $2T$, and so forth. With this interpretation, we can recover a sampled sequence from its z transform. For example, the transform

$$Y(z) = 5 + \frac{3}{z} + \frac{1}{z^2} + \frac{-1}{z^3} + \frac{-4}{z^4} + \frac{-5}{z^5} + \cdots$$

corresponds to the sample values $y(t) = 5, 3, 1, -1, -4, -5, \ldots$ for $t = 0, T, 2T, 3T, 4T, 5T, \ldots$.

Commonly Occurring Transforms

If $y^*(t)$ is a unit impulse at $t = 0$, then $y(0) = 1$ and $y(it) = 0$ for $i \neq 0$. The z transform of this function is 1, and it is identical to its Laplace transform. Unfortunately, the simple relationship between a sampled sequence and its z transform series does not always provide a convenient form to handle, because the series is usually infinite. What is needed is a closed-form expression for the transform. Fortunately, they are available for commonly occurring functions like the step, ramp, and exponential. Here, we take all signals to be zero for $t < 0$ and thus confine our attention to the one-sided z transform. If $y(t)$ is nonzero for $t < 0$, the two-sided transform is needed (Reference 3).

Example 10.4

Find the z transform of the sequence $y = \{1, a, a^2, a^3, \ldots\}$ with $y = 0$ for $t < 0$.

The sequence values are the coefficients of the transform's series expression; that is,

$$Y(z) = 1 + \frac{a}{z} + \frac{a^2}{z^2} + \frac{a^3}{z^3} + \cdots = \sum_{k=0}^{\infty} a^k z^{-k} \tag{10.4-4}$$

This is the *geometric* series, and from most calculus texts, we know that its sum can be expressed as

$$Y(z) = \frac{1}{1 - az^{-1}} = \frac{z}{z - a} \tag{10.4-5}$$

if $|z| > |a|$. The series is said to have a radius of absolute convergence of $|z| = |a|$.

If $a = 1$, the preceding sequence corresponds to the sample sequence of a unit-step function that starts just before sampling begins. The z transform of this function is

$$Y(z) = \frac{z}{z-1} \tag{10.4-6}$$

Similarly, the exponential $y(t) = \exp(-bt)$ has the transform

$$Y(z) = 1 + \frac{e^{-bT}}{z} + \frac{e^{-2bT}}{z^2} + \cdots \tag{10.4-7}$$

This is the form of (10.4-4) with $a = \exp(-bT)$. Thus, from (10.4-5)

$$\mathscr{Z}(e^{-bt}) = \frac{z}{z - e^{-bT}}$$

Example 10.5

Compute the z transform of the sampled unit ramp $y(t) = t$.

The sampled series representation of the unit ramp is

$$\begin{aligned} Y(z) &= 0 + \frac{T}{z} + \frac{2T}{z^2} + \cdots \\ &= Tz\left(\frac{1}{z^2} + \frac{2}{z^3} + \cdots\right) \\ &= -Tz\frac{d}{dz}\left(\frac{1}{z} + \frac{1}{z^2} + \cdots\right) \end{aligned}$$

With (10.4-6), this can be written as

$$Y(z) = -Tz\frac{d}{dz}\left(\frac{z}{z-1}\right) = \frac{Tz}{(z-1)^2} \tag{10.4-8}$$

The method of differentiation used to obtain the preceding transform can be generalized. In summation notation, the z transform is defined as

$$\mathscr{Z}[y(kT)] = Y(z) = \sum_{k=0}^{\infty} y(kT)z^{-k} \tag{10.4-9}$$

or equivalently as

$$\mathscr{Z}[y(k)] = Y(z) = \sum_{k=0}^{\infty} y(k)z^{-k} \tag{10.4-10}$$

Differentiation and multiplication by z gives

$$\begin{aligned} -z\frac{d}{dz}Y(z) &= \sum_{k=0}^{\infty} ky(kT)z^{-k} \\ &= \mathscr{Z}[ky(kT)] \end{aligned}$$

The resulting property is

$$\mathscr{Z}[ky(kT)] = -z\frac{d}{dz}Y(z) \tag{10.4-11}$$

with a similar result obtained for the form given by (10.4-10). This property can be used to obtain new z transform pairs.

Table 10.2 lists the z transforms of common sampled functions along with their Laplace transforms. The z transforms are obtained by manipulating series summations in a manner similar to that used in the previous examples. The time period T does not appear in the transform if the time indexing is referenced solely by k, as in (10.4-10). This was the case in Example 4.4. The conversion between the two forms is easily done by letting $T=1$ in the sampled transform. For example, $y(k)=k$ represents the unit ramp expressed without reference to an absolute sampling time. Its transform is $z/(z-1)^2$, which can be obtained either from (10.4-10) or from (10.4-8)

TABLE 10.2 Laplace and z Transforms for Sampled Functions

	$y(t)$ $t \geq 0$	$Y(s) = \int_0^\infty y(t)e^{-st}\,dt$	$Y(z) = \sum_{k=0}^{\infty} y(kT)z^{-k}$
1.	1	$\frac{1}{s}$	$\frac{z}{z-1}$
2.	t	$\frac{1}{s^2}$	$\frac{zT}{(z-1)^2}$
3.	$\frac{t^2}{2}$	$\frac{1}{s^3}$	$\frac{z(z+1)T^2}{2(z-1)^3}$
4.	e^{-at}	$\frac{1}{s+a}$	$\frac{z}{z-e^{-aT}}$
5.	te^{-at}	$\frac{1}{(s+a)^2}$	$\frac{zTe^{-aT}}{(z-e^{-aT})^2}$
6.	$\sin \omega t$	$\frac{\omega}{s^2+\omega^2}$	$\frac{z \sin \omega T}{z^2 - 2z\cos \omega T + 1}$
7.	$\cos \omega t$	$\frac{s}{s^2+\omega^2}$	$\frac{z(z-\cos \omega T)}{z^2 - 2z\cos \omega T + 1}$
8.	$e^{-at}\sin \omega t$	$\frac{\omega}{(s+a)^2+\omega^2}$	$\frac{ze^{-aT}\sin \omega T}{z^2 - 2ze^{-aT}\cos \omega T + e^{-2aT}}$
9.	$e^{-at}\cos \omega t$	$\frac{s+a}{(s+a)^2+\omega^2}$	$\frac{z^2 - 2e^{-aT}\cos \omega T}{z^2 - 2ze^{-aT}\cos \omega T + e^{-2aT}}$

TABLE 10.3 z Transform Pairs

	$y(k)$ for $k \geqslant 0$	$Y(z) = \sum_{k=0}^{\infty} y(k) z^{-k}$
1.	1	$\dfrac{z}{z-1}$
2.	a^k	$\dfrac{z}{z-a}$
3.	k	$\dfrac{z}{(z-1)^2}$
4.	k^2	$\dfrac{z(z+1)}{(z-1)^3}$
5.	ka^k	$\dfrac{az}{(z-a)^2}$
6.	$a^k \sin \omega k$	$\dfrac{az \sin \omega}{z^2 - 2az \cos \omega + a^2}$
7.	$a^k \cos \omega k$	$\dfrac{z(z - a \cos \omega)}{z^2 - 2az \cos \omega + a^2}$

with $T = 1$. Note that this does not imply that the sampling period actually is unity. For convenience, this alternative form is shown in Table 10.3.

10.5 THE PULSE TRANSFER FUNCTION AND SYSTEM RESPONSE

Our primary interest in the z transform is that it helps us to compute the forced response of difference equation models and allows a transfer function description of discrete-time linear systems. Both of these applications can be demonstrated with the following first-order model:

$$y[(k+1)T] = ay(kT) + bu(kT) \tag{10.5-1}$$

If T is specified, this is more simply expressed as

$$y(k+1) = ay(k) + bu(k) \tag{10.5-2}$$

where $u(k)$ is a specified input or forcing function; a and b are constants.

It is easily established from the definition (10.4-10) that the z transform is a linear operator. Therefore, transforming both sides of (10.5-2) gives

$$\begin{aligned} \mathscr{Z}[y(k+1)] &= \mathscr{Z}[ay(k) + bu(k)] \\ &= a\mathscr{Z}[y(k)] + b\mathscr{Z}[u(k)] \end{aligned} \tag{10.5-3}$$

Equation (10.4-10) also implies that

$$\begin{aligned}\mathscr{Z}[y(k+1)] &= y(1) + y(2)z^{-1} + y(3)z^{-2} + \cdots \\ &= z[y(0) + y(1)z^{-1} + y(2)z^{-2} + \cdots] - zy(0)\end{aligned}$$

or

$$\mathscr{Z}[y(k+1)] = zY(z) - zy(0) \tag{10.5-4}$$

This is the *left-shifting property* of the transform and analogous to the Laplace transform of a derivative. Application of the shifting property to (10.5-3) gives

$$zY(z) - zy(0) = aY(z) + bU(z)$$

or

$$Y(z) = \frac{z}{z-a}y(0) + \frac{b}{z-a}U(z) \tag{10.5-5}$$

This shows that the complete solution consists of the sum of two terms: a free response and a forced response. As with the first-order linear differential equation, the free response depends on only the initial condition, while the forced response results from the input function. Thus, the two effects can be treated separately.

With the linearity property and the results of Example 10.4, the free response is found to be

$$\mathscr{Z}^{-1}\left[\frac{z}{z-a}y(0)\right] = y(0)a^k \tag{10.5-6}$$

where $\mathscr{Z}^{-1}$ represents the inverse transformation. We have seen this before in (10.2-4).

The Pulse Transfer Function

If $y(0)$ is zero, the transform of the solution is

$$Y(z) = \frac{b}{z-a}U(z)$$

The ratio $Y(z)/U(z)$ of the output and input transforms is the transfer function of the system. It is also called the *pulse transfer function* or *discrete transfer function* to distinguish it from its Laplace transform counterpart for continuous-time systems.

The transfer function of (10.5-2) is

$$\frac{Y(z)}{U(z)} = \frac{b}{z-a} \tag{10.5-7}$$

The characteristic root of the model is the root of the denominator of the transfer function after it is normalized to a ratio of two polynomials in z. Here, the root is $z = a$, and it tells much about the system behavior, as can be seen from Figure 10.2.

Example 10.6

Investigate the behavior of the system for which $a = 0$, $b = 1$, and $y(0) = 0$. Interpret its transfer function.

The transfer function is $1/z$ and

$$Y(z) = \frac{1}{z} U(z) \tag{10.5-8}$$

The original difference equation shows that

$$y(k+1) = u(k) \tag{10.5-9}$$

This model represents a simple time shift of the input such that $y(1) = u(0)$, $y(2) = u(1)$, and so forth. In this sense, the transfer function $1/z$ represents a shift of one unit forward in time.

Systems Diagrams

One advantage of the transfer function concept is that its input-output structure provides a basis for a graphic description of system behavior, such as a block diagram. The three basic elements of block diagrams for discrete-time systems are shown in Figure 10.7. The multiplier and comparator are identical to their continuous-time counterparts. The new element in the *delay* unit, which delays the input signal by one time unit to produce the output. From Example 10.6 the transfer function of the delay unit is seen to be $1/z$. In discrete-time diagrams the delay unit plays a central role similar to that played by the integrator element whose transfer function is $1/s$.

These elements are sufficient to describe the first-order system (10.5-2), as shown in Figure 10.8*a*. The corresponding signal flow diagram is shown in Figure 10.8*b*. For clarity the time variables have been shown instead of the transformed variables required by the strict interpretation of the diagram. The block diagram with transformed variables is shown in Figure 10.8*c*.

The discrete-time system equations become algebraic equations after being transformed. Then they obey the same rules of diagram algebra as Laplace-transformed continuous-time equations. Figure 4.4 contains the common reduction formulas. For example, the loop reduction formula in Figure 4.4*b* can be applied to Figure 10.8*c* to obtain the transfer function given by (10.5-7).

A pulse transfer function for an analog system can be found by approximating its differential equation model with a difference equation representation such as that given by the Euler method. However, several different approximations are in common use (such as the trapezoidal rule), and each will give a different pulse transfer function.

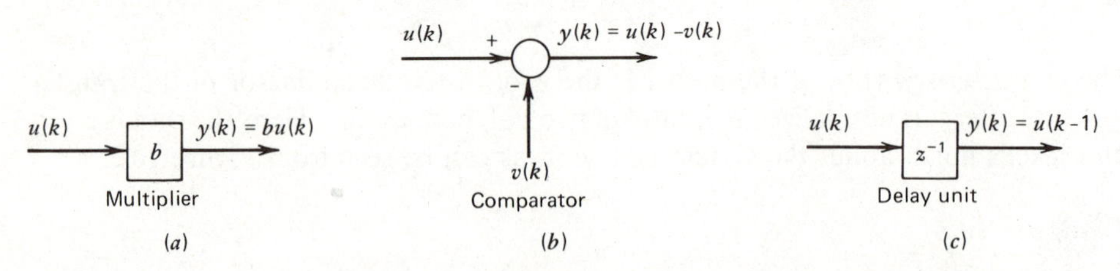

FIGURE 10.7 Three basic elements of block diagrams for discrete-time systems. (*a*) Multiplier. (*b*) Comparator. (*c*) Delay unit.

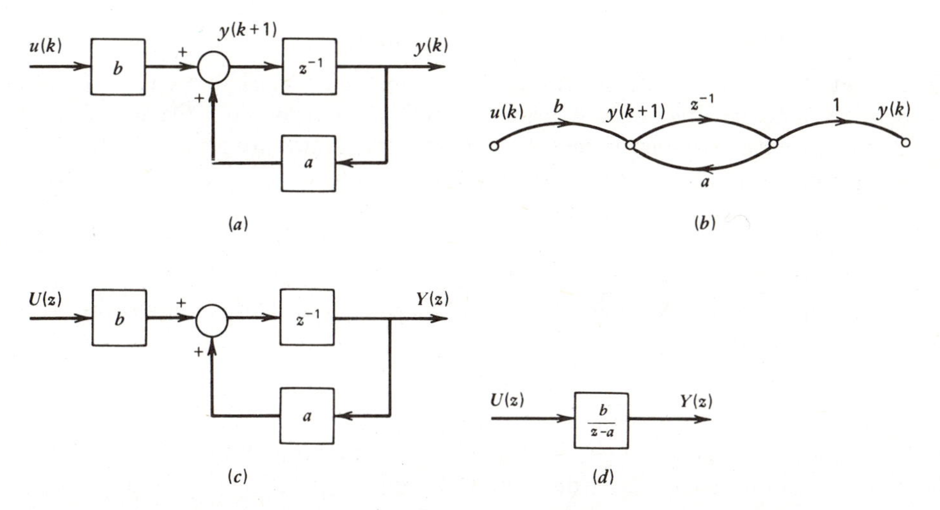

FIGURE 10.8 Representations of a first-order discrete-time system. (*a*) Block diagram in the time domain. (*b*) Signal flow graph in the time domain. (*c*) Block diagram in terms of *z*-transforms. (*d*) Reduced block diagram showing the system transfer function.

Example 10.7

Use the Euler method to find the pulse transfer function of a series RC circuit. The input voltage is denoted here by $u(t)$, and the output voltage across the capacitor by $y(t)$.

The differential equation relating y to u is

$$RC\frac{dy}{dt} = u - y \tag{10.5-10}$$

The Euler approximation assumes the right-hand side of this equation is constant over the step size, denoted here by T. Using this approximation and the expression (10.1-2) for dy/dt, we obtain

$$y(k+1) = y(k) + \frac{T}{RC}[u(k) - y(k)] \tag{10.5-11}$$

For zero initial conditions, the z transform of this equation is

$$zY(z) = \left(1 - \frac{T}{RC}\right)Y(z) + \frac{T}{RC}U(z)$$

or

$$T(z) = \frac{Y(z)}{U(z)} = \frac{b}{z+b-1} \tag{10.5-12}$$

where $b = T/RC$. This is the desired pulse transfer function.

System Response

The system transfer function is useful in computing the response to a given input. In most cases, a partial fraction expansion is required to decompose the transformed function into a set of simpler terms whose inverse transforms are given in Table 10.3. For functions with a simple root, the transform has the form

$$\frac{Cz}{z+d}$$

Thus, the expansion is made in terms of similar expressions. This is more easily done by expanding $Y(z)/z$ as a series of terms of the form $1/(z+d)$.

Example 10.8

Find the impulse response of (10.5-2) for $y(0)=0$ and an impulse of strength A.

For the impulse, $U(z)=A$ from Table 10.3, and from (10.5-7),

$$Y(z)=\frac{b}{z-a}A=\frac{bz}{z-a}\frac{A}{z}$$

The expansion of $Y(z)/z$ is

$$\frac{Y(z)}{z}=\frac{b}{z-a}\frac{A}{z}=\frac{C_1}{z-a}+\frac{C_2}{z}$$

Cross multiplication gives

$$C_1 z+C_2(z-a)=bA$$

or

$$(C_1+C_2)z-aC_2=bA$$

From

$$\begin{aligned} C_1+C_2&=0\\ -aC_2&=bA \end{aligned}$$

the solution is

$$C_1=-C_2=\frac{bA}{a}$$

Thus,

$$Y(z)=\frac{bA}{a}\left(\frac{z}{z-a}-1\right)$$

From the transform table, we obtain

$$y(k)=\frac{bA}{a}[a^k-\delta(k)]$$

This satisfies the initial condition $y(0)=0$, because $\delta(0)=1$. For $k>0$, the solution reduces to

$$y(k)=\frac{bA}{a}a^k=bAa^{k-1} \tag{10.5-13}$$

This is equivalent to the free response with $y(0) = bA/a$. However, if we consider the start of the problem to be at $k = 1$, the impulse response is equivalent to the free response with initial condition $y(1) = bA$. With this interpretation, the result resembles that for the continuous-time model.

Example 10.9

Compute the unit-step response of (10.5-2) if $y(0) = 0$.

Here, $u(k) = \{1, 1, 1, \ldots\}$ for $k \geqslant 0$, and from Table 10.3 and (10.5-7),

$$U(z) = \frac{z}{z-1}$$

$$Y(z) = \frac{b}{z-a}\frac{z}{z-1}$$

The expansion is

$$\frac{Y(z)}{z} = \frac{b}{(z-a)(z-1)} = \frac{b}{a-1}\left(\frac{z}{z-a} - \frac{z}{z-1}\right)$$

From the transform table, we obtain

$$y(k) = \frac{b}{a-1}(a^k - 1) = \frac{b}{1-a}(1 - a^k) \qquad \textbf{(10.5-14)}$$

The response normalized by the factor $b/(1-a)$ is shown in Figure 10.9 for stable values of the root $z = a$. Note the occurrence of oscillations for $a < 0$.

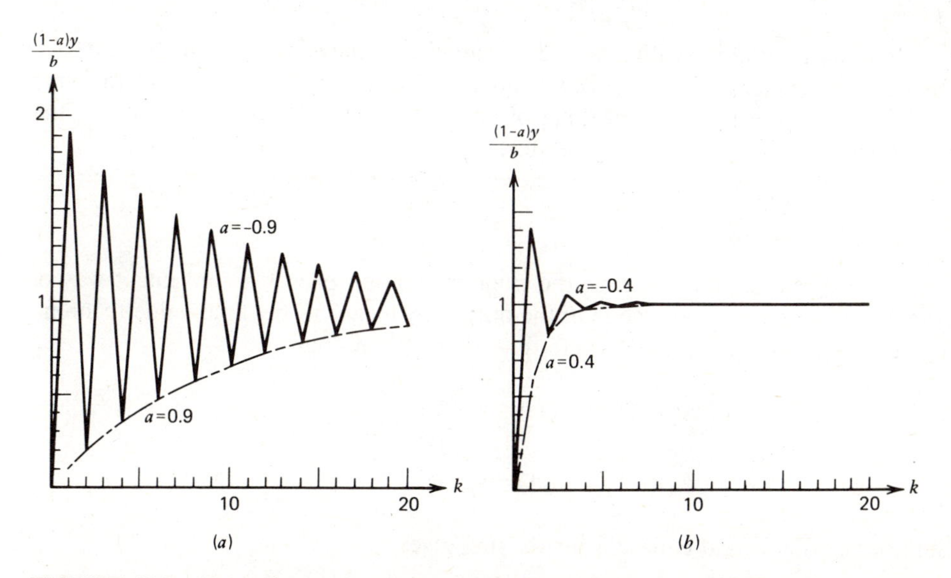

FIGURE 10.9 Normalized step response of the first-order system (10.5-2). The response for $a > 0$ provides the envelope of the oscillatory response for $a < 0$.

Example 10.10

Compute the ramp response of (10.5-2) with $y(0)=0$.

For the ramp function with a slope m,

$$u(k)=mk$$

$$U(z)=\frac{mz}{(z-1)^2}$$

and

$$\frac{Y(z)}{z}=\frac{b}{z-a}\frac{m}{(z-1)^2}$$

Thus,

$$\frac{Y(z)}{z}=\frac{b}{z-a}\frac{m}{(z-1)^2}=\frac{C_1}{z-a}+\frac{C_2}{(z-1)^2}+\frac{C_3}{z-1}$$

where the expansion for repeated roots is performed similarly to that in Example 4.3.

The result is

$$C_1=\frac{mb}{(a-1)^2}=-C_3$$

$$C_2=(1-a)C_1$$

and

$$Y(z)=\frac{mb}{(a-1)^2}\left[\frac{z}{z-a}+\frac{(1-a)z}{(z-1)^2}-\frac{z}{z-1}\right]$$

The response is

$$y(k)=\frac{mb}{(a-1)^2}[a^k+(1-a)k-1] \tag{10.5-15}$$

If uniform sampling with a period T is applied to the *unit*-ramp function $u(t)=t$, the sampled sequence is $u(kT)=kT$. This is not a *unit* ramp in terms of the time index k unless $T=1$. When speaking of unit functions, such as the unit ramp, the time variables (t or k) should be specified.

Final Value Theorem

As with continuous-time systems, the transfer function proves a convenient way of computing the steady-state value of the output by means of the final value theorem. To derive this theorem for a discrete-time system, note that

$$\mathscr{Z}[y(k+1)]=zY(z)-zy(0)=\sum_{k=0}^{\infty}y(k+1)z^{-k}$$

$$\mathscr{Z}[y(k)]=Y(z)=\sum_{k=0}^{\infty}y(k)z^{-k}$$

Subtracting the second equation for the first gives

$$(z-1)Y(z)-zy(0)=\sum_{k=0}^{\infty}[y(k+1)-y(k)]z^{-k}=y(\infty)-y(0)$$

Thus,

$$\frac{(z-1)}{z}Y(z) = y(0) + \frac{y(\infty) - y(0)}{z}$$

and

$$\lim_{z\to 1}\frac{z-1}{z}Y(z) = y(\infty)$$

The theorem is valid only when the preceding limit exists. This occurs when the system is stable and the input is such that the output can approach a definite value.

TABLE 10.4 Properties of the z Transform

Property	$y(t)$ or $y(k)$	$\mathscr{Z}[y(t)]$ or $\mathscr{Z}[y(k)]$
1. Linearity	$ay(t) + bx(t)$	$aY(z) + bX(z)$
2. Right-shifting	$y(t - mT)$ $y(k - m)$	$z^{-m}Y(z)$
3. Left-shifting	(a) $y(t + T)$ $y(k + 1)$	$zY(z) - zy(0)$
	(b) $y(t + 2T)$ $y(k + 2)$	$z^2Y(z) - z^2y(0) - zy(T)$ $z^2Y(z) - z^2y(0) - zy(1)$
	(c) $y(t + mT)$	$z^mY(z) - \sum_{i=0}^{m-1} y(iT)z^{m-i}$
	$y(k + m)$	$z^mY(z) - \sum_{i=0}^{m-1} y(i)z^{m-i}$
4. Differentiation	$ty(t)$	$-Tz\frac{d}{dz}[Y(z)]$
	$ky(k)$	$-z\frac{d}{dz}[Y(z)]$
5. Convolution	$\sum_{i=0}^{h} x(iT)y(kT - iT)$	$X(z)Y(z)$
6. Summation	$\sum_{i=0}^{h} y(iT)$	$\frac{z}{z-1}Y(z)$
7. Multiplication by an exponential	(a) $e^{-at}y(t)$	$Y(ze^{aT})$
	(b) $a^k y(k)$	$Y\left(\frac{z}{a}\right)$
8. Initial value theorem	$y(0) = \lim_{\lvert z\rvert\to\infty} Y(z)$	
9. Final value theorem	$y(\infty) = \lim_{z\to 1}\frac{(z-1)}{z}Y(z)$ if $\frac{(z-1)}{z}Y(z)$ is analytic for $\lvert z\rvert \geqslant 1$	

Mathematically, these conditions are satisfied if the product $(z-1)Y(z)/z$ has no poles z such that $|z| \geqslant 1$. In this case, the product is said to be *analytic* for $|z| \geqslant 1$.

This and other properties of the z transform are listed in Table 10.4.

10.6 SAMPLED-DATA SYSTEMS

In most engineering applications, there is at least one component of the system that is analog in nature. The motion of a mechanical element is an example and thus can be described by differential equations. If a digital device is used to measure or control the mechanical element, the resulting system is a discrete-continuous hybrid. Such systems are called *sampled-data* systems. The mathematical modeling of sampled-data systems must be treated with special care because of their mixed nature. Here, we present pertinent modeling and analysis techniques for such applications.

Zero-Order Hold Circuit

The binary form of the output from the digital device is first converted to a sequence of short voltage pulses. However, it is usually not possible to drive a load, such as a motor, with short pulses. In order to deliver sufficient energy, the pulse amplitude might have to be so large that it is infeasible to be generated. Also, large voltage pulses might saturate or even damage the system being driven.

The solution to this problem is to smooth the output pulses to produce a signal in analog (continuous) form. Thus the digital-to-analog converter performs two functions: first, generation of the output pulses from the digital representation produced by the machine, and second, conversion of these pulses to analog form.

The simplest way of converting an impulse sequence into a continuous signal is to hold the value of the impulse until the next one arrives. Since the impulses are really short-duration pulses, the net effect is to extend their duration to T, the sampling period (Figure 10.10). The device that accomplishes this is called a *zero-order hold* or *boxcar hold*, from the output signal shape. The term *zero order* refers to the zero-order polynomial used to extrapolate between the sampling times. A first-order hold uses a first-order polynomial (a straight line with nonzero slope) for extrapolation. The zero-order hold is by far the easiest to construct physically and program, and it is the most widely used.

If the input to the hold is a unit impulse, its resulting output will reveal the transfer function. For such an input at $t=0$, the output is a pulse with a unit amplitude from $t=0$ to $t=T$. The Laplace transform of this pulse is obtained with the shifting property and is

$$G(s) = \frac{1}{s}(1 - e^{-Ts}) \tag{10.6-1}$$

This is the transfer function of the zero-order hold. Referring to Figure 10.10*a*, we see that

$$Y(s) = G(s)U^*(s) \tag{10.6-2}$$

Each impulse at the input to the zero-order hold is converted into a rectangular pulse of width T and a height $u(kT)$ equal to the sample value at that time. If these

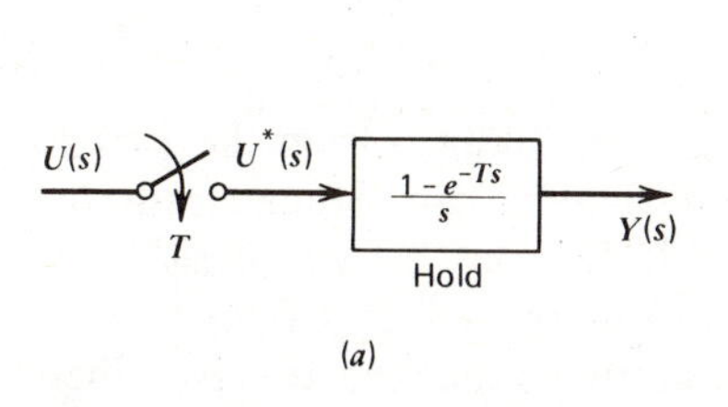

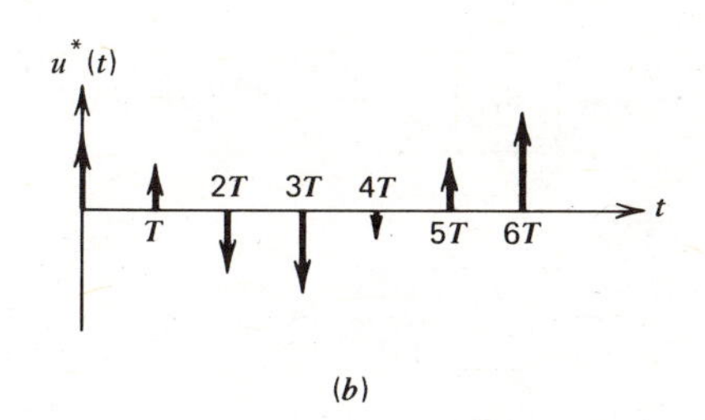

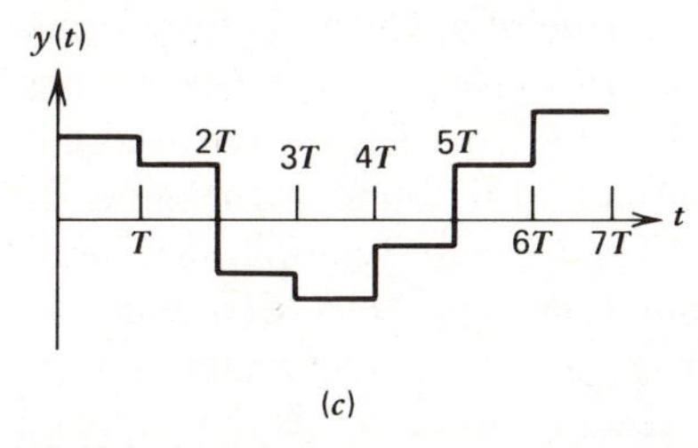

FIGURE 10.10 Zero-order hold. (*a*) Block diagram representation of the hold with a sampler. (*b*) Sampled input sequence. (*c*) Resulting analog output from the hold for the input sequence shown in (*b*).

pulses are applied to an element whose time constants are large compared to T, the pulses can be considered to be impulses with a strength $Tu(kT)$. Thus, the hold not only converts the impulses into analog form, but also supplies a gain T.

Pulse Transfer Functions of Analog Elements

When discrete-time signals are used with a hold to drive an analog device, such as a motor, the block diagram representation naturally consists of a mixture of such discrete elements as samplers, and continuous elements representing the transfer functions of the analog devices. The alternative to this hybrid diagram is to describe all the system elements by difference equations and pulse transfer functions and to use the discrete-time block diagram elements shown in Figure 10.7. For example, the differential equation describing the dynamics of a motor and its load might be converted into a difference equation by the Euler method and its discrete-time transfer function obtained as in Section 10.5. This alternative approach avoids a mixed description of the system but has the disadvantages associated with approximations required to obtain the difference equations that describe the system. The optimal choice is not always clear. In this section, we present the mixed description.

We know that for the system shown in Figure 10.11*b*

$$Y(s) = G(s)U^*(s)$$

and for Figure 10.11*c*

$$Y(z) = G(z)U(z)$$

The question to be answered is: Given $G(s)$, what is $G(z)$? Since the transfer functions $G(s)$ and $G(z)$ represent the unit-impulse response $g(t)$ in the s and z domains,

$$g(t) = \mathscr{L}^{-1}[G(s)] \qquad \textbf{(10.6-3)}$$

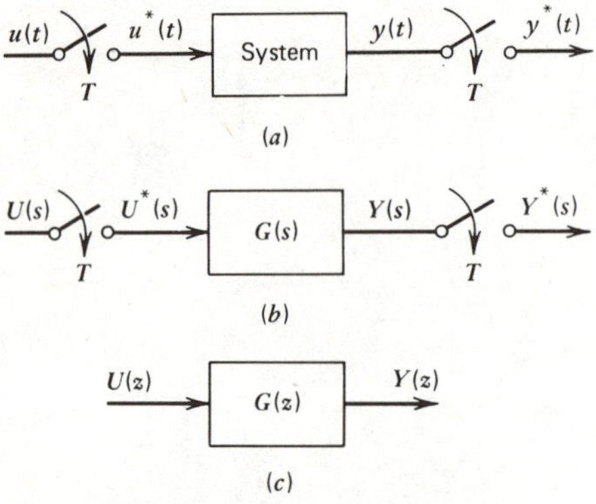

FIGURE 10.11 Equivalent representations of a sampled-data system. (*a*) Time-domain diagram. (*b*) *s*-domain diagram. (*c*) *z*-domain diagram (automatically implies the existence of an input sampler and of either a real or a fictitious output sampler).

and

$$G(z) = \mathscr{Z}[g(t)] \qquad \textbf{(10.6-4)}$$

Thus, $G(z)$ is obtained from the z transform of $g(t)$, which can be found from $G(s)$.

Table 10.2 has been constructed to facilitate this procedure. Commonly encountered time functions are given there along with their Laplace and z transforms. These entries give $G(s)$ and $G(z)$ for the corresponding $g(t)$. Thus, given $G(s)$, the appropriate $G(z)$ is found on the same line in the table. For more complicated forms not found in the table, $G(s)$ must be written as a partial fraction expansion. The preceding procedure is then applied to each term in the expansion.

Example 10.11

Find the pulse transfer function and difference equation for the series RC circuit shown in Figure 10.12*a*. The input voltage $u(t)$ is sampled before being applied to the circuit. The output is the voltage y. The block diagram is shown in Figure 10.12*b*.

From the diagram and Figure 10.11*b*, we see that

$$G(s) = \frac{1}{RCs+1} = \frac{1/RC}{s+1/RC} \qquad \textbf{(10.6-5)}$$

From Table 10.2, no. 4,

$$G(z) = \frac{1}{RC}\frac{z}{z-e^{-T/RC}} = \frac{Y(z)}{U(z)} \qquad \textbf{(10.6-6)}$$

and

$$y(k+1) - e^{-T/RC}y(k) = \frac{1}{RC}u(k+1) \qquad \textbf{(10.6-7)}$$

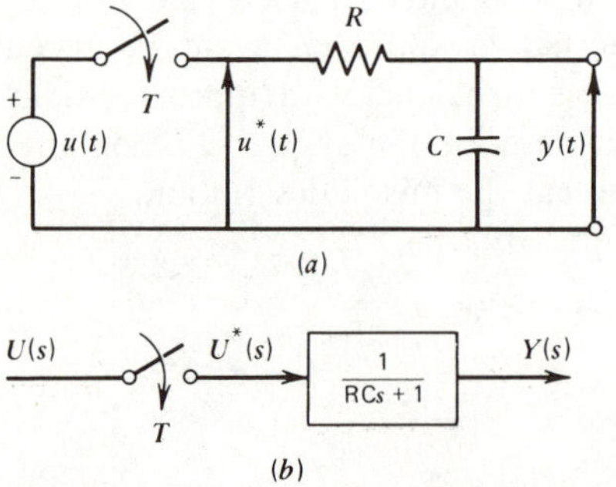

FIGURE 10.12 Series *RC* circuit with a sampled input voltage and its block diagram.

This pulse transfer function and difference equation are quite different from those found in Example 10.7 via the Euler method. One reason for the difference is that the Euler method assumes the input u to be constant over the sampling interval T, whereas the present method treats the input as a train of impulses. The second reason is that the z-transform entries $G(z)$ in Table 10.2 represent the exact solution for the impulse response corresponding to $G(s)$. Equation (10.6-7) represents this solution for $y(t_{k+1})$ with initial condition $y(t_k)$ and an impulse input $u(t_{k+1})$ occurring at time t_{k+1} (not at t_k). It involves no approximation for the derivative, as does the Euler method.

The diagrams in Figures 10.11*a* and 10.11*b* both have samplers at their outputs, whereas the one in Figure 10.12*b* does not. Although the output voltage of the *RC* circuit in Figure 10.12*b* is analog, its pulse transfer function relates the output to the input at only the sample times. That is, imagine a "fictitious" output sampler that is synchronized with the input sampler. The output of this fictitious sampler is $y^*(t)$, and it is the output that can be computed from the pulse transfer function. Thus, Figure 10.11 portrays the general situation, in that the output samplers shown there may be real or fictitious, depending on the application.

If the sampler is fictitious (i.e., an analog output) and we desire to compute the output values between the sample times, a more complicated analysis is required. The essence of one technique is to choose the fictitious sampling period to be a submultiple of the real period T.

Figure 10.13 summarizes the commonly-used symbols for diagrams in the time domain, the s domain, and the z domain. Note that a sampler never appears in a z domain diagram because its presence must be assumed. The output of an A/D

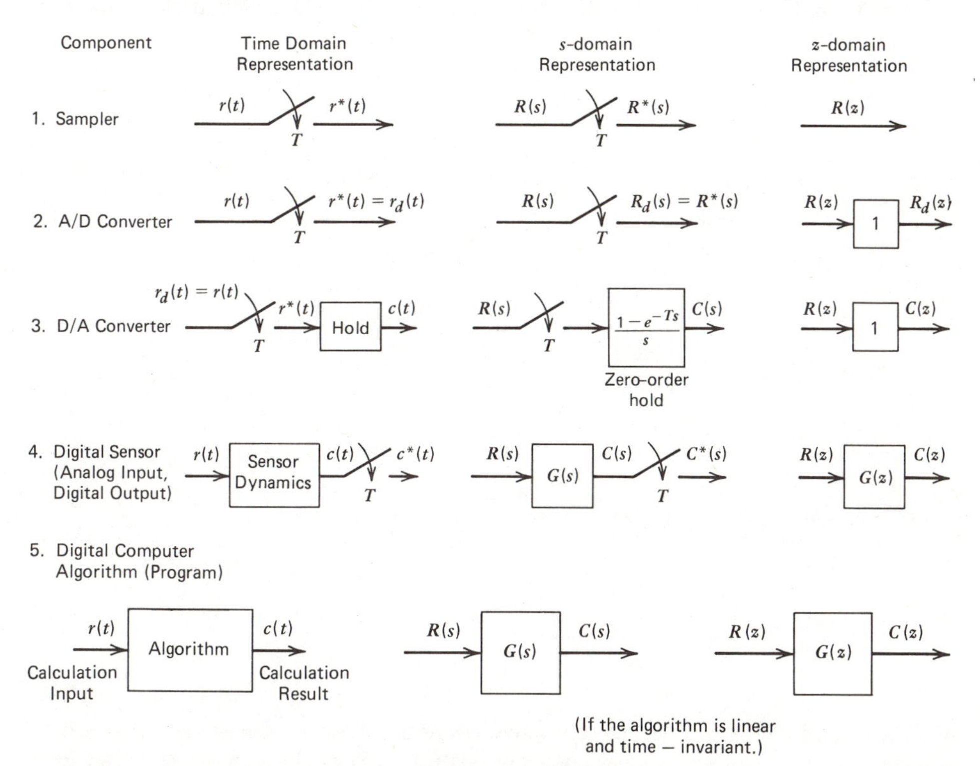

FIGURE 10.13 Representations of common digital elements.

converter or the input to a D/A converter is a number that has been sampled, quantized, and digitized (see Figure 10.3*b*). We denote this digital sensor by $r_d(t)$, but it is important to note that the specific numerical coding used to represent $r_d(t)$ is irrelevant for block diagram purposes. That is, the distinction between $r^*(t)$ and $r_d(t)$ is irrelevant for diagramming purposes. Similarly, $R_d(s)$ is equivalent to $R^*(s)$, $R_d(z)$ is equivalent to $R(z)$, and the subscript d is not used unless it is necessary to distinguish between A/D and D/A converters on a z domain diagram.

A digital sensor is used to provide a digital representation of the analog variable being sensed. Thus, an A/D function is built in. Also, the sensor can have its own dynamic characteristics. For example, a tachometer will have inertia. Thus, a digital sensor model will be of the form given in Figure 10.13, item 4.

Item 5 in Figure 10.13 shows how a digital computer and its algorithm (its program) is represented in a block diagram. Not shown is the A/D and D/A converters that must be present at the input and the output of the computer in order to interface it to the outside analog environment. If the algorithm is linear and time-invariant, we can represent it by a transfer function, $G(s)$ or $G(z)$.

Cascaded Elements

Care is needed in obtaining the pulse transfer function for cascaded elements. The pulse transfer function between U and Y in Figure 10.14*a* is not the same as that between U and Y in Figure 10.14*b*. If no sampler is present between elements, the usual cascade formula applies. Thus, for Figure 10.14*a*,

$$Y(s) = G_1(s)G_2(s)U^*(s) = G_a(s)U^*(s) \tag{10.6-8}$$

and

$$Y(z) = G_a(z)U(z)$$

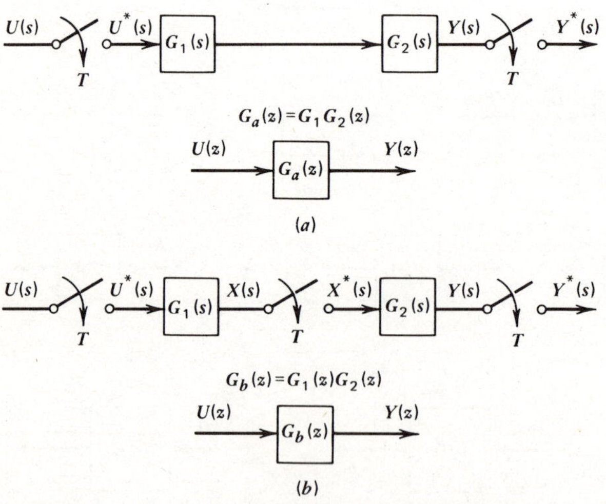

FIGURE 10.14 Transfer functions of cascaded elements. (*a*) Analog elements not separated by a sampler and their cascaded pulse-transfer function $G_a(z)$. (*b*) Analog elements separated by a sampler and their cascaded pulse-transfer function $G_b(z)$.

where

$$G_a(z) = \mathscr{Z}[g_a(t)]$$
$$g_a(t) = \mathscr{L}^{-1}[G_1(s)G_2(s)] \tag{10.6-9}$$

For Figure 10.14*b*,

$$Y(s) = G_2(s)X^*(s)$$
$$X(s) = G_1(s)U^*(s)$$
$$X(z) = G_1(z)U(z)$$
$$Y(z) = G_2(z)X(z) = G_2(z)G_1(z)U(z) = G_b(z)U(z) \tag{10.6-10}$$

where

$$G_i(z) = \mathscr{Z}[g_i(t)] \qquad i = 1, 2 \tag{10.6-11}$$
$$g_i(t) = \mathscr{L}^{-1}[G_i(s)] \qquad i = 1, 2$$

So the pulse transfer functions $G_a(z)$ and $G_b(z)$ are not equal in general, because $G_a(z)$ does not equal $G_2(z)G_1(z)$ in general. This is most easily proved by a counterexample.

Example 10.12

Refer to Figure 10.14. Let $G_1(s) = 1/(s + c)$ and $G_2(s) = 1/(s + d)$. Find $G_a(z)$ and $G_b(z)$.

(a) Find $G_a(z)$. From (10.6-8),

$$G_a(s) = \frac{1}{s+c}\,\frac{1}{s+d}$$
$$= \frac{1}{d-c}\left(\frac{1}{s+c} - \frac{1}{s+d}\right)$$

Table 10.2 gives

$$G_a(z) = \frac{1}{d-c}\left(\frac{z}{z-e^{-cT}} - \frac{z}{z-e^{-dT}}\right)$$
$$= \frac{z}{d-c}\,\frac{e^{-cT} - e^{-dT}}{(z-e^{-cT})(z-e^{-dT})} \tag{10.6-12}$$

(b) Find $G_b(z)$. From (10.6-10),

$$G_b(z) = G_1(z)G_2(z)$$

where from Table 10.2,

$$G_1(z) = \frac{z}{z-e^{-cT}}$$

$$G_2(z) = \frac{z}{z-e^{-dT}}$$

Thus,

$$G_b(z) = \frac{z^2}{(z-e^{-cT})(z-e^{-dT})} \tag{10.6-13}$$

$G_a(z)$ is obviously not equal to $G_b(z)$.

In light of the preceding observation, we now introduce a new notation for use with cascaded elements. In Figure 10.14*a*, $G_a(z)$ is found by operating on the product $G_1(s)G_2(s)$. We will denote this operation by

$$G_a(z) = G_1G_2(z) = \mathscr{Z}\{\mathscr{L}^{-1}[G_1(s)G_2(s)]\} \tag{10.6-14}$$

Note that $G_1G_2(z) = G_2G_1(z)$. In this notation, the z operator is meant to apply to the product, and Example 10.12 shows that in general

$$G_1(z)G_2(z) \neq G_1G_2(z) \tag{10.6-15}$$

Thus, it is important to look for a sampler between elements (the preceding analysis assumes that all samplers are synchronized with the same period).

We can use these results to obtain the transfer function for a system with a zero-order hold. If the hold element is cascaded with an analog element with no sampler in between, write the transfer function product as (see Table 10.5)

$$G(s) = G_1(s)G_2(s) \tag{10.6-16}$$

where $G_1(s) = 1 - e^{-Ts}$ and $G_2(s)$ is the remainder of $G(s)$. This gives

$$G(s) = G_2(s) - e^{-Ts}G_2(s)$$

and

$$g(t) = g_2(t) - g_2(t - T)$$

TABLE 10.5 Transfer Functions and the Zero-Order Hold

s domain

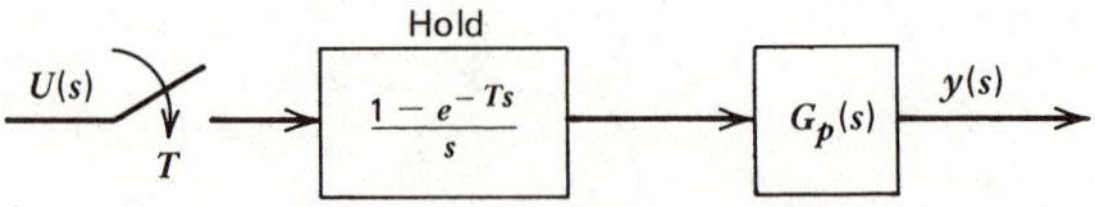

$$\frac{Y(s)}{U(s)} = G(s) = \frac{1 - e^{-Ts}}{s}G_p(s) = (1 - e^{-Ts})G_2(s)$$

where

$$G_2(s) = \frac{G_p(s)}{s}$$

z domain

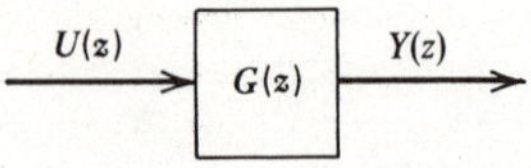

$$\frac{Y(z)}{U(z)} = G(z) = \frac{z - 1}{z}G_2(z)$$

From the third property in Table 10.4,

$$G(z) = G_2(z) - z^{-1}G_2(z)$$
$$= (1 - z^{-1})G_2(z) = \frac{z-1}{z}G_2(z) \qquad \textbf{(10.6-17)}$$

This identity simplifies the analysis.

Example 10.13

The series RC circuit is now shown in Figure 10.15 with a sample-and-hold acting on the input voltage. Find the pulse transfer function and difference equation.

From (10.6-16),

$$G_2(s) = \frac{1}{s}\frac{1/RC}{s + 1/RC} = \frac{1}{s} - \frac{1}{s + 1/RC}$$

and

$$G_2(z) = \frac{z}{z-1} - \frac{z}{z - e^{-T/RC}}$$
$$= \frac{z(1 - e^{-T/RC})}{(z-1)(z - e^{-T/RC})}$$

The required transfer function is obtained from (10.6-17).

$$G(z) = \frac{z-1}{z}G_2(z) = \frac{1 - e^{-T/RC}}{z - e^{-T/RC}} = \frac{Y(z)}{U(z)} \qquad \textbf{(10.6-18)}$$

This implies that

$$y(k+1) = e^{-T/RC}y(k) + (1 - e^{-T/RC})u(k) \qquad \textbf{(10.6-19)}$$

Because of the zero-order hold's effect, (10.6-19) represents the pulse response of the differential equation (10.5-10) for the circuit, just as (10.6-7) represents the impulse response. Even with a piecewise constant input, (10.6-19) still differs from (10.5-11), because the latter involves a derivative approximation.

Block Diagram Algebra for Sampled-Data Systems

Given the block diagram in terms of Laplace transforms, the usual rules of block diagram algebra can be applied to reduce the diagram and obtain the pulse transfer function. Here, we must be careful whenever a sampler is present within a loop. The

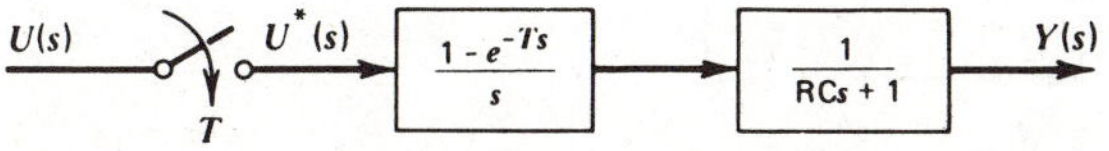

FIGURE 10.15 Series *RC* circuit block diagram with a sample-and-hold acting on the input voltage.

procedure is to check all signal paths. If a signal passes from one element to another without going through a sampler, the transfer functions for these elements must be combined according to (10.6-14). The algebraic relations are then written in terms of z-transformed variables and the reduction carried out.

For example, consider the diagram shown in Figure 10.16*a*. From this we can write the following relations.

$$E(s) = R(s) - B(s) \tag{10.6-20}$$

$$C(s) = G(s)E^*(s) \tag{10.6-21}$$

$$B(s) = H(s)C^*(s) \tag{10.6-22}$$

Equation (10.6-20) implies that

$$E^*(s) = R^*(s) - B^*(s)$$

Because the z transform specifies quantities only at the sample times, we have $E^*(z) = E(z)$, and so on. Thus,

$$E(z) = R(z) - B(z)$$

Equations (10.6-21) and (10.6-22) give

$$C(z) = G(z)E(z)$$

$$B(z) = H(z)C(z)$$

Combining the last three equations gives the input–output relation

$$C(z) = \frac{G(z)}{1 + G(z)H(z)} R(z) \tag{10.6-23}$$

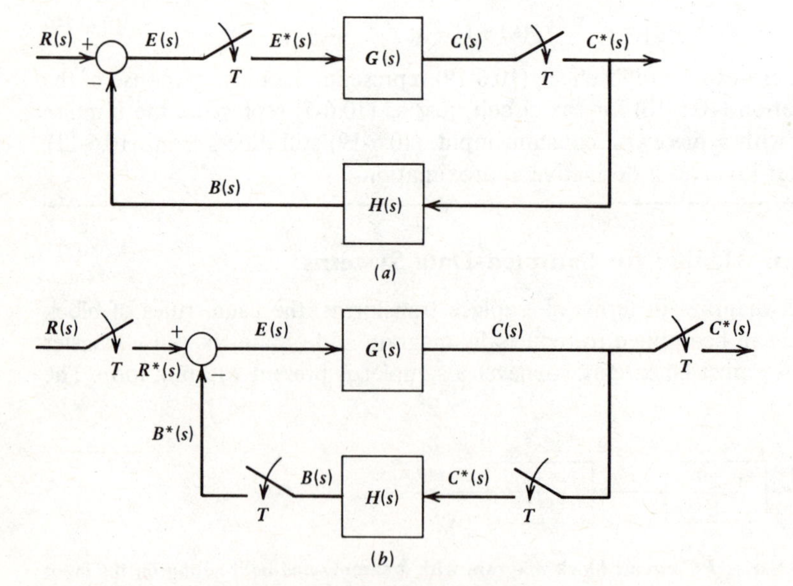

FIGURE 10.16 Equivalent locations for samplers.

Because comparators and take-off points are not dynamic elements, we can move samplers across them to obtain equivalent block diagrams. With reference to Figure 10.16*b*, note that $E(s) = E^*(s)$. With this observation it is easy to see that (10.6-20) and (10.6-21) also apply, and that the system input–output relation is identical to (10.6-23).

Another common diagram form is given by Figure 10.17*a*, which is a representation of a digital control system. The command input is assumed to be given as an analog signal, which must be converted to digital form for use by the computer running the control algorithm. Figure 10.17*b* gives the *s*-domain diagram. The transfer function $P(s)$ represents the product of the zero-order hold's transfer function with the plant's transfer function, as summarized in Table 10.5. The sensor's dynamics are described by $H(s)$, and the control algorithm is described by $G(s)$. For example, $G(s) = K_p + K_I/s$ for *PI* control.

We can simplify the diagram by moving the samplers across the comparator as shown in Figure 10.17*c*. The following equations can then be written.

$$E^*(s) = R^*(s) - B^*(s) \tag{10.6-24}$$

$$F(s) = G(s)E^*(s) \tag{10.6-25}$$

$$C(s) = P(s)F^*(s) \tag{10.6-26}$$

$$B(s) = H(s)P(s)F^*(s) \tag{10.6-27}$$

Equations (10.6-24)–(10.6-26) give

$$E(z) = R(z) - B(z)$$

$$F(z) = G(z)E(z)$$

$$C(z) = P(z)F(z)$$

Equation (10.6-27) involves the product of two unstarred quantities, $H(s)$ and $P(s)$, because there is no sampler between their blocks. Therefore, we must write

$$B(z) = HP(z)F(z)$$

Combining the last four equations gives

$$C(z) = \frac{G(z)P(z)}{1 + G(z)HP(z)} R(z) \tag{10.6-28}$$

Figure 10.18 illustrates a phenomenon that we have not yet encountered. When an input signal is acted upon by a dynamic element before being sampled, it is impossible to obtain a transfer function for the system. The system in Figure 10.18 differs from that in Figure 10.17 in that the analog error signal $e(t)$ is first amplified before being converted to digital form for the control computer. The amplifier's dynamics are given by $G_1(s)$. Also, the feedback sensor must now be analog to be compatible with the amplifier $G_1(s)$. From Figure 10.18 we can write

$$F(s) = G_1(s)[R(s) - H(s)G_3(s)A^*(s)] \tag{10.6-29}$$

$$A(s) = G_2(s)F^*(s) \tag{10.6-30}$$

$$C(s) = G_3(s)A^*(s) \tag{10.6-31}$$

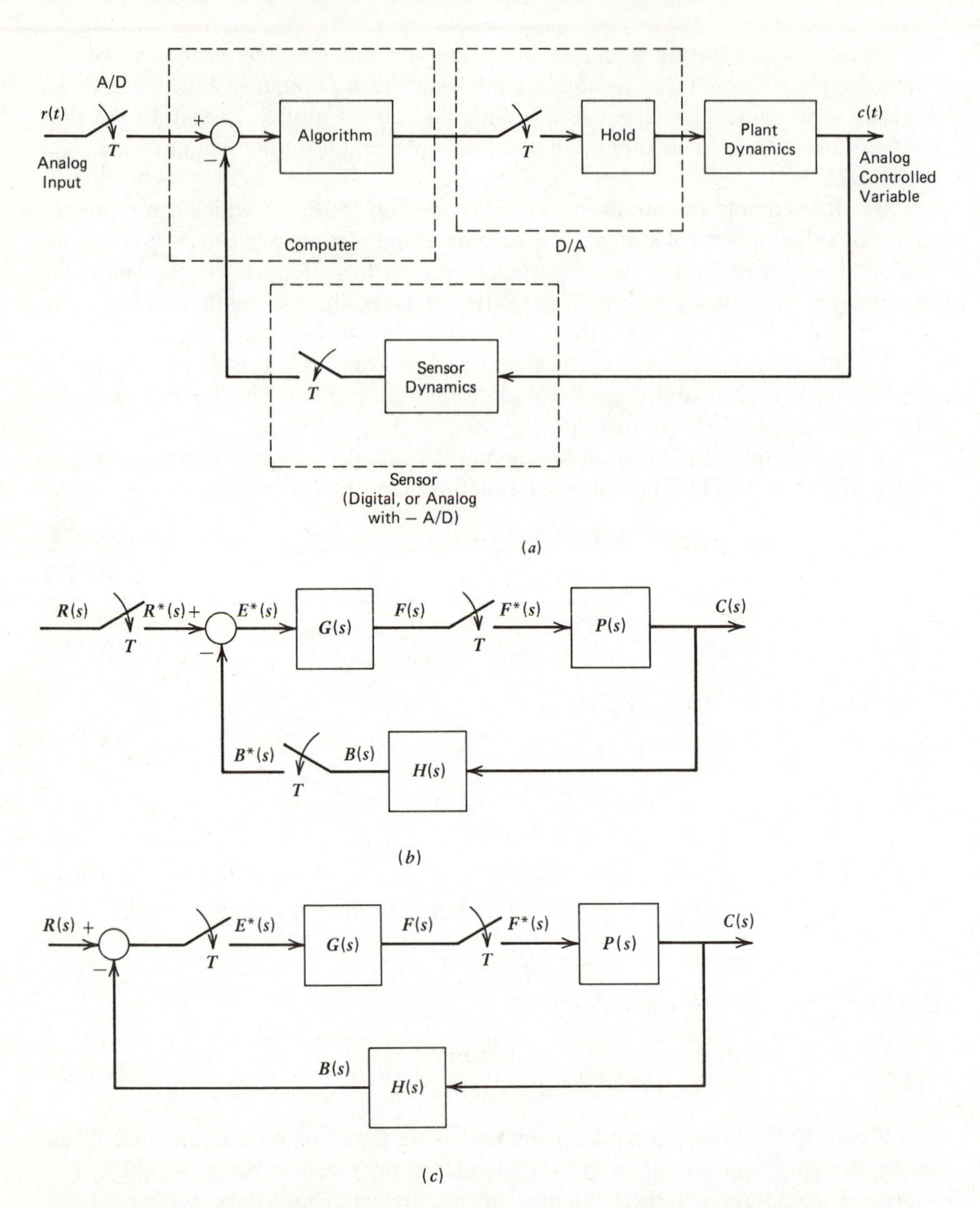

FIGURE 10.17 Three representations of a digital control system. (*a*) Component diagram. (*b*) Block diagram in the *s*-domain. (*c*) Equivalent representation with fewer samplers.

Thus

$$F(z) = G_1R(z) - G_1HG_3(z)A(z)$$
$$A(z) = G_2(z)F(z)$$
$$C(z) = G_3(z)A(z)$$

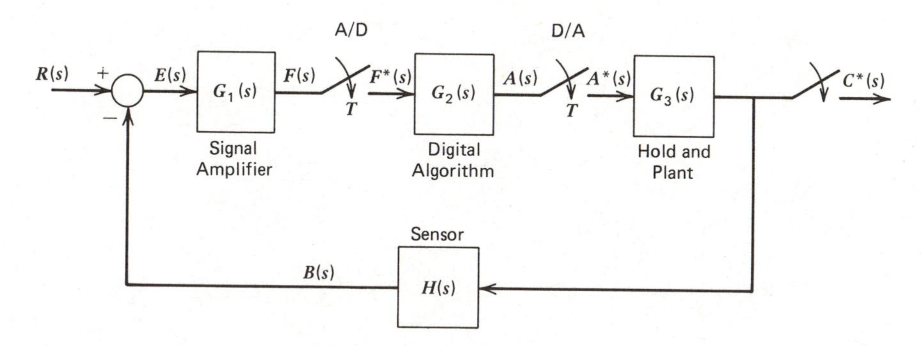

FIGURE 10.18 A digital system for which a pulse transfer function does not exist.

This gives

$$C(z) = \frac{G_2(z)G_3(z)G_1R(z)}{1 + G_2(z)G_3(z)G_1H(z)} \tag{10.6-32}$$

Since it is impossible to form the ratio $C(z)/R(z)$ from (10.6-32), we cannot obtain a transfer function for this system, and we cannot analyze it further without specifying a functional form for the input $r(t)$.

Note that any starred (sampled) Laplace quantity is *not* combined in product form in the manner of (10.6-14). The only variables to which (10.6-14) applies are those that have never been subjected to sampling or have passed through a dynamic operation after being sampled (the dynamic operation actually creates a new, unsampled variable).

It is not permissible to create a pulse transfer function for the system in Figure 10.18 by inserting a fictitious sampler at the input, because this would change the physics of the situation represented by the diagram (the analog input would be replaced by a train of impulses). Fictitious samplers are permissible at only the output, because they are simply a means of selecting the values of the output at the times of interest to us, namely, the sample times.

It is apparent that an analog input's point of entry into the diagram determines whether or not a transfer function can be found for that input. Figure 10.19*a* shows a digital control system with a disturbance input. To find the relation between $C(z)$ and $R(z)$, set $D(s) = 0$. This gives the diagram shown in Figure 10.19*b*. Using Table 10.5, we obtain Figure 10.19*c* and the desired transfer function:

$$\frac{C(z)}{R(z)} = \frac{G_c(z)G(z)}{1 + G_c(z)G(z)} \tag{10.6-33}$$

To find the relation between $C(z)$ and $D(z)$, set $R(s) = 0$ in Figure 10.19*a*; then note that

$$C(s) = G_p(s)\left[-D(s) + \frac{1 - e^{-Ts}}{s}B^*(s)\right]$$

$$B(s) = -G_c(s)C^*(s)$$

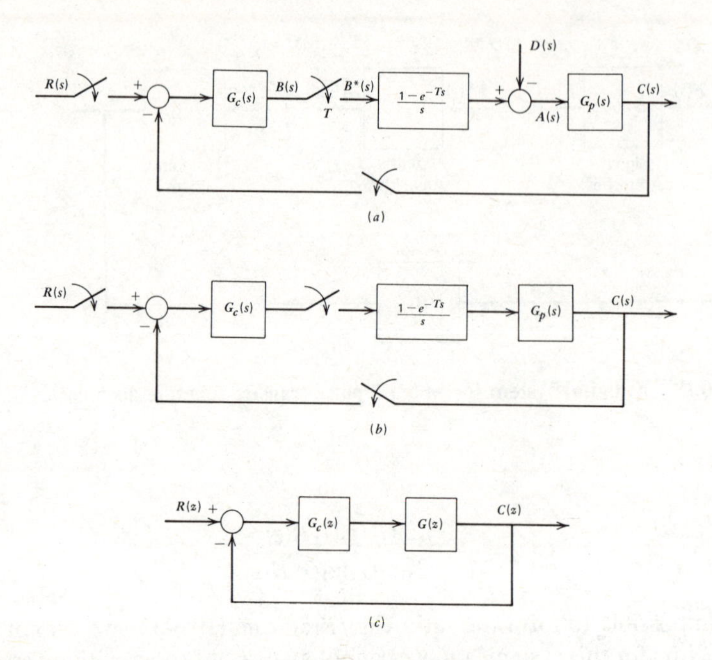

FIGURE 10.19 Digital control system with a disturbance input.

From Table 10.5, $G_2(s) = G_p(s)/s$. Therefore

$$C(z) = \frac{-zG_pD(z)}{z + (z-1)G_2(z)G_c(z)} \tag{10.6-34}$$

and thus no transfer function can be found for the disturbance input.

Sampling and Stability

The introduction of sampling can destabilize a system, as shown by Example 10.14.

Example 10.14

Compare the stability properties of the system shown in Figure 10.20 with and without a sample-and-hold on the error signal. Assume $K > 0$.

The analog system in Figure 10.20 has the transfer function

$$T(s) = \frac{K}{s + K + 3}$$

This system is stable for all $K > 0$. For the system in Figure 10.20*b*,

$$G(s) = (1 - e^{-sT})\frac{K}{s(s+3)} = (1 - e^{-sT})\frac{K}{3}\left(\frac{1}{s} - \frac{1}{s+3}\right)$$

$$G(z) = \frac{K}{3}\frac{z-1}{z}\left(\frac{z}{z-1} - \frac{z}{z - e^{-3T}}\right) = \frac{K}{3}\frac{1 - e^{-3T}}{z - e^{-3T}}$$

$$GH(z) = G(z)$$

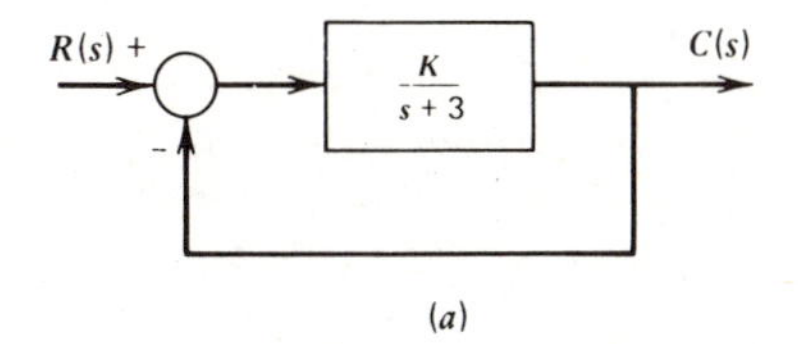

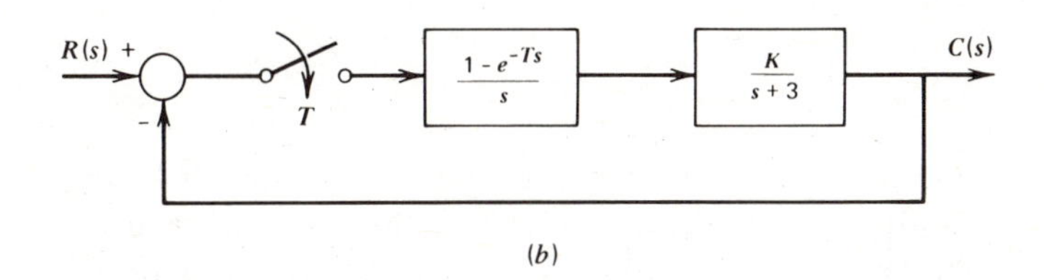

FIGURE 10.20 Instability caused by sampling. (*a*) Original analog system. (*b*) Analog system with a sample and hold inserted.

Thus, its transfer function is

$$T(z) = \frac{K}{3} \frac{1 - e^{-3T}}{z - e^{-3T} + K(1 - e^{-3T})/3}$$

For an unstable system, $|z| > 1$. This occurs if either

$$e^{-3T} > \frac{K+3}{K+3} = 1$$

or

$$e^{-3T} < \frac{K-3}{K+3}$$

The first condition is never true for $T > 0$, and we now consider the implications of the second condition. As the sampling frequency is increased ($T \to 0$), the stability of the system is improved (its behavior approaches that of the analog system). For example, with $T = 1$, the system is unstable for $K > 3.315$. For $T = 0.5$, instability occurs if $K > 4.723$.

10.7 DISCRETE-TIME MODELS OF HIGHER ORDER

We have thus far used first-order models to provide simple illustrations of the chapter's methods, but of course, higher order models often occur in digital control systems. For example, designing digital control systems sometimes requires a model of a dc motor whose input voltage is produced by a sample-and-hold device. Assume the motor is field controlled. Thus, no back emf is present. For now, we neglect the

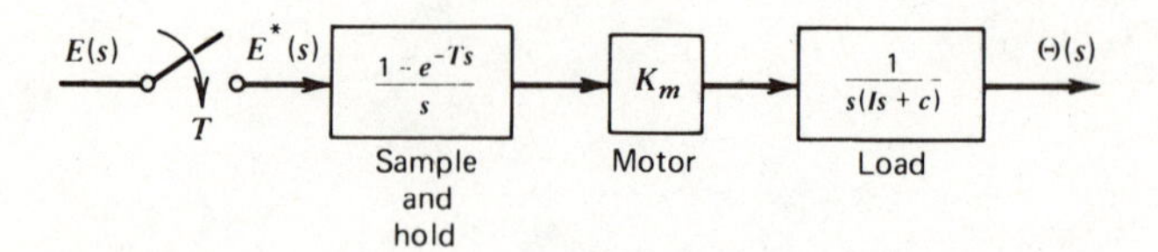

FIGURE 10.21 Block diagram of a dc motor-amplifier system with a sampled voltage input and an angular displacement as the output.

armature inductance of the motor, and lump its inertia with that of the load. If viscous friction is present and the output is angular displacement, the resulting model is shown in Figure 10.21. From the property stated in Table 10.5 for a zero-order hold,

$$G(z) = \frac{z-1}{z} G_2(z) = \frac{\Theta(z)}{E(z)}$$

where

$$G_2(s) = \frac{K_m}{s^2(Is+c)} = \frac{K}{s^2(s+b)}$$

The new constants are $K = K_m/I$, $b = c/I$. Expand $G_2(s)$ to compute $G_2(z)$ as follows:

$$G_2(s) = \frac{K}{b}\left[\frac{1}{s^2} - \frac{1}{b}\left(\frac{1}{s} - \frac{1}{s+b}\right)\right]$$

From Table 10.2,

$$G_2(z) = \frac{K}{b}\left[\frac{zT}{(z-1)^2} - \frac{1}{b}\left(\frac{z}{z-1} - \frac{z}{z-a}\right)\right]$$

where $a = e^{-bT}$. The overall transfer function is

$$G(z) = \frac{K}{b^2} \frac{(bT - 1 + a)z + 1 - a - bTa}{(z-1)(z-a)} \tag{10.7-1}$$

The denominator indicates a second-order model.

In addition to the transfer function form given by (10.7-1), discrete-time models can also appear in the reduced form; for example,

$$a_2 y(k+2) + a_1 y(k+1) + a_0 y(k) = b_1 v(k+1) + b_0 v(k) \tag{10.7-2}$$

or in state variable form, as

$$x_1(k+1) = a_{11}x_1(k) + a_{12}x_2(k) + b_1 u(k) \tag{10.7-3}$$

$$x_2(k+1) = a_{21}x_1(k) + a_{22}x_2(k) + b_2 u(k) \tag{10.7-4}$$

The techniques for converting from one model form to another are similar to those developed for the continuous-time case. For example, to construct a state model from a transfer function, use the same procedure as in Section 5.4, but replace $1/s$ with $1/z$.

Consider the discrete model

$$T(z) = \frac{X(z)}{U(z)} = \frac{1}{\alpha z^2 + \beta z + \gamma} = \frac{1/\alpha z^2}{1 + \beta/\alpha z + \gamma/\alpha z^2} \tag{10.7-5}$$

Cross multiply and rearrange to obtain

$$X(z) = \frac{1}{z}\left\{-\frac{\beta}{\alpha}X(z) + \frac{1}{z}\left[\frac{1}{\alpha}U(z) - \frac{\gamma}{\alpha}X(z)\right]\right\}$$

Define $X_1(z) = X(z)$, $X_2(z) = [U(z) - \gamma X(z)]/\alpha$. This gives the state equations

$$\left.\begin{aligned} x_1(k+1) &= -\frac{\beta}{\alpha}x_1(k) + x_2(k) \\ x_2(k+1) &= -\frac{\gamma}{\alpha}x_1(k) + \frac{1}{\alpha}u(k) \end{aligned}\right\} \tag{10.7-6}$$

To obtain a reduced model, cross multiply (10.7-5) directly, and revert to the time domain. The result is

$$\alpha x(k+2) + \beta x(k+1) + \gamma x(k) = u(k) \tag{10.7-7}$$

The process is identical to that for Laplace transforms. The characteristic equation for the second-order case can be written as

$$\alpha z^2 + \beta z + \gamma = 0 \tag{10.7-8}$$

As should be expected, different types of response occur depending on whether the roots of this equation are real or complex. In the first-order case treated earlier in this chapter, we saw that the basic solution form is

$$x(k) = Az^k$$

For two distinct roots z_1 and z_2, the free response in (10.7-7) is

$$x(k) = A_1 z_1^k + A_2 z_2^k \tag{10.7-9}$$

where A_1 and A_2 depend on two initial conditions, say, $x(0)$ and $x(1)$. The concept of a dominant root applies in discrete time as well. For example, if $|z_1| > |z_2|$, the z_1 term dominates the behavior. If either root is negative, the solution oscillates. If any root has a magnitude greater than unity, the system is unstable.

For repeated roots, the solution form becomes

$$x(k) = A_1 z_1^k + kA_2 z_1^k \tag{10.7-10}$$

Complex Roots

We have seen that oscillations can occur in the free response of a first-order discrete system, but these oscillations are restricted to a period of two time steps, because they are caused by a negative number raised to increasing integer powers. Free oscillations with a period different from two are possible for only higher order models with complex roots. Assume that these are of the form $z = -a \pm ib$. In this case, (10.7-9) is not convenient, so we seek a simpler form. This is obtained by using *DeMoivre's theorem* for a complex number raised to a power. The complex number $r = z_1 = -a + ib$ can be written in terms of its magnitude and phase angle as

$$r = |r|e^{i\theta}$$

Thus,

$$\begin{aligned} r_k &= |r|^k e^{ik\theta} \\ &= |r|^k(\cos k\theta + i \sin k\theta) \end{aligned} \tag{10.7-11}$$

where, in this case, $|r| = \sqrt{a^2 + b^2}$ and $\tan \theta = -b/a$. For $z_2 = -a - ib$, a conjugate relation is the result. Substituting these expressions into (10.7-9), we find that

$$x(k) = |r|^k[(A_1 + A_2) \cos k\theta + i(A_1 - A_2) \sin k\theta]$$

From the same logic as in the continuous-time case, the result is

$$x(k) = |r|^k B \sin (k\theta + \phi) \tag{10.7-12}$$

where r is the magnitude of the upper root $z_1 = -a + ib$, θ is its phase angle, and the constants B and ϕ depend on the initial conditions $x(0)$ and $x(1)$.

The response for the complex-roots case has a frequency of θ rad/time step. The model is stable if $|r| < 1$, neutrally stable if $|r| = 1$, and unstable otherwise. This completes all the cases that can occur, and the general stability criterion for linear discrete-time models can now be stated in terms of a unit circle in the z plane (Figure 10.22). In an nth-order system, if any one of the n roots lies outside the unit circle, the system is unstable. A root lies outside the circle if its magnitude is greater than 1.

The roots of the denominator of the transfer function (10.7-1) are $z = 1$ and $z = a$. Thus, the sampled-data motor system is neutrally stable, since one root lies on the

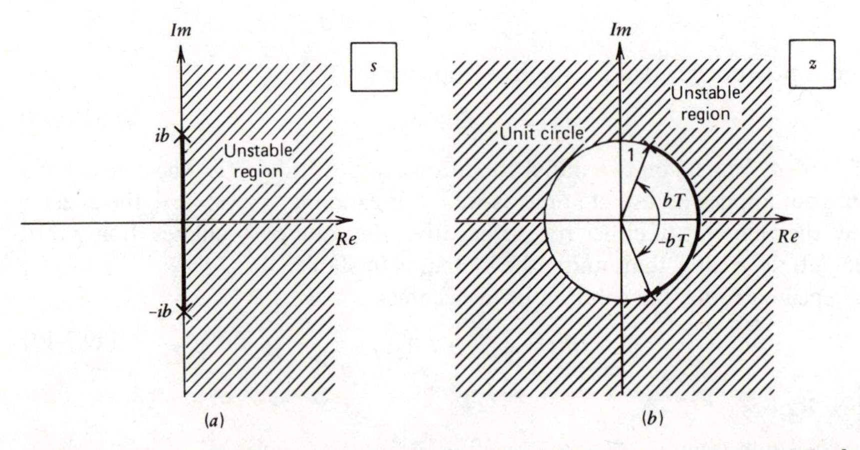

FIGURE 10.22 Stability criteria for continuous and discrete time systems. (*a*) In the *s*-plane, the unstable region is the entire right-half plane. (*b*) In the *z*-plane, the unstable region is that outside the unit circle at the origin. Equivalent locations of a pair of neutrally stable roots are also shown.

circumference of the unit circle while the other lies within the circle ($a < 1$ if $bT > 0$). The root $z = 1$ produces a constant term proportional to 1^k in the free response, while the root $z = a$ produces the term a^k, which decays with time.

Stability Tests

The Routh–Hurwitz criterion provides a convenient stability test that can be applied to the characteristic polynomial without solving for the roots. Originally developed to detect the occurrence of a root in the right half of the s plane, it can be converted for use with discrete models by employing the following transformation:

$$z = \frac{s+1}{s-1} \tag{10.7-13}$$

This transformation maps the inside of the unit circle in the z plane onto the entire left half of the s plane. Substitute z from (10.7-13) into the characteristic equation in terms of z. This gives a polynomial in s to which the Routh–Hurwitz criterion can be applied.

Example 10.15

Determine the stability conditions for the characteristic equation

$$\alpha z^2 + \beta z + \gamma = 0$$

With (10.7-13), the characteristic equation becomes

$$\alpha\left(\frac{s+1}{s-1}\right)^2 + \beta\left(\frac{s+1}{s-1}\right) + \gamma = 0$$

or

$$(\alpha + \beta + \gamma)s^2 + 2(\alpha - \gamma)s + (\alpha - \beta + \gamma) = 0$$

The result of the Routh–Hurwitz criterion says that no root lies in the right half of the s plane if $(\alpha + \beta + \gamma)$, $(\alpha - \gamma)$ and $(\alpha - \beta + \gamma)$ all have the same sign. Thus, the second-order discrete-time system is stable if and only if

$$\frac{\alpha - \gamma}{\alpha + \beta + \gamma} > 0 \tag{10.7-14}$$

and

$$\frac{\alpha - \beta + \gamma}{\alpha + \beta + \gamma} > 0 \tag{10.7-15}$$

The above method becomes tedious for higher-order systems. However, there are several other methods available for such situations. One of the most popular is *Jury's stability test* (Reference 3). It is given in general form in Appendix F. Jury's results for low-order systems are given in Table 10.6. The results for the second-order case are

TABLE 10.6 Jury's Stability Results for First-, Second-, and Third-Order Systems

Characteristic Equation	Stability Requirements	
1. $F(z) = b_1 z + b_0 = 0$	$b_1 > \lvert b_0 \rvert$	(1.1)
$b_1 > 0$		
2. $F(z) = b_2 z^2 + b_1 z + b_0 = 0$	$b_2 + b_1 + b_0 > 0$	(2.1)
$b_2 > 0$	$b_2 - b_1 + b_0 > 0$	(2.2)
	$b_2 > \lvert b_0 \rvert$	(2.3)
3. $F(z) = b_3 z^3 + b_2 z^2 + b_1 z + b_0 = 0$	$b_3 + b_2 + b_1 + b_0 > 0$	(3.1)
$b_3 > 0$	$b_3 - b_2 + b_1 - b_0 > 0$	(3.2)
	$b_3 > \lvert b_0 \rvert$	(3.3)
	$\lvert b_0^2 - b_3^2 \rvert > \lvert b_0 b_2 - b_1 b_3 \rvert$	(3.4)

equivalent to those obtained in Example 10.15. It is easier to show this equivalence if the equation $\alpha z^2 + \beta z + \gamma = 0$ is first normalized so that $\alpha = 1$.

Transient Performance Specifications

The Laplace transform of a sampled time function in Section 10.4 led naturally to the definition of the variable z as

$$z = e^{sT} \tag{10.7-16}$$

This formula is useful in relating the behavior of a time function as specified by its roots in the s plane to the location of the corresponding roots in the z plane. In Chapter 3, we saw how transient behavior can be characterized by the damping ratio, natural frequency, and time constant of the dominant root in the s plane. Therefore, (10.7-16) gives the root locations in the z plane required to produce the same transient behavior. The transformation (10.7-16) is not one-to-one. If $z = e^{s_1 T}$, then $z = e^{s_2 T}$ also, where

$$s_2 = s_1 + i\frac{2\pi n}{T}$$

and n is integer. However, if guard filters are used to prevent aliasing, we need not be concerned with the solutions for $n > 0$.

Consider a stable root pair $s = -a \pm ib$. The corresponding z values are

$$z = e^{-aT} e^{\pm ibT} = e^{-aT}(\cos bT \pm i \sin bT) \tag{10.7-17}$$

We have seen that purely imaginary roots $s = \pm ib$ correspond to z roots on the unit circle $z = \cos bT \pm i \sin bT$ (Figure 10.22). Horizontal lines of constant frequency and variable time constant in the s plane correspond to radial lines in the z plane (Figure 10.23*a*). These make an angle of $\theta = \pm bT$ with the positive real axis. For small time constants, the z roots lie close to the origin (these roots are denoted by ○ in both plots). As the time constant increases, the roots move in the direction of the arrows until the stability limit is reached (denoted by X). From this, we see that the *dominant* root in the z plane is that root lying closest (radially) to the unit circle.

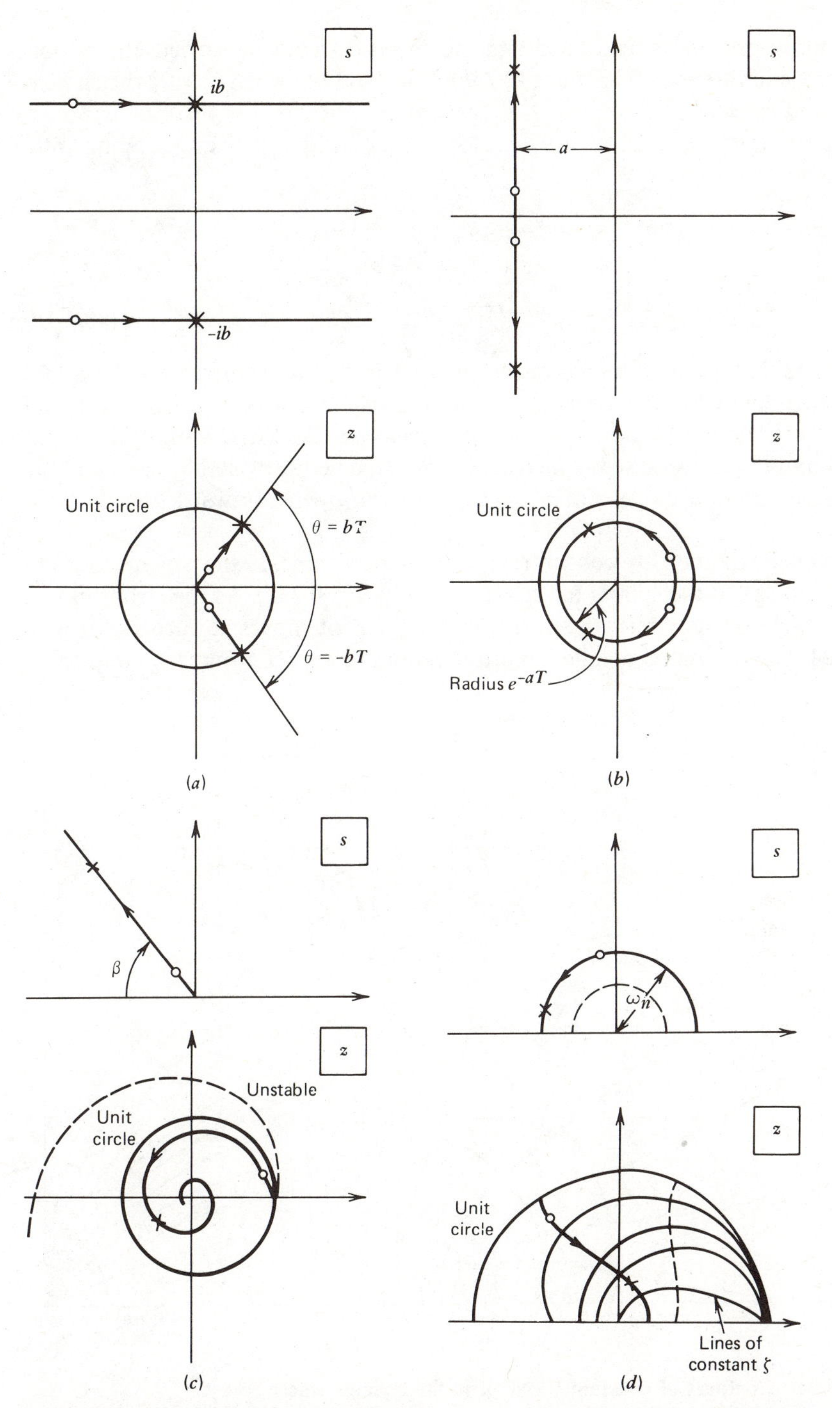

FIGURE 10.23 Equivalent root paths in the s plane and z plane. (*a*) Roots with the same oscillation frequency b. (*b*) Roots with the same time constant $\tau = 1/a$. (*c*) Roots with the same damping ratio ζ. (*d*) Roots with the same natural frequency ω_n.

If the time constant is held fixed and the damped frequency varied, the z roots move in a circle of radius e^{-aT} (Figure 10.23*b*). For a stable system ($a > 0$), the circle is within the unit circle.

To see the result of a root moving along a radial line of fixed damping ratio $\zeta = \cos \beta$, write the upper root as

$$s = -a + ib = -\zeta\omega_n + i\omega_n\sqrt{1-\zeta^2}$$

The z root is

$$z = e^{-\zeta\omega_n T} e^{i\omega_n T\sqrt{1-\zeta^2}} \tag{10.7-18}$$

For fixed ζ and T, the s root moves out from the origin as ω_n is increased; the z root rotates counterclockwise with decaying amplitude. The result is a logarithmic spiral (Figure 10.23*c*). The lower root $s = -a - ib$ produces a clockwise spiral that is not shown. As damping is decreased, β increases to 90° (the neutrally stable case), and the spiral becomes the unit circle. For $\beta \geqslant 90°$, the spiral opens outward (the unstable case).

Curves of constant ω_n are concentric circles in the s plane. In the z plane, they are lines perpendicular to the spirals for constant ζ (Figure 10.23*d*). As the s root moves counterclockwise on its circle, the corresponding z root moves toward the origin. Figure 10.24 gives a more detailed representation of the contours of constant ζ and ω_n.

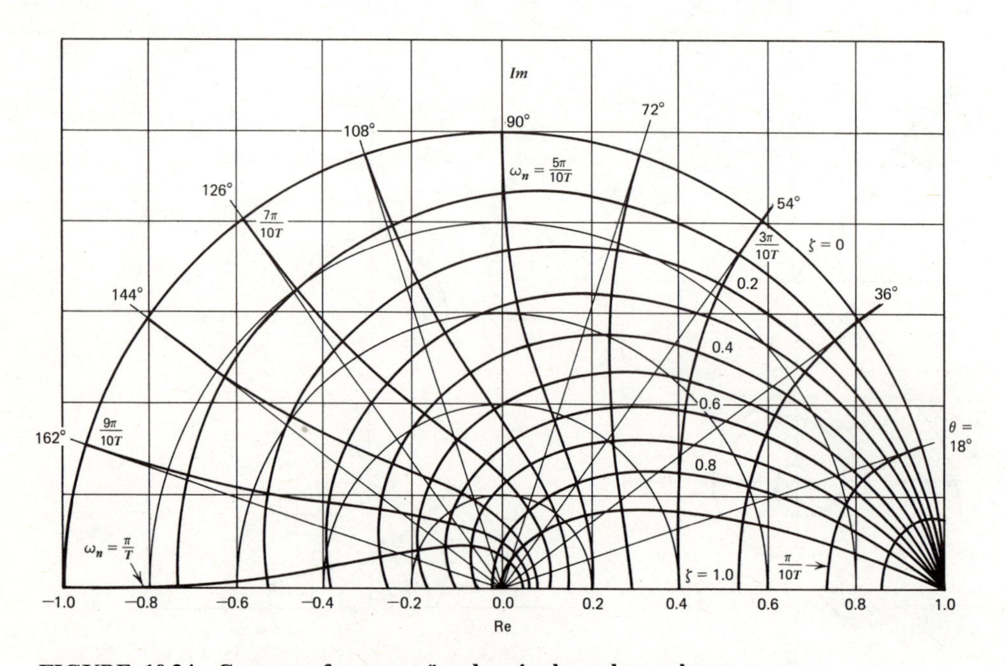

FIGURE 10.24 Contours of constant ζ and ω_n in the z-plane, where

$$z = e^{-\zeta\omega_n T} e^{i\omega_n T\sqrt{1-\zeta^2}} = re^{i\theta}$$

The radial lines are lines of constant θ. The semicircular lines are lines of constant r.

Equation (10.7-18) can be used to derive formulas for ζ, ω_n, and τ as functions of the z roots. First write z in polar form as

$$z = e^{sT} = re^{i\theta}, \qquad r > 0 \tag{10.7-19}$$

Comparison with (10.7-18) shows that

$$r = e^{-\zeta\omega_n T} \tag{10.7-20}$$

$$\ln r = -\zeta\omega_n T \tag{10.7-21}$$

and

$$\theta = \omega_n T\sqrt{1-\zeta^2} \tag{10.7-22}$$

Divide (10.7-21) by (10.7-22) to obtain

$$\frac{\ln r}{\theta} = -\frac{\zeta}{\sqrt{1-\zeta^2}}$$

Since ζ is defined only for stable systems ($\zeta > 0$), we must require that $r < 1 (\ln r < 0)$. With this in mind,

$$\zeta = -\frac{\ln r}{\sqrt{\theta^2 + \ln^2 r}} \tag{10.7-23}$$

Using this relation with (10.7-21) gives

$$\omega_n = -\frac{\ln r}{\zeta T} = \frac{1}{T}\sqrt{\theta^2 + \ln^2 r} \tag{10.7-24}$$

Finally, if $\zeta \leqslant 1$, the time constant is

$$\tau = \frac{1}{\zeta\omega_n}$$

which gives

$$\tau = -\frac{T}{\ln r}, \qquad \zeta \leqslant 1 \tag{10.7-25}$$

Figure 10.25 shows how the free response varies with root locations in the z-plane. Note how the rate of decay with time is faster for roots located closer to the origin. Note also that the oscillations due to negative real roots are fundamentally different from those generated by complex roots. For negative real roots, the frequency of oscillation is always the same; namely, the response changes sign every sample period. For complex roots the frequency depends on the root location (see Figure 10.23*b* for this effect).

To illustrate how the relations are applied, consider a system described by the following transfer function:

$$T(z) = \frac{z + p}{\alpha z^2 + \beta z + \gamma}$$

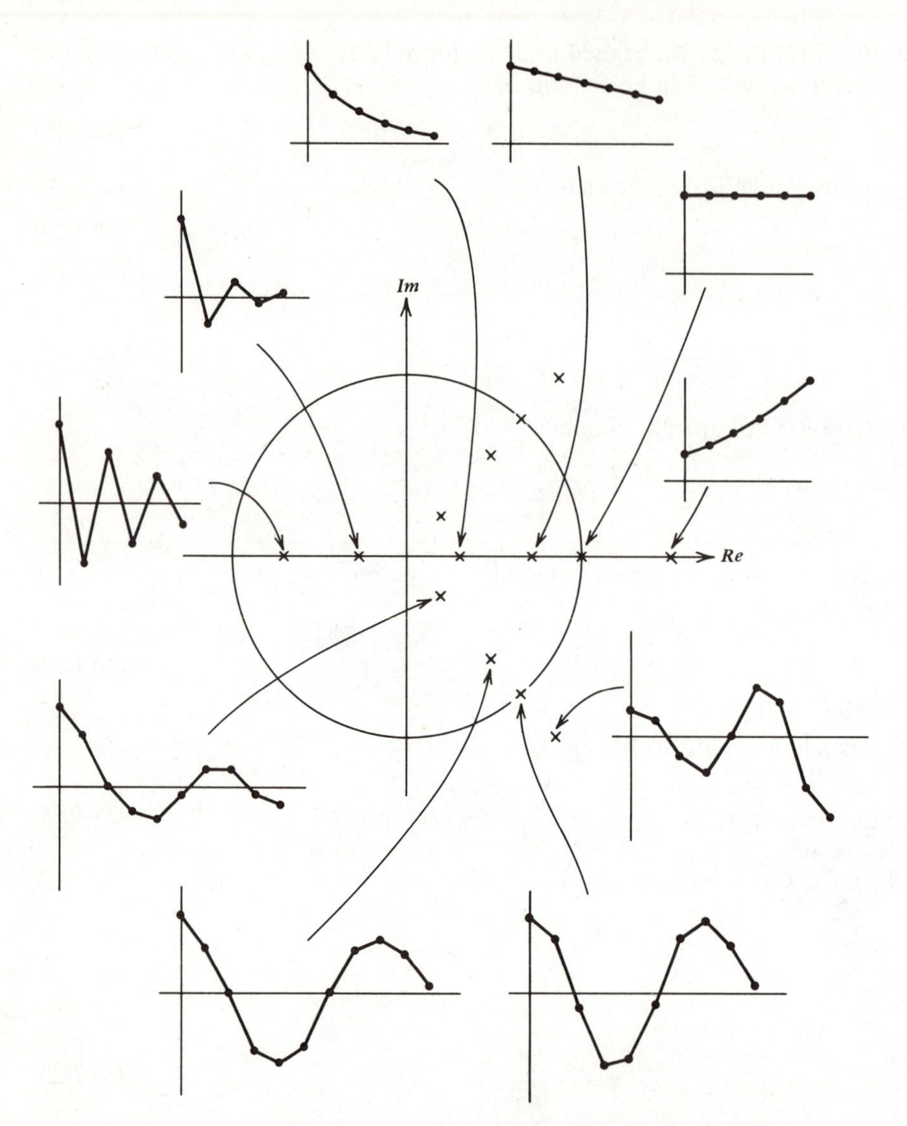

FIGURE 10.25 Free response as a function of root location in the z plane.

This could be the transfer function of a sampled-data system whose output is a continuous function of time. Thus, it would be natural to describe the desired transient response in terms of the specifications given in Section 3.5. The sampling time T will appear in some if not all of the parameters α, β, γ, and p. We assume that these can be selected to some extent by the designer.

Suppose that the step response of the preceding system is to have the following properties: (1) maximum percent overshoot $\leqslant 10\%$; (2) 100% rise time $\leqslant 5$; and (3) 2% settling time $\leqslant 20$. From the appropriate relations in Section 3.5, we can see that these are equivalent to the following:

1. $\zeta \geqslant 0.6$
2. $\omega_n \geqslant 3/5$
3. $\zeta\omega_n \geqslant 4/20$

We note in passing that the results are for a second-order system without any finite zeros, whereas the transfer function has a zero at $z = -p$. In Example 4.15, the effect of such a zero was shown to increase the overshoot. We temporarily neglect this effect here; the resulting design should be checked analytically or by simulation to see if an adjustment is necessary.

The stability requirement is implicit. For $\zeta \geqslant 0.6$, both roots of the denominator of $T(z)$ must be within the region enclosed by the spirals shown in Figure 10.26*a*. These can be plotted from (10.7-18) for a fixed T with $\zeta = 0.6$ and ω_n as a parameter. The lower spiral results from the other conjugate root for $\zeta = 0.6$. For $\omega_n \geqslant 3/5$, the z roots must lie to the left of the curve obtained from (10.7-18) with $\omega_n = 3/5$ and ζ as a parameter (Figure 10.26*b*). Finally, the third specification is a restriction on the real part of the s roots. This requires the z roots to lie within a circle of radius $R = \exp(-4T/20)$ (Figure 10.26*c*).

The design is achieved by selecting a suitable sampling period T, sketching the boundaries, and choosing the design parameters to place the z roots within the region allowed by all the boundaries. A technique that greatly eases this task is the root locus method. We will see an example of this later.

Forced Response

The forced response of discrete-time models can be computed by direct substitution if the solution form is known or by the systematic method of the z transform, as in Section 10.5. The only difference occurs when complex characteristic roots exist. In this case, the computations can be simplified by introducing a complex factor similar to the factor $R(-a+ib)$ in Table 4.6 and by using DeMoivre's theorem (10.7-11).

For example, consider a single-input, single-output second-order system of the form

$$\alpha x(k+2) + \beta x(k+1) + \gamma x(k) = \delta u(k) \tag{10.7-26}$$

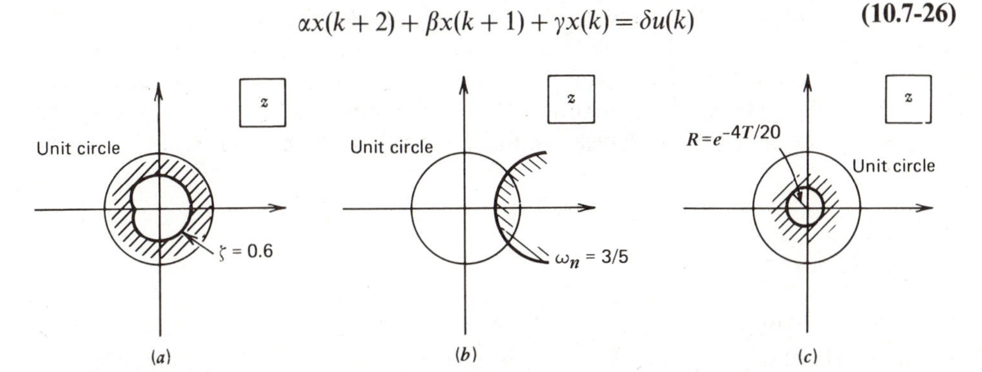

FIGURE 10.26 Transient response specifications in the z plane. (*a*) Overshoot constraint. (*b*) Rise-time constraint. (*c*) Settling-time constraint.

with complex conjugate roots $z = -a \pm ib = |r|e^{\pm i\theta}$. The unit-step response is form

$$x(k) = A + B|r|^k \sin(k\theta + \phi) \tag{10.7-27}$$

The unknown constants are A, B, and ϕ. They can be found by substituting $x(k)$ into (10.7-26) and using the initial conditions $x(0)$ and $x(1)$.

10.8 SUMMARY

Analysis of discrete-time models resembles that of continuous-time models in many ways. The Laplace and z transforms allow linear models to be reduced to algebraic problems in terms of the s and z variables, respectively, and the concept of a transfer function can be applied to both model types. The response can be obtained by partial fraction expansion and inverting the transform. The only difference occurs when we return to the time domain. For example, the unit-step function corresponds to $1/s$ in the Laplace domain but to $z/(z-1)$ in the z domain. The stability regions reflect this difference [$|z| < 1$ versus $Re(s) < 0$]. Thus, facility for continuous-time analysis is easily transferred to discrete-time analysis, and vice versa.

Important applications of discrete-time modeling and analysis occur where an analog system is coupled to a digital device to produce a sampled-data system. No natural sampling period exists in such cases, and the designer must select it unless it has been specified by some other source. Of course, no comparable decision must be made in continuous-time systems. Also, the location of the samplers in the system must be carefully considered before reducing the block diagram to find the pulse transfer function. This is because cascaded elements combine differently depending on whether or not a sampler is located between them. We have also seen that a transfer function will not exist for configurations that do not contain an input sampler (see Figures 10.18 and 10.19). Finally, the presence of sampling with some type of hold, usually a zero-order hold, can alter the stability characteristics of the system. This phenomenon is not seen in continuous-time systems.

In addition to altering stability characteristics, the presence of sampling changes the general dynamic behavior of a system. This can be examined with the important relation: $z = \exp(sT)$, which shows that the sampling time T greatly influences the discrete-time dynamics. This relation was also used to show how transient performance specifications in the s plane can be used to determine the required root locations in the z plane. These methods will be very important for understanding the performance of digital control systems, to be treated in Chapter 11.

REFERENCES

1. J. B. Peatman, *The Design of Digital Systems*, McGraw-Hill, New York, 1972.

2. R. W. Hamming, *Digital Filters*, Prentice-Hall, Englewood Cliffs, NJ, 1977.

3. E. I. Jury, *Theory and Application of the z-Transform Method*, John Wiley & Sons, New York, 1964.

PROBLEMS

10.1 Numerical algorithms preprogrammed into a calculator's memory are often in the form of difference equations. An example of this is the equation

$$x(k) = \tfrac{1}{2}\left[x(k-1) + \frac{u}{x(k-1)}\right]$$

where u is a constant (for example, a number entered from the keyboard).

(a) What function does this equation implement? (*Hint.* Find the equilibrium solution.)

(b) Simulate the algorithm for a value of u, say, $u = 4$, with several initial values $x(0)$, and investigate its dynamics.

10.2 One way of computing the integral

$$x(t) = \int_0^t u(\lambda)\, d\lambda$$

is to convert it into a difference equation. Let

$$x(kT) = \int_0^{kT} u(\lambda)\, d\lambda$$

where T is the discretization interval.

(a) Use the rectangular approximation

$$\int_{kT-T}^{kT} u(\lambda)\, d\lambda = Tu(kT - T)$$

to derive a difference equation for $x(kT)$ in terms of $x(kT - T)$ and $u(kT - T)$. What is the initial condition?

(b) Suppose $u(t) = t$ and we wish to find $x(t = 1)$. Select an interval T and simulate the equation found in (a) to obtain $x(1)$. Compare with the exact value.

10.3 Suppose we apply a piecewise-constant torque $m(kT)$ to an inertia I. Use the step response of the continuous-time model $I\dot{\omega} = m$ to obtain a difference equation model of the speed ω at the discrete times 0, $2T$, $3T$, ...

10.4 Determine the Fourier series representation of the function shown in Figure P10.4 and plot its spectrum.

10.5 For the signal whose spectrum was found in Problem 10.4, compute the frequency range that contains 99% of the signal's energy and estimate the required sampling frequency.

10.6 Compute the z transforms for the following functions:

(a) $\frac{1}{2}t^2$
(b) e^{-at}
(c) te^{-at}
(d) $\sin \omega t$

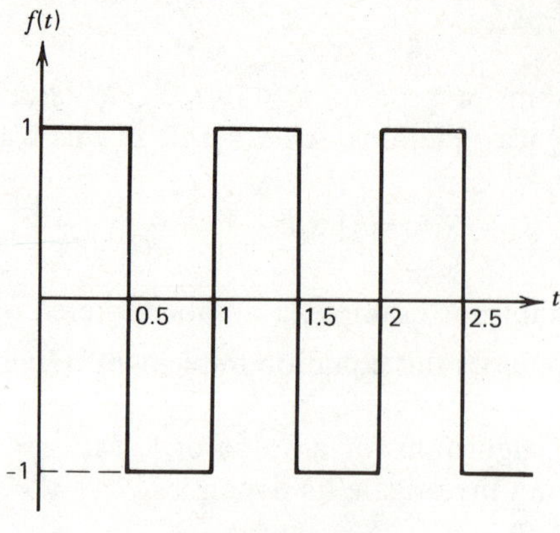

FIGURE P10.4

(e) $\cos \omega t$
(f) $e^{-at} \sin \omega t$
(g) A unit amplitude pulse of duration L.

10.7 Use the Euler method to compute the pulse transfer function of the models with dead time.

(a) $\dot{y} = ay(t) + by(t - D) + u(t)$
(b) $\dot{y} = ay(t) + bu(t - D)$

Assume that the dead time D is an integral multiple of the sampling period T, so that $D = nT$, $n =$ integer.

10.8 Use the Euler method to find the pulse transfer function of the model with numerator dynamics.

$$\dot{y} = ay + b\dot{u} + cu$$

10.9 Compute the forced response of the system $y(k + 1) = 0.5y(k) + u(k)$ for $u(k)$ as:

(a) k^2
(b) ka^k
(c) $a^k \sin \omega k$
(d) $a^k \cos \omega k$

10.10 Use the convolution summation (Table 10.4) to compute the forced response of the model $y(k + 1) = 0.1y(k) + u(k)$ for $0 \leqslant k \leqslant 5$, where

k	$u(k)$
0	2
1	3
2	4
3	2
4	1

10.11 Write a calculator or computer program to simulate the response of the model $y(k+1) = ay(k) + u(k)$, with $u(k)$ a given set of values.

10.12 Draw the block diagram and signal flow graph for the model

$$y(k+1) = ay(k) + bu(k) + cu(k+1)$$

10.13 Compute the step response of the model

$$y(k+1) = 0.1y(k) + bu(k) + cu(k+1)$$

10.14 Use the final value theorem to compute the steady-state error e between the input and the output for the system shown in Figure P10.14 with the input functions:

(a) $r(k) = 1$
(b) $r(k) = k$

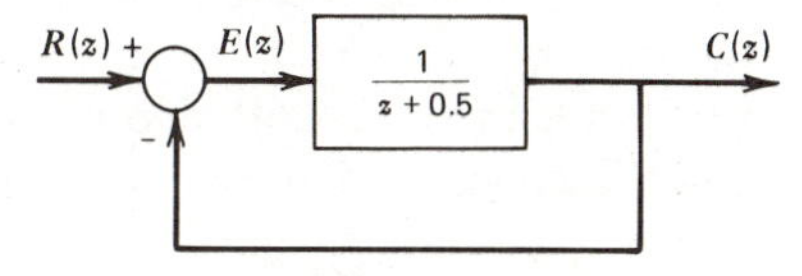

FIGURE P10.14

10.15 Find the pulse transfer function and difference equation for the following models where u is a sampled input (no hold is present):

(a) $\dot{y} = ay(t) + bu(t-D) \qquad D = nT$
(b) $\dot{y} = ay + b\dot{u} + cu$

10.16 Find the pulse transfer function and difference equation for the following systems when a sample-and-hold is applied to the input u:

(a) $G(s) = \dfrac{Y(s)}{U(s)} = K\dfrac{s+c}{s+a}$

(b) $G(s) = \dfrac{Y(s)}{U(s)} = K\dfrac{s}{s+a}$

10.17 Without solving for the roots, determine whether or not the following characteristic equations represent stable discrete-time systems.

(a) $10z^2 + 5z + 1 = 0$
(b) $10z^2 - 5z + 1 = 0$
(c) $10z^3 + 7z^2 + 5z - 3 = 0$

10.18 Determine the stability properties of the system with negative unity feedback around the element $G(s)$, where

$$G(s) = \frac{K}{s+5}$$

(a) with, and (b) without a sample-and-hold on the error signal.

10.19 The relationship between the Laplace variable s and the z variable is $z = e^{sT}$. Suppose that a first-order continuous-time system is required to have a time constant in the range $2 \leqslant \tau \leqslant 5$. Find the corresponding range for the characteristic root z that a sampled data system must have to meet this requirement. Consider two cases:

(a) $T = 1$.
(b) $T = 0.1$.

10.20 (a) Consider the analog element shown in Figure P10.20*a*. We wish to use this element to create a system with a time constant of 0.5. To do this, we place a gain K in series with the element and a unity feedback loop around the combination. This is shown in Figure P10.20*b*. Find the value of K required to give a time constant of 0.5.

(b) We now wish to implement the design in (a) in digital form. This is shown in Figure P10.20*c*. Find the closed-loop transfer function $T(z) = C(z)/R(z)$. Determine the equivalent root location for z that corresponds to a time constant of 0.5, using a sample period of $T = 0.4$. Determine the value of K that will give this response.

(c) For $T = 0.4$, compute the system's unit-step response for the values of K found in (a) and (b). Which value of K best meets the specification of a time constant of 0.5?

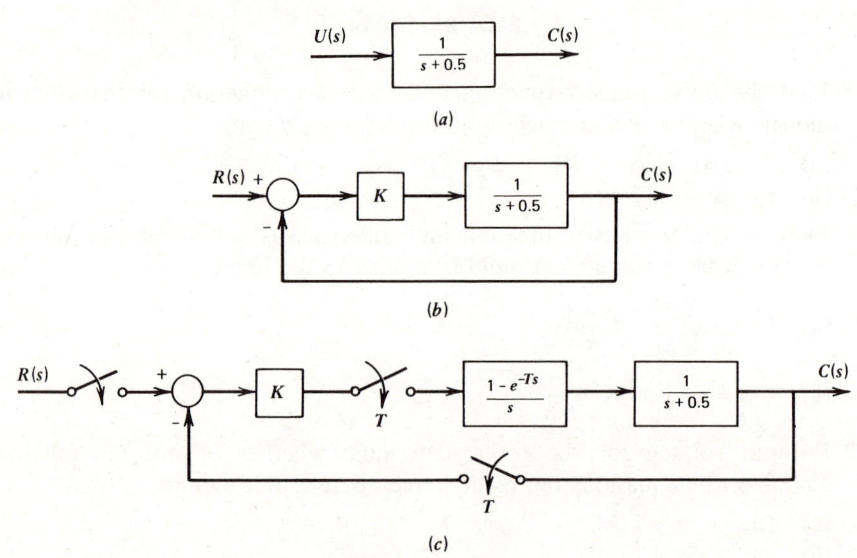

FIGURE P10.20

10.21 (a) Repeat parts (b) and (c) in Problem 10.20 using $T = 0.2$.

(b) Compare the two design methods for digital systems.

(1) Calculating K from s-plane specifications, followed by discretization.
(2) Discretization followed by calculating K from equivalent z-plane root locations.

10.22 Consider the system shown in Figure P10.22. Find the transfer function $C(z)/R(z)$ and determine the range of values of K for which the system will be stable for

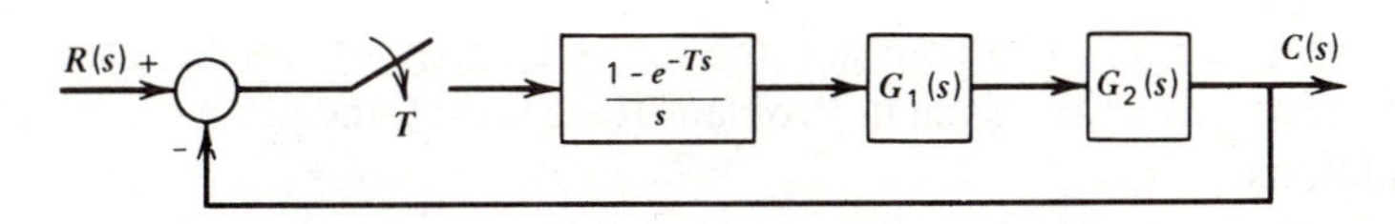

FIGURE P10.22

the following cases:

(a) $G_1(s) = K \qquad G_2(s) = \dfrac{1}{s(s+4)}$ for $T = 0.25$ and $T = 0.025$.

(b) $G_1(s) = \dfrac{K}{s+1} \qquad G_2(s) = \dfrac{1}{s+2}$ for $T = 0.1$

(c) $G_1(s) = K \qquad G_2(s) = \dfrac{1}{s^2 + 2s + 5}$ for $T = 0.1$

10.23 Consider the armature-controlled dc motor treated in Example 3.8.

(a) For the parameter values given in the example, find the pulse transfer function between the motor speed ω and the motor voltage v. Assume that the voltage v is applied through a sample and zero-order hold.

(b) The characteristic roots of the analog system are $s = -102.8 \pm i48.94$. Let the sampling period be $T = 0.001$ sec. Find the corresponding characteristic z roots and compare.

10.24 Put the transfer function models found in Problems 10.22 and 10.23 into state variable form.

10.25 Given the transfer function

$$\frac{C(z)}{R(z)} = T(z) = \frac{z+3}{2z(z^2-4)}$$

obtain a state variable model.

10.26 For the following transfer functions $T(z) = C(z)/R(z)$, find the free response. The initial conditions are $c(0) = 1$, $c(1) = 2$.

(a) $T(z) = \dfrac{1}{(z-0.5)(z+0.3)}$

(b) $T(z) = \dfrac{1}{(z-0.5)^2(z-0.1)}$

(c) $T(z) = \dfrac{z+0.2}{z^2 + 0.2z + 0.26}$

(d) $T(z) = \dfrac{1}{z^2 + 0.2z + 0.26}$

10.27 Develop a procedure to handle complex roots for discrete-time systems, using a factor similar to $R(-a+ib)$ in Table 4.6.

10.28 Use the z transform to compute the unit-step response for the systems whose transfer functions are given in Problem 10.26 (a), (c), and (d). Assume zero initial conditions.

10.29 Use the z transform to compute the unit-ramp response for the systems whose transfer functions are given in Problem 10.26 (a) and (c).

10.30 Plot the locations in the z plane that correspond to the following s-plane locations. Assume that $T = 0.01$.

(a) $s = -a \pm i3 \quad 0 \leqslant a \leqslant 2$
(b) $s = -a \pm i \quad 0 \leqslant a \leqslant 2$
(c) $s = -2 \pm ib \quad 0 \leqslant b \leqslant 2$
(d) $s = -3 \pm ib \quad 0 \leqslant b \leqslant 2$
(e) $\zeta = 0.707 \quad 0.1 \leqslant \tau \leqslant 1$
(f) $\omega_n = 2 \quad 0.5 \leqslant \zeta \leqslant 1$
(g) $\omega_n = 1 \quad 0.5 \leqslant \zeta \leqslant 1$

10.31 Compute the values of ζ, ω_n, and τ that correspond to each of the following characteristic roots in the z plane. The sampling time is 10 secs. (*Hint:* first put the roots into polar form.)

(a) $z = 0.1414 \pm i0.1414$
(b) $z = -3, -3$

CHAPTER ELEVEN

Digital Control Systems

Now that we have developed the prerequisite analytical techniques in Chapter 10, we can use them to design digital control systems. Section 11.1 discusses discrete-time versions of the analog PID family of control laws developed in Chapter 6 and shows that several variants are possible due to the approximations inherent in the discretization process. Section 11.2 develops the corresponding pulse transfer functions and illustrates how to compute the control gains. Many familiar design methods can also be applied to digital controllers, and Section 11.3 gives an example of a root-locus application. Since digital control algorithms are implemented in software, they may take many forms. Section 11.4 shows how to take advantage of this by deriving the control law directly from the desired response. Design methods based on frequency response are also useful, as shown in Section 11.5. They can be used for designing digital filters, which are algorithms for extracting the maximum information from noisy feedback measurements. These are introduced in Section 11.6. The chapter concludes with an example of a digital controller implemented with a microprocessor (Section 11.7).

11.1 DIGITAL IMPLEMENTATION OF CONTROL ALGORITHMS

There are two types of applications of digital computers to control problems. The first is *supervisory control*, where analog controllers are directly involved with the plant to be controlled, while the digital computer provides command signals to the controllers. The second application is *direct digital control* (*DDC*), where the digital device acts at the lowest levels of the system, in direct control of the plant.

There are several advantages to using digital computers in control systems. Complex control algorithms can be implemented easily, because the algorithm is created in software. Thus, the design difficulties that exist in electronic, pneumatic, and hydraulic implementation of control laws do not occur. Nonlinear algorithms to prevent saturation of the final control elements and ease start-up problems can be easily programmed (see Section 7.3). Since the software can be easily created, it can be easily changed either to alter the algorithm itself or change gain values. The latter is especially useful when the command signal drives the system far away from the linearization reference point and new gains are required for the new operating condition. Finally, the computer is valuable for keeping records of energy consumption, downtime, and so forth, for administrative purposes. The sometimes higher cost of the digital system can often be justified because of these advantages. Development of inexpensive microcomputer systems is making this justification less necessary all the time.

Because digital-control algorithms are not so limited by hardware as analog

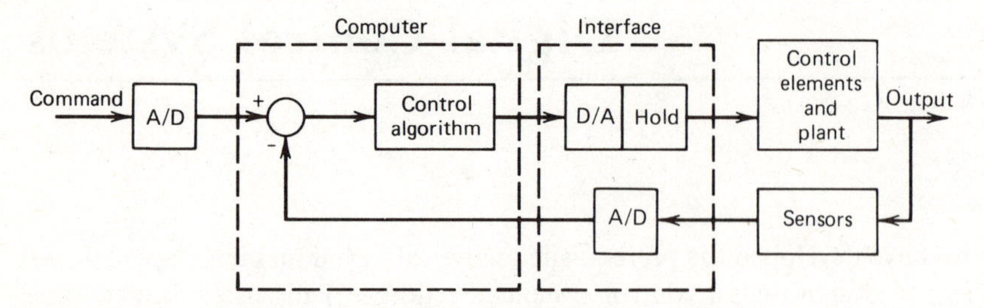

FIGURE 11.1 Structure of a digital control system.

controllers, the algorithm is essentially limited by only the designer's imagination. One of the reasons for the widespread use of the PI-control algorithm is that it can be physically implemented relatively easily. While this algorithm is often implemented in DDC systems with good reason, some of the justifications for its use historically are no longer relevant, and the designer should feel free to consider algorithms that were previously out of the question; some of these are considered in Section 11.4. Here, we limit ourselves to DDC versions of the PID family of algorithms.

The basic structure of a single-loop DDC controller is shown in Figure 11.1. The computer with its internal clock drives the D/A and A/D converters. It compares the command signals with the feedback signals and generates the control signals to be sent to the final control elements. These control signals are computed from the control algorithm stored in the memory. Slightly different structures exist, but Figure 11.1 shows the important aspects. For example, the comparison between the command and feedback signals can be done with analog elements and the A/D conversion made on the resulting error signal.

Digital Control of a Motor

An example of a DDC system is obtained by using a displacement sensor and digital controller with the dc motor system shown in Figure 10.21. The resulting position controller is shown in Figure 11.2*a*. The transform of the control algorithm is $G_c(z)$.

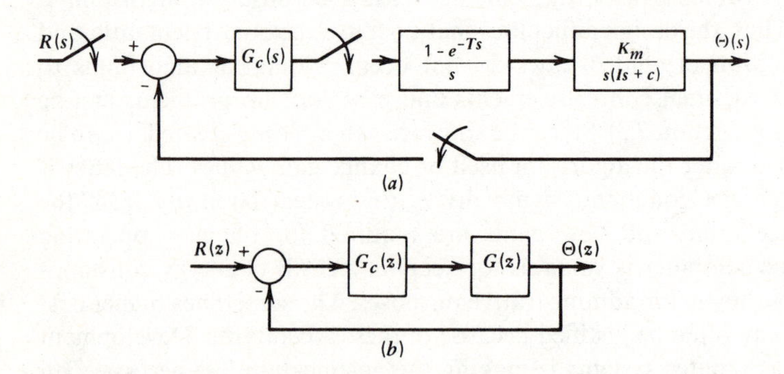

FIGURE 11.2 Digital control of a dc motor. (*a*) Sampled data diagram in the *s* domain. (*b*) Diagram in the *z* domain.

The transfer function of the sample-and-hold, motor-and-load combination was obtained in Section 10.7 (10.7-1) and is repeated here.

$$G(z) = \frac{K}{b^2} \frac{(bT - 1 + a)z + 1 - a - bTa}{(z-1)(z-a)} \tag{11.1-1}$$

where $K = K_m/I$, $b = c/I$, $a = e^{-bT}$, and T is the sampling period. This can be further simplified to

$$G(z) = K \frac{b_1 z + b_2}{(z-1)(z-a)} \tag{11.1-2}$$

where the identification of b_1 and b_2 is easily made.

With (11.1-2), we can analyze the system in the z domain entirely without further consideration of the sampling. The system diagram is shown in Figure 11.2*b*. The transfer function is

$$T(z) = \frac{\Theta(z)}{R(z)} = \frac{G_c(z)G(z)}{1 + G_c(z)G(z)}$$

or

$$T(z) = \frac{KG_c(z)(b_1 z + b_2)}{(z-1)(z-a) + KG_c(z)(b_1 z + b_2)} \tag{11.1-3}$$

Design of the control algorithm can proceed in a manner similar to that used in Sections 6.5, 6.6, and 6.7 for analog controllers. For example, we can try proportional control $G_c(z) = K_p$. In this case, (11.1-3) gives

$$T(z) = \frac{KK_p(b_1 z + b_2)}{z^2 + (KK_p b_1 - 1 - a)z + a + KK_p b_2} \tag{11.1-4}$$

In the analog case, only one parameter, K_p, has to be selected. However, in the digital case, we must often choose more parameters, because the sampling period and the machine's word length might be under the designer's control. For the sake of simplicity, let us assume that these quantities have been specified for this example and only the control gains are to be found.

For given values of K_m, I, c, and T, (11.1-4) is completely determined except for the value of K_p. The question now becomes *how* to determine K_p. The results of Example 10.15 can be used to determine the values of K_p that will give a stable system. The conditions that must be satisfied are (10.7-14) and (10.7-15). For the present system, these stability requirements are

$$\frac{1 - a - KK_p b_2}{KK_p(b_1 + b_2)} > 0 \tag{11.1-5}$$

$$\frac{2(1 + a) + KK_p(b_2 - b_1)}{KK_p(b_1 + b_2)} > 0 \tag{11.1-6}$$

With all parameters except K_p specified, these conditions are easy to use. However, the reader is urged to consider how the situation changes if T must also be selected.

In addition to the stability requirement, K_p must also be selected to yield the desired response characteristics, which are frequently difficult to visualize in the

z domain. For example, the damping ratio provides a simple test in the s domain for the existence of oscillatory behavior. No simple method exists in the z domain, because such behavior can be caused by negative as well as complex roots. Our conclusion is that it is often preferable to specify the response characteristics in terms of continuous-time features and to convert the root locations in the s domain into corresponding root locations in the z domain. The procedure for doing this was presented in Section 10.7.

Position and Velocity Algorithms

Extending this line of reasoning a bit further, we see that there are two ways of describing the form of the digital control law. One way is to select a controller transfer function $G_c(s)$ in the s domain, such as a PID-type law, and convert it into an equivalent $G_c(z)$ in the z domain. We will present methods for doing this shortly. The other way is to select $G_c(z)$ directly, as was done for the proportional controller in the previous example. With this approach, the question now arises: What are discrete-time equivalents of the integral and derivative actions? Two commonly used forms are the *proportional-plus-sum* and the *proportional-plus-difference* algorithms. They are

$$f(k) = K_p e(k) + K_I T \sum_{i=0}^{k} e(i) \tag{11.1-7}$$

$$f(k) = K_p e(k) + \frac{K_D}{T}[e(k) - e(k-1)] \tag{11.1-8}$$

where $f(k)$ and $e(k)$ are the control and error signals and T is the sampling period. The analogy to PI and PD control is apparent. In transform notation, they become

$$F(z) = \left(K_p + K_I T \frac{z}{z-1}\right) E(z) \tag{11.1-9}$$

$$F(z) = \left[K_p + \frac{K_D}{T}(1 - z^{-1})\right] E(z) \tag{11.1-10}$$

Generalization of PID control is made in the obvious way. These algorithms are the *position* versions of the PI and PD laws.

The *incremental* or *velocity* versions of the algorithms determine the *change* in the control signal $f(k) - f(k-1)$. To obtain them, decrement k by 1 in (11.1-7) and (11.1-8), and subtract the results from (11.1-7) and (11.1-8). This gives

$$f(k) = f(k-1) + K_p[e(k) - e(k-1)] + K_I T e(k) \tag{11.1-11}$$

$$f(k) = f(k-1) + K_p[e(k) - e(k-1)] + \frac{K_D}{T}[e(k) - 2e(k-1) + e(k-2)] \tag{11.1-12}$$

or

$$F(z) = \frac{(K_p + K_I T)z - K_p}{z-1} E(z) \tag{11.1-13}$$

$$F(z) = \frac{(K_p T + K_D)z^2 - (K_p T + 2K_D)z + K_D}{Tz(z-1)} \tag{11.1-14}$$

Suppose that the control signal $f(k)$ is a valve position. The position version of the algorithm is so named because it specifies the valve position directly as a function of the error signal. The incremental algorithm, on the other hand, specifies the change in valve position. The incremental version has the advantage that the valve will maintain its last position in the event of failure or shutdown of the control computer. Also, the valve will not "saturate" at start-up if the controller is not matched to the current valve position. In addition to having these safety features, the incremental algorithm is also well suited for use with incremental output devices, such as stepper motors. These differences between the position and velocity algorithms do not show up in a transform analysis, for (11.1-13) reduces to (11.1-9), and (11.1-14) reduces to (11.1-10).

Some Practical Considerations

Some practical aspects of implementing these digital algorithms that do not appear in analog design are related to the step changes that occur in the digital signals between the sample times and to the finite word length of the control computer.

Consider the integral term in (11.1-7). The change in this term is given in (11.1-11) and is $K_I Te(k)$. If T and $e(k)$ are small, the finite word length of the machine can result in a zero change in the integral output. For example, assume that the word length of the controller is 12 bits. If the full-scale value is 4095, the smallest nonzero value that can be represented is 1. If $K_I = 0.001$ and $T = 1$ second, any value of the error $e(k)$ less than 1000 (24% of full scale) will result in the term $K_I Te(k)$ being less than 1. The change in the output of the integral mode will be zero. The fact that this nonzero error causes no change in the control signal results in a steady-state offset error, which never occurs with I action in analog systems.

Two remedies are available. The first is to improve the resolution by increasing the word length of the computer, but this can be expensive. An alternative solution is modifying the software. Before the term $K_I Te(k)$ is computed, any "ineffective" portion of $e(k)$ is removed and saved to be added to the next error sample. In the preceding example, if $e(k) = 1200$, the additional value of 200 will not influence the controller's output [the output will be 1 both for $e(k) = 1000$ and $e(k) = 1200$]. So the ineffective portion of value 200 is removed and saved. If the next error sample is 1900, it is modified by the remainder term to become 2100. The controller output is then 2, not 1.

The output from D action in analog controllers is constant if the error signal increases at a constant rate. However, D action in digital controllers can produce a fluctuating output for such an error signal. The effect again results from the round-off required by the finite word length of the machine. The derivative term from (11.1-8) is $K_D[e(k) - e(k-1)]/T$. Suppose $K_D = 1000$ and $T = 1$ sec, with a 12-bit controller. If the error signal is increasing at a rate of 0.5 parts/sec, the sampled error series looks like 0, 1, 1, 2, 2, 3, 3, ... The resulting value of the derivative term will be 1000, 0, 1000, 0, ... This behavior is referred to as *derivative-mode kick*. It can be reduced by improving the resolution or using an improved approximation to the derivative.

Increasing the sampling period T will not eliminate the preceding problems with the integral and derivative terms. The dynamics of the discrete-time system are heavily influenced by the value of T; thus, the values of K_I and K_D depend on T. Increasing T in order to increase the product $K_I T$ and to decrease the ratio K_D/T might not work, because the correct values of K_I and K_D will shift accordingly. Finally, the larger T is, the less accurate the discrete approximation.

The approximation to the derivative can be improved by using values of the sampled error signal at more instants. For example, in the velocity algorithm (11.1-12), we can replace the D-action term with one obtained from a four-point central-difference technique. Let m be the mean of the previous four error samples.

$$m = \frac{e(k) + e(k-1) + e(k-2) + e(k-3)}{4} \tag{11.1-15}$$

For $\hat{e}(k) = e(k) - m$, the new D-action term is

$$\begin{aligned} f_D(k) &= \frac{K_D}{4T}\left[\frac{\hat{e}(k)}{1.5} + \frac{\hat{e}(k-1)}{0.5} + \frac{\hat{e}(k-2)}{0.5} + \frac{\hat{e}(k-3)}{1.5}\right] \\ &= \frac{K_D}{6T}[e(k) + 3e(k-1) - 3e(k-2) - e(k-3)] \end{aligned} \tag{11.1-16}$$

This requires slightly more programming and storage for two additional values of the error sample.

The I-action term in (11.1-7) represents the rectangular integration formula, and its accuracy can be improved by substituting a more sophisticated algorithm, such as the trapezoidal rule. With this, the I-action term becomes

$$f_I(k) = \sum_{i=0}^{k} \tfrac{1}{2}[e(i) + e(i-1)]K_I T \tag{11.1-17}$$

The improvements given by (11.1-16) and (11.1-17) can also be applied to the velocity algorithm. The steps are similar to those used to derive (11.1-11) and (11.1-12).

Another form of derivative kick occurs when the command input is a step function. The D action is the most sensitive to the resulting rapid change in the error samples. This effect can be eliminated by reformulating the control algorithm as follows (Reference 1, chapter 11). To do this, I action must be included. The velocity algorithm for PID control is

$$\begin{aligned} f(k) &= f(k-1) + K_p[e(k) - e(k-1)] + K_I T e(k) \\ &\quad + \frac{K_D}{T}[e(k) - 2e(k-1) + e(k-2)] \end{aligned} \tag{11.1-18}$$

The error is $e(k) = r(k) - c(k)$, where r and c are the set point and output. The key step is to treat r as a constant temporarily and to write the algorithm in terms of r and c. This gives

$$\begin{aligned} f(k) &= f(k-1) + K_p[c(k-1) - c(k)] + K_I T[r - c(k)] \\ &\quad + \frac{K_D}{T}[-c(k) + 2c(k-1) - c(k-2)] \end{aligned} \tag{11.1-19}$$

TABLE 11.1 Common Discrete-Time Approximations to Analog PID Controllers

$c(k)$ = measured output	$f(k)$ = control signal
$r(k)$ = command input	$e(k)$ = error signal

1. PID: position version [see (11.1-7) and (11.1-8)].

$$f(k) = K_p e(k) + K_I T \sum_{i=0}^{k} e(i) + \frac{K_D}{T}[e(k) - e(k-1)]$$

2. PID: velocity version [see (11.1-11) and (11.1-12)].

$$f(k) = f(k-1) + K_I T e(k) + \frac{K_D}{T}[e(k) - 2e(k-1) + e(k-2)]$$

3. PID with no derivative kick [see (11.1-19)].

$$f(k) = f(k-1) + K_p[c(k-1) - c(k)] + K_I T[r(k) - c(k)]$$
$$+ \frac{K_D}{T}[-c(k) + 2c(k-1) - c(k-2)] \qquad K_I \neq 0$$

4. Improved D action using four-point central differences [see (11.1-16)].

$$f_D(k) = \frac{K_D}{6T}[e(k) + 3e(k-1) - 3e(k-2) - e(k-3)]$$

5. Improved I action using trapezoidal integration [see (11.1-17)].

$$f_I(k) = \sum_{i=0}^{k} \frac{1}{2}[e(i) + e(i-1)]K_I T$$

The set point r can now be replaced with $r(k)$. Note that integral action is required since r now appears only in this term. These control laws are summarized in Table 11.1.

11.2 PULSE TRANSFER FUNCTIONS FOR DIGITAL CONTROL LAWS

Equations (11.1-7) through (11.1-14) are digital versions of the PID family of control laws that were obtained by mimicking the attributes of the analog versions. For example, the sum in the I-action term in (11.1-7) is supposed to represent the behavior of an analog integrator.

It is instructive to see how a digital control algorithm can be obtained from the analog law by other means. Two common ways of converting a continuous-time model into a discrete-time form are the Euler and the z-transform methods. The PID algorithm in analog form is

$$f(t) = K_p e(t) + K_I \int_0^t e\,dt + K_D \frac{de}{dt} \tag{11.2-1}$$

or its equivalent

$$\frac{F(s)}{E(s)} = K_p + \frac{K_I}{s} + K_D s \tag{11.2-2}$$

Differentiating (11.2-1) gives

$$\frac{df}{dt} = K_p \frac{de}{dt} + K_I e + K_D \frac{d^2 e}{dt^2} \tag{11.2-3}$$

Applying the Euler method with a step size T results in a difference equation that is identical to the velocity algorithm given by (11.1-18). Thus, this algorithm suffers from the same inaccuracies as the Euler method *unless* the algorithm is applied in the manner to be indicated. For later reference, the transfer function of (11.1-18) is

$$\frac{F(z)}{E(z)} = \frac{a_1 z^2 + a_2 z + a_3}{z(z-1)} \tag{11.2-4}$$

where

$$a_1 = K_p + K_I T + a_3 \tag{11.2-5}$$

$$a_2 = -(K_p + 2a_3) \tag{11.2-6}$$

$$a_3 = \frac{K_D}{T} \tag{11.2-7}$$

The line of reasoning is easier to follow if we assume that no D action is present. With $K_D = 0$, (11.2-4) becomes

$$\frac{F(z)}{E(z)} = \frac{a_1 z + a_2}{z-1} \tag{11.2-8}$$

with $a_3 = 0$ and

$$a_2 = -K_p = K_I T - a_1 \tag{11.2-9}$$

Finally, with $K_D = 0$, use Table 10.2 to find the pulse transfer function equivalent to (11.2-2). This is

$$\begin{aligned}\frac{F(z)}{E(z)} &= K_p + \frac{K_I z}{z-1} \\ &= \frac{(K_p + K_I)z - K_p}{z-1}\end{aligned} \tag{11.2-10}$$

which is of the form (11.2-8) with

$$a_1 = K_p + K_I \tag{11.2-11}$$

$$a_2 = -K_p \tag{11.2-12}$$

Thus, for PI control, the Euler and z-transform operations lead to the same *form* of the digital control law transform (11.2-8). The only difference is in the relation of the coefficients a_1 and a_2 to T, K_I, and K_p. We will show that this difference is irrelevant for many applications and that only the form of (11.2-8) is important.

If D action is used in the algorithm, the Euler and z-transform methods yield the same discrete-time form. Following the same procedure as before, it can be shown that the Euler method and Table 10.2 both give a control law of the form (11.2-4). Of course, the values of a_1, a_2, and a_3 will be different for the two methods. Table 11.2 summarizes the commonly used pulse transfer function equivalents for the PID family of control laws.

Example 11.1

Consider PI control of the first-order plant whose transfer function is $1/(2s+1)$. The block diagram in Figure 6.39 applies, with $I=2$, $c=1$. For analog control, the appropriate transfer function is (6.6-9). Let us suppose that no oscillations are wanted with a step input, so that $\zeta=1$ is specified. Assume that the desired time constant is $\tau=1$. This gives $K_p=3$, $K_I=2$.

Design a digital control law for this system as shown in Figure 11.3*a*. Assume that the sampling rate is one-tenth of the desired time constant, so that $T=0.1$. Neglect any disturbances.

We demonstrate two approaches to the problem.

(a) Use the procedure in Example 10.13 to convert the plant's transfer function with a zero-order hold into one in the z domain. The result is given in Figure 11.3*b*. $G_c(z)$ is the control law, which is designed by converting the performance specifications in the s-domain into equivalent ones in the z-domain. Refer to (10.7-18), and note that for $\zeta=1$, the corresponding z roots are repeated and equal to

$$z = e^{-\omega_n T} \tag{11.2-13}$$

where ω_n is arbitrary. Its value can be obtained by noting that $\zeta=1$ and $\tau=1$ imply that $s=-1,-1$. Thus, $\omega_n=1$. For $T=0.1$, the two required roots are both located at $z=\exp[-0.1]=0.905$.

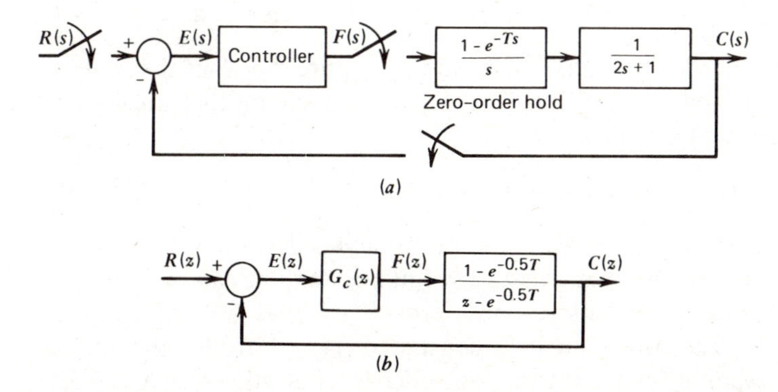

FIGURE 11.3 Digital control of a first-order plant. (*a*) s domain. (*b*) z domain.

TABLE 11.2 Pulse Transfer Function Approximations for PID Control Laws

$e(k)$ = error signal $\qquad$ $f(k)$ = control signal

1. P action.

$$\frac{F(z)}{E(z)} = a_1$$

2. I action.

$$\frac{F(z)}{E(z)} = \frac{a_1 z}{z-1}$$

3. PI using the Euler approximation or Table 10.2 [see (11.2-8)].

$$\frac{F(z)}{E(z)} = \frac{a_1 z + a_2}{z-1}$$

4. PID using the Euler approximation [see (11.2-4)].

$$\frac{F(z)}{E(z)} = \frac{a_1 z^2 + a_2 z + a_3}{z(z-1)}$$

5. PID using Tustin's approximation (see Problems 11.7 and 11.8).

$$\frac{F(z)}{E(z)} = \frac{a_1 z^2 + a_2 z + a_3}{z^2-1}$$

Using the form of PI control from (11.2-8), we obtain

$$G_c(z) = \frac{a_1 z + a_2}{z-1} = \frac{F(z)}{E(z)} \tag{11.2-14}$$

$$\frac{C(z)}{R(z)} = \frac{(1-b)(a_1 z + a_2)}{z^2 + (a_1 - 1 - b - a_1 b)z + b + a_2 - a_2 b} \tag{11.2-15}$$

where $b = e^{-0.5T}$. With $T = 0.1$, the values of a_1 and a_2 required to place the roots at $z = 0.905$ are $a_1 = 2.8996$ and $a_2 = -2.716$. The difference equation that the control computer must implement is, from (11.2-14),

$$f(k) = f(k-1) + a_1 e(k) + a_2 e(k-1) \tag{11.2-16}$$

The response to a unit-step input can be computed analytically from (11.2-15). The analog controller gives a response that crosses $c = 1$ at $t = 2$ and has an overshoot of 2.5% at $t = 3$. The digital controller's response crosses $c = 1$ at $t = 2$ and has an overshoot of 2.2% at $t = 2.9$. After four time constants, $c(4) = 1.018$ for the analog controller, while the response for the digital controller reaches this value at $t = 3.5$. Thus, the digital system responds similarly to the analog system. The overshoot in both cases results from the numerator dynamics. The slight differences in response result from the discrete approximations for the transfer function of the PI-control law and from the effects of the sample-and-hold operation.

(b) The second approach is to convert the acceptable analog control law with $K_p = 3$, $K_I = 2$ into discrete form. Inserting these values into the Euler coefficients (11.2-9), we obtain $a_1 = 3.2$ and $a_2 = -3$. From (11.2-15), the characteristic roots are $z = 0.926, 0.870$. The system is approaching the instability region where $|z| > 1$.

For the equivalent z-transform method, (11.2-11) and (11.2-12) give $a_1 = 5$, $a_2 = -3$, and $z = 0.854 \pm i0.276$. The system is stable but oscillatory.

The response for the Euler approximation resembles that for the design in (a). However, for the equivalent z-transform method, the oscillatory response has a maximum overshoot of 43%. The performance obtained with this method obviously does not meet the specifications.

It is impossible to draw general conclusions about the relative merits of the three approximation methods on the basis of one example. However, the example clearly shows the undesirable response that can be generated when using a discrete approximation with the gain values computed for an analog control law. Decreasing the sample period T improves the approximation, but it is not always possible to do so. For example, it is not unusual to have an A/D conversion time greater than 0.1 sec. The sample period must be large enough to accommodate this time as well as that required for D/A conversion and processing the control algorithm. At present, there are significant application areas, like robotics, in which these times are large enough to limit the performance of the control system in a serious way. A good design requires a balance between the complexity of the algorithm and the speed of the computer.

Comparison of Methods

We can categorize the design methods that could be used as follows:

Method 1

The controller design is done in the s domain, and the gain values K_p, K_I, K_D, are computed using the continuous-time methods Chapters 3, 4, 6, and 7. The resulting analog control law $G_c(s)$ must then be converted to discrete-time form with one of the approximation techniques given in Table 11.1. (This was the approach in Example 11.1*b*.)

Method 2

The performance specifications are given in terms of the desired continuous-time response and/or desired root locations in the s plane. From these the corresponding root locations in the z plane are found, and a discrete control law $G_c(z)$ is designed using one of the forms given in Table 11.2. (This was the method used in Example 11.1*a*.)

Method 3

The performance specifications are given in terms of the desired discrete-time response and/or desired root locations in the z plane. The rest of the procedure follows Method 2.

Obviously, Method 3 is the most direct, since the s plane is bypassed entirely. However, since most of our applications involve analog plants, it is difficult to state specifications in the z domain. Therefore, the approach described in Method 2 is the most practical one to use. It avoids the approximation errors that are inherent in Method 1.

If the sampling period is small, the approximations used in Method 1 can be successfully applied. The technique is widely used for two reasons. When existing analog controllers are converted to digital control, the form of the control law and the values of its associated gains are known to have been satisfactory. Therefore, a digital version is designed. Second, because analog design methods are well established, many engineers prefer to take this route and then convert the design into a discrete-time equivalent. Other methods for developing discrete equivalents from analog transfer functions can be developed with frequency response techniques (these are treated in advanced works on digital control; see Reference 2, for example).

11.3 ROOT LOCUS FOR DISCRETE-TIME SYSTEMS

The root locus simply presents the locations of the roots of an equation in terms of some parameter. Thus, the root locus plotting guides apply to the characteristic equation of a discrete-time system without any modifications. The variable s is simply replaced with the variable z. However, the interpretation of the resulting root locus plot is different for discrete-time systems because of the different relationships between the s and z variables and the time response. These relationships are summarized in Section 10.7 (Figure 10.23). Example 11.2 illustrates the method.

Example 11.2

Design a digital version of integral control to be used with a first-order plant whose transfer function is

$$G_p(s) = \frac{1}{\tau s + 1} \tag{11.3-1}$$

where $\tau = 2$ sec. When a unit-step input is applied, the minimum overshoot should be less than 20%, the 100% rise time should be less than 5 sec, and the 2% settling time less than 20 sec.

With a sample-and-hold device, the pulse transfer function for it combined with the plant can be found with the results in Table 10.5. It is

$$G_p(z) = \frac{1 - a}{z - a} \tag{11.3-2}$$

$$a = e^{-T/\tau} \tag{11.3-3}$$

The discrete form of the analog I action K_I/s is $K_1 z/(z-1)$. See Table 11.2. The relation between K_1 and K_I is irrelevant for our purposes, since we will determine K_1 directly from the specifications. Thus, the digital controller transfer function is

$$D(z) = \frac{M(z)}{E(z)} = \frac{K_1 z}{z-1} \tag{11.3-4}$$

The characteristic equation is thus

$$1 + \frac{K_1(1-a)z}{(z-1)(z-a)} = 0 \tag{11.3-5}$$

The block diagram and root locus of this system are shown in Figure 11.4. The root locus parameter is $K = K_1(1-a)$. The plot is obtained from the guides in Table 8.1, with z replacing s. Off the real axis, the locus is a circle of radius $\sqrt{a}$. From (11.3-3), this radius is always less than unity. The system becomes unstable if the gain K is large enough to make the locus in the left half-plane cross the unit circle. This will occur for a combination of large K_1 and small a (that is, for a sampling time T large relative to τ).

Select the sampling period T to be a fraction of the plant's time constant. Thus, try $T = 2/4 = 0.5$ sec. In this case, $a = \exp(-0.5/2) = 0.78$. The radius of the locus is $\sqrt{a} = 0.88$.

Referring to Figure 3.14*a*, we see that the overshoot specification requires that $\zeta \geqslant 0.45$. From Figure 3.14*c*, this implies that $\omega_n t_r \geqslant 2.3$. Therefore, the rise time specification calls for $\omega_n \geqslant 2.3/5 = 0.46$. Finally, the settling time from (3.5-22) requires that $\zeta\omega_n \geqslant 4/20 = 0.2$. Note that if $\zeta \geqslant 0.45$ and $\omega_n \geqslant 0.46$, the last requirement is always satisfied.

The z-plane relation (10.7-18) is repeated here.

$$z = e^{-\zeta\omega_n T} e^{i\omega_n T\sqrt{1-\zeta^2}} \tag{11.3-6}$$

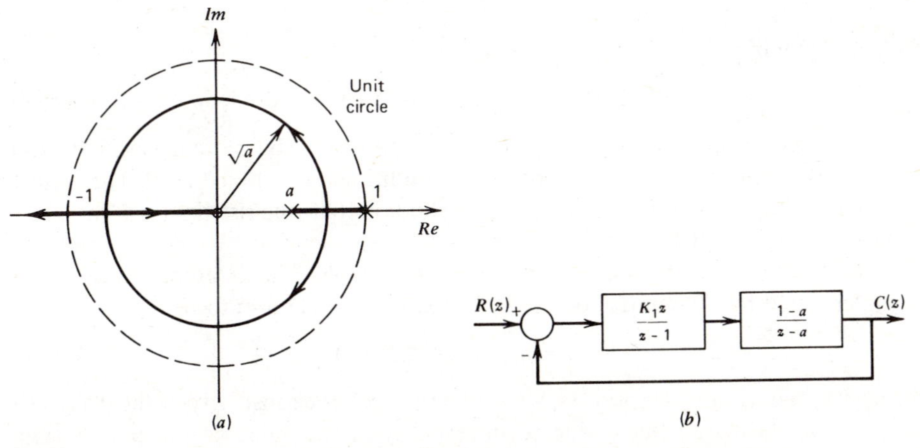

FIGURE 11.4 Root locus and block diagram for the system of Example 11.2.

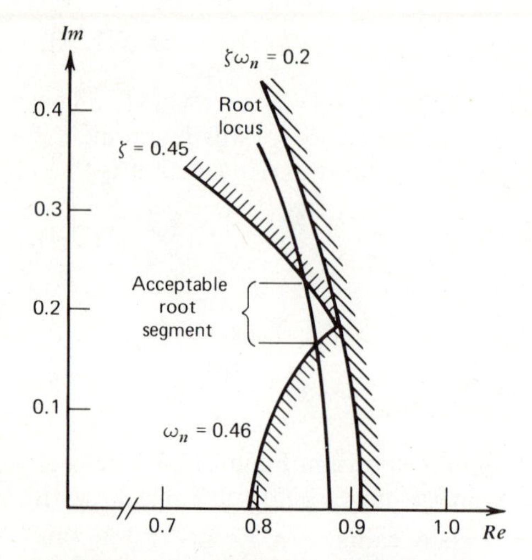

FIGURE 11.5 Root-locus segment and boundaries corresponding to the specifications of Example 11.2.

For $\omega_n = 0.46$ and $T = 0.5$, (11.3-6) gives

$$z = e^{-0.23\zeta} e^{i0.23\sqrt{1-\zeta^2}} \tag{11.3-7}$$

This specifies two roots. The upper root is plotted in Figure 11.5 for $\zeta \geqslant 0.45$.

For $\zeta = 0.45$, (11.3-6) gives

$$z = e^{-0.225\omega_n} e^{i0.4465\omega_n} \tag{11.3-8}$$

This curve is plotted in Figure 11.5 for $\omega_n \geqslant 0.46$. Also shown is the line corresponding to $\zeta\omega_n = 0.2$, or

$$|z| = e^{-0.1} = 0.905$$

The characteristic roots of the system must not lie on the hatched side of these three curves. The appropriate segment of the root locus is also displayed. From this, we can see that the only acceptable upper roots lie in the region indicated. The value of K_1 can be obtained in the usual manner with the magnitude criterion. For example, with $z = 0.86 + i0.2$,

$$K = K_1(1-a) = \frac{|z-1||z-a|}{|z|} = \frac{0.24(0.22)}{0.88} = 0.06$$

or

$$K_1 = \frac{0.06}{0.22} = 0.27$$

With this value for K_1, the closed-loop transfer function is

$$T(z) = \frac{C(z)}{R(z)} = \frac{0.06z}{z^2 - 1.72z + 0.78} \tag{11.3-9}$$

The corresponding difference equation is

$$c(k) = 1.72c(k-1) - 0.78c(k-2) + 0.06r(k-1) \tag{11.3-10}$$

For a unit-step input, $r(i) = 0$, $i < 0$ and $r(i) = 1$, $i \geqslant 0$. The system (11.3-10) was simulated with this input and the zero initial conditions $c(-1) = c(0) = 0$. The results are shown in Figure 11.6. The maximum overshoot is 18%, the rise time is 4.3 sec, and the settling time is approximately 15 sec.

Our choice of $T = 0.5$ thus appears to be acceptable. The difference equation to be implemented by the control computer is found from (11.3-4) to be

$$m(k) = m(k-1) + 0.27e(k) \tag{11.3-11}$$

The next step in the design process, would be to check the sensitivity of the design to variations in the parameter a due to uncertainty in the value of the plant's time constant.

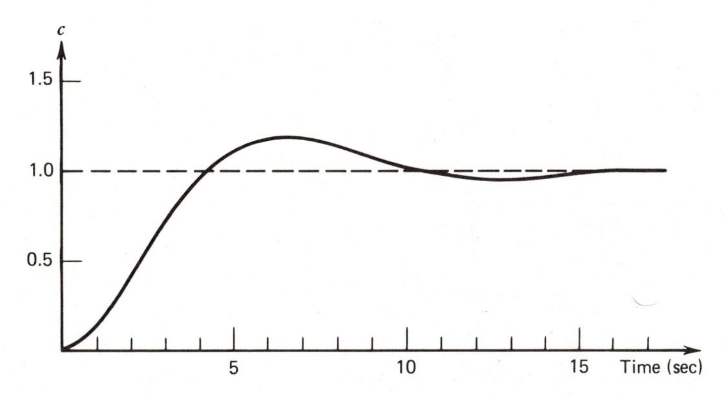

FIGURE 11.6 Response of (11.3-10) for a unit step input.

11.4 DIRECT DESIGN OF DIGITAL CONTROL ALGORITHMS

The approach to the design of analog controllers has been governed to a great extent by the limitations of the physical realization of the control algorithm. The PID family of algorithms is realizable with electronic, hydraulic, and pneumatic devices because of some very inventive designs. However, we should try to envision the difficulties that would be encountered if a more complicated control law were to be implemented, especially if the number of active elements such as op amps, is limited for reliability purposes.

When implementing the control law with a computer, such difficulties vanish. Limiting the controller output to avoid saturating the control elements and the reset windup phenomenon is easily accomplished with digital control. Also, the objection that true derivative action is not obtainable physically is not relevant with digital control in which the corresponding difference action is easily implemented. Because almost any algorithm can be implemented digitally, we can specify the desired response and work backward to find the required control algorithm. This is the *direct-design* method.

Consider the control system shown in Figure 11.7*a*. The pulse transfer function diagram is shown in Figure 11.7*b*. The transfer function is

$$T(z) = \frac{C(z)}{R(z)} = \frac{G(z)D(z)}{1 + G(z)D(z)} \tag{11.4-1}$$

With the direct design method, $T(z)$ is specified, the plant-hold transfer function $G(z)$ is given, and the control algorithm $D(z)$ is found from (11.4-1). This gives

$$D(z) = \frac{T(z)}{G(z)[1 - T(z)]} \tag{11.4-2}$$

Alternatively, the input $R(z)$ and desired output $C(z)$ can be specified. These, of course, determine $T(z)$, and (11.4-2) can still be used.

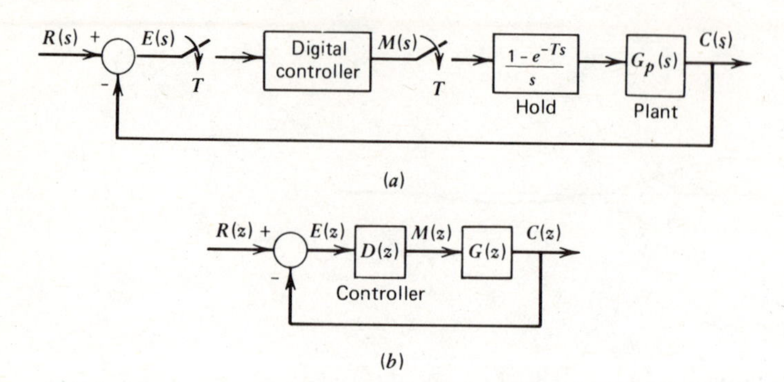

FIGURE 11.7 Digital controller configurations in the *s* domain and the *z* domain.

Example 11.3

Find the control algorithm $D(z)$ so that the response to a unit-step function will be $c(t) = 1 - e^{-t}$. The plant transfer function is $G_p(s) = 1/(10s + 1)$. Assume that the sampling time is $T = 2$ sec.

From the results in Example 10.13 (10.6-18), we obtain the plant-hold transfer function.

$$G(z) = \frac{1 - e^{-2/10}}{z - e^{-2/10}} = \frac{0.18}{z - 0.82} \tag{11.4-3}$$

From the input information, $R(s) = 1/s$, so $R(z) = z/(z - 1)$. Also, the specified output form implies that

$$C(s) = \frac{1}{s} - \frac{1}{s + 1}$$

Therefore, with $T = 2$,

$$C(z) = \frac{z}{z - 1} - \frac{z}{z - e^{-2}} = \frac{0.86z}{(z - 1)(z - 0.14)} \tag{11.4-4}$$

and

$$T(z) = \frac{C(z)}{R(z)} = \frac{z - 1}{z} \frac{0.86z}{(z - 1)(z - 0.14)}$$

$$= \frac{0.86}{z - 0.14} \tag{11.4-5}$$

Substituting this into (11.4-2) gives

$$D(z) = \frac{z - 0.82}{0.18} \frac{0.86}{z - 1}$$

$$= \frac{4.8 - 3.9z^{-1}}{1 - z^{-1}} = \frac{M(z)}{E(z)} \tag{11.4-6}$$

The difference equation that must be implemented in the software of the computer control system is

$$m(k) = m(k - 1) + 4.8e(k) - 3.9e(k - 1) \tag{11.4-7}$$

Several considerations are important when specifying the transfer function $T(z)$. If $T(z)$ has the form

$$T(z) = \frac{C(z)}{R(z)} = \frac{b_0 + b_1 z^{-1} + b_2 z^{-2} + \cdots}{a_0 + a_1 z^{-1} + a_2 z^{-2} + \cdots} \tag{11.4-8}$$

causality requires $b_0 = 0$. This can be seen by writing the relation as a difference equation.

$$\begin{aligned} a_0 c(k) = &-a_1 c(k-1) - a_2 c(k-2) - \cdots \\ &+ b_0 r(k) + b_1 r(k-1) + b_2 r(k-2) + \cdots \end{aligned} \tag{11.4-9}$$

The output of a physical system will require at least one sampling period to react to the input. Thus, $r(k)$ cannot affect $c(k)$ in a real system, and b_0 must be zero. Given the denominator of (11.4-8), the same argument shows that the numerator cannot contain terms such as z^i, $i \geqslant 0$.

The controller transfer function $D(z)$ resulting from (11.4-2) will produce the desired output $C(z)$ with the specified input $R(z)$, but how will it perform with other input functions or disturbances? These questions will be important for some applications and should be answered before accepting the design as final.

The algorithm designed in Example 11.3 might be useful for starting up the process according to the prescribed function $c(t)$. The control software can be programmed to switch when the process reaches $c = 1$ to another control algorithm more suitable for regulation in the presence of disturbances.

Compensation with Finite-Time Setting Algorithms

The lead and lag compensation methods in Section 9.4 can be applied to digital systems by either performing the design in the s domain and converting it into difference form or converting the specifications into root locations in the z domain. In the latter case, lead and lag compensation simply consists of introducing additional poles and zeros to the open-loop transfer function in terms of z. The techniques are similar to those used with analog systems.

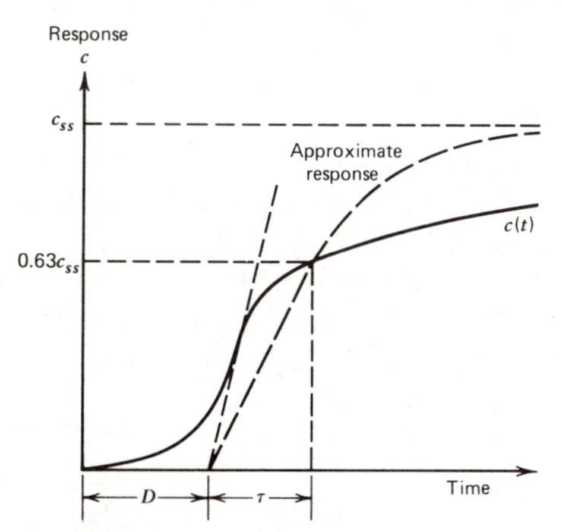

FIGURE 11.8 Approximation of process response with a first-order lag τ and a deadtime element D. The time D is found from the maximum slope of the response curve.

We now indicate how a digital controller can be designed to compensate for the effects of dead time. With the direct-design method, the required controller transfer function $D(z)$ can be found in terms of that of the plant-and-hold $G(z)$ and the desired closed-loop transfer function $T(z)$ (see Figure 11.7 and (11.4-2).

Suppose that the plant has a step response like the S-shaped signature shown in Figure 11.8. This is similar to that shown in Figure 7.1 for the Ziegler–Nichols process reaction method. Thus, a simple model of the plant consists of a first-order lag with a dead-time element; that is,

$$G_p(s) = \frac{e^{-Ds}}{\tau s + 1} \tag{11.4-10}$$

The pulse transfer function of the first-order lag with a sample-and-hold is given by (11.3-2). The dead time results in a term z^{-n}, where n relates the dead time to the sampling period, as

$$D = nT \tag{11.4-11}$$

We assume that the sampling period is chosen so that n is an integer, for simplicity. Thus,

$$G_p(z) = z^{-n}\frac{1-a}{z-a} \tag{11.4-12}$$

where $a = \exp(-T/\tau)$.

Consider the closed-loop transfer function

$$T(z) = \frac{C(z)}{R(z)} = z^{-(n+1)} \tag{11.4-13}$$

Then

$$C(z) = z^{-(n+1)}R(z) \tag{11.4-14}$$

From Tables 10.2 and 10.4, (11.4-14) implies that

$$c(k) = r(k-n-1) \tag{11.4-15}$$

If the input $r(k)$ is a unit step, $r(k) = 1$, $k \geqslant 0$; $r(k) = 0$, $k < 0$, then

$$c(k) = 0 \qquad k \leqslant n$$

$$c(k) = 1 \qquad k \geqslant n+1$$

That is, the output lags the input signal by $(n+1)$ sampling periods, one more sampling period than is in the dead time D. We assume that the extra sample period is required, because no system can respond instantaneously to an input.

The preceding response is desirable for start-up operations, and the controller transfer function required to obtain it is given by substituting (11.4-13) and (11.4-12) into (11.4-2). This yields

$$\frac{M(z)}{E(z)} = D(z) = \frac{1}{1-a}\frac{1-az^{-1}}{1-z^{-(n+1)}} \tag{11.4-16}$$

Such a fast response might require a value for the manipulated variable beyond the capability of the final control element. To check this, obtain $M(z)$ in terms of $R(z)$. From block diagram manipulation, the required relation is found to be

$$M(z) = \frac{T(z)}{G_p(z)}R(z) = \frac{1-az^{-1}}{1-a}R(z) \tag{11.4-17}$$

If $R(z)$ represents a unit-step input,

$$M(z) = \frac{1 - az^{-1}}{1 - a} \frac{z}{z - 1} = \frac{a}{1 - a} + \frac{z}{z - 1}$$

and

$$m(k) = \frac{a}{1 - a} \delta(k) + 1$$

where $\delta(k) = 1$ when $k = 0$ and $\delta(k) = 0$ otherwise. The maximum value of the manipulated variable occurs at $k = 0$ and is $1 + a/(1 - a) = 1/(1 - a)$. For a fixed dead time D, an increase in the sampling period T causes an increase in this maximum value.

The algorithm for $D(z)$, (11.4-16), is called a *finite-settling time* algorithm, because the response reaches the desired value in a finite, prescribed time. Such algorithms produce a closed-loop transfer function that is a finite polynomial in z^{-1}, such as (11.4-13). The principle can be applied to other plant types. It should be noted, however, that the controller transfer function cancels the plant pole at $z = a$, and the design is therefore subject to the criticism given in Section 9.5, page 499.

11.5 FREQUENCY RESPONSE METHODS

The response of a discrete-time system to a sinusoidal input is determined in a manner similar to that used for continuous-time systems. If a sinusoid of circular frequency ω and amplitude A is sampled with a sampling period T, the resulting sequence is

$$u(k) = A \sin k\omega T \tag{11.5-1}$$

If this sequence is applied as an input to a stable system whose transfer function is $T(z)$, the steady-state output is

$$y(k) = B \sin (k\omega T + \phi) \tag{11.5-2}$$

where

$$M = \frac{B}{A} = |T(e^{i\omega T})| \tag{11.5-3}$$

$$\phi = \measuredangle T(e^{i\omega T}) \tag{11.5-4}$$

Thus, the steady-state output is also sinusoidal with the same frequency ω. The amplitude ratio M and phase shift ϕ are found by substituting $z = \exp(i\omega T)$ into the transfer function. [Recall that the substitution $s = i\omega$ is made in $T(s)$ for continuous-time systems]. As in Chapter 4, polar or rectangular plots can be constructed for M and ϕ as functions of ω.

Both M and ϕ are periodic functions of ω with period $2\pi/T$. This is true because $\exp(i\omega T)$ is periodic in ω with period $2\pi/T$, as can be shown from the Euler identity relating $\exp(i\omega T)$ to $\sin \omega T$ and $\cos \omega T$. Also, from the sampling theorem, the highest significant frequency can be no greater than π/T if T has been chosen correctly. Thus, we must consider the behavior of M and ϕ only over the frequency range $0 \leqslant \omega \leqslant \pi/T$.

Digital Compensation

As we have seen, analog system performance can be improved by introducing new poles and zeros with a compensator. The same is true for digital systems, and the design can be made with either the root locus or the frequency response plots. The Nyquist stability criterion remains essentially the same for discrete-time systems, and phase and gain margins can be used as before. Bode showed that for a minimum phase system with a rational transfer function, the amplitude curve must have a slope of no more than about -20 db/decade at the gain crossover in order for the phase to stay above $-180°$ (the stability boundary).

The difficulty with discrete-time systems is that the frequency transfer functions are obtained by substituting $z = \exp(i\omega T)$. Thus, the transfer functions are not rational, and Bode's technique does not apply. An alternative is to transform the variable to obtain rational transfer functions in the new plane. Franklin and Powell (Reference 2) have proposed the transformation

$$w = \frac{2}{T}\frac{z-1}{z+1} \tag{11.5-5}$$

or

$$w = \frac{2}{T}\frac{e^{sT}-1}{e^{sT}+1} = \frac{2}{T}\tanh\frac{sT}{2} \tag{11.5-6}$$

From (10.7-16), we see that if $s = i\omega$ (the stability boundary), then $z = \exp(i\omega T)$. The latter expression specifies the unit circle as ω varies from 0 to infinity. Equation (11.5-6) in this case gives

$$w = i\frac{2}{T}\tan\frac{\omega T}{2} = iv \tag{11.5-7}$$

Thus, as z goes around the unit circle, v, the frequency in the w plane, goes from 0 to ∞.

Given the open-loop transfer function $G(z)H(z)$, we substitute

$$z = \frac{1 + wT/2}{1 - wT/2} \tag{11.5-8}$$

The open-loop transfer function is now $G(w)H(w)$, and the Bode design procedure is the same as before. However, introducing the sample-and-hold operation can introduce additional zeros whose locations depend on the sampling period. As the sampling period is decreased, these additional zeros become less important in the design analysis. If they occur in the right-half plane, they make the system nonminimum phase (see Section 9.1), and Bode's relationship between gain and phase no longer applies. A comprehensive design example with this effect is given in Reference 2 (pp. 114–117).

11.6 DIGITAL FILTERING

Over the years, technological developments have caused an evolution in the meaning of the term *filter*. The primitive concept referred to a physical device that removes high-frequency components from a signal. These components are referred to generally as "noise". An example of such a device is the low-pass *RC* circuit. The signal variable is not always electrical, however. For example, an accumulator (Section 2.6) filters out high-frequency pressure fluctuations. With the improvement and miniaturization of op amp circuitry, it became possible to build devices that operated on sensor signals in a closed-loop manner to reduce the uncertainties inherent in the measurements. In the sense that sensor measurement uncertainties constituted "noise" in the measurement signal, these devices came to be called filters, because they removed the noise to yield a better estimate of the measured variable. One of the most powerful and popular algorithms for designing such circuits is the *Kalman filter* algorithm, named after its developer.

Filtering took on a new meaning with the introduction of digital devices. Now, the term *filter* also includes any algorithm or formula that acts on a sequence of input *numbers* to produce another sequence. Such an algorithm is called a *digital filter.* Its function is the same as that of an analog filter; namely, to remove unwanted components from a signal or number sequence. Perhaps the most common example is the running average of a sequence of numbers $u(k)$. The two-point average is expressed by the formula

$$\hat{y}(k+1) = \tfrac{1}{2}[u(k+1) + u(k)] \qquad \textbf{(11.6-1)}$$

where $\hat{y}$ is the estimate.

Modern systems theory terminology distinguishes between *filtering*, *smoothing*, and *prediction.* If the independent variable is time, smoothing refers to the process of obtaining the best estimate of a previous value of the measured variable, using past measurements up to the present time. A common example of a smoothing algorithm is the least-squares criterion for fitting a curve to data. Filtering is the process of estimating the current value of the variable, and prediction refers to estimating a future value. The term *estimation* refers to all three processes. The techniques of digital filtering can be extended to develop methods for smoothing and prediction.

Filter Design in the Time Domain

The two-point running average formula given by (11.6-1) can be analyzed by difference equation techniques. It contains the implicit assumption that y, the value to be estimated, is a constant and the measurements $u(k)$ fluctuate about the true value of y. If the fluctuation amplitude is large, an average over a longer string of data would be used. The n-point average is

$$\hat{y}(k+1) = \frac{1}{n}[u(k+1) + u(k) + u(k-1) + \cdots + u(k-n+2)] \qquad \textbf{(11.6-2)}$$

The larger n is, the more digital storage is required. This might be a critical consideration for a small special-purpose device.

Equation (11.6-2) is an example of a *nonrecursive* filter. This implies that the data $u(k)$ must be processed in batch form and the algorithm does not take advantage of previous calculations as new data arrive. However, the algorithm can be rearranged to alleviate this problem somewhat by noting that

$$\hat{y}(k) = \frac{1}{n}[u(k) + u(k-1) + \cdots + u(k-n+1)] \qquad \textbf{(11.6-3)}$$

Thus,

$$u(k) + u(k-1) + \cdots + u(k-n+1) = n\hat{y}(k) - u(k-n+1)$$

Substituting this into (11.6-2) gives

$$\hat{y}(k+1) = \hat{y}(k) + \frac{1}{n}[u(k+1) - u(k-n+1)] \qquad \textbf{(11.6-4)}$$

This says that the estimate of y at time $(k+1)$ consists of the sum of the previous estimate and a correction term proportional to the difference between the latest measurement and the measurement n time periods earlier. The proportional factor $1/n$ weights the correction term relative to the previous estimate $\hat{y}(k)$. If n is large, the filter places more emphasis on the previous estimate. This makes sense, because this estimate results from a long string of data for large n and presumably contains more information than in the two measurements $u(k+1)$ and $u(k-n+1)$. This form of the running-average algorithm is said to be *recursive*, because it uses the previous average to reduce the number of required calculations.

While the number of computations required by (11.6-4) is less than those for (11.6-2), the same number of data points must be stored even though (11.6-4) uses only two at a time. If the sampling rate is high, the number of storage locations (n) will be large. Thus, if we can find a filter algorithm that uses only the current measurement and the previous estimate, we will have considerably reduced both the storage requirements and the number of required calculations. With the additional insight that linear equations are easier to work with, we are led to try the following filter form:

$$\hat{y}(k+1) = w_1\hat{y}(k) + w_2u(k+1) \qquad \textbf{(11.6-5)}$$

where w_1 and w_2 are weighting factors measuring the relative importance of the previous estimate and the current measurement $u(k+1)$.

Up to now, we have been considering measuring a constant y. A more general case is the one where the quantity to be estimated is a variable y described by the linear model:

$$y(k+1) = ay(k) \qquad \textbf{(11.6-6)}$$

We emphasize that the variable y, which presumably represents a physical process, usually cannot be described exactly by such a simple model. On the other hand, our measurements of y are not exact either. The coefficients w_1 and w_2 can be chosen to express the relative degrees of confidence we have in our model of the process and our measurements. If we believe our model completely and do not trust our measurements, we choose $w_1 = a$ and $w_2 = 0$. This choice tells the filter not to use the measurements at all and to estimate $y(k+1)$ using the model given by (11.6-6). The other extreme is complete faith in our measurements and none in the model which leads to the choice $w_1 = 0$ and $w_2 = 1$. The current estimate of y is then taken to be the value of the last measurement.

The preceding interpretation of the filter weights suggests the following rearrangement of the algorithm (11.6-5):

$$\hat{y}(k+1) = a\hat{y}(k) + K[u(k+1) - a\hat{y}(k)] \qquad \textbf{(11.6-7)}$$

where $w_1 = a(1-K)$ and $w_2 = K$. The first term on the right of the equation is the prediction of the current value of y based on the model (11.6-6). The second term is a correction term consisting of the difference between the current measurement and the current value of y as predicted by the model. This difference is weighted by the filter gain K. From the two extreme cases discussed previously, we see that a choice of $K=0$ reflects complete faith in the model, and $K=1$, complete faith in the measurement.

The filter's dynamics are revealed by its transfer function. The z transform of (11.6-7) is

$$z\hat{Y}(z) - z\hat{y}(0) = a\hat{Y}(z) + KzU(z) - Kzu(0) - aK\hat{Y}(z) \tag{11.6-8}$$

or

$$\hat{Y}(z) = \frac{z[\hat{y}(0) - Ku(0)]}{z - a + aK} + \frac{Kz}{z - a + aK} U(z) \tag{11.6-9}$$

The denominator of the transfer function has the single root $z = a(1-K)$, and the filter estimate is inherently nonoscillatory if $a(1-K) > 0$. [An oscillating data sequence $u(k)$ could cause oscillations, however.] From the preceding discussion, we see that K lies in the range $0 < K < 1$. Thus, no inherent oscillation of the estimate occurs unless $a < 0$. This is acceptable, because $a < 0$ implies that the process being estimated is itself oscillatory. If the quantity to be estimated is modeled as a constant, then $a = 1$, and the filter is stable if $0 \leqslant K < 1$.

Example 11.4

Assume a process $y(t)$ is to be modeled as the decaying exponential $y(0)\exp(-t/\tau)$, where τ is assumed to be known at least approximately. The variable y might represent the temperature of an object being cooled. Develop a linear recursive filter to estimate $y(t)$ using only the current measurement.

The given process is continuous time, so a suitable sampling period T must be selected. The sampling theorem states that $T = 0.049\tau$ is sufficient for this process. With a safety factor of ten, we use $T = 0.0049\tau$. The discrete-time model of the process is

$$y[(k+1)T] = y(kT)e^{-T/\tau} = y(kT)e^{-0.0049} = 0.9951y(kT)$$

With the filter form given by (10.13-7), $a = 0.9951$, and the desired algorithm is

$$\hat{y}(k+1) = 0.9951\hat{y}(k) + K[u(k+1) - 0.9951\hat{y}(k)]$$

where the time between discrete instants is T, and $u(k)$ is the error-contaminated measurement of the process variable $y(k)$. (If the measurements did not contain error, we would not need a filter.) Our choice of a value for K depends on how confident we are of (1) the form of the process model (exponential), (2) our knowledge of the model parameter τ, and (3) our measurements.

The filter algorithm given in (11.6-7) requires an initial value $\hat{y}(0)$ in order to start. Frequently, this value is taken to be the value given by the first measurement $u(1)$; that is, $\hat{y}(0) = u(1)$. In view of the additional requirement of choosing a value for the gain K, we might wonder if this filter has any advantages over traditional estimation methods, such as averaging or least-squares techniques. If y is taken to be a constant, an estimate can be obtained by averaging the measurements; that is,

$$\hat{y}(k) = \frac{1}{k} \sum_{i=1}^{k} u(i) \tag{11.6-10}$$

When the next measurement becomes available at time $k+1$, the new estimate is

$$\begin{aligned}\hat{y}(k+1) &= \frac{1}{k+1} \sum_{i=1}^{k+1} u(i) \\ &= \frac{k}{k+1}\left[\frac{1}{k}\sum_{i=1}^{k} u(i)\right] + \frac{1}{k+1}u(k+1)\end{aligned}$$

or

$$\hat{y}(k+1) = \frac{k}{k+1}\hat{y}(k) + \frac{1}{k+1}u(k+1) \tag{11.6-11}$$

This is the recursive form of the simple averaging process. It eliminates the need to store all the past measurements. Note that in this case, a starting value of $\hat{y}$ at $k=0$ need not be provided, because the coefficient of $\hat{y}(k)$ vanishes when $k=0$. The algorithm can be rewritten as

$$\hat{y}(k+1) = \hat{y}(k) + \frac{1}{k+1}[u(k+1) - \hat{y}(k)] \tag{11.6-12}$$

In this form, a resemblance can be seen to the filter form given in (11.6-7) with $a=1$. However, here the filter gain $1/(k+1)$ is time varying. The measurement is weighted less and less as time goes on.

Note that both methods (11.6-7) and (11.6-12) require a model of the process being estimated. The averaging algorithm (11.6-12) implicitly assumes that y is constant. Similarly, estimation techniques based on least-squares criteria also require a process model. For example, if we fit a straight line to the data, we are assuming that the process can be modeled by a straight line. The so-called *weighted* least-squares method also requires selecting weighting factors for every measurement as indicators of their reliability.

The filter algorithm given by (11.6-7) also requires a selection of values for the gain K and initial condition $\hat{y}(0)$. In spite of this, the filter has advantages, because the rate of convergence to the true value can be faster than for the other methods. The algorithm can be extended to higher order models, and systematic methods are available for determining the filter gains in terms of the statistical properties of the process model uncertainties and measurement errors (see Reference 3).

Example 11.5

Investigate the frequency response characteristics of the filter algorithm given by (11.6-7).

The transfer function from (11.6-9) is

$$T(z) = \frac{Kz}{z - a + aK} \tag{11.6-13}$$

Substituting $z = \exp(i\omega T)$ and using Euler's identity gives

$$T(e^{i\omega T}) = \frac{Ke^{i\omega T}}{e^{i\omega T} - a + ak} = \frac{K}{1 + a(K-1)e^{-i\omega T}} = \frac{K}{1 + a(K-1)\cos\omega T - ia(K-1)\sin\omega T}$$

After some manipulation, the magnitude and phase angle can be shown to be

$$M = |T(e^{i\omega T})| = K[a^2(K-1)^2 + 2a(K-1)\cos\omega T + 1]^{-1/2} \qquad \textbf{(11.6-14)}$$

$$\phi = \tan^{-1}\left[\frac{a(K-1)\sin\omega T}{a(K-1)\cos\omega T + 1}\right]$$

Two cases are possible: $a \geqslant 0$ and $a < 0$ (we assume here that $0 \leqslant K \leqslant 1$). The frequency range of interest is $0 \leqslant \omega \leqslant \pi/T$. The values of M at the two extreme frequencies are also of interest. At $\omega = 0$,

$$M = \frac{K}{|a(K-1)+1|}$$

and at $\omega = \pi/T$,

$$M = \frac{K}{|a(K-1)-1|}$$

If $a > 0$, the maximum value of M occurs at $\omega = 0$ and the minimum at $\omega = \pi/T$. The opposite is true for $a < 0$. Thus, the algorithm defined by (11.6-7) is a *low-pass* filter if $a > 0$, and a *high-pass* filter if $a < 0$. This is consistent with our previous interpretation of the algorithm. For $a < 0$, the filter is trying to estimate the value of a variable assumed to be alternating in sign at every time step (and hence possesses a circular frequency of $\omega = \pi/T$). We would expect the filter to "look for" such fluctuations and to pass the high-frequency components.

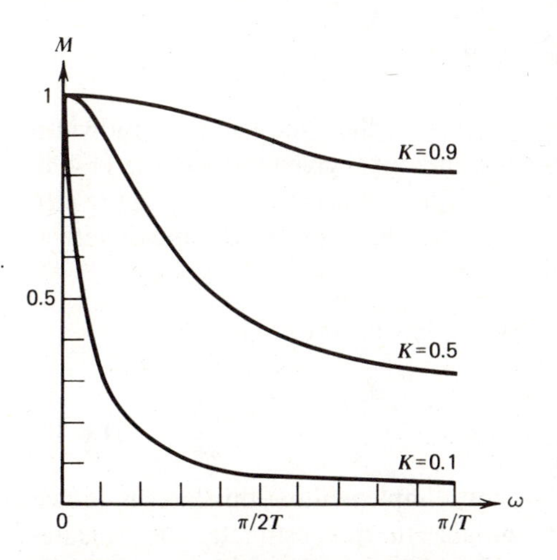

FIGURE 11.9 Magnitude ratio M versus circular frequency ω for the digital filter (11.6-13) with $a = 1$.

Figure 11.9 shows the magnitude of ratio M versus frequency plotted for the case $a = 1$ and several values of K. For $K = 0.9$, very little attenuation occurs at any frequency. This is the case where heavy weight is given to the measurement. For $K = 0.1$, heavy weight is given to the model; the filter algorithm expects to be estimating a constant and therefore tends to ignore measurement fluctuations with higher frequencies. From Table 10.1, the time constant for the filter is $1/\ln(1-K)$. The smaller value of K produces a larger time constant and hence a more sluggish system. For small K, the filter estimate generally takes longer to converge to the true value, but also does a better job of filtering out high-frequency components resulting from, for example, measurement noise.

The general linear algorithm given by

$$y(k+1) = ay(k) + bu(k+1) \tag{11.6-15}$$

is a low-pass filter if $a > 0$ and a high-pass filter if $a < 0$. Define the system bandwidth in the same way as for continuous-time systems; that is, the bandwidth is the range of frequencies for which the magnitude ratio M satisfies

$$M \geqslant \frac{M_{max}}{\sqrt{2}} \tag{11.6-16}$$

where M_{max} is the maximum value of M. In general, as the designer reduces the filter's bandwidth to suppress measurement noise, the response of the filter becomes sluggish. If a satisfactory compromise cannot be reached, a higher order algorithm often works.

Example 11.6

Write the software required to implement a digital filter for estimating an exponentially decaying voltage whose time constant is thought to be approximately 0.5 sec. The electrical instrumentation has line noise due to nearby equipment operating on 60-Hz ac current.

From the results in Example 11.4 we require a sampling period of $T = 0.0049(0.5) = 0.00245$ sec. Thus, the highest circular frequency the digital filter will see is

$$\omega = \frac{\pi}{T} = 1282 \text{ rad/sec}$$

Since 60 Hz is 377 rad/sec, the noise frequency is seen by the filter, and the bandwidth should be made less than this frequency. For the filter form given in Example 11.4, in terms of the notation of (11.6-7), $a = 0.9951$. Using (11.6-14) and the bandwidth definition (11.6-16), we find that the noise frequency lies outside the bandwidth if $K \leqslant 0.58$. For $K = 0.58$, the filter algorithm is

$$\hat{y}(k+1) = 0.9951\hat{y}(k) + 0.58[u(k+1) - 0.9951\hat{y}(k)]$$

or

$$\hat{y}(k+1) = 0.4179\hat{y}(k) + 0.58u(k+1) \tag{11.6-17}$$

The general outline of a FORTRAN program implementing this filter is shown on page 607. The first measurement is used to initialize the filter estimate. The measurements will be read by the program every 2.45×10^{-3} sec. Thus, the time for the program to execute one cycle must be less than this time if the filter is to run in "real time."

```
J=0
1 READ U
IF (J.EQ.0)  YJ=U
J=J+1
Y=0.4179*YJ+0.58*U
OUTPUT Y
YJ=Y
GO TO 1
```

The frequency response curve used to design this filter is a representation of the filter's steady-state response and does not include the effect of transients. That is, the filter might perform well at steady state but give poor estimates before the steady state is reached. An estimate of how long this takes is provided by the time constant. Referring to Table 10.1, the time constant is

$$k = -\frac{1}{\ln 0.4179} = 1.15 \text{ steps}$$

Thus, the transients will essentially be gone after $4(1.15) \cong 5$ steps, or $5(2.45 \times 10^{-3}) = 0.012$ sec. This time is short compared to the estimated time constant of the input (0.5 sec).

Higher Order Digital Filters

Numerical algorithms used for digital filtering and processing data often require a model of higher order than the first-order algorithms just presented. For example, the frequency response of the low-pass filter can be improved by cascading more than one such filter. For n filters in cascade, the overall filter transfer function is

$$T(z) = \left(\frac{Kz}{z - a + aK} \right)^n \tag{11.6-18}$$

Cascading allows the constants a and K to be selected with more freedom to improve the signal rejection properties (a sharper falloff with frequency) while preserving the transient-response characteristics. Similar statements apply to the design of high-pass and band-pass filters, and higher order models are common in such applications.

The following second-order filter is easily derived and is useful for estimating displacement and velocity, given only displacement measurements contaminated with error (Ref. 4). Define the following variables:

$u(k)$ = displacement measurement at time k

$p(k)$ = prediction of the displacement at time k, after processing the measurement $u(k-1)$

$y(k)$ = estimate of the displacement at time k, after processing the measurement $u(k)$

$v(k)$ = estimate of the velocity at time k, after processing k measurements

The filter consists of three stages. The first is a prediction of the current displacement at time k based on the estimates obtained from the previous measurement at time $k-1$. Noting that velocity is the derivative of displacement, we apply the Euler approximation to obtain

$$p(k) = y(k-1) + Tv(k-1)$$

where T is the time between measurements. The second stage corrects this prediction by using the difference between the measurement and the prediction at time k; that is,

$$y(k) = p(k) + a[u(k) - p(k)]$$

where a is a weighting factor that indicates the reliability of the measurement. The final stage of the filter is a correction of the velocity estimate, with a weighted velocity error correction term.

$$v(k) = v(k-1) + \frac{b}{T}[u(k) - p(k)]$$

Using the first equation to eliminate $p(k)$, we obtain the filter algorithm.

$$y(k) = (1-a)y(k-1) + (1-a)Tv(k-1) + au(k) \tag{11.6-19}$$

$$v(k) = -\frac{b}{T}y(k-1) + (1-b)v(k-1) + \frac{b}{T}u(k) \tag{11.6-20}$$

The filter's characteristic equation is

$$z^2 + (a+b-2)z + 1 - a = 0$$

For the stability criteria (10.7-14) and (10.7-15),

$$\alpha + \beta + \gamma = b$$
$$\alpha - \beta + \gamma = 4 - 2a - b$$
$$\alpha - \gamma = a$$

For $a > 0$ and $b > 0$, the filter will be stable if and only if $4 - 2a - b > 0$. If this stability condition is not satisfied, the displacement and velocity estimates $y(k)$ and $v(k)$ will never settle down to the correct values. Thus, the effects of the initial (and probably incorrect) estimates of displacement and velocity will never be eliminated, regardless of the accuracy of the measurements $u(k)$.

Figure 11.10 shows an example of how this displacement-velocity filter can be used in a digital control system. A radar is used to measure the aircraft's altitude h, and the elevator angle ϕ is controlled to keep the aircraft at the desired altitude r. If the digital equivalent of tachometer compensation is used (see Section 6.7), we need an estimate of the time rate of change of the altitude to feed back to the control law (see Figure 11.10*b*). Let u denote the radar's measurement of the altitude, y the altitude estimate, and v the altitude rate estimate. Then the filter described by (11.6-19) and (11.6-20) can be used. The designer must select K_1, K_2, a, b, and possibly T.

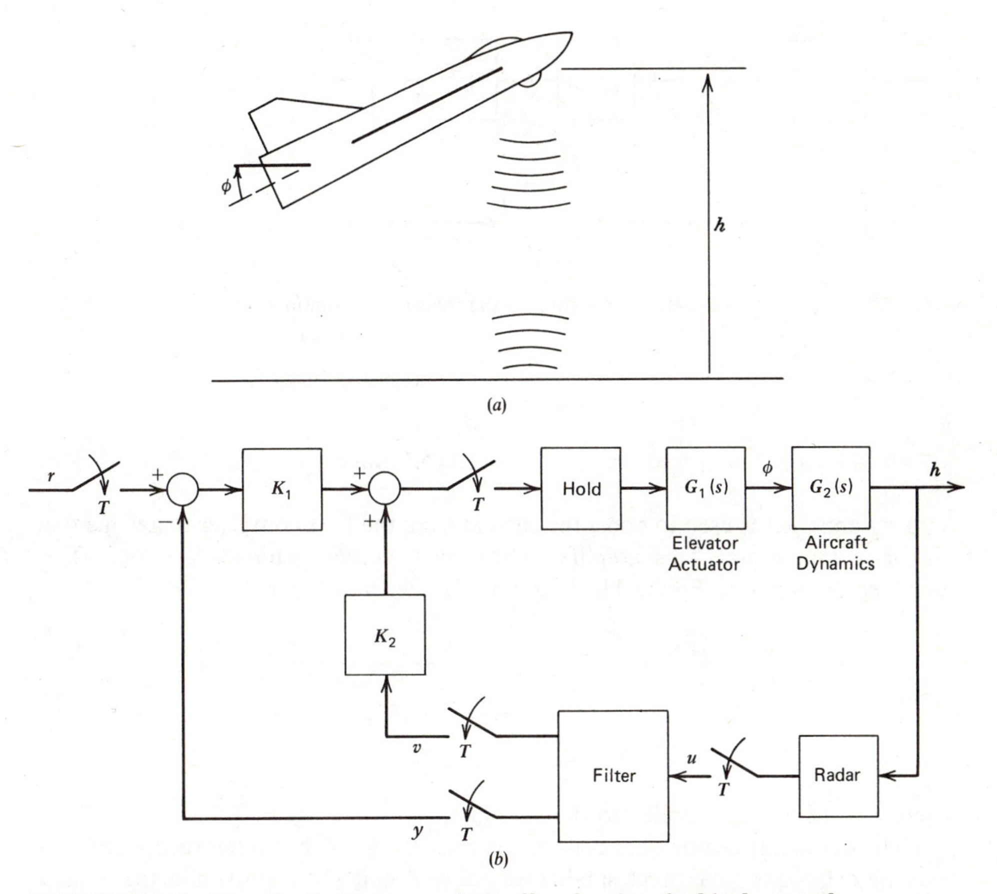

FIGURE 11.10 Use of a displacement-velocity filter for aircraft altitude control.

11.7 MICROPROCESSOR IMPLEMENTATION OF DIGITAL CONTROL: A CASE STUDY

So far we have concentrated on the modeling of the plant, the sensors, and the actuators, and on the design of the algorithms necessary to control them. We now consider how the digital hardware can be configured to produce a working control system. This requires a treatment of *interfaces*, which are devices that allow digital computers to communicate with A/D and D/A converters as well as other types of equipment such as printers, modems, and disk drives. A complete treatment of interfacing is beyond the scope of this text (see Reference 5 for details). However, our treatment will be sufficient to understand most digital controller configurations.

Modeling the Plant and Sensor

We will use speed control of a motor as a specific example. The motor is a dc permanent magnet type with the following parameter values as given in the manufacturer's literature (see Example 2.8 for the motor model).

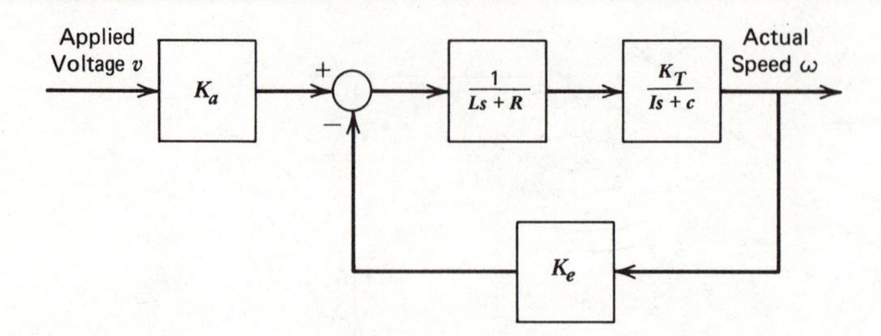

FIGURE 11.11 The subsystem consisting of the motor and amplifier.

$K_T = 5.8$ oz-in/amp $I = 0.0055$ oz-in-sec^2

$K_e = 0.05185$ volts/rad/sec $R = 1.55$ ohms

$c = 9.549 \times 10^{-4}$ oz-in/rad/sec $L = 5.12$ mH

A power amplifier is used to boost the current from the D/A converter to a value that can drive the motor. This amplifier also has a voltage gain of $K_a = 9.62$. This subsystem is shown in Figure 11.11, and its transfer function is

$$\frac{\Omega(s)}{V(s)} = \frac{K_a K_T}{LIs^2 + (RI + cL)s + Rc + K_e K_T}$$
$$= \frac{184.69}{(0.0244s + 1)(0.00381s + 1)} \qquad \textbf{(11.7-1)}$$

where v is the voltage applied to the amplifier.

The dominant motor time constant is $\tau = 0.0244$ sec. Since the secondary time constant of 0.00381 sec is much smaller we will neglect it and approximate the transfer

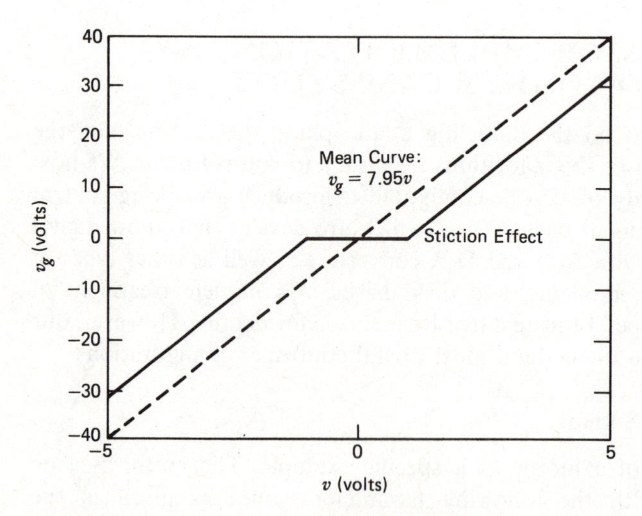

FIGURE 11.12 Steady-state tachometer voltage v_g as a function of applied voltage v.

function by

$$\frac{\Omega(s)}{V(s)} = \frac{184.69}{0.0244s + 1} \tag{11.7-2}$$

The tachometer is incorporated in the motor unit, and its damping and inertia are included in the previous parameter values. Its voltage versus speed characteristics were obtained experimentally by applying a voltage v to the amplifier and measuring the tachometer voltage v_g after the motor speed reached steady state. The results are shown in Figure 11.12. The stiction effect occurs because the motor needs a minimum applied voltage to rotate. This nonlinearity could be compensated for with the digital control algorithm, but for simplicity we will use the mean curve shown in Figure 11.12. The tachometer gain K_g was found as follows. Since $v_g = K_g \omega_{ss}$, the steady-state motor relation gives

$$v_g = K_g \omega_{ss} = K_g(184.69v) = 7.85v \tag{11.7-3}$$

Thus, $K_g = 7.95/184.69 = 0.043$ volt/rad/sec.

Selecting the Control Law

Chapter 6 shows that proportional control with feedforward compensation eliminates the offset error with a step input applied to a first-order plant like (11.7-2). This control algorithm is shown in Figure 11.13, in which the pulse transfer function of the plant has been obtained as the series combination of the analog plant (11.7-2) with a zero-order hold (see Table 10.5 and Example 10.13). The constant a is

$$a = e^{-T/0.0244} \tag{11.7-4}$$

The gains G_1 and G_2 are associated with the system's interfaces and will be determined subsequently. The closed-loop transfer function obtained from Figure 11.13 is

$$\frac{\Omega(z)}{\Omega_d(z)} = \frac{G_3(K_f + K_p)}{z - a + G_3 K_p} \tag{11.7-5}$$

where

$$G_3 = 184.69(1 - a)G_1 G_2 \tag{11.7-6}$$

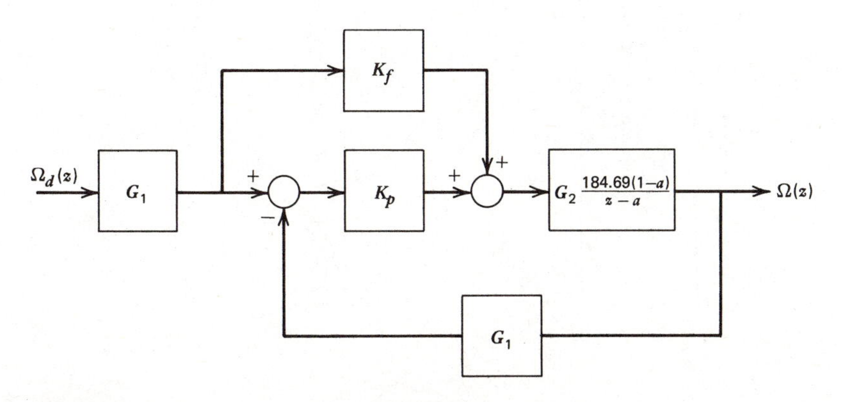

FIGURE 11.13 Control system diagram in the z domain.

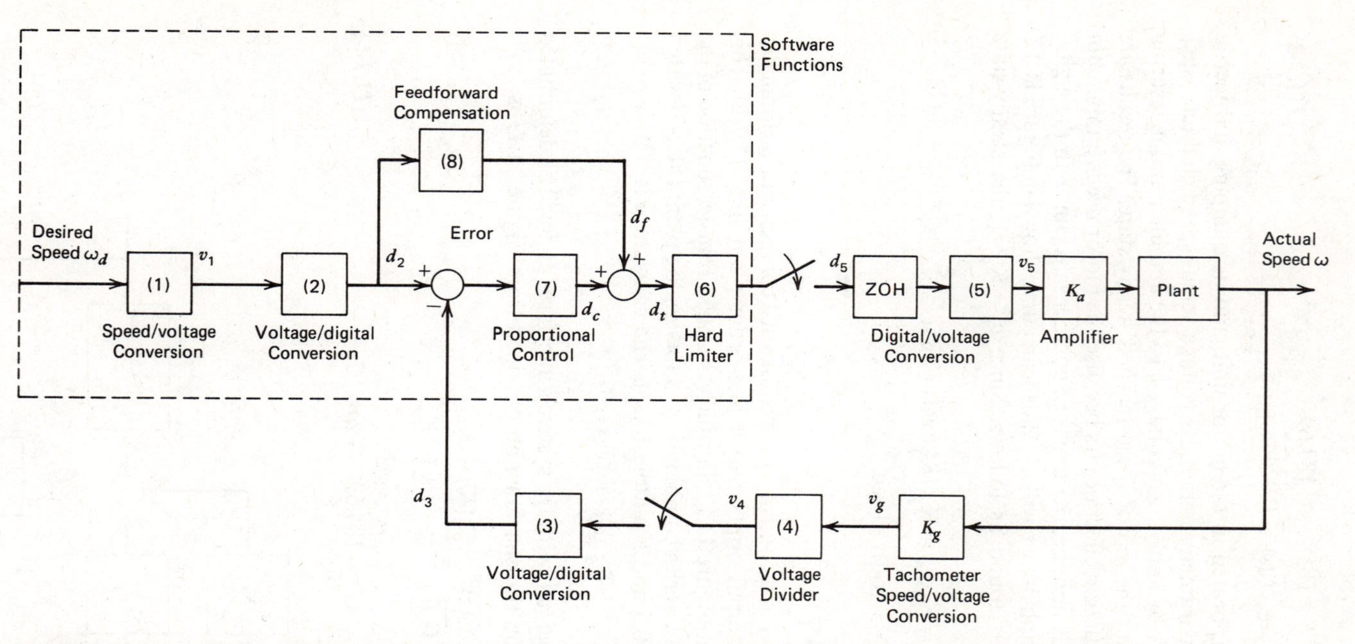

FIGURE 11.14 Control system diagram showing the digital conversion operations and the software functions.

For a zero steady-state error with a step input, we set K_f as follows.

$$K_f = 1/184.69 G_1 G_2 \tag{11.7-7}$$

The gain K_p can then be selected along with the sampling period T to achieve the desired characteristic root $z = a - G_3 K_p$.

Modeling the Interfaces

The meaning of the gains G_1 and G_2 can be explained with Figure 11.14, which is a more detailed representation of the operation of the system. First consider the feedback loop. The tachometer produces the voltage v_g. Let us assume that the A/D converter can only detect voltages in the range $r_1 \leqslant v_4 \leqslant r_2$. Therefore the tachometer voltage v_g must be reduced to this range. This is done by applying v_g to a *voltage divider*, as shown in Figure 11.15. The voltage v_4 is proportional to v_g as follows.

$$v_4 = K_4 v_g \tag{11.7-8}$$

$$K_4 = R_2/(R_1 + R_2) \tag{11.7-9}$$

Since $v_g = K_g \omega$, we have

$$v_4 = K_4 K_g \omega \tag{11.7-10}$$

Let us also assume that the A/D converter uses a *8-bit parallel* interface to transmit data to the computer. This means that eight electrical lines connect the A/D to a specific memory location in the computer. In this memory location (which is specified by the interface manufacturer) there resides an 8-bit number in the decimal range 0 to 255 (recall that $2^8 = 256$, so that we can represent 256 values with an 8-bit number). This number is denoted d_3 and represents the measured voltage v_4 as follows (see Figure 11.16).

$$d_3 = 255(v_4 - r_1)/(r_2 - r_1) \tag{11.7-11}$$

We have assumed that the A/D is linear (this can be checked by experiment), and that $d_3 = 255$ corresponds to the maximum voltage r_2 (in some devices, $d_3 = 0$ corresponds to r_2).

From (11.7-10) and (11.7-11) we can see that the number d_3 represents the

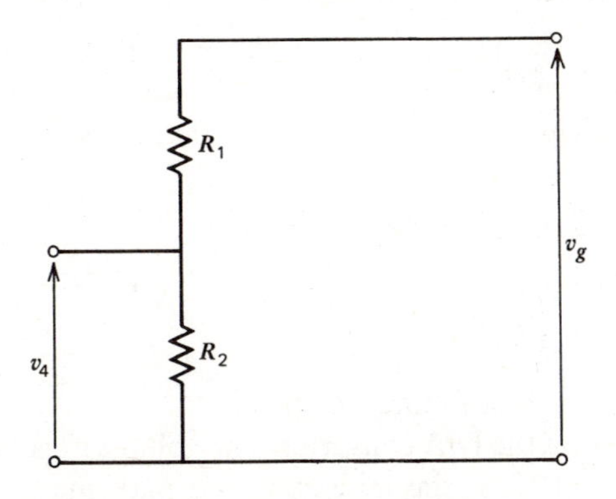

FIGURE 11.15 The voltage divider.

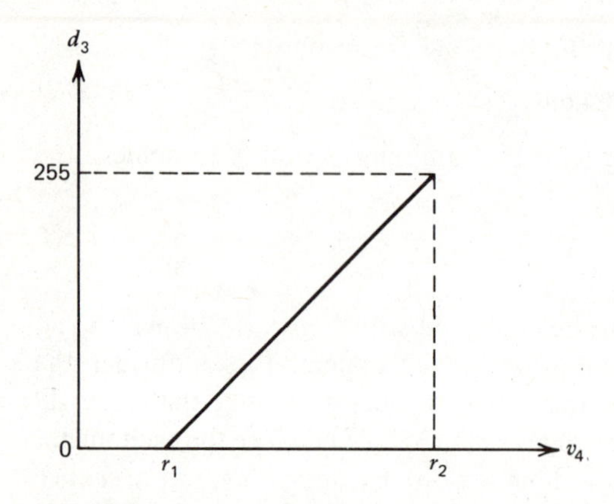

FIGURE 11.16 A/D digital output d_3 as a function of A/D input voltage v_4.

measured motor speed ω. We assume that the desired speed ω_d is entered from a keyboard in rad/sec. In order that it can be compared with d_3 to generate an error signal, this number must be converted to a number d_2 in the range 0 to 255. This conversion must be done in a meaningful way so that comparing d_2 and d_3 is equivalent to comparing ω_d and ω. In light of (11.7-10) and (11.7-11), the only way this can be done is to compute d_2 as follows.

$$d_2 = 255(v_1 - r_1)/(r_2 - r_1) \tag{11.7-12}$$

$$v_1 = K_4 K_g \omega_d \tag{11.7-13}$$

Thus v_1 is not a voltage, but a number that plays the role of the voltage v_4. The conversions specified by (11.7-12) and (11.7-13) are implemented entirely in software, and are represented by blocks 1 and 2 in Figure 11.14.

The final portion of Figure 11.14 to be considered is the D/A converter. Assuming that the converter has an 8-bit parallel interface, a number d_5 in the range 0 to 255 must be generated by the control software and placed into the memory location connected to the converter. From Figure 11.14 we see that the number d_t computed by the proportional control/feedforward compensation algorithm is

$$d_t = K_p(d_2 - d_3) + K_f d_2 \tag{11.7-14}$$

Since d_5 must be in the range 0 to 255, we implement a software function that keeps d_5 in this range, as follows

$$d_5 = \begin{cases} 0 & \text{if } d_t \leqslant 0 \\ d_t & \text{if } 0 < d_t < 255 \\ 255 & \text{if } d_t \geqslant 255 \end{cases} \tag{11.7-15}$$

This function is the *hard limiter* shown in block 6 of Figure 11.14.

Finally we must relate d_5 to the voltage v_5 of the D/A converter. The voltage v_5 is the same as voltage v in Figures 11.11 and 11.12. If the minimum and maximum possible voltage outputs of the D/A are r_3 and r_4, and if we assume that v_5 is a linear

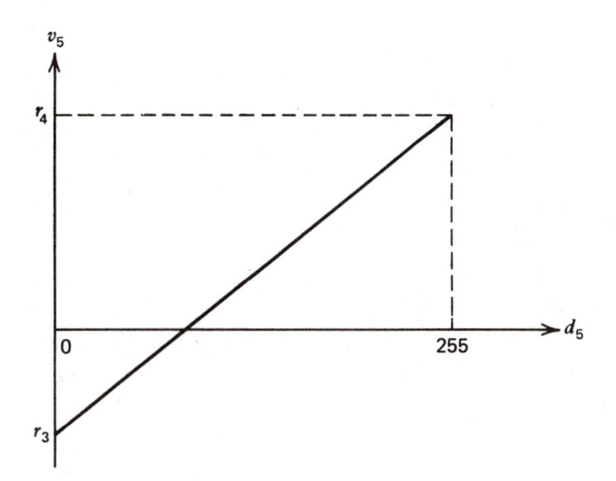

FIGURE 11.17 D/A voltage output v_5 as a function of D/A digital input d_5.

function of d_5 (see Figure 11.17), then

$$v_5 = \frac{r_4 - r_3}{255} d_5 + r_3 \tag{11.7-16}$$

Figure 11.14 can be used with the previous relations to obtain the diagram shown in Figure 11.18. The quantities r_1 and r_3 are simply constant inputs to the system, and they can be set to zero temporarily to obtain the relation between $\Omega(z)$ and $\Omega_d(z)$. Assume the hard limiter is operating in the linear range, set $r_1 = r_3 = 0$, and compare Figure 11.18 with Figure 11.13 to see that

$$G_1 = 255K_4K_g/(r_2 - r_1) \tag{11.7-17}$$

$$G_2 = (r_4 - r_3)/255 \tag{11.7-18}$$

Computing the Gains

Let us take a specific numerical example at this point. Assume that the A/D and D/A converters operate in the range of 0 to 5 volts (a typical range). Thus, $r_1 = r_3 = 0$ and $r_2 = r_4 = 5$. From Figure 11.12 we see that the magnitude of the tachometer voltage v_g is less than 50 volts, so a voltage divider gain K_4 of less than 0.1 will be sufficient to reduce v_g to the 0–5 volt range required by the A/D. Available resistors of $R_1 = 47\ k\Omega$ and $R_2 = 5.1\ k\Omega$ gave $K_4 = 0.098$. These values give $G_1 = 0.2149$, $G_2 = 0.0196$, and $K_f = 1.285$.

Suppose we desire to have a closed-loop time constant of $\tau_d = 0.5$ sec, and that we use a sampling time of $T = 0.05$ sec. Then the characteristic root must be $z = \exp(-T/\tau_d) = \exp(-0.1) = 0.9048$. From (11.7-4), $a = 0.1289$ and the proportional gain must be

$$K_p = \frac{a - z}{G_3} = \frac{a - z}{184.69(1 - a)G_1G_2} = -1.14446$$

A difference equation model of the system can be developed using Figure 11.18 as a

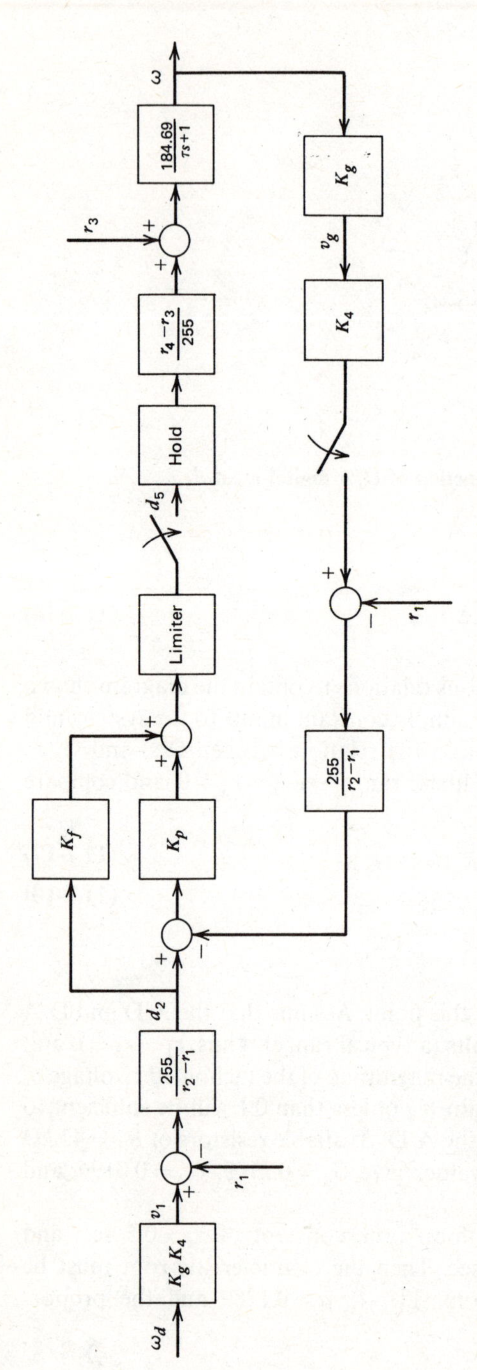

FIGURE 11.18 Block diagram with the equivalent inputs due to the constants r_1 and r_3.

guide. A simulation should be performed with this model, using the gain values just computed, in order to see if the system is operating in the linear region of the hard limiter. If the limiter is being saturated, the controller is demanding more power than the actuator can supply (see Section 7.3). In that case, we should review our performance specification for τ_d, and perhaps increase τ_d to allow the gains to be made smaller. Another factor is the sampling time T. If a smaller value of T is obtainable with the hardware, we might be able to improve the performance by using this smaller value. How this is done is discussed later.

Control Software

From Figures 11.14 and 11.18 and the preceding relations, we can draw the flowchart for the control software (Figure 11.19). The program starts when the desired speed ω_d is entered from the keyboard. After computing the constant d_2, the program queries the memory location to which the A/D converter is interfaced, and returns with a value for d_3. This is used to compute d_t. The hard limiter produces d_5, which is then placed in the memory location to which the D/A converter is interfaced. This step sends the required voltage to the motor amplifier. The program then enters a "wait loop" before repeating the process. This loop insures that the program's cycle time matches the value of the sampling time T used to compute the control gains. That is, the sampling time T is actually the total time required for the program to execute one cycle, if the A/D and D/A conversion times are less than the program's execution time.

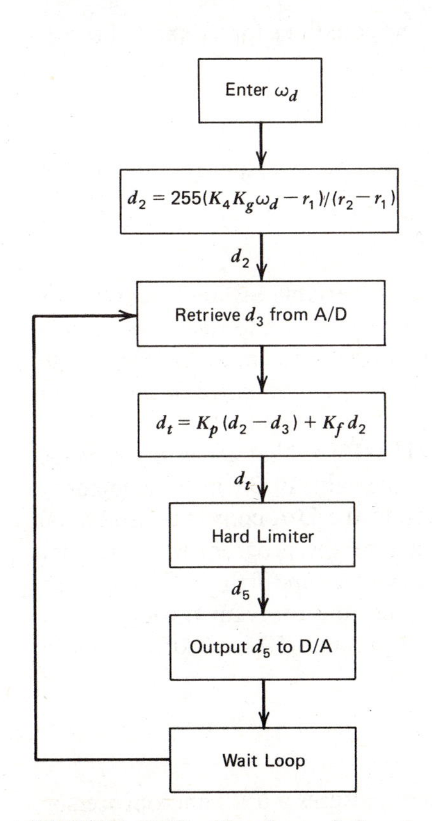

FIGURE 11.19 Flowchart of the control program.

For simplicity we have not shown additional details that are normally required for a working control program. Such a program should have provision for *interrupts*, which allow the program to jump to another segment of code in response to external events. For example, we might incorporate an interrupt that allows the user to enter a new value of the desired speed ω_d. Such an interrupt would then return the program to the "Enter ω_d" portion of the program.

A listing of the control program in BASIC is given in Figure 11.20. BASIC is used here as an illustration because it is probably the language familiar to most readers (Note that this program will not run on all machines because of slight differences in BASIC versions).

```
 10 REM N DETERMINES THE CYCLE TIME OF THE WAIT LOOP
 20 N = the required value
 30 REM SD IS THE DESIRED SPEED
 40 INPUT SD
 50 D3 = PEEK(5000)
 60 DT = 0.030202* SD + 1.14446* D3
 70 IF DT < 0 THEN DT = 0
 80 IF DT > 255 THEN DT = 255
 90 POKE 6000, DT
100 FOR J = 1 TO N
110 NEXT J
120 GO TO 50
130 END
```

FIGURE 11.20 A BASIC listing of the control program.

However, BASIC programs can usually not be made to run fast enough for control purposes. Most control programs are written in assembly language, FORTRAN, Pascal, or FORTH.

With the parameter values given previously, the equations for d_2 and d_t become

$$d_2 = 0.214914\omega_d$$
$$d_t = -1.14446(d_2 - d_3) + 1.285d_2$$

To increase the program's speed, we have combined these two equations as follows.

$$d_t = 0.030202\ \omega_d + 1.14446 d_3$$

This equation is programmed in line 60.

The desired speed ω_d is represented by the BASIC variable SD and is entered by the user at line 40. We have assumed that the A/D is interfaced to memory location 5000. The line 50 statement D3 = PEEK(5000) causes the program to set D3 equal to the value of the number stored in location 5000. Since this is assumed to be an 8-bit number, D3 will be in the range 0 to 255.

Lines 70 and 80 implement the hard limiter (11.7-15) so that DT is in the range 0–255. The line 90 statement POKE 6000,DT puts the value of DT into the memory location 6000, which we have assumed is interfaced to the D/A converter. Lines 100 and 110 implement the wait loop. The user must time the program with different values of N until the time for one cycle equals the sample time T (here T = 0.05 sec). This can be done with an oscilloscope or by running the program through many cycles (say 1000), and adjusting N until it takes the program 50 sec (0.05 × 1000) to complete 1000 cycles.

Test Results

The program was timed on an Apple II+ computer that uses a 6502 microprocessor with a 1 MHz clock. With N = 1, one pass through the program running in *interpreted* BASIC took 0.08 sec to execute. Since this is larger than the assumed sample time of

T = 0.05 sec, we cannot use interpreted BASIC unless we are willing to recompute the gains using T = 0.08 sec. This of course will tend to degrade the system's performance.

When the program was run in *compiled* BASIC, one pass took 0.014 sec. This is acceptable and a value of N must be found to make up the difference of 0.05–0.014 = 0.036 sec. Alternatively, we could recompute the gains using T = 0.014 to improve the performance.

For applications requiring a sampling time of less than 0.014 sec, we must write the program in assembly language. This program takes 130 machine cycles (this can be computed from the assembly code itself; see Reference 6). With a 1 MHz clock the assembly program would then take 0.00013 sec for one pass. Thus T could be set as low as 0.00013 sec if the A/D and D/A conversion times are fast enough. If a smaller value is required, a faster microprocessor must be used.

Implementing Other Control Laws

For ease of illustration we chose a simple control law. However, other control laws can be implemented with only a slight change in the software. For example, the proportional-plus-sum law (11.1-7) would be programmed as follows, using our current notation.

```
E = D2 - D3
SUM = SUM + E
DT = KP*E + KI*T*SUM
```

The variable SUM is

$$\text{SUM} = \sum_{i=0}^{k} e(i)$$

and is set to zero at the start of the program. Other digital control laws can be programmed just as easily.

11.8 SUMMARY

The concepts of analog control theory provide a good foundation for understanding digital control. Commonly-used digital control laws are similar to those of analog control and include variations of the PID family of algorithms. Design methods, such as the root locus plot, can be easily adapted for designing digital systems. Because of the presence of sampling in digital systems, one must be more careful in developing a system model and in deriving transfer functions. Also, the designer must select the control gains in light of the sampling period T, and might even have to specify T as part of the design.

Digital filters are proving very useful in extracting the maximum information from sensors, and applications for them continue to expand. Many techniques for designing filters are based on control theory, and the modern systems engineer should be familiar with them.

There are several advantages to using a digital controller instead of an analog element.

- Digital implementation readily allows the use of time-varying coefficients—in the simple averaging process (11.6-12), for example.
- Algorithms can be implemented digitally even when analog implementation is impossible because of unrealizable component values (for example, extremely large capacitance values).
- Nonlinear algorithms can easily be implemented in digital form. Designing nonlinear analog elements is often difficult.
- Greater accuracy is possible with digital systems.
- Digital systems are less sensitive to such environmental factors as humidity, temperature, pressure, and noise.

On the other hand, digital systems have some disadvantages.

- Digital systems are active and thus require power (analog filters like the *RC* circuit are often passive).
- Input signals with a wide range of values cannot be handled unless the number of bits is sufficient for accurate quantization.
- Processing is limited to signals whose highest frequency content is less than the logic circuitry speed.

For these reasons, there are applications for both analog and digital systems.

REFERENCES

1. Y. Takahashi, M. Rabins, and D. Auslander, *Control*, Addison-Wesley, Reading, Mass., 1970.
2. G. F. Franklin and J. D. Powell, *Digital Control of Dynamic Systems*, Addison-Wesley, Reading, Mass., 1980.
3. A. Gelb, ed., *Applied Optimal Estimation*, MIT Press, Cambridge, 1974.
4. T. R. Benedict and G. W. Bordner, "Synthesis of an Optimal Set of Radar Track-While-Scan Smoothing Equations", *IEEE Transactions On Automatic Control*, Vol. AC-7, pp. 27–32, July, 1962.
5. J. B. Peatman, *Microprocessor-Based Design*, McGraw-Hill, New York, 1977.
6. L. J. Scanlon, *6502 Software Design*, H. W. Sams, Indianapolis, 1980.

PROBLEMS

11.1 Each of the following problems refers to a previous problem in which an analog controller was developed. For each problem, assume that a sampler and zero-order hold is inserted between the controller and the plant $G_p(s)$ as shown in Figure P11.1. Assume also that the command input and any feedback measurements are sampled. The disturbance input is not sampled.

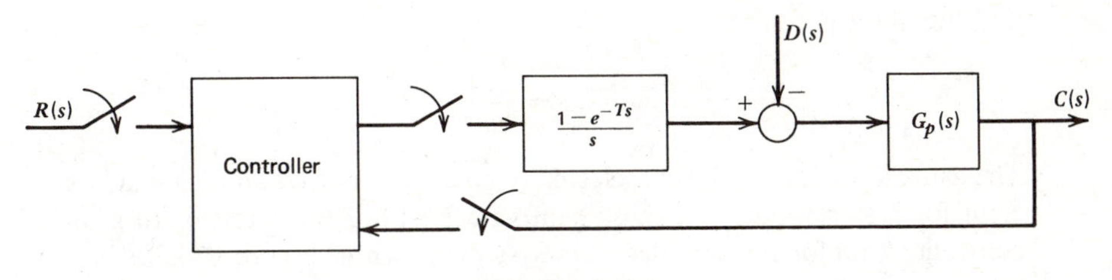

FIGURE P11.1

For each problem, design a digital controller to meet the specifications given for the analog system in the problem cited. Evaluate the resulting design to see if it meets the continuous-time specifications. Use the sampling period T given for each problem. This value corresponds to 1/10 or less of the smallest relevant time constant of the plant or as required by the specifications. If a satisfactory design cannot be achieved with the given sampling period, find a suitable value for T.

(a) Problem 6.9*c*, $T = 0.01$.
(b) Problem 6.15*a*,*b*,*c*, $T = 0.04$.
(c) Problem 6.21, $T = 0.1$.
(d) Problem 6.23, $T = 0.08$.
(e) Problem 6.24, $T = 0.2$.
(f) Problem 6.26*e*, $T = 0.01$.
(g) Problem 6.28*a*, $T = 0.01$.
(h) Problem 6.32, $T = 0.01$.
(i) Problem 7.17*b*. $T = 0.01$.

11.2 The diagram shown in Figure P11.2 represents digital control of a dc motor system with a disturbance torque included. Obtain the relation for the output $\Theta(z)$ in terms of $D(z)$ for:

(a) Proportional control $G_c(s) = K_p$.
(b) Proportional-plus-integral control $G_c(s) = K_p + K_I/s$.

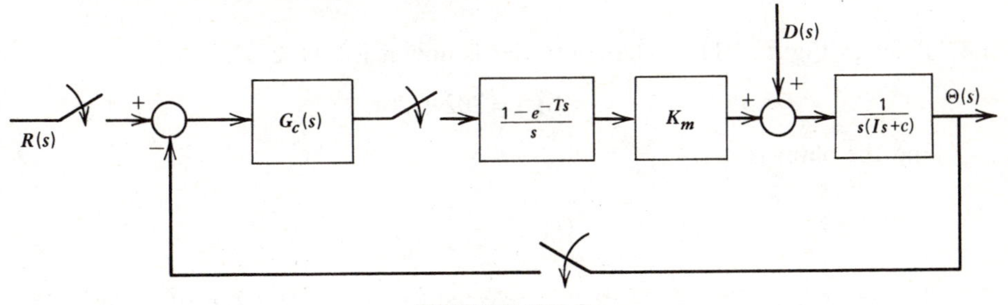

FIGURE P11.2

11.3 In Figure P11.33*a* the controller is a PI type with

$$G(s) = 3 + \frac{2}{s}$$

and the plant is

$$G_p(s) = \frac{1}{2s+1}$$

The gains $K_p = 3$, $K_I = 2$ were selected to give $\zeta = 1$, $\tau = 1$, and zero steady-state error for a step input. Use these gain values with $T = 0.1$ to obtain a digital controller $G(z)$ for the sampled data system shown in Figure P11.3*b*.

(a) Use the rectangular integration formula (11.1-7) for the I action.

(b) Use the trapezoidal integration form in (11.1-17) for the I action.

Find the unit-step response of each design. Compare them with each other and with the analog controller's response.

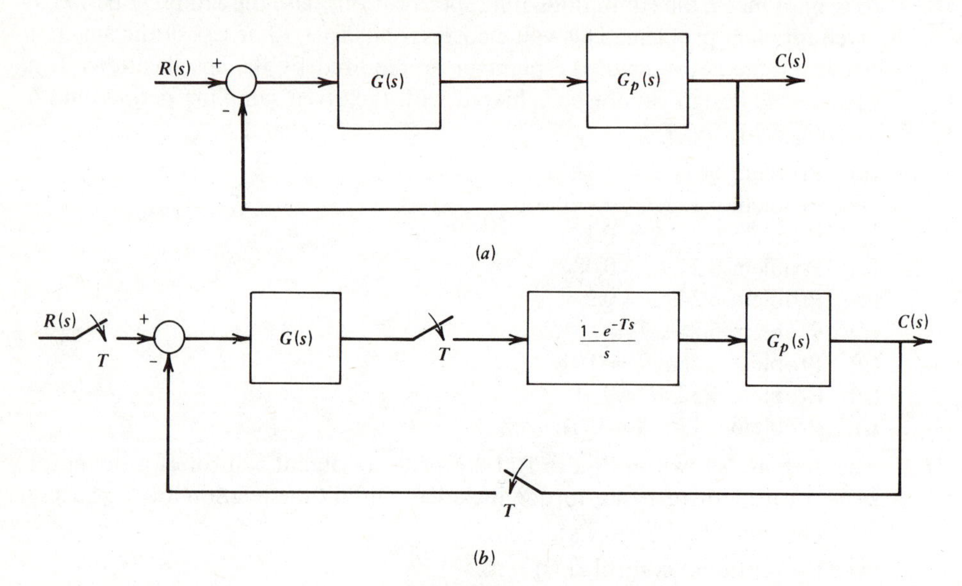

FIGURE P11.3

11.4 Refer to Figure P11.3. The controller is now a PD type with

$$G(s) = 202 + 20s$$

and the plant is

$$G_p(s) = \frac{1}{s^2 - 2}$$

The gains $K_p = 202$, $K_D = 20$ were selected to give $\zeta = 0.707$, $\tau = 0.1$, and a 1% steady-state error with a step input. Use these gain values with $T = 0.01$ to obtain a digital controller $G(z)$ for the sampled data system shown in Figure P11.3*b*.

(a) Use the proportional-plus difference law (11.1-8).

(b) Use the four-point central difference formula (11.1-16) for the D action.

Find the unit-step response of each design. Compare them with each other and that of the analog controller.

11.5 Refer to Figure P11.3. The controller is a PID type with

$$G(s) = 120 + \frac{100}{s} + 20s$$

and the plant is

$$G_p(s) = \frac{1}{s(s+1)}$$

The gains $K_p = 120$, $K_I = 100$, $K_D = 20$ were selected to give a dominant root at $s = -1$, with the remaining roots at $s = -10$, and a zero steady-state error for a step input. Use these gain values with $T = 0.1$ to obtain a digital controller $G(z)$ for the sampled data system shown in Figure P11.3*b*.

(a) Use (11.1-18).
(b) Use (11.1-19).

Find the unit-step response of each design. Compare the responses with each other and that of the analog controller.

11.6 Repeat Example 11.1 with (a) $T = 0.2$ and (b) $T = 0.05$. Discuss the results.

11.7 The *Tustin* approximation for differentiation is

$$\mathscr{Z}\left(\frac{dy}{dt}\right) = D(z)Y(z)$$

where

$$D(z) = \frac{2}{\Delta t}\frac{1 - z^{-1}}{1 + z^{-1}}$$

(a) Show that the Tustin approximation is equivalent to integration of the following equation using the trapezoidal rule:

$$\frac{dy}{dt} = v$$

[*Hint:* Apply the trapezoidal rule to the preceding equation and show that $Y(z) = V(z)/D(z)$.]

(b) Use the Tustin approximation to obtain a difference equation approximation for

$$\dot{y} = ry + bv$$

[*Hint:* Apply the z transform to both sides of the preceding equation, and use the Tustin approximation for $\mathscr{Z}(dy/dt)$.]

(c) Show that

$$\mathscr{Z}\left(\frac{d^n y}{dt^n}\right) = [D(z)]^n Y(z)$$

(d) Apply the results in (c) to find a difference equation approximation for the following equation:

$$\frac{d^2y}{dt^2} + 5\frac{dy}{dt} + 3y = v$$

11.8 The Tustin approximation introduced in Problem 11.7 can also be used to derive discrete-time control algorithms from their analog equivalents.

(a) Show that the Tustin method when applied to the PI-control law (11.2-1) with $K_D = 0$ results in a transfer function like (11.2-8). Derive the values of a_1 and a_2 in terms of the analog gains K_p, K_I, and T.

(b) Apply the results of (a) to the problem given in Example 11.1*b*. Find the values of a_1 and a_2, and the characteristic roots. Compare the unit step response with that of the analog controller designed in Example 11.1*a*.

(c) Show that the Tustin approximation, when applied to the full PID control law (11.2-1) with $K_D \neq 0$, gives the following transfer function:

$$\frac{F(z)}{E(z)} = \frac{a_1 z^2 + a_2 z + a_3}{z^2 - 1}$$

Derive the expressions for the coefficients a_1, a_2, and a_3 in terms of the analog gains K_p, K_I, K_D, and T.

11.9 The system shown in Figure P11.9*a* represents the digital equivalent of I action, while in Figure P11.9*b*, the controller is a delayed I action. Plot the root locus for each system and determine the stability limit.

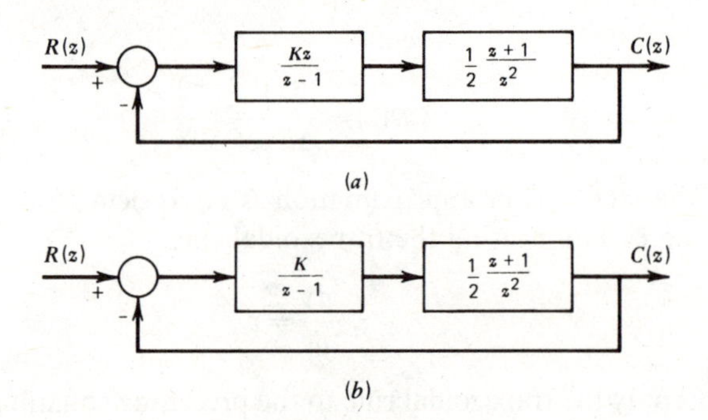

FIGURE P11.9

11.10 Consider the proportional controller shown in Figure P11.10, where

$$G(s) = \frac{1}{s(s+1)}$$

Let $T = 1$.

(a) Plot the root locus in terms of K.

(b) Find the stability limit and compare with that of the analog system.

(c) Find the value of K required to give a damping ratio of 0.5. What is the

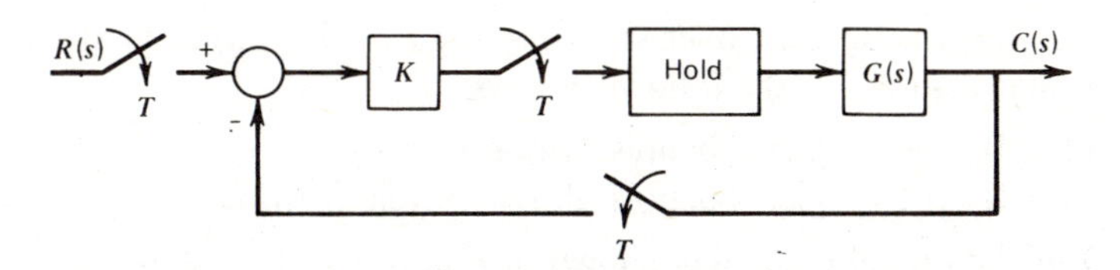

FIGURE P11.10

corresponding time constant? Compare the unit-step response with that of the analog system.

(d) Is it possible to find a value of K to give a damping ratio of 1 with a time constant of 2? Explain.

11.11 Consider the system shown in Figure P11.10 where

$$G(s) = \frac{1}{s(s+1)(s+3)}$$

and $T = 0.1$. Plot the root locus in terms of K.

11.12 Repeat Example 11.3 using a sampling period of $T = 1$ sec. Compare the responses for the two values of T.

11.13 Suppose that the plant in the control system shown in Figure 11.7 is

$$G_p(s) = \frac{1}{2s+1}$$

with $T = 1$.

(a) Find the control algorithm $D(z)$ so that the closed-loop transfer function will be $T(z) = z^{-1}$.

(b) Determine $c(k)$, $e(k)$, and $m(k)$ for a unit-step input.

11.14 We desire the open-loop system shown in Figure P11.14 to exhibit the following response to a unit-step input:

$$c(0) = 0 \qquad c(1) = 0.5 \qquad c(2) = c(3) = \cdots = 1$$

(a) Find the required transfer function $G(z)$.

(b) The given unit-step response suggests a first-order system. Is this correct? Explain.

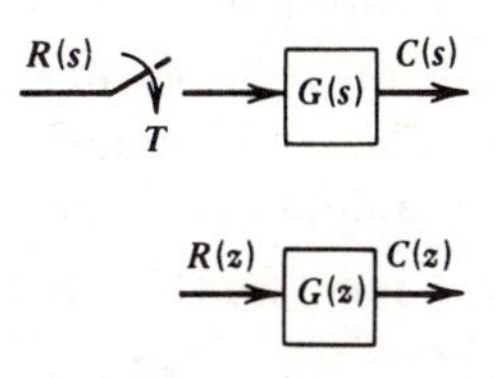

FIGURE P11.14

11.15 The P control of a pure integrator with dead time is considered. Suppose that the dead time is $D = 1$.

(a) For the analog controller shown in Figure P11.15*a*, compute the proportional gain K_p in the following ways:

(1) By the Ziegler–Nichols response method.
(2) By the Ziegler–Nichols ultimate-cycle method.

(b) For the digital controller shown in Figure P11.15*b*, compute the stability limit for K_p, and compare with that found for the analog system.

(c) Determine the characteristic root locations using the two values of K_p found in (a), assuming the following:

(1) $T = D = 1$
(2) $T = 0.5D = 0.5$

Compare the results.

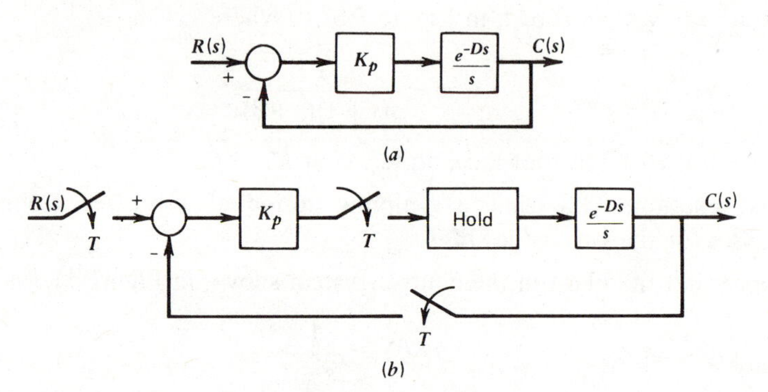

FIGURE P11.15

11.16 Consider the system shown in Figure P11.10, where

$$G(s) = \frac{e^{-Ds}}{4s + 1}$$

Suppose that the dead time is $D = 1$.

(a) Plot the root locus in terms of K for the following:

(1) $T = D = 1$
(2) $T = 0.5D = 0.5$
(3) $T = 0.25D = 0.25$

(b) What is the effect on the root locus of the choice of $n = D/T$?

11.17 Write a program to simulate the behavior of the filter in Example 11.4 for a set of exponentially decaying data contaminated with a random component. Investigate the choice of K and the filter's initial condition $\hat{y}(0)$.

11.18 Plot the frequency response curve for (11.6-9) for $a = -1$ and $K = 0.1$, 0.5, and 0.9.

11.19 Redo the filter design in Example 11.6 for the case in which the voltage time constant is 1 sec instead of 0.5 sec.

11.20 Simulate the behavior of the filter in Example 11.6 for the case in which the measured voltage is $u(t) = 10e^{-2t} + 0.5 \sin(377t)$.

11.21 A bandpass filter centered on the frequency ω_c can be obtained by cascading two first-order filters with roots at $z = r\exp(\pm i\omega_c T)$, where r is a parameter. The filter transfer function is

$$T(z) = \frac{bz^2}{(z - re^{ia})(z - re^{-ia})}$$

where $a = \omega_c T$.

(a) What restrictions must be placed on r for the filter to be stable?
(b) What value must the parameter b have in order for the filter gain to be unity at $\omega = \omega_c$?

11.22 Consider the displacement velocity filter (11.6-19) and (11.6-20), for the situation where the velocity $v(t) = 10$ ft/sec and the initial displacement is $y(0) = 10$ ft. That is, $y(t) = 10 + 10t$. Assume that the measurements are perfect and the filter's estimates are initialized to zero. Assume that the sampling period is $T = 0.1$ sec.

(a) Suppose that the filter's behavior in the continuous-time domain is specified to be critically damped $\zeta = 1$. Find the requirements on the filter's coefficient a in terms of b in order to achieve this. The filter must be stable.
(b) Compute the response of the filter in terms of b, given the previously specified inputs $y(t)$ and $v(t)$. Discuss the performance of the filter and the effect of b on this performance.

11.23 Figure P11.23*a* shows a representation of a missile whose orientation θ from horizontal can be controlled by the elevator angle ϕ. The desired orientation angle is r; the error is e. Part (b) of the figure shows a control system based on the concept of tachometer feedback compensation (see Section 6.7). The rate ω is measured by an analog sensor, but the angle θ is measured by a sensor that incorporates a digital filter algorithm. The output of this filter is converted to analog form with the sample-and-hold. The resulting analog estimate of θ is denoted $\hat{\theta}$ and is fed to the error detector for use with the analog controller $G_1(s)$.

The time required for the filter algorithm to make its calculation is T, so we set the sampling time to this value. Because the calculations are done in the z domain, we need to be given only $H_2(z)$ and not $H_2(s)$.

(a) Find the input–output relation for this system, given that

$$G_1(s) = K_1 \qquad G_2(s) = K_2$$
$$H_1(s) = K_3 \qquad H_2(z) = Kz/(z + b)$$

(b) Investigate the stability properties of the proposed system for the transfer functions given in (a).

11.24 For the system shown in Figure 11.10, assume that $T = 0.1$ sec and that

$$G_1(s) = 1 \qquad G_2(s) = 10/s^2$$

The filter shown in the figure is described by (11.6-19) and (11.6-20).

(a) Select the filter coefficients a and b so that the filter will have a time constant of 1 sec and a damping ratio $\zeta = 1$.

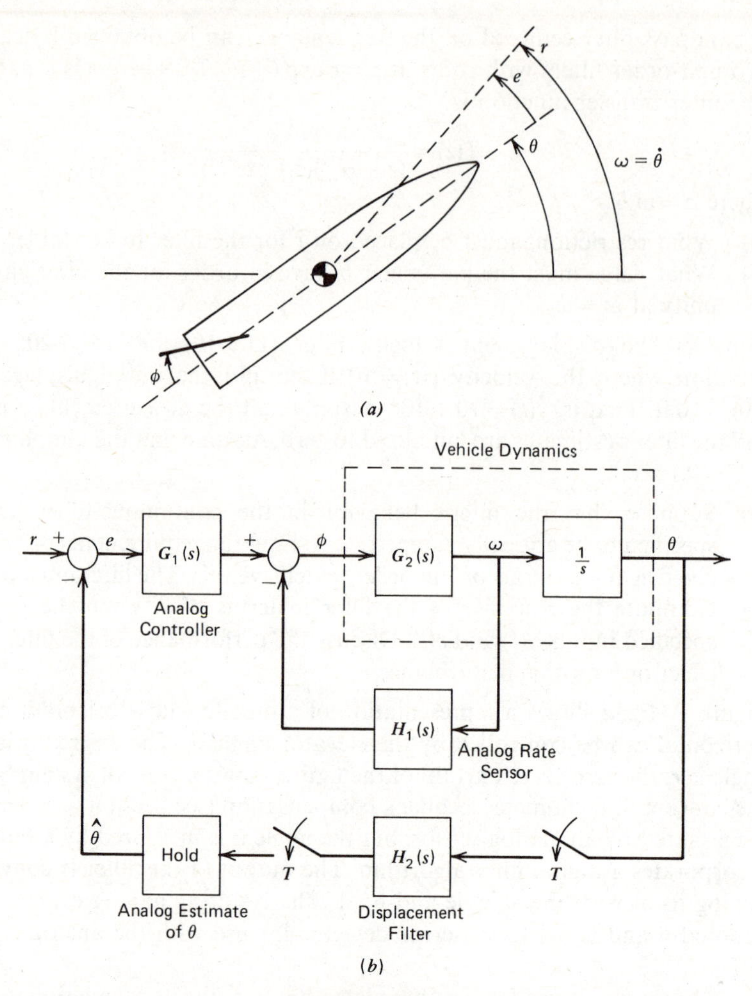

FIGURE P11.23

(b) Select the gains K_1 and K_2 so that the system will have a time constant of 10 sec and a damping ratio $\zeta = 1$. To do this assume that the radar measurement is exact and that the filter is working perfectly, so that $y(k) = h(k)$ and $v(k) = [h(k) - h(k-1)]/T$.

(c) Using the values of a, b, K_1, and K_2 found in parts a and b, find the transfer function $H(z)/R(z)$. Compute the time constant and damping ratio of the dominant roots.

11.25 For the motor control system discussed in Section 11.7, suppose we use the PID control law (11.1-19). The specifications require that $\zeta = 1$ and that the dominant time constant be 0.5 sec. Compute the gains K_p, K_I, K_D, and make the necessary modifications to the BASIC program shown in Figure 11.20.

11.26 For the motor control system discussed in Section 11.7, suppose that the A/D and D/A converters operate in the range -10 to $+10$ volts. Obtain the new equation for d_t that must replace line 60 in Figure 11.20.

APPENDIX A

Some Useful Analytical Techniques

A.1 THE TAYLOR SERIES EXPANSION

Taylor's theorem states that a function $f(y)$ can be represented in the vicinity of $y = a$ by the expansion

$$f(y) = f(a) + f'(a)(y - a) + \frac{1}{2!} f''(a)(y - a)^2 + \cdots + \frac{1}{k!} f^{(k)}(a)(y - a)^k + \cdots + R_n \tag{A.1-1}$$

where the kth derivative evaluated at $y = a$ is denoted by

$$f^{(k)}(a) = \left. \frac{d^k f}{dy^k} \right|_{y=a} \tag{A.1-2}$$

The term R_n is the remainder and is given by

$$R_n = \frac{f^{(n)}(b)(y - a)^n}{n!} \tag{A.1-3}$$

where b lies between a and y. These results hold if $f(y)$ has continuous derivatives through order n. If R_n approaches zero for large n, (A.1-1) is called the Taylor series for $f(y)$ about $y = a$. If $a = 0$, the series is sometimes called the Maclaurin series.

Some common examples of the Taylor series are

$$\sin y = y - \frac{y^3}{3!} + \frac{y^5}{5!} - \frac{y^7}{7!} + \cdots \tag{A.1-4}$$

$$e^y = 1 + y + \frac{y^2}{2!} + \frac{y^3}{3!} + \cdots \tag{A.1-5}$$

where $a = 0$ in both examples.

Generally, the series converges for all values of y in some interval called the interval of convergence and diverges for all values of y outside this interval. The accuracy of the expansion depends on how close y is to a and on how many terms are used in the expansion, in addition to the nature of the function in question. If only the first-order term is retained, the function is approximated by a straight line.

$$f(y) \cong f(a) + f'(a)(y - a) \tag{A.1-6}$$

The error is given by the remainder term

$$R_2 = \frac{f''(b)(y - a)^2}{2} \tag{A.1-7}$$

This result is the basis of the technique used to linearize a nonlinear function or nonlinear differential equation. If we define the deviations

$$g = f(y) - f(a) \tag{A.1-8}$$

$$x = y - a \tag{A.1-9}$$

then (A.1-6) can be expressed more simply as

$$g(x) \cong mx \qquad m = f'(a) \tag{A.1-10}$$

where m is the slope of the straight-line approximation at the reference point $y = a$.

Multivariable Case

The Taylor series can be generalized to any number of variables. For two variables, it is

$$\begin{aligned} f(y_1, y_2) = {} & f(a_1, a_2) + f_{y_1}(a_1, a_2)(y_1 - a_1) + f_{y_2}(a_1, a_2)(y_2 - a_2) \\ & + \frac{1}{2!} f_{y_1 y_1}(a_1, a_2)(y_1 - a_1)^2 + \frac{1}{2!} f_{y_1 y_2}(a_1, a_2)(y_1 - a_1)(y_2 - a_2) \\ & + \frac{1}{2!} f_{y_2 y_2}(a_1, a_2)(y_2 - a_2)^2 + \cdots \end{aligned} \tag{A.1-11}$$

where

$$f_{y_i y_j}(a_1, a_2) = \left. \frac{\partial^2 f}{\partial y_i \partial y_j} \right|_{\substack{y_1 = a_1 \\ y_2 = a_2}} \tag{A.1-12}$$

The linearizing approximation is

$$f(y_1, y_2) \cong f(a_1, a_2) + f_{y_1}(a_1, a_2)(y_1 - a_1) + f_{y_2}(a_1, a_2)(y_2 - a_2) \tag{A.1-13}$$

or

$$g(x_1, x_2) \cong m_1 x_1 + m_2 x_2 \tag{A.1-14}$$

Where

$$g = f(y_1, y_2) - f(a_1, a_2) \tag{A.1-15}$$

$$x_1 = y_1 - a_1 \tag{A.1-16}$$

$$x_2 = y_2 - a_2 \tag{A.1-17}$$

$$m_1 = f_{y_1}(a_1, a_2) \tag{A.1-18}$$

$$m_2 = f_{y_2}(a_1, a_2) \tag{A.1-19}$$

The linearizing approximation for two variables is thus seen to be a plane passing through the point (a_1, a_2) and having the slopes m_1 and m_2 in the y_1 and y_2 directions, respectively.

The two-variable case can easily be generalized to the case with any number of variables. This is represented most easily in matrix notation (see Section A.4 for a review of matrix methods).

A.2 NEWTON ITERATION

A useful application of the Taylor series occurs in the numerical solution of nonlinear equations. The problem is as follows. Find the value y such that

$$f(y) = 0 \tag{A.2-1}$$

where $f(y)$ is some given function. Let y^* be an initial guess. If it is assumed that y is close enough to y^* so that the linearized form (A.1-6) is valid, then

$$f(y) \cong f(y^*) + f'(y^*)(y - y^*) \tag{A.2-2}$$

Using (A.2-1), we can solve for the estimate of the solution

$$y \cong y^* - \frac{f(y^*)}{f'(y^*)} \tag{A.2-3}$$

The solution y from (A.2-3) gives a new y^*, and the process is repeated until (A.2-1) is satisfied to the desired degree of accuracy. In difference equation form, the algorithm is

$$y_k \cong y_{k-1} - \frac{f(y_{k-1})}{f'(y_{k-1})} \tag{A.2-4}$$

The initial guess occurs at $k = 0$ and is y_0. The usual criterion for termination is

$$|f(y_k)| < \varepsilon \tag{A.2-5}$$

where ε is a small number chosen by the analyst. Often in practice, however, the iteration is stopped when the solution stops changing in the number of significant figures required of the answer.

When the initial guess is reasonably close to the correct value, Newton's method converges very rapidly. It is possible that convergence will not occur, however. We cannot overestimate the importance of sketching the function $f(y)$ in order to obtain a reasonable idea of its shape and the number and locations of its roots. For example, consider the function

$$f(y) = 10ye^{-y} - 1 \tag{A.2-6}$$

Assuming that only positive values of y are of interest, we see that $f(0) = -1$, $f(1) = 2.7$ and $f(y) \to 0$ as $y \to \infty$. From this, we can tell that a root lies in the range $0 < y < 1$ and no other roots occur. Thus, a reasonable starting guess would be in this range. For a starting value of $y_0 = 0$, the solution sequence is

$$\begin{aligned} y_0 &= 0 \\ y_1 &= 0.1 \\ y_2 &= 0.1117 \\ y_3 &= 0.1118 \end{aligned}$$

where more significant figures have been retained in the calculation than are shown. The last value gives $f(0.1118) = -0.000258$. If we are satisfied with this accuracy, we can stop. If the guess is farther away, say, $y_0 = 0.5$, the convergence will take longer.

These results are

$$y_0 = 0.5$$
$$y_1 = -0.1703$$
$$y_2 = 0.0473$$
$$y_3 = 0.1077$$
$$y_4 = 0.1118$$

However, there is a peak in the function at $y = 1$. If we guess $y_0 > 1$, the iteration will not converge but will approach $y = \infty$. The reason is that the method approximates the function with a straight line whose slope is the slope of $f(y)$ at the current value of y_k. Here, for $y > 1$, the slope is always negative, and the method is deceived into thinking that the curve is approaching an axis crossing (a root) when, in fact, the function is approaching zero asymptotically. Near the peak at $y = 1$, the slope is very small. Thus, for y_0 close to but less than 1, the method will at first head for large negative values of y before converging, and more iterations will be required.

Of course, we will not always be able to determine the shape of the function as well as in the previous example, but the insights generated from it are important to keep in mind. The two most important applications of Newton's method for us are (1) the solution for polynomial roots required to determine the location of breakaway and break-in points on the root locus plot (Chapter 8), and (2) the solution of nonlinear equations that specify the equilibrium points of a nonlinear differential or difference equation. In the root locus application, we know where the locus can exist on the real axis and in general where the breakaway and break-in points will occur. Thus, we can usually make a good initial guess, and there is no difficulty due to multiple or complex roots. (Newton iteration cannot find complex roots and might converge to the wrong root if more than one exists. We must then begin again with a new guess.)

A.3 FOURIER SERIES

The Taylor series is useful for representing a function with an approximate straight line, while the Fourier series is a technique for representing a periodic function as a sum of sines, cosines, and a constant. With such a description, the sinusoidal response of a given linear system can be used with the principle of superposition to obtain the system's response to the periodic function. The concept of system bandwidth is helpful in reducing the number of terms to be considered.

Consider a periodic function $f(y)$ of period $2L$ such that $f(y+2L) = f(y)$. The Fourier series representation of this function defined on the interval $c \leqslant y \leqslant c + 2L$, where c and $L > 0$ are constants, is defined as

$$f(y) = \frac{a_0}{2} + \sum_{n=1}^{\infty} \left(a_n \cos \frac{n\pi y}{L} + b_n \sin \frac{n\pi y}{L} \right) \tag{A.3-1}$$

where

$$a_n = \frac{1}{L} \int_c^{c+2L} f(y) \cos \frac{n\pi y}{L} \, dy \tag{A.3-2}$$

$$b_n = \frac{1}{L}\int_c^{c+2L} f(y)\sin\frac{n\pi y}{L}\,dy \tag{A.3-3}$$

If $f(y)$ is defined outside the specified interval by a periodic extension of period $2L$ and $f(y)$ and $f'(y)$ are piecewise continuous, then the series converges to $f(y)$ if y is a point of continuity, and to the average value $[f(y_+)+f(y_-)]/2$ otherwise.

For example, a train of unit pulses of width π and alternating in sign is described by

$$f(y) = \begin{cases} 1 & 0 < y < \pi \\ -1 & -\pi < y < 0 \end{cases} \tag{A.3-4}$$

Here, the period is 2π, so that $L = \pi$. We can take the constant c to be zero. From (A.3-2) and (A.3-3), with the aid of a table of integrals, the coefficients are found to be

$$a_n = 0 \quad \text{for all } n$$

$$b_n = \frac{4}{n\pi} \quad \text{for } n \text{ odd}$$

$$b_n = 0 \quad \text{for } n \text{ even}$$

The resulting Fourier series is

$$f(y) = \frac{4}{\pi}\left(\frac{\sin y}{1} + \frac{\sin 3y}{3} + \frac{\sin 5y}{5} + \cdots\right) \tag{A.3-5}$$

The cosine terms and the constant a_0 will not appear in the series if the function is odd; that is, if $f(-y) = -f(y)$. If the function is even, $f(-y) = f(y)$, and no sine terms will appear. If the function is neither even nor odd, both sines and cosines, and perhaps the constant a_0, will appear in the series. For some applications, it is convenient to define the function outside its basic interval to be either odd or even in order to obtain a special series form, but this technique is of little interest here.

A.4 VECTORS AND MATRICES

The use of vector matrix notation allows simplified representation of complex mathematical expressions. For example, a set of many equations can be represented as one vector matrix equation. This has advantages in presenting analytical concepts and developing techniques applicable to models with an arbitrary number of variables.

The set of linear algebraic equations

$$\left.\begin{aligned} a_{11}x_1 + a_{12}x_2 + \cdots + a_{1n}x_n &= b_1 \\ a_{21}x_1 + a_{22}x_2 + \cdots + a_{2n}x_n &= b_2 \\ \cdots\cdots\cdots\cdots\cdots\cdots\cdots\cdots \\ a_{n1}x_1 + a_{n2}x_2 + \cdots + a_{nn}x_n &= b_n \end{aligned}\right\} \tag{A.4-1}$$

can be represented in compact form as

$$\mathbf{Ax} = \mathbf{b} \tag{A.4-2}$$

The *matrix* **A** is an array of numbers that corresponds in an ordered fashion to the coefficients a_{ij}. It is ordered as follows:

$$\mathbf{A} = \begin{bmatrix} a_{11} & a_{12} & \cdots & a_{1n} \\ a_{21} & a_{22} & \cdots & a_{2n} \\ \cdot & \cdot & \cdots & \cdot \\ a_{n1} & a_{n2} & \cdots & a_{nn} \end{bmatrix} \tag{A.4-3}$$

This matrix has n rows and n columns, so it is a *square* matrix. Its dimension is expressed as $(n \times n)$.

The matrix should not be confused with the *determinant*, which is a single number derived from the rows and columns of (A.4-3) by a prescribed set of reduction rules. A determinant is denoted by a pair of parallel lines; square brackets denote matrices and vectors. The determinant of the matrix **A** and also of the set of equations (A.4-1) is

$$\det(\mathbf{A}) = |\mathbf{A}| = \begin{vmatrix} a_{11} & a_{12} & \cdots & a_{1n} \\ a_{21} & a_{22} & \cdots & a_{2n} \\ \cdot & \cdot & \cdots & \cdot \\ a_{n1} & a_{n2} & \cdots & a_{nn} \end{vmatrix} \tag{A.4-4}$$

Determinant dimensions are expressed in the same way as matrices.

The (2×2) determinant is evaluated as follows:

$$|\mathbf{A}| = \begin{vmatrix} a_{11} & a_{12} \\ a_{21} & a_{22} \end{vmatrix} = a_{11}a_{22} - a_{12}a_{21} \tag{A.4-5}$$

Higher order determinants are defined in a similar manner. Let M_{ij} denote the *minor* of the element a_{ij} and C_{ij} the *cofactor*. The minor is the determinant obtained by crossing out the row and column of the element a_{ij}. The cofactor is defined as

$$C_{ij} = (-1)^{i+j} M_{ij} \tag{A.4-6}$$

A determinant is evaluated by selecting a row or column (preferably the one with the most zero elements), computing the product $a_{ij}C_{ij}$ for each element, and summing the results. In equation form, this may be expressed as follows. For expansion about row i,

$$\det(\mathbf{A}) = \sum_{k=1}^{n} a_{ik}C_{ik} \tag{A.4-7}$$

For expansion about column j,

$$\det(\mathbf{A}) = \sum_{k=1}^{n} a_{kj}C_{kj} \tag{A.4-8}$$

These results are independent of which row or column is selected. A determinant is always square and is said to be *singular* if its value is zero. If it is the determinant of a matrix, the matrix is also said to be singular.

For example, consider the determinant

$$|\mathbf{A}| = \begin{vmatrix} 1 & 0 & -1 \\ 0 & 2 & 5 \\ 1 & 2 & 4 \end{vmatrix} \tag{A.4-9}$$

For expansion about the second row, the minors are

$$M_{21} = \begin{vmatrix} 0 & -1 \\ 2 & 4 \end{vmatrix} \qquad M_{22} = \begin{vmatrix} 1 & -1 \\ 1 & 4 \end{vmatrix} \qquad M_{23} = \begin{vmatrix} 1 & 0 \\ 1 & 2 \end{vmatrix}$$

or $M_{21} = 2$, $M_{22} = 5$, $M_{23} = 2$. The cofactors are $C_{21} = -2$, $C_{22} = 5$, $C_{23} = -2$. The value of the determinant is

$$\det(\mathbf{A}) = 0C_{21} + 2C_{22} + 5C_{23} = 0$$

The determinant is singular.

Types of Matrices

A *vector* is a special case of a matrix; it has only one row or one column. Thus, a *row vector* has the dimension $(1 \times n)$, since it has one row and n columns. A *column vector* has one column and is of dimension $(n \times 1)$. In this text, a vector is taken to be a column vector unless otherwise specified. Examples of column vectors are the vectors **x** and **b** in (A.4-1), where

$$\mathbf{x} = \begin{bmatrix} x_1 \\ x_2 \\ \vdots \\ x_n \end{bmatrix} \qquad \mathbf{b} = \begin{bmatrix} b_1 \\ b_2 \\ \vdots \\ b_n \end{bmatrix} \tag{A.4-10}$$

A matrix with only one element is a *scalar*.

A column vector can be converted into a row vector by the *transpose* operation, in which the rows and columns are interchanged. Here, we denote this operation by the superscript T. For an $(n \times m)$ matrix **A** with n rows and m columns, $\mathbf{A}^T$ (read "*A* transpose") is an $(m \times n)$ matrix. If **A** is given by

$$\mathbf{A} = \begin{bmatrix} 2 & 4 & 1 \\ 5 & 3 & 0 \end{bmatrix}$$

its transpose is

$$\mathbf{A}^T = \begin{bmatrix} 2 & 5 \\ 4 & 3 \\ 1 & 0 \end{bmatrix}$$

If $\mathbf{A}^T = \mathbf{A}$, the matrix is *symmetric*. Only a square matrix can be symmetric.

It is sometimes convenient to consider a matrix as consisting of submatrices or vectors. The matrix is said to be *partitioned*. For example, if the matrix given by (A.4-4) is partitioned into column vectors $\mathbf{a}_i$,

$$\mathbf{A} = [\mathbf{a}_1 \quad \mathbf{a}_2 \quad \dots \quad \mathbf{a}_n] \tag{A.4-11}$$

where the column vectors are

$$\mathbf{a}_1 = \begin{bmatrix} a_{11} \\ a_{21} \\ \vdots \\ a_{n1} \end{bmatrix} \qquad \mathbf{a}_2 = \begin{bmatrix} a_{12} \\ a_{22} \\ \vdots \\ a_{n2} \end{bmatrix} \qquad \dots \qquad \mathbf{a}_n = \begin{bmatrix} a_{1n} \\ a_{2n} \\ \vdots \\ a_{nn} \end{bmatrix} \tag{A.4-12}$$

A similar procedure is used to partition by rows.

A *diagonal* matrix is one where all the elements $a_{ij} = 0$ for $i \neq j$. The only nonzero elements can be the elements a_{ii}. A diagonal matrix must be square and is automatically symmetric. An example is

$$\mathbf{A} = \begin{bmatrix} 1 & 0 & 0 \\ 0 & -2 & 0 \\ 0 & 0 & 5 \end{bmatrix}$$

The *identity* or *unity* matrix is a square matrix of any order that is diagonal with all the diagonal elements equal to 1. The symbol **I** is reserved for the identity matrix.

$$\mathbf{I} = \begin{bmatrix} 1 & 0 & 0 & \dots & 0 \\ 0 & 1 & 0 & \dots & 0 \\ \cdot & \cdot & \cdot & \dots & \cdot \\ 0 & \cdot & \cdot & 0 & 1 \end{bmatrix} \tag{A.4-13}$$

The *null* matrix is a matrix of any dimension (not necessarily square) whose elements are all zero. Its symbol is **0**.

Matrix Addition

Two matrices **A** and **B** are equal to one another if they have the same dimensions and the corresponding elements are equal; that is, if $a_{ij} = b_{ij}$ for every i and j.

Two or more matrices can be added if they have all the same dimensions. The resulting matrix is obtained by adding all the corresponding elements; that is,

$$\mathbf{A} + \mathbf{B} = \mathbf{C} \tag{A.4-14}$$

implies that

$$c_{ij} = a_{ij} + b_{ij} \tag{A.4-15}$$

The matrix **C** has the same dimensions as **A** and **B**.

For example,

$$\begin{bmatrix} 2 & 5 \\ 4 & 1 \end{bmatrix} + \begin{bmatrix} 3 & 8 \\ 6 & 7 \end{bmatrix} = \begin{bmatrix} 5 & 13 \\ 10 & 8 \end{bmatrix}$$

Matrix subtraction is defined in a similar way, so that

$$\begin{bmatrix} 2 & 5 \\ 4 & 1 \end{bmatrix} - \begin{bmatrix} 3 & 8 \\ 6 & 7 \end{bmatrix} = \begin{bmatrix} -1 & -3 \\ -2 & -6 \end{bmatrix}$$

Matrix addition and subtraction are both associative and commutative. This means that for addition

$$(\mathbf{A} + \mathbf{B}) + \mathbf{C} = \mathbf{A} + (\mathbf{B} + \mathbf{C}) \tag{A.4-16}$$

$$\mathbf{A} + \mathbf{B} + \mathbf{C} = \mathbf{B} + \mathbf{C} + \mathbf{A} = \mathbf{A} + \mathbf{C} + \mathbf{B} \tag{A.4-17}$$

Matrix Multiplication

Equation (A.4-2), the matrix form of (A.4-1), implies a multiplication rule for matrices. Consider an $(n \times p)$ matrix **A** and a $(p \times q)$ matrix **B**. The $(n \times q)$ matrix **C** is the product **AB** where

$$c_{ij} = \sum_{k=1}^{p} a_{ik} b_{kj} \tag{A.4-18}$$

for all $i = 1, \ldots, n$ and $j = 1, \ldots, q$. In order for the product to be defined, the matrices **A** and **B** must be *conformable*; that is, the number of rows in **B** must equal the number of columns in **A**.

The algorithm defined by (A.4-18) is easily remembered. Each element in the ith row of **A** is multiplied by the corresponding element in the jth column of **B**. The sum of the products is the element c_{ij}. For example,

$$\begin{bmatrix} 2 & 5 \\ 3 & 6 \\ 8 & 9 \end{bmatrix} \begin{bmatrix} 1 & 4 \\ 0 & 7 \end{bmatrix} = \begin{bmatrix} 2 & 43 \\ 3 & 54 \\ 8 & 95 \end{bmatrix}$$

The element c_{11} is obtained from $2(1) + 5(0) = 2$. The element c_{32} is $8(4) + 9(7) = 95$.

Matrix multiplication does not commute; that is, in general, $\mathbf{AB} \neq \mathbf{BA}$. A simple counterexample is all that is needed to be convinced of this fact. Overlooking this property is a common mistake of beginning students.

Since vectors are also matrices, they are multiplied in the fashion shown. You should convince yourself of the equivalence of the two forms (A.4-1) and (A.4-2) at this point.

The only time two matrices need not be conformable for multiplication is when one of them is a scalar. In this case, the multiplication is defined as

$$w\mathbf{A} = [wa_{ij}] \tag{A.4-19}$$

where w is a scalar. Multiplying **A** by a scalar w produces a matrix whose elements are found by multiplying each element of **A** by w.

The general exceptions to the noncommutative property are the null matrix and the identity matrix. For the latter,

$$\mathbf{IA} = \mathbf{AI} = \mathbf{A} \tag{A.4-20}$$

Verify this fact.

The associative and distributive laws hold for matrix multiplication. The former states that

$$\mathbf{A}(\mathbf{B}+\mathbf{C}) = \mathbf{AB}+\mathbf{AC} \tag{A.4-21}$$

while for the latter

$$(\mathbf{AB})\mathbf{C} = \mathbf{A}(\mathbf{BC}) \tag{A.4-22}$$

The transpose operation affects multiplication and addition in the following ways:

1. $(w\mathbf{A})^T = w\mathbf{A}^T$, where w is a scalar (A.4-23)
2. $(\mathbf{A}+\mathbf{B})^T = \mathbf{A}^T + \mathbf{B}^T$ (A.4-24)
3. $(\mathbf{AB})^T = \mathbf{B}^T\mathbf{A}^T$ (A.4-25)

The last property is particularly important to remember.

Inverse of a Matrix

The division operation of scalar algebra has an analogous operation in matrix algebra. For example, to solve (A.4-2) for $\mathbf{x}$ in terms of $\mathbf{b}$ and $\mathbf{A}$, we must somehow "divide" $\mathbf{b}$ by $\mathbf{A}$. The procedure for doing this is developed from the concept of a *matrix inverse*. The inverse of a matrix $\mathbf{A}$ is defined only if $\mathbf{A}$ is square and nonsingular; it is denoted by $\mathbf{A}^{-1}$ and has the property that

$$\mathbf{A}^{-1}\mathbf{A} = \mathbf{A}\mathbf{A}^{-1} = \mathbf{I} \tag{A.4-26}$$

Using this property, we multiply both sides of (A.4-2) from the left by $\mathbf{A}^{-1}$ to obtain

$$\mathbf{A}^{-1}\mathbf{A}\mathbf{x} = \mathbf{A}^{-1}\mathbf{b}$$

or

$$\mathbf{x} = \mathbf{A}^{-1}\mathbf{b} \tag{A.4-27}$$

Evidently, the solution for $\mathbf{x}$ is given by (A.4-27). We now need to know how to compute $\mathbf{A}^{-1}$.

We state without proof that $\mathbf{A}^{-1}$ is found as follows. The *cofactor* matrix cof($\mathbf{A}$) is the matrix consisting of the cofactors of the matrix $\mathbf{A}$ [see (A.4-6).] The *adjoint* matrix adj($\mathbf{A}$) is the transpose of cof($\mathbf{A}$).

$$\text{adj}(\mathbf{A}) = [\text{cof}(\mathbf{A})]^T = [C_{ij}]^T \tag{A.4-28}$$

The inverse is computed from

$$\mathbf{A}^{-1} = \frac{\text{adj}(\mathbf{A})}{|\mathbf{A}|} \tag{A.4-29}$$

For example, the inverse of

$$\mathbf{A} = \begin{bmatrix} a_{11} & a_{12} \\ a_{21} & a_{22} \end{bmatrix}$$

is

$$\mathbf{A}^{-1} = \frac{1}{a_{11}a_{22} - a_{12}a_{21}} \begin{bmatrix} a_{22} & -a_{12} \\ -a_{21} & a_{11} \end{bmatrix} \tag{A.4-30}$$

Consider another example:

$$\mathbf{A} = \begin{bmatrix} 1 & 1 & 0 \\ -1 & 0 & 3 \\ 2 & 0 & 1 \end{bmatrix}, \qquad |\mathbf{A}| = 7$$

The inverse is

$$\mathbf{A}^{-1} = \tfrac{1}{7} \begin{bmatrix} 0 & 7 & 0 \\ -1 & 1 & 2 \\ 3 & -3 & 1 \end{bmatrix}^T = \begin{bmatrix} 0 & -1/7 & 3/7 \\ 1 & 1/7 & -3/7 \\ 0 & 2/7 & 1/7 \end{bmatrix}$$

Since calculations are error-prone, you should check the result to see if $\mathbf{A}^{-1}\mathbf{A} = \mathbf{I}$.

Some important properties of the inverse are

$$(\mathbf{A}^{-1})^{-1} = \mathbf{A} \tag{A.4-31}$$

$$(\mathbf{AB})^{-1} = \mathbf{B}^{-1}\mathbf{A}^{-1} \tag{A.4-32}$$

if the inverses exist. Also, if $\mathbf{A}^{-1}$ exists, then

$$\mathbf{AB} = \mathbf{AC} \tag{A.4-33}$$

implies that $\mathbf{B} = \mathbf{C}$. Otherwise, this is not true in general.

A.5 COMPLEX NUMBER ALGEBRA

Complex numbers are those that have a real part and an imaginary part. The primary way of representing a complex number is in the form

$$s = a + ib \tag{A.5-1}$$

where s is the complex number, a is its real part, b is the imaginary part (b is a real number), and $i = \sqrt{-1}$. The abbreviations Re and Im are used to denote real and imaginary parts, respectively. Thus, $a = Re(s)$ and $b = Im(s)$. The number $\bar{s} = a - ib$ is the *complex conjugate* of $a + ib$.

Operations on Complex Numbers

One advantage of the representation (A.5-1) is that it allows the rules of ordinary algebra to be applied to complex numbers. To do this, consider $a + ib$ to be like any algebraic quantity. Let $s_1 = a_1 + ib_1$, and $s_2 = a_2 + ib_2$. Adding s_1 and s_2 is performed as follows:

$$\begin{aligned} s_1 + s_2 &= a_1 + ib_1 + a_2 + ib_2 \\ &= (a_1 + a_2) + i(b_1 + b_2) \end{aligned} \tag{A.5-2}$$

where it is the usual practice to group the resulting real and imaginary parts. Subtraction is done in a similar way.

The product of s_1 and s_2 is

$$\begin{aligned} s_1 s_2 &= (a_1 + ib_1)(a_2 + ib_2) \\ &= (a_1 a_2 - b_1 b_2) + i(a_1 b_2 + a_2 b_1) \end{aligned} \tag{A.5-3}$$

where we have used the fact that $i^2 = -1$.

The ratio obtained from division can be simplified by multiplying top and bottom by the complex conjugate of the denominator.

$$\frac{s_1}{s_2} = \frac{a_1 + ib_1}{a_2 + ib_2} \frac{a_2 - ib_2}{a_2 - ib_2} = \frac{(a_1 a_2 + b_1 b_2) + i(a_2 b_1 - a_1 b_2)}{a_2^2 + b_2^2} \tag{A.5-4}$$

From the denominator result, note that the product of a number with its conjugate is the sum of the squares of the real and imaginary parts.

Two complex numbers are equal if and only if their real parts are equal and their imaginary parts are equal; that is, $s_1 = s_2$ if and only if $a_1 = a_2$ and $b_1 = b_2$.

Alternative Representations

The representation given in (A.5-1) is the *rectangular* form. Two other forms are commonly used. Each has its own advantages.

The *polar* form treats the complex number as a two-dimensional vector whose components are the real and imaginary parts. This is shown in Figure A.1. The *modulus* r is the length of the vector and the magnitude of the complex number. Its angle is ϕ. From the figure, we see that

$$r = \sqrt{a^2 + b^2} \tag{A.5-5}$$

$$\phi = \tan^{-1} \frac{b}{a} \tag{A.5-6}$$

The polar representation is

$$s = r(\cos \phi + i \sin \phi) \tag{A.5-7}$$

Note that the complex conjugate is the mirror image of s about the real axis. It has the same modulus, but its angle is opposite in sign. Thus,

$$\bar{s} = r(\cos \phi - i \sin \phi) \tag{A.5-8}$$

The reciprocal of s is

$$\begin{aligned} \frac{1}{s} &= \frac{1}{a + ib} = \frac{\bar{s}}{r^2} = \frac{1}{r^2}(a - ib) \\ &= \frac{1}{r}(\cos \phi - i \sin \phi) \end{aligned} \tag{A.5-9}$$

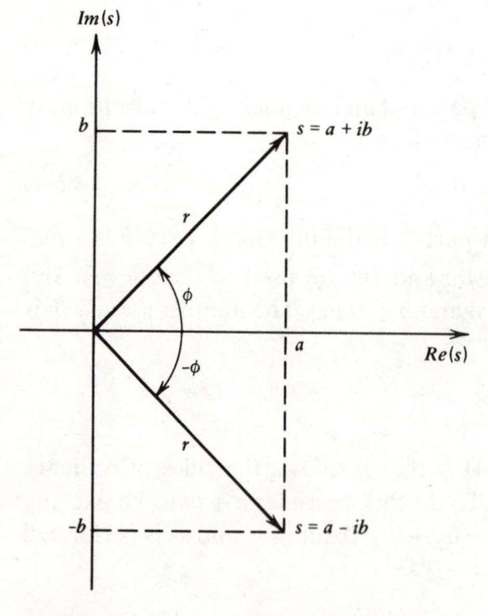

FIGURE A.1 Polar representation of complex numbers.

Thus, the reciprocal lies along the complex conjugate but has a magnitude of $1/r$.

The Taylor series expansion can be applied to functions of complex numbers. For example,

$$e^s = e^{(a+ib)} = e^a e^{ib}$$

Since a is real, we can evaluate e^a, but as yet, we do not know how to evaluate e^{ib}. To see how this is done, use (A.1-5).

$$\begin{aligned} e^{ib} &= 1 + ib + \frac{1}{2}(ib)^2 + \frac{1}{3!}(ib)^3 + \frac{1}{4!}(ib)^4 + \cdots \\ &= 1 - \frac{1}{2}b^2 + \frac{1}{4!}b^4 - \cdots + i\left(b - \frac{1}{3!}b^3 + \cdots\right) \end{aligned}$$

From (A.1-4), the imaginary part can be recognized as sin b. The real part is the Taylor series for cos b. Thus,

$$e^{ib} = \cos b + i \sin b \tag{A.5-10}$$

This is *Euler's identity*, and it is frequently used in the text. Letting $(-b)$ replace b, we obtain

$$e^{-ib} = \cos b - i \sin b \tag{A.5-11}$$

With Euler's identity, we can derive the *complex exponential* representation of a complex number. From (A.5-7),

$$s = re^{i\phi} \tag{A.5-12}$$

From (A.5-8) and (A.5-9),

$$\bar{s} = re^{-i\phi} \tag{A.5-13}$$

$$\frac{1}{s} = \frac{1}{r}e^{-i\phi} \tag{A.5-14}$$

The exponential form greatly simplifies multiplication and division.

$$s_1 s_2 = r_1 e^{i\phi_1} r_2 e^{i\phi_2} = r_1 r_2 e^{i(\phi_1 + \phi_2)} \tag{A.5-15}$$

$$\frac{s_1}{s_2} = \frac{r_1 e^{i\phi_1}}{r_2 e^{i\phi_2}} = \frac{r_1}{r_2} e^{i(\phi_1 - \phi_2)} \tag{A.5-16}$$

Vector Interpretation

A concept that is useful in the root locus method is the vector addition and subtraction of complex numbers. Equation (A.5-2) shows that two complex numbers can be added in component form. The vector interpretation is shown in Figure A.2, where $s_3 = s_1 + s_2$. This is the parallelogram law of vector addition. The law for subtraction is shown in Figure A.3 and can be obtained from Figure A.2. From this figure note that

$$s_2 = s_3 - s_1 = r_2 e^{i\phi_2} \tag{A.5-17}$$

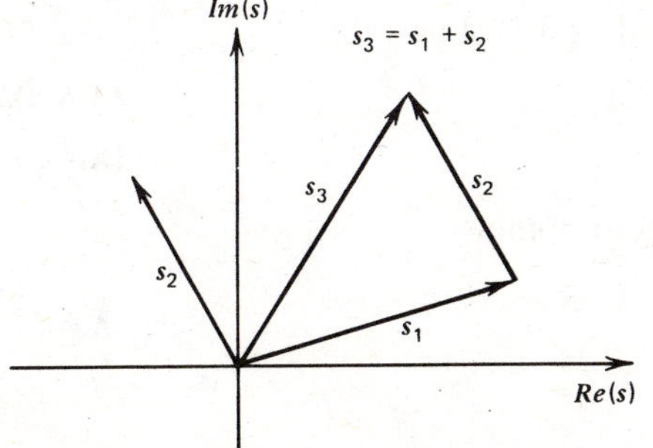

FIGURE A.2 Vector addition of two numbers.

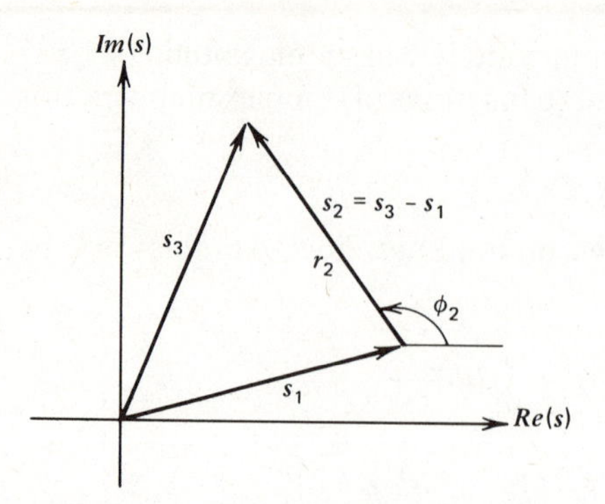

FIGURE A.3 Vector subtraction of two numbers.

DeMoivre's Theorem

The analysis of discrete-time systems requires raising complex numbers to a power. DeMoivre's theorem provides the means. It states that

$$s^p = r^p e^{ip\phi} = r^p(\cos p\phi + i \sin p\phi) \tag{A.5-18}$$

and is a consequence of Euler's identity.

A.6 A USEFUL TRIGONOMETRIC IDENTITY

Several applications require converting the form

$$x(t) = A_1 \sin \omega t + A_2 \cos \omega t \tag{A.6-1}$$

into

$$x(t) = B \sin (\omega t + \phi) \tag{A.6-2}$$

This is because (A.6-2) is easier to interpret than (A.6-1).

We now show how to obtain B and ϕ given A_1, A_2, and ω. Use the identity

$$B \sin (\omega t + \phi) = B \cos \phi \sin \omega t + B \sin \phi \cos \omega t \tag{A.6-3}$$

Comparing the right-hand sides of (A.6-1) and (A.6-3) shows that

$$B \cos \phi = A_1 \tag{A.6-4}$$

$$B \sin \phi = A_2 \tag{A.6-5}$$

If $\cos \phi \neq 0$, (A.6-5) can be divided by (A.6-4) to obtain

$$\frac{\sin \phi}{\cos \phi} = \tan \phi = \frac{A_2}{A_1} \tag{A.6-6}$$

Thus,

$$\phi = \tan^{-1}\left(\frac{A_2}{A_1}\right) \tag{A.6-7}$$

TABLE A.1

		A_1		
	ϕ	$-$	0	$+$
A_2	$+$	2d quadrant	$\pi/2$	1st quadrant
	0	π	---	0
	$-$	3d quadrant	$-\pi/2$	4th quadrant

When formulas such as (A.6-7) are encountered, additional information is required to establish the proper quadrant, because the arctangent is a double-valued function. Its principal value is defined to be in the range $(-90°, 90°)$. Thus, most calculators and computers will return an arctangent value in the fourth quadrant $(-90°, 0°)$ if a negative number is entered and a value in the first quadrant $(0°, 90°)$ if a positive number is entered.

Here, the quadrant of ϕ is determined as follows. For ease of interpretation, we define B to be positive and force ϕ to absorb the necessary sign. Thus, from (A.6-4) and (A.6-5), the sign of A_1 determines the sign of $\cos\phi$, and the sign of A_2 determines the sign of $\sin\phi$. Therefore, the quadrant of ϕ can be found from the signs of A_1 and A_2. This is shown in Table A.1. Note that when $A_1 = 0$, $\cos\phi = 0$, and (A.6-7) cannot be used to find ϕ. In this case, $\phi = \pi/2$ or $-\pi/2$ depending on the sign of A_2. Finally, when $A_2 = 0$, $\sin\phi = 0$, and $\phi = 0$ or π, depending on the sign of A_1. These cases are also shown in Table A.1.

The B is found by squaring (A.6-4) and (A.6-5), adding the results, and using the identity $\sin^2\phi + \cos^2\phi = 1$. Thus,

$$B^2(\sin^2\phi + \cos^2\phi) = A_1^2 + A_2^2$$

or

$$B = +\sqrt{A_1^2 + A_2^2} \tag{A.6-8}$$

Equation (A.6-8) can be used to find B.

Finally, note that the units of ωt in (A.6-2) are radians. Therefore, ϕ should also be expressed in radians.

APPENDIX B
The Laplace Transform

The Laplace transform is a technique for converting a functional dependence on one variable, say, time, into a functional dependence on another variable, denoted by s. Just as the logarithmic transformation makes the operations of multiplication and division easier to perform, so the Laplace transform makes such operations as differentiation and integration easier to perform in the s domain. Here we give a brief introduction to the topic sufficient for the purposes of this text. The application of Laplace transforms to the solution of differential equations is treated in Chapter 4.

The definition of the Laplace transform of the function $f(t)$ is

$$\mathscr{L}[f(t)] = F(s) = \int_0^\infty f(t)e^{-st}\,dt \tag{B-1}$$

where s is in general a complex number. The definition presumes that the integral will exist for some range of values of s. The question of existence is usually no problem for our purposes; the functions of interest to us have well-defined transforms. Equation (B-1) is the definition of the *one-sided* transform. It assumes that the function $f(t)$ is zero for $t < 0$. This assumption is satisfied for our applications, and we will not need the two-sided transform, which has a lower limit of $t = -\infty$.

B.1 TRANSFORMS OF COMMON FUNCTIONS

A table of Laplace transforms for common functions can easily be generated by applying the definition (B-1). Consider the transform of a constant $f(t) = A$ for $t \geqslant 0$. From (B-1),

$$\mathscr{L}(A) = F(s) = \int_0^\infty Ae^{-st}\,dt = A\int_0^\infty e^{-st}\,dt = \left.\frac{-Ae^{-st}}{s}\right|_{t=0}^{t=\infty}$$

or,

$$\mathscr{L}(A) = F(s) = \frac{A}{s} \tag{B.1-1}$$

if we assume that the real part of s is greater than zero so that the limit of e^{-st} exists as $t \to \infty$.

For the *pulse function*, defined as

$$f(t) = \begin{cases} A & \text{for } 0 \leqslant t \leqslant T \\ 0 & \text{elsewhere} \end{cases} \tag{B.1-2}$$

the transform is

$$F(s) = \int_0^\infty Ae^{-st}\,dt = -\left.\frac{Ae^{-st}}{s}\right|_{t=0}^{t=T}$$

$$= A\frac{(1 - e^{-sT})}{s} \tag{B.1-3}$$

TABLE B.1 Laplace Transform Pairs

$F(s)$	$f(t),\ t \geqslant 0$
1. 1	$\delta(t)$, unit impulse at $t=0$
2. $\dfrac{1}{s}$	1, unit step
3. $\dfrac{n!}{s^{n+1}}$	t^n
4. $\dfrac{1}{s+a}$	e^{-at}
5. $\dfrac{1}{(s+a)^n}$	$\dfrac{1}{(n-1)!}t^{n-1}e^{-at}$
6. $\dfrac{a}{s(s+a)}$	$1-e^{-at}$
7. $\dfrac{1}{(s+a)(s+b)}$	$\dfrac{1}{(b-a)}(e^{-at}-e^{-bt})$
8. $\dfrac{s+p}{(s+a)(s+b)}$	$\dfrac{1}{(b-a)}[(p-a)e^{-at}-(p-b)e^{-bt}]$
9. $\dfrac{1}{(s+a)(s+b)(s+c)}$	$\dfrac{e^{-at}}{(b-a)(c-a)}+\dfrac{e^{-bt}}{(c-b)(a-b)}+\dfrac{e^{-ct}}{(a-c)(b-c)}$
10. $\dfrac{s+p}{(s+a)(s+b)(s+c)}$	$\dfrac{(p-a)e^{-at}}{(b-a)(c-a)}+\dfrac{(p-b)e^{-bt}}{(c-b)(a-b)}+\dfrac{(p-c)e^{-ct}}{(a-c)(b-c)}$
11. $\dfrac{b}{s^2+b^2}$	$\sin bt$
12. $\dfrac{s}{s^2+b^2}$	$\cos bt$
13. $\dfrac{b}{(s+a)^2+b^2}$	$e^{-at}\sin bt$
14. $\dfrac{s+a}{(s+a)^2+b^2}$	$e^{-at}\cos bt$
15. $\dfrac{\omega_n^2}{s^2+2\zeta\omega_n s+\omega_n^2}$	$\dfrac{\omega_n}{\sqrt{1-\zeta^2}}e^{-\zeta\omega_n t}\sin \omega_n\sqrt{1-\zeta^2}\,t,\quad \zeta<1$
16. $\dfrac{\omega_n^2}{s(s^2+2\zeta\omega_n s+\omega_n^2)}$	$1+\dfrac{1}{\sqrt{1-\zeta^2}}e^{-\zeta\omega_n t}\sin(\omega_n\sqrt{1-\zeta^2}\,t+\phi)$ $\phi=\tan^{-1}\dfrac{\sqrt{1-\zeta^2}}{\zeta}+\pi$ (3d quadrant)

The area under the pulse function is AT. If we let this area remain constant at the value B and let the pulse width T approach zero, we obtain the *impulse function*. Its transform is thus

$$F(s) = \lim_{T \to 0} \left[\frac{B(1 - e^{-sT})}{sT} \right]$$

From L'Hôpital's limit rule,

$$F(s) = \lim_{T \to 0} \left[\frac{Bse^{-sT}}{s} \right] = B \qquad \textbf{(B.1-4)}$$

If $B = 1$, the function is a *unit impulse*.

Transforms for other common functions are obtained similarly, with the help of integration theorems and a table of integrals. Table B.1 lists the results.

B.2 PROPERTIES OF THE TRANSFORM

The Laplace transform has several properties that are useful. These are listed in Table B.2. The transformation is a linear operation, because

$$\mathscr{L}[af_1(t) + bf_2(t)] = a\mathscr{L}[f_1(t)] + b\mathscr{L}[f_2(t)] \qquad \textbf{(B.2-1)}$$

if a and b are constants. This property follows from the linearity property of the integral in (B-1). Note that the property has two elements; namely, the transform of a sum equals the sum of the transforms, and the transform of a constant times a function equals the constant times the function's transform.

The initial and final value theorems are useful for evaluating the initial and steady-state values of a function. If the limit of $f(t)$ as $t \to \infty$ exists, the final value theorem states that

$$f(\infty) = \lim_{s \to 0} sF(s) \qquad \textbf{(B.2-2)}$$

For a function whose limit exists as t approaches zero from positive values of time, the initial value theorem gives

$$f(0+) = \lim_{s \to \infty} sF(s) \qquad \textbf{(B.2-3)}$$

The proofs are straightforward applications of integration theorems and given in most texts dealing with advanced engineering mathematics.

Solving differential equations with Laplace transforms requires the following differentiation and integration properties. Let $f(t) = dx/dt$, where $x = x(t)$. From (B-1),

$$\mathscr{L}\left(\frac{dx}{dt}\right) = \int_0^\infty \frac{dx}{dt} e^{-st}\, dt = x(t)e^{-st}\Big|_{t=0}^{t=\infty} + s\int_0^\infty x(t)e^{-st}\, dt$$

where we have used integration by parts. If $x(t)$ is transformable, this gives

$$\mathscr{L}\left(\frac{dx}{dt}\right) = sX(s) - x(0) \qquad \textbf{(B.2-4)}$$

TABLE B.2 Properties of the Laplace Transform

$f(t)$	$F(s) = \int_0^\infty f(t)e^{-st}\,dt$
1. $af_1(t) + bf_2(t)$	$aF_1(s) + bF_2(s)$
2. $\dfrac{df}{dt}$	$sF(s) - f(0)$
3. $\dfrac{d^2f}{dt^2}$	$s^2F(s) - sf(0) - \left.\dfrac{df}{dt}\right\|_{t=0}$
4. $\dfrac{d^nf}{dt^n}$	$s^nF(s) - \sum_{k=1}^{n} s^{n-k}g_{k-1}$ $g_{k-1} = \left.\dfrac{d^{k-1}f}{dt^{k-1}}\right\|_{t=0}$
5. $\int_0^t f(t)\,dt$	$\dfrac{F(s)}{s} + \dfrac{h(0)}{s}$ $h(0) = \left.\int f(t)\,dt\right\|_{t=0}$
6. $\begin{Bmatrix} 0, & t < D \\ f(t-D), & t \geq D \end{Bmatrix}$	$e^{-sD}F(s)$
7. $e^{-at}f(t)$	$F(s+a)$
8. $f\left(\dfrac{t}{a}\right)$	$aF(as)$
9. $f(t) = \int_0^t x(t-\tau)y(\tau)\,d\tau$ $= \int_0^t y(t-\tau)x(\tau)\,d\tau$	$F(s) = X(s)Y(s)$
10. $f(\infty) = \lim_{s\to 0} sF(s)$	
11. $f(0+) = \lim_{s\to\infty} sF(s)$	

Extending this procedure to the second derivative, we obtain

$$\mathcal{L}\left(\frac{d^2x}{dt^2}\right) = s^2X(s) - sx(0) - \left.\frac{dx}{dt}\right|_{t=0} \tag{B.2-5}$$

The initial value of the derivative is commonly abbreviated as $\dot{x}(0)$. In general,

$$\mathcal{L}\left(\frac{d^nx}{dt^n}\right) = s^nX(s) - \sum_{k=1}^{n} s^{n-k}g_{k-1} \tag{B.2-6}$$

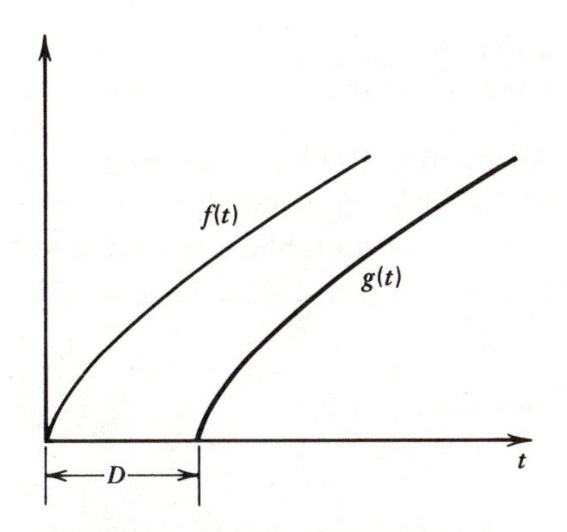

FIGURE B.1 Translated function.

where

$$g_{k-1} = \left. \frac{d^{k-1}x}{dt^{k-1}} \right|_{t=0}$$

The following property can be derived by applying integration by parts to the transform of $\int f(t)\,dt$. The result is

$$\mathscr{L}\left[\int f(t)\,dt\right] = \frac{F(s)}{s} + \frac{h(0)}{s} \tag{B.2-7}$$

where $h(0) = \int f(t)\,dt$ evaluated at $t = 0+$.

The transform of a time-shifted function is useful for discrete-time systems analysis and systems with dead-time elements. The shifted, or translated, function is defined as

$$g(t) = \begin{cases} 0 & \text{for } t < D \\ f(t-D) & \text{for } t \geqslant D \end{cases} \tag{B.2-8}$$

where $f(t)$ is some arbitrary but transformable function. This is shown in Figure B.1. The transform is

$$\begin{aligned} \mathscr{L}[g(t)] &= \int_0^\infty f(t-D)e^{-st}\,dt = \int_D^\infty f(t-D)e^{-st}\,dt \\ &= \int_0^\infty f(q)e^{-s(q+D)}\,dq = e^{-sD}\int_0^\infty f(q)e^{-sq}\,dq \end{aligned}$$

where the integration variable was changed to $q = t - D$. Thus,

$$\mathscr{L}[g(t)] = e^{-sD}F(s) \tag{B.2-9}$$

If $u(t)$ is the unit-step function,

$$u(t) = \begin{cases} 0 & \text{for } t < 0 \\ 1 & \text{for } t \geqslant 0 \end{cases} \tag{B.2-10}$$

we note that the translated function $g(t)$ can be expressed as $f(t-D)u(t-D)$.

The final property of use to us is the convolution theorem. If $F(s) = X(s)Y(s)$ and the time function $f(t)$ corresponding to $F(s)$ is written as

$$f(t) = \mathscr{L}^{-1}[F(s)] \tag{B.2-11}$$

the theorem states that

$$\begin{aligned} f(t) &= \int_0^t x(t-\tau)y(\tau)\,d\tau \\ &= \int_0^t y(t-\tau)x(\tau)\,d\tau \end{aligned} \tag{B.2-12}$$

B.3 THE INVERSE TRANSFORMATION AND PARTIAL FRACTION EXPANSION

When the transform of the solution to a differential equation has been obtained, it is necessary to invert the transformation to obtain the solution as a function of time. Unless the transform is a simple one appearing in the transform table, it will have to be represented as a combination of simple transforms. Most transforms occur in the form of a ratio of two polynomials, such as

$$F(s) = \frac{N(s)}{D(s)} = \frac{b_m s^m + b_{m-1}s^{m-1} + \cdots + b_1 s + b_0}{s^n + a_{n-1}s^{n-1} + \cdots + a_1 s + a_0} \quad \textbf{(B.3-1)}$$

If this transform results from a dynamic system, integral causality (Chapter 1) requires that $m \leqslant n$. If $F(s)$ is of the form in (B.3-1), the method of partial fraction expansion can be used. We assume that the coefficient a_n has been absorbed into the b_i coefficients. The first step is to solve for the n roots of the denominator (several methods for doing this are discussed in the text). If the a_i coefficients are real, the roots will fall into one of the following categories:

1. Real and distinct.
2. Repeated.
3. Complex conjugate pairs.

If all the roots are real and distinct, we can express $F(s)$ in factored form as follows.

$$F(s) = \frac{N(s)}{(s+r_1)(s+r_2)\ldots(s+r_n)} \quad \textbf{(B.3-2)}$$

where the roots are $s = -r_1, -r_2, \ldots, -r_n$. This form can be expanded as

$$F(s) = \frac{C_1}{(s+r_1)} + \frac{C_2}{(s+r_2)} + \cdots + \frac{C_n}{(s+r_n)} \quad \textbf{(B.3-3)}$$

where

$$C_i = \lim_{s \to -r_i} [F(s)(s+r_i)] \quad \textbf{(B.3-4)}$$

Multiplying by the factor $(s + r_i)$ cancels that term in the denominator before the limit is taken. This is a good way of remembering (B.3-4). Each factor corresponds to an exponential function of time, and the solution is

$$f(t) = C_1 e^{-r_1 t} + C_2 e^{-r_2 t} + \cdots + C_n e^{-r_n t} \quad \textbf{(B.3-5)}$$

Now, suppose that p of the roots have the same value $s = -r_1$ and the remaining $(n-p)$ roots are real and distinct. Then $F(s)$ is of the form

$$F(s) = \frac{N(s)}{(s+r_1)^p(s+r_{p+1})(s+r_{p+2})\ldots(s+r_n)} \quad \textbf{(B.3-6)}$$

The expansion is

$$F(s) = \frac{C_1}{(s+r_1)^p} + \frac{C_2}{(s+r_1)^{p-1}} + \cdots + \frac{C_p}{s+r_1} + \cdots + \frac{C_{p+1}}{s+r_{p+1}} + \cdots + \frac{C_n}{s+r_n} \tag{B.3-7}$$

The coefficients for the repeated roots are found from

$$C_1 = \lim_{s \to -r_1} [F(s)(s+r_1)^p] \tag{B.3-8}$$

$$C_2 = \lim_{s \to -r_1} \left\{ \frac{d}{ds} [F(s)(s+r_1)^p] \right\} \tag{B.3-9}$$

$$\vdots$$

$$C_i = \lim_{s \to -r_1} \left\{ \frac{1}{(i-1)!} \frac{d^{i-1}}{ds^{i-1}} [F(s)(s+r_1)^p] \right\} \qquad i = 1, 2, \ldots, p \tag{B.3-10}$$

The coefficients for the distinct roots are found from (B.3-4). The solution for the time function is

$$\begin{aligned} f(t) = {} & C_1 \frac{t^p}{p!} e^{-r_1 t} + C_2 \frac{t^{p-1}}{(p-1)!} e^{-r_1 t} + \cdots + C_p e^{-r_1 t} + \cdots \\ & + C_{p+1} e^{-r_{p+1} t} + \cdots + C_n e^{-r_n t} \end{aligned} \tag{B.3-11}$$

When the roots are complex conjugates, the expansion has the same form as (B.3-3), since the roots are in fact distinct. The coefficients C_i can be found from (B.3-4). However, these will be complex numbers, and the solution form given by (B.3-5) will not be convenient to use. In this case, the following method can be used to give the complex-roots response in the form of a sine function with a phase shift and an exponential amplitude. The derivation of the following relationships is given in Section 4.9. Write $F(s)$ as

$$F(s) = \frac{P(s)}{Q(s)} \tag{B.3-12}$$

where $Q(s)$ is the quadratic polynomial corresponding to the complex roots $s = -a \pm ib$.

$$Q(s) = (s+a)^2 + b^2 \tag{B.3-13}$$

The term $P(s)$ represents the remaining terms in $F(s)$. The time response due to the complex roots is

$$\frac{1}{b} |R(-a+ib)| e^{-at} \sin(bt + \phi) \tag{B.3-14}$$

where

$$R(-a+ib) = \lim_{s \to -a+ib} [P(s)] \tag{B.3-15}$$

$$\phi = \measuredangle R(-a+ib) \tag{B.3-16}$$

The complete response is found by finishing the partial fraction expansion for all the factors of $F(s)$ and adding all of the resulting response terms. When the roots contain a mixture of the three root types, the partial fraction expansion is a combination of the preceding forms, as was the case in (B.3-7). Examples are given in Chapter 4.

When repeated complex roots occur, the solution form is given by (B.3-11) where r_1 is now a complex number. However, (B.3-11) will contain complex exponentials, and will be difficult to interpret. In this case, a procedure similar to that used to derive (B.3-14) can be applied. This is left as an exercise for the reader, but we note that repeated complex root pairs require at least a fourth-order system, and rarely occur in control systems models.

APPENDIX C

A Runge–Kutta Subroutine

The fourth-order Runge–Kutta method is widely used in engineering applications because of its accuracy and the ease with which it can be applied. It does not require a starting solution, unlike some other powerful methods. Runge–Kutta algorithms are currently available in magnetic card and solid-state memory modules for programmable hand-held calculators. These are generally limited to a system of third order or less. For higher order systems, or for applications in which the output is to be plotted, the following FORTRAN subroutine can be used. Many computer installations have a graphic plotting routine. If not, one such program is given in Reference C.1.

The following subroutine was adapted by F. M. White of the University of Rhode Island from a similar routine used with the MAD language at the University of Michigan.* It is an implementation of Gill's method and has been found to be very accurate for a wide variety of problems.

The user is required only to write the expressions for the derivatives, with the model expressed in state variable form. The user must select the step size *H*, and this value can be varied by the programmer during the run. The subroutine is called by the statement

```
CALL RUNGE(N, Y, F, T, H, M, K)
```

The arguments N to K may be described in order of appearance.

N = Number of differential equations to be solved (set by the programmer).

Y = Array of N dependent variables (with initial values set by the programmer).

F = Array of the N derivatives of the variables Y [the programmer must provide an expression in the main program for the calculation of each F(J)].

T = Independent variable (initialized by the programmer).

H = Step size ΔT (set by the programmer).

M = Index used in the subroutine that must be set equal to zero by the programmer before the first CALL.

K = Integer from the subroutine that is used as the argument of a computed GO TO statement in the main program, such as GO TO (10, 20), K. Statement 10 calculates the derivatives F(J) and statement 20 prints the answers, T and Y(J).

For example, a schematic main program that makes use of subroutine RUNGE might look like this.

---

*F. M. White, *Viscous Fluid Flow*, McGraw-Hill, New York, 1974. Used with permission.

```
      DIMENSION Y(10), F(10)
    4 FORMAT(3F10.0, 2I5/(8F10.0))
    5 FORMAT(1X, 6E15.8)
      READ (5, 4) T, TLIM, H, M, N, (Y(J), J=1, N)
    8 IF(T-TLIM) 6, 6, 7
    6 CALL RUNGE (N, Y, F, T, H, M, K)
      GO TO (10, 20), K
   10 F(1) = the derivative of Y(1)
      F(2) = the derivative of Y(2)
      ................................
      F(N) = the derivative of Y(N)
      GO TO 6
   20 WRITE (6, 5) T, (Y(J), J=1, N)
      GO TO 8
    7 STOP
      END
```

It is only necessary to place the proper expressions for the derivatives in the statements for F(J) = … .

Note that it is not necessary to increment T in the main program. This is automatically done correctly in the subroutine. We could change H at any time by inserting a logic expression for changing H underneath the WRITE statement (statement 20).

In this example, Y and F are dimensioned to only a size of 10. There is no inherent reason why more equations could not be handled. The only limitations would be accuracy (round-off error) and computer memory space. It would be necessary only to redimension Y and F in both the main program and subroutine and to redimension Q in the subroutine. It is not necessary to have the same dimensions in both the main program and subroutine as long as dimensions in the subroutine equal or exceed those in the main program.

A listing of the subroutine follows:

```
      SUBROUTINE RUNGE (N, Y, F, T, H, M, K)
C     THIS ROUTINE PERFORMS RUNGE–KUTTA CALCULATION
C     BY GILLS METHOD
      DIMENSION Y(10), F(10), Q(10)
      M=M+1
      GO TO (1, 4, 5, 3, 7), M
    1 DO 2 J=1, N
    2 Q(J)=0
      A=.5
      GO TO 9
    3 A=1.707107
C     IF YOU NEED MORE ACCURACY, USE
C     A=1.7071067811865475244
    4 T=T+.5*H
```

```
    5   DO 6 J = 1, N
        Y(J) = Y(J) + A*(F(J)*H - Q(J))
    6   Q(J) = 2.*A*H*F(J) + (1. - 3.*A)*Q(J)
        A = .2928932
C  00   IF YOU NEED MORE ACCURACY SET
C       A = .2928932188134524756
        GO TO 9
    7   DO 8 J = 1, N
    8   Y(J) = Y(J) + H*F(J)/6. - Q(J)/3.
        M = 0
        K = 2
        GO TO 10
    9   K = 1
   10   RETURN
        END
```

The preceding listings can easily be converted into BASIC as follows. The statement numbers must be replaced by the appropriate line numbers and the DO loops replaced with FOR-NEXT loops. The statement GO TO (10, 20) K in the main program must be replaced by an IF-THEN test on the value of K. A GOSUB statement must be substituted for the CALL to subroutine RUNGE. Finally, the formatted READ and WRITE statements should be replaced by their BASIC equivalents.

Some examples will illustrate the use of the program.

Example C.1

Consider the nth-order equation

$$\frac{d^n y}{dt^n} = f\left(t, y, \frac{dy}{dt}, \frac{d^2 y}{dt^2}, \dots, \frac{d^{n-1} y}{dt^{n-1}}\right) \tag{C-1}$$

One way of converting into state variable form is to define

$$y_1 = \frac{d^{n-1} y}{dt^{n-1}}$$

$$y_2 = \frac{d^{n-2} y}{dt^{n-2}}$$

$$\vdots$$

$$y_{n-1} = \frac{dy}{dt}$$

$$y_n = y$$

Then with the dot notation for derivatives, (C-1) becomes

$$\dot{y}_1 = f(t, y_n, y_{n-1}, \dots, y_1)$$

$$\dot{y}_2 = y_1$$

$$\dot{y}_3 = y_2$$
$$\vdots$$
$$\dot{y}_n = y_{n-1}$$

The functions to be inserted into the main program are:

```
F(1) = f[T, Y(N), Y(N-1), Y(N-2), ..., Y(2), Y(1)]
F(2) = Y(1)
F(3) = Y(2)
   :
F(N-1) = Y(N-2)
F(N) = Y(N-1)
```

Of course, any other choice of state variables can be used.

Example C.2

The subroutine can also be used to compute integrals. For example,

$$z = \iint f(t)\,dt$$

is equivalent to

$$\ddot{z} = f(t)$$

The programming looks like

```
F(1) = f(T)
F(2) = Y(1)
```

Example C.3

Example C.2 shows that the subroutine can be used to evaluate a performance index (IAE, ISE, etc., from Chapter 7) while solving the system's differential equations. For example, the system model

$$\dot{y} = -y + u(t)$$

and the index

$$J = \int_0^t y^2\,dt$$

can be reduced to the system

$$\dot{y}_1 = -y_1 + u(t)$$
$$\dot{y}_2 = y_1^2$$

by defining $y_1 = y$ and $y_2 = J$. For most indices, the upper limit is $t = \infty$. Here, we can handle this by using a value for TLIM that is large enough to allow J to become approximately constant. Several trial values might have to be used before a suitable one is found.

REFERENCE

C.1 J. L. Melsa and S. K. Jones, *Computer Programs for Computational Assistance in the Study of Linear Control Theory*, McGraw-Hill, New York, 1973.

APPENDIX D

Mason's Rule for Diagram Reduction

When the number of loops or forward paths in a diagram is large, repeated application of the loop reduction and cascaded element formulas becomes tedious. In these instances, using Mason's rule simplifies the task of finding the transfer function. The rule applies to both block diagrams and signal flow graphs, but we present it here in terms of the former.

To establish the terminology, consider the single-loop path shown in Figure D.1. The transfer function of the forward path is $G(s)$. The loop transfer function is $G(s)H(s)$. The closed-loop transfer function is that of the forward path divided by one minus the loop transfer function; that is,

$$\frac{C(s)}{R(s)} = \frac{G(s)}{1 - G(s)H(s)} \tag{D-1}$$

Note that the sign of the feedback loop is positive. It is wise when using Mason's rule to absorb any minus signs into the feedback element $H(s)$.

A block diagram is *tightly connected* if each forward path touches every loop path, as is shown in Figure D.2. Mason's rule may be stated as follows (Reference D.1): *The transfer function of a tightly connected block diagram is the sum of the forward-path*

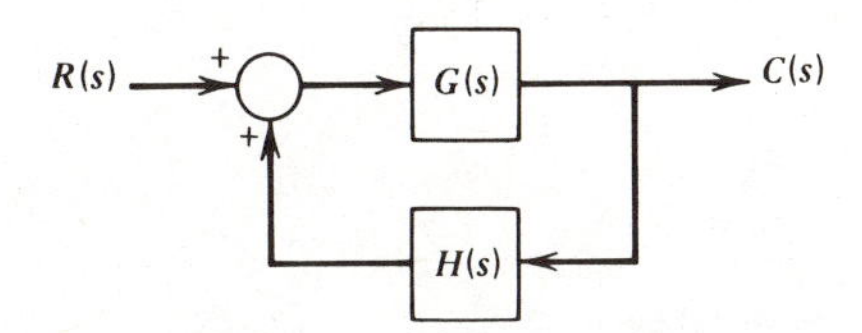

FIGURE D.1 Single-loop system.

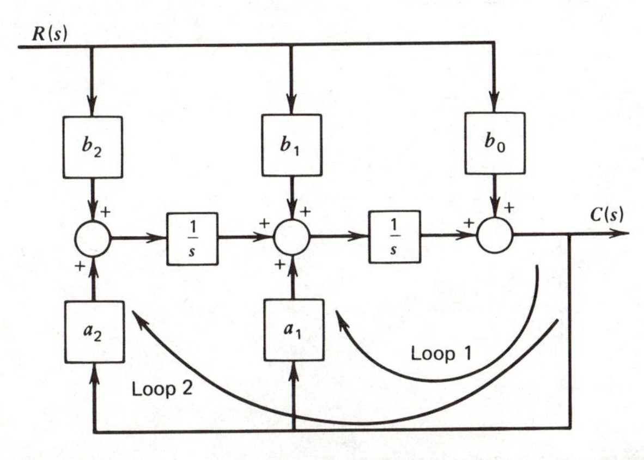

FIGURE D.2 Second-order system in observer canonical form.

transfer functions divided by one minus the sum of all the loop transfer functions. Applying this rule to Figure D.2, we obtain

$$\frac{C(s)}{R(s)} = \frac{b_0 + b_1 s^{-1} + b_2 s^{-2}}{1 - (a_1 s^{-1} + a_2 s^{-2})} \qquad \textbf{(D-2)}$$

or in the usual form

$$\frac{C(s)}{R(s)} = \frac{b_0 s^2 + b_1 s + b_2}{s^2 - a_1 s - a_2} \qquad \textbf{(D-3)}$$

Note that the input signal can reach the output in three ways while traveling in only the forward direction. The product of the transfer functions along such a path is the forward-path transfer function. For example, the path from $R(s)$ through b_2 also passes through two integrators to give a forward-path transfer function of $b_2(s^{-1})(s^{-1}) = b_2 s^{-2}$.

A loop transfer function is found by ignoring the other loops and the input and following the signal around the loop. For example, ignore the input in Figure D.2 and separate loop 1 from loop 2 to give Figure D.3. From this figure, it is clear that the loop transfer function for loop 2 is $a_2(s^{-1})(s^{-1}) = a_2 s^{-2}$. Similarly, that for loop 1 is $a_1 s^{-1}$.

Now, consider the system shown in Figure D.4. Mason's rule is easily applied to show that the system transfer function is that given by (D-2).

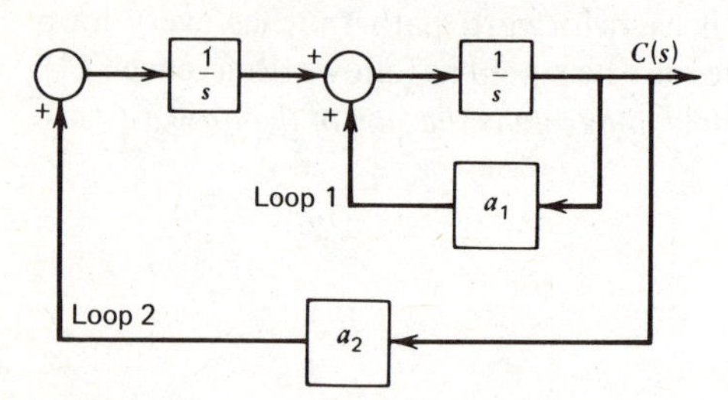

FIGURE D3. Zero-input version of Figure D.2 with loops separated.

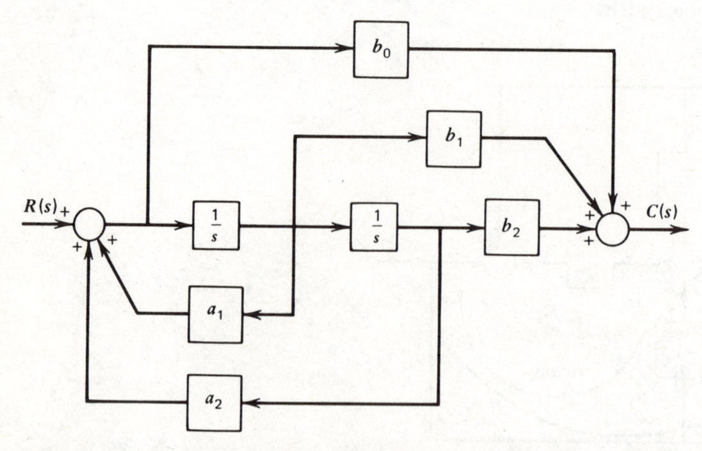

FIGURE D.4 Second-order system in control canonical form.

The diagram in Figure D.2 is the *observer canonical* form because all the feedback loops come from the output or "observed" signal. Figure D.4 shows the *control canonical* form, because all the feedback loops return to the comparator where the actuating or error signal is formed for use by the controller. Both forms are called *direct canonical* forms, because the gains in the diagram are coefficients in the numerator and denominator polynomials of the system transfer function. In *cascade canonical* form, the poles and zeros of the transfer function appear in the diagram, and the transfer function appears as a ratio of the products of these factors.

A diagram that is not tightly connected can be treated with Mason's rule by reducing the diagram until subdiagrams appear that are tightly connected.

REFERENCE

D.1. S. J. Mason, "Feedback Theory: Further Properties of Signal Flow Graphs," *Proceedings of the IRE*, 44, 1956, 970-976.

APPENDIX E

Simulation with an Analog Computer

The operational amplifier can be used to solve the differential equation model of a dynamic system. The essence of the technique is to construct an op amp circuit whose governing equations are the same as those of the system under study except for magnitude and time scale factors. The resulting circuit is said to be an *analog computer*.

Analog computation has lost some of its importance because of the increasing availability of digital computers. Nevertheless, we can identify two advantageous features of analog simulation.

1. Selecting a time step is not required, unlike digital methods. This has an advantage for fast systems, which require small time steps and many iterations in order to achieve accurate results with digital methods.
2. Often, the actual components to be used in the real system can be employed in the analog simulation. This provides an excellent way of ensuring that the simulation includes the important characteristics of the system.

E.1 ANALOG COMPUTER ELEMENTS

The basic element of most analog computers is the op amp. In Section 6.8, the characteristics of this device were explored, and it was shown how to implement such functions as multiplication by a constant, summation, and integration. These constitute all the operations necessary to develop an analog simulation of a linear system. In fact, the electronic implementation of the PID control laws are analog computer circuits. The only additional considerations required to develop a general procedure for analog simulation are those relating to magnitude and time scale factors.

Table E.1 lists the relevant op amp circuits and their functions. The potentiometer is useful for multiplying by a fraction without using an amplifier. The computing diagram symbols are helpful for setting up the analog circuit. The block symbols represent active elements (those containing an amplifier). Another commonly used set of symbols is shown in Table E.2. They do not display the sign inversion associated with each amplifier, and we must be careful not to forget its existence when using these symbols.

Some commercially available analog computers use standard resistance and capacitance values, so that only certain multiplication factors are available. In such cases, a potentiometer can be used in series with the multiplier to obtain any other desired factor.

TABLE E.1 Analog Computing Elements

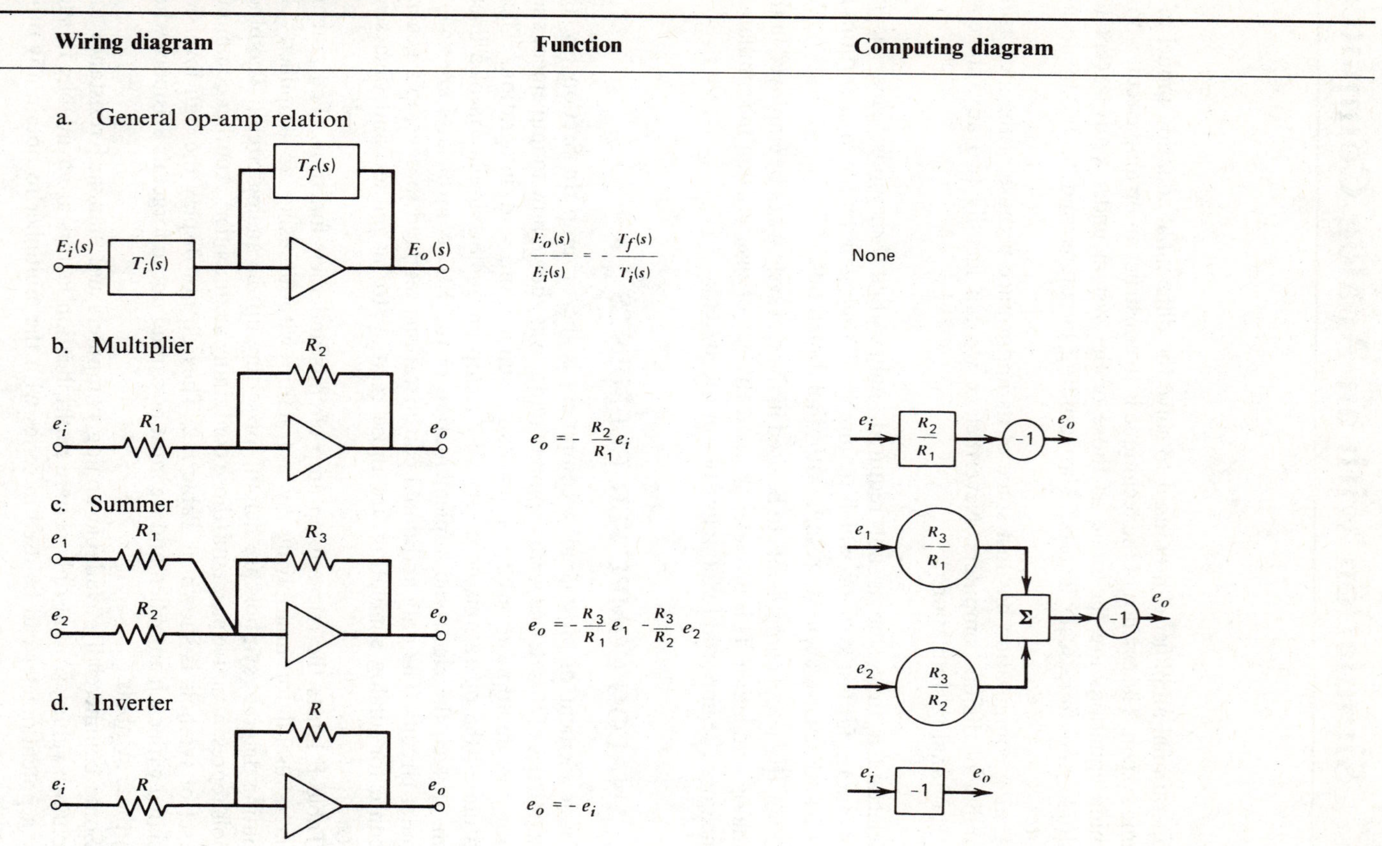

Wiring diagram	Function	Computing diagram
a. General op-amp relation	$\frac{E_o(s)}{E_i(s)} = -\frac{T_f(s)}{T_i(s)}$	None
b. Multiplier	$e_o = -\frac{R_2}{R_1} e_i$	
c. Summer	$e_o = -\frac{R_3}{R_1} e_1 - \frac{R_3}{R_2} e_2$	
d. Inverter	$e_o = -e_i$	

e. Integrator

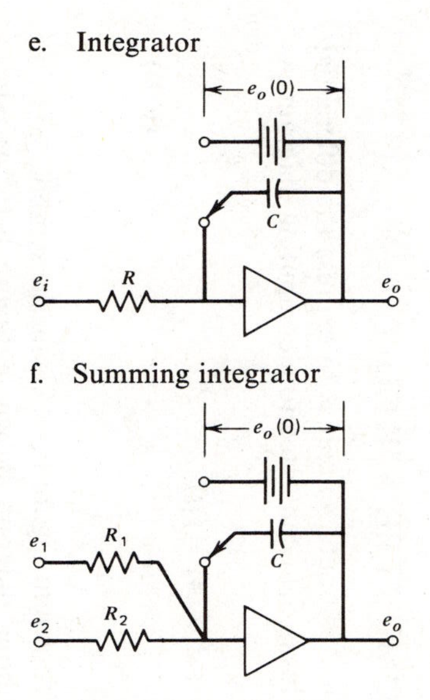

$$e_o = e_o(0) - \frac{1}{RC}\int_0^t e_i \, dt$$

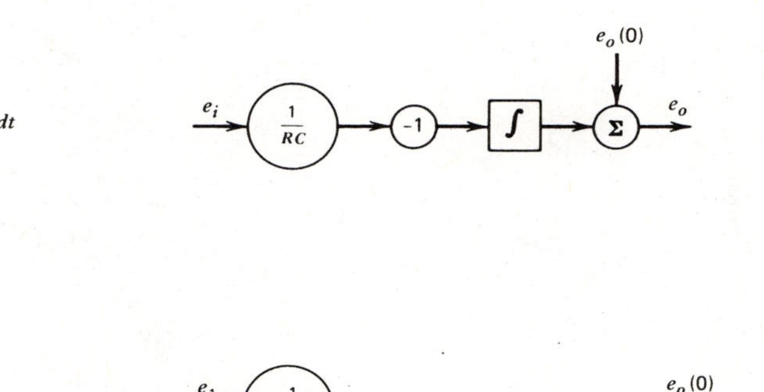

f. Summing integrator

$$e_o = e_o(0) - \int_0^t \left(\frac{1}{R_1C} e_1 + \frac{1}{R_2C} e_2 \right) dt$$

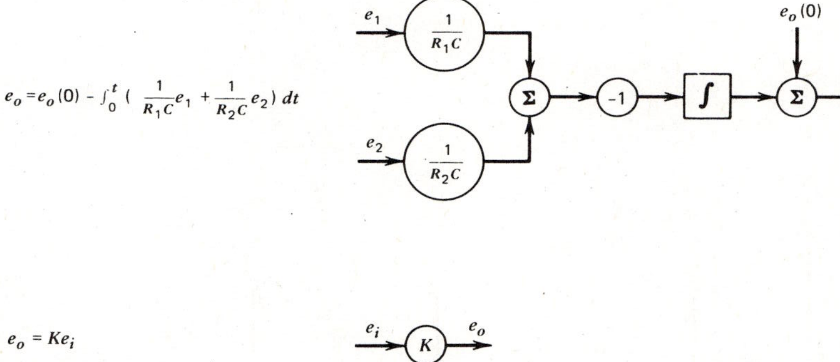

g. Fractional multiplier (pot)

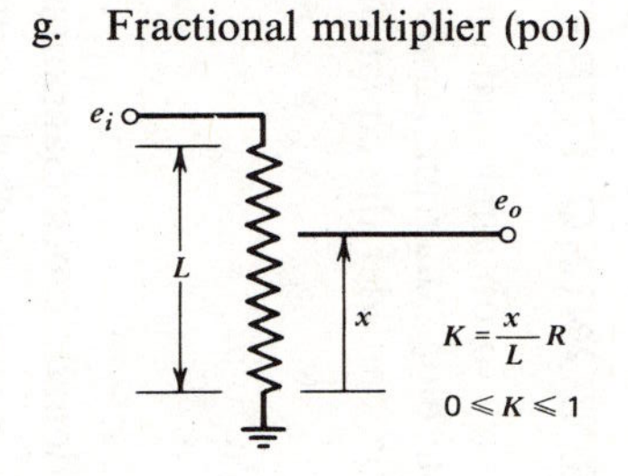

$$e_o = Ke_i$$

TABLE E.2 Simplified Analog Computing Symbols

Symbol	Function
1. Multiplier	$e_o = -Ke_i$
2. Integrator 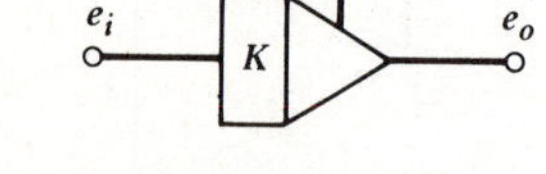	$e_o = e_o(0) - K\int_0^t e_i dt$
3. Summer	$e_o = -K_1e_1 - K_2e_2$
4. Summing integrator 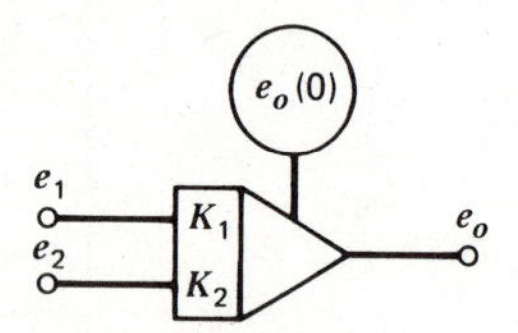	$e_o = -\int_0^t (K_1e_1 + K_2e_2)\, dt$

E.2 DESIGNING THE COMPUTER CIRCUIT

The first step in obtaining an analog computer circuit is to express the differential equation in block diagram form. This is then converted to a computing diagram. As an example, consider the equation

$$\begin{aligned} \ddot{y} + 2\dot{y} + 4y &= f(t) \\ y(0) &= 1 \\ \dot{y}(0) &= 3 \end{aligned} \tag{E.2-1}$$

The time domain block diagram is shown in Figure E.1*a*. Using the symbols given in Table E.1, we construct the computing diagram shown in Figure E.1*b*. An alternative diagram is shown in Figure E.1*c*. Note that the sign of $\dot{y}(0)$ must be reversed in Figures E.1*b* and E.1*c* because of where it enters the circuit.

The main differences between the block diagrams and the computing diagrams

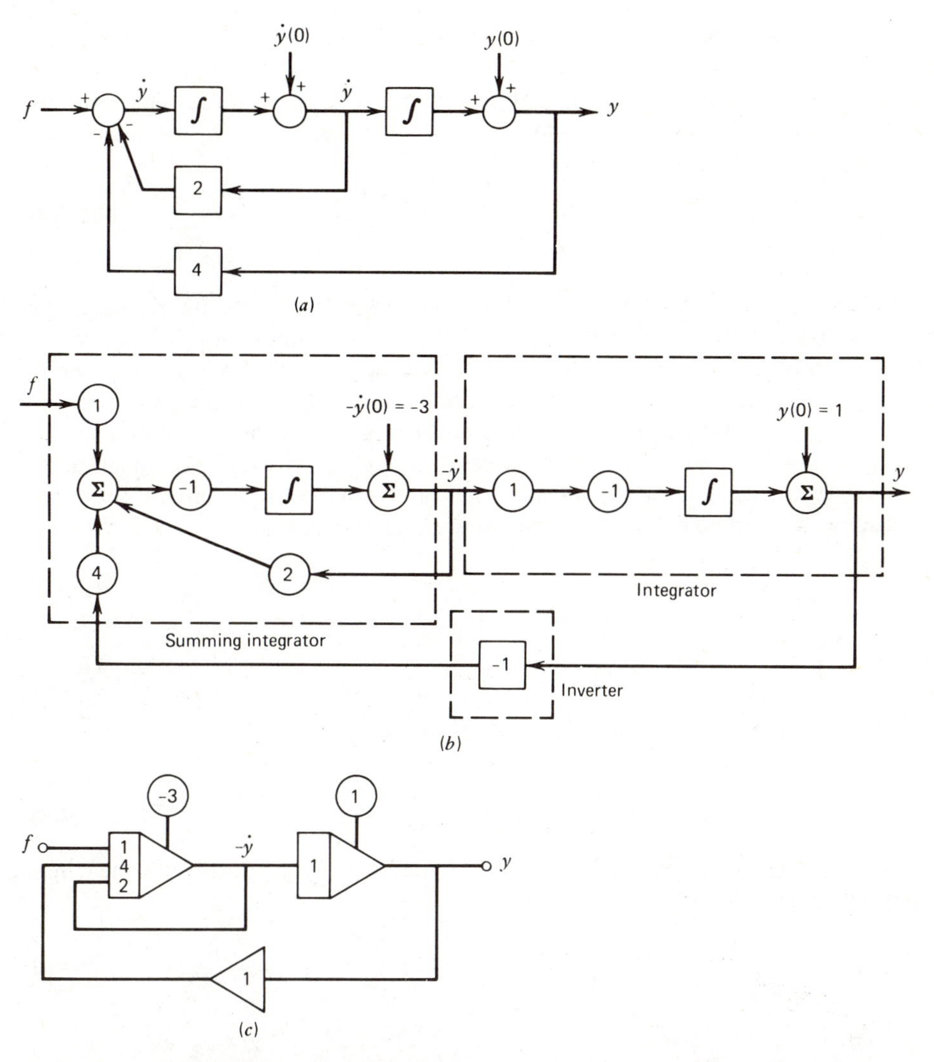

FIGURE E.1 Analog computing diagrams for the equation $\ddot{y} + 2\dot{y} + 4y = f$, $y(0) = 1$, $\dot{y}(0) = 3$. (*a*) Block diagram. (*b*) Computation diagram. (*c*) Simplified computation diagram.

are (1) the computing diagram accounts for the sign inversions due to the op amps, and (2) the computing diagram does not allow subtraction. Subtraction must be developed with a summer and an inverter.

E.3 MAGNITUDE SCALE FACTORS

The circuit shown in Figure E.1 will implement the solution in (E.2-1) provided the voltage saturation limits of the amplifiers are not exceeded. These limits are typically either ± 10 or ± 100V for most commercial units. In addition, the voltages should be large enough to prevent noise from introducing errors. For these reasons, magnitude scale factors are usually required. These relate the voltage levels in the circuit to the corresponding values of the variables f, y, and $\dot{y}$. For example, a large enough value of f will produce values of $|y|$ and $|\dot{y}|$ that exceed the saturation limits. Therefore, we represent these variables by the voltages v_1, v_2, and v_f according to the relations

$$\begin{aligned} y &= k_1 v_1 \\ \dot{y} &= k_2 v_2 \\ f &= k_f v_f \end{aligned} \tag{E.3-1}$$

where k is the magnitude scale factor. These factors are determined by the maximum expected magnitudes of y, $\dot{y}$, and f and by the saturation limits of the machine.

In order to see how the scale factors are chosen, we redraw the computing diagram in Figure E.1b except that we now leave the gains in terms of the resistance and capacitance values. We denote the diagram variables as v_1, v_2, and v_f, rather than y, $\dot{y}$, and f, as shown in Figure E.2. The voltages v_1 and v_2 are the result of integrations, which suggests that they can be taken as the state variables of the circuit. From the diagram, we can immediately write the voltage equations as

$$\dot{v}_1 = -\frac{1}{R_4 C_2} v_2 \tag{E.3-2}$$

$$\dot{v}_2 = -\frac{1}{R_1 C_1} v_f + \frac{1}{R_2 C_1} v_1 - \frac{1}{R_3 C_1} v_2 \tag{E.3-3}$$

Next, write (E.2-1) in state variable form as

$$\dot{x}_1 = x_2 \tag{E.3-4}$$

$$\dot{x}_2 = f - 4x_1 - 2x_2 \tag{E.3-5}$$

where $x_1 = y$ and $x_2 = \dot{y}$. From (E.3-1), $x_1 = k_1 v_1$ and $x_2 = k_2 v_2$. Substituting them into (E.3-4) and (E.3-5), we obtain

$$\dot{v}_1 = \frac{k_2}{k_1} v_2 \tag{E.3-6}$$

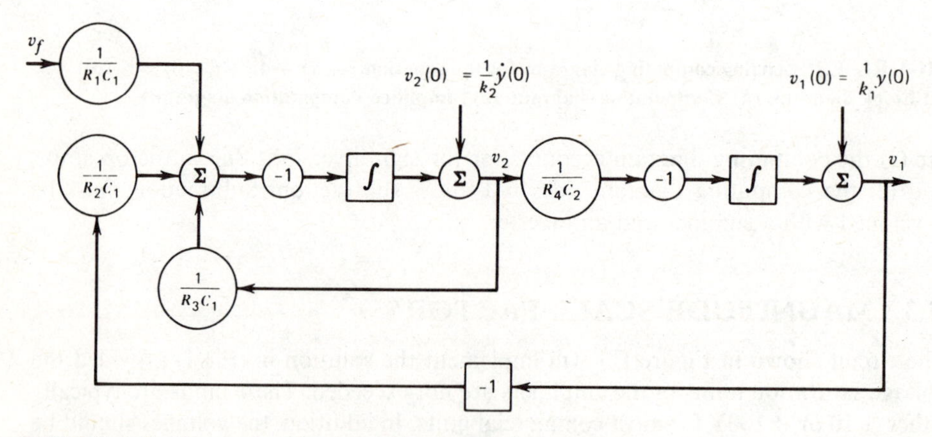

FIGURE E.2 Computing diagram with magnitude scale factors.

$$\dot{v}_2 = \frac{1}{k_2}(k_f v_f - 4k_1 v_1 - 2k_2 v_2) \quad \textbf{(E.3-7)}$$

Comparing (E.3-2), (E.3-3), (E.3-6), and (E.3-7) shows that

$$\left.\begin{array}{ll} \dfrac{k_2}{k_1} = -\dfrac{1}{R_4 C_2} & \dfrac{k_f}{k_2} = -\dfrac{1}{R_1 C_1} \\ \dfrac{-4k_1}{k_2} = \dfrac{1}{R_2 C_1} & -2 = -\dfrac{1}{R_3 C_1} \end{array}\right\} \quad \textbf{(E.3-8)}$$

Examining Figures E.1*b* and E.2 shows that if $\dot{y} > 0$, then $v_2 < 0$. Thus, k_2 must be negative. In a similar way, we can see that $k_1 > 0$ and $k_f > 0$.

Let us suppose that our machine has voltage saturation limits of ± 10v. In order to have a safety margin, we will thus attempt to keep the voltage magnitudes less than, say, 9V but well above zero to maintain an adequate signal-to-noise ratio. Therefore, from (E.3-1) we have (noting that $k_2 < 0$),

$$\left.\begin{array}{l} |y|_{max} = k_1 |v_1|_{max} = 9k_1 \\ |\dot{y}|_{max} = k_2 |v_2|_{max} = -9k_2 \\ |f|_{max} = k_f |v_f|_{max} = 9k_f \end{array}\right\} \quad \textbf{(E.3-9)}$$

Suppose that we can estimate the maximum values as follows:

$$|y|_{max} = 6 \qquad |\dot{y}|_{max} = 3 \qquad |f|_{max} = 5 \quad \textbf{(E.3-10)}$$

This information might come from insight concerning the physical process being simulated. Estimating the maximum values of the model's variables is the trickiest aspect of analog simulation, since it requires knowledge of the very solution being sought. But with physical insight and some trial and error, a useful set of scale factors can usually be found.

Substituting (E.3-10) into (E.3-9) gives

$$k_1 = \tfrac{2}{3} \qquad k_2 = -\tfrac{1}{3}, \qquad k_f = \tfrac{5}{9} \quad \textbf{(E.3-11)}$$

These values are then substituted into (E.3-8) to obtain

$$\left.\begin{array}{ll} \dfrac{1}{R_4 C_2} = \dfrac{1/3}{2/3} = \dfrac{1}{2}, & \dfrac{1}{R_1 C_1} = \dfrac{5/9}{1/3} = \dfrac{5}{3} \\ \dfrac{1}{R_2 C_1} = \dfrac{4(2/3)}{1/3} = 8 & \dfrac{1}{R_3 C_1} = 2 \end{array}\right\} \quad \textbf{(E.3-12)}$$

These are the gains in the computing diagram of Figure E.2. We are fortunate here, because the gains given by (E.3-12) are reasonably close in value. Large differences in gain values should be avoided, because they require disproportionate sizes for the resistances and capacitances. Such wide ranges are not available on commercial analog computers. Gains on such machines are generally obtainable in the range from 0.1 to 10.

Equations (E.3-12) contain six unknowns, so there is some flexibility in the resulting circuit. Choosing $C_1 = C_2 = 10^{-6}F$ as a commonly available value, we obtain

$$\left.\begin{aligned} R_1 &= \tfrac{3}{5}\times 10^6\Omega \qquad R_2 = \tfrac{1}{8}\times 10^6\Omega \\ R_3 &= \tfrac{1}{2}\times 10^6\Omega \qquad R_4 = 2\times 10^6\Omega \end{aligned}\right\} \tag{E.3-13}$$

These values are roughly the same size, so the design would seem to be acceptable. However, before drawing the final wiring diagram, we will investigate some additional considerations.

E.4 CHOICE OF TIME SCALE

We often desire to speed up the dynamics of the process under study so that many simulations can be made within a reasonable period. On the other hand, it is sometimes necessary to slow down the simulation to be compatible with the response of an xy plotter. Denote the analog computer's time scale by T and that of the process by t. The two scales are related by the scale factor k_t as follows:

$$t = k_t T \tag{E.4-1}$$

Then the rate of change of voltage in the two scales are related as

$$\frac{dV}{dT} = \frac{dV}{dt}\frac{dt}{dT} = k_t\frac{dV}{dT} = k_t\dot{V} \tag{E.4-2}$$

In terms of the new scale T, the voltage equations (E.3-2) and (E.3-3) become

$$\left.\begin{aligned} \frac{dv_1}{dT} &= -\frac{1}{R_4C_2}v_2 \\ \frac{dv_2}{dT} &= -\frac{1}{R_1C_1}v_f + \frac{1}{R_2C_1}v_1 - \frac{1}{R_3C_1}v_2 \end{aligned}\right\} \tag{E.4-3}$$

With (E.4-2) we obtain

$$\left.\begin{aligned} \frac{dv_1}{dt} &= -\frac{1}{k_t}\frac{1}{R_4C_2}v_2 \\ \frac{dv_2}{dt} &= -\frac{1}{k_t}\left(\frac{1}{R_1C_1}v_f - \frac{1}{R_2C_1}v_1 + \frac{1}{R_3C_1}v_2\right) \end{aligned}\right\} \tag{E.4-4}$$

Comparing (E.4-4) with (E.3-6) and (E.3-7) gives the following requirements:

$$\left.\begin{aligned} \frac{k_2}{k_1} &= -\frac{1}{k_tR_4C_2} \qquad \frac{k_f}{k_2} = -\frac{1}{k_tR_1C_1} \\ \frac{-4k_1}{k_2} &= \frac{1}{k_tR_2C_1} \qquad -2 = -\frac{1}{k_tR_3C_1} \end{aligned}\right\} \tag{E.4-5}$$

Note the similarity to (E.3-8).

Suppose that we wish to slow down the simulation by a factor of two. Then $k_t = 1/2$. Using the previous values for k_1, k_2, and k_f, we obtain the required resistances as before. From (E.4-5),

$$\left.\begin{aligned} \frac{1}{R_4C_2} &= \frac{1}{4} \qquad \frac{1}{R_1C_1} = \frac{5}{6} \\ \frac{1}{R_2C_1} &= 4 \qquad \frac{1}{R_3C_1} = 1 \end{aligned}\right\} \tag{E.4-6}$$

With $C_1 = C_2 = 10^{-6}F$, we have $R_1 = 6/5\ M\Omega$, $R_2 = 1/4\ M\Omega$, $R_3 = 1\ M\Omega$, and $R_4 = 4\ M\Omega$.

E.5 USE OF POTENTIOMETERS

Thus far, we have made the implicit assumption that a continuous range of values are available for the resistances and capacitances of the multipliers, summers, and integrators of the computer. Usually, this is not the case, and only discrete values are available. If so, the required gain values can still be obtained by choosing an available resistance or capacitance value lower than the computed value and using a

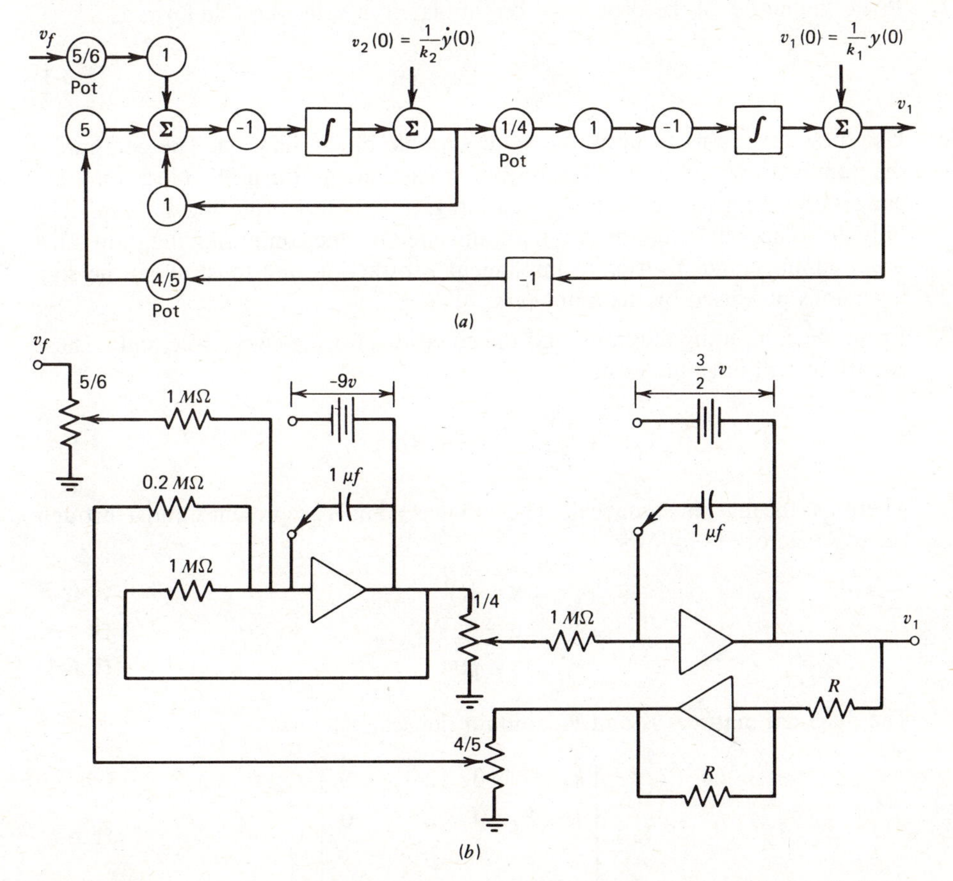

FIGURE E.3 Computing and wiring diagrams including magnitude and time-scale factors.

potentiometer in series to reduce the gain to the required value. The technique also has the advantage of producing gains that are adjustable. This is convenient if we wish to modify the scale factors somewhat.

For example, suppose that $10^{-6}F$ is an available capacitance and the available resistances are integer multiples of $100k\Omega$. Then the gain $1/R_4C_2 = 1/4$ can be realized by choosing $R_4 = 1M\Omega$ (instead of $1/4\ M\Omega$) and by using a series potentiometer with a fractional gain of 1/4. The other resistances are chosen similarly. The resulting computing and wiring diagrams are shown in Figure E.3. This circuit simulates (E.2-1) with the scale factors $k_1 = 2/3$, $k_2 = -1/3$, $k_f = 5/9$, and $k_t = 1/2$, and with the given initial conditions.

E.6 A GENERAL PROCEDURE

The preceding approach to designing a simulation circuit can be concisely stated in terms of state variable notation, as follows:

1. Write the model of the process to be simulated in state variable form as

$$\frac{d\mathbf{x}}{dt} = \mathbf{Ax} + \mathbf{Bu} \tag{E.6-1}$$

2. Use these equations to draw the state variable block diagram. The computing diagram is drawn from this. The output of each integrator in the block diagram corresponds to a voltage output of an integrator in the computing diagram. The only differences between the two diagrams are that the computing diagram must use a summer and inverter to implement subtraction and must show the sign inversions produced by the amplifiers.
3. From the computing diagram, find the equations for the electrical circuit. These equations will have the form

$$\frac{d\mathbf{V}}{dT} = \mathbf{A}_E\mathbf{V} + \mathbf{B}_E\mathbf{V}_u \tag{E.6-2}$$

where T is the machine time scale. The voltages $\mathbf{V}$ and $\mathbf{V}_u$ represent $\mathbf{x}$ and $\mathbf{u}$ through the scale factors as follows:

$$\mathbf{x} = \mathbf{KV} \tag{E.6-3}$$

$$\mathbf{u} = \mathbf{K}_u\mathbf{V}_u \tag{E.6-4}$$

$$t = k_tT \tag{E.6-5}$$

The diagonal matrices $\mathbf{K}$ and $\mathbf{K}_u$ contain the scale factors.

$$\mathbf{K} = \begin{bmatrix} k_1 & 0 & 0 & \cdots & . & 0 \\ 0 & k_2 & 0 & \cdots & . & 0 \\ . & . & . & \cdots & . & . \\ 0 & . & . & \cdots & 0 & k_n \end{bmatrix} \tag{E.6-6}$$

$$\mathbf{K}_u = \begin{bmatrix} k_{1u} & 0 & 0 & \dots & . & 0 \\ 0 & k_{2u} & 0 & \dots & . & 0 \\ . & . & . & \dots & . & . \\ 0 & . & . & \dots & 0 & k_{mu} \end{bmatrix} \tag{E.6-7}$$

The elements of the matrices $\mathbf{A}_E$ and $\mathbf{B}_E$ are functions of the circuit's resistances and capacitances.

4. Substituting (E.6-3)–(E.6-5) into (E.6-1), and comparing the result with (E.6-2) shows that

$$\mathbf{A}_E = k_t \mathbf{K}^{-1} \mathbf{A} \mathbf{K} \tag{E.6-8}$$

$$\mathbf{B}_E = k_t \mathbf{K}^{-1} \mathbf{B} \mathbf{K}_u \tag{E.6-9}$$

These are the scaling relations that relate the resistances and capacitances to the scale factors. Since $\mathbf{K}$ and $\mathbf{K}_u$ are diagonal, their inverses are easily computed.

5. Pick the scale factors so that the given specifications are satisfied. These include the given saturation limits, the desired time scale, and the limits on the available coefficient ranges on the machine. The latter usually requires the magnitudes of the elements of $\mathbf{A}_E$ and $\mathbf{B}_E$ to lie within the range 0.1 to 10.
6. Once $\mathbf{K}$, $\mathbf{K}_u$, and k_t have been selected, use the scaling relations (E.6-8) and (E.6-9) to obtain the resistances and capacitances. Some flexibility occurs here, since there are usually more unknowns than constraints.
7. If only discrete resistance and capacitance values are available, use potentiometers to obtain the desired gains.

For the example problem given by (E.2-1), the state equations (E.3-4) and (E.3-5) with $u = f$, give the following matrices:

$$\mathbf{A} = \begin{bmatrix} 0 & 1 \\ -4 & -2 \end{bmatrix} \qquad \mathbf{B} = \begin{bmatrix} 0 \\ 1 \end{bmatrix} \tag{E.6-10}$$

From the computing diagram shown in Figure E.2, we obtained the circuit equations (E.4-3). These give the matrices $\mathbf{A}_E$ and $\mathbf{B}_E$.

$$A_E = \begin{bmatrix} 0 & -\dfrac{1}{R_4 C_2} \\ \dfrac{1}{R_2 C_1} & -\dfrac{1}{R_3 C_1} \end{bmatrix} \qquad B_E = \begin{bmatrix} 0 \\ -\dfrac{1}{R_1 C_1} \end{bmatrix} \tag{E.6-11}$$

where

$$\mathbf{V} = \begin{bmatrix} v_1 \\ v_2 \end{bmatrix} \qquad V_u = v_f$$

$$\mathbf{x} = \begin{bmatrix} k_1 & 0 \\ 0 & k_2 \end{bmatrix} \mathbf{V} \qquad u = k_u v_f \qquad k_u = k_f \tag{E.6-12}$$

The scaling equations (E.6-8) and (E.6-9) become

$$\mathbf{A}_E = k_t \begin{bmatrix} 1/k_1 & 0 \\ 0 & 1/k_2 \end{bmatrix} \begin{bmatrix} 0 & 1 \\ -4 & -2 \end{bmatrix} \begin{bmatrix} k_1 & 0 \\ 0 & k_2 \end{bmatrix}$$

$$= k_t \begin{bmatrix} 0 & k_2/k_1 \\ -4k_1/k_2 & -2 \end{bmatrix} \tag{E.6-13}$$

$$\mathbf{B}_E = k_t \begin{bmatrix} 1/k_1 & 0 \\ 0 & 1/k_2 \end{bmatrix} \begin{bmatrix} 0 \\ 1 \end{bmatrix} k_f = k_t k_f \begin{bmatrix} 0 \\ 1/k_2 \end{bmatrix} \tag{E.6-14}$$

Comparing the right-hand sides with the expressions for $\mathbf{A}_E$ and $\mathbf{B}_E$ in (E.6-11) gives the scaling equations obtained in (E.4-5). In steps 5–7, the scale factors, resistances, capacitances, and potentiometer settings are found as before.

Even with the assumption that we can roughly estimate the maximum values of each variable, the preceding procedure might still require some trial-and-error adjustment of the scale factors in order to obtain the proper magnitudes for the elements of $\mathbf{A}_E$ and $\mathbf{B}_E$. If this fails to work, the formulation of the problem should be reviewed. Such a failure can occur if the model to be simulated has very fast and very slow response modes (some small and some large time constants). A solution is to separate the system model into two parts, one for the fast response and one for the slow response. Each model is then simulated separately.

E.7 ANALOG SIMULATION OF NONLINEAR SYSTEMS

Simulating nonlinear systems is possible with analogs, but it requires a special circuit for each type of nonlinearity encountered. For example, a saturation nonlinearity can be simulated by using diodes to limit the op amp multiplier's output, as in Figure 6.52. Many nonlinear functions can be represented well enough by a series of straight-line segments, each generated by a multiplier–diode combination. References 7 and 10 in chapter 6 contain examples of such circuits.

APPENDIX F

The Routh–Hurwitz and Jury Stability Criteria

Two powerful stability tests are available for linear time-invariant models. These are the Routh–Hurwitz criterion for continuous-time models, and the Jury criterion for discrete-time models. Here we state both criteria without proof.

Consider the following characteristic equation:

$$b_n s^n + b_{n-1}s^{n-1} + b_{n-2}s^{n-2} + \cdots + b_2 s^2 + b_1 s + b_0 = 0 \tag{F-1}$$

We assume that this equation has been normalized so that $b_n > 0$. The Routh–Hurwitz criterion consists of several rules applied to the coefficients in (F-1). The first rule is as follows.

Rule 1

If any of the coefficients b_i, $i = 0, 1, 2, \ldots, n-1$ are zero or negative, the system is not stable. It can be either unstable or neutrally stable.

For example, the equation $s^2 + 1 = 0$ has two imaginary roots and represents a neutrally stable system. The equation $s^2 - 1 = 0$ represents an unstable system, as does the equation $s^2 - 3s + 1 = 0$. The reader is cautioned that Rule 1 cannot be used to prove stability, because its converse is not necessarily true. A polynomial with all positive coefficients can represent an unstable or neutrally stable system. For example, the polynomial $s^3 + 2s^2 + s + 2$ has the roots $s = -2, \pm i$, and so it represents a neutrally stable system.

F.1 THE ROUTH ARRAY

The remaining rules can be expressed in terms of the following array, suggested by Routh. Arrange the coefficients in rows and columns as follows:

$$\begin{array}{c|ccccc}
s^n & b_n & b_{n-2} & b_{n-4} & b_{n-6} & \cdots \\
s^{n-1} & b_{n-1} & b_{n-3} & b_{n-5} & b_{n-7} & \cdots \\
s^{n-2} & c_1 & c_2 & c_3 & c_4 & \cdots \\
s^{n-3} & d_1 & d_2 & d_3 & d_4 & \cdots \\
\cdots & \cdots & \cdots & \cdots & \cdots & \cdots \\
s^3 & e_1 & e_2 & 0 & & \\
s^2 & f_1 & f_2 & 0 & & \\
s^1 & g_1 & 0 & & & \\
s^0 & h_1 & 0 & & &
\end{array} \tag{F.1-1}$$

The powers of $s(s^n, s^{n-1}, \ldots, s^0)$ are used for indexing purposes. The array has $(n+1)$ rows. The entries in the third and succeeding rows are obtained from the following procedure:

$$c_1 = \frac{-\begin{vmatrix} b_n & b_{n-2} \\ b_{n-1} & b_{n-3} \end{vmatrix}}{b_{n-1}} = \frac{b_{n-1}b_{n-2} - b_n b_{n-3}}{b_{n-1}}$$

$$c_2 = \frac{-\begin{vmatrix} b_n & b_{n-4} \\ b_{n-1} & b_{n-5} \end{vmatrix}}{b_{n-1}} = \frac{b_{n-1}b_{n-4} - b_n b_{n-5}}{b_{n-1}} \qquad \textbf{(F.1-2)}$$

The entry c_3 is found by forming a determinant whose first column consists of b_n and b_{n-1} and whose second column consists of the two b_i entries above and immediately to the right of c_3. The negative of this determinant is then divided by the pivotal element b_{n-1}, the element above the first entry in the c_i row. The result is

$$c_3 = \frac{-\begin{vmatrix} b_n & b_{n-6} \\ b_{n-1} & b_{n-7} \end{vmatrix}}{b_{n-1}} = \frac{b_{n-1}b_{n-6} - b_n b_{n-7}}{b_{n-1}}$$

This process is easily generalized to produce the remaining entries. Moving down a row, the d_i entries are obtained as

$$d_1 = \frac{-\begin{vmatrix} b_{n-1} & b_{n-3} \\ c_1 & c_2 \end{vmatrix}}{c_1} = \frac{c_1 b_{n-3} - c_2 b_{n-1}}{c_1}$$

$$d_2 = \frac{-\begin{vmatrix} b_{n-1} & b_{n-5} \\ c_1 & c_3 \end{vmatrix}}{c_1} = \frac{c_1 b_{n-5} - c_3 b_{n-1}}{c_1} \qquad \textbf{(F.1-3)}$$

and so forth until the $(n+1)$ row is obtained.

Consider the polynomial

$$s^4 + 2s^3 + s^2 + 4s + 2$$

Rule 1 gives no information, so we form the array

$$\begin{array}{c|rrr} s^4 & 1 & 1 & 2 \\ s^3 & 2 & 4 & 0 \\ s^2 & -1 & 4 & \\ s^1 & 12 & 0 & \\ s^0 & 4 & & \end{array}$$

Zeros may be included to fill in missing elements in any determinant used to find the entries in the array.

Rule 2

The number of roots of (F-1) with positive real parts is equal to the number of sign changes in the entries in the first column of the array.

Rule 3

The necessary and sufficient condition for all roots of (F-1) to lie in the left-half plane (a stable system) is that Rule 1 is satisfied and the first column of the array contain no sign changes.

Example F.1

Determine the stability criteria for the second-order system whose characteristic equation is

$$b_2 s^2 + b_1 s + b_0 = 0$$

Assume that $b_2 > 0$. (If not, multiply the equation by -1.)

Rule 1 requires that $b_1 > 0$ and $b_0 > 0$. To apply Rule 2, form the array

$$\begin{array}{c|cc} s^2 & b_2 & b_0 \\ s^1 & b_1 & 0 \\ s^0 & b_0 & \end{array}$$

Since $b_2 > 0$, no sign changes occur if Rule 1 is satisfied. The necessary and sufficient condition for a stable second-order system is that all three coefficients must be positive.

Example F.2

Derive the stability conditions for the third-order polynomial

$$b_3 s^3 + b_2 s^2 + b_1 s + b_0 = 0 \qquad \textbf{(F.1-4)}$$

where $b_3 > 0$.

Rule 1 requires that $b_2 > 0$, $b_1 > 0$ and $b_0 > 0$. The Routh array is

$$\begin{array}{c|cc} s^3 & b_3 & b_1 \\ s^2 & b_2 & b_0 \\ s^1 & \dfrac{b_1 b_2 - b_0 b_3}{b^2} & \\ s^0 & b_0 & \end{array}$$

Given the conditions already imposed on the signs of b_i, no sign changes will occur in the first column if $(b_1 b_2 - b_0 b_3) > 0$. Thus, the necessary and sufficient conditions for stability are

$$b_3 > 0 \qquad b_2 > 0 \qquad b_1 > 0 \qquad b_0 > 0 \qquad b_1 b_2 - b_0 b_3 > 0 \qquad \textbf{(F.1-5)}$$

F.2 SPECIAL CASES

Several situations can occur that require additional interpretation.

Rule 4

If a zero occurs in the first column, and the remaining elements in that row are not all zero, replace the zero element with a small positive number ε, and continue forming the array. When completed, let $\varepsilon \to 0$.

Rule 5

If the entries above and below a zero in the first column have the same sign, a pair of purely imaginary roots exists. If the signs are opposite, there is one sign change (and therefore at least one root in the right-half plane).

To illustrate Rules 4 and 5, consider the polynomial

$$s^3 + 2s^2 + s + 2$$

The array has a zero in the third row. This is replaced with ε, and the rest of the array is constructed, which gives

$$\begin{array}{c|cc} s^3 & 1 & 1 \\ s^2 & 2 & 2 \\ s^1 & \varepsilon & \\ s^0 & 2\varepsilon/\varepsilon = 2 & \end{array}$$

As $\varepsilon \to 0$, we see $+2$ occurs above and below the zero. Thus, the polynomial has a pair of imaginary roots, while the third root must lie in the left-half plane, since no sign changes occur in the first column. We can use this information to determine all three roots easily. Let the imaginary root factors be written as $s^2 + b$, where $b > 0$. Let the real root factor be $s + a$, $a > 0$. Then

$$\begin{aligned}(s + a)(s^2 + b) &= s^3 + as^2 + bs + ab \\ &= s^3 + 2s^2 + s + 2\end{aligned}$$

Comparing the two forms shows that $a = 2$, $b = 1$. The roots are $s = -2, \pm i$.

The polynomial

$$s^3 + 2s^2 - s - 2$$

has the array

$$\begin{array}{c|cc} s^3 & 1 & -1 \\ s^2 & 2 & -2 \\ s^1 & \varepsilon & \\ s^0 & -2 & \end{array}$$

The polynomial has one root in the right-half plane. It must therefore be a real root. However, the array gives no other information in this case.

The existence of one or more roots at the origin is obvious by the absence of the term b_0, or both terms b_0 and b_1, and so forth. This is also indicated by the array if the last row is zero or if the last two rows are zero, and so forth. However, the occurrence of a zero row followed by a nonzero row is another matter. We now consider this case. (Note that Rules 4 and 5 do not apply to identically zero rows of more than one element. "Filler" zeros are not counted as zero elements.)

Rule 6

If any derived row is identically zero, the system is not stable. It may be unstable or neutrally stable. The zero row corresponds to roots located symmetrically about the origin. These can be pairs of real roots with the same magnitude but opposite sign ($\pm r$), purely imaginary complex pairs ($\pm i\omega$), or two pairs of complex conjugate roots ($a \pm ib$, $-a \pm ib$). The auxiliary polynomial containing these roots is obtained from the row above the zero row. The coefficients of the derivative of the auxiliary polynomial are used to replace the zero row, and the rest of the array is then constructed.

If the auxiliary polynomial is even and of order $2m$, there will be m pairs of equal and opposite roots.† Knowledge of these auxiliary roots, and the use of synthetic division often enables us to determine all the characteristic roots. For example, consider the polynomial

$$s^6 + 8s^5 + 18s^4 + 24s^3 + 41s^2 - 32s - 60$$

From Rule 1, this represents an unstable system, but sometimes we wish to know the degree of instability; that is, how many roots lie in the right-half plane, and how close are all of the roots to the imaginary axis. For the first four rows, the Routh array is

$$\begin{array}{c|cccc} s^6 & 1 & 18 & 41 & -60 \\ s^5 & 8 & 24 & -32 & \\ s^4 & 15 & 45 & -60 & \\ s^3 & 0 & 0 & & \end{array}$$

The appearance of the zero row requires the use of Rule 6. The auxiliary equation from the s^4 row is

$$A(s) = 15s^4 + 45s^2 - 60 = 0$$

This indicates the usefulness of the s^i terms for indexing each row. The derivative gives

$$\frac{dA(s)}{ds} = 60s^3 + 90s$$

†The auxiliary polynomial is not always even. For example, the characteristic equation $s(s^2 - 1)(s^2 + 2s + 4) = 0$ produces the auxiliary polynomial $A(s) = 4s^3 - 4s$. The author is grateful to a reviewer for pointing this out.

The zero row is now replaced with the preceding coefficients, and the array becomes

$$
\begin{array}{c|rrrr}
s^6 & 1 & 18 & 41 & -60 \\
s^5 & 8 & 24 & -32 & \\
s^4 & 15 & 45 & -60 & \\
s^3 & 60 & 90 & & \\
s^2 & 22.5 & -60 & & \\
s^1 & 250 & & & \\
s^0 & -60 & & &
\end{array}
$$

In this example, we need not complete the array in order to determine the roots. Four of them are given by the roots of the auxiliary equation (after dividing by 15)

$$s^4 + 3s^2 - 4 = 0$$

or

$$s^2 = -4, +1$$

The auxiliary roots are $s = \pm 2i, \pm 1$. The remaining two characteristic roots can be found by synthetic division. Dividing the auxiliary polynomial into the characteristic polynomial gives

$$
\begin{array}{r}
s^2 + 8s + 15 \\
s^4 + 3s^2 - 4 \overline{\big) s^6 + 8s^5 + 18s^4 + 24s^3 + 41s^2 - 32s - 60} \\
\underline{s^6 \qquad + 3s^4 \qquad\qquad - 4s^2 \qquad\qquad} \\
8s^5 + 15s^4 + 24s^3 + 45s^2 - 32s - 60 \\
\underline{8s^5 \qquad\quad + 24s^3 \qquad\quad - 32s \qquad} \\
15s^4 \qquad + 45s^2 \qquad - 60 \\
\underline{15s^4 \qquad + 45s^2 \qquad - 60}
\end{array}
$$

The other factor in the characteristic polynomial is seen to be $s^2 + 8s + 15$. This gives the roots $s = -3, -5$. We have thus found the roots of a sixth-order polynomial, without recourse to numerical methods.

Note in the preceding example that if we divide $A(s)$ by 15 and then compute the derivative, we obtain $4s^3 + 6s$. The entries to replace the zero row are now 4 and 6, not 60 and 90. Both are correct because of Rule 7.

Rule 7

Any derived row in the Routh array can be multiplied or divided by any nonzero positive number without changing the results.

This can be used to simplify the calculations by reducing the size of the numbers to be manipulated.

F.3 RELATIVE STABILITY

The Routh–Hurwitz criterion provides a yes-or-no answer to the stability question. Thus, it gives information about the *absolute stability* of the system. However, for design purposes, we would also like to know "how close" the system is to being unstable if the criterion predicts a stable system. That is, since the numerical values of the coefficients of the characteristic equation are not known exactly, then what appears to be a stable system mathematically might in reality be an unstable one. Thus, we are also interested in the *relative stability* of the system, which is indicated by the proximity of the roots to the imaginary axis.

We have seen that all roots lying on a vertical line in the complex plane have the same time constant, which is the reciprocal of the distance from the line to the imaginary axis (Figure 3.8*d*). Any root lying to the right of this line has a larger time constant and is closer to being an unstable root than those roots to its left. Thus, a measure of relative stability is given by the system's largest time constant; that is, the time constant of the dominant root. The Routh–Hurwitz criterion can be used to estimate the relative stability and sometimes to find the location of the dominant root. If we create a new polynomial by shifting the imaginary axis to the left, the Routh–Hurwitz criterion can be applied to this new polynomial to determine how many roots lie to the right of the new imaginary axis. This technique is stated in Rule 8.

Rule 8

The number of roots lying to the right of the vertical line $s = -\sigma$ can be determined by substituting $s = p - \sigma$ into (F-1) and applying the Routh–Hurwitz criteria to the resulting polynomial in p.

Example F.3

The characteristic polynomial of a third-order system is given to be

$$s^3 + 9s^2 + 26s + K$$

Find the value of K required so that the dominant time constant is no larger than 1/2. We want to design the system to be slightly underdamped. Determine whether or not this is possible, and if so, find the value of K needed and the resulting damping ratio.

The largest time constant can be 1/2. This means that no root can lie to the right of the line $s = -2$. From Rule 8, we have $\sigma = 2$, and we substitute $s = p - 2$. This gives

$$(p-2)^3 + 9(p-2)^2 + 26(p-2) + K$$

or

$$p^3 + 3p^2 + 2p + K - 24$$

We see immediately from Rule 1 that $K > 24$ is required but does not guarantee that no root lies to the right of $s = -2$. To obtain the complete set of requirements, construct the array from the polynomial in p.

$$
\begin{array}{c|cc}
p^3 & 1 & 2 \\
p^2 & 3 & (K-24) \\
p^1 & (10-K/3) & \\
p^0 & (K-24) &
\end{array}
$$

No sign changes will occur in the first column if $24 \leqslant K \leqslant 30$. This condition guarantees that the dominant time constant will be no larger than 1/2.

If $K = 24$, the last row is zero, and there exists a single root at the origin of the p coordinate system (at $p = 0$) or, equivalently, a root at $s = -2$. If $K = 30$, the third row is zero, with positive entries above and below. From Rule 5, a pair of purely imaginary roots exists (in the p coordinates). Therefore, if $K = 30$, two roots are $s = -2 \pm ib$, where b is unknown at this point. Note that $K = 30$ is the only value that will give underdamped behavior given the restriction on the dominant time constant. To compute the damping ratio for $K = 30$, write the p polynomial in factored form (we now know the third root must be real).

$$
\begin{aligned}
p^3 + 3p^2 + 2p + 6 &= (p + a)(p^2 + b^2) \\
&= p^3 + ap^2 + b^2 p + ab^2
\end{aligned}
$$

Comparing the two forms, we see that

$$
\begin{aligned}
a &= 3 \\
b^2 &= 2 \\
ab^2 &= 6
\end{aligned}
$$

or

$$
\begin{aligned}
a &= 3 \\
b &= \sqrt{2}
\end{aligned}
$$

In terms of s, this gives $s = p - 2 = -a - 2$, $\pm ib - 2$, or $s = -5$, $-2 \pm i\sqrt{2}$. The damping ratio of the dominant root is found from Figure 3.7 to be $\zeta = 0.82$.

In practical problems, the coefficients of the characteristic equation rarely produce an exactly zero entry or row in the Routh array. However, as Example F.3 shows, it is important to be able to interpret the significance of zeros when considered as limiting cases.

The Routh–Hurwitz criterion is intimately related to the geometry of the complex plane, as shown by Rule 8. Some other transformations are sometimes useful, and we briefly mention these before concluding our discussion. If we let $s = -q$ in the original polynomial, the Routh–Hurwitz criteria applied to the resulting polynomial in q will tell how many roots lie in the left-half plane. This is useful for determining the number of purely imaginary roots. The transformation $s = re^{-i\theta}$ rotates the axes through an angle θ. The resulting polynomial in r will reveal how many roots of the original polynomial lie to the right of the lines making angles of $\pm(90° + \theta)$ relative to the positive real axis in the s plane. This can be used to determine the minimum value of the damping ratio $\zeta = \sin\theta$, but the required algebra can sometimes be prohibitive.

F.4 JURY'S STABILITY CRITERION

Jury's criterion can be used to determine the stability of a discrete-time model whose characteristic equation is of the form

$$b_n z^n + b_{n-1} z^{n-1} + \cdots + b_1 z + b_0 = 0, \quad b_n > 0 \tag{F.4-1}$$

The criterion uses the following array, which is formed from the b_i coefficients [Reference F.1].

Row	z^0	z^1	z^2	...	z^{n-j}	...	z^{n-1}	z^n
1	b_0	b_1	b_2	...	b_{n-j}	...	b_{n-1}	b_n
2	b_n	b_{n-1}	b_{n-2}	...	b_j	...	b_1	b_0
3	c_0	c_1	c_2	...	c_{n-j}	...	c_{n-1}	
4	c_{n-1}	c_{n-2}	c_{n-3}	...	c_{j-1}	...	c_0	
5	d_0	d_1	d_2	...	d_{n-j}	...		
6	d_{n-2}	d_{n-3}	d_{n-4}	...	d_{j-2}	...		
	.	.	.	...	.			
	.	.	.	...	.			
	.	.	.	...	.			
$2n-5$	f_0	f_1	f_2	f_3				
$2n-4$	f_3	f_2	f_1	f_0				
$2n-3$	g_0	g_1	g_2					

The array's elements are computed as follows.

$$c_j = \begin{vmatrix} b_0 & b_{n-j} \\ b_n & b_j \end{vmatrix} \tag{F.4-2}$$

$$d_j = \begin{vmatrix} c_0 & c_{n-1-j} \\ c_{n-1} & c_j \end{vmatrix} \tag{F.4-3}$$

$$e_j = \begin{vmatrix} d_0 & d_{n-2-j} \\ d_{n-2} & d_j \end{vmatrix} \tag{F.4-4}$$

$$\vdots$$

The array is continued until a row with only three elements: g_0, g_1, g_2 is obtained.

The model is stable if and only if (F.4-1) has no roots on or outside of the unit circle in the z plane. The Jury criterion says that the system is stable if and only if the following $n+1$ conditions are satisfied.

$$
\begin{aligned}
F(1) &> 0 \\
(-1)^n F(-1) &> 0 \\
|b_0| &< b_n \\
|c_0| &> |c_{n-1}| \\
|d_0| &> |d_{n-2}| \\
|e_0| &> |e_{n-3}| \\
&\vdots \\
|g_0| &> |g_2|
\end{aligned}
\qquad \textbf{(F.4-5)}
$$

REFERENCE

F.1. E. I. Jury, *Theory and Application of the z-Transform Method*, John Wiley & Sons, New York, 1964.

APPENDIX G

Typical Units and Physical Constants

TABLE G.1 Conversion Factors

Length	1 m = 3.281 ft	1 ft = 0.3048 m
Force	1 N = 0.2248 lb	1 lb = 4.4482 N
Mass	1 kg = 0.06852 slug	1 slug = 14.594 kg
Energy	1 J = 0.7376 ft-lb	1 ft-lb = 1.3557 J
Power	1 W = 1.341×10^{-3} hp	1 hp = 745.7 W
Temperature	$T°C = \frac{5}{9}(T°F - 32)$	
	$T°F = \frac{9}{5}\,T°C + 32$	

TABLE G.2 Typical Units in Mechanical Systems

Quantity and Symbol	British Engineering	SI	Conversion[a] Factor	Other Common Units
Translational displacement x	ft	m	0.3048	mile[b]
Translational velocity v or $\dot{x}$	ft/sec	m/sec	0.3048	mile/hr (mph)[c]
Angular displacement θ	radian (rad)	radian	—	degrees[d]
Angular velocity ω or $\dot{\theta}$	rad/sec	rad/sec	—	revolutions/min[e] cycles/sec (cps) = hertz(Hz)[f]
Force f	lb	N	4.4482	ounce (oz)[g]
Torque T	lb-ft[h]	N-m[h]	1.3557	ounce-inch (oz-in.)[i]
Translational impulse M or M_x	lb-sec	N-sec	4.4482	—
Angular impulse M or M_θ	lb-ft-sec	N-m-sec	1.3557	—

[a] To convert from British engineering units to SI units, multiply the British units by this factor.

[b] 1 mile = 5280 ft.

[c] 1 mph = 1.46667 ft/sec.

[d] 1 degree = 0.017453 rad.

[e] 1 rpm = 0.1047198 rad/sec.

[f] 1 cps = 6.283185 rad/sec.

[g] 1 oz = 1/16 lb = 0.0625 lb.

[h] The dimensions of torque and energy are identical even though they are different physical quantities.

[i] 1 oz-in. = 0.005208 lb-ft.

TABLE G.3 Mechanical Properties of Steel and Aluminum

	Steel[a]	Aluminum
Mass density ρ[b]	15.2 slug/ft^3	5.25 slug/ft^3
Modulus of elasticity E[b]	4.32×10^9 lb/ft^2	1.44×10^9 lb/ft^2
Shear modulus of elasticity G[b]	1.73×10^9 lb/ft^2	5.76×10^8 lb/ft^2

[a]Steel is the collective name for the family of alloys containing iron, carbon, and frequently other substances. These properties vary with composition, but the values given here are representative.
[b]Factors for converting to SI units are given in Table G.2.

TABLE G.4 Units and Representations for Common Electrical Quantities[a]

Quantity and Symbol	Units	Representation
Voltage v	volt (V)	Voltage source v (+, −)
Charge Q	coulomb (C) = N-m/volt	
Current i	ampere (A) = coulomb/second	Current source i
Flux ϕ	volt-second	
Resistance R	ohm (Ω) = volt/ampere	R
Capacitance C	farad (F) = coulomb/volt	C
Inductance L	henry (H) = volt-second/ampere	L
Battery (voltage source plus internal resistance)	—	+ −
Ground (zero-voltage reference)	—	

[a]British Engineering and SI units are identical for electrical systems.

TABLE G.5 Typical Units in Fluid Mechanics

Quantity and Symbol	British Engineering[a]	SI[a]	Conversion Factor[b]	Other Common Units[a]
Pressure p	lb/ft^2 (psf)	N/m^2 (pascal)	47.88	$lb/in.^2$ (psi) atmospheres (atm)[c] millimeters of mercury (mm Hg)[d]
Volume flow rate q	ft^3/sec (cfs)	m^3/sec	0.02832	gal./min (gpm)[e]
Mass density ρ[f]	$slugs/ft^3$	kg/m^3	515.4	—
Viscosity μ	$lb\text{-}sec/ft^2$	$N\text{-}sec/m^2$	47.88	Centipoise[g]

[a]Other names or abbreviations are shown in parentheses.

[b]To convert from British Engineering Units to SI units, multiply the British units by this factor.

[c]1 atmosphere = 1.01325×10^5 N/m^2 = 14.6959 psi = 760 mm Hg.

[d]1 mm Hg at 0°C = 133.322 N/m^2.

[e]1 U.S. liquid gallon = 0.13368 ft^3.

[f]Weight density $\gamma = \rho g$, where $g = 32.174$ ft/sec^2 = 9.8066 m/sec^2.

[g]1 centipoise = 10^{-3} $N\text{-}sec/m^2$.

TABLE G.6 Properties of Water and Fuel Oil

Fluid	Mass Density ρ ($slugs/ft^3$)	Viscosity μ ($lb\text{-}sec/ft^2$)[a]	Specific Gravity[b]
Water at 68°F	1.94	2×10^{-5}	1
Fuel oil at 68°F	1.88	2×10^{-2}	0.97

[a]Viscosity is usually highly temperature dependent.

[b]Specific gravity is the ratio of the density to that of pure water at 4°C and a pressure of 1 atm.

TABLE G.7 Properties of Air at Standard Temperature and Pressure[d]

Gas constant R_g[a]	1715 ft-lb/slug-°R
Specific heats[b]	
c_p	7.72 Btu/slug-°R
c_v	5.5 Btu/slug-°R
Specific heat ratio $\gamma = c_p/c_v$	1.40
Mass density ρ[c]	0.0025 $slug/ft^3$
Viscosity μ[c]	3.3×10^{-7} $lb\text{-}sec/ft^2$

[a]To convert the gas constant to N-m/kg-K, multiply the British Units by 0.1672.

[b]To convert specific heat to J/kg-K, multiply the British units by 130.1.

[c]Conversion factors for mass density and viscosity are given in Table G.5.

[d]Standard temperature is 32°F. Standard pressure is 14.7 psia.

TABLE G.8 Typical Units in Heat Transfer

Quantity and Symbol	British Engineering	SI Metric	Conversion Factor[a]	Other Common Units
Heat energy Q_h	ft-lb	J	1.3557	Btu,[b] calorie (cal)[c]
Heat flow rate q_h	ft-lb/sec	W	1.3557	Btu/hr[b]
Thermal conductivity k	ft-lb/sec-ft-°F	W/m-K	8.0061	Btu/hr-ft-°F[b]
Specific heat c, c_p, or c_v	ft-lb/slug-°F	J/kg-K	0.16721	Btu/slug-°F[d]
Film (convection) coefficient h	ft-lb/sec-ft^2-°F	W/m^2-K	26.267	Btu/hr-ft^2-°F[e]

[a]To convert from British Engineering to SI units, multiply the British units by this factor.

[b]Conversion factors are given in Table G.1.

[c]1 cal = 4.18585 J.

[d]The conversion factor is given in Table G.1.

[e]To convert h from ft-lb/sec-ft^2-°F to Btu/hr-ft^2-°F, multiply by 4.6272.

TABLE G.9 Thermal Conductivity, Specific Heat, and Mass Density of Some Common Materials

Material[a]	Thermal Conductivity k Btu/hr-ft-°F[b]	Specific Heat Btu/slug-°F	Mass Density ρ slug/ft^3 [d]
Copper 32°F	224	2.93	17.3
Concrete 68°F	0.65	6.8	4.04
Fiber insulating board 70°F	0.028	—	0.46
Plate glass 70°F	0.44	6	5.25
Air 80°F, atmos. pressure	0.01516	7.728 (c_p)	0.0023
Water 68°F	0.345	32.14	1.942

[a]Values given for insulating board, concrete, and glass are typical, but can vary somewhat with composition.

[b]To convert thermal conductivity to *W/m-K*, multiply the British units by 1.7302.

[c]Conversion factor for specific heat is given in Table G.7.

[d]Conversion factor for mass density is given in Table G.5.

Index